Food Process Engineering and Technology

Second Edition

Food Science and Technology
International Series

Series Editor

Steve L. Taylor
University of Nebraska – Lincoln, USA

Advisory Board

Ken Buckle
The University of New South Wales, Australia

Mary Ellen Camire
University of Maine, USA

Roger Clemens
University of Southern California, USA

Hildegarde Heymann
University of California – Davis, USA

Robert Hutkins
University of Nebraska – Lincoln, USA

Ron S. Jackson
Quebec, Canada

Huub Lelieveld
Bilthoven, The Netherlands

Daryl B. Lund
University of Wisconsin, USA

Connie Weaver
Purdue University, USA

Ron Wrolstad
Oregon State University, USA

A complete list of books in this series appears at the end of this volume.

Food Process Engineering and Technology

Second Edition

Zeki Berk

Professor (Emeritus)
Department of Biotechnology and
Food Engineering
TECHNION
Israel Institute of Technology
Israel

AMSTERDAM • BOSTON • HEIDELBERG • LONDON
NEW YORK • OXFORD • PARIS • SAN DIEGO
SAN FRANCISCO • SINGAPORE • SYDNEY • TOKYO
Academic Press is an imprint of Elsevier

Academic Press is an imprint of Elsevier
32 Jamestown Road, London NW1 7BY, UK
525 B Street, Suite 1800, San Diego, CA 92101-4495, USA

Second edition

Copyright © 2013, 2009 Elsevier Inc. All rights reserved

No part of this publication may be reproduced, stored in a retrieval system or transmitted in any form or by any means electronic, mechanical, photocopying, recording or otherwise without the prior written permission of the publisher.

Permissions may be sought directly from Elsevier's Science & Technology Rights Department in Oxford, UK: phone (+44) (0) 1865 843830; fax (+44) (0) 1865 853333; email: permissions@elsevier.com.

Alternatively, visit the Science and Technology Books website at www.elsevierdirect.com/rights for further information.

Notice
No responsibility is assumed by the publisher for any injury and/or damage to persons or property as a matter of products liability, negligence or otherwise, or from any use or operation of any methods, products, instructions or ideas contained in the material herein. Because of rapid advances in the medical sciences, in particular, independent verification of diagnoses and drug dosages should be made

British Library Cataloguing-in-Publication Data
A catalogue record for this book is available from the British Library

Library of Congress Cataloging-in-Publication Data
A catalog record for this book is available from the Library of Congress

ISBN: 978-0-12-415923-5

For information on all Academic Press publications
visit our website at elsevierdirect.com

Typeset by MPS Ltd, Chennai, India
www.adi-mps.com

Printed and bound by CPI Group (UK) Ltd, Croydon, CR0 4YY

To my students

Contents

Introduction

"Food is life"

"Food is Life" was the theme of the 13th World Congress of the International Union of Food Science and Technology (IUFoST), held in Nantes, France, in September 2006. We retain this phrase as the motto of this book, in recognition of the vital role of food and food processing in our lives. The need to subject natural food materials to some kind of treatment before consumption was apparently realized very early in prehistory. Some of these operations, such as the removal of inedible parts, cutting, grinding, and cooking, were aimed at rendering the food more palatable, easier to consume and to digest. Others had as their objective prolongation of the useful life of food, by retarding or preventing spoilage. To this day, *transformation* and *preservation* are still the two basic objectives of food processing. While transformation is the purpose of the manufacturing industry in general, the objective of preservation is specific to the processing of foods.

The food process

Literally, a "process" is defined as a set of actions in a specific sequence, to a specific end. A manufacturing process starts with *raw materials* and ends with *products* and *by-products*. The number of actually existing and theoretically possible processes in any manufacturing industry is enormous. Their study and description individually would be nearly impossible. Fortunately, the "actions" that constitute a process may be grouped into a relatively small number of operations governed by the same basic principles and serving essentially similar purposes. Early in the 20th century, these operations, called *unit operations*, became the backbone of chemical engineering studies and research (Loncin and Merson, 1979). Since the 1950s, the unit operation approach has also been extensively applied by teachers and researchers in food process engineering (Bimbenet *et al.*, 2002; Bruin and Jongen, 2003; Fellows, 1988). Some of the unit operations of the food processing industry are listed in Table I.1.

In recent years, the concept of "food processing" has been expanded beyond the manufacturing plant to include operations associated with the ultimate consumption of foods. Methods of engineering thinking and research have been applied to the behavior of foods in the human digestive track, from mastication to gastric digestion and mass transfer in the small intestine (see, for example,

Table I.1 Unit Operations of the Food Processing Industry by Principal Groups

Group	Unit Operation	Examples of Application
Cleaning	Washing	Fruits, vegetables
	Peeling	Fruits, vegetables
	Removal of foreign bodies	Grains
	Cleaning-in-place (CIP)	All food plants
Physical separation	Filtration	Sugar refining
	Screening	Grains
	Sorting	Coffee beans
	Membrane separation	Ultrafiltration of whey
	Centrifugation	Separation of milk
	Pressing, expression	Oilseeds, fruits
Molecular (diffusion-based) separation	Adsorption	Bleaching of edible oils
	Distillation	Alcohol production
	Extraction	Vegetable oils
Mechanical transformation	Size reduction	Chocolate refining
	Mixing	Beverages, dough
	Emulsification	Mayonnaise
	Homogenizing	Milk, cream
	Forming	Cookies, pasta
	Agglomeration	Milk powder
	Coating, encapsulation	Confectionery
Chemical transformation	Cooking	Meat
	Baking	Biscuits, bread
	Frying	Potato fries
	Fermentation	Wine, beer, yogurt
	Aging, curing	Cheese, wine
	Extrusion cooking	Breakfast cereals
Preservation (Note: Many of the unit operations listed under "Preservation" also serve additional purposes, such as cooking, volume and mass reduction, improving the flavor, etc.)	Thermal processing (blanching, pasteurization, sterilization)	Pasteurized milk
		Canned vegetables
	Chilling	Fresh meat, fish
	Freezing	Frozen dinners
		Ice cream
		Frozen vegetables
	Concentration	Tomato paste
		Citrus juice concentrate
		Sugar
	Addition of solutes	Salting of fish
		Jams, preserves

(*Continued*)

Table I.1 (Continued)

Group	Unit Operation	Examples of Application
	Chemical preservation	Pickles
		Salted fish
		Smoked fish
	Dehydration	Dried fruit
		Dehydrated vegetables
		Milk powder
		Instant coffee
		Mashed potato flakes
	Freeze-drying	Instant coffee
Packaging	Filling	Bottled beverages
	Sealing	Canned foods
	Wrapping	Fresh salads

Ferrua and Singh, 2010; Kong and Singh, 2010; Sun and Xu, 2010; Tharakan *et al.*, 2010).

While the type of unit operations and their sequence vary from one process to another, certain features are common to all food processes:

- Material balances and energy balances are based on the universal principle of the conservation of matter and energy.
- Practically every operation involves exchange of material, momentum and/or heat between the different parts of the system. These exchanges are governed by rules and mechanisms, collectively known as transport phenomena.
- In any manufacturing process, adequate knowledge of the properties of the materials involved is essential. The principal distinguishing peculiarity of food processing is the outstanding complexity of the materials treated and of the chemical and biological reactions induced. This characteristic reflects strongly on issues related to process design and product quality, and calls for the extensive use of approximate models. Mathematical – physical modeling is indeed particularly useful in food engineering. Of particular interest are the physical properties of food materials and the kinetics of chemical reactions.
- One of the distinguishing features of food processing is the concern for food safety and hygiene. This aspect constitutes a fundamental issue in all the phases of food engineering, from product development to plant design, from production to distribution.

- The importance of packaging in food process engineering and technology cannot be over-emphasized. Research and development in packaging is also one of the most innovative areas in food technology today.
- Finally, common to all industrial processes, regardless of the materials treated and the products made, is the need to control. The introduction of modern measurement methods and control strategies is, undoubtedly, one of the most significant advances in food process engineering of recent years.

Accordingly, the first part of this book is devoted to basic principles, common to all food processes, and includes chapters on the physical properties of foods, momentum transfer (flow), heat and mass transfer, reaction kinetics, and elements of process control. The second part deals with transformation and separation processes. The third and final part of the book deals with food preservation.

Batch and continuous processes

Processes may be carried out in batch, continuous or mixed fashion.

In *batch processing*, a portion of the materials to be processed is separated from the bulk and treated separately. The conditions, such as temperature, pressure, composition, etc., usually vary during the process. The batch process has a definite duration, and after its completion a new cycle begins, with a new portion of material. The batch process is usually less capital-intensive, but may be more costly to operate and involves costly equipment dead time for loading and unloading between batches. It is easier to control, and lends itself to intervention during the process. It is particularly suitable for small-scale production, and when there are frequent changes in product composition and process conditions. A typical example of a batch process would be the mixing of flour, water, yeast and other ingredients in a bowl mixer to make bread dough. After having produced one batch of dough for white bread, the same mixer can be cleaned and used to make a batch of dark dough.

In *continuous processing*, the materials pass through the system continuously, without separation of a part of the material from the bulk. The conditions *at a given point in the system* may vary for a while at the beginning of the process, but ideally they remain constant during the best part of the process. In engineering terms, a continuous process is ideally run at *steady state* for most of its duration. Continuous processes are more difficult to control and require higher capital investment, but provide better utilization of production capacity and at lower operational cost. They are particularly suitable for lines producing large quantities of one type of product for a relatively long duration. A typical example of a continuous process would be the continuous pasteurization of milk.

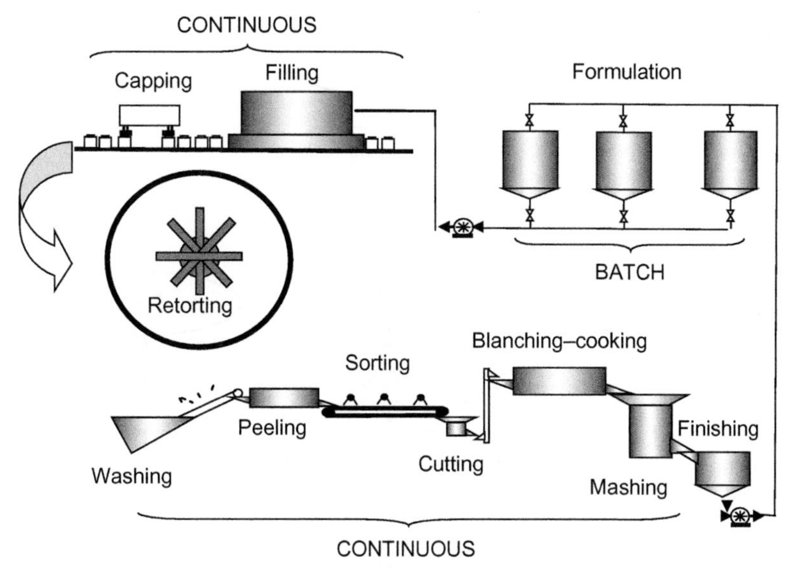

FIGURE I.1

Flow diagram for the production of strained vegetable (e.g., carrots) baby food as an example of a mixed process.

Mixed processes are composed of a sequence of continuous and batch processes. An example of a mixed process would be the production of strained infant food (Figure I.1). In this example, the raw materials are first subjected to a continuous stage consisting of washing, sorting, continuous blanching or cooking, mashing, and finishing (screening). Batches of the mashed ingredients are then collected in *formulation tanks* where they are mixed according to formulation. Usually, at this stage a sample is sent to the quality assurance laboratory for evaluation. After approval, the batches are pumped, one after the other, to the continuous homogenization, heat treatment and packaging line. Thus, this mixed process is composed of one batch phase between two continuous phases. To run smoothly, mixed processes require that *buffer storage* capacity be provided between the batch and continuous phases.

Process flow diagrams

Flow diagrams, also called *flow charts* or *flow sheets*, serve as the standard graphical representation of processes. In its simplest form, a flow diagram shows the major operations of a process in their sequence, the raw materials, the products and the by-products. Additional information, such as flow-rates and process

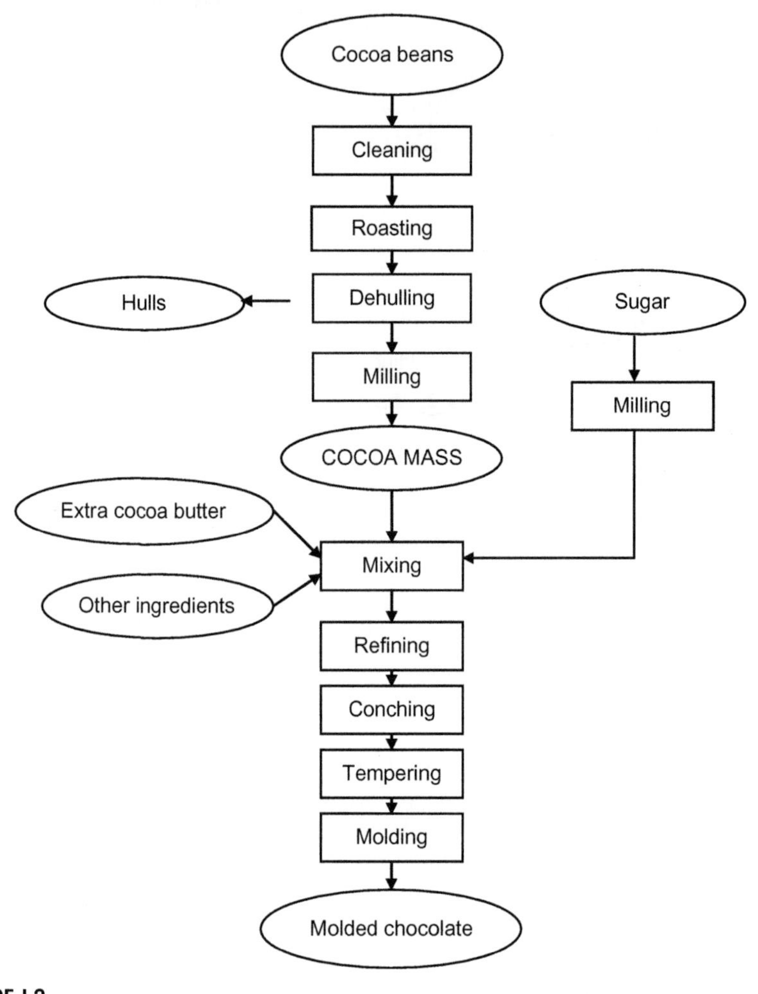

FIGURE I.2

Simplified block diagram for chocolate production.

conditions such as temperatures and pressures, may be added. Because the operations are conventionally shown as rectangles or "blocks", flow charts of this kind are also called *block diagrams*. Figure I.2 shows a simplified block diagram for the manufacture of chocolate.

A more detailed description of the process provides information on the main pieces of equipment selected to perform the operations. Standard symbols are used for frequently utilized equipment items such as pumps, vessels, conveyors, centrifuges, filters, etc. (Figure I.3).

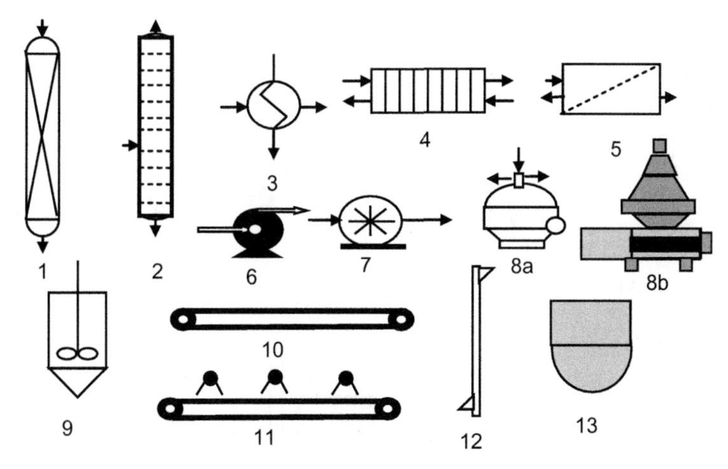

FIGURE I.3

Some equipment symbols used in process flow diagrams: 1, Reactor; 2, Distillation column; 3, Heat exchanger; 4, Plate heat exchanger; 5, Filter or membrane; 6, Centrifugal pump; 7, Rotary positive displacement pump; 8a, 8b, Centrifuges; 9, Stirred tank; 10, Belt conveyor; 11, Sorting table; 12, Elevator; 13, Kettle.

Other pieces of equipment are represented by custom symbols, resembling the actual equipment or identified by a legend. Process piping is schematically included. The resulting drawing is called an *equipment flow diagram*. A flow diagram is not drawn to scale and has no meaning whatsoever concerning the location of the equipment in space. A simplified equipment flow diagram for the chocolate production process is shown in Figure I.4.

The next step of process development is the creation of an *engineering flow diagram*.

In addition to the items shown in the equipment flow diagram, auxiliary or secondary equipment items, measurement and control systems, utility lines and piping details such as traps, valves, etc., are included. The engineering flow diagram serves as a starting point for the listing, calculation and selection of all the physical elements of a food plant or production line and for the development of a *plant layout*.

The design of a food process is a time-consuming and costly operation. The cost of process design is said to represent 10% of the total cost of plant design. (Diefes *et al.*, 2000). A number of software packages for the computer-aided design of food processes are available (Datta, 1998; Diefes *et al.*, 2000). These tools contain usually additional modules for the design of quality and food safety assurance systems, economic evaluation, environmental characteristics of the process, and many other relevant issues.

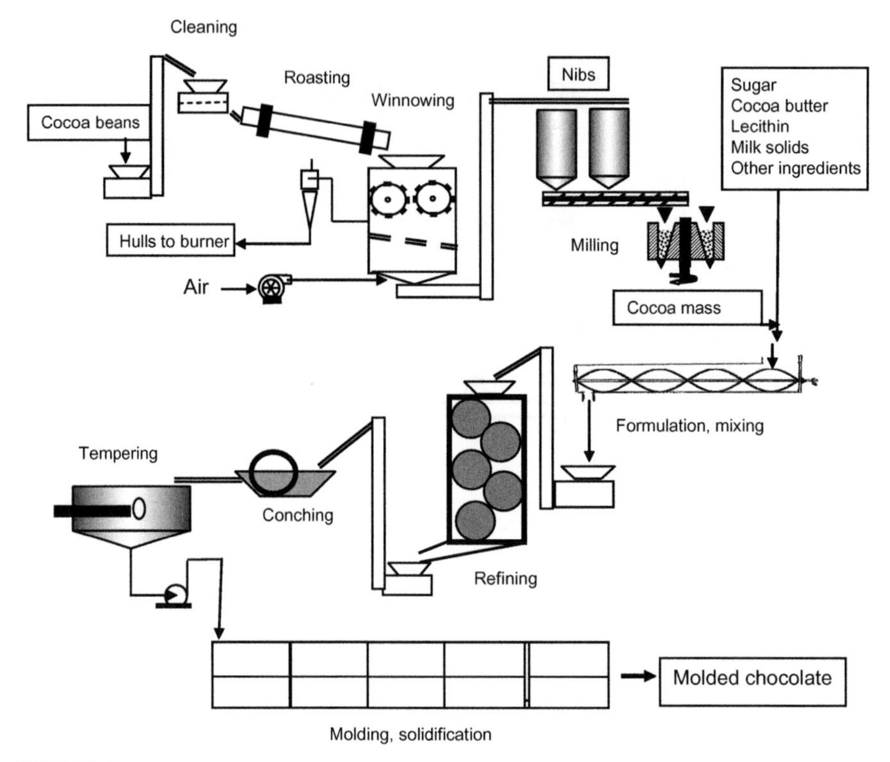

FIGURE I.4

Simplified equipment flow diagram for chocolate production.

References

Bimbenet, J.J., Duquenoy, A., Trystram, G., 2002. Génie Des Procédés Alimentaires. Dunod, Paris, France.

Bruin, S., Jongen, Th.R.G., 2003. Food process engineering: the last 25 years and challenges ahead. Compr. Rev. Food Sci. Food Saf. 2, 42–54.

Datta, A.K., 1998. Computer-aided engineering in food process and product design. Food Technol. 52 (10), 44–52.

Diefes, H.A., Okos, M.R., Morgan, M.T., 2000. Computer-aided process design using food operations oriented design system block library. J. Food Eng. 46 (2), 99–108.

Fellows, P.J., 1988. Food Processing Technology. Ellis Horwood Ltd, New York, NY.

Ferrua, M.J., Singh, R.P., 2010. Modeling the fluid dynamics in a human stomach to gain insight of food digestion. J. Food Sci. 75 (7), R151–R162.

Kong, F., Singh, R.P., 2010. A human gastric simulator (HGS) to study food digestion in human stomach. J. Food Sci. 76 (5), E627–E635.

Loncin, M., Merson, R.L., 1979. Food Engineering, Principles and Selected Applications. Academic Press, New York, NY.

Sun, Y., Xu, W.L., 2010. Simulation of food mastication based on discrete elements method. J. Int. Comput. Appl. Technol. 39 (1–3), 3–11.

Tharakan, A, Norton, I.T., Fryer, P.J., Bakalis, S., 2010. Mass transfer and nutrient absorption in a simulated model of small intestine. J. Food Sci. 75 (6), E339–E346.

CHAPTER

Physical Properties of Food Materials

1

1.1 Introduction

Dr Alina Szczesniak defined the physical properties of foods as "those properties that lend themselves to description and quantification by physical rather than chemical means" (Szczesniak, 1983). This seemingly obvious distinction between physical and chemical properties reveals an interesting historical fact. Indeed, until the 1960s, the chemistry and biochemistry of foods were by far the most active areas of food research. The systematic study of the physical properties of foods (often considered a distinct scientific discipline called *food physics* or *physical chemistry of foods*) is of relatively recent origin.

The physical properties of foods are of utmost interest to the food engineer, for many reasons:

1. Many of the characteristics that define the quality (e.g., texture, structure, appearance) and stability (e.g., water activity) of a food product are linked to its physical properties.
2. One of the most active areas of cutting-edge food research deals with the development of foods with novel physical structures. The incorporation of man-made nanoscale elements is one example of application in this area that requires in-depth understanding of physical structure.
3. Quantitative knowledge of many of the physical properties, such as thermal conductivity, density, viscosity, specific heat, enthalpy and many others, is essential for the rational design and operation of food processes and for the prediction of the response of foods to processing, distribution and storage conditions. These are sometimes referred to as *engineering properties*, although most physical properties are significant both from the product quality and process engineering points of view.

In recent years, the growing interest in the physical properties of foods has become manifest. A number of books dealing specifically with the subject have been published (e.g., Balint, 2001; Belton, 2007; Figura and Teixeira, 2007; Jowitt, 1983; Lewis, 1990; Lillford and Aguilera, 2008; Mohsenin, 1980; Peleg and Bagley, 1983; Rahman, 2009; Sahin and Sumnu, 2006; Scanlon, 2001; Walstra, 2003). The number of scientific meetings on related subjects held every year is considerable. Specific courses on the subject are being included in most food science, engineering and technology curricula.

Food Process Engineering and Technology. DOI: http://dx.doi.org/10.1016/B978-0-12-415923-5.00001-0
© 2013 Elsevier Inc. All rights reserved.

Some of the "engineering" properties will be treated in connection with the unit operations where such properties are particularly relevant (e.g., viscosity in fluid flow, particle size in size reduction, thermal properties in heat transfer, diffusivity in mass transfer, etc.). Properties of more general significance and wider application are discussed in this chapter.

1.2 Mass, volume, density

Some of the most important characteristics of foods are physical attributes such as size, shape, volume, weight, density and appearance (color, turbidity, gloss). Most of these properties are immediately perceivable and easily measurable.

Density is the mass per unit volume. Its symbol is the Greek letter ρ and its units in the SI system are $kg \cdot m^{-3}$. If m (kg) is the mass and V (m^3) is the volume, then:

$$\rho = \frac{m}{V} \tag{1.1}$$

Specific gravity (also known as *relative density*) is the ratio of the density of a substance to the density of another substance taken as reference. Nearly always, the reference material is water. Unlike density, specific gravity is dimensionless.

Density is an important engineering property as it appears in almost every equation describing a process where mass and gravity are involved. At the same time, it is a factor that must be taken into account in technological considerations. For example, the density of a finished product must be known and controlled, as food packages are most often filled volumetrically but sold by weight.

In *porous foods*, one distinguishes between the *apparent density* (density of the food including the pores) and the *true density* (density of the solid matrix without the pores). *Bulk density* is used to characterize powders, and depends on the relative position of the powder particles inside the inter-particle void space.

Density is usually determined experimentally. Methods for the prediction of density, based on three-dimensional (3D) scanning of the food volume, have been proposed (Kelkar *et al.*, 2011; Uyar and Erdoğdu, 2009). Density data for various materials are given in Table A.3 (see Appendix).

1.3 Mechanical properties

1.3.1 Definitions

By mechanical properties we mean those properties that determine the behavior of food materials when subjected to external forces. As such, mechanical properties are relevant both to processing (e.g., conveying, size reduction) and to consumption (texture, mouth feel).

The forces acting on the material are usually expressed as *stress* – i.e., intensity of the force per unit area ($N \cdot m^{-2}$, or Pa). The dimensions and units of stress are like those of pressure. Very often, but not always, the response of materials to stress is deformation, expressed as *strain*. Strain is usually expressed as a dimensionless ratio, such as the elongation as a percentage of the original length. The relationship between stress and strain is the subject matter of the science known as *rheology* (Steffe, 1996).

We define three ideal types of deformation (Szczesniak, 1983):

Elastic deformation. This kind of deformation appears instantly with the application of stress and disappears instantly with the removal of stress. For many materials the strain is proportional to the stress, at least for moderate values of the deformation. The condition of linearity, called Hooke's Law (Robert Hooke, 1635–1703, English scientist), is formulated in Eq. (1.2):

$$E = \frac{\text{stress}}{\text{strain}} = \frac{F/A_0}{\Delta L/L_0} \tag{1.2}$$

where:

E = Young's Modulus (after Thomas Young, 1773–1829, English scientist), Pa
F = force applied, N
A_0 = original cross-sectional area, m^2
ΔL = elongation, m
L_0 = original length, m.

Plastic deformation. This kind of deformation does not occur as long as the stress is below a limit value known as *yield stress*. Deformation is permanent – i.e., the body does not return to its original size and shape when the stress is removed.

Viscous deformation. This kind of deformation (flow) occurs instantly with the application of stress, and is permanent. The *rate* of strain is proportional to the stress (see Chapter 2, Fluid Flow).

The types of stress are classified according to the direction of the force in relation to the material. *Normal stresses* are those that act in a direction perpendicular to the material's surface. Normal stresses are *compressive* if they act towards the material and *tensile* if they act away from it. *Shear stresses* act in a direction parallel (tangential) to the material's surface (Figure 1.1).

The increase in the deformation of a body under constant stress is called *creep*. The decay of stress with time, under constant strain, is called *relaxation*.

Many of the important quality attributes, collectively known as *texture*, actually reflect the mechanical properties of foods. The objective study of food texture is therefore based on the quantitative determination of basic mechanical properties with the help of appropriate instruments (Steffe, 1996).

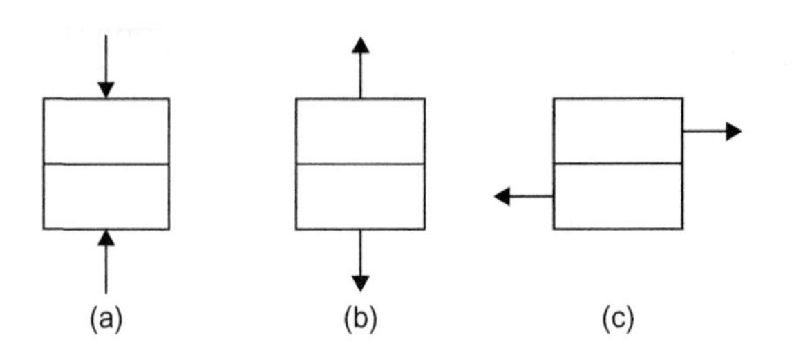

FIGURE 1.1

Types of stress: (a) compression; (b) tension; (c) shear.

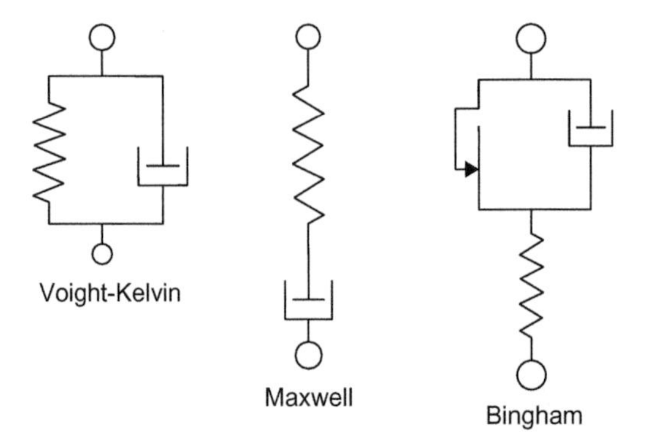

FIGURE 1.2

Three rheological models.

1.3.2 Rheological models

The stress–strain relationship in food materials is usually complex. It is therefore useful to describe the real rheological behavior of foods with the help of simplified mechanical analogs. These analogs are constructed by connecting ideal elements (elastic, viscous, friction, rupture, etc.) in series, in parallel, or in combinations of both. The ideal solid (elastic) behavior is graphically represented as a spring, while the ideal fluid (viscous) behavior is introduced as a dashpot. Models consisting of a combination of springs and dashpots represent *visco-elastic* behavior. Some of these models are shown in Figure 1.2. The mechanical models are useful in the development of mathematical models (equations) for the description and prediction of the complex rheological behavior of foods. The Maxwell model and the Kelvin model are used to simulate stress relaxation and creep, respectively (Figura and Teixeira, 2007). The rheological characteristics of fluids are discussed in some detail in a subsequent section (Chapter 2, Section 2.3).

1.4 Thermal properties

Almost every process in the food industry involves thermal effects such as heating, cooling, or phase transition. The thermal properties of foods are therefore of considerable relevance in food process engineering. The following properties are of particular importance: thermal conductivity, thermal diffusivity, specific heat, latent heat of phase transition, and emissivity. A steadily increasing volume of information on experimental values of these properties is available in various texts (e.g., Choi and Okos, 1986; Mohsenin, 1980; Rahman, 2009) and electronic databases. In addition, theoretical or empirical methods have been developed for the prediction of these properties in the light of the chemical composition and physical structure of food materials.

Specific heat, C_p ($kJ \cdot kg^{-1} \cdot K^{-1}$), is among the most fundamental of thermal properties. It is defined as the quantity of heat (kJ) needed to increase the temperature of 1 unit mass (kg) of the material by 1 Kelvin (K) at constant pressure. The specification of "at constant pressure" is relevant to gases where the heat input needed to cause a given increase in temperature depends on the process. It is practically irrelevant in the case of liquids and solids. A short survey of the methods for the prediction of specific heat is included below. Most of the other thermal properties of foods are defined and discussed in detail in Chapter 3, dealing with transport phenomena.

The definition of specific heat can be formulated as follows:

$$C_p = \frac{1}{m}\left(\frac{dQ}{dT}\right)_P \tag{1.3}$$

The specific heat of a material can be determined experimentally by static (adiabatic) calorimetry or differential scanning calorimetry, or calculated from measurements involving other thermal properties. It can be also predicted quite accurately with the help of a number of empirical equations.

The simplest model for solutions and liquid mixtures assumes that the specific heat of the mixture is equal to the sum of the pondered contribution of each component. The components are grouped in classes: water, salts, carbohydrates, proteins and lipids. The specific heat, *relative to water*, is taken as: salts = 0.2; carbohydrate = 0.34; proteins = 0.37; lipids = 0.4; water = 1. The specific heat of water is $4.18\ kJ \cdot kg^{-1} \cdot K^{-1}$. The specific heat of a solution or liquid mixture is therefore:

$$C_p = 4.18(0.2X_{salt} + 0.34X_{carbohyd} + 0.37X_{prot} + 0.4X_{lip} + X_{water}) \tag{1.4}$$

where X represents the mass fraction of each of the component groups (Rahman, 2009).

For mixtures that approximate solutions of sugar in water (e.g., fruit juices), Eq. (1.4) becomes:

$$C_p = 4.18[0.34X_{sugar} + 1(1 - X_{sugar})] = 4.18(1 - 0.66X_{sugar}) \qquad (1.5)$$

Another frequently used model assigns to the total dry matter of the mixture a single relative specific value of $0.837\ \mathrm{kJ \cdot kg^{-1}}$. The resulting approximate empirical expressions for temperatures above and below freezing are given in Eq. (1.6):

$$\begin{aligned} C_p &= 0.837 + 3.348X_{water} \quad \text{for temperatures above freezing} \\ C_p &= 0.837 + 1.256X_{water} \quad \text{for temperatures below freezing} \end{aligned} \qquad (1.6)$$

EXAMPLE 1.1

Meat loaf contains 21% protein, 12% fat, 10% carbohydrates, 1.5% minerals (salt), 0.555% water. Estimate the specific heat of the loaf.

Solution

We substitute the data in Eq. (1.4):

$$C_P = 4.18(0.2 \times 0.015 + 0.34 \times 0.1 + 0.37 \times 0.21 + 0.4 \times 0.12 + 0.555) = 3.00\ \mathrm{kJ \cdot g^{-1} \cdot K^{-1}}$$

EXAMPLE 1.2

Estimate the specific heat of 65-Bx orange juice concentrate.

Solution

We consider the concentrate as a solution of sugar in water and we read 65 Bx as 65% sugar w/w. We apply Eq. (1.5):

$$C_p = 4.18(1 - 0.66 \times 0.65) = 2.39\ \mathrm{kJ \cdot kg^{-1} \cdot K^{-1}}$$

EXAMPLE 1.3

Raw salmon contains 63.4% water. Estimate, approximately, the specific heat of salmon at temperatures above freezing and below freezing.

Solution

Since an approximate solution is acceptable, we shall apply Eq. (1.6):

$$C_p = 0.837 + 3.348 \times 0.634 = 2.96\ \mathrm{kJ \cdot kg^{-1} \cdot K^{-1}} \text{ for temperatures above freezing}$$
$$C_p = 0.837 + 1.256 \times 0.634 = 1.63\ \mathrm{kJ \cdot kg^{-1} \cdot K^{-1}} \text{ for temperatures below freezing.}$$

1.5 Electrical properties

The electrical properties of foods are particularly relevant to microwave and ohmic heating of foods, and to the effect of electrostatic forces on the behavior of powders. The most important properties are electrical conductivity and the dielectric properties. These are discussed in Chapter 3, in relation to ohmic heating and microwave heating, respectively.

1.6 Structure

Very few foods are truly homogeneous systems. Most foods consist of mixtures of distinct physical phases, in close contact with each other. The heterogeneous nature of foods may be visible to the naked eye, or perceived only when examined under a microscope or electron microscope. In foods, the different phases are seldom in complete equilibrium with each other, and many of the desirable properties of "fresh" foods, such as the crispiness of fresh bread crust, are due to the lack of equilibrium between phases. The structure, microstructure and, lately, nanostructure of foods are extremely active areas of research (see, for example, Aguilera, 2011; Anon. 2008; Chen *et al.*, 2006; Garti *et al.*, 2005; Graveland-Bikker and de Kruif, 2006; IFT, 2006; Kokini, 2011; Markman and Livney, 2011; Morris, 2004).

The following are some of the different structural elements in foods.

1. *Cellular structures*. Vegetables, fruits and muscle foods consist in large part of cellular tissue. The characteristics of the cells and, more particularly, of the cell walls determine the rheological and transport properties of cellular foods. One of the characteristics particular to cellular foods is *turgidity* or *turgor pressure*. Turgor is the intracellular pressure resulting from osmotic differences between the cell content and the extracellular fluid (Taiz and Zeiger, 2006). This is the factor responsible for the crisp texture of fruits and vegetables, and for the "fleshy" appearance of fresh meat and fish. Cellular food structures may also be created artificially. Wheat bread consists of gas-filled cells with distinct cell walls. The numerous puffed snacks and breakfast cereals produced by extrusion owe their particular crispiness to their cellular structure with brittle cell walls (see *Foams*, below).
2. *Fibrous structures*. In this context we refer to physical fibers (i.e., solid structural elements with one dimension much larger than the other two) and not to "dietary fiber", which is a notion related to dietary function. The most obvious of the fibrous foods is meat. Indeed, protein fibers are responsible for the chewiness of muscle foods. The creation of a man-made fibrous structure is the main challenge of the meat analog developer.

3. *Gels*. Gels are macroscopically homogeneous colloidal systems, where dispersed particles (generally polymeric constituents such as polysaccharides or proteins) have combined with the solvent (generally water) to create a semi-rigid solid structure. Gels are usually produced by first dissolving the polymer in the solvent, than changing the conditions (cooling, concentration, cross-linking) so that the solubility is decreased. Gelation is particularly important in the production of set yogurt, dairy deserts, custard, tofu, jams and confectionery. The structural stability of food gels subjected to shear and certain kinds of processing (e.g., freezing–thawing) is an important consideration in product formulation and process design.
4. *Emulsions*. Emulsions (Dickinson, 1987; McClements, 2005) are intimate mixtures of two mutually immiscible liquids, where one of the liquids is dispersed as fine globules in the other (Figure 1.3). In the case of foods, the two liquid media are, in most cases, fats and water.

   Two possibilities exist for emulsions consisting of oil and water:
   a. The dispersed phase is oil (oil-in-water, o/w emulsions). This is the case in milk, cream, sauces and salad dressings.
   b. The dispersed phase is water (water-in-oil, w/o emulsions). Butter and margarine are w/o emulsions.

   Emulsions are not thermodynamically stable systems. They do not form spontaneously. Emulsification requires energy input (mixing, homogenization) in order to shear one of the phases into small globules and disperse them in the continuous phase (see Chapter 7, Section 7.6, on homogenization). Emulsions tend to break apart as the result of coalescence (fusion of the dispersed droplets into larger ones) and creaming (separation of the original emulsion into a more concentrated emulsion or cream, and some free continuous phase). Emulsions are stabilized with the help of surface active agents known as emulsifiers.
5. *Foams*. Foams are cellular structures consisting of gas (air)-filled cells and liquid cell walls. Due to surface forces, foams behave like solids. Ice cream is essentially frozen foam, since almost half of its volume is air. Highly porous solid foods, such as many cereal products, may be considered as solid foams (Guessasma *et al.*, 2011). Foams with specific characteristics (bubble size distribution, density, stiffness, stability) are important in milk-containing

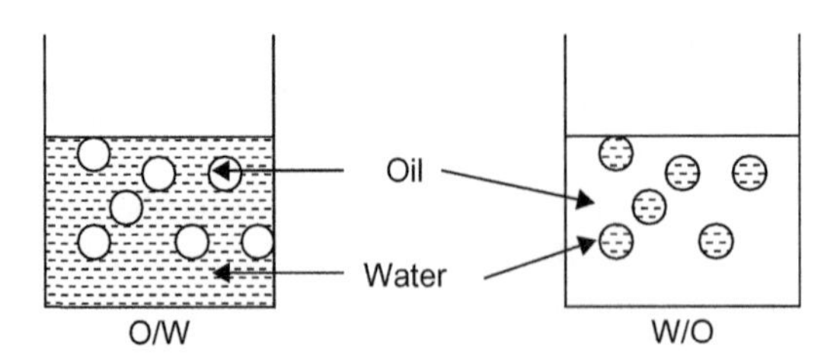

FIGURE 1.3

Schematic structure of oil-in-water and water-in-oil emulsions.

beverages and in beer. On the other hand, the spontaneous excessive foaming of some liquid products (e.g., skim milk) during transportation and processing may create serious engineering problems. Undesired foaming is controlled by proper design of the equipment, by mechanical foam breakers, or through the use of food-grade chemical antifoaming (prevention) and defoaming (foam abatement) agents such as oils and certain silicone-based compounds.

6. *Powders*. Solid particles, 10–1000 micrometers in size, are defined as powders. Smaller particles are conventionally called *dust* and larger particles *granules*. Some food products and many of the raw materials of the food industry are powders. Powders are produced by size reduction, precipitation, crystallization or spray drying. One of the main issues related to powders in food engineering is the flow and transportation of particulate materials, discussed in Chapter 3.
7. *Nanostructure*. Natural structural/functional elements measuring a few nanometers occur abundantly in all foods (Magnuson *et al*., 2011). Nano-emulsions, casein micelles and very thin films measuring a few nanometers in thickness are but a few examples of nanomaterials naturally present in foods. Although the study of the occurrence, structure and function of natural nano-materials is of considerable importance, *food nanotechnology* deals primarily with the creation of man-made nano-elements that can perform, because of their size and structure, specific useful functions in foods (Weiss *et al*., 2006). Many of these elements can be prepared from food bio-polymer systems, taking advantage of the capability of these systems to form nano-particles of specific shape through mechanisms of *self-assembly*. The following are a few of the potential applications of engineered nano-elements in the food industry:
   - Nano-sizing of antimicrobial agents to enhance their effectiveness (Weiss *et al*., 2009). One of the most promising applications is the incorporation of nano-sized silver (Chen and Schluesener, 2008).
   - Nano-emulsification to improve dispersibility/solubility and bio-availability (Magnuson *et al*., 2011).
   - Controlled and efficient delivery of nutrients and nutraceuticals through nano-encapsulation (Chen *et al*., 2006; McClements, 2010).
   - Improved protection of sensitive nutrients through nano-encapsulation and coating (Weiss *et al*., 2006).
   - Use of nanostructures as biosensors for detecting pathogens and providing information related to food traceability (Charych *et al*., 1996).
   - Incorporation of specific nano-elements (nanofibers, nanotubes, nanoclays) in polymers in order to improve mechanical, barrier and delivery characteristics as coating or packaging materials (Arora and Padua, 2010; Brody, 2006; Kumar *et al*., 2011; Sanchez-Garcia *et al*., 2010).

   It should be pointed out that the technical/economic feasibility of these and similar applications of nanotechnology to foods at industrial scale remains to be demonstrated. Furthermore, extensive further research is needed regarding the safety of the inclusion of engineered nano-elements in foods.

1.7 Water activity

1.7.1 The importance of water in foods

Water is the most abundant constituent in most foods. Indicative values of water content in a number of food products are shown in Table 1.1. Classification of foods into three groups according to their water content (high-, intermediate- and low-moisture foods), has been suggested (Franks, 1991). Fruits, vegetables, juices, raw meat, fish and milk belong to the high-moisture category. Bread, hard cheeses and sausages are examples of intermediate-moisture foods, while the low-moisture group includes dehydrated vegetables, grains, milk powder and dry soup mixtures.

The functional importance of water in foods goes far beyond its mere quantitative presence in their composition. On one hand, water is essential for the good texture and appearance of fruits and vegetables. In such products, loss of water usually results in lower quality. On the other hand, water, being an essential requirement for the occurrence and support of chemical reactions and microbial growth, is often responsible for the microbial, enzymatic and chemical deterioration of food.

It is now well established that the effect of water on the stability of foods cannot be related solely to the quantitative water content. As an example, honey containing 23% water is perfectly shelf-stable, while dehydrated potato will undergo rapid spoilage at a moisture content half as high. To explain the influence of water, a parameter that reflects both the quantity and the "effectiveness" of water is needed. This parameter is *water activity*.

Table 1.1 Typical Water Content of Some Foods

Food	Water %
Cucumbers	95–96
Tomatoes	93–95
Cabbage	90–92
Orange juice	86–88
Apples	85–87
Cow milk	86–87
Eggs, whole	74
Chicken, broiled	68–72
Hard cheese	30–50
White bread	34
Jams, preserves	30–35
Honey	15–23
Wheat	10–13
Nuts	4–7
Dehydrated onion	4–5
Milk powder	3–4

1.7.2 Water activity, definition, and determination

Water activity, a_w, is defined as the ratio of the water vapor pressure of the food to the vapor pressure of pure water at the same temperature.

$$a_w = \frac{p}{p_0} \tag{1.7}$$

where p = partial pressure of water vapor of the food at temperature T and p_0 = equilibrium vapor pressure of pure water at temperature T.

The same type of ratio also defines the relative humidity of air, RH (usually expressed as a percentage):

$$RH = \frac{p'}{p_0} \times 100 \tag{1.8}$$

where p′ = partial pressure of water vapor in air

If the food is in equilibrium with air, then p = p′. It follows that the water activity of the food is equal to the relative humidity of the atmosphere in equilibrium with the food. For this reason, water activity is sometimes expressed as the *equilibrium relative humidity*, ERH:

$$a_w = \frac{ERH}{100} \tag{1.9}$$

Many of the methods and instruments for the determination of water activity are based on Eq. (1.9). A sample of the food is equilibrated with a small headspace of air in a close chamber, and the relative humidity of the headspace is then measured by an appropriate hygrometric method such as the "chilled mirror" technique (Figure 1.4).

Typical water activity values of some food products are given in Table 1.2.

1.7.3 Water activity: Prediction

The principal mechanisms responsible for the depression of vapor pressure of water in foods are solvent–solute interaction, binding of water molecules to the polar sites of polymer constituents (e.g. polysaccharides and proteins), adsorption of water on the surface of the solid matrix, and capillary forces (Le Maguer, 1987). In high-moisture foods, such as fruit juices, the depression may be attributed entirely to water–solute interaction. If such foods are assumed to behave as "ideal solutions", then their water vapor pressure obeys Raoult's Law (see Chapter 13, Section 13.2, on vapor–liquid equilibrium). In this case:

$$p = x_w p_0 \Rightarrow x_w = \frac{p}{p_w} = a_w \tag{1.10}$$

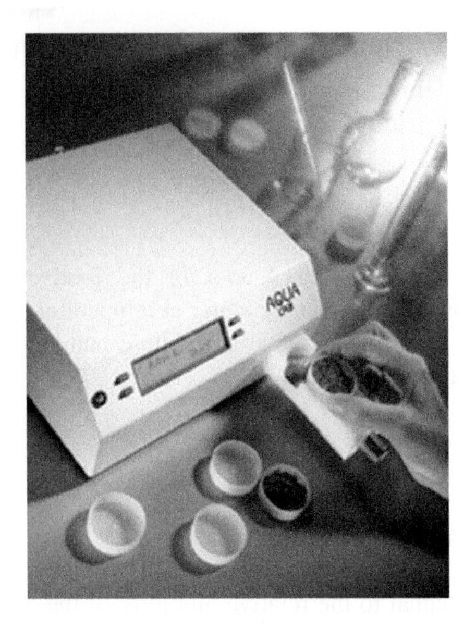

FIGURE 1.4

Measurement of water activity.

Table 1.2 Typical Water Activities of Selected Foods

a_w Range	Product Examples
0.95 and above	Fresh fruits and vegetables, milk, meat, fish
0.90–0.95	Semi-hard cheeses, salted fish, bread
0.85–0.90	Hard cheese, sausage, butter
0.80–0.85	Concentrated fruit juices, jelly, moist pet food
0.70–0.80	Jams and preserves, prunes, dry cheeses, legumes
0.50–0.70	Raisins, honey, grains
0.40–0.50	Almonds
0.20–0.40	Nonfat milk powder
<0.2	Crackers, roasted ground coffee, sugar

where x_w is the water content (in molar fraction) of the food. It follows that the water activity of an ideal aqueous solution is equal to the molar concentration of water x_w. The water activity of high-moisture foods (with a_w of 0.9 or higher) can be calculated quite accurately by this method.

EXAMPLE 1.4

Estimate the water activity of honey. Consider honey as an 80% w/w aqueous solution of sugars (90% hexoses, 10% disaccharides).

Solution

The composition of 100 g of honey is:

Water	20 g	$20/18 = 1.11\ g \cdot mol^{-1}$
Hexoses	72 g	$72/180 = 0.40\ g \cdot mol^{-1}$
Disaccharides	8 g	$8/342 = 0.02\ g \cdot mol^{-1}$

Assuming Raoult's Law, the water activity of honey is equal to the molar fraction of water:

$$a_w = x_w = \frac{1.11}{1.11 + 0.40 + 0.02} = 0.725$$

As the water content is reduced, water binding by the solid matrix and capillary forces becomes an increasingly significant factor and overshadows water–solute interaction. Furthermore, the assumption of ideal solution behavior can no longer be applied because of the elevated concentration of the liquid phase. The relationship between water content and water activity, $a_w = f(X)$, becomes more complex. This is discussed in the next section.

Water activity is temperature dependent. Considering the definition of water activity, as given in Eq. (1.8), one would be tempted to conclude the opposite. Temperature affects both p and p_0 in the same manner (the Law of Clausius–Clapeyron); therefore, their ratio should not be affected by temperature. This is true for the liquid phase, and, indeed, the water activity of high-moisture foods is affected by temperature very slightly, if at all. The situation is different at lower levels of water content. Temperature affects not only the water molecules but also the solid matrix interacting with water. Therefore, temperature affects water activity at intermediate and low moisture levels where adsorption and capillary effects are strong. The direction and intensity of temperature effects are not predictable.

1.7.4 Water vapor sorption isotherms

The function representing the relationship between water content (e.g., as grams of water per gram of dry matter) and water activity at constant temperature is called the *water vapor sorption isotherm* or *moisture sorption isotherm* of a food. The general form of a hypothetical sorption isotherm is shown in Figure 1.5. Sorption isotherms of a large number of foods have been compiled by Iglesias and Chirife (1982).

Sorption isotherms are determined experimentally. Basically, samples of the food are equilibrated at constant temperature with atmospheres at different known relative humidities. After equilibration, the samples are analyzed for water (moisture) content. Each pair of ERH–moisture content data gives one point on the isotherm. The experimental methods for the determination of sorption isotherms fall into two groups, namely static and dynamic procedures. In static methods, weighed samples of food are placed in jars over saturated aqueous solutions of different salts and left to equilibrate at constant temperature. At constant temperature, the concentration of saturated solutions is constant and so is their water vapor pressure. The relative humidity of the atmosphere in equilibrium with saturated solutions of some salts is given in Table 1.3.

In dynamic methods, the sample is equilibrated with a gas stream, the relative humidity of which is continuously changed. The quantity of moisture adsorbed or desorbed is determined by recording the change in the weight of the sample.

The two curves shown in Figure 1.5 indicate the phenomenon of *hysteresis*, which is often encountered. One of the curves consists of experimental data points at which the food sample came to equilibrium by losing moisture (desorption). The other curve represents points obtained by the opposite path–i.e., by gain of moisture (adsorption). The physical explanation of sorption hysteresis has been the subject of many studies. Generally, hysteresis is attributed to the condensation of some of the water in the capillaries (deMann 1990; Kapsalis, 1987; Labuza, 1968). The observation that, depending on the path of sorption, food can have two different values of water activity at the same moisture content casts doubt on the thermodynamic validity of the concept of sorption equilibrium (Franks, 1991).

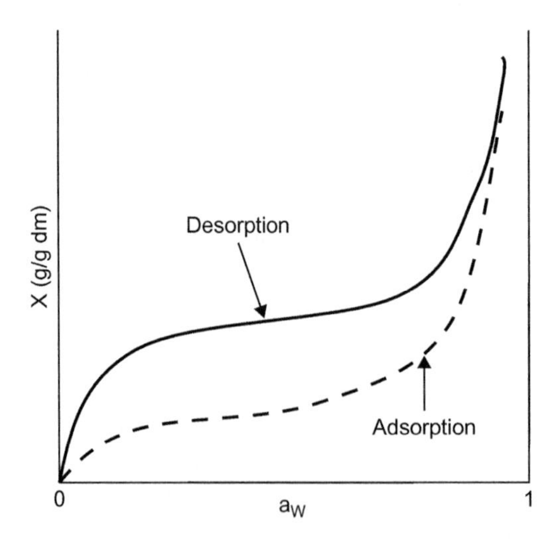

FIGURE 1.5

General form of a sorption isotherm.

Table 1.3 Saturated Salt Solutions Commonly Used in the Determination of Sorption Isotherms

Salt	a_w at 25°C
Lithium chloride	0.11
Potassium acetate	0.22
Magnesium chloride	0.33
Potassium carbonate	0.44
Sodium chloride	0.75
Potassium sulfate	0.97

Numerous attempts have been made to develop mathematical models for the prediction of sorption isotherms (Chirife and Iglesias, 1978). Some of the models developed are based on physical theories of adsorption (see Chapter 12). Others are semi-empirical expressions developed by curve-fitting techniques. A model based on the distinction of "surface absorption water" and "solution water" has been suggested (Yanniotis and Blahovec, 2009). A distributive algorithm, based on the sorption characteristics of the main constituents (glucose, fructose, sucrose, salt, protein, fiber), has been reported to predict successfully the water sorption isotherms of vegetables and fruits (Moreira *et al.*, 2009).

One of the best-known models is the Brunauer–Emmett–Teller (BET) equation. The basic assumptions on which the BET model is based are discussed in Chapter 12, Section 12.2, on adsorption. Applied to water vapor sorption, the BET equation is written as follows:

$$\frac{X}{X_m} = \frac{C\,a_w}{(1-a_w)[1+a_w(C-1)]} \tag{1.11}$$

where:

X = water content, grams water per gram of dry matter
X_m = a parameter of the equation, interpreted as the value of X for the saturation of one monomolecular layer of water on the adsorbing surface (the BET monolayer)
C = A constant, related to the heat of adsorption.

To find X_m and C from experimental sorption data, the BET equation is written as follows:

$$\frac{a_w}{(1-a_w)X} = \Phi = \frac{1}{X_m C} + \frac{C-1}{X_m C}a_w \tag{1.12}$$

If the group Φ is plotted against a_w, a straight line is obtained (Figure 1.6). X_m and C are calculated from the intercept and the slope.

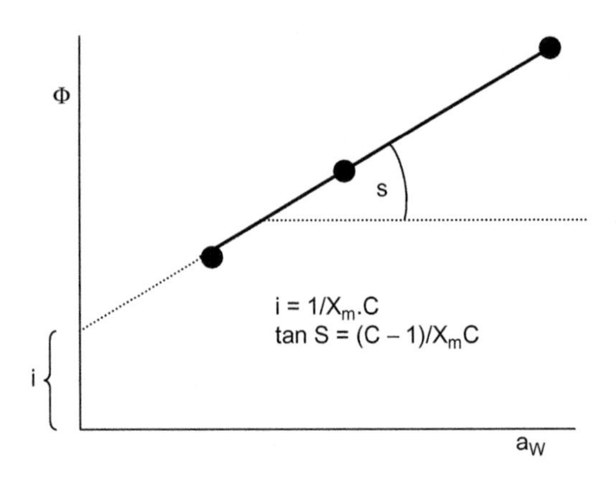

FIGURE 1.6

Linearization of the BET equation with three experimental points.

The BET model has been found to fit well with sorption isotherms, up to water activity values of about 0.45.

EXAMPLE 1.5

The following are three points from the sorption isotherm of potato at 20°C:

a_w	X (g Water per g Dry Matter)
0.12	0.05
0.47	0.11
0.69	0.18

Estimate the monolayer value of potato, based on the data.

Solution

We calculate the group Φ in Eq. (1.11), as a function of a_w. We find:

a_w	Φ
0.12	2.73
0.47	8.06
0.69	12.36

The equation of the linear plot of Φ versus a_w is found to be: $\Phi = 16.74\ a_w + 0.57$

$$\frac{C-1}{X_m C} = 16.74 \quad \frac{1}{X_m C} = 0.57$$

Solving for C and X_m we find $C = 30.36$ and $X_m = 0.058$.

Another equation that is often used to predict sorption isotherms is the GAB (Guggenheim–Anderson–de Boer) model shown below:

$$\frac{X}{X_m} = \frac{CK\,a_w}{(1 - Ka_w)[1 - Ka_w + CKa_w]} \tag{1.13}$$

where C and K are constants, both related to the temperature and heat of adsorption. The range of applicability of the GAB equation is wider than that of the BET model.

1.7.5 Water activity: Effect on food quality and stability

Bacterial growth does not occur at water activity levels below 0.9. With the exception of osmophilic species, the water activity limit for the growth of molds and yeasts is between 0.8 and 0.9. Most enzymatic reactions require water activity levels of 0.85 or higher. The relationship between water activity and chemical reactions (Maillard browning, lipid oxidation) exhibits more complex behavior with maxima and minima (Figure 1.7).

1.8 Phase transition phenomena in foods

1.8.1 The glassy state in foods

With few exceptions, foods should be regarded as metastable systems capable of undergoing change. Stability is a consequence of the rate of change. In turn, the rate of change depends on molecular mobility. In recent years, molecular mobility has become a subject of strong interest among food scientists. The subject is

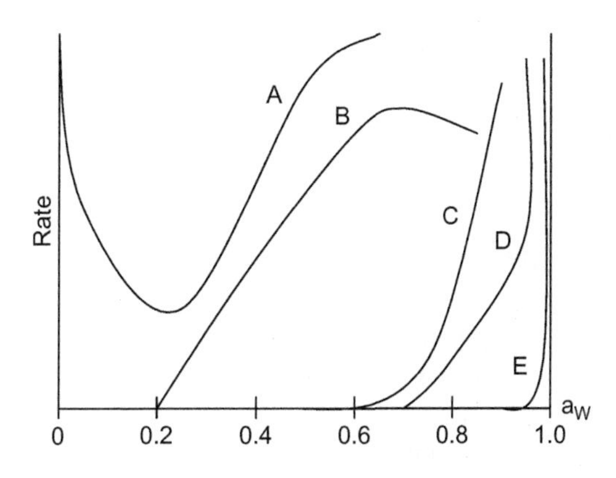

FIGURE 1.7

Relative rate of deterioration mechanisms as affected by water activity. A, lipid oxidation; B, Maillard browning; C, enzymatic activity; D, mold growth; E, bacteria growth.

particularly important in solid and semi-solid foods with low to intermediate water content. In the majority of foods belonging to this category, the interaction between polymeric constituents, water, and solutes is the key issue in connection with molecular mobility, diffusion and reaction rates. Accordingly, concepts and principles developed by polymer scientists are now being applied to foods (Le Meste *et al.*, 2002; Slade and Levine, 1991, 1995).

Consider a liquid food product, such as honey, consisting of a concentrated aqueous solution of sugars. The physical properties and stability of such a solution depend on two variables: concentration and temperature. If the concentration is increased by slowly removing some of the water and the temperature is lowered gradually, solid crystals of sugar will be formed. If the process of concentration and cooling is carried out under different conditions, crystallization will not take place but the viscosity of the solution will increase until a rigid, transparent, glass-like material is obtained. The familiar transparent hard candy is an example of glassy (vitreous) food. The glassy state is not limited to sugar–water systems. Intermediate- and low-moisture foods often contain glassy regions consisting of polymer materials (e.g., gelatinized starch) and water. The phenomenon of passage from the highly viscous, rubbery semi-liquid to the rigid glass is called *glass transition*, and the temperature at which that occurs is the *glass transition temperature*, T_g.

Physically, a glass is an *amorphous solid*. It is also sometimes described as a supercooled liquid of extremely high viscosity. Conventionally, the viscosity assigned to a glass is in the order of 10^{11} to 10^{13} Pa·s, although it is practically impossible to verify this convention experimentally. The molecules of a glass do not have an orderly arrangement as in a solid crystal, but they are sufficiently close and sufficiently immobile to possess the characteristic rigidity of solids. Because of the negligible molecular mobility, the rate of chemical and biological reactions in glassy material is extremely low. Therefore the glassy state in foods represents a condition of maximum stability (Akköse and Aktaş, 2008; Roos, 2010; Sablani *et al.*, 2007; Syamaladevi *et al.*, 2011). The rigidity of the glassy regions affects the texture of the food. Staling of bread is due to the transition of the starch–water system from a rubbery to a glassy state. The crunchiness of many snack products is due to their glassy structure.

1.8.2 Glass transition temperature

The different physical states of an aqueous solvent–solute system capable of forming an amorphous solid and the processes of passage from one state to another have been described by Roos and Karel (1991a) in a frequently cited diagram (Figure 1.8).

Boiling of a liquid or melting of a crystal are *thermodynamic* phase transitions, also known as *first-order transitions* (Roos, 1995). They occur at a fixed, definite set of conditions (temperature, pressure), independent of rate. The phases in transition are mutually in equilibrium. In contrast, glass transition is of kinetic nature. It does not involve large step changes in properties, and does not require a considerable latent heat of transition. The glass transition temperature of a given rubbery material is not a fixed point; it varies somewhat with the rate and direction of the change (e.g., rate of heating or cooling), and therefore the procedure for its determination has to be specified exactly.

Glass transition temperatures of pure dry sugars are given in Table 1.4.

Glass transition temperature is strongly dependent on concentration. Dilute solutions have lower T_g. This led Roos and Karel (1991b) to conclude that water acts on the amorphous food as a *plasticizer* in a polymer system. The effect of concentration on T_g is shown in Figure 1.9 for a solution of sucrose.

It has been suggested that the glass transition temperature of a binary blend can be predicted using the Gordon–Taylor equation, borrowed from polymer science and shown in Eq. (1.14):

$$T_g = \frac{w_1 T_{g1} + k w_2 T_{g2}}{w_1 + k w_2} \tag{1.14}$$

where:

T_g = glass transition temperature of the mixture
T_{g1}, T_{g2} = absolute glass transition temperatures (K) of components 1 and 2, respectively, where the subscript 2 refers to the component with the higher T_g
w_1, w_2 = weight fractions of components 1 and 2, respectively
k = a constant.

A simpler approximate expression is the Fox equation (Schneider, 1997):

$$\frac{1}{T_g} = \frac{w_1}{T_{g1}} + \frac{w_2}{T_{g2}} \tag{1.15}$$

According to Johari *et al.* (1987), the T_g of water is 138 K or −135°C. Glass transition temperatures of some carbohydrates are shown in Table 1.4.

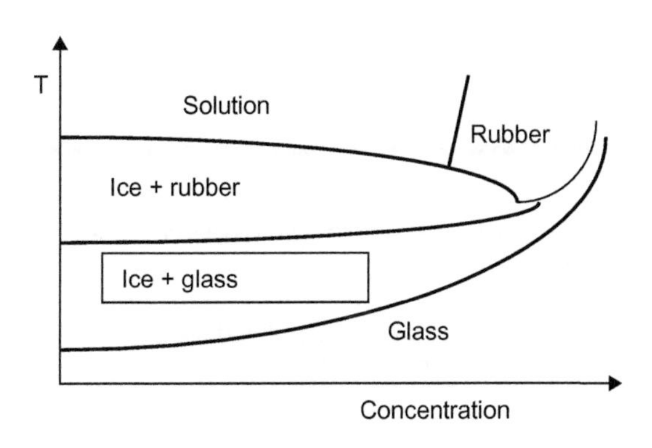

FIGURE 1.8

State diagram of a carbohydrate solution.

Adapted from Roos and Karel (1991a).

Table 1.4 Glass Transition Temperatures of Sugars

Substance	T_g(°C)
Sorbitol	−2
Xylose	9.5
Glucose	31
Fructose	100
Maltose	43
Sucrose	52

Data from Belitz et al., *(2004).*

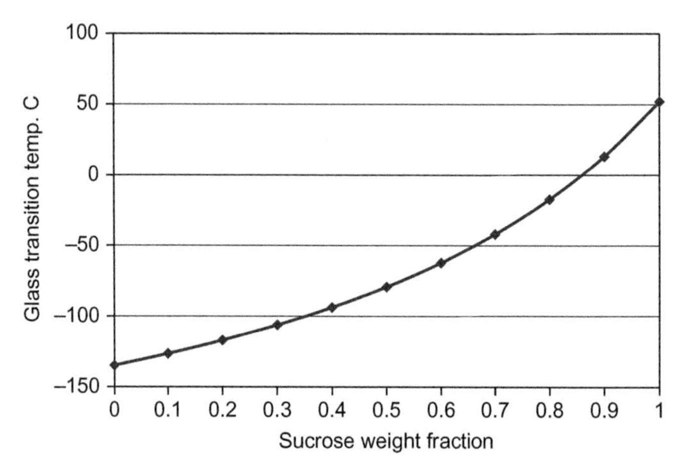

FIGURE 1.9

Effect of temperature on T_g of sucrose solutions.

EXAMPLE 1.6

Estimate the glass transition temperature of honey. Consider honey to be an 80% w/w aqueous solution of dextrose.

Solution

We use the Fox equation, and index water as 1 and dextrose as 2. We take the T_g of water to be 138 K and the T_g of dextrose to be 304 K (31°C).

$$\frac{1}{T_g} = \frac{w_1}{T_{g1}} + \frac{w_2}{T_{g2}} = \frac{0.2}{138} + \frac{0.8}{304} = 0.00408$$

$$T_g = 245 \text{ K} = -28°\text{C}$$

The viscosity of solutions increases sharply as T_g is approached. Near T_g, the effect of temperature on viscosity does not comply with the Arrhenius Law (Chapter 4, Section 4.2.3) but follows the Williams–Landel–Ferry (WLF) relationship (Roos and Karel, 1991c), shown in Eq. (1.15):

$$\log \frac{\mu}{\mu_g} = \frac{-17.44\,(T - T_g)}{51.6 + (T - T_g)} \tag{1.16}$$

where μ and μ_g stand for the viscosity at the temperatures T and T_g respectively.

As with water activity, glass transition temperature has become a key concept in food technology, with applications in quality assessment and product development. Since the glassy state is considered as a state where molecular mobility is at a minimum, it has been suggested that food properties and stability can be studied not as a function of the temperature T but as a function of the difference $T - T_g$ (Simatos *et al.*, 1995).

Several methods exist for the determination of T_g (Otles and Otles, 2005). The results may vary somewhat, depending on the method used. The most commonly applied method is differential scanning calorimetry (DSC). DSC measures and records the heat capacity (i.e. the amount of heat necessary to increase the temperature by 1 degree Celsius) of a sample and of a reference as a function of temperature. A sharp increase or decrease in heat capacity indicates an endothermic or exothermic phase transition at the temperature where this occurs. In the case of first-order transitions, such as melting, the amplitude of the change is considerable. In contrast, glass transition is detected as a subtle inflexion in the heat capacity curve (Figure 1.10).

Since the change occurs over a temperature range and not sharply at one temperature, the decision regarding where to read T_g is subject to interpretation. The two most common conventions are the mid-point of the step and the point representing the onset of the transition (Simatos *et al.*, 1995).

Other techniques, such as a rheological method applicable to spray-dried dairy powders, have also been proposed (Hogan *et al.*, 2010).

1.9 Optical properties

The term *optical properties* refers to the reaction of the food to electromagnetic radiation, and particularly to visible light. Practically, the important optical properties are *transparency*, *turbidity* and *color*. As a quality attribute, high transparency is expected in drinking water, wine, most beers, some juices, jellies, oils, etc. Turbidity is a consequence of light scattering by dispersed particles or droplets in emulsions and suspensions. A dispersed system will be turbid if the dispersed phase is opaque or has a refractive index different from that of the

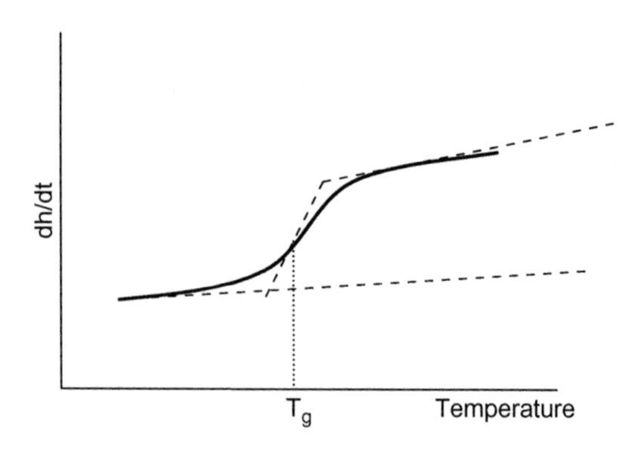

FIGURE 1.10

Glass transition temperature from DSC plot.

continuous phase *and* if the dispersed particles are larger than the wavelength of the light.

Light is *refracted* when it passes from one transparent medium to another. Refraction is a consequence of the difference in velocity of the light in the two media. The amplitude of refraction is a function of the *index of refraction* of the two media, as shown in Figure 1.11 and Eq. (1.17), known as Snell's Law.

$$\frac{\sin\alpha}{\sin\beta} = \frac{n_2}{n_1} \tag{1.17}$$

where α, β = incidence and refraction angles respectively, and n_1, n_2 = refractive indexes of the two media against air (or vacuum).

If the light penetrates from air into a liquid, $n_1 = 1$ and Eq. (1.17) becomes

$$\frac{\sin\alpha}{\sin\beta} = n \tag{1.18}$$

where n is the refractive index of the liquid.

The *refractive index* of solutions depends on their concentration. Indeed, refractometry is extensively used for the rapid determination of concentration, particularly in solutions of sugar and sugar-like solutes.

Another important area related to the optical properties of foods is the reaction of foods to electromagnetic radiation in the wavelength range of 800–2500 μm, known as the near-infrared (NIR) range. The principal components of foods (water, proteins, fats, carbohydrates) have characteristic absorption spectra when exposed to NIR radiation. In recent years, *NIR spectroscopy*, NIRS (Christy *et al.*, 2004) has become a valuable tool for the rapid approximate analysis of foods. More recently, the potential use of NIRS for process control, through monitoring of material properties, has been demonstrated (Bock and Connelly, 2008).

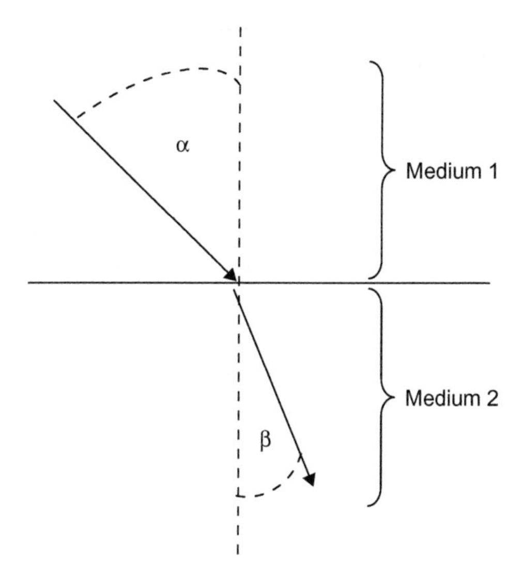

FIGURE 1.11

Refraction of light.

1.10 Surface properties

The molecules at the surface of a liquid or solid body are in a state different from that of the molecules inside the body. While the molecules in the bulk are subject to balanced attraction by all the surrounding molecules, the molecules at the free surface of the body are attracted on one side only. While the molecules in the bulk are arranged at random, the molecules at the surface assume a more or less orderly arrangement approaching that of a solid film under tension. Energy has to be supplied in order to increase the free surface of a liquid or the interface surface between two immiscible liquids. The energy increment ΔE needed to increase the free surface by 1 unit is named the *surface tension* of the liquid. It is designated by the symbol γ (sometimes σ), and its units in the SI system are $J \cdot m^{-2}$ or $N \cdot m^{-1}$:

$$\gamma = \frac{\Delta E}{\Delta A} \tag{1.19}$$

Surface tension is relevant in many areas of food technology, such as capillary phenomena, emulsions and suspensions, homogenization, foams, bubble formation, spraying, coating, wetting, washing, cleaning of solid surfaces, adsorption, etc.

The surface tension of water at room temperature is approximately $73 \times 10^{-3}\ N \cdot m^{-1}$. Surface tension decreases with increasing temperature. For pure substances, it becomes 0 at the critical temperature. Surface tension data for a number of additional liquids are listed in Table A.18, in the Appendix to this volume.

1.11 Acoustic properties

Solid foods emit audible sounds when fractured (Lewicki *et al.*, 2009). The sound generated in the mouth cavity when solid foods are fractured during mastication is an important quality factor, particularly in crunchy/crispy foods. Loss of the ability to emit sound when fractured (for example, as a result of moisture adsorption) is detected by the consumer as a loss of quality. Analysis of the emitted sound has become an important tool in the study of crunchiness/crispiness, as well as a useful technique in quality assessment (Castro-Parada *et al.*, 2007).

Feeling the acoustic response (echo) of solid foods to contact or knocking is a well-established popular method for the evaluation of texture quality in some foods, such as melons and hard cheeses. It has been shown that the analysis of the acoustic response can be used for the evaluation of fruit texture (Zdunek *et al.*, 2011).

References

Aguilera, J.M., 2011. Where is the "nano" in foods? In: Taoukis, P.S., Stoforos N.G., Karathanos, V.T., Saravacos, G.D. (Eds.), Food Process Engineering in a Changing World. vol. 1. ICEF 11 Congress Proceedings, Athens, Greece.

Akköse, A., Aktaş, N., 2008. Determination of glass transition temperature of beef and effects of various cryoprotective agents on some chemical changes. Meat. Sci. 80, 875–878.

Anon, 2008. The nanoscale food science, engineering, and technology section. J. Food Sci. 73, VII.

Arora, A., Padua, G.W., 2010. Review: nanocomposites in food packaging. J. Food Sci. 75 (1), R43–R49.

Balint, A., 2001. Prediction of physical properties of foods for unit operations. Periodica Polytechnica Ser. Chem. Eng. 45 (1), 35–40.

Belitz, H.D., Grosch, W., Schieberle, P., 2004. Food Chemistry, third ed. Springer-Verlag, Berlin.

Belton, P.S., 2007. The Chemical Physics of Foods. Blackwell Publishing Co., Oxford, UK.

Bock, J.E., Connelly, R.K., 2008. Innovative uses of near-infrared spectroscopy in food processing. J. Food Sci. 73 (7), R91–R98.

Brody, A.L., 2006. Nano and food packaging technologies converge. Food Technol. 60 (3), 92–94.

Castro-Parada, E.M., Luyten, H., Lichtendonk, W., Hamer, R.J., Van Vliet, T., 2007. An improved instrumental characterization of mechanical and acoustic properties of crispy cellular solid food. J. Texture Stud. 38, 698–724.

Charych, D., Cheng, Q., Reichert, A., Kuziemko, G., Stroh, N., Nagy, J., et al., 1996. A "litmus test" for molecular recognition using artificial membranes. Chem. Biol. 3, 113–120.

Chen, H., Weiss, J., Shahidi, F., 2006. Nanotechnology in nutraceuticals and functional foods. Food Technol. 60 (3), 30–36.

Chen, X., Schluesener, H.J., 2008. Nanosilver: a nano product in medical application. Toxicol. Lett. 176 (1), 1–12.

Chirife, J., Iglesias, H.A., 1978. Equations for fitting sorption isotherms of foods. J. Food Technol. 13, 159–174.

Choi, Y., Okos, M.R.,1986. Effect of temperature and composition on the thermal properties of foods. In: Le Maguer, L., Jelen, P. (Eds.), Food Engineering and Process Applications. Transport Phenomena, vol. 1, Elsevier, New York, NY.

Christy, A.A., McClure, W.F., Ozaki, Y. (Eds.), 2004. Near-Infrared Spectroscopy in Food Science and Technology. Wiley, New York.

deMann, J.M., 1990. Principles of Food Chemistry, second ed. Van Nostrand Reinhold, New York, NY.

Dickinson, E. (Ed.), 1987. Food Emulsions and Foams. Woodhead Publishing Ltd, Cambridge, UK.

Figura, L.O., Teixeira, A.A., 2007. Food Physics. Springer-Verlag, Berlin, Germany.

Franks, F., 1991. Hydration phenomena: an update and implications for the food processing industry. In: Levine, H., Slade, L. (Eds.), Water Relationships in Foods. Plenum Press, New York, NY.

Garti, N., Spernath, A., Aserin, A., Lutz, R., 2005. Nano-sized self-assemblies of nonionic surfactants as solubilization reservoirs and microreactors for food systems. Soft Matter 1, 206–218.

Graveland-Bikker, J.F., de Kruif, C.G., 2006. Unique milk protein based nanotubes. Trends Food Sci. Techn. 17, 196–203.

Guessasma, S., Chaunier, L., Della Valle, G., Lourdin, D., 2011. Mechanical modelling of cereal solid foods. Trends Food Sci. Technol. 22 (4), 142–153.

Hogan, S.A., Famelart, M.H., O'Callaghan, D.J., Schuck, P., 2010. A novel technique for determining glass-rubber transition in dairy powders. J. Food Eng. 99 (1), 76–82.

IFT, 2006. Functional materials in food nanotechnology. A scientific status summary of the Institute of Food Technologists. J. Food Sci. 71 (9), R107–R116.

Iglesias, H.A., Chirife, J., 1982. Handbook of Food Isotherms. Academic Press, New York, NY.

Johari, G.P., Hallbrucker, A., Mayer, E., 1987. The glass transition of hyperquenched glassy water. Nature 330, 552–553.

Jowitt, R. (Ed.), 1983. Physical Properties of Foods. Elsevier, Amsterdam, The Netherlands.

Kapsalis, J.G., 1987. Influences of hysteresis and temperature on moisture sorption isotherms. In: Rockland, L.B., Beuchat, L.R. (Eds.), Water Activity: Theory and Applications to Food. Marcel Dekker, New York, NY.

Kelkar, S., Stella, S., Boushey, C., Okos, M., 2011. Development of novel 3D measurement techniques and prediction method for food density determination. In: Taoukis, P. S., Stoforos N.G., Karathanos, V.T., Saravacos, G.D., (Eds), Food Process Engineering in a Changing World. vol. 1. ICEF 11 Congress Proceedings, Athens, Greece.

Kokini J.I., 2011. Advances in nanotechnology as applied to food systems. In: Taoukis, P. S., Stoforos N.G., Karathanos, V.T., Saravacos, G.D., (Eds), Food Process Engineering in a Changing World. vol. 1. ICEF 11 Congress Proceedings, Athens, Greece.

Kumar, P., Sandeep, K., Alavi, S., Truong, V., 2011. A review of experimental and modeling techniques to determine properties of biopolymer-based nanocomposites. J. Food Sci. 76 (1), E2–E14.

Labuza, T.P., 1968. Sorption phenomena in foods. Food Technol. 22 (3), 263–266.
Le Maguer, M., 1987. Mechanics and influence of water binding on water activity. In: Rockland, L.B., Beuchat, L.R. (Eds.), Water Activity: Theory and Applications to Food. Marcel Dekker, New York, NY.
Le Meste, M., Champion, D., Roudaut, G., Blond, G., Simatos, D., 2002. Glass transition and food technology: a critical appraisal. J. Food Sci. 67 (7), 2444–2458.
Lewicki, P.P., Marzec, A., Ranachowski, Z., 2009. Acoustic properties of foods. In: Rahman, M.S. (Ed.), Food Properties Handbook, second ed. CRC Press, New York, NY.
Lewis, M.J., 1990. Physical Properties of Foods and Food Processing Systems. Woodhead Publishing Ltd, Cambridge, UK.
Lillford, P., Aguilera, J.M., 2008. Food Materials Sciences. Springer, New York, NY.
Magnuson, B.A., Jonaitis, T.S., Card, J.W., 2011. A brief review of the occurrence, use and safety of food-related nanomaterials. J. Food Sci. 76 (6), R126–R133.
Markman, G., Livney, T.D., 2011. Maillard-reaction based nano-capsules for protection of water-insoluble neutraceuticals in clear drinks. In: Taoukis, P.S., Stoforos N.G., Karathanos, V.T., Saravacos, G.D. (Eds.), Food Process Engineering in a Changing World. vol. 1. ICEF 11 Congress Proceedings, Athens, Greece.
McClements, D.J., 2005. Food Emulsions, Principles, Practice and Technologies, second ed. CRC Press, New York, NY.
McClements, D.J., 2010. Design of nano-laminated coatings to control bioavailability of lipophilic food components. J. Food Sci. 75 (1), R30–R42.
Mohsenin, N.N., 1980. Physical Properties of Plant and Animal Materials. Gordon & Breach Science Publishers, New York, NY.
Moreira, R., Chenlo, F., Torres, M.D., 2009. Simplified algorithm for the prediction of water sorption isotherms of fruits, vegetables and legumes based upon chemical composition. J. Food Eng. 94 (3–4), 334–343.
Morris, V., 2004. Probing molecular interactions in foods. Trends Food Sci. Technol. 15, 291–297.
Otles, S., Otles, S., 2005. Glass transition in food industry–Characteristic properties of glass transition and determination techniques. Elect J. Polish Agric. Univ. 8 (4), 69.
Peleg, M., Bagley, E.B. (Eds.), 1983. Physical Properties of Foods. Avi Publishing Company, Westport, CT.
Rahman, M.S. (Ed.), 2009. Food Properties Handbook. second ed. CRC Press, New York, NY.
Roos, Y., 1995. Phase Transitions in Foods. Academic Press, San Diego, CA.
Roos, Y., Karel, M., 1991a. Applying state diagrams to food processing and development. Food Technol. 45 (12), 66 (68–71 & 107).
Roos, Y., Karel, M., 1991b. Plasticizing effect of water on thermal behavior and crystallization of amorphous food models. J. Food Sci. 56, 38–43.
Roos, Y., Karel, M., 1991c. Phase transition of mixtures of amorphous polysaccharides and sugars. Biotechnol. Prog. 7, 49–53.
Roos, Y.H., 2010. Glass transition temperature and its relevance in food processing. Annu. Rev. Food Sci. Technol. 1, 469–496.
Sablani, S.S., Al-Belushi, K., Al-Marhubi, I., Al-Belushi, R., 2007. Evaluating stability of vitamin C in fortified formula using water activity and glass transition. Intl. J. Food Prop. 10, 61–71.

Sahin, S., Sumnu, S.G., 2006. Physical Properties of Foods. Springer, New York, NY.
Sanchez-Garcia, M.D., Lopez-Rubio, A., Lagaron, M.J., 2010. Natural micro and nanobiocomposites with enhanced barrier properties and novel functionalities for food biopackaging applications. Trends Food Sci. Technol. 21 (11), 528–536.
Scanlon, M.G., 2001. Physical properties of foods, special issue. Food Res. Intl. 34 (10), 839.
Schneider, H.A., 1997. Conformational entropy contribution to the glass temperature of blends of miscible polymers. J. Res. Natl. Ins. Stand. Technol. 102, 229–232.
Simatos, D., Blond, G., Perez, J., 1995. Basic physical aspects of glass transition. In: Barbosa-Canovas, G.V., Welti-Chanes, J. (Eds.), Food Preservation by Moisture Control. Technomic Publishing Co., Lancaster, PA.
Slade, L., Levine, H., 1991. Beyond water activity: recent advances based on an alternative approach to the assessment of food quality and safety. Crit. Rev. Food Sci. Nutr. 30 (2), 115–360.
Slade, L., Levine, H., 1995. Polymer science approach to water relationships in foods. In: Barbosa-Canovas, G.V., Welti-Chanes, J. (Eds.), Food Preservation by Moisture Control. Technomic Publishing Co., Lancaster, PA.
Steffe, J.F., 1996. Rheological Methods in Food Process Engineering, second ed. Freeman Press, East Lansing, MI.
Syamaladevi, R.M., Sablani, S.S., Tang, J., Powers, J., Swanson, B.G., 2011. Stability of anthocyanins in frozen and freeze-dried raspberries during long-term storage: in relation to glass transition. J. Food Sci. 76 (6), E414–E421.
Szczesniak, A.S., 1983. Physical properties of foods: what they are and their relation to other food properties. In: Peleg, M., Bagley, E.B. (Eds.), Physical Properties of Foods. Avi Publishing Co. Inc., Westport, CT.
Taiz, L., Zeiger, E., 2006. Plant Physiology, fourth ed. Sinauer Associates, Inc., Sunderland, MA.
Uyar, R., Erdoğdu, F., 2009. Potential use of 3-dimensional scanners for food process modeling. J. Food Eng. 93 (3), 337–343.
Walstra, P., 2003. Physical Chemistry of Foods. Marcel Dekker, New York, NY.
Weiss, J., Takhistov, P., McClements, D.J., 2006. Functional materials in food nanotechnology. J. Food Sci. 71 (9), R107–R116.
Weiss, J., Gaysinsky, S., Davidson, M., McClements, J., 2009. Nanostructured encapsulation systems: food Antimicrobials. In: Barbosa-Canovas, G. (Ed.), Global Issues in Food Science and Technology. Academic Press, London, UK.
Yanniotis, S., Blahovec, J., 2009. Model analysis of sorption isotherms. Food Sci. Technol. (LWT) 42 (10), 1688–1695.
Zdunek, A., Cybulska, J., Konopacka, D., Rutkowski, K., 2011. Evaluation of apple texture with contact acoustic emission detector: a study on performance of calibration models. J. Food Eng. 106 (1), 80–87.

CHAPTER

Fluid Flow

2

2.1 Introduction

The majority of industrial food processes involve fluid movement. Liquid foods such as milk and juices have to be pumped through processing equipment or from one container to another. In a blast freezer, a rapid stream of cold air is blown over the food. In a wheat mill, the grain, the milled intermediates and the final products are most often conveyed in a stream of air (pneumatic conveying). Essential process service media (utilities) such as water, steam and various gases have to be distributed about the plant in properly designed pipelines. A number of important unit operations, such as filtration, pressing and mixing, are, essentially, particular applications of fluid flow. The mechanism and rate of energy and mass transfer are strongly dependent on flow characteristics. Finally, the sensory quality of many liquid and semi-liquid foods depends, to a large extent, on the flow properties of the product.

This chapter consists of four parts. The first part deals with the study of fluids in motion. This is the realm of a discipline known as *fluid mechanics*. The second part is about the flow and deformation properties of fluids. This is the subject matter of the science called *rheology*. Technical elements such as pumps and piping used for *conveying* fluids are discussed in the third part. The fourth part deals with flow and flow-related phenomena involving *particulate solids*. Techniques for the measurement and control of flow are described in Chapter 5.

2.2 Elements of fluid mechanics

2.2.1 Viscosity

Consider a mass of fluid confined between two flat plates (Figure 2.1). The lower plate is held stationary, and the upper plate moves in the x direction at a constant velocity v_x. Assume that the liquid layer in immediate contact with each plate moves at the velocity of that plate (no slippage). What has been described is clearly a *shearing* action on the fluid.

Newton's law states that the shearing force F_x required to maintain the upper plate in movement is proportional to the area of the plate A and to the velocity gradient $-dv_x/dz$. It is assumed that there is no movement other than in the x direction.

$$F_x = -\mu A \frac{dv_x}{dz} \tag{2.1}$$

Food Process Engineering and Technology. DOI: http://dx.doi.org/10.1016/B978-0-12-415923-5.00002-2
© 2013 Elsevier Inc. All rights reserved.

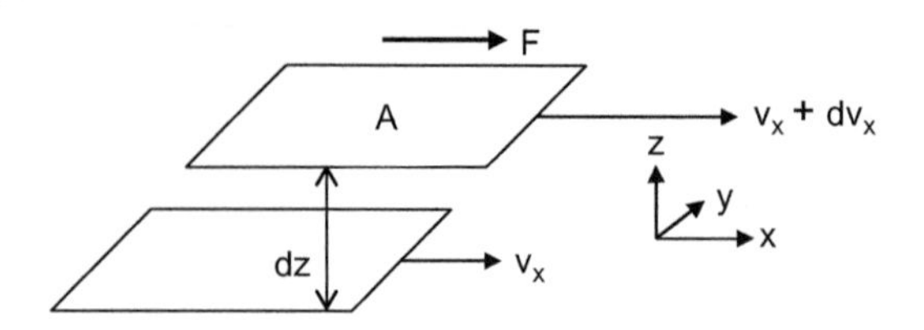

FIGURE 2.1

Definition of viscosity.

The proportionality factor μ is called *viscosity*. Viscosity is a property of the fluid, and represents the resistance of the fluid to shearing action. Its units in the SI system are *Pascal · second* (Pa · s). The traditional c.g.s unit is the *poise* (and its subdivision, the centipoise, cP) after the French physicist Poiseuille (1799–1869). The conversion factor is:

$$1\text{Pa}\cdot\text{s} = 10\text{poise} = 1000\text{cP}$$

The viscosity of liquids is strongly temperature-dependent and almost pressure-independent. The viscosity of gases increases with pressure and decreases slightly with increasing temperature.

Viscosity data for various materials of interest in food process engineering are given in the Table A.3 in the Appendix to this volume.

The shearing force per unit area is the *shear stress*, shown by the symbol τ. The velocity gradient $-(dv_x/dz)$ is called the *shear rate*, represented by the symbol γ. Eq. (2.1) can now be written as:

$$\tau = \mu\gamma \tag{2.2}$$

For many fluids, the viscosity is independent of the shear rate and Eq. (2.2) is linear. Such fluids are qualified as "Newtonian". Gases, water, milk and dilute solutions of low molecular weight solutes are practically Newtonian. Other fluids such as solutions of polymers and concentrated suspensions are non-Newtonian. Their viscosity depends on the shear rate. For non-Newtonian liquids Eq. (2.2) is not linear. Some liquid foods may be Newtonian under given conditions and non-Newtonian under others. Thus, milk cream is reported to behave as a Newtonian fluid at moderate fat content and ambient temperature, and as a non-Newtonian fluid at high fat content and low temperature (Flauzino *et al.*, 2010). A more detailed discussion of non-Newtonian flow behavior is to be found in Section 2.3.

2.2.2 Fluid flow regimes

There are two types or *regimes* of fluid flow. The first is called *laminar* or *streamlined* flow; the other is *turbulent* flow. The type of flow depends on the mass flow rate, the density and viscosity of the fluid, and the geometry of the flow channel. These variables are combined in a dimensionless group known as the *Reynolds number* (Re), named after Osborne Reynolds (1842–1912), one of the most important pioneers of engineering science.

$$\mathrm{Re} = \frac{Dv\rho}{\mu} \tag{2.3}$$

where:

D = a linear dimension characterizing the geometry of the channel (in the case of flow in a full pipe, D is the pipe diameter), m
v = average linear velocity of the fluid, $m \cdot s^{-1}$
ρ = density of the fluid, $kg \cdot m^{-3}$
μ = viscosity of the fluid, Pa · s.

Physically, the Reynolds number represents the balance between inertial forces (the nominator) and the viscous constraints (the denominator).

In the case of flow in a duct with a non-circular cross-section, or in a partially filled duct, D is replaced by the *hydraulic diameter*, defined as follows:

$$D_{hydr.} = \frac{4A}{p_w}$$

where A = cross-section of the fluid in a direction perpendicular to the direction of flow, and p_w = wetted perimeter.

In laminar flow, movement is only in one direction at all points of the fluid. This type of flow may be visualized as if the fluid were flowing in distinct, parallel layers sliding over one another with no passage between the layers. Physically, the flow is laminar if the viscous constraints overcome inertia (i.e., at low Re). In the case of flow inside cylindrical pipes, laminar regime prevails for $\mathrm{Re} < 2300$ (approximately). Laminar regime is common in food processes where the velocities are relatively low and the viscosities relatively high.

EXAMPLE 2.1

Milk ($\rho = 1032\ kg \cdot m^3$, $\mu = 2 \times 10^{-3}$ Pa · s), is flowing in a smooth stainless steel pipe, 0.025 m in diameter. What is the maximum flow rate at which the flow can be still considered laminar?

Solution

We shall assume that the condition for laminar flow is $\mathrm{Re} < 2000$. Applying the data:

$$\mathrm{Re}_{max} = D \cdot v_{max} \cdot \rho/\mu = 0.025 \cdot v_{max} \cdot 1032/0.002 = 2000$$

Hence, $v_{max} = 0.155\ m \cdot s^{-1}$.

$$Q_{max} = 0.155 \cdot \pi \cdot (0.025)^2/4 = 76 \times 10^{-6}\ m^3 \cdot s^{-1} = 0.274\ m^3 \cdot h^{-1}$$

In turbulent flow, the local instantaneous velocity in a given point of the fluid varies at random in magnitude and in direction. Eddies are created and mixing occurs. In food processes, turbulent flow is encountered when high velocities are imparted to fluids of relatively low viscosity – for example, pumping of water, milk or diluted solutions through pipes at high mass flow rate, air blasts used in freezers and dryers, etc.

Laminar flow lends itself to simple mathematical analysis. On the other hand, mathematical modeling of turbulent flow is much more complex due to the random nature of the movement. Calculation of turbulent flow therefore makes extensive use of experimental and semi-experimental techniques.

2.2.3 Typical applications of Newtonian laminar flow

Several flow configurations represent situations that are frequently encountered in industry. Some of these configurations are treated in the following analysis, assuming Newtonian behavior of the fluid and laminar flow.

2.2.3.1 Laminar flow in a cylindrical channel (pipe or tube)

Consider a fluid flowing in laminar fashion in a pipe of radius R (Figure 2.2). We take an annular layer (shell) of the fluid at a distance r from the central axis of the pipe, moving at a constant linear velocity v_r. Let L be the length of the shell and dr its thickness.

The forces acting on the shell are:

- The force resulting from pressure on the rear annular area of the layer: $[2\pi r(dr)P_0]$
- The force resulting from pressure on the front annular area of the layer: $[-2\pi r(dr)P_L]$
- Shearing force acting on the inner cylindrical surface of the shell: $[2\pi rL\tau_r]$
- Shearing force acting on the outer cylindrical surface of the shell: $[-2\pi(r+dr)L\tau_{r+dr}]$

Since the shell is moving at constant velocity, the sum of forces acting on it is zero:

$$2\pi r(dr)P_0 - 2\pi r(dr)P_L + 2\pi rL\tau_r - 2\pi(d+dr)L\tau_{r+dr} = 0 \tag{2.4}$$

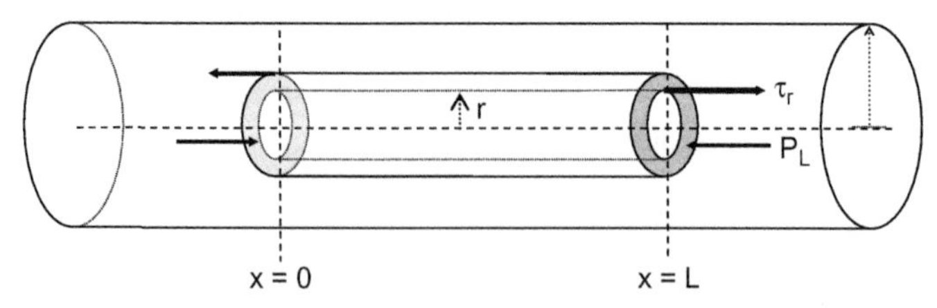

FIGURE 2.2

Laminar flow in a tube or pipe.

Setting $P_0 - P_L = \Delta P$ and rearranging, we can write:

$$d(r\tau) = \frac{r\Delta P}{L}dr \tag{2.5}$$

Integration gives the relationship between the shear stress and the distance from the central axis r:

$$\tau = \frac{r}{2L}\Delta P \tag{2.6}$$

If the fluid is Newtonian, then:

$$\tau = -\mu\frac{dv}{dr}$$

Substitution in Eq. (2.6) gives:

$$dv = -\frac{\Delta P}{2L\mu}rdr \tag{2.7}$$

Integration of Eq. (2.7) gives the velocity profile as a function of r:

$$v = -\frac{\Delta P}{4L\mu}r^2 + C \tag{2.8}$$

The integration constant C is calculated by setting the velocity at the wall to 0. The final result is:

$$v = \frac{\Delta P}{4L\mu}(R^2 - r^2) \tag{2.9}$$

Equation (2.9) leads to the following conclusions:

- The velocity profile is parabolic (Figure 2.3), varying from 0 at the wall ($r = R$) to a maximum at the center axis of the pipe ($r = 0$). The maximum velocity is:

$$v_{max} = \frac{\Delta P}{4L\mu}R^2 \tag{2.10}$$

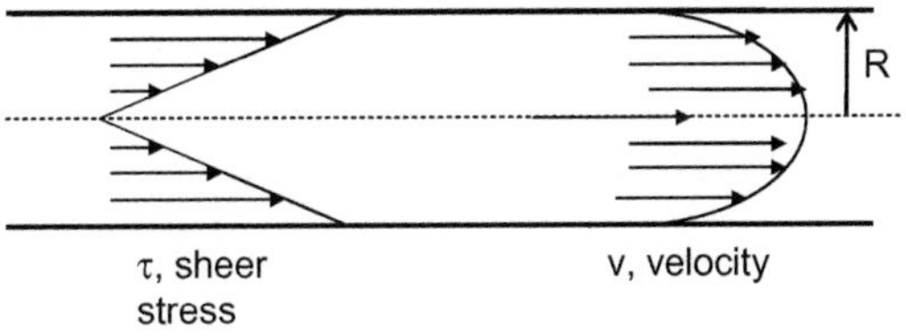

FIGURE 2.3

Shear stress and velocity profile in laminar flow in a tube.

- The shear stress profile is linear (Figure 2.3), varying from 0 at the center axis to a maximum at the wall. The maximum shear stress is:

$$\tau_{max} \equiv \tau_{wall} = \frac{R}{2L}\Delta P \tag{2.11}$$

- The volumetric flow rate Q, $m^3 \cdot s^{-1}$, is calculated from the velocity profile by integration:

$$Q = \int_0^R \frac{\Delta P}{4L\mu}(R^2 - r^2)2\pi r dr = \frac{\pi R^4 \Delta P}{8L\mu} \tag{2.12}$$

This is known as the *Hagen–Poiseuille* equation.

- The mean velocity is calculated by dividing the volumetric flow rate by the cross-sectional area of the tube:

$$v_{average} = \frac{Q}{A} = \frac{\pi \Delta P R^4}{8L\mu} \times \frac{1}{\pi R^2} = \frac{\Delta P R^2}{8L\mu} \tag{2.13}$$

It follows that the mean velocity is half the maximum velocity:

$$v_{average} = \frac{v_{max}}{2} \tag{2.14}$$

EXAMPLE 2.2

In "capillary viscosimetry", the fluid is forced to flow through a capillary of known diameter and length, under the action of a known applied pressure. The viscosity of the fluid is calculated from the relationship between flow rate and pressure drop, using the Hagen–Poiseuille equation.

Our "capillary viscosimeter" consists of a stainless steel capillary (internal diameter = 2 mm, length = 60 mm) attached to a syringe, 50 mm in diameter (Figure 2.4). The syringe is filled with fluid, and the force necessary to move the piston at a certain velocity is measured.

A sample of honey (assumed Newtonian) was tested. The force necessary to move the piston at a speed of $0.05\ mm \cdot s^{-1}$ was found to be 176.6 N. What is the viscosity of the honey?

Solution

The mean velocity of the honey in the capillary is:

$$V_m = 0.05 \times (50/2)^2 = 31.25\ mm \cdot s^{-1}$$

The pressure applied is:

$$P = 176.6/[\pi \cdot (0.05)^2/4] = 89\,987\ Pa$$

We assume that the pressure is totally used to compensate the pressure drop in the capillary. Substituting in the Hagen–Poiseuille equation:

$$\mu = \frac{\Delta P R^2}{8 L v_m} = \frac{89987 \times (10^{-3})^2}{8 \times 60 \cdot (10^{-3}) \times 31.25(10)^{-3}} = 6 Pa \cdot s$$

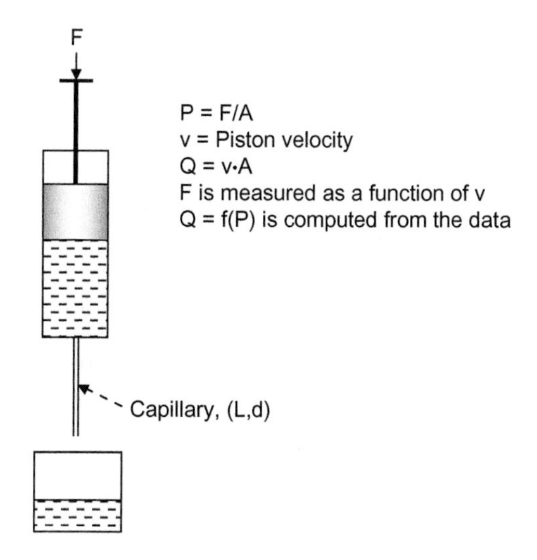

FIGURE 2.4

Capillary viscosimeter.

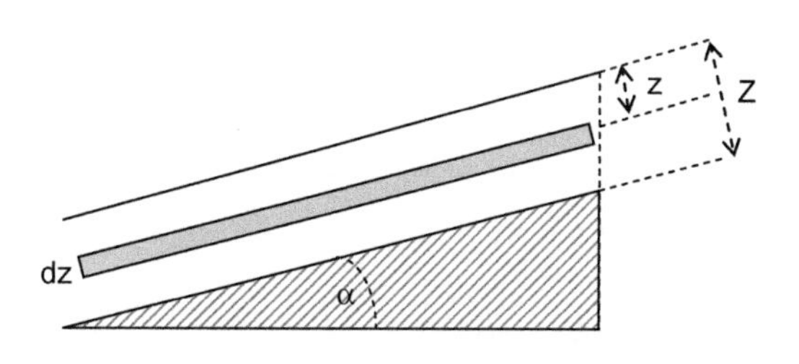

FIGURE 2.5

Laminar flow on an inclined plane.

2.2.3.2 Laminar fluid flow on flat surfaces and channels

Fluid flow on flat surfaces is of interest in many practical situations, such as fluid transport in channels and processes whereby a liquid flows on a solid surface as a thin film (e.g., film evaporators, coating of food items with a liquid film). In such cases, flow is generally induced by gravity.

Consider a mass of fluid moving along an inclined flat surface in laminar flow at steady state (Figure 2.5). A fluid layer of thickness dz, at a depth z from the free surface of the liquid, is taken. The angle of inclination is α.

The forces acting on the layer are identified and the sum of these forces is set to 0, as in the case of the analysis of laminar flow in a tube. Assuming

Newtonian behavior, the following expressions for velocity, shear stress and flow rate are developed:

For the velocity profile:

$$v_z = \frac{\rho g}{2\mu} \sin\alpha (Z^2 - z^2) \tag{2.15}$$

where g = gravitational acceleration, $m \cdot s^{-2}$.

For the shear stress profile:

$$\tau_z = \rho g z \sin\alpha \tag{2.16}$$

For the volumetric flow rate Q, in a channel of width W, m:

$$Q = \frac{Z^3 W \rho g \sin\alpha}{3\mu} \tag{2.17}$$

Equation (2.17) is frequently utilized for the calculation of the thickness Z of a falling film, corresponding to a given flow rate Q.

EXAMPLE 2.3

Water at 60°C is slowly flown as a uniform film over a 0.5-m wide vertical wall, at a flow rate of 60 liters per hour. What is the thickness of the film?

Solution

$$Q = 60/(3600 \times 1000) = 16.7 \times 10^{-6} m^3 \cdot s^{-1}$$

For water at 60°C,

$$\rho = 983\, kg \cdot m^3 \text{ and } \mu = 0.466 \times 10^{-3} Pa \cdot s.$$

Since the wall is vertical, sin $\alpha = 1$.

$$z^3 = \frac{3Q\mu}{W\rho g} = \frac{3 \times 16.7 \times 10^{-6} \times 0.466 \times 10^{-3}}{0.5 \times 983 \times 9.81} = 0.00482 \times 10^{-9} \Rightarrow z = 0.169 \times 10^{-3} m$$

The thickness of the film is 0.169 mm.

2.2.3.3 Laminar fluid flow around immersed particles

Fluid flow around individual particles is of interest in many applications, such as settling of particles in a suspension, centrifugation, shear disintegration of droplets in an emulsion, fluidization, pneumatic transport, etc. In the following discussion we shall consider the flow of a fluid around a stationary particle, but the same analysis would be valid for a particle moving in a stagnant fluid. For simplicity, the shape of the particle will be assumed to be spherical.

The Reynolds number for this flow geometry is defined as in Eq. (2.3), but with the diameter of the particle d_p as the linear parameter. The velocity v is the *approach velocity*, i.e. the velocity of the fluid at considerable distance from the sphere. For this geometry the regime is found to be laminar (i.e., no eddies are induced) for Re < 2.

As the fluid flows around the sphere, the particle experiences a force, due to friction, in the direction of the flow. This is the force that tends to drag the particle along with the fluid in the direction of the current, and is therefore called the *drag force*, F_D, or simply the *drag*. Dimensional analysis (see Section 2.2.4) leads to the definition of a dimensionless expression C_D, called the *drag coefficient*, defined as follows:

$$C_D = \frac{2F_D}{v^2 \rho A_p} \tag{2.18}$$

where A_p is the projected area of the particle, on a plane normal to the direction of flow. For a sphere, $A_p = \pi d_p/4$. The drag coefficient is a function of the Reynolds number. George Gabriel Stokes (1819–1903) found that for laminar flow, the drag coefficient is given by:

$$C_D = \frac{24}{\mathrm{Re}} \tag{2.19}$$

Substitution of Eq. (2.19) and $A_p = \pi d_p/4$ in Eq. (2.18) gives the magnitude of the drag acting on a sphere, as a function of fluid viscosity, velocity and particle size:

$$F_D = 3\pi v \mu d_p \tag{2.20}$$

Consider a spherical solid particle immersed in a Newtonian fluid. Assume that the density of the solid ρ_S is higher than that of the fluid ρ_L (i.e., the particle sinks). The forces acting on the sphere are:

- A downwards force due to gravity, $F_g = \pi d_p^3(\rho_S - \rho_L)g/6$
- The drag force, opposing the downwards movement, as per Eq. (2.20).

The net vertical force downwards is:

$$F_{net\downarrow} = \frac{\pi\ d_p^3}{6}(\rho_S - \rho_L)g - 3\pi\ v\ \mu d_p \tag{2.21}$$

At first, the velocity (and hence the drag) is low and the particle experiences vertical downward acceleration. As the velocity increases, the drag becomes stronger, so the net force (and hence the acceleration) decreases. When the drag becomes equal to the gravity force, the net force acting on the particle is zero, acceleration ceases and the particle continues to move downwards at a constant velocity, known as *terminal velocity* or *settling velocity*, v_t, calculated by setting the net force to zero in Eq. (2.21):

$$v_t = \frac{d_p^2(\rho_S - \rho_L)g}{18\mu} \tag{2.22}$$

Equation (2.22) is known as *Stokes' Law*.

EXAMPLE 2.4

A suspension contains particles of two different insoluble solid substances, A and B, in water. It is desired to obtain, by elutriation with water at 20°C, a sample of pure A for analytical purposes. The densities of A and B are 3400 and 1500 kg·m^{-3}, respectively. The particle diameter range of both substances is 20–60 μm, and the particles are assumed to be spherical. Calculate the minimal water velocity needed in order to obtain a precipitate of pure A. What is the particle size range in the precipitate?

Solution

In order to obtain a precipitate of pure A, the velocity of the water should be sufficient to prevent precipitation of the largest particle of B.

$$v_{min} = \frac{d_p^2(\rho_s - \rho_L)g}{18\mu} = \frac{(60 \times 10^{-6})^2(1500 - 1000) \times 9.8}{18 \times 1 \times 10^{-3}} = 0.98 \times 10^{-3}\,\text{m}\cdot\text{s}^{-1}$$

The diameter of the smallest particle of A in the precipitate will be:

$$d_{min,A} = \sqrt{\frac{(18 \times 10^{-3}) \times (0.98 \times 10^{-3})}{(3400 - 1000) \times 9.8}} = 27.4 \times 10^{-6}\,\text{m}$$

The particle size range will be 27.4 to 60 μm.

2.2.3.4 *Fluid flow through porous media*

Fluid flow through porous media, such as a bed of solid particles, a porous membrane or a fibrous mat, has many applications in food processes. Filtration, membrane processes, and flow through a packed column are examples of such applications. The fundamental principles of flow through porous media are discussed in Section 2.5.3 and in Chapter 8.

2.2.4 Turbulent fluid flow

Turbulence occurs if the Reynolds number is high – i.e., when inertial forces predominate over viscous constraints (see Section 2.2.2). In fact, in the case of turbulent flow, friction can be neglected everywhere in the fluid except in a thin layer of fluid adjacent to the wall, where flow is assumed to be laminar. The concept of a *boundary layer*, first proposed by Ludwig Prandtl (1875–1953), has widespread applications in the analysis of transport phenomena. By this model, the totality of the velocity gradient and hence the totality of the friction occurs within the laminar boundary layer (Figure 2.6).

As stated in Section 2.2.2, purely analytical modeling of turbulent flow is difficult. The laws of turbulent flow are based, to a large extent, on experimentation. The treatment of experimental data is considerably simplified through the use of a technique known as *dimensional analysis*. First, a list is made of all the variables capable of affecting a certain process. These variables are then grouped into dimensionless groups, and the experimental data are used to determine the quantitative relationship between the *dimensionless groups* rather than between the individual variables.

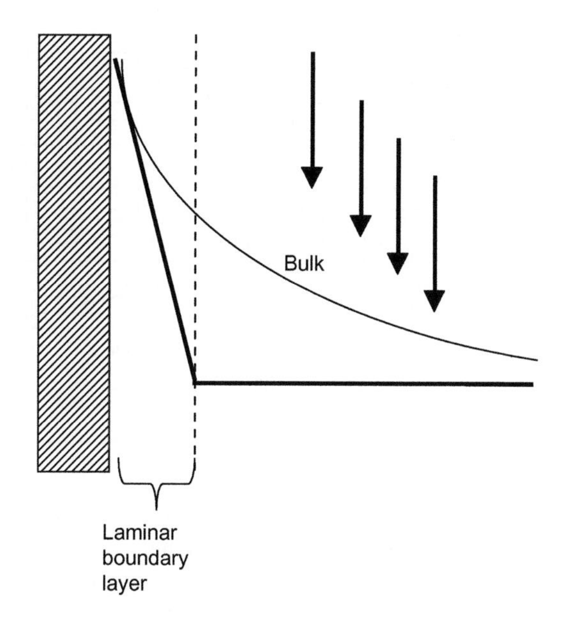

FIGURE 2.6

Boundary layer in turbulent flow.

2.2.4.1 Turbulent Newtonian fluid flow in a cylindrical channel (tube or pipe)

It has been found experimentally that the onset of turbulence in pipes or tubes occurs at a Reynolds number above 2000 but, often, fully turbulent flow does not develop below 4000. The range between 2000 and 4000 is the *transition range* where flow may be laminar or turbulent. The value of 2100 is conventionally set as a criterion for turbulent flow in pipes.

In process engineering, the most common calculation in connection with turbulent flow in pipes is the estimation of the pressure drop due to friction. Dimensional analysis leads to the definition of a dimensionless *friction factor*, *f*, as follows:

$$f = \frac{2\Delta P D}{L\,v^2\,\rho} \tag{2.23}$$

where:

ΔP = pressure drop, Pa
D = diameter of the channel, m
L = length of the channel, m
v = average velocity of the fluid, $m \cdot s^{-1}$
ρ = density of the fluid, $kg \cdot m^{-3}$.

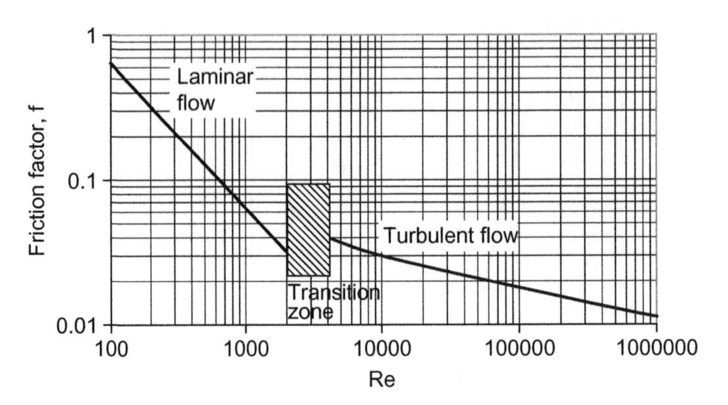

FIGURE 2.7

Friction factor chart for smooth pipes.

The numerical coefficient of 2 is included conventionally in order to form the familiar expression of $v^2/2$, from the definition of kinetic energy.

The friction factor depends on the Reynolds number and on the roughness of the channel walls. For smooth channels, such as stainless steel piping used in the food industry, the friction factor is a function of Re alone. The relationship between f and Re has been determined experimentally, and is usually presented as log–log plots known as *friction-factor charts* (see, for example, Fox and McDonald, 2005; Figure 2.7).

The relationship between f and Re in laminar flow, calculated by comparing Eq. (2.23) with Eq. (2.12) (Hagen–Poiseuille Law), is found to be:

$$f_{la\min ar} = \frac{64}{\text{Re}}$$

The procedure for calculating the pressure drop in a straight pipe of diameter D and length L, carrying a fluid of known viscosity and density at a given volumetric flow rate Q, is as follows:

- From Q and the cross-sectional area of the pipe, calculate the average flow velocity v.
- Calculate Re from D, v, ρ and μ. If $\text{Re} < 2100$, assume laminar flow and calculate the pressure drop with the help of Poiseuille's equation. If $\text{Re} > 2100$, assume turbulent flow and consult a friction-factor chart to find the value of f corresponding to Re.
- Calculate the pressure drop ΔP from Eq. (2.23), rearranged as follows:

$$\Delta P = f \frac{L}{D} \frac{v^2}{2} \rho \tag{2.24}$$

Fittings and valves installed on the pipeline cause additional pressure drop. The usual method of including the contribution of such elements to the total pressure drop is to assign to each type of fitting or valve an empirical *equivalent length of pipe*,

Table 2.1 Equivalent Pipe Length Factors of Fittings and Valves

Fitting	Equivalent L/D
90° Elbow	20–30
45° Elbow	16
Tee, flow through run	20
Tee, flow through branch	60
Globe valve, fully open	340
Gate valve, fully open	8–14
Gate valve, half open	160

expressed as number of pipe diameters. Standard values for some common fittings are given in Table 2.1. These values are simply added to the L/D term in Eq. (2.24).

In engineering terminology, pressure is often expressed as *fluid head*, H_f, and pressure drop as *loss of fluid head*. Fluid head is the height of a column of fluid exerting on its base a pressure equal to the pressure in question. The conversion of pressure to fluid head is shown in Eq. (2.25):

$$H_f = \frac{P}{\rho g} \tag{2.25}$$

EXAMPLE 2.5

Skim milk at 20°C is being pumped through a 25-mm internal diameter smooth pipe, at a volumetric flow rate of $0.001\ m^3 \cdot s^{-1}$. The pipeline is 26 m long, and horizontal. Calculate the pressure drop due to friction.

Data: For skim milk at 20°C, $\rho = 1035\ kg \cdot m^{-3}$, $\mu = 0.002\ Pa \cdot s$.

Solution

We first calculate the Reynolds number to find out if flow is laminar or turbulent. The mean velocity is:

$$v_m = \frac{0.001}{\pi(0.025)^2/4} = 2.04\ m \cdot s^{-1}$$

$$Re = \frac{dv\rho}{\mu} = \frac{0.025 \times 2.04 \times 1035}{0.002} = 26400$$

The flow is turbulent. On the friction-factor chart, for a smooth pipe at Re = 26400 we find $f = 0.024$.

Substituting in Eq. (2.24):

$$\Delta P = f \frac{L}{D} \frac{v^2}{2} \rho = \frac{0.024 \times 26 \times (2.04)^2 \times 1035}{2 \times 0.025} = 53\ 750\ Pa$$

The loss of fluid head, in meters of milk:

$$H_f = \frac{P}{\rho g} = \frac{53750}{1035 \times 9.8} = 5.3\ m$$

2.2.4.2 Turbulent fluid flow around immersed particles

Flow around a sphere becomes turbulent for $Re > 2$. The relationship between the drag coefficient and Re has to be determined empirically. Note that the drag coefficient is exactly similar to the friction factor discussed in Section 2.2.4.1. Empirical charts and approximate expressions are available for the calculation of C_D as a function of Re. For Reynolds numbers between 2 and 500, the following empirical equation is often used:

$$C_D = \frac{18.5}{\mathrm{Re}^{0.6}} \tag{2.26}$$

At higher Re, the drag coefficient becomes nearly constant (independent of Re). A good approximation for this range is $C_D = 0.44$.

2.3 Flow properties of fluids

2.3.1 Types of fluid flow behavior

With respect to flow properties, fluids are classified according to the relationship between shear stress τ and shear rate γ (see Section 2.2.1). Shear stress–shear rate curves for typical classes of fluids are shown in Figure 2.8.

- *Newtonian fluids*: The $\tau - \gamma$ relationship is linear with a zero intercept. The constant slope of the straight line is the viscosity. Gases, low molecular weight liquids and dilute solutions behave like Newtonian fluids.
- *Bingham fluids*: The $\tau - \gamma$ relationship is linear with a non-zero positive intercept called "yield stress", τ_0. The fluid does not flow if the shear stress is less than τ_0. Some paste-like concentrated suspensions behave like Bingham fluids (also called Bingham plastic materials).
- *Shear thinning* or *pseudoplastic fluids*: The $\tau - \gamma$ relationship is non-linear with a zero intercept. The curve is concave, i.e., the viscosity decreases with

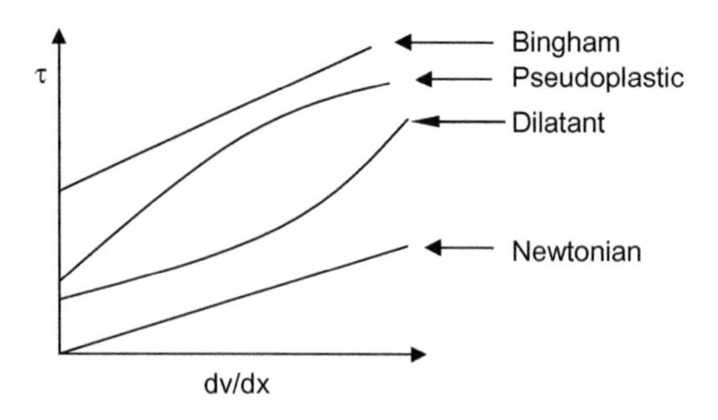

FIGURE 2.8

Types of rheological behavior of fluids.

increasing shear, hence the qualification of shear thinning. Fruit juice concentrates and solutions of proteins or hydrocolloid gums exhibit pseudoplastic behavior.

- *Shear thickening* or *dilatent fluids*: The $\tau - \gamma$ relationship is non-linear with a zero intercept but the curve is convex. The viscosity increases with increasing shear. Some types of honey have been reported to have dilatent behavior (Steffe, 1996).
- *Herschel-Bulkley* fluid: This type is non-linear with a positive yield stress.

The models of flow behavior listed above are *time-independent* because the response of shear stress to a change in shear rate is immediate. It can be said that these models have *no memory*. In contrast, the flow behavior of certain fluids is time-dependent. At constant shear, their viscosity changes with time. If shear is applied to certain fluids, qualified as *thixotropic*, their viscosity decreases gradually with time. If shear is stopped, their viscosity increases gradually. Tomato ketchup and some gels are thixotropic. Fluids that show the opposite behavior are called *rheopectic*. Rheopectic behavior is not common (Steffe, 1996).

Several mathematical models have been proposed for the representation of the types of flow behavior described above. The Herschel–Bulkley model, shown in Eq. (2.27), known as the *power law model*, is most commonly used.

$$\tau = \tau_0 + b(\gamma)^s \tag{2.27}$$

For a Newtonian fluid, $\tau_0 = 0$, $s = 1$
For a Bingham fluid, $\tau_0 > 0$, $s = 1$
For a shear thinning fluid, $\tau_0 = 0$, $0 < s < 1$
For a shear thickening fluid, $\tau_0 = 0$, $s > 1$
For a Herschel–Buckley fluid, $\tau_0 > 0$, $s > 0$.

Another expression, known as the Casson equation, is given in Eq. (2.28):

$$\sqrt{\tau} = \sqrt{\tau_0} + b\sqrt{\gamma} \tag{2.28}$$

Apparent viscosity is τ/γ. In a Newtonian fluid the apparent viscosity is also the true (Newtonian) viscosity, independent of shear rate. For all the types of non-Newtonian fluids, apparent viscosity depends on the shear rate. For shear thinning and shear thickening fluids (so-called *Power Law fluids*), the apparent viscosity is:

$$\mu_{app.} = b(\gamma)^{s-1} \tag{2.29}$$

2.3.2 Non-Newtonian fluid flow in pipes

The relationship between the shear stress and the distance from the wall, in the case of laminar flow in a pipe or tube was developed in Section 2.2.3.1, Eq. (2.6):

$$\tau = \frac{r}{2L}\Delta P \tag{2.6}$$

Assume that the flow behavior of the fluid fits the Herschel–Bulkley model [Eq. (2.27)]

$$\tau = \tau_0 + b(\gamma)^s \tag{2.27}$$

Substitution, followed by integration, gives the velocity profile as follows:

$$v = \frac{2L}{\Delta P b^{1/s}\left(\frac{1}{s}+1\right)}\left[\left(\frac{\Delta PR}{2L}-\tau_0\right)^{1/s+1} - \left(\frac{\Delta Pr}{2L}-\tau_0\right)^{1/s+1}\right] \tag{2.30}$$

The velocity profile is clearly not parabolic as in the case of Newtonian flow ($s = 1$). Furthermore, if there is a positive yield stress, Eq. (2.30) is valid only if the local shear stress exceeds the yield stress. Equation (2.6) indicates that the shear stress is at a maximum at the wall and zero at the central axis. At all points where $\tau < \tau_0$ the velocity gradient is 0 – i.e., the velocity profile is flat. It follows that throughout the region between the axis and a distance at which $\tau = \tau_0$ the fluid flows as a *plug* (Figure 2.9). If this region extends up to the tube wall, the fluid does not flow at all. The condition for non-flow is then:

$$\frac{R}{2L}\Delta P < \tau_0 \tag{2.31}$$

EXAMPLE 2.6

What is the minimum pressure drop needed to make peach puree flow through 20 m of pipe of 4 cm inside diameter? Peach puree is a pseudoplastic fluid with a yield stress of 60dyne · cm^{-2}.

Solution

We apply Eq. (2.31).

The yield stress is converted to Pa:

$$\tau_0 = 60 \text{ dyne}\cdot\text{cm}^{-2} = 60\times10^{-5}\ \text{N}\cdot\text{cm}^{-2} = 60\times10^{-5}\times10^{4}\ \text{N}\cdot\text{m}^{-2} = 6\ \text{Pa}$$

$$\Delta P_{min} = 2L\tau_0/R = 2\times20\times6/0.04 = 6000\ \text{Pa}$$

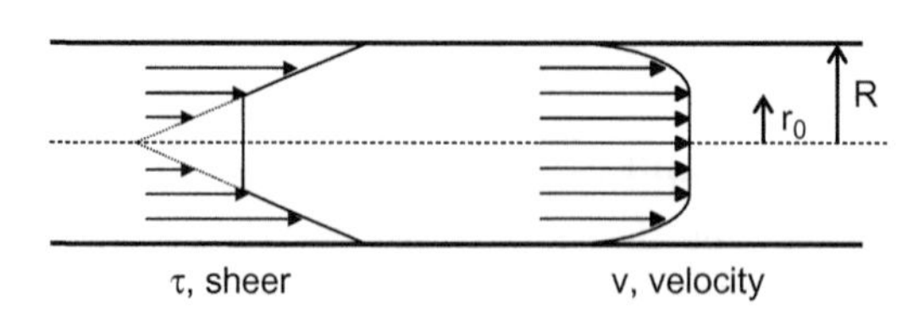

FIGURE 2.9

Non-Newtonian flow with yield stress in tube.

EXAMPLE 2.7

Small cakes are coated with melted chocolate at 50°C. At that temperature, chocolate behaves as a Bingham fluid, with a yield stress of 12 Pa. Its density is 1011 $kg \cdot m^{-2}$. Calculate the thickness of the chocolate coating on the vertical face of the cakes.

Solution

The chocolate will not flow over the vertical face, as long as its weight is less than the yield force. The maximum thickness z of coating is then calculated from:

$$Az\rho g = A\tau_0 \quad \Rightarrow \quad z = \frac{\tau_0}{\rho g} = \frac{12}{1011 \times 9.8} = 0.0012\ \text{m} = 1.2\ \text{mm}$$

2.4 Transportation of fluids

2.4.1 Energy relations: The Bernoulli equation

The *Bernoulli equation* or Bernoulli Law, first enounced in 1738 by the Swiss physicist–mathematician Daniel Bernoulli (1700–1782), is, essentially, the law of the conservation of energy applied to fluid flow.

The First Law of thermodynamics is often stated as follows:

$$Q - W = \Delta E \tag{2.32}$$

where:

Q = heat transferred to the system from outside
W = work transferred from the system to the outside
ΔE = Increase of the total internal energy of the system.

Consider a fluid mass moving from point 1 to point 2 (Figure 2.10).

In the absence of heat exchange (Q = 0), an energy balance can be written as follows:

$$E_2 - E_1 = \sum_1^2 W \tag{2.33}$$

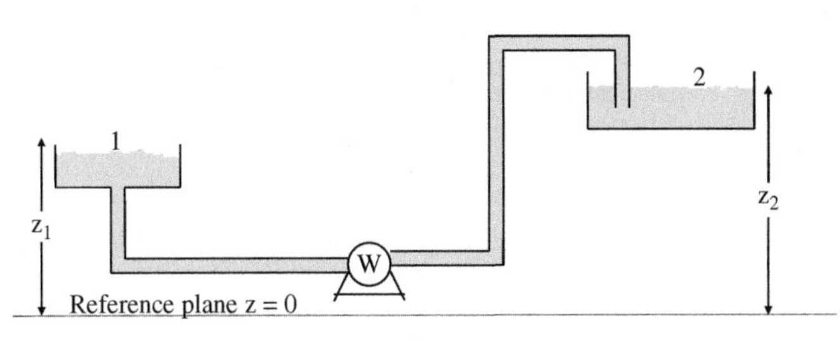

FIGURE 2.10

Simple case of application of Bernoulli's Law.

The total mechanical energy of the fluid consists of its kinetic energy (due to velocity) and potential energy (resulting from its position in the gravity field, i.e., its height relative to a plane of reference). The work transferred along the 1–2 path consists of mechanical work done *on the fluid* by a pump, work done *by the fluid* by expansion by virtue of its pressure, and work done *by the fluid* by moving against friction. It is customary to bring all these energy elements to a common length dimension expressed in meters as "head" (see Eq. (2.25)), which is essentially the energy divided by the weight, mg. In terms of head, the potential energy becomes the height z and kinetic energy $mv^2/2$ becomes $v^2/2g$. In the case of an incompressible fluid (constant volume), the expansion work becomes $V(P_2 - P_1)$. The friction work per unit weight is the "friction head", defined in Eq. (2.25).

Remembering that, conventionally, work done by the system is assigned a negative sign, Eq. (2.33) can be expanded as follows:

$$\left(z_2 + \frac{v_2^2}{2g}\right) - \left(z_1 + \frac{v_1^2}{2g}\right) = -\left(\frac{P_2 - P_1}{\rho g}\right) + \Delta H_{pump} - \Delta H_{friction} \quad \text{or:}$$

$$(z_2 - z_1) + \left(\frac{v_2^2 - v_1^2}{2g}\right) + \left(\frac{P_2 - P_1}{\rho g}\right) + \Delta H_{friction} = \Delta H_{pump} \tag{2.34}$$

where:

z = relative height
v = velocity
$\Delta H_{pump} = W_{pump}/mg$ = Pump head = Pump work per unit weight of fluid
$\Delta H_{friction}$ = Pressure drop due to friction, divided by ρg = Friction head.

Equation (2.34) is just one form of the Bernoulli equation. It is widely used for the calculation of the pump work necessary to transport a fluid from one point of the process to another.

EXAMPLE 2.8

A large vertical vessel of cross-section A_1 contains a non-compressible fluid that is allowed to flow out through a small opening at the bottom of the tank. The cross-sectional area of the opening is A_2. Calculate the time needed to empty the vessel.

Solution

We first find the relationship between the instantaneous velocity of the fluid at the exit and its height in the vessel. Neglecting friction loss and assuming that the velocity in the vessel is negligible compared to the velocity at the outlet (because the ratio A_1/A_2 is very large), Eq. (2.34) is reduced to:

$$z = \frac{v^2}{2g} \quad \Rightarrow \quad v = \sqrt{2gz}$$

Material balance gives:

$$-A_1\frac{dz}{dt} = vA_2 = A_2\sqrt{2gz} \quad \Rightarrow \quad \frac{-dz}{\sqrt{z}} = \frac{A_2}{A_1}\sqrt{2g}\ dt$$

Integration between z = z and z = 0 leads to:

$$2\sqrt{z} = \frac{A_2}{A_1}t\sqrt{2g}$$

Application: Given z = 2 m, $A_2/A_1 = 0.001$, find t:

$$t = \frac{2\sqrt{2}}{0.001\sqrt{2\times 9.8}} = 638.7\ \text{s}$$

EXAMPLE 2.9

Concentrated orange juice has to be pumped continuously from a vacuum evaporator to a storage tank open to the atmosphere (Figure 2.11). Calculate the head to be developed by the pump.

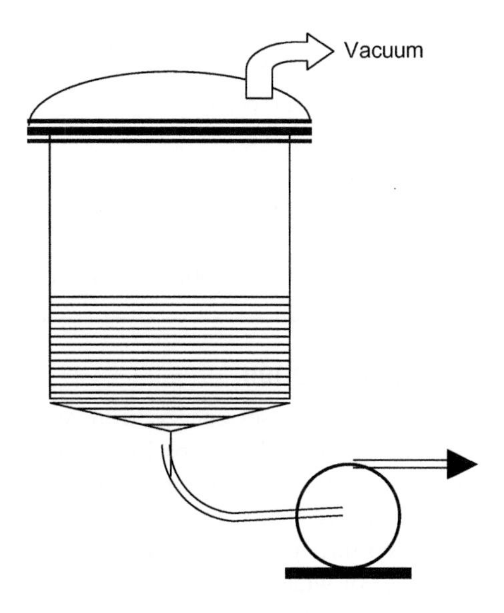

FIGURE 2.11

Pumping product from a vacuum vessel.

Data:

Height of evaporator outlet above the pump	0.6 m
Height of storage tank inlet above the pump	7.8 m
Length of pipe from evaporator to pump	1.6 m
Length of pipe from pump to storage tank	7.8 m
Pipe diameter (uniform)	4 cm
Concentrate mass flow rate	2400 kg · h – 1
Pressure in the evaporator	7500 Pa
Atmospheric pressure	100 000 Pa
Concentrate properties at the temperature of operation:	$\rho = 1150\ kg \cdot m^{-3}$
	$\mu = 0.11\ Pa \cdot s$

Solution

We calculate the volumetric flow rate Q and the flow velocity v:

$$Q = \frac{2400}{1150 \times 3600} = 0.58 \times 10^{-3} m^3 \cdot s^{-1} \quad v = \frac{0.58 \times 10^{-3}}{\pi \times (0.04)^2/4} = 0.46\ m \cdot s^{-1}$$

We test for turbulence:

$$Re = \frac{0.04 \times 0.46 \times 1150}{0.11} = 192$$

The flow is laminar. We calculate the friction pressure drop per unit length of pipe, using the Hagen–Poiseuille equation:

$$\frac{\Delta P_{frict.}}{L} = \frac{8Q\mu}{\pi R^4} = \frac{8 \times 0.58 \times 10^{-3} \times 0.11}{\pi \times (0.02)^4} = 1020.8\ Pa \cdot m^{-1}$$

We write the Bernoulli equation in terms of pressure:

$$(z_2 - z_1)\rho g \; + \; (P_2 - P_1) + \Delta P_{frict.} = \Delta P_{pump}$$

We have neglected the kinetic energy term since the velocity is the same at both ends of the line.

$$\Delta P_{pump} = (3 - 0.6)1150 \times 9.8 + (100\,000 - 7500) + 1020.8(1.6 + 7.8) = 129\,143\ Pa$$

2.4.2 Pumps: Types and operation

A pump is, essentially, a device that uses energy to increase the total head of a fluid. By this general definition, not only pumps for moving liquids but also gas compressors, vacuum pumps, blowers and fans are considered to be pumps. Only mechanical pumps (pumps in which the energy input is in the form of *shaft work*, i.e. mechanical work delivered through a shaft) are discussed in this section. One of the non-mechanical devices of widespread industrial application (the ejector) is considered in a subsequent paragraph.

One of the main engineering characteristics of a pump is the relationship between the volumetric flow rate (capacity) and the total head increment (usually expressed as the increase in pressure). The graphical presentation of this relationship is called a *characteristic curve*. Characteristic curves usually provide information on additional interrelated values such as power, pump efficiency and *net suction pressure* (to be defined later).

One of the important factors of pump performance is the mechanical power required to operate the pump. Theoretically, at a volumetric flow rate Q ($m^3 \cdot s^{-1}$), the net energy input to the fluid as its pressure is increased by ΔP (Pas) is:

$$W_{th} = Q\,\Delta P \tag{2.35}$$

where W_{th} (watt) is the theoretical power requirement.

The actual power requirement W may be considerably higher than the theoretical value. The ratio of the theoretical to the actual value is the *mechanical efficiency* η_m of the pump:

$$\eta_m = \frac{W_{th}}{W} \tag{2.36}$$

Pumps are classified into two major groups: kinetic pumps and positive displacement pumps.

Kinetic pumps

Kinetic pumps impart to the fluid a velocity, hence kinetic energy, which is then converted to pressure according to Bernoulli's Law. The most widespread type of kinetic pump is the *centrifugal pump*. The mode of operation of a centrifugal pump is shown in Figure 2.12. The fluid is admitted at the center of the pump housing. A rapidly rotating *impeller* (rotor) imparts to the fluid a rotational movement at high velocity. The fluid moves in a radial direction away from the center to the periphery under the action of centrifugal forces. The velocity is gradually reduced in the *volute* (the space between the impeller and the pump housing) as the fluid advances from the tip of the impeller towards the pump outlet. Most of the velocity head is converted almost quantitatively to pressure head. The impeller consists of curved vanes. Impellers can be *open*, *semi-open* or *closed* (*shrouded)* (Figure 2.13).

The theoretical performance of a centrifugal pump can be calculated in function of pump geometry and operational conditions, but the actual pump performance parameters may deviate considerably from the theoretical values. The main reasons for deviation are:

- Hydraulic losses due to sudden changes in flow direction, friction between the fluid and the solid surfaces, vortexes, etc.

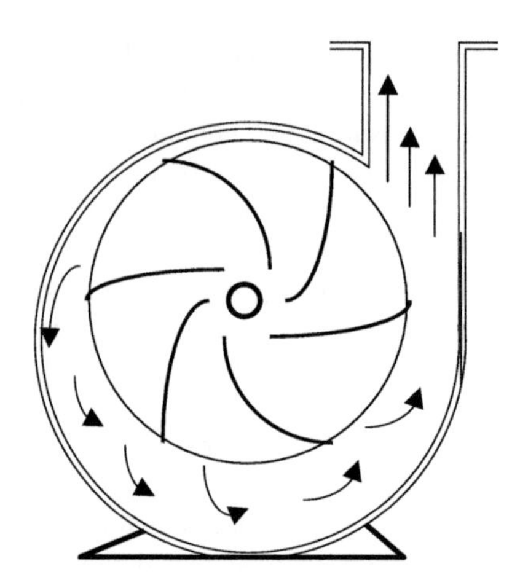

FIGURE 2.12

Principle of operation of a centrifugal pump.

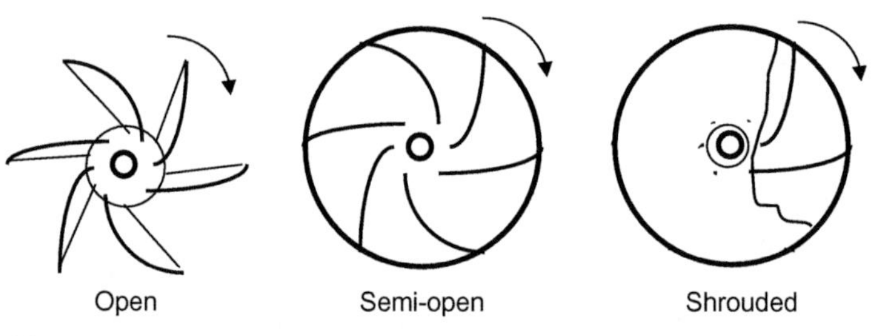

FIGURE 2.13

Centrifugal pump impellers.

- Leakage losses due to back-flow of the fluid from the high-pressure zones back towards the suction, as a result of poor fitting between the impeller and the housing. This kind of loss is less serious in closed impellers.
- Mechanical loss due to friction in bearings, stuffing boxes, etc.
- Recirculation losses due to differences in local velocity in the space between two adjacent vanes.

As a result of these and other losses, the total fluid head at the pump discharge is less and the energy expenditure of the pump is greater than the values predicted by theory. The basic form of a characteristic curve is shown in Figure 2.14. Detailed actual characteristic curves, obtained experimentally, are usually available from pump manufacturers.

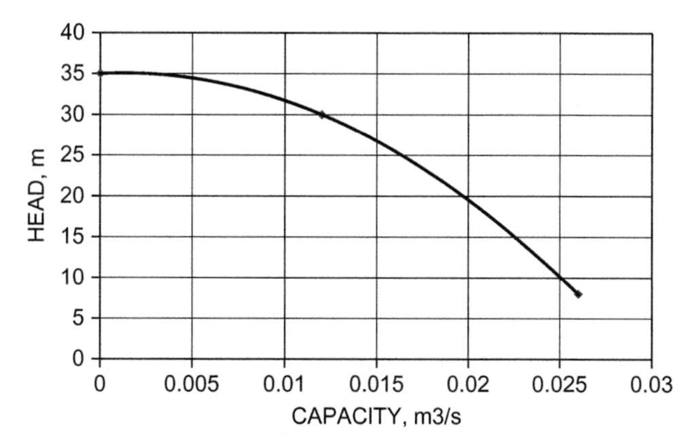

FIGURE 2.14

Centrifugal pump, typical characteristic curve.

Examination of the characteristic curves leads to a number of important practical conclusions:

- The discharge head (pressure) and volumetric flow rate are interrelated. Reducing the volumetric flow rate (e.g., by partially closing a valve at the discharge) while the pump operates at normal speed results in an increase in pressure. If the flow rate is reduced to 0 (e.g., by closing the valve completely), the pressure does not increase indefinitely but reaches a finite, maximum value.
- For a given pump, the flow rate, the discharge head and the power requirement all increase with increasing speed of revolution. An approximate rule-of-thumb, known as the *rule of one–two–three* states that:

$$Q \propto N \quad \Delta P \propto N^2 \quad W \propto N^3 \tag{2.37}$$

where N is the speed of revolution (s^{-1}).

The pressure (or head) increase is, of course, the difference between the pressure at the pump outlet and that at the pump inlet. Unless pressure-fed, the pump must create a sufficiently low pressure (suction head) at the inlet in order to draw liquid from a lower location, or through a resistance to flow or from a vessel at low pressure (e.g., a vacuum vessel, an evaporator operating under vacuum). In the case of centrifugal pumps, suction is created mainly by the movement of the liquid from the center to the periphery. Consequently, no significant suction can be created if the pump is empty of fluid. Centrifugal pumps that require suction at their inlet cannot be started unless first filled with fluid (primed or self-primed). Not all of the hydrostatic head developed by the pump at the inlet is available for suction. The *net positive suction head*, NPSH, is the hydrostatic head at the inlet less the vapor pressure of the liquid. The suction capacity of a pump is therefore considerably lower for liquids at relatively high temperature, and drops to zero if the temperature of the liquid is above its boiling point at the hydrostatic pressure at the inlet.

Positive displacement pumps

In a positive displacement pump, a portion of the fluid is confined in a chamber and mechanically moved (displaced) forward. The main types of positive displacement pumps are reciprocating (piston, diaphragm) pumps, rotary (lobe, gear) pumps, progressing cavity (Moyno) pumps and peristaltic pumps.

The reciprocating (piston) pump consists of a stationary cylindrical chamber containing a reciprocating piston or plunger (Figure 2.15). Two no-return check valves are fixed on the cylinder head. The inlet valve can open only inwards and the outlet valve only outwards. As the piston retreats, the inlet valve opens and draws fluid into the expanding chamber. As the piston advances, the inlet valve closes, the outlet valve opens, and the fluid is pushed

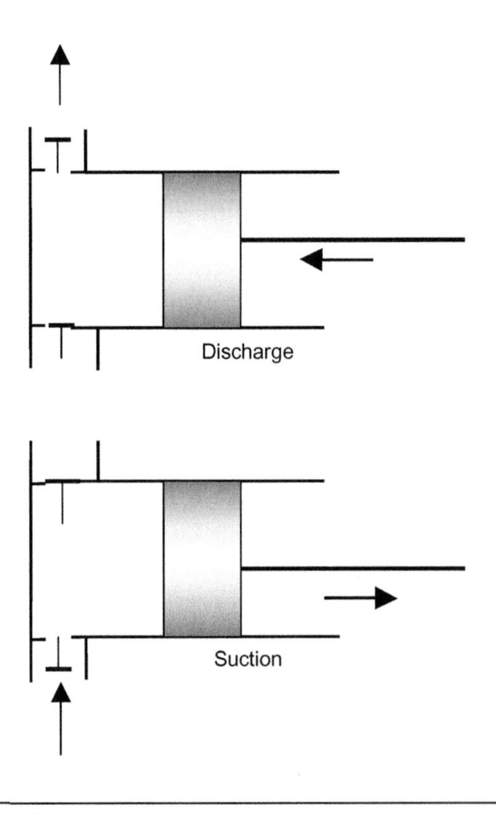

FIGURE 2.15

Operation of reciprocating piston pump.

out against the discharge pressure. For an incompressible fluid, the volumetric capacity is:

$$Q = N\left(\frac{\pi D^2}{4}L\right)\eta_v \tag{2.38}$$

where:

Q = volumetric flow rate (capacity), $m^3 \cdot s^{-1}$
N = pump speed, strokes per second
D = diameter of the cylinder (bore), m
L = length of the piston movement (stroke), m
η_v = volumetric efficiency.

In pump and compressor terminology, D and L are respectively called *bore* and *stroke*. In the case of non-compressible fluids, the volumetric efficiency is high (usually around 0.95) and takes into account the inertia of the valves and the back-flow due to less than perfect fitting between the chamber wall and the piston. In the case of compressible fluids (e.g., air compressors), the volumetric efficiency is lower and decreases with increasing compression ratio. For this reason, compression of gases to very high pressures or creating a very high vacuum can only be achieved by multi-stage compressors or multi-stage vacuum pumps.

Equation (2.38) indicates that the volumetric capacity of a reciprocating piston pump for liquids, operating at constant speed, is nearly constant and independent of the pressure. The volumetric flow rate does not change considerably if the discharge pressure is increased by increasing the resistance to flow after the pump (e.g., closing a valve partially). If the discharge is totally blocked, the pressure rises tremendously (in theory, infinitely) and may cause serious damage to the pump. The characteristic curve of a piston pump (Figure 2.16) is very different from that of a centrifugal pump.

The diaphragm pump (Figure 2.17) is another type of reciprocating positive displacement pump. The reciprocating piston is replaced by a flexible diaphragm made of rubber or another elastomer. The diaphragm seals one side of the chamber and is mechanically flexed to and fro, thus alternately expanding and reducing the free volume of the chamber as in a piston pump. The chamber is equipped with check valves as inlet and outlet ports. The use of a diaphragm eliminates the problem of wear as a result of friction between the piston and the cylinder wall. For the same reason, diaphragm pumps can be used for fluids containing erosive particles. The diaphragms are replaceable and relatively inexpensive.

Rotary pumps are positive displacement pumps whereby a "moving chamber" is formed between a pair of rotating elements (rotors) or between a single rotor and the pump housing. The fluid moves with the chamber from the suction end to the discharge. There are many kinds of rotary pumps. One type (the lobe pump) is shown in Figure 2.18.

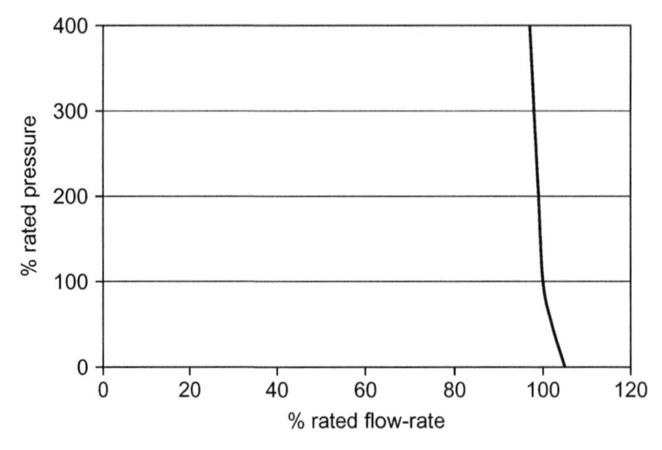

FIGURE 2.16

Characteristic curve of a reciprocating piston pump.

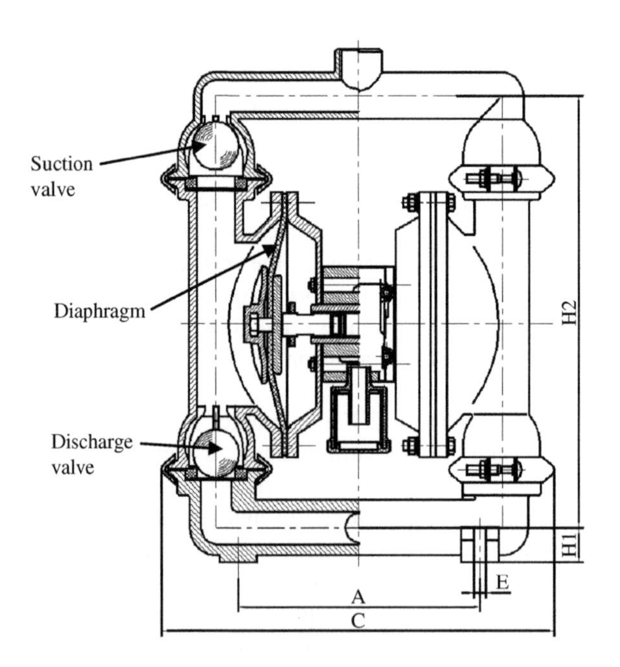

FIGURE 2.17

Air-operated double diaphragm valve.

Progressing cavity pumps, better known by their commercial name of *Moyno pumps* (Figure 2.19), are a special type of rotary positive displacement pumps. They consist of a helical shaft (rotor) turning inside a helical sleeve (stator). A cavity is formed between the sleeve and the rotor, and this cavity advances from

FIGURE 2.18

Lobe pump.

Photograph courtesy of Viking Pumps.

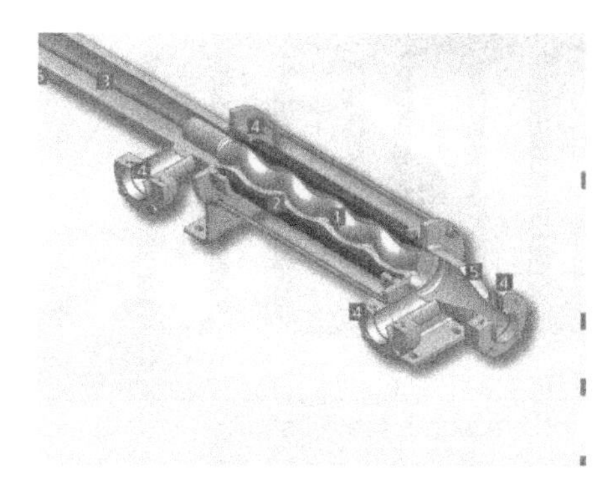

FIGURE 2.19

Progressing cavity pump.

Photograph courtesy of NETZSCH.

the suction end to the discharge as the shaft rotates. The fluid trapped in the cavity is thus transported at constant volume and with almost no shear. This type of pump is therefore especially suitable for high-viscosity fluids and for liquids containing particles. One of the ways to visualize the mode of operation of the progressing cavity pump is to consider the radial cross-section of the stator–rotor assembly (Figure 2.20). The cross-section of the stator is nearly elliptical, while that of the rotor is a circle. As the rotor turns, the circle moves up and down the static ellipse, drawing fluid into the expanding side and pushing fluid forward from the shrinking side. The stator sleeve is usually made of rubber or another suitable polymer, and the rotor is generally made of stainless steel. In a new pump, the stator–rotor fit is tight and friction is reduced by the lubricating effect of the transported fluid itself. When the fitting becomes less tight with usage, the sleeve is replaced by a new one.

Peristaltic pumps are mainly used in low-capacity, low-pressure applications. They consist of a flexible tube, progressively squeezed by external rotating rollers or linear fingers (Figure 2.21). The liquid comes into contact only with the tube material. There are no valves, shafts or bearings. The tube is replaceable and even disposable when needed (e.g., when sterility is to be maintained). Peristaltic pumps are extensively used in the laboratory, in medical applications, and in some industrial operations requiring fluid dispensing at low but precisely controllable flow rates.

2.4.3 Pump selection

From the short and necessarily partial description above, it is obvious that the number of different kinds of pumps is very large (Philby and Stewart, 1984;

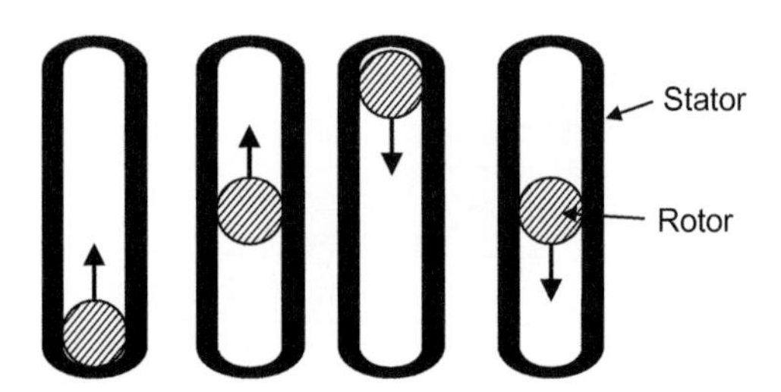

FIGURE 2.20

Principle of operation of the progressing cavity pump.

FIGURE 2.21

Peristaltic pump.

Photograph courtesy of the Department of Biotechnology and Food Engineering, Technion, I.I.T.

Nesbitt, 2006). Pumps are selected in the light of process requirements (volumetric capacity and its controllability, inlet and discharge pressure), process conditions (viscosity, temperature, corrosiveness and erosiveness of the fluid to be pumped), power requirement, cost, and additional conditions dictated by the specific usage. For pumping food materials, compliance with sanitary requirements (ease of cleaning in-place, no stagnant regions, building materials suitable for contact with food, no risk of contamination with lubricants, no leakage through seals, etc.) is essential.

Generally, centrifugal pumps are used for high-volume, low-head applications with liquids of relatively low viscosity. Positive displacement pumps, on the contrary, are more frequently used in high-pressure, low-capacity applications and can handle highly viscous liquids. As a rule, centrifugal pumps are simpler and less expensive. Pumps suitable for use in food processes are available in both categories (Figures 2.22–2.24).

Excessive agitation, high viscosities, impact and shear may not be acceptable in the case of fluids with a structure, materials prone to foaming, and, of course, liquids carrying fragile solid particles. Special pumps are available for such applications. Among the most remarkable examples are the *fish pumps* capable of pumping fairly large live fish in water without causing damage to the animals.

When high discharge pressure is required, the best choice is a reciprocating piston pump. This is the type of pump used, for example, in high-pressure homogenizers for forcing fluids through the homogenizing head at pressures of 10–70 MPas.

Variable (controllable) flow rate is often a process requirement. With centrifugal pumps this is easily achieved by installing a valve, preferably after the discharge. Centrifugal pumps usually operate at constant speed, which is generally the speed of the motor directly coupled with the pump. Positive displacement

FIGURE 2.22

Sanitary centrifugal pump.

Photograph courtesy of Alva-Laval.

FIGURE 2.23

Impellers and rotors for sanitary pumps.

Photograph courtesy of Alva-Laval.

FIGURE 2.24

Sanitary rotating lobe pump.

Photograph courtesy of Alva-Laval.

pumps run at lower speed, and their shaft is connected to the motor through a speed-reducing gearbox. Variable flow rate is achieved by using a variable-speed motor or a drive transmission with speed variability. A device for changing the length of piston travel is found in some reciprocating piston pumps.

2.4.4 Ejectors

Ejectors (also known as jet ejectors) fit the definition of pumps and compressors (see Section 2.4.2) but they have no moving parts. The energy required to increase the pressure head of a fluid is provided by another fluid (motive fluid) moving at high velocity. The principal application of ejectors in the food industry is their use for drawing a vacuum and for thermal vapor recompression in evaporators (Chapter 19). For this application steam is used as the motive fluid, and therefore

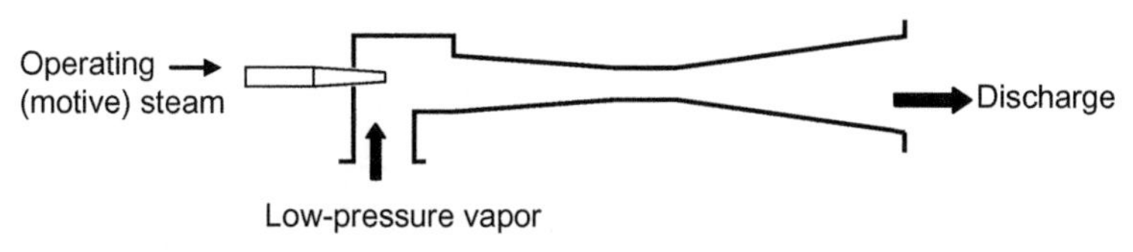

FIGURE 2.25

Steam-jet ejector.

the device is called a steam-jet ejector (Figure 2.25). The moved fluid is low-pressure water vapor from the evaporator. High-pressure steam is introduced through a nozzle, as a high-velocity jet, into a mixing chamber where it meets the low pressure vapor to be compressed. By virtue of its kinetic energy, the motive steam entrains the low-pressure vapor into a converging–diverging conduct known as the diffuser. The mixture expands as it moves through the diverging portion of the diffuser whereby its velocity head (kinetic energy) is converted to pressure head, hence the compressor action of the ejector. At the discharge from the ejector the gases may have a pressure up to 10 times higher than their pressure at the entrance, but such compression ratios require a high consumption of motive fluid per unit mass of compressed fluid and may not be economical. The most economical compression ratio depends on the cost of high-pressure steam. If higher compression ratios (e.g., a higher vacuum) are required, two or three ejectors may be installed in series. A two-stage ejector is shown in Figure 2.26.

Ejectors are remarkably simple yet extremely useful devices. They require almost no maintenance for long periods. Steam-jet ejectors for vacuum duty or vapor recompression represent only one area of ejector application. The familiar water jet "pump" found in chemistry laboratories is, in essence, an ejector that uses water as the motive fluid. Small ejectors are used for various purposes in food service and dispensing systems. In some systems, the ejector principle is used for moving waste water and sewage.

2.4.5 Piping

The pipelines of a food plant fall into two categories:

1. Pipelines for the transportation of utility fluids such as cold and hot water (except water to be used in product formulation), steam, compressed air, liquid or gas fuels, etc.
2. Pipelines for the transportation of products, semi-products and raw materials used in product formulation, including water.

The pipes and fittings of utility pipelines are usually made of steel. Their dimensions (inside and outside diameter, wall thickness) and modes of connection are specified by official standards and listed in detailed tables. Pipelines for steam and hot water are thermally insulated. For esthetic and sanitary reasons, the utility pipelines are laid out following mutually perpendicular planes.

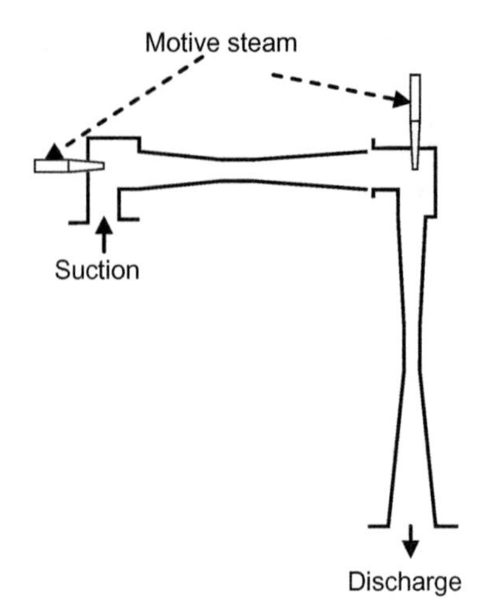

FIGURE 2.26

Two-stage steam jet ejector.

Product pipelines consist of tubes and fittings made of stainless steel. The design diameters of the tubes are calculated according to the flow rate, pressure drop and fluid properties, but tubing with a diameter below about 20 mm is seldom used for foods. The connections between tubes and fittings may be dismantleable or permanent (welded). The former can be taken apart for cleaning and inspection and provide the flexibility required for changing layouts. A considerable number of different and mutually incompatible systems of stainless steel piping connections are available. This lack of standardization must be taken into consideration in designing a dismantleable product pipeline and in specifying the type of connections between the piping and equipment items.

Permanent product pipelines are less expensive and particularly suitable for plants with a fixed layout, such as large dairy plants. With the development of reliable clean-in-place (CIP) systems (see Chapter 25), the use of welded stainless steel pipelines is becoming more widespread.

2.5 Flow of particulate solids (powder flow)

2.5.1 Introduction

The flow behavior of powders and granulated solids differs radically from that of liquids and gases (Jaeger *et al.*, 1996). The fundamental difference between the two kinds of flow stems from the fact that the flow properties of particulate solids depend on their state of compression and sometimes on their history of compression.

Note that in this section, the term *powder flow* is used for *flow of particulate solids*, for briefness. Most of the discussion, however, is applicable also to particles outside the powder range, such as crystals and granules.

Powder flow is important in a number of areas of food engineering and technology, including:

- Charging and discharging of storage bins, hoppers and vehicles for bulk transport
- Reliable operation of feeders
- Flow of powders in weighing, portioning and packaging equipment
- Dosing powder products in automatic dispensing and vending equipment
- Powder mixing
- Compaction (caking) of powder products in storage
- Pneumatic conveying
- Fluidization.

2.5.2 Flow properties of particulate solids

Powder flow is the displacement of powder particles in relation to each other, under the effect of some directional force (gravity, entrainment in a fluid stream, mechanical force of an auger, scraper, vibrator, agitator, mixer, etc., and, occasionally, electrostatic forces). The resistance of the powder to such displacement is called "powder strength". Powder strength is a consequence of inter-particle forces. The main types of adhesive forces between particles (or between particles and solid surfaces) are:

- Van der Waals forces
- Liquid bridge (capillary) forces
- Electrostatic forces.

In an absolutely dry powder and under conditions found in common food applications, van der Waals forces are the main cause of particle–particle adhesion. However, even in an apparently dry powder a thin film of liquid may often exist on the surface of particles. In this case, liquid bridge forces become at least as important as van der Waals forces.

Van der Waals forces (named after Johannes Diderik van der Waals, 1837–1923, Dutch physicist, 1910 Nobel Prize in Physics) are universal forces of attraction between material bodies. They are dependent on the distance between the bodies, and are significant only when that distance is very small (say <100 nm). Their origin is electromagnetic.

The general expression for van der Waals forces between two spheres of equal size has the following form:

$$F_{vdW} = \frac{Cd}{x^2} \tag{2.39}$$

where d is the particle diameter, x is the inter-particle distance and C is a constant.

Liquid bridge forces (Figure 2.27) are related to capillary forces. For two equal spheres of diameter d, bound by a liquid bridge, the adhesion force is approximately:

$$F_{lb} \approx \pi \gamma d \tag{2.40}$$

where γ is the surface tension of the liquid.

Analysis of powder flow in terms of the forces acting between individual particles is difficult. An alternative approach to powder flow considers the powder bulk to be a continuous solid, and applies the concepts and methods of solid mechanics to powder flow. At a "macro" scale, inter-particle forces are treated as friction, since friction is the force that opposes movement of two touching bodies with respect to each other.

Consider a solid body resting on a horizontal plane (Figure 2.28). A constant force N (which includes the weight of the solid) acts in the direction normal to the plane and a gradually increasing force T is applied in the direction parallel to the plane. The body starts to move when T overcomes the static friction force F. According to Coulomb's Law of friction (Charles Augustin Coulomb, French physicist, 1736–1806), that value of T (and therefore F) is proportional to the normal force N:

$$F = kN \tag{2.41}$$

where k is the coefficient of friction.

FIGURE 2.27

Liquid bridge binding together powder particles.

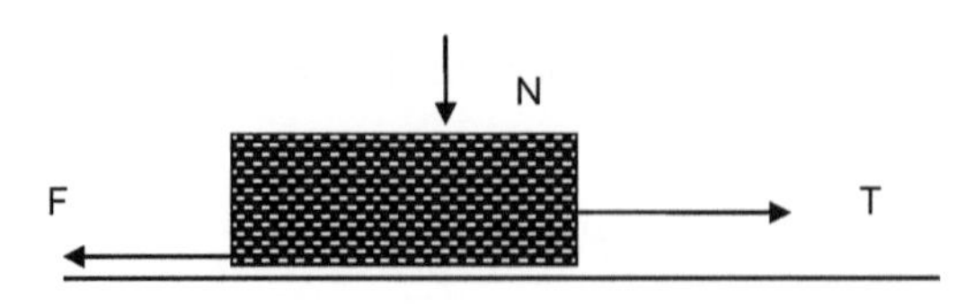

FIGURE 2.28

Friction on a horizontal plane.

Now consider a solid body of weight G resting on a slope (Figure 2.29). Assume that the angle of the slope α is increased gradually until the body just starts to slide down. At this moment, the force T parallel to the repose plane is equal to the friction F. The force T can be written as $T = G \sin \alpha$. Applying Coulomb's Law, F can be written as: $F = kN$ or $F = k\, G \cos \alpha$. The coefficient of friction is then related to the base angle α:

$$k = \tan \alpha$$

The use of an angle as an expression of friction is common in powder flow studies.

In solid mechanics, it is often more convenient to deal with stresses (i.e., force per unit area) rather than forces. Referring to Figures 2.28 or 2.29, with A as the contact area, we define:

- Normal (compression) stress: $\sigma = N/A$
- Tangential (shear) stress: $\tau = T/A$.

Then, Coulomb's Law of friction can be written as follows:

$$\tau = k\sigma \tag{2.42}$$

The curve describing the relationship between τ and σ, in an ideal Coulomb system, is then a straight line, with a slope k, passing through the origin (Figure 2.30). The slope of the curve is equal to the coefficient of friction.

As explained above, one way of dealing with powder flow is to describe flow as a result of shear, and the resistance of a powder to flow as friction. When applied to powder flow, the plot of τ versus σ is known as the *yield locus* of the powder. Experimental determination of the yield locus is one of the approaches for the definition of the flow properties of powders. The application of this approach is illustrated below for the case of the *Jenike Shear Cell.*

The Jenike Shear Cell, developed by A.W. Jenike of the University of Utah (Jenike, 1964) consists of a shallow cylindrical box composed of a base and a ring (Figure 2.31). The base and ring assembly is filled with the powder sample to be tested. After a standard procedure of conditioning and consolidation of the sample, the powder is covered with a lid and a known normal force N (usually a weight) is

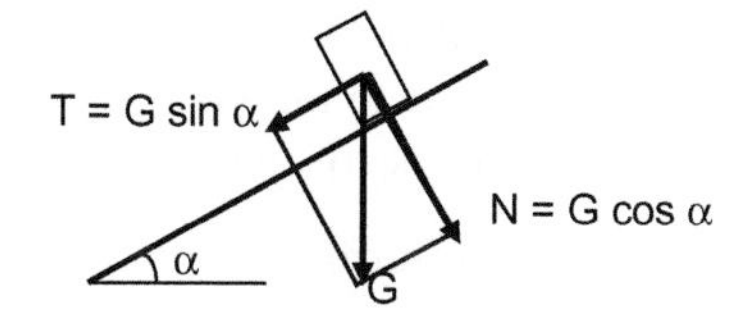

FIGURE 2.29

Friction on an inclined plane.

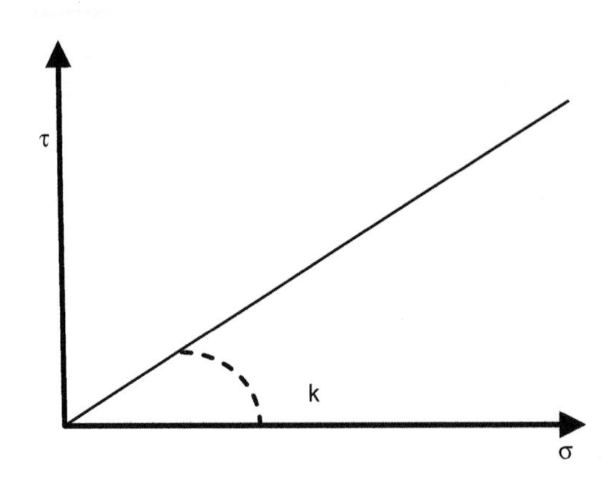

FIGURE 2.30

A "yield locus".

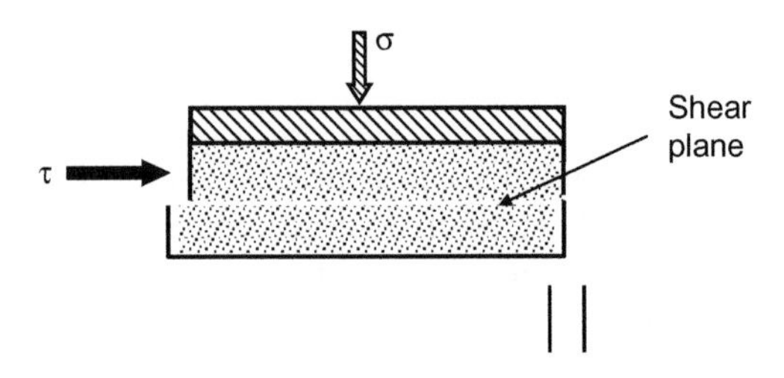

FIGURE 2.31

The Jenike Shear Cell.

applied. Now the ring is pushed horizontally and the horizontal steady-state force T required to move the ring (i.e., to shear the powder along the plane separating the base from the lid) is recorded. This gives one point on the τ versus σ plot. The yield locus is constructed by repeating the test with different values of the normal force.

Two hypothetical yield loci are shown in Figure 2.32:

1. Curve A (straight line through the origin, Coulomb's Law) is characteristic of free-flowing, non-cohesive powders such as dry crystals of salt or sugar.
2. Curve B is typical of cohesive powders. Most food powders exhibit this kind of yield loci. The intercept τ_0 (similar to the yield stress of non-Newtonian fluids) is a measure of the cohesiveness of the powder.

The yield locus furnishes useful information on the flow properties of the powder:

- It delimits the flow and no-flow regions of the w−s (or $\tau-\sigma$) space
- It gives the intrinsic cohesiveness
- The slope of the locus gives the "angle of internal friction" of the powder.

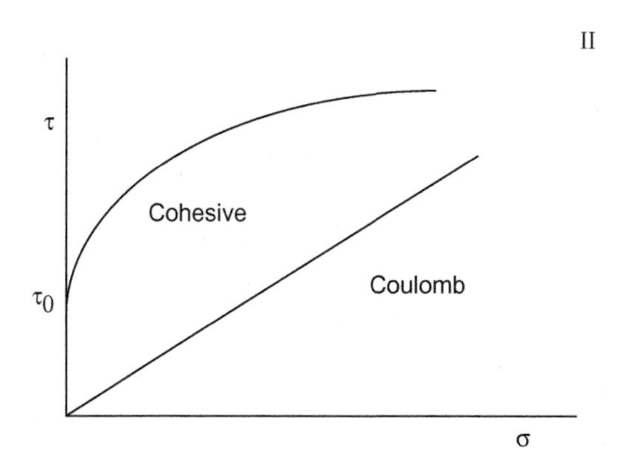

FIGURE 2.32

Yield loci of two types of powders.

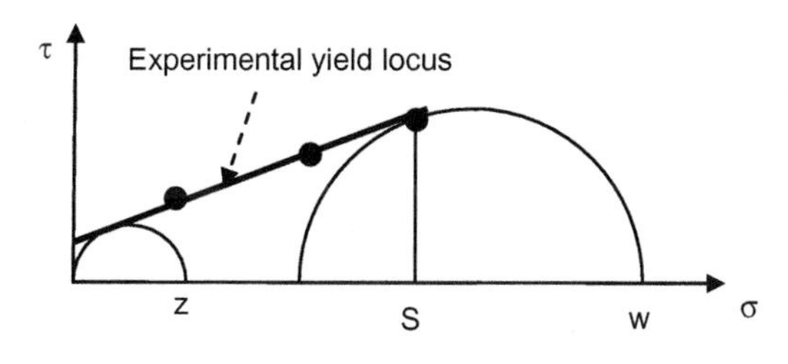

FIGURE 2.33

Analysis of yield tests using Mohr diagrams.

The test results can be analyzed further (Barbosa-Cánovas *et al.*, 2005), following Jenike's graphical method (Figure 2.33), to determine a number of additional parameters that are used in the design of storage bins as well as for the classification of powders according to their flowability:

- On the graph showing the yield locus, a half-circle with the center on the σ axis, passing through the consolidation point (w,S depending on the conditions of consolidation of the sample before the shearing tests) and tangential to the yield locus is constructed. This is the *large Mohr circle.* (Note: A circle on the $\sigma-\tau$ plane with its center on the σ axis is known as a Mohr circle. It is a graphical representation of the state of stress, extensively used in solid mechanics). The value w at the farthest point of the large circle on the σ axis (consolidation stress) is recorded.
- A second Mohr circle, passing through the origin and tangential to the yield locus is drawn. This is the *small Mohr circle.* The value z at the farthest point of the small circle on the σ axis is recorded. z is the *unconfined yield strength.*
- Jenike's *flow function*, ffc is defined as the ratio w/z.

The Flow Function is frequently used to classify and compare powders with regard to their flowability (Svarovsky, 1987). A higher value of ffc indicates better flowability. Jenike (1964) suggested the following classification:

ffc = 1–2	Poor flowability (cohesive)
ffc = 2–4	Moderate flowability
ffc = 4–10	Good flowability
ffc > 10	Excellent flowability (free flowing)

In addition to tests based on the shear cell principle, other experimental methods are available for the evaluation of the flow properties of powders:

- *Funnels*. The time required for a given volume of powder to flow out of a standard funnel is used as an indication of flowability.
- *Angle of repose*. When a bulk powder falls freely on a plane, it forms a conical heap (Figure 2.34). The base angle of the heap is "the angle of repose" of the powder. In analogy to the solid on a slope discussed before, the angle of repose is related to the friction between particles. It is therefore a useful parameter related to powder flowability (Tenou *et al.*, 1995). A small angle of repose indicates good flowability.
- *Mechanical methods*. An impeller, similar to a paddle mixer, is rotated inside a mass of bulk powder. The torque developed, or the power necessary to maintain the rotation at constant speed, is a measure of the resistance of the powder to flow.
- *Compressibility/compactability*. The bulk density of a powder depends on the degree of compaction of the bed. Consequently, depending on the mode of measurement, two values are obtained: free (loose, aerated) bulk density and tapped (consolidated) bulk density. The ratio of tapped to loose bulk density is known as the Hausner ratio. It has been suggested that the Hausner ratio may provide an indication as to the flowability of a powder (Geldart *et al.*, 1984; Harnby *et al.*, 1987).

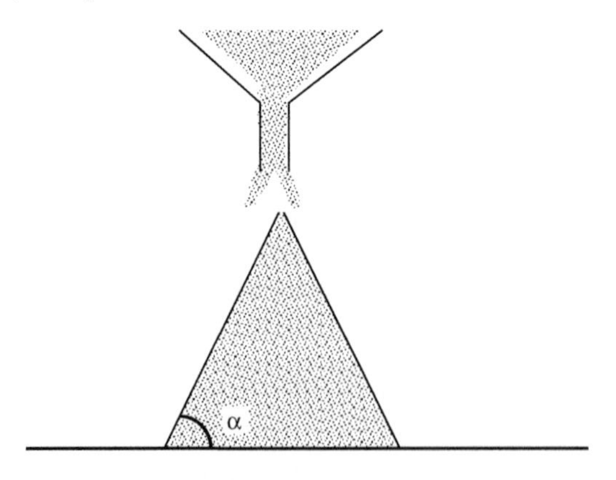

FIGURE 2.34

Angle of repose.

2.5.3 Fluidization

Consider a bed of solid particles resting on the perforated bottom of a tall container (Figure 2.35). Assume that a fluid is forced upward through the bed. As long as the fluid flow rate is low, the particles do not move and the volume of the bed does not change (fixed bed). As the flow rate is increased, the pressure drop through the bed also rises (Figure 2.36). When the flow rate reaches the point where the pressure drop through the bed exceeds the weight of the solid bed, the bed starts to expand. This point is called the "fluidization point". If the flow rate is increased further, the bed continues to expand. The particles of the bed begin to move freely and at random. The pressure drop remains nearly constant, despite the increasing flow rate of the fluid, because of the increased porosity of the bed (Figure 2.36).

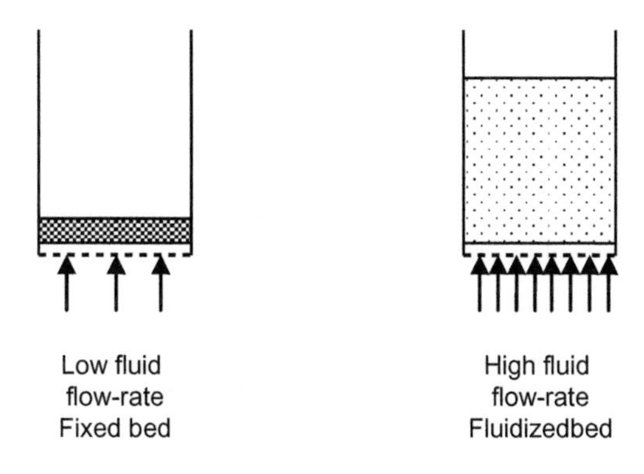

FIGURE 2.35

Fluidization of a powder bed.

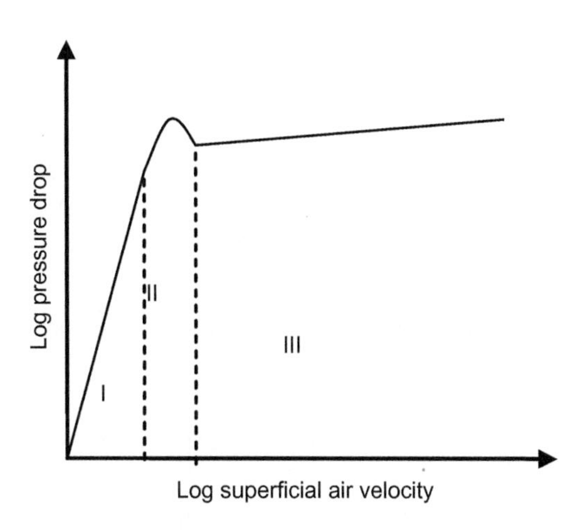

FIGURE 2.36

Pressure drop through a powder bed: I, fixed bed; II, intermediate region; III, fluidized bed.

At this stage, the bed is said to be *fluidized*. Indeed the fluid–solid mixture in a fluidized bed behaves in many respects as a liquid. It can be poured from one container to another, it has a well-defined level, and it follows the principle of communicating vessels.

Fluidized bed combustion of solid fuel is used in power plants. In the food industry, processes based on *fluidized beds* are finding increasing use. By virtue of the large fluid–solid contact area and the vigorous agitation created by the random movement of the particles, the fluidized bed is an ideal system for rapid heat and mass transfer. In addition, through fluidization it is possible to treat particulate solids in continuous processes, like liquids. Another advantage of the fluidized bed is uniformity of conditions throughout the material. A fluidized bed is practically isothermal.

Some of the applications of the fluidized bed in food process technology include:

- Fluidized bed drying
- Individual quick freezing (IQF)
- Agglomeration of particulate solids
- Coating of particulate solids
- Rapid heating or cooling of particulate solids.

The minimum fluid velocity required for fluidization, v_{mf}, can be calculated by equating the pressure drop with the apparent weight of the bed. Various expressions have been proposed for pressure drop as a result of flow through a porous bed. One of the most commonly cited is the Ergun equation (Sabri Ergun, Turkish chemical engineer, 1918–2006). Ergun defines a Reynold's number for a bed of spherical particles of equal size as follows:

$$\mathrm{Re}_p = \frac{d_p v \rho}{\mu(1-\varepsilon)} \tag{2.43}$$

where:

d_p = particle diameter, m
v = superficial velocity, $m \cdot s^{-1}$ = the fluid velocity that would exist for the same volumetric flow rate if the vessel were empty ($v = Q/A$)
ρ, μ = density and viscosity of the fluid, $kg \cdot m^{-3}$ and $Pa \cdot s$ respectively
ε = porosity (void fraction) of the bed.

Ergun's equation (Ergun and Orning, 1952) for the pressure drop per unit bed height is:

$$\frac{\Delta P}{h} = v\frac{1-\varepsilon}{\varepsilon^3}\left[150\frac{\mu}{d_p^2}(1-\varepsilon)\right] + 1.75\frac{\rho v}{d_p} \tag{2.44}$$

The first term of Ergun's equation represents the pressure drop due to friction between the fluid and the particles. The second term is the pressure drop due to sudden changes in direction. For $Re_p < 10$, the second term is negligible and Ergun's equation is reduced to its first term:

$$\frac{\Delta P}{h} = v\frac{1-\varepsilon}{\varepsilon^3}\left[150\frac{\mu}{d_p^2}(1-\varepsilon)\right] \tag{2.45}$$

Eq. (2.45) is identical to the Kozeny–Carman equation discussed in Chapter 8. The apparent weight of the solid, per unit bed height is:

$$\frac{G}{h} = (1-\varepsilon)(\rho_p - \rho)g \tag{2.46}$$

where ρ_p stands for the density of the particles.

Equating Eq. (2.44) with Eq. (2.46) and setting $v = v_{mf}$, the following quadratic expression for the minimum velocity for fluidization is obtained:

$$\frac{1.75}{\varepsilon^3}\left(\frac{d_p v_{mf}\rho}{\mu}\right)^2 + \frac{150(1-\varepsilon)}{\varepsilon^3}\left(\frac{d_p v_{mf}\rho}{\mu}\right) = \frac{d_p^3\rho(\rho_p-\rho)g}{\mu 2} \tag{2.47}$$

In most food applications of fluidization, the particles are small and light and the condition of $Re_p < 10$ is largely met. In this case, the Kozeny–Carman approximation is used. The minimum velocity for fluidization is then calculated with the help of Eq. (2.48):

$$v_{mf} \approx d_p^2\frac{(\rho_p-\rho)}{150\mu}\cdot\frac{\varepsilon^3}{1-\varepsilon}\cdot g \tag{2.48}$$

EXAMPLE 2.10

For the design of a fluidized bed freezer, it is desired to calculate the minimum fluidization velocity of peas in air at −40 °C. Suppose the peas are uniform spheres.

Data:

Diameter of the peas	5 mm
Particle density of peas	$1080\ kg\cdot m^{-3}$
Porosity of the pea bed at rest	0.42
For air at −40°C	$\rho = 1.4\ kg\cdot m^{-3}$
	$\mu = 16\times10^{-6}\ Pa\cdot s$

Solution

The approximate value of the minimal velocity for fluidization will be calculated from Eq. 2.48:

$$v_{mf} \approx d_p^2\frac{(\rho_p-\rho)}{150\mu}\cdot\frac{\varepsilon^3}{1-\varepsilon}\cdot g \approx (0.005)^2\frac{(1080-1.4)}{150\times16\times10^{-6}}\cdot\frac{(0.42)^3}{(1-0.42)}\cdot 9.8 = 14\ m\cdot s^{-1}$$

Note: For equal spheres in random arrangement the porosity ε is about 0.40–0.45.

As long as the fluid velocity is not much higher than the minimum velocity for fluidization, the expanded bed is uniform and stable. This stage is called *particulate fluidization*. Process wise, this is the preferred state of the fluidized bed. If the velocity is increased, the uniformity of the bed is disrupted. The fluid passes through the solid particles as large bubbles and the bed takes the aspect of a boiling liquid. This is the state of *aggregative fluidization*. Finally, if the velocity is further increased, the bed disintegrates and the particles are carried away with the fluid, out of the vessel (Figure 2.37). This occurs when the velocity of the fluid exceeds the settling velocity of the particles, v_s, which can be calculated with the help of Eq. (2.22). Pneumatic transport (see Section 2.5.4) makes use of fluid velocities above the settling velocity.

The ratio between the minimum fluidization velocity and the settling velocity is indicative of the velocity range in which a fluidized bed can be operated. For the low Reynolds condition explained above, this ratio is given by Eq. (2.49):

$$\frac{v_s}{v_m} = \frac{25(1-\varepsilon)}{3\varepsilon^3} \tag{2.49}$$

For a porosity value of 0.45, for example, Eq. (2.49) gives a ratio of 50. In practice, fluidized beds are operated at superficial velocities up to 30–35 times higher than the minimum velocity of fluidization.

In most of the applications of fluidized bed in food processes (e.g., drying, freezing, agglomeration, coating) the fluidizing fluid is air. In rare cases, such as ion exchange in fluidized columns, the solids are fluidized in a liquid stream.

2.5.4 Pneumatic transport

Particulate solids in bulk can be transported in a stream of gas (pneumatic conveying) or in a flowing liquid (hydraulic conveying). Both methods are used in

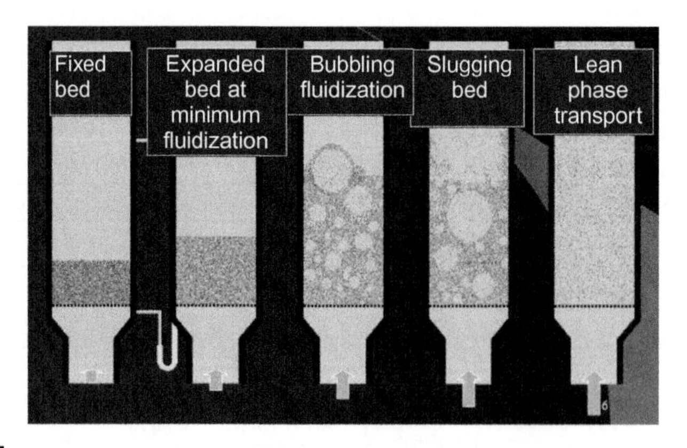

FIGURE 2.37

Stages of fluidization.

Figure reproduced courtesy of GEA-NIRO.

the food industry. Grain, flour and other mill products, sugar, etc. are often transported pneumatically. Hydraulic conveying is common in vegetable processing (peas, tomatoes). Pumping fish with sea water from submerged nets up to the fishing boat may be considered hydraulic transport. The basic principles of the two processes are the same, but the physical set-ups and equipment used are obviously different. Only pneumatic conveying will be discussed in this section.

The advantages of pneumatic transport, over other types of conveying are:

- The transport takes place in a closed system, protecting the product against contamination and protecting the environment against pollution.
- The product is conveyed in a pipeline that can be easily disassembled and reassembled and can even contain flexible sections as hoses. Pneumatic conveyors are particularly suitable for serving changing points of pick-up and delivery.
- Solids can be transported over long distances, through complicated pathways.
- The pneumatic conveyor is compact.

The disadvantages are:

- Energy consumption per unit weight of solids transported is considerably higher than in mechanical conveyors.
- Due to the intimate contact with air, drying and oxidation of the product may occur.

Systems for pneumatic conveying in pipes are classified into two categories: pressure systems and vacuum systems. In the pressure system, the air blower is installed upstream and blows air into the system. The pipeline is under positive pressure. In the vacuum system, the blower is installed downstream and sucks air through the system. The pipeline is under negative pressure.

Pressure (or positive pressure) conveyors are suitable for systems with a single pick-up point and multiple delivery points (Figure 2.38). Pressure systems can be used with higher solid-to-gas ratios. Introducing the solid into a system at positive

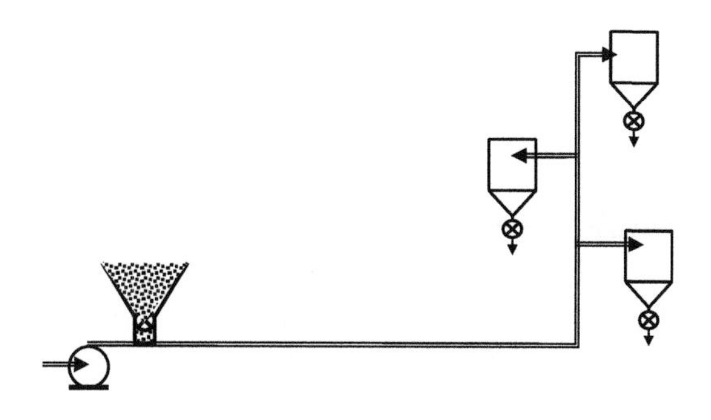

FIGURE 2.38

Pneumatic transport: positive pressure system.

pressure requires special devices such as locks. Escape of dust through leakages in the system may be a problem.

Vacuum systems are suitable for multiple pick-up–single delivery set-ups (Figure 2.39). Because of the limited pressure difference available for transport, vacuum systems cannot convey solids for long distances.

The pressure drop resulting from flow of a mixture of gas and solid particles cannot be calculated by the methods used for air alone. In addition to the usual variables (pipe diameter, velocity, gas viscosity, gas density), parameters of the solid (particle size and shape, solid to gas ratio, and the gravity acceleration g) must be included in the dimensional analysis. In one of the methods of calculation described by McCabe and Smith (1956), a ratio β is defined as follows:

$$\beta = \frac{(\Delta P/L)_s}{(\Delta P/L)_g} \tag{2.50}$$

where $(\Delta p/L)_s$ and $(\Delta p/L)_g$ are the pressure drop per unit length for the suspension and for the gas alone, flowing at the same velocity. An empirical equation for the calculation of β is proposed, as follows:

$$\beta - 1 = a\left(\frac{D}{d_p}\right)^2 \left(\frac{\rho\, r}{\rho_p \mathrm{Re}}\right)^K \tag{2.51}$$

where:

D = diameter of the pipe
r = mass ratio of solids to fluid
a, K = parameters of the equation, functions of the other variables of the equation and of the direction of flow (vertical or horizontal).

The most important design variable in pneumatic transport is the velocity of the air. Insufficient air velocity may result in settling of particles in horizontal sections, and in elbows, tees, and sudden expansion. Usual air velocities are in the range of 20–30 $\mathrm{m \cdot s^{-1}}$.

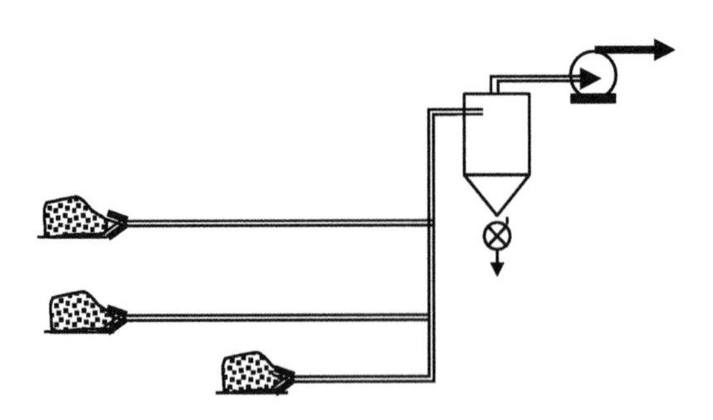

FIGURE 2.39

Pneumatic transport: vacuum system.

2.5.5 Flow of powders in storage bins

The food industry makes extensive use of vertical bins or silos for the storage of powders in bulk. Irregular flow of powders out of storage bins and hoppers is a major cause of loss in production capacity and product deterioration. Three types of powder flow patterns in bins are known: mass flow, funnel flow and expanded flow (Barbosa-Cánovas *et al.*, 2005).

Mass flow (Figure 2.40) is observed in bins with smooth surfaces and a long, steep converging bottom, and with free-flowing powders. In this type of flow, every part of the powder is in movement while the bin is discharged. It is said that the entire bin is "live". The first portion of powder entering the bin is also the first to be discharged. The bin can be emptied completely.

Funnel flow (Figure 2.40) occurs with cohesive powders, in bins with rough surfaces, a short converging bottom and a discharge outlet that is not sufficiently large. The powder flows out through a central vertical channel. The powder outside of the channel is stagnant, but may collapse into the channel and cause irregular flow. A portion of the powder is permanently stagnant and remains in the bin at the end of the discharge. If the powder is susceptible to chemical or microbial spoilage, funnel flow may result in product quality loss. Funnel flow is also more favorable to segregation by particle size, resulting in heterogeneity in the discharged material. The major advantages of bins for funnel flow are the lower cost of production and lower height requirement (Barbosa-Cánovas *et al.*, 2005).

Expanded flow is achieved in hoppers with a converging bottom consisting of a less steep upper part followed by a steeper lower part. Thus, funnel flow occurs in the upper part and mass flow above the discharge opening.

Irregularity of powder discharge from bins and silos is a major problem in industries handling bulk solids. Two of the most common causes for such irregularities are *ratholes* and *arching* (Figure 2.41). Both are due to compression of the powder as a result of flow in the bin, and both are more frequent with fine and cohesive powders. The phenomenon of ratholes refers to the flow of the powder

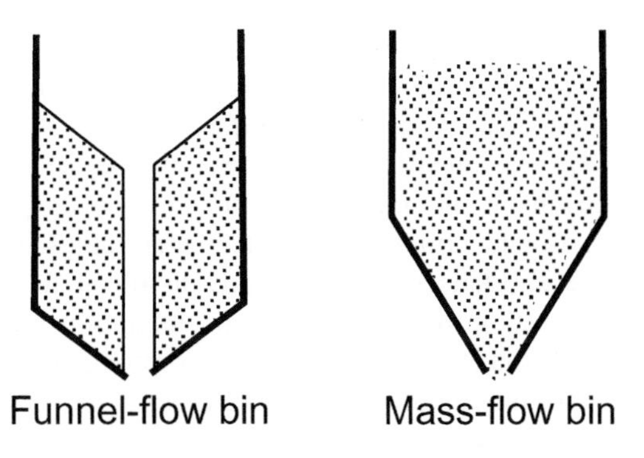

FIGURE 2.40

Mass-flow and funnel-flow bins.

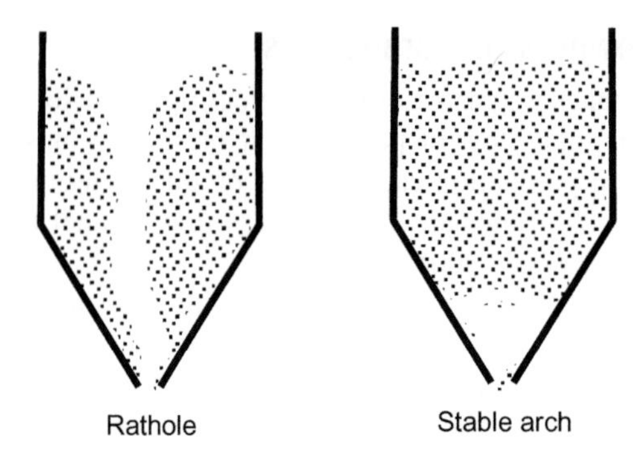

FIGURE 2.41

Ratholing and arching in powder bins.

through channels (ratholes), while the mass outside the channels remains stationary. Occurrence of ratholes causes a decrease in the rate of discharge and irregularities in flow due to the cycles of formation and collapse of ratholes. Arching refers to the tendency of cohesive powders to form a stable arch of compressed material above the discharge outlet The strength of an arch may be sufficient to support the weight of the powder remaining in the hopper, in which case the only way to restore flow is to destroy the arch with the help of vibrators or knocks on the hopper wall. The tendency of powders to form stable arches depends on their moisture content and state of compression (Guan and Zhang, 2009).

2.5.6 Caking

Caking is a change of state whereby a free-flowing powder is transformed to large solid chunks upon storage under certain conditions. With the exception of intentional agglomeration of powders (see Chapter 22, Section 22.7), caking represents a serious loss of quality in particulate materials. The relative humidity and the temperature of storage, in addition to the chemical composition, particle size and particle shape of the powder, are the main factors connected with caking.

Most frequently, caking starts with the adsorption of moisture by the powder. The adsorbed moisture condenses on the surface of the powder particles and forms the *liquid bridge* explained in Section 2.5.2. Caking often occurs also as a result of internal movement of moisture due to temperature gradients (Billings and Paterson, 2008). Liquid bridging in itself creates lumps, but if the powder contains water-soluble substances and these are dissolved in the liquid bridge, subsequent loss of moisture by evaporation or migration converts the interparticle bond to even stronger *solid bridges*. Adsorption of moisture may also cause caking without forming liquid bridges (Aguilera *et al.*, 1995). If the storage temperature is sufficiently high the powder particles may undergo phase transition to the rubbery state, at which cohesion of the particles occurs due to the stickiness of the rubbery surfaces (see Chapter 1, Section 1.8).

Prevention of caking in granular materials requires avoiding exposure to high relative humidity and high temperature. Furthermore, additives called *anti-caking agents* are sometimes added to powders, to facilitate flow. Water insoluble, non-hygroscopic substances, such as silica (silicon dioxide), magnesium carbonate, calcium carbonate, dicalcium phosphate, sodium aluminosilicate and calcium stearate are anti-caking agents added to some food powders, at the level of 1–2% or less. Anti-caking agents are very fine powders that act, primarily, by adhering to the surface of the food particles, thus preventing inter-particle contact.

References

Aguilera, J.M., del Valle, J.M., Karel, M., 1995. Caking phenomena in amorphous food powders. Trends. Food Sci. Techn. 6 (5), 149–155.

Barbosa-Cánovas, G.V., Ortega-Rivas, E., Juliano, P., Yan, H., 2005. Food Powders. Kluwer Academic/Plenum Publishers, New York, NY.

Billings, S.W., Paterson, A.H.J., 2008. Prediction of the onset of caking in sucrose from temperature induced moisture movement. J. Food Eng. 88 (4), 466–473.

Ergun, S., Orning, A.A., 1952. Fluid flow through packed columns. Chem. Eng. Prog. 48, 89–94.

Flauzino, R.D., Gut, J.A.W., Tadini, C.C., Telis-Romero, J., 2010. Flow properties and tube friction factor of milk cream: influence of temperature and fat content. J. Food Proc. Eng. 33 (5), 820–836.

Fox, R.W., McDonald, A.T., 2005. Introduction to Fluid Mechanics. Wiley, New York, NY.

Geldart, D., Harnby, N., Wong, A.C.Y., 1984. Fluidization of cohesive powders. Powder Technol. 37, 25–37.

Guan, W., Zhang, Q., 2009. The effect of moisture content and compaction on the strength and arch formation of wheat flour in a model bin. J. Food Eng. 94 (2–3), 227–232.

Harnby, N., Hawkins, A.E., Vandame, D., 1987. The use of bulk density determination as a means of typifying the flow characteristics of loosely compacted powders under conditions of variable relative humidity. Chem. Eng. Sci. 42, 879–888.

Jaeger, H.M., Nagel, S.R., Behringer, R.P., 1996. The physics of granular materials. Rev. Mod. Phys. 68, 1259–1273.

Jenike, A.W. ,1964. Storage and flow of solids. Bulletin No. 123. Utah Engineering Experiment Station, Salt Lake City, UT.

McCabe, W., Smith, J.C., 1956. Unit Operations of Chemical Engineering. McGraw-Hill, New York, NY.

Nesbitt, B., 2006. Handbook of Pumps and Pumping. Elsevier Science, London, UK.

Philby, T., Stewart, H.L., 1984. Pumps. MacMillan Publishing Co., London, UK.

Steffe, J.F., 1996. Rheological Methods in Food Process Engineering, second ed. Freeman Press, East Lansing, MI.

Svarovsky, L., 1987. Powder Testing Guide. Methods of Measuring the Physical Properties of Bulk Powders. British Materials Handling Board, UK.

Tenou, E., Vasseur, J., Krawczyk, M., 1995. Measurement and interpretation of bulk solids angle of repose for industrial process design. Powder Handl. Process. 7 (3), 2003–2227.

CHAPTER

Heat and Mass Transfer: Basic Principles

3

3.1 Introduction

Most food processes, such as cooking, pasteurization, sterilization, drying, evaporation, distillation, chilling, freezing, etc., involve some sort of heat transfer. Important industrial processes such as membrane separation, drying, salting, candying, humidification, adsorption, extraction, etc., involve exchange of materials between different parts (phases) of the system, often combined with heating or cooling. Transport of moisture and oxygen through the package often determines the shelf life of the product within. Our chemical senses (taste, odor) can function only if molecules of the taste and odor components can be transported to the sensing bodies. Most importantly, life itself depends on the exchange of material through biological membranes.

Heat transfer and mass transfer are based on essentially similar physical principles. Both processes obey laws that are, in principle, identical. Therefore, the basic principles of heat and mass transfer will be described together while their applications will be treated separately.

3.2 Basic relations in transport phenomena

3.2.1 Basic laws of transport

All transport phenomena (fluid flow, heat and mass transfer, electric current, etc.) are the result of lack of equilibrium between parts of the system. In principle, they all obey a universal law, similar to the familiar *Ohm's law*, which can be expressed, in general terms, as follows: *The rate of transport (i.e., the quantity transported per unit time) is proportional to the driving force and inversely proportional to the resistance of the medium to the transport.* Applying the law to the particular case of heat transfer, one can write:

$$q = \frac{dQ}{dt} = \frac{F}{R} = kF \quad (k = 1/R) \tag{3.1}$$

Food Process Engineering and Technology. DOI: http://dx.doi.org/10.1016/B978-0-12-415923-5.00003-4
© 2013 Elsevier Inc. All rights reserved.

where:

$q = dQ/dt$ = rate of heat transfer (the quantity of heat transported per unit time)
F = driving force
R = resistance of the medium to heat transfer
k = compliance or conductance of the medium to heat transfer.

In other cases of transport, the quantity of whatever is being transported (mass, electric charge, momentum...) would replace Q, in appropriate units (kilograms, coulombs, Newton-meters, etc.).

The driving force F is always a gradient, representing the deviation from equilibrium: a temperature gradient in the case of heat transfer; a gradient of concentration in the case of mass transfer; a gradient of pressure in the case of fluid flow; voltage (which is a gradient of field intensity) in the case of electric current, etc. By definition, at equilibrium $F = 0$.

It is sometimes convenient to decompose the complex notion of resistance into its physical components. Thus, as the rate of transport can logically be expected to be proportional to the area A available to the transport, it is often customary to deal with the *rate of transport per unit area*. This dimension is known as *flux* and denoted by the symbol J.

3.2.2 Mechanisms of heat and mass transfer

Heat transfer occurs via three fundamental mechanisms: *conduction, convection* and *radiation* (McAdams, 1954).

Conduction refers to the transfer of heat through a stationary medium. The mass transfer equivalent of conduction (conductive mass transfer) is *molecular diffusion* through a stationary medium.

Convection occurs when heat travels along with a moving fluid. In mass transfer, convection (*convective mass transfer*) refers to a situation whereby molecular diffusion occurs simultaneously with bulk flow.

Radiation is the transfer of heat in the form of electromagnetic energy. Unlike the former two mechanisms, radiative heat transfer does not require the presence of a material medium between the two points.

In practice, more than one mechanism may be involved in a transfer process.

3.3 Conductive heat and mass transfer

3.3.1 The Fourier and Fick laws

In media with no considerable internal mobility (e.g., solids), heat travels by conduction and mass travels by molecular diffusion. These transfers are governed by Fourier's and Fick's Laws, respectively.

For heat transfer:

$$\frac{dQ}{Adt} = -k\frac{\partial T}{\partial z} \quad \text{(Fourier First Law)} \tag{3.2}$$

(after Jean-Baptiste Joseph Fourier, French mathematician–physicist, 1768–1830)

For mass transfer:

$$\frac{dm_B}{Adt} = J_B = -D_B\frac{\partial C_B}{\partial z} \quad \text{(Fick First Law)} \tag{3.3}$$

(after Adolf Eugen Fick, German physiologist, 1829–1901) where:

Q = heat transferred, J
T = temperature, K
t = time, s
k = thermal conductivity of the medium, $J \cdot s^{-1} \cdot m^{-2} \cdot K^{-1} \cdot m = W \cdot m^{-1} \cdot K^{-1}$
z = distance in the direction of the transport, m
m = mass of substance B transferred, mol
C_B = concentration of substance B, $mol \cdot m^{-3}$
D_B = diffusivity (coefficient of diffusion) of molecular species B through the medium, $m^2 \cdot s^{-1}$.

The minus sign before the gradients serves to indicate that heat flows in the direction of decreasing temperatures, and mass travels in the direction of decreasing concentration.

3.3.2 Integration of Fourier's and Fick's laws for steady state conductive transport

At steady state, all the properties that define the "state" of the system (temperature, pressure, chemical composition, etc.) remain constant with time. They may vary with location within the system. Keeping in mind that, at steady state, the temperature and the concentration depend only on the location (z), Eqs (3.2) and (3.3) can be rewritten as ordinary differential equations:

$$\frac{dQ}{Adt} = \frac{q}{A} = -k\frac{dT}{dz} \tag{3.4}$$

$$\frac{dm_B}{Adt} = J_B = -D_B\frac{dC_B}{dz} \tag{3.5}$$

The boundary conditions for the integration of Eqs (3.4) and (3.5) are:

$T = T_1$ and $C = C_1$ at $z = z_1$
$T = T_2$ and $C = C_2$ at $z = z_2$.

Assuming that the thermal conductivity k does not change considerably with the temperature, and that the diffusivity is independent of concentration, integration gives:

$$q = \frac{Q}{t} = kA\frac{T_2 - T_1}{z} \tag{3.6}$$

$$\dot{m}_B = \frac{m_B}{t} = D_B A\frac{C_{2B} - C_{1B}}{z} \tag{3.7}$$

Unsteady state heat and mass transfer will be discussed later in this chapter.

EXAMPLE 3.1

Calculate the rate of heat transfer through a 3×4 m concrete wall. One face of the 0.2-m thick wall is at 22°C and the other face is at 35°C. The thermal conductivity of the concrete is $1.1\ \mathrm{W \cdot m^{-1} \cdot K^{-1}}$.

Solution

Steady state is assumed and Eq. (3.6) is applied:

$$q = kA\frac{T_2 - T_1}{z} = 1.1 \times (3 \times 4)\frac{35 - 22}{0.2} = 858\ \mathrm{W}$$

EXAMPLE 3.2

The diffusivity of the vapor of a volatile aroma in air is measured by a simple procedure known as the Winkelman method. The volatile liquid is placed in a vessel connected to a tube, through which a stream of air is passed (Figure 3.1). The air flow-rate is sufficiently large to carry away the vapors completely from the zone of connection between the vessel and the tube. The distance z from the connection to the level of the fluid is measured as a function of time.

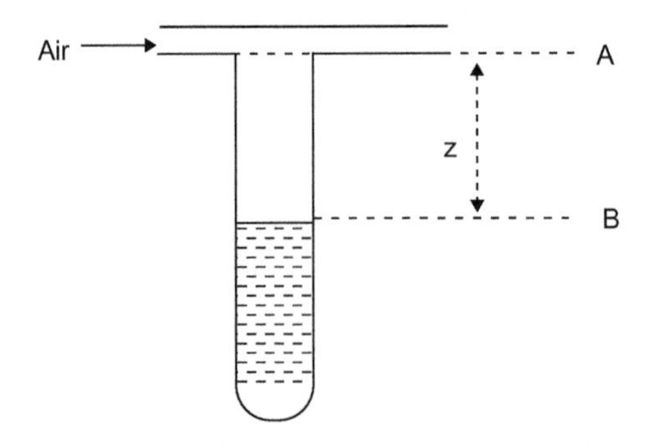

FIGURE 3.1

Schema of the system for measuring the diffusivity of vapors by the Winkelman method.

In a test with a volatile aroma at 25°C, the following results were obtained:

Time (h)	z (mm)
0	30
6	95

Data:

Molecular weight of the aroma	110
Density of the aroma liquid	940 $kg \cdot m^{-3}$
Vapor pressure of the aroma at 25°C	34 mmHg
Atmospheric pressure	100 kPa

What is the diffusivity of the aroma vapor in air?

Solution

The method is based on the assumption that the concentration of the vapor in air is zero at plan A and at saturation at plan B. Both are constant with time, therefore the steady-state equation is applicable, albeit the system is not truly in steady state (z changes with time).

$$\frac{dm}{dt} = DA\frac{C_B - C_A}{z} = A\rho\frac{dz}{dt}$$

But $C_A = 0$, therefore:

$$\frac{D \cdot C_B}{\rho} dt = z \cdot dz$$

Integration gives:

$$\frac{D \cdot C_B}{\rho} \cdot t = \frac{z^2}{2} \Rightarrow D = \frac{\rho \cdot z^2}{2 \cdot C_B \cdot t} = \frac{940 \times (0.095 - 0.030)^2}{2 \times C_B \times 6 \times 3600} = \frac{91}{C_B} \times 10^{-9}$$

The saturation concentration in air ($kg \cdot m^{-3}$) is easily calculated from the vapor pressure data.

3.3.3 Thermal conductivity, thermal diffusivity and molecular diffusivity

3.3.3.1 Thermal conductivity and thermal diffusivity

Thermal conductivity is a property of the material. It varies with temperature, and is strongly pressure-dependent in the case of gases. In the SI system its units are $W \cdot m^{-1} \cdot K^{-1}$. Within a narrow range of temperatures, the temperature dependence of the thermal conductivity is approximated by the linear equation $k = k_0(1 + aT)$, provided that no phase change (melting of fats, gelation, etc.) occurs.

Table 3.1 Thermal Conductivity and Thermal Diffusivity of some Materials (Approximate Representative Values)

Material	T (°C)	k ($W \cdot m^{-1} \cdot K^{-1}$)	α($m^2 \cdot s^{-1}$)
Air	20	0.026	21×10^{-6}
Air	100	0.031	33×10^{-6}
Water	20	0.599	0.14×10^{-6}
Water	100	0.684	0.17×10^{-6}
Ice	0	2.22	1.1×10^{-6}
Milk	20	0.56	0.14×10^{-6}
Edible oil	20	0.18	0.09×10^{-6}
Apple	20	0.5	0.14×10^{-6}
Meat (lamb leg)	20	0.45	0.14×10^{-6}
Stainless steel	20	17	4×10^{-6}
Glass	20	0.75	0.65×10^{-6}
Copper	20	370	100×10^{-6}
Concrete	20	1.2	0.65×10^{-6}

Thermal diffusivity α is a useful concept in heat transfer analysis. It is defined as *the ratio of thermal conductivity to the "volumetric heat capacity" of the material.* Volumetric heat capacity is obtained by multiplying the mass specific heat C_p by the density ρ.

$$\alpha = \frac{k}{\rho C_p} \tag{3.8}$$

Physically, thermal diffusivity can be interpreted as the ratio of the material's ability to transfer heat to its capacity to store heat. The SI units for thermal diffusivity are $m^2 \cdot s^{-1}$.

A novel instrumental method for the simultaneous determination of thermal conductivity and thermal diffusivity is available (Huang and Liu, 2009).

Approximate representative values of the thermal conductivity and thermal diffusivity of some materials are given in Table 3.1. One of the methods proposed for the evaluation of the thermal conductivity of a food material from its composition is illustrated in Example 3.3.

EXAMPLE 3.3

Based on experimental data, Sweat (1986), (cited by Singh and Heldman, 2003), suggested the following equation for the calculation of the thermal conductivity of foods:

$$k = 0.25X_c + 0.155X_p + 0.16X_f + 0.135X_a + 0.58X_w$$

where X represents mass fraction, c = carbohydrate, p = protein, f = fat, a = ash, w = water.

Calculate the thermal conductivity of meat loaf containing 21% protein, 12% fat, 10% carbohydrates, 1.5% ash, 55.5% water and no air bubbles.

Solution

$$k = (0.25 \times 0.1) + (0.155 \times 0.21) + (0.16 \times 0.12) + (0.135 \times 0.015) + (0.58 \times 0.555)$$
$$= 0.401 \; \mathrm{W \cdot m^{-1} \cdot K^{-1}}$$

3.3.3.2 Molecular (mass) diffusivity, diffusion coefficient

The diffusion coefficient D in Fick's law depends on the diffusing molecular species, the medium in which it is diffusing, and the temperature. In the SI system, it is expressed in units of $m^2 \cdot s^{-1}$, just as thermal diffusivity.

Diffusivity in gases can be predicted quite accurately, with the help of the kinetic theory of gases. The diffusivity of gases in binary mixtures, at room temperature and atmospheric pressure, is in the order of 10^{-5} to 10^{-4} $m^2 \cdot s^{-1}$.

Several models have been proposed for the prediction of diffusivity in liquids. One of the best known is the Einstein–Stokes equation for the Brownian diffusion of a solute. According to this model, the diffusivity D of a solute molecule (assumed spherical), with a molecular radius r, in a liquid of viscosity μ is given by:

$$D = \frac{\kappa T}{6\pi r \mu} \tag{3.9}$$

where κ is the Boltzmann's constant (1.38×10^{-23} $J \cdot K^{-1}$) and r is the radius of the particle. T is the absolute temperature. The assumption of sphericallity is, in many cases, a serious deviation from reality. While its use for the quantitative prediction of diffusivity is problematic, the model is instructive regarding, for example, the effect of viscosity and molecule size on the diffusivity.

The diffusivity of solutes in water, at room temperature, ranges from 10^{-9} for small molecules to 10^{-11} $m^2 \cdot s^{-1}$ for large ones (e.g., proteins). Diffusivities of some substances in air and water are given in Table 3.2. Diffusivities of additional substances are given in Tables A.6 and A.7 (see Appendix).

Diffusion in true solids is extremely slow. In crystals and metals molecular transport occurs mainly through defects (holes) in the crystal lattice, via a "single-jump" process. The diffusion coefficients for small ions in solid glasses may be as low as 10^{-25} $m^2 \cdot s^{-1}$. In porous solids, the bulk of mass transfer occurs through the gas or liquid filling the pores and not through the solid matrix. The *effective diffusivity*, D_{eff}, through a porous solid is related to the diffusivity D through the medium in the pores as follows:

$$D_{eff} = D\frac{\varepsilon}{\tau} \tag{3.10}$$

where ε (dimensionless) is the porosity (i.e., the void volume fraction of the porous solid) and τ (dimensionless) is the "tortuosity factor", a consequence of the tortuous path of the diffusing molecule through the porosity of the solid.

Table 3.2 Diffusivity of Some Substances at 25°C

Substance	D in Air ($m^2 \cdot s^{-1}$)	D in Water (Dilute Solutions) ($m^2 \cdot s^{-1}$)
Water	25×10^{-6}	–
CO_2	16×10^{-6}	1.98×10^{-9}
Ethanol	12×10^{-6}	1.98×10^{-9}
Acetic acid	13×10^{-6}	1.98×10^{-9}
H Cl		2.64×10^{-9}
Na Cl		1.55×10^{-9}
Glucose		0.67×10^{-9}
Sucrose		0.52×10^{-9}
Soluble starch		0.1×10^{-9}

Data from Bimbenet et al. *(2002).*

EXAMPLE 3.4

The diffusivity of ethanol vapor in air at room temperature is $12 \times 10^{-6}\ m^2 \cdot s^{-1}$. Estimate the diffusivity of ethanol vapor through a bed of inert powder having a void volume fraction of 45%. A tortuosity factor of 2.2 is assumed.

Solution

Eq. (3.10) is applied:

$$D_{eff} = D\frac{\varepsilon}{\tau} = \frac{12 \times 10^{-6} \times 0.45}{2.2} = 2.45 \times 10^{-6}\ m^2 \cdot s^{-1}$$

3.3.4 Examples of steady-state conductive heat and mass transfer processes

3.3.4.1 Steady-state conduction through a single slab

Consider a slab of homogeneous material between two parallel planes, a distance z apart (Figure 3.2). Let T_1 and T_2 be the temperature at the two planes. Both temperatures are assumed constant with time.

According to Eq. (3.6) the rate of heat transfer through the slab is given by:

$$Q/t = q = kA(T_1 - T_2)/z$$

At steady state, the heat flux q/A is the same at any plane perpendicular to the direction of heat flow. It follows that the temperature distribution across the slab is linear.

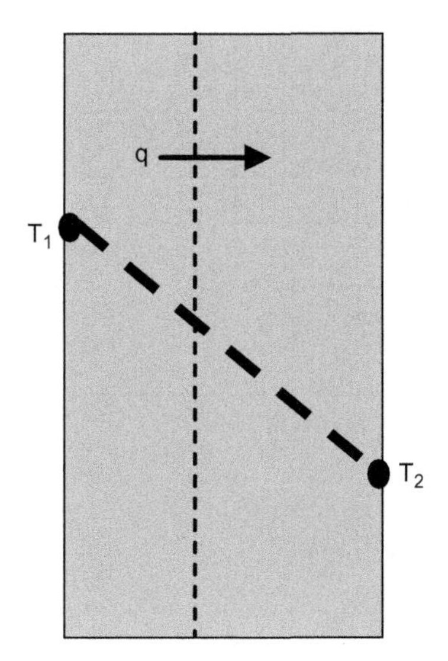

FIGURE 3.2

Steady-state conduction heat transfer in a slab.

3.3.4.2 Steady-state conduction through a multi-layer slab; total resistance of resistances in series

Consider a composite slab (Figure 3.3) formed by several layers of material, each with a different thermal conductivity. Let their thicknesses be $z_1, z_2, z_3 \ldots z_n$ and their thermal conductivities be $k_1, k_2, k_3 \ldots k_n$.

At steady state, the heat flux through any plane perpendicular to the direction of heat transfer must be equal:

$$q/A = \frac{k_1(T_1 - T_2)}{z_1} = \frac{k_2(T_2 - T_3)}{z_2} = \frac{k_3(T_3 - T_4)}{z_3} = \cdots = \frac{k_n(T_n - T_f)}{z_n} \quad (3.11)$$

What would be the thermal conductivity K of a homogeneous slab of thickness Z, equal to that of the multi-layer slab that would produce the same heat flux when submitted to the same overall temperature difference $T_1 - T_f$?

We can write heat transfer equations for each slab, as follows:

$$\frac{qz_1}{Ak_1} = T_1 - T_2 = (\Delta T)_1$$

$$\frac{qz_2}{Ak_2} = T_2 - T_3 = (\Delta T)_2 \quad \ldots.$$

$$\frac{qz_n}{Ak_n} = T_n - T_f = (\Delta T)_n$$

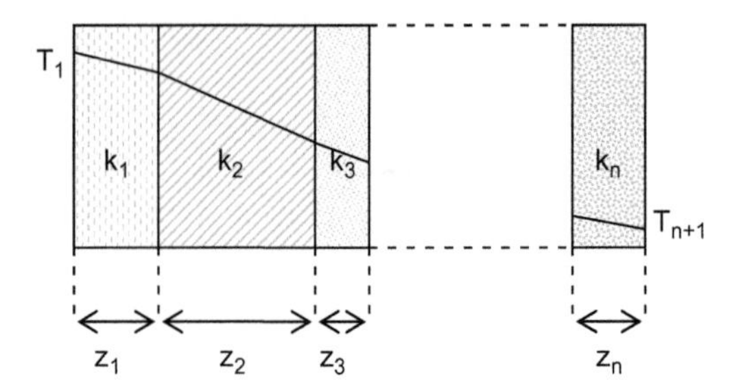

FIGURE 3.3

Conduction in multi-layer slab.

Summing the equations we obtain:

$$\frac{q}{A}\sum_{1}^{n}\frac{z_i}{k_i} = \sum \Delta T = T_1 - T_f \tag{3.12}$$

For the equivalent homogeneous slab we can write:

$$\frac{q}{A}\frac{Z}{K} = T_1 - T_2$$

It follows that:

$$\frac{Z}{K} = \frac{z_1}{k_1} + \frac{z_2}{k_2} + \ldots + \frac{z_n}{k_n} = \sum_{1}^{n}\frac{z_i}{k_i} \tag{3.13}$$

We define the ratio z/k as thermal resistance. Equation (3.13) then means that the total thermal resistance of a composite conductor consisting of several conductors in series is equal to the sum of the individual thermal resistances. The same conclusion is valid for mass transfer or, for that matter, for any kind of transfer (*cf.* electricity). In the same manner, resistance to mass transfer is defined as z/D.

EXAMPLE 3.5

The external wall of a cold storage room is made of three layers, as follows, from the inside out:

1. Stainless steel sheet, 2 mm thick ($k = 15\ W \cdot m^{-1} \cdot K^{-1}$)
2. Thermal insulation, 80 mm thick ($k = 0.03\ W \cdot m^{-1} \cdot K^{-1}$)
3. Concrete, 150 mm thick ($k = 1.3\ W \cdot m^{-1} \cdot K^{-1}$).

The inside face is at −18°C, the outside face is at 20°C.

Calculate the heat flux through the wall and the temperature at the insulator–concrete interface.

Solution

The total thermal resistance of the wall is:

$$\frac{Z}{K} = \frac{z_1}{k_1} + \frac{z_2}{k_2} + \frac{z_3}{k_3} = \frac{0.002}{15} + \frac{0.08}{0.03} + \frac{0.15}{1.3} = 0.13 \times 10^{-3} + 2.67 + 0.11$$
$$\approx 2.78\ \mathrm{m^2 \cdot K \cdot W^{-1}}$$

$$J = \frac{\Delta T_{Total}}{(Z/K)} = \frac{20-(-18)}{2.78} = 13.67\ \mathrm{W \cdot m^{-2}}$$

It is seen that the contribution of the stainless steel sheet to the thermal resistance is negligible. We can safely assume, therefore, that the temperature at the steel–insulator interface will be very nearly $-18°C$. Since the same flux crosses all the layers, the temperature T′ at the insulator–concrete interface can be calculated from:

$$J = 13.67 = \frac{0.03 \cdot \{T' - (-18)\}}{0.08} \Rightarrow T' = 0.08 \times 13.67/0.03 - 18 = 18.5°C$$

3.3.4.3 Steady-state transfer through varying area

A common application of this case is heat transfer through a thick cylindrical shell. Consider radial convective transfer through such a shell (Figure 3.4). Let r_1 and r_2 be the inner and outer radii of the shell, and L its length.

The heat flux through an annular element of radius z and thickness dz is:

$$\frac{q}{A} = \frac{q}{2\pi z L} = k\frac{dT}{dz} \Rightarrow \int_{r_1}^{r_2} \frac{dz}{z} = \frac{2\pi L}{q}\int_{T_1}^{T_2} dT$$

Integration gives:

$$q = 2\pi L k \frac{T_1 - T_2}{\ln\left(\frac{r_2}{r_1}\right)} \tag{3.14}$$

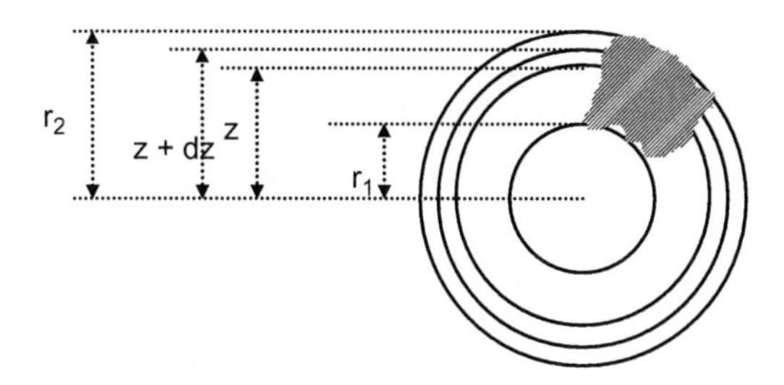

FIGURE 3.4

Heat transfer through thick cylindrical shell.

EXAMPLE 3.6

A 5-cm outside diameter (od) steel pipe is carrying steam at 150°C. The pipe is insulated with a cylindrical shell of insulator, 3 cm thick ($k = 0.03\ W \cdot m^{-1} \cdot K^{-1}$).

Calculate the rate of heat loss per meter of pipe length, if the temperature of the outer surface of the insulation is 35°C.

Solution

We shall assume that the outer wall of the steel pipe is at the temperature of the steam. Applying Eq. (3.14):

$$q/L = 2\pi k \frac{T_1 - T_2}{\ln\left(\frac{r_2}{r_1}\right)} = 2\pi \times 0.03 \times \frac{150 - 35}{\ln\left(\frac{5.5}{2.5}\right)} = 27.5\ W \cdot m^{-1}$$

3.3.4.4 Steady-state mass transfer of gas through a film

A common practical application of this example is the transfer of water vapor or oxygen through a film of packaging material. Consider a film of thickness z (Figure 3.5). Let p_1 and p_2 be the partial pressures of a gaseous component G at each side of the film, respectively. Assume $p_1 > p_2$. The difference in partial pressure (or concentration) causes G to penetrate through the film. The transfer of G through the film occurs in three consecutive steps:

Step a: Adsorption (dissolution) of G in the film material on plane 1
Step b: Molecular diffusion of G from plane 1 to plane 2 by virtue of the concentration gradient
Step c: Desorption of G from the film material at plane 2.

Assume that the equilibrium concentration of the component G in the film material is proportional to its partial pressure (Henry's Law). Concentration C is then related to the partial pressure p as follows:

$$C = sp \tag{3.15}$$

The proportionality constant s is the *solubility* of the gas G in the film material.

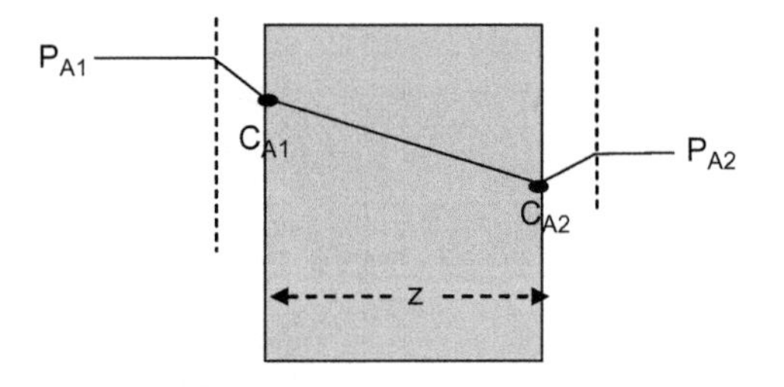

FIGURE 3.5

Steady-state mass transfer of a gas through a film.

At steady state, the flux is given by Eq. (3.7):

$$J_G = D_G \frac{C_1 - C_2}{z}$$

Writing the concentrations in terms of equilibrium partial pressure we obtain:

$$J_G = D_G s_G \frac{p_1 - p_2}{z} = \Pi \frac{p_1 - p_2}{z} \tag{3.16}$$

The expression D · s (diffusivity multiplied by solubility) is called *permeability* (Π). The permeability to different gases is an important characteristic of packaging materials. Its SI units are $kg \cdot m^{-1} \cdot s^{-1} \cdot Pa^{-1}$, but it is usually expressed in various practical units. More information on the permeability of packaging materials to various gases is included in Chapter 27.

EXAMPLE 3.7

The permeabilities of two polymers (A and B) to oxygen, in cm^3 STP/24 h, $ft^2 \cdot atm \cdot mil$, are:

Polymer A, 550
Polymer B, 25
(Note: 1 mm = 0.001 inch.)

a. Calculate the permeability to oxygen of a laminate made of 6 mm of A and 1 mm of B.

b. Calculate the total quantity of oxygen (in $g \cdot mol^{-1}$) that penetrates in 100 days into a pouch made of the said laminate. The total surface area of the pouch is 400 cm^2. The pouch contains a food rich in unsaturated fat, in an atmosphere of pure nitrogen. It is assumed that all the oxygen that penetrates combines quickly with the fat. The storage temperature is 25°C.

Solution

a. The permeability units given in this example are one of the many "practical" units in use (see Chapter 27, Section 27.2.3). The "resistance" of the laminate is equal to the sum of the individual resistances of the two films:

$$\frac{Z}{\Pi} = \frac{z_1}{\Pi_1} + \frac{z_2}{\Pi_2} = \frac{6}{550} + \frac{1}{25} = \frac{7}{\Pi} \Rightarrow \Pi = 137.5 \frac{cm^3 \cdot mil}{ft^2 \cdot atm \cdot 24\ h}$$

b.

$$V = \Pi \cdot (\Delta P) \cdot A \cdot t / z$$

where V = volume of oxygen (STP), A = surface area, t = time.

The partial pressure of oxygen in air outside the pouch is 21% of the total pressure, and 0 inside the pouch.

$\Delta P = 0.21$ atm.

$$A = 400\ cm^2 = 400/929 = 0.43\ ft^2 \quad (1\ ft^2 = 929\ cm^2)$$

$$V = 137.5 \times 0.21 \times 0.43 \times 100/7 = 177.3\ cm^3\ \text{oxygen} = 7.25 \times 10^{-3} g \cdot mol^{-1}\ \text{oxygen}$$

3.4 Convective heat and mass transfer

Heat and mass transfer in fluids almost always occur simultaneously with bulk movement of the medium. We define two kinds of convection.

1. *Natural (or free) convection*: The movement is caused by heat or mass transfer itself, usually by virtue of density differences. Consider a hot stove in a cold room. Air in contact with the stove surface is heated, expands, becomes less dense, moves upwards, and is replaced by colder, heavier air. Natural circular "convection currents" continue to move the air, as long as temperature differences exist in the room. The same type of density-driven currents would occur and be observed in an unstirred cup of tea to which a spoonful of sugar has been added.
2. *Forced convection*: Fluid movement (flow) is caused by factors independent of the transfer. Consider the stove-room model, this time with a fan blowing the air over the stove or the cup of tea, stirred with a spoon.

In contrast to conduction, it is extremely difficult to predict convective heat or mass transfer rates analytically, particularly in the case of turbulent flow. One of the simplifying models most commonly used in convective transfer between a surface and a moving fluid is the *film* or *contact layer* model, already introduced in connection with turbulent flow.

3.4.1 Film (or surface) heat and mass transfer coefficients

Consider heat transfer from a surface (e.g., a wall) at temperature T_1 to a fluid in turbulent flow and in contact with it (Figure 3.6). The film theory makes the following hypotheses:

a. There exists, in contact with the surface, a stationary layer (film) of fluid, of thickness δ.
b. The bulk of the fluid beyond the film is perfectly mixed, by virtue of the turbulence. The temperature in the bulk T_2 is therefore uniform.

(Note: As far as heat transfer is concerned, a laminar film would also be treated as stationary, since there is no fluid movement in the direction of heat flow).

The rate of heat transfer is:

$$q = kA(T_1 - T_2)/\delta$$

The division of the fluid into a stationary film and a turbulent bulk is, of course, a purely theoretical model and does not correspond to reality. The thickness δ is, therefore, a notion rather than a measurable physical value. A *coefficient of convective heat transfer* h is defined as:

$$h = \frac{k}{\delta} \tag{3.17}$$

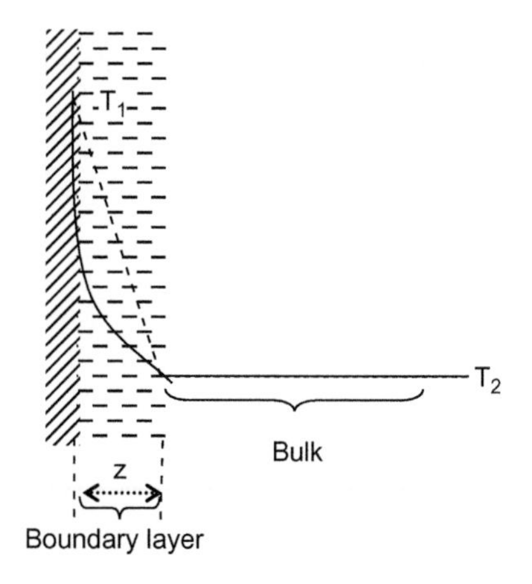

FIGURE 3.6

Boundary layer in turbulent convection.

The rate of heat transfer by convection is then expressed by the following equation:

$$q = hA\Delta T \tag{3.18}$$

The heat transfer coefficient h is a fundamental notion in heat transfer. Its SI units are $W \cdot m^{-2} \cdot K^{-1}$. It is determined experimentally. Its value depends on the properties of the fluid (specific heat, viscosity, density and thermal conductivity), the turbulence (average velocity), and the geometry of the system. As usual, these parameters are grouped in dimensionless expressions. The principal dimensionless groups used in heat transfer are:

$$\text{Reynolds} \quad Re = \frac{dv\rho}{\mu}$$

$$\text{Nusselt} \quad Nu = \frac{hd}{k}$$

$$\text{Prandl} \quad Pr = \frac{c_P\mu}{k}$$

$$\text{Grashof} \quad Gr = \frac{d^3\rho(\Delta\rho)\,g}{\mu^2}$$

d is a linear dimension (diameter, length, height, etc.) characterizing the geometry of the system. Note that Prandl number consists only of properties of the fluid. The Grashof number will be discussed later, in relation with natural convection.

Table 3.3 Approximate Order of Magnitude of Convective Heat Transfer Coefficients for Various Applications

System	h ($W \cdot m^{-2} \cdot K^{-1}$)
Natural convection (gas)	10
Natural convection (liquid)	100
Flowing gas	50–100
Flowing liquid (low viscosity)	1000–5000
Flowing liquid (high viscosity)	100–500
Boiling liquid	20 000
Condensing steam	20 000

Representative values of h for different applications are given in Table 3.3. Experimental correlations for the calculation of h are usually expressed in the general form:

$$\text{Nu} = f(\text{Re}, \text{Pr}, \text{Gr}, \ldots)$$

The convective mass transfer coefficient k_m is defined, in the same manner, as:

$$k_m = \frac{D}{\delta} \tag{3.19}$$

If the driving force for mass transfer is expressed in terms of a difference in concentration, then the steady state flux is written as:

$$J = k_C \Delta C \tag{3.20}$$

The SI units of k_c are $m \cdot s^{-1}$.

In the case of mass transfer in gases, the driving force is usually given in terms of a difference in partial pressures rather than concentrations. The coefficient of mass transfer is then indexed as k_g. The flux is then:

$$J = k_g \Delta C \tag{3.21}$$

The SI units of k_g are $kg \cdot m^{-2} \cdot s^{-1} \cdot Pa^{-1}$.

Two additional dimensionless groups are useful in mass transfer calculations:

$$\text{Sherwood} \quad \text{Sh} = \frac{k_C}{dD}$$

$$\text{Schmidt} \quad \text{Sc} = \frac{\mu}{D\rho}$$

As in the Prandl number, the Schmidt number contains only properties of the matter. The Sherwood number contains the transfer coefficient, as does the Nusselt number in heat transfer.

3.4.2 Empirical correlations for convection heat and mass transfer

Engineering literature is replete with empirical or semi-empirical data relating transfer coefficients to material properties and operation conditions, in the form of graphs, charts, tables or correlation equations. Only a few of the most commonly used correlations will be discussed here.

For natural (free) convection, which is essentially based on differences in density, and hence on thermal expansion of the fluid, the correlations contain the Grashof number Gr. As shown in Section 3.4.1, this dimensionless group contains the term $\Delta\rho$, the difference in the density of the fluid, which in turn is related to the differences in temperature ΔT and the coefficient of thermal expansion β.

In engineering applications, heat transfer by natural convection is particularly important for the calculation of heat or refrigeration "losses" from the surface of buildings, equipment, vessels, etc. The following correlation is often recommended for the calculation of natural convection heat transfer from vertical surfaces:

$$\mathrm{Nu} = 0.59\ \mathrm{Gr}^{0.25}\mathrm{Pr}^{0.25} \tag{3.22}$$

In this case, the linear dimension d in both dimensionless groups is the height of the surface.

For a sphere immersed in fluid, the following equation is proposed:

$$\mathrm{Nu} = 2 + 0.6\ \mathrm{Gr}^{0.25}\mathrm{Pr}^{0.33} \tag{3.23}$$

EXAMPLE 3.8

Estimate the rate of heat loss from the vertical wall of an oven by natural convection. The temperature of the wall is 50°C; the temperature of the ambient air is 20°C. The wall is 1.2 m high and 3 m wide.

Solution

First, the convective heat transfer coefficient will be calculated from Eq. (3.22). The properties of air will be taken at the average temperature of the film at the wall–air interface.

$$T_{average} = (50 + 20)/2 = 35°\mathrm{C}$$

The properties of air at 35°C (Table A.17, by interpolation) are as follows:

$\rho = 1.14\ \mathrm{kg \cdot m^{-3}}$
$\mu = 8.9 \times 10^{-6}\ \mathrm{Pa \cdot s}$
$C_p = 1014\ \mathrm{J \cdot kg^{-1} \cdot K^{-1}}$
$k = 0.026\ \mathrm{W \cdot m^{-1} \cdot K^{-1}}$
$\Delta\rho$ = The difference between the densities at 20°C and 50°C = 1.16 − 1.06 = 0.10 $\mathrm{kg \cdot m^{-3}}$.

$$\mathrm{Gr} = \frac{d^3\rho(\Delta\rho)g}{\mu^2} = (1.2)^3 \times 1.14 \times 0.1 \times 9.8/(18.9 \times 10^{-6})^2 = 4.5 \times 10^9$$

$$\mathrm{Pr} = \frac{C_p\mu}{k} = (1014 \times 18.9 \times 10^{-6})/0.026 = 0.757$$

$$\mathrm{Nu} = \frac{hd}{k} = 0.59\ \mathrm{Gr}^{0.25}\mathrm{Pr}^{0.25} = 0.59\cdot(4.5\times10^{9})^{0.25}\cdot(0.757)^{0.25} = 142$$

$$h = 142\cdot\frac{k}{d} = 142\cdot\frac{0.026}{1.2} = 3.08\ \mathrm{W\cdot m^{-2}\cdot K^{-1}}$$

$$q = 308\cdot(1.2\times3)(50-20) = 332.6\ \mathrm{W}$$

Forced convective transfer is extremely common in engineering applications such as heat exchangers, evaporation, convective drying, blast freezing, agitated vessels, etc. One of the most widely quoted correlations is the Dittus–Boelter equation, also known as Sieder–Tate Equation (3.24) for *turbulent* heat transfer between the walls of a cylindrical duct (pipe) and a fluid flowing inside.

$$\mathrm{Nu} = 0.023\mathrm{Re}^{0.8}\mathrm{Pr}^{0.3-0.4}\left(\frac{\mu}{\mu_0}\right)^{0.14} \tag{3.24a}$$

The exponent of Pr is 0.3 for heating and 0.4 for cooling. The dimension d is the inside diameter of the pipe. The fluid properties are taken at the average temperature, except for μ_0, which is the viscosity at the temperature at the exchange surface.

For transfer between a sphere and a fluid in turbulent flow, the following correlations are used.

For heat transfer:

$$\mathrm{Nu} = 2 + 0.6\mathrm{Re}^{0.5}\cdot\mathrm{Pr}^{0.33} \tag{3.24b}$$

For mass transfer:

$$\mathrm{Sh} = 2 + 0.6\mathrm{Re}^{0.5}\cdot\mathrm{Sc}^{0.33} \tag{3.24c}$$

For other correlations, covering heat and mass transfer in different geometries and flow regimes, see Bimbenet and Loncin (1995) and Singh and Heldman (2003). For the evaluation of the heat transfer rate in turbulent flow, using numerical methods, see Nitin et al. (2006).

EXAMPLE 3.9

Estimate the convective heat transfer coefficient between orange juice flowing inside a pipe and the pipe wall. The juice is being cooled.

Data:

Pipe diameter	0.05 m
Volumetric flow-rate	$4\ \mathrm{m^3\cdot h^{-1}}$
Properties of orange juice at the temperature of the process:	$\rho = 1060\ \mathrm{kg\cdot m^{-3}}$
	$\mu = 3\times10^{-3}\ \mathrm{Pa\cdot s}$
	$C_p = 3900\ \mathrm{J\cdot kg^{-1}\cdot K^{-1}}$
	$k = 0.54\ \mathrm{W\cdot m^{-1}\cdot K^{-1}}$

Solution

The flow velocity is:

$$v = q/A = \frac{4}{3600 \times \pi \times (0.05)^2/4} = 0.57 \ \text{m}\cdot\text{s}^{-1}$$

The Reynolds number is:

$$\text{Re} = \frac{dv\rho}{\mu} = \frac{0.05 \times 0.57 \times 1060}{0.003} = 10\,070$$

The Prandl number is:

$$\text{Pr} = \frac{C_p\mu}{k} = \frac{3900 \times 0.003}{0.54} = 21.7$$

The flow is turbulent, therefore the Dittus–Boelter correlation is applicable. The viscosity correction term will be neglected.

$$\text{Nu} = 0.023\text{Re}^{0.8}\text{Pr}^{0.4} = 0.023\cdot(10\,070)^{0.8}\cdot(21.7)^{0.4} = 125.4$$

$$h = 125.4.\frac{0.54}{0.05} = 1354 \ \text{W}\cdot\text{m}^{-2}\cdot\text{K}^{-1}$$

EXAMPLE 3.10

Cooked wheat grits (couscous) are being dried by hot air in turbulent flow, in a fluidized bed dryer. Estimate the convective heat transfer coefficient, assuming that the grit particles are spherical.

Data:

Air: velocity	$0.5 \ \text{m}\cdot\text{s}^{-1}$
Temperature	80°C
Grits: diameter	1 mm

Solution

From Table A.17 (see Appendix to this volume), we read the properties of air at 80°C, as follows:

$\rho = 0.968 \ \text{kg}\cdot\text{m}^{-3}$
$\mu = 20.8 \times 10^{-6} \ \text{Pa}\cdot\text{s}$
$\text{Pr} = 0.71.$

$$\text{Re} = \frac{0.001 \times 0.5 \times 0.968}{0.0000208} = 23.3$$

$$\text{Nu} = \frac{hd}{k} = 2 + 0.6.(23.3)^{0.5}(0.71)^{0.33} = 4.586$$

$$h = 4.586\cdot\frac{k}{d} = \frac{4.586 \times 0.0293}{0.001} = 134.4 \ \text{W}\cdot\text{m}^{-1}\cdot\text{K}$$

When using empirical correlations for the calculation of transfer coefficients, the following points are noteworthy:

1. Each correlation is valid only for a certain geometry and a specific set of operating conditions (e.g., a specific range of Re and/or Nu).
2. Empirical correlations are approximations. Their degree of accuracy is often specified in the literature source proposing their use.
3. In their general form, heat transfer correlations are usually valid also for mass transfer (and *vice versa*), provided that the dimensionless groups are replaced by their proper counterpart (Nu by Sh, Pr by Sc, etc.)

3.4.3 Steady-state interphase mass transfer

Mass transfer between physical phases is of central importance in many processes. The process of drying is based on the transfer of water molecules from a liquid or a wet material to a gas (usually air). De-aeration of fruit juices consists of the transfer of oxygen from a liquid where it is dissolved to a gaseous phase. Liquid–liquid extraction is based on the transport of a solute from one liquid solvent to another.

A model that is useful in the analysis of interphase mass transfer phenomena is the *two-film model*, formulated by Lewis and Whitman in 1924 in connection with gas adsorption. The model assumes the existence of two stagnant or laminar films, each on one side of the boundary (interface) between the two phases. The two phases may be a gas and a liquid, or two immiscible liquids. A substance is transferred from one phase to the other by virtue of a difference in concentration or partial pressure. (Rigorously speaking, the transfer is driven by a difference in "activity", which can be replaced by a difference in concentration only in the case of ideal mixtures.) The following assumptions are made:

- The total resistance to transfer resides in the two films; therefore the total drop in concentration or partial pressure occurs in the two films
- The two phases are in equilibrium at the interface
- Accumulation at the interface is zero.

Figure 3.7 shows the concentration or partial pressure profiles across two phases in contact, according to the two-film model.

Consider the case of gas–liquid contact in Figure 3.7. Assume that a substance A is transported from the gas to the liquid. Since there can be no accumulation at the interface, the flux of A from gas to interface must be equal to the flux from the interface to the liquid:

$$k_g(\overline{p}_{A,g} - \overline{p}_{A,i}) = k_L(C_{A,i} - C_{A,L}) \tag{3.25}$$

where:

k_g, k_L = convective mass transfer coefficients of the gas and liquid films, respectively

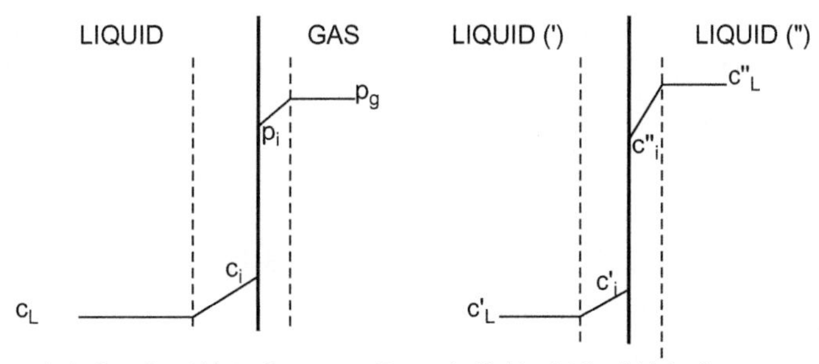

FIGURE 3.7

Interphase mass transfer.

$p_{A,g}$, $p_{A,i}$ = partial pressures of A, in the gas bulk and at the interface, respectively
$C_{A,L}$, $C_{A,i}$ = concentrations of A, in the liquid bulk and at the interface, respectively.

Because of equilibrium at the interface, $C_{A,i}$ and $p_{A,I}$ are interrelated by the relevant equilibrium function. If, for example, Henry's Law is assumed, the interrelation is:

$$C_{A,i} = s\,\overline{p}_{A,i}$$

(see Eq. (3.15)).

We define an *overall coefficient of mass transfer* K_L for the transfer driven by the total drop in concentration:

$$J_A = K_L(s\,\overline{p}_{A,g} - C_{A,L}) \tag{3.26}$$

Equally, we can define an overall coefficient of mass transfer K_g for the transfer in terms of the total drop in partial pressure:

$$J_A = K_g\left(\overline{p}_{A,g} - \frac{C_{A,L}}{s}\right) \tag{3.27}$$

The total resistance is equal to the sum of the resistances of the two individual films:

$$\frac{1}{K_L} = \frac{1}{k_L} + \frac{s}{k_g} \quad \text{or} \quad \frac{1}{K_g} = \frac{1}{k_g} + \frac{1}{sk_L} \tag{3.28}$$

Often, one of the resistances is much larger than the other. Assume, for example, that the resistance to the transport of A in the gas is much smaller than that in the liquid, i.e., $k_g >>> k_L$. In this case:

$$K_L \approx k_L \quad \text{and} \quad K_g \approx sk_L \tag{3.29}$$

The equations developed above are only valid in the case of liquid–gas couples obeying Henry's Law. Otherwise, more precise equilibrium functions or experimental data such as sorption isotherms have to be used.

3.5 Unsteady state heat and mass transfer

Within a piece of vegetable subjected to drying in a current of hot air, a cut of meat undergoing smoking, a piece of fruit immersed in a sugar syrup for candying, or a can of solid food heated in a retort for sterilization, the conditions (temperature, concentration) at a given point change with time. The heat or mass flux at a given location also varies with time. These are examples of transient or unsteady-state heat or mass transfer in food processing.

3.5.1 The second Fourier and Fick laws

Consider a slab of material (Figure 3.8) in which transient heat transfer is taking place. Assume that the flow of heat is unidirectional, in the direction perpendicular to the surface of the slab. Consider a slice of the slab at a distance z from one surface. Let the thickness of the slice be dz.

At unsteady state, the rate of heat input to the slice (q_i) is not equal to the rate of heat output (q_o). The difference is the accumulated heat q_a (which can be

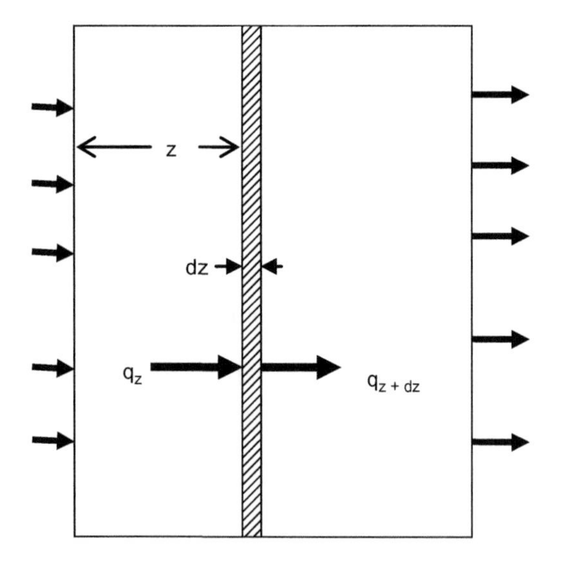

FIGURE 3.8

Unsteady-state heat transfer in a slab.

positive or negative). If there is no heat generation (no internal source of energy) in the slab, heat balance requires that:

$$\text{Input} - \text{Output} = \text{Accumulation}$$

(An internal source or an internal sink of energy must be included in the balance in certain cases, such as post-harvest respiration of fruit or a phase change. For an application with an internal source of energy, see Cuesta and Lamúa, 2009.)

In the case of heat transfer by conduction:

$$q_i = -A(k\frac{\partial T}{\partial z})_z \tag{3.30}$$

$$q_o = -A(k\frac{\partial T}{\partial z})_{z+dz} \tag{3.31}$$

The rate of heat accumulation is related to the rate of temperature change, through the mass of the slice element and its specific heat, as follows:

$$q_a = A\frac{\partial(\rho C_P T)}{\partial t}dz \tag{3.32}$$

Assuming that the properties k, C_p and ρ are fairly temperature-independent and uniform throughout the slice, substituting in the heat balance equation, eliminating A from all the terms and remembering the definition of thermal diffusivity α, we obtain:

$$\frac{\partial T}{\partial t} = \left(\frac{k}{\rho C_p}\right)\frac{\partial^2 T}{\partial z^2} = \alpha\frac{\partial^2 T}{\partial z^2} \tag{3.33}$$

For transient molecular diffusion, we would obtain in the same manner:

$$\frac{\partial C}{\partial T} = D\frac{\partial^2 C}{\partial z^2} \tag{3.34}$$

The above equations can be expanded for three-dimensional transfer, by analyzing the transfers in and out a three-dimensional volume element. For heat transfer, for example, such analysis would yield:

$$\frac{\partial T}{\partial t} = \alpha\frac{\partial^2 T}{\partial x^2} + \alpha\frac{\partial^2 T}{\partial y^2} + \alpha\frac{\partial^2 T}{\partial z^2} \tag{3.35}$$

(Note: If the material is not isotropic, the thermal diffusivity may be different in the x, y and z directions. It is known, for example, that the thermal diffusivity of meat in a direction parallel to the fibers differs from that in a direction normal to the fibers. The same is true for molecular diffusivity.)

Equations (3.33) and (3.34) are known as Fourier's and Fick's Second Laws, respectively. The solution of these equations gives the temperature or concentration distribution within the solid, and its variation with time. Analytical solutions of these equations are available for regular geometries (slab, sphere, cylinder) and simplified boundary conditions. One of these classical solutions is discussed below. For more complex situations, numerical methods are used.

3.5.2 Solution of Fourier's second law equation for an infinite slab

Consider an infinite solid slab of thickness 2L. Assume that initially the temperature everywhere in the slab is T_0. Both surfaces of the slab are suddenly brought to contact with a fluid at temperature T_∞. The fluid is kept at that temperature indefinitely. What is the temperature distribution within the slab at time t?

First, the variables are transformed to dimensionless expressions. The dimensionless temperature expression θ is defined as:

$$\theta = \frac{T_\infty - T}{T_\infty - T_0} \tag{3.36}$$

Physically, θ represents the *unachieved fraction* of the temperature change that would be achieved at $t = t_\infty$.

Next, we define a dimensionless expression for the relationship between the surface resistance and the internal resistance to transfer. Let h be the coefficient of convective heat transfer from the fluid to the slab surface, and k be the thermal conductivity of the solid. The dimensionless *Biot number*, named after the French physicist Jean-Baptiste Biot (1774–1862), is defined as:

$$N_{Bi} = \frac{hL}{k} \tag{3.37}$$

We place the origin for distance z on the central plane of the slab, and define the dimensionless distance z′ as follows:

$$z' = \frac{z}{L} \tag{3.38}$$

The dimensionless expression of time t is the Fourier number (Fo), defined as follows (with α = thermal diffusivity):

$$Fo = \frac{\alpha t}{L^2} \tag{3.39}$$

The general solution of Fourier's second law equation, in terms of the dimensionless groups, for the boundary conditions given above is an infinite series:

$$\theta = \sum_{i=1}^{\infty} \left[\frac{2\sin\beta_i}{\beta_i + \sin\beta_i \cos\beta_i} \cos(\beta_i z') \right] \exp(-\beta_i^2 Fo) \tag{3.40}$$

The parameter β_i is related to the Biot number as follows:

$$N_{Bi} = \beta_i \tan\beta_i \tag{3.41}$$

If time is sufficiently long (say $Fo > 0.1$), the sum converges approximately to the first term of the series. Furthermore, if the surface resistance to heat transfer is negligible in comparison to the internal resistance ($N_{Bi} = \infty$, hence $\beta = \pi/2$), the solution becomes:

$$\theta = \frac{4}{\pi}\left(\cos\frac{\pi z'}{2}\right)\exp\left(-\frac{\pi^2}{4}Fo\right) \tag{3.42}$$

In practice, the assumption of $N_{Bi} = \infty$ is justifiable in many situations, such as heating a can of solid food by vapor condensing on the surface of the can, a large chunk of solid food heated or cooled in a turbulent stream of air, etc.

Equation (3.42) represents the temperature distribution in the slab, as a function of location (z') and time (Fo). The location function is trigonometric. The time function is exponential.

At a given location in the slab (a given z'), the time–temperature relationship is approximated by an equation in the following general form:

$$\ln\frac{T - T_\infty}{T_0 - T_\infty} = \ln j - \left(\frac{\pi^2\alpha}{4L^2}\right)t \tag{3.43}$$

The temperature T_c at the central axis of the slab is found by setting $z = 0$.

A slab of thickness L, exchanging heat only from one of its faces, is treated as one half of a slab of thickness 2L, exchanging heat at both faces.

Equations of the same general form are obtained for other geometries (infinite cylinder, sphere). Graphically, all these equations are represented as a curve that becomes approximately a straight line on the log θ vs t plane if the time is sufficiently long (Figure 3.9).

Complete charts for the temperature distribution for different values of Biot number and various geometries are available in the literature (see, for example, Carslaw and Jaeger, 1986). The general form of these charts is shown in Figures A.4–A.6 (see Appendix).

Exactly the same equations are also valid for mass transfer, by substituting the thermal variables by their mass transfer equivalent: concentration instead of temperature, molecular diffusivity instead of thermal diffusivity, etc.

3.5.3 Transient conduction transfer in finite solids

Solutions are available for unsteady-state transfer in finite solids such as a finite cylinder (a round can) or a brick-shaped body (a rectangular can, a loaf of bread). These shapes can be considered as the intersection of infinite bodies. The brick is seen as the intersection of three mutually perpendicular infinite slabs. If θ_x, θ_y, θ_z are the solutions for unidirectional heat transfer in the three infinite slabs, respectively, then the solution for the brick is:

$$\theta_{brick} = (\theta_x) \times (\theta_y) \times (\theta_z) \tag{3.44}$$

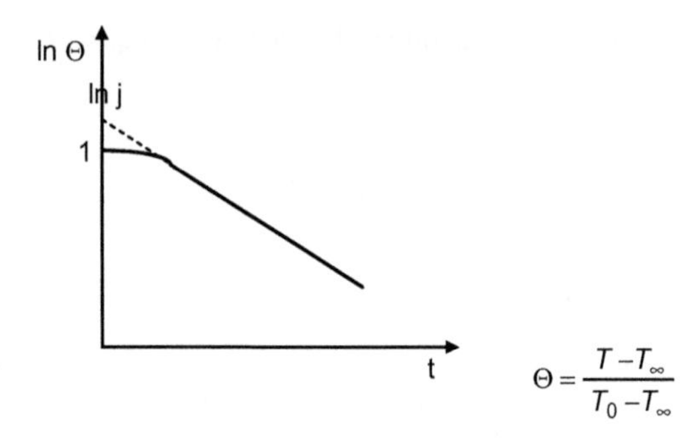

FIGURE 3.9

General form of unsteady-state heat transfer plot.

In the same manner, the finite cylinder is considered as the intersection of an infinite cylinder and an infinite slab. The solution for the finite cylinder is:

$$\theta_{finite\ cyl.} = (\theta_{slab}) \times (\theta_{\text{infinite}\ cyl.}) \tag{3.45}$$

This rule, also known as Newman's law, is applicable to mass transfer as well.

EXAMPLE 3.11

A can contains meat loaf, solidly packed, at 70°C. The can is placed in an autoclave where it is heated with steam at 121°C. What will be the temperature at the can center after 30 minutes? How long will it take to reach a temperature of 120°C at the center? Assume that the temperature of the can surface is 121°C from t = 0 on.

Data:

Can dimensions	Diameter = 8 cm, height = 6 cm
Thermal diffusivity of the loaf	$0.2 \times 10^{-6}\ m^2 \cdot s^{-1}$

Solution

We treat the can as an intersection of an infinite plate and an infinite cylinder (Newman's Law).

For the cylinder:

$$\frac{\alpha t}{R^2} = \frac{0.2 \times 10^{-6} \times 30 \times 60}{0.04^2} = 0.225$$

From Figure A.5, for Bi = ∞ E(cyl) = 0.44

For the slab:

$$\frac{\alpha t}{d^2} = \frac{0.2 \times 10^{-6} \times 30 \times 60}{0.03^2} = 0.400$$

From Figure A.4, for Bi = ∞, E(slab) = 0.48

$$E(can) = E(cyl) \times E(slab) = 0.44 \times 0.48 = 0.2112$$

$$E(can) = \frac{T - T_\infty}{T_0 - T_\infty} = \frac{T - 121}{70 - 121} = 0.2112 \quad \Rightarrow \quad T = 110.2°C$$

The simplest way to find the time taken to reach 120°C at the center is to proceed by trial and error. We calculate the temperature at the center after, say, 40, 60, 80, 100 minutes, plot the results, and estimate the time corresponding to 120°C. The result is 62 minutes.

EXAMPLE 3.12

A manufacturer of sausages has decided to produce spherical sausages instead of the traditional cylindrical shape. The traditional sausage has a diameter of 4 cm and a length of 30 cm. The spherical sausage will have the same weight as the cylindrical one. Both types of sausage are smoked.

If the smoking time for the cylinders is 14 hours, how long will it take to smoke the spheres to reach the same average concentration of smoke compound in the product?

The diffusion coefficient of the smoke substance in sausage is $1.65 \times 10^{-9}\ m^2 \cdot s$.

Solution

It will be assumed that mass transfer in the cylindrical sausage is only in the radial direction (the sausage is relatively slim and long) and governed by the internal resistance ($Bi = \infty$).

The volume of the cylindrical sausage is $\pi(0.02)^2 \cdot 0.3 = 0.000377\ m^3$.

The volume of the sphere will be the same:

$$\frac{4\pi R^3}{3} = 0.000377 \quad \Rightarrow \quad R = 0.045\ m$$

For the cylindrical sausage:

$$\left(\frac{Dt}{a^2}\right)_{inf.cyl.} = \left(\frac{Dt}{R^2}\right)_{inf.cyl.} = \frac{(1.65 \times 10^{-9})(14 \times 3600)}{(0.02)^2} = 0.207$$

From the chart (Figure A.7):

$$E_{inf.cyl} = 0.19$$

Since the spheres must reach the same final average concentration of smoke substance, while being exposed to the same smoke, this is also the value of E for the spheres:

$$E_{sphere} = 0.19$$

The corresponding value of Dt/a^2 for spheres, from the chart is:

$$\left(\frac{Dt}{a^2}\right) = \left(\frac{Dt}{R^2}\right) = 0.11$$

$$t = \frac{0.11R^2}{D} = \frac{0.11 \times (0.045)^2}{1.65 \times 10^{-9}} = 135\,000\ s = 37.5\ \text{hours}$$

Smoking time has increased 2.6-fold! It seems that there is a good reason to form smoked sausages as slim cylinders.

EXAMPLE 3.13

Peas, initially at a temperature of 10°C, are blanched in steam at atmospheric pressure. Calculate the blanching time needed to bring the temperature at the center of the peas to 88°C. Consider the peas to be spheres, 6 mm in diameter. Assume that the entire surface of the pea is in contact with steam, that the surface resistance to heat transfer is negligible, and that the average thermal diffusivity of peas is $0.16 \times 10^{-6}\ m^2 \cdot s^{-1}$.

Solution

$$\theta = \frac{T_\infty - T}{T_\infty - T_0} = \frac{100 - 88}{100 - 10} = 0.12$$

On the unsteady heat transfer chart for spheres (Figure A.7), with $\Theta = 0.12$ and $N_{Bi} = \infty$ we find $\alpha t/R^2 = 0.2 \cdot R = 6/2 = 3$ mm.

Hence:

$$t = \frac{0.2 \times (0.003)^2}{0.16 \times 10^{-6}} = 11.2\ \text{s}$$

3.5.4 Transient convective transfer in a semi-infinite body

A semi-infinite body is one that is delimited by one plane only. A very deep body of water comes close to the description of a semi-infinite body.

As an example, consider molecular diffusion of a pollutant from air to a very deep, essentially stagnant mass of water. What is the concentration of the pollutant in the water, as a function of time and location (depth)?

Setting as a boundary condition that the concentration at the surface remains constant at c_∞ from t = 0 on (equivalent to $N_{Bi} = \infty$), the following solution has been developed:

$$\frac{c - c_\infty}{c_0 - c_\infty} = erf \frac{z}{2\sqrt{Dt}} \tag{3.46}$$

Error function of x, (erf x), is defined as follows:

$$erf\ x = \frac{2}{\sqrt{\pi}} \int_0^x e^{-x^2} dx \tag{3.47}$$

Often, the flux of transfer and the total quantity transferred as a function of time, rather than the concentration profile, are of interest. These values are calculated from the following expressions.

Flux through the free surface:

$$J\Big|_{z=0} = (C_\infty - C_0)\sqrt{\frac{D}{\pi t}} \tag{3.48}$$

The total mass transferred from t = 0 to t = t:

$$m = 2(C_\infty - C_0)\sqrt{\frac{Dt}{\pi}} \tag{3.49}$$

3.5.5 Unsteady-state convective transfer

Let us now consider the case where the main resistance to transfer is at the interface, while the resistance inside the body is negligible, i.e., $N_{Bi} \approx 0$. In practice, this would be the case of a liquid heated in a well-stirred jacketed kettle. Since the liquid is well mixed, the temperature (or the concentration) at any moment would be the same everywhere but would vary with time. Since the temperature is now a function of time alone and not of location, an ordinary differential equation suffices to describe the energy (or material) balance. For heat transfer, the energy balance gives:

$$hA(T_\infty - T) = V\rho\, C_P \frac{dT}{dt} \tag{3.50}$$

where V is the volume of the liquid, ρ is its density and A is the area of heat transfer.

Integration between t = 0 and t = t gives:

$$\ln\frac{T_\infty - T}{T_\infty - T_0} = -\frac{hA}{V\rho\, C_P}t \tag{3.51}$$

EXAMPLE 3.14

A sauce must be heated from 30°C to 75°C before bottling. Heating will be done in batches, in a jacketed kettle equipped with a stirrer. The jacket is heated with steam condensing at 110°C. The batch size is 500 kg. The specific heat of the sauce is 3100 $J \cdot kg^{-1} \cdot K^{-1}$. The production rate requires that heating time be 30 minutes or less. The estimated overall convective heat transfer coefficient is 300 $W \cdot m^{-2} \cdot K^{-1}$. What is the minimum heat transfer area required?

Solution

We apply Eq. (3.51):

$$\ln\frac{T_\infty - T}{T_\infty - T_0} = \ln\frac{110-75}{110-30} = -\frac{hA}{V\rho\, C_P}t = -\frac{300 \times 30 \times 60 \times A}{500 \times 3100}$$

$$-0.8267 = --0.348A \quad \Rightarrow \quad A = 2.38\ \text{m}^2$$

3.6 Heat transfer by radiation

The term *radiation* covers a vast array of phenomena that involve energy transport in the form of waves. In this section we deal only with a particular kind of radiation, called *thermal radiation*. Thermal radiation refers to electromagnetic radiation in the wavelength range of 10^{-7} to 10^{-4} m and encompasses mainly the range of infrared radiation. It is so called because it's practically sole effect is thermal – i.e., cooling of the emitting body and heating of the receiving body. Above the absolute temperature of 0 K, all substances emit electromagnetic radiation. The intensity and the “color” (wavelength distribution) of the radiation depend strongly on the temperature of the source. In contrast with conduction and convection, heat transfer by radiation does not require the presence of a material medium.

3.6.1 Interaction between matter and thermal radiation

Radiation energy impinging on matter is partly transmitted, partly reflected and partly absorbed. The relative proportion of each part is called *transmissivity*, *reflectivity* and *absorptivity*, respectively. Only the absorbed portion causes heating. If none of the radiation received is transmitted, the body is called “opaque”. A body that absorbs the incident radiation totally is called a *black body*. The absorptivity of a black body is unity, and its reflectivity and transmissivity are both zero. A black body is also a body that emits the maximum quantity of thermal radiation at a given temperature. The emissive power of a black body depends only on its temperature. The quantitative relationship between the emissive power of a black body and its temperature is given by the Stefan–Boltzmann equation:

$$E_b = \sigma T^4 \tag{3.52}$$

where:

E_b = emissive power of a black body, $W \cdot m^{-2}$
σ = Stefan's Constant = $5.669 \times 10^{-8}\ W \cdot m^{-2} K^{-4}$
T = absolute temperature, K.

(Note: Throughout this section, T refers to the absolute temperature.)

The term “black body” refers to an ideal physical model as described above, and not necessarily to the visual color of the body. Thus, the face of the Earth and even white snow are nearly black bodies with respect to most of the thermal energy spectrum.

By definition, a real surface absorbs and emits less energy than a black surface at the same temperature. The ratio of the emissive power of a real body to that of a black body at the same temperature is called *emissivity*, ε, and is always less than unity.

A body is *gray* if its emissivity is independent of wavelength. The gray body is also an ideal physical model with no reference to perceptible color. The emissive power of a gray body, E_g, is given in Eq. (3.49):

$$E_g = \varepsilon \sigma T^4 \tag{3.53}$$

The emissivity of real surfaces does depend on wavelength. The *average emissivity* over the relevant wavelength range is used in calculations. The average emissivity of various surfaces is given in Table 3.4.

Table 3.4 Total Emissivity of Various Surfaces

Material	Temperature (°C)	Emissivity
Aluminum, polished	100	0.095
Brass, polished	100	0.06
Steel, polished	100	0.066
Stainless steel, polished	100	0.074
Building brick	1000	0.45
Fireclay brick	1000	0.75
Carbon black	50–1000	0.96
Water	0–100	0.95–0.96

Table adapted from McAdams (1954).

3.6.2 Radiation heat exchange between surfaces

In this section we shall examine the net heat exchange between surfaces as a result of radiation. Before going into the quantitative aspects of the subject, it is important to note a few qualitative points:

- Radiation, in itself, is electromagnetic energy but not heat. It is converted to heat and causes thermal effects only after it is absorbed by a body.
- Thermal radiation has very low penetration power. Thermal radiant energy is absorbed by a thin layer on the surface of a body. From thereon, further penetration of heat occurs by convection and conduction. Therefore, in bodies with high internal resistance to heat transfer, radiation creates a large difference of temperature between the surface and the interior of the body. This property is advantageously utilized in many instances, such as the formation of crust in bread baking.
- Since the interaction between radiation and matter occurs at the surface, absorptivity and emissivity are properties of the surface. The treatment and state of the surface (e.g., polishing) strongly affect these properties.

The net rate of heat exchange between radiating bodies depends on two kinds of variables:

1. The properties and state of the radiating surfaces – namely, their temperature, emissivity and absorptivity.
2. The position in space of the surfaces relative to each other, including the distance between them. For simplicity, we shall assume that the space between the surfaces does not contain radiation-absorbing or -emitting bodies.

We shall first consider heat exchange between two black surfaces. The simplest case is that of two black surfaces 1 and 2, of equal areas, that see only each other – i.e., all the radiation emitted by one surface is captured by the other and *vice versa*. For this case, the net heat transfer from 1 to 2 is:

$$q_{1\to 2} = A\sigma(T_1^4 - T_2^4) \tag{3.54}$$

EXAMPLE 3.15

Frost in orchards occurs in clear, windless nights, as a result of radiation heat loss of the fruit. "Clear nights" mean that the fruit radiates to an infinite, empty space. "Windless" means that there is no other heat transfer (e.g., from air) to the fruit.

Calculate the time elapsed to cool an orange from 20°C to −1°C (frost) exposed to "clear night" conditions. Consider the orange as a sphere, 7 cm in diameter, with a density of 1000 $kg \cdot m^{-3}$ and a specific heat of 3800 $J \cdot kg^{-1} \cdot K^{-1}$. Assume that the internal resistance to heat transfer is negligible (there is no temperature gradient inside the fruit).

Solution

The absolute temperature of the infinite empty space is 0 K. Let T be the absolute temperature of the orange. The rate of heat loss is:

$$q = \sigma A T^4 = m\ C_p\ dT/dt$$

$$dt = \frac{mC_p}{\sigma A}\int_{273+20}^{273-1} \frac{dT}{T^4}$$

The area and the mass of the orange are: $A = 0.0154\ m^2$, $m = 0.179$ kg.

$$t = \frac{0.179 \times 3800}{(5.67 \times 10^{-8}) \times 0.0154} \cdot \frac{1}{3}\left[\frac{1}{(272)^3} - \frac{1}{(293)^3}\right] = 2600\ \text{s}$$

Note: Taking the average temperature of the orange instead of calculating the integral would give a slightly shorter time.

Situations where the radiating surfaces see only each other are highly improbable. In practice, only part of the radiation leaving one surface reaches the other. We shall define as "view factor m − n" (F_{m-n}) the fraction of radiant energy emitted by surface m which impinges on surface n. Adding the spatial considerations to Eq. (3.50) and omitting the condition of equal areas we obtain:

$$q_{1\to 2} = A_1\sigma F_{1-2}T_1^4 - A_2\sigma F_{2-1}T_2^4 \tag{3.55}$$

It can be mathematically proven (reciprocity theorem) that:

$$A_n F_{n-m} = A_m F_{m-n} \tag{3.56}$$

Equation (3.52) can then be written as:

$$q_{1-2} = A_1\sigma F_{1-2}(T_1^4 - T_2^4) = A_2\sigma F_{2-1}(T_1^4 - T_2^4) \tag{3.57}$$

Of the two options offered by Eq. (3.57), it is preferable to choose the one with the simplest or best-known view factor. View factors for some simple cases

Table 3.5 View Factor for Some Simple Geometries

Surfaces Interchanging Radiation	F_{1-2}
Infinite parallel planes	1
A_1 completely enclosed by A_2; A_1 cannot see any part of itself	1
Element dA and parallel circular disc of radius a with its center directly above dA at a distance L	$a^2/(a^2 + \times L^2)$
Two parallel discs, distance L apart, centers aligned, radius a smaller than radius b	$\frac{1}{2a^2}\left[(L^2+a^2+b^2)-\sqrt{(L^2+a^2+b^2)-4a^2b^2}\right]$

Data from Kreith (1973).

are listed in Table 3.5. View factors for more complex geometries have been calculated, and are generally presented as charts.

The expressions developed above for radiation heat transfer between black surfaces need to be corrected for radiation heat exchange between real surfaces (assumed gray). An interchange emissivity, ε_{1-2}, is defined. The calculation of this factor takes into account not only the individual emissivity of each surface but also the geometry. For example, the interchange emissivity for two very large parallel gray plane surfaces is:

$$\varepsilon_{1-2} = \frac{1}{\frac{1}{\varepsilon_1} + \frac{1}{\varepsilon_2} - 1} \tag{3.58}$$

For two gray surfaces, the overall radiation heat exchange equation becomes:

$$q_{1-2} = A_1 \sigma F_{1-2} \varepsilon_{1-2}(T_1^4 - T_2^4) = A_2 \sigma F_{2-1} \varepsilon_{1-2}(T_1^4 - T_2^4) \tag{3.59}$$

3.6.3 Radiation combined with convection

In practice, heat transfer often involves more than one mechanism. Heat transfer by a combination of convection and radiation is very common. For example, in a baking oven, heat is transferred to the product both from hot air by convection and from hot bodies by radiation (Sakin et al., 2009). Direct summing of the heat transferred by the two mechanisms is difficult due to the fact that radiation transfer involves the fourth power of temperature, while in expressions for convection and conduction the heat transfer temperature is at power 1. To overcome this difficulty a "pseudo" heat transfer coefficient for radiation, h_r, is defined, so as to express radiation heat transfer in terms of a temperature difference, as follows:

$$q = A h_r (T_1 - T_2) \tag{3.60}$$

For exchange between two black bodies, h_r is obtained by comparing Eq. (3.60) with Eq. (3.54). The result is:

$$h_r = \sigma \frac{T_1^4 - T_2^4}{T_1 - T_2} \tag{3.61}$$

The combined convection–radiation heat transfer coefficient is the sum of the convection heat transfer coefficient h_c and the radiation heat transfer coefficient h_r. Since both transfers occur on the surface of the solid, the combined coefficient is also called the "combined surface heat transfer coefficient".

3.7 Heat exchangers

Heat exchangers are devices for the exchange of heat between two fluids separated by a heat-conducting partition. Heat exchangers are extensively used in the food industry for heating (e.g., pasteurizers), cooling (chilled water generators) and heat-induced phase change (freezing, evaporation). Each one of the two fluids may be confined or unconfined (free), stagnant or flowing. The partition is a heat-conducting solid wall, usually made of metal.

The design of a heat exchanger usually involves two main domains, namely thermal analysis and hydraulic calculations. This section will deal only with the thermal performance of heat exchangers.

3.7.1 Overall coefficient of heat transfer

Consider heat exchange between two fluids A and B, separated by a heat-conducting wall (Figure 3.10).

Assume $T_A > T_B$. Heat travels from A to B, through three thermal resistances in series:

1. The resistance to convective heat transfer from fluid A to the contact area of the wall
2. The resistance of the wall to conductive heat transfer
3. The resistance to convective heat transfer from the wall surface to fluid B.

The rate of heat transfer, driven by the overall temperature difference $T_A - T_B$ is:

$$q = A \frac{T_1 - T_2}{\frac{1}{h_1} + \frac{x}{k} + \frac{1}{h_2}} \tag{3.62}$$

where:

A = heat transfer area, m^2
h_1, h_2 = convection heat transfer coefficients on the side of A and B respectively, $W \cdot m^{-2} \cdot K^{-1}$
T_1, T_2 = bulk temperatures of A and B respectively, K

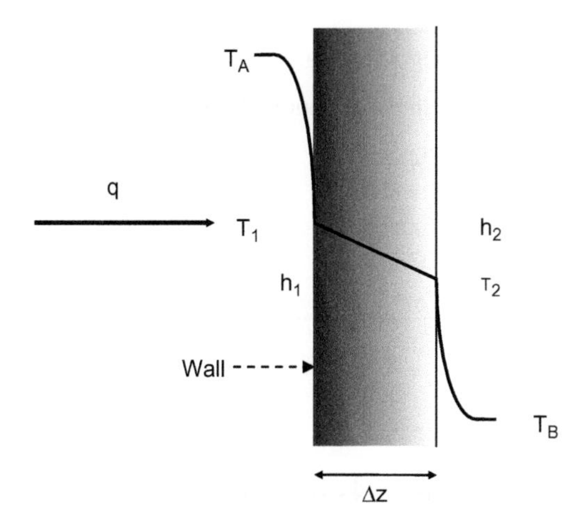

FIGURE 3.10

Resistances in overall heat transfer.

x = thickness of the wall, m
k = thermal conductivity of the wall, $W \cdot m^{-1} \cdot K^{-1}$.

An overall coefficient of heat transfer, U ($W \cdot m^{-2} \cdot K^{-1}$), is defined as follows:

$$\frac{1}{U} = \frac{1}{h_1} + \frac{x}{k} + \frac{1}{h_2} \tag{3.63}$$

Substituting in Eq. (3.62):

$$q = UA\Delta T \tag{3.64}$$

The overall coefficient U is a fundamental concept in heat exchange. It is affected by a large number of variables, such as the properties of the two fluids, flow conditions, flow pattern, geometry, physical dimensions, the thickness and the thermal conductivity of the wall, etc. The three resistances mentioned above are usually not of equal significance. If, for example, the medium on one side of the exchanger is a condensing vapor (steam) and on the other side is a viscous liquid, the thermal resistance on the steam side may be neglected. If the partition consists of a thin metal of high thermal conductivity, such as copper, the resistance of the wall may be neglected. The magnitude of U for different common cases of heat exchange is given in Table 3.6.

3.7.2 Heat exchange between flowing fluids

In continuous heat exchangers, both fluids are in movement. There are three main types of flow patterns: parallel, countercurrent and cross-flow (Figure 3.11).

Table 3.6 Typical Order of Magnitude Values of U

Fluid A	Fluid B	Heat Exchanger Type	Typical U, $W \cdot m^{-2} \cdot K^{-1}$
Gas	Gas	Tubular	5–50
Gas	Liquid	Tubular	100–400
Liquid	Liquid	Tubular	200–800
Liquid	Steam	Tubular	500–1200
Liquid	Liquid	Plate	1000–3000

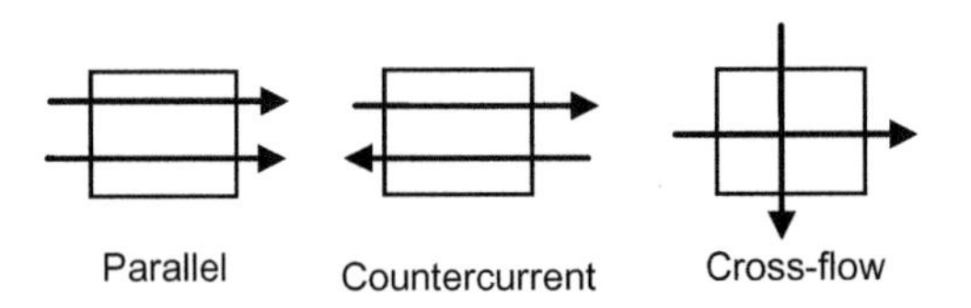

FIGURE 3.11

Types of flow patterns in heat exchangers.

Parallel and countercurrent flow are most common in liquid-to-liquid and liquid-to-condensing vapor heat exchange. Cross-flow exchange is particularly common for heating or cooling air.

As a result of flow, the temperature of each fluid and therefore the temperature drop for heat transfer may change from one point in the exchanger to another. Consider the simple shell and tube countercurrent heat exchanger and the element dA of the exchange surface, shown in Figure 3.12.

The hot and cold fluids will be indexed h and c, respectively.

By virtue of energy balance:

$$dq = G_c(cP)_c dT_c \tag{3.65}$$

$$dq = -G_h(cP)_h dT_h \tag{3.66}$$

where G represents the mass flow-rate, $kg \cdot s^{-1}$.

By virtue of rate of heat transfer (Eq. (3.60)):

$$dq = U\, dA\, (T_h - T_c) = U\, dA\, \Delta T \tag{3.67}$$

Equations (3.65) and (3.66) indicate that if the flow rates and the specific heats are constant, then both T_h and T_c change linearly with q. Therefore, their difference ΔT also changes linearly with q:

$$\frac{d(\Delta T)}{dq} = \frac{d(\Delta T)}{U\, dA\, \Delta T} = \frac{(\Delta T)_2 - (\Delta T)_1}{q} \tag{3.68}$$

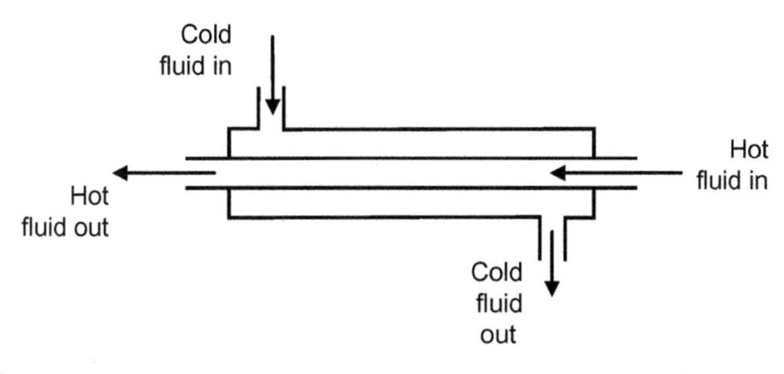

FIGURE 3.12

Heat transfer in a shell-and-tube heat exchanger.

After separation of the variables and integration between points 1 and 2 for the entire exchanger assuming constant U, we obtain:

$$q = UA\left[\frac{(\Delta T)_2 - (\Delta T)_1}{\ln\frac{(\Delta T)_2}{(\Delta T)_1}}\right] \tag{3.69}$$

The expression in brackets is the logarithmic mean temperature difference, abbreviated as LMTD or ΔT_{ml}. The total rate of heat exchange for the entire exchanger is, therefore:

$$q = UA\Delta T_{ml} \tag{3.70}$$

In the analysis above, the partition has been assumed to be flat and thin. Therefore the size of dA and the total area A were assumed to be the same for the cold and the hot sides. For a thick and curved surface (e.g., a thick-walled tube), the area would have to be averaged as well.

Equation (3.70) is valid for countercurrent and co-current (parallel) flow but not for mixed flow, such as a shell and tube exchanger with two passes in the tube side (Figure 3.12). Correction factors for various flow configurations are available in the literature (see, for example, Kreith, 1965, p. 493).

EXAMPLE 3.16

A tubular heat exchanger will be used to heat tomato paste from 60°C to 105°C, prior to holding, cooling, and aseptic filling into drums. The exchanger consists of two concentric tubes. Tomato paste is pumped through the inner tube. Steam condenses at 110°C in the annular space. Calculate the required heat transfer area.

Data:

Tomato paste mass flow-rate	$5000\ kg \cdot h^{-1}$
Tomato paste specific heat	$3750\ J \cdot kg^{-1} \cdot K^{-1}$
Convective heat transfer coefficient, tomato side	$200\ W \cdot m^{-2} \cdot K^{-1}$
Convective heat transfer coefficient, steam side	$20\,000\ W \cdot m^{-2} \cdot K^{-1}$
Wall of tube (stainless steel):	Thickness = 2 mm
	Thermal conductivity = $15\ W \cdot m^{-2} \cdot K^{-1}$

Solution

We apply Eq. (3.70):

$$q = UA\Delta T_{ml}$$

$$q = \frac{dm}{dt} C_P(T_2 - T_1) = \frac{5000}{3600} \times 3750 \times (105 - 60) = 234\,375 \ \mathrm{J \cdot s^{-1}}$$

$$\frac{1}{U} = \frac{1}{h_i} + \frac{1}{h_o} + \frac{z}{k} = \frac{1}{200} + \frac{1}{20000} + \frac{0.002}{15} \cong \frac{1}{200} \quad \Rightarrow \quad U = 200 \ \mathrm{W \cdot m_2{}^{-1} \cdot K^{-1}}$$

$$\Delta T_{ml} = \frac{(\Delta T)_2 - (\Delta T)_1}{\ln \frac{(\Delta T)_2}{(\Delta T)_1}} = \frac{(110 - 105) - (110 - 60)}{\ln \frac{110 - 105}{110 - 60}} = 16.6°\mathrm{C}$$

$$A = \frac{q}{U\Delta T_{ml}} = \frac{234\,375}{200 \times 16.6} = 70.6 \ \mathrm{m^2}$$

3.7.3 Fouling

The efficiency of heat exchangers may be considerably reduced during operation, as a result of the deposition of various kinds of materials with relatively low thermal conductivity on the exchange surface (Fryer and Belmar-Beiny, 1991). The nature of these materials depends on the fluids treated: denatured proteins in the case of milk, burnt pulp in the case of tomato juice, caramelized sugar in the case of syrups, scale in the case of hard water, etc. In the food industry, fouling may also result in loss of product quality (burnt flavor, dark specks, etc.). Fouling often determines the maximum operation time between two interruptions for cleaning.

If the rate of deposition of the fouling material and its thermal conductivity are known, the effect of fouling can be taken into account in heat exchanger calculations. The resistance of the fouling film to heat transfer is simply added to the overall resistance. In the absence of exact data, it is assumed that the resistance of the fouling film to heat transfer increases linearly with time:

$$\frac{1}{U} = \frac{1}{U_0} + \beta t \tag{3.71}$$

where:

U = overall heat transfer coefficient of the "dirty" exchanger
U_0 = overall heat exchange coefficient of the "clean" exchanger
β = fouling factor $\mathrm{W^{-1} \cdot m^{-2} \cdot K^{-1} \cdot s^{-1}}$.

It should be noted that Eq. (3.71) provides only an approximate estimate of the fouling resistance. The build-up of fouling is often not uniform over the entire exchange surface. For example, the rate of deposition is higher in locations where flow is slower.

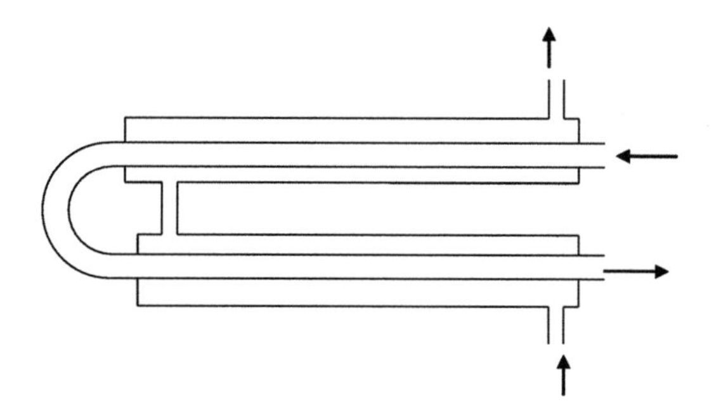

FIGURE 3.13

Basic structure of a tubular heat exchanger.

FIGURE 3.14

Tubular heat exchanger assembly in aseptic processing plant.

Photograph courtesy of Rossi & Catelli.

3.7.4 Heat exchangers in the food process industry

Although heating vessels and cooking kettles are, by definition, heat exchangers, only continuous in-flow heat exchangers will be discussed in this section. Because of the strict sanitary requirements, only a few of the many heat exchanger types utilized in the process industry are suitable for food applications.

- *Tubular heat exchangers*. The simplest representative of this group consists of a pair of concentric tubes (Figures 3.13 and 3.14). For ease of cleaning, the

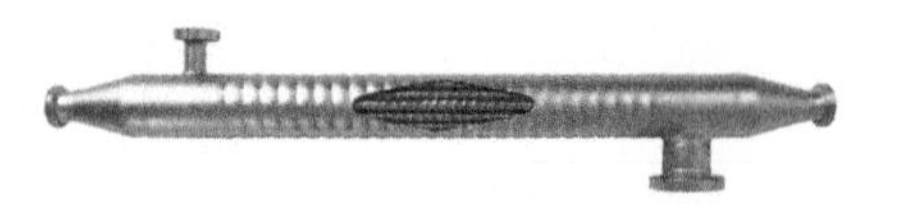

FIGURE 3.15

Shell-and-tube heat exchanger.

Photograph courtesy of Alva-Laval.

food product usually flows in the inner tube and the heating or cooling medium in the outer annular space. A variation of this type is the *triple tube* (or tube-in-tube-in tube) exchanger, consisting of three concentric tubes. The product is fed to the middle tube and the heating/cooling medium to the inner and outer tubes, thus providing heat transfer areas on both sides of the middle tube. The calculation of the overall heat transfer coefficient for this type of equipment is more complex than for a double tube exchanger (Batmaz and Sandeep, 2005). Asteriadou et al. (2010) have studied the performance of the triple tube exchanger in cooling mode with non-Newtonian fluids.

- *Shell-and-tube exchangers* (Figure 3.15). These are tubular exchangers consisting of bundles of parallel tubes inside a larger cylindrical jacket (shell). Again, the product is fed to the tube side. In a type of tubular exchanger known as a *Joule effect heater*, the tube wall is electrically heated (Fillaudeau et al., 2009). Tubular heat exchangers are particularly suitable for heating or cooling highly viscous products and where relatively high pressures must be applied. They are therefore utilized for the bulk in-flow sterilization of products containing solid particles, or for the heat treatment of cooling of tomato paste prior to aseptic packaging. Tubular heat exchangers are also the heat transfer component in tubular evaporators (see Chapter 21).
- *Plate heat exchangers*. Originally developed for the pasteurization of milk, plate heat exchangers (Figure 3.16) are now used for a vast variety of heating, cooling and evaporation applications in the food industry. They consist of a stack of corrugated thin metal plates, pressed together so as to form two continuous flow channels for the fluids exchanging heat. Gaskets are placed between the plates to prevent leakage.

  The advantages of the plate heat exchangers are:

  1. Flexibility – the capacity can be increased or decreased by adding or removing plates
  2. Sanitation – by opening the stack, both sides of the entire exchange area are made accessible for cleaning and inspection
  3. A high heat transfer coefficient, due to increased turbulence in the narrow flow channel
  4. Compactness – a high exchange surface to volume ratio.

  On the other hand, the narrow size of the flow channels results in high pressure drop and limits its use to low-viscosity fluids not containing large suspended particles. The need for gaskets is also a disadvantage.

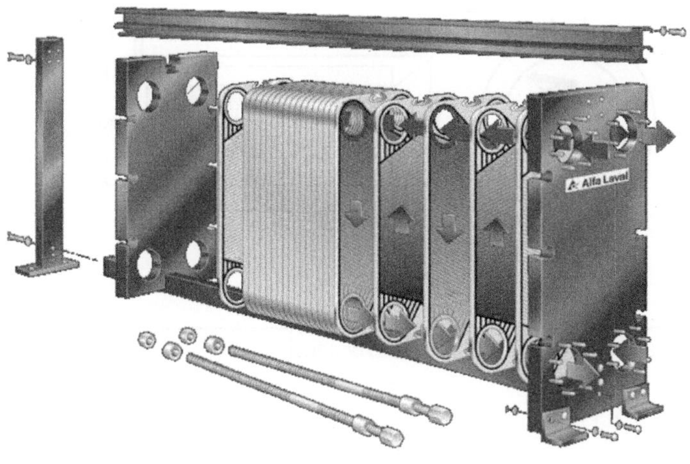

FIGURE 3.16

Plate heat exchanger.

Photograph courtesy of Alfa-Laval.

- *Scraped surface heat exchangers*. These consist of a jacketed cylinder equipped with a central rotating dasher with scraping blades (Figure 3.17). They can be horizontal or vertical. The product is fed into the cylinder, and the rapidly (600–700 rpm) rotating dasher spreads, scrapes and moves the product as a film over the wall. The heating or cooling fluid is fed into the jacket. Scraped surface heat exchangers are used for heating and cooling highly viscous fluids, and for slush-freezing. Continuous ice cream freezers and slush freezers are, essentially, scraped surface heat exchangers with a refrigerant evaporating in the jacket. The scraped surface exchanger is an expensive piece of equipment, both in price and in operating costs (moving parts).

3.8 Microwave and radio frequency (RF) heating

Microwave and radio frequency heating are special forms of radiative heat transfer, but they differ in many ways from heat transfer by thermal radiation discussed in Section 3.6. Both are based on the transfer of energy in the form of electromagnetic waves. The term *microwaves* applies to electromagnetic radiation in the wavelength range of 0.1–1 m in air, corresponding to a frequency range of 0.3–3 GHz. Radio frequency heating uses lower frequencies, in the range of 3 kHz to 300 MHz. For communication safety reasons, only certain frequencies are allowed by the authorities. In the USA, the permissible frequencies are 915 and 2450 MHz for microwaves, and 13.56, 27.12 and 40.68 MHz for RF (Piyasena et al., 2003). Microwaves and RF waves are, like any other radiation,

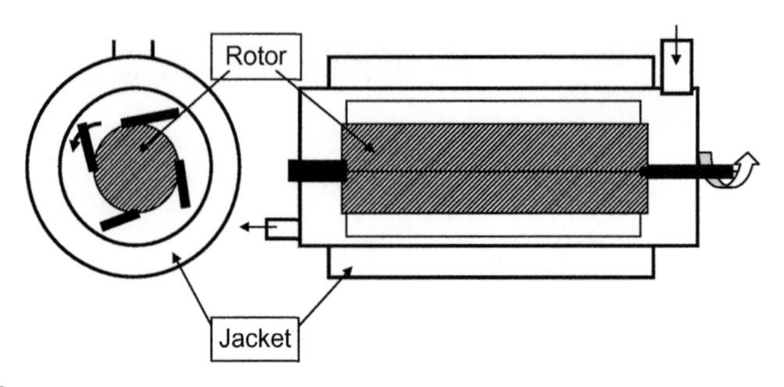

FIGURE 3.17

Basic structure of a scraped-surface heat exchanger.

partly reflected, partly transmitted and partly absorbed. The absorbed fraction generates heat. Unlike infrared radiation, however, microwaves and RF waves penetrate deep into the receiving body (see below regarding effect of frequency on depth of penetration). In microwave and RF heating, heat is generated deep inside the irradiated material and the rate of energy input does not depend on thermal conductivity or surface resistance.

Microwave heating is mainly used in household and food service (e.g., restaurants, airlines) ovens, to thaw frozen products, to heat food, to cook, to bake and even to boil water. First introduced in the 1950s, the microwave oven has become the main food heating device in modern homes and in many classes of food service establishments (Datta and Ananesthwaran, 2001; Meredith, 1998). Consequently, the food processing industry dedicates considerable effort to developing products and packages compatible with the capabilities and limitations of the domestic microwave oven. As to industrial applications, the use of microwave heating in the food industry is still limited to a relatively small but increasing number of cases where this method offers clear technological and economic advantages over conventional heating. One of the more successful industrial applications of microwave heating is in the tempering (partial thawing) of frozen meat and fish prior to cutting (Schiffmann, 2001). Microwave drying is also used in special cases, such as the dehydration of hydrocolloid gums.

Radio frequency heating is much less widespread. It has been also much less intensely studied. It is not used in homes and its use in industry has been limited. Reported applications include post-baking drying of cookies, tempering and thawing of frozen meat, cooking of ham and roasting of cocoa beans (Piyasena et al., 2003). It has also been suggested as a technique for the disinfestation of dry cereals, flours and nuts, and its use for the disinfestation of legumes has been studied by Wang et al. (2010). According to Tiwari et al. (2011), heating non-uniformity is a major problem in the application of RF heating in dry foods.

3.8.1 Basic principles of microwave and RF heating

Unlike thermal radiation, the heating action of microwaves strongly depends on the chemical composition of the material being irradiated. Microwaves interact mainly with polar molecules and charged particles. In the case of foods, by far the most important interaction is with the molecules of water. As the electromagnetic field of the microwaves alternates at high frequency, the dipoles of the material rotate in their attempt to align themselves with the field. Part of the energy of the waves provides the increase in kinetic energy associated with these rotational oscillations and is converted to intermolecular friction work and finally to heat. Thus, heat is generated *inside the material* as the microwaves penetrate it. This is, in principle, the phenomenon of dielectric heating. This phenomenon involves two characteristics of the dielectric material, namely:

- The relative dielectric permittivity ε', linked to the concentration of polar molecules
- The "loss factor" ε'' , representing the fraction of the microwave energy converted to heat.

(Note: In electrical devices such as capacitors the ε'' represents a "loss", since the generation of heat is an undesirable side effect. In microwave heating the ε'' represents a "gain", since conversion to heat is the desirable effect.)

The ratio of the loss factor to the relative permittivity is the "loss tangent", tan δ.

$$\tan\delta = \frac{\varepsilon''}{\varepsilon'} \tag{3.72}$$

The rate of heat generation per unit volume of material W_V ($W \cdot m^{-3}$) is given by:

$$W_V = 2\pi\varepsilon_0 f E^2 \varepsilon'' \tag{3.73}$$

where:

ε_0 = permittivity of vacuum = 8.86×10^{-12} farad/m
f = frequency of the field, s^{-1}
E = strength of the electric field (potential gradient), volt $\cdot$ m^{-1}.

It follows that the rate of temperature increase in microwave heating is (Orfeuil, 1987):

$$\frac{dT}{dt} = \frac{2\pi f\,\varepsilon_0\,\varepsilon'' E^2}{c_P\,\rho} \tag{3.74}$$

Equation (3.73) indicates that if E and f are constant, the rate of heat generation per unit volume is proportional to the loss factor. Table 3.7 gives the loss factor of a number of materials. Water, with a loss factor of about 12 at 3 GHz and 25°C is by far the most important component of foods with respect to

Table 3.7 Dielectric Properties of Some Foods

		ε′/ε″	
Product	Temperature (°C)	1 GHz	2.5 GHz
Apple	19	61/9	57/12
Beef (raw)	25	56/22	52/17
Bread dough	15	24/12	20/10
Fish (cod)	−25	4/0/7	
Fish (cod)	10	66/40	
Potato	22	66/30	61/19

Data from Meredith (1998).

microwave heating. In contrast, polyethylene with a loss factor four orders of magnitude lower does not absorb any appreciable amount of microwave energy.

The effect of the field frequency is more complicated. On one hand, the rate of heat generation is proportional to the frequency f, according to Eq. (3.73). On the other hand, the loss factor itself is frequency-dependent (Metaxas and Meredith, 1983). It increases to a maximum at a critical frequency. Furthermore, the loss factor is temperature-dependent and therefore changes in the course of microwave heating. Nevertheless, the overall practical effect of an increase in frequency is an increase in heating rate. Consequently, RF heating is considerably slower than MW heating. The remarkable rapidity of microwave heating is the main reason for the success of the microwave oven as a home device.

Another important factor to be considered is the *depth of penetration*. Depth of penetration is conventionally defined as the distance over which the wave energy is reduced to 1/e of its initial level. The depth of penetration is inversely proportional to the frequency. The penetration power of MW is less than that of RF. Consequently, microwave heating is not suitable for large items because of serious temperature heterogeneity.

3.9 Ohmic heating

3.9.1 Introduction

In general, ohmic heating is a technique whereby a material is heated by passing an electric current through it (Knirsch et al., 2010). In a more restricted sense, ohmic heating refers to processes whereby an electric current is passed through fluids with the specific purpose of rising their temperature rapidly. This technique has been in some use for many years for rapid heating of water in some household appliances and for on-demand heating of water in food service outlets. Its application to food is fairly recent.

In ohmic heating, as in microwave heating, heat as such is not delivered to the food through its surface but is generated inside the food by conversion of another type of energy to heat. A temperature gradient is not required for heating to occur. The rate of heating does not depend on thermal conductivity or heat transfer coefficients. Consequently, heating is rapid and uniform, at least theoretically. Since there is no heat transfer surface, there is practically no fouling (Stancl and Zitny, 2010). These characteristics give ohmic heating a distinct advantage over conventional heat transfer, particularly in the following cases:

1. Heating of liquid foods containing relatively large (2- to 3-cm) particles, such as fruits in syrup (Sastry and Palaniappan, 1992).
2. Thermal processes where even moderate temperature gradients (i.e., even light overheating of parts of the food) cannot be tolerated. A typical example of this type of application is the pasteurization of liquid egg, where temperatures slightly higher than the pasteurization temperature would cause coagulation of the product.
3. Heating of highly viscous but pumpable foods.

Basically, the ohmic heater consists of one or several pairs of electrodes to which voltage is applied. The food flows in the space between the electrodes and acts as a moving electrical resistance. (Figure 3.18). Several commercial systems for ohmic heating of foods are available but, generally, the industrial application of ohmic heating has not been extensive, mainly for economic reasons and because of some technical problems (to be discussed later). The commercial applications are almost exclusively in the area of continuous thermal processing (i.e., heating for the destruction of microorganisms and enzymes; see Chapter 17).

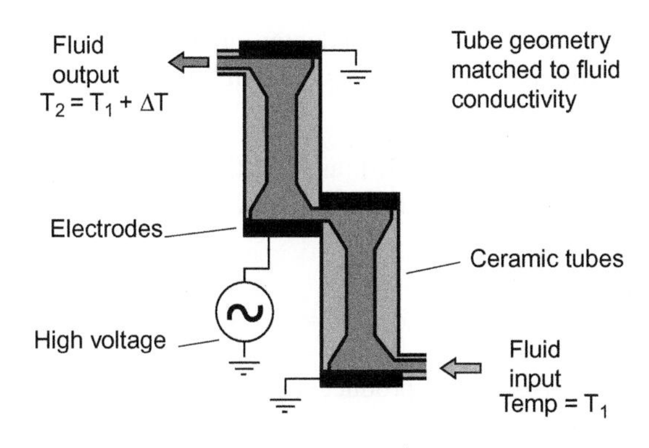

FIGURE 3.18

Principle of ohmic heating.

Figure Courtesy of Raztec Ltd.

3.9.2 Basic principles

The rate of heat generation in a purely resistive conductor traversed by an electric current is governed by Joule's Law (James Prescott Joule, English physicist, 1818–1889), formulated in Eq. (3.75):

$$q = I^2R = \frac{E^2}{R} \tag{3.75}$$

where:

q = rate of heat dissipation, W
I = current intensity, amp
R = electrical resistance of the conductor, ohm
E = voltage applied, volt.

The electric current applied in ohmic heating is alternating current, in order to avoid electrolysis. Most systems now use common commercial line frequency (50 or 60 Hz), although higher frequencies (in the range of kHz and MHz) have been found to result in a higher heating rate and less texture damage in some products (Shynkaryk et al., 2010). The voltage is regulated so as to reach the desired final temperature, despite fluctuations in feed composition and flow-rate.

The electrical resistance of a conductor depends on the geometry of the conductor and its electrical conductivity as shown in Eq. (3.76):

$$R = \frac{L}{A \cdot k_e} \tag{3.76}$$

where:

L = length of the path of the current = length of the conductor, m (in practice, L is the distance between the electrodes)
A = cross-sectional area of the conductor, normal to the direction of the current, m^2
k_e = electrical conductivity of the conductor, $ohm^{-1} \cdot m^{-1} = S \cdot m^{-1}$
(S = Siemens = 1/ohm, which replaces the unit "mho" still used to express the conductance of insulators and water).

Combining Eqs (3.74) and (3.75) and writing q in terms of temperature increase in the food as it passes through the heater at a mass flow-rate G, we obtain:

$$q = G \cdot c_p \cdot \Delta T = \frac{E^2 \cdot A \cdot k_e}{L} \tag{3.77}$$

where G = mass flow-rate of the product, $kg \cdot s^{-1}$ and C_p = specific heat of the food, $J \cdot kg^{-1} \cdot K^{-1}$.

Equation (3.77) indicates that in a heater with a fixed geometry (fixed A and L), operating at constant mass flow-rate, the temperature increment ΔT is controlled by adjusting the voltage (Sastry and Li, 1996).

The electrical conductivity of foods depends strongly on their composition and particularly on their content of electrolytes (salt) and moisture (Fryer and Li, 1993). It is also temperature-dependent. Contrary to metals, where the conductivity decreases with temperature, the conductivity of foods generally increases with temperature (Szczesniak, 1983; Reznick, 1996). Therefore it can be said that, in most cases, ohmic heating is an auto-accelerating process.

Table A.14 in the Appendix shows electrical conductivity data for some materials.

In practice, the theoretical uniformity of temperature is not achieved for a number of reasons:

1. If the food is not homogeneous, the different phases may have different electrical conductivities. Thus, for example, in a product containing solid particles, the heating rate of the particle may be faster or slower than that of the sauce or syrup (Sarang et al., 2007). In a piece of meat, the fat would be colder than the lean parts. Thus, the process is not entirely insensitive to variables such as viscosity, flow rate, geometry and thermal conductivity, as internal heat transfer by convection and conduction does occur by virtue of local temperature differences.
2. In continuous ohmic heating, the temperature rise of any part of the food depends on its residence time between the electrodes. Therefore, temperature uniformity requires even residence time (i.e., plug flow), and this is not always easy to achieve. In fact, considerable residence time non-uniformity has been observed in continuous ohmic heating of foods containing suspended particles (Sarang et al., 2009).

3.9.3 Applications and equipment

Potentially, ohmic heating could be used in a large number of thermal operations such as cooking, drying, blanching, thawing, etc. (Bozkurt and Icier, 2010; Icier et al., 2010; Knirsch et al., 2010). However, most commercially available systems use ohmic heating for pasteurization or sterilization (e.g., Leizerson and Shimoni, 2005). The technical problems associated with this particular application and their solutions are the most important aspects of ohmic heating system design and operation (Larkin and Spinak, 1996). The limitations and advantages of ohmic heating as a particular method of thermal processing will be discussed in Chapter 17.

The equipment for ohmic heating is relatively simple and extremely compact. The optimal field frequency to be used is an issue. Working on the ohmic heating of peaches, Shynkaryk et al. (2010) observed that greater damage to tissue occurred at lower frequency. Nevertheless, industrial units now use alternating current at normal line frequency. Consequently, the bulky and energetically wasteful frequency converters formerly used in older systems are no longer necessary. In a particular commercial system, the flow channel is made of ceramic material and the electrodes are made of graphite, to reduce electrode erosion. Graphite electrodes also seem to be less prone to fouling when heating milk (Stancl and Zitny, 2010).

References

Asteriadou, K., Hasting, A., Bird, M., Melrose, J., 2010. Modeling heat exchanger performance for non-Newtonian fluids. J. Food Proc. Eng. 33 (2), 1010–1035.

Batmaz, E., Sandeep, K.P., 2005. Overall heat transfer coefficients and axial temperature distribution in a triple tube heat exchanger. J. Food Proc. Eng. 31 (2), 260–279.

Bimbenet, J.J., Loncin, M., 1995. Bases du Génie des Procédés Alimentaires. Masson, Paris.

Bozkurt, H., Icier, F., 2010. Ohmic cooking of ground beef: effects on quality. J. Food Eng. 96 (4), 481–490.

Carslaw, H.S., Jaeger, J.C., 1986. Conduction of Heat in Solids, second ed. Oxford Science Publications, Oxford, UK.

Cuesta, F.J., Lamúa, M., 2009. Fourier series solution to the heat conduction equation with an internal heat source linearly dependent on temperature: application to chilling of fruit and vegetables. J. Food Eng. 90 (2), 291–299.

Datta, A., Anantheswaran, R.C. (Eds.), 2001. Handbook of Microwave Technology for Food Applications. Marcel Dekker, New York, NY.

Fillaudeau, L., Le-Nguyen, K., André, C., 2009. Influence of flow regime and thermal power on residence time distribution in tubular joule effect heaters. J. Food Eng. 95 (3), 489–498.

Fryer, P., Li, Z., 1993. Electrical resistance heating of foods. Trends Food Sci. Technol. 4 (11), 364–369.

Fryer, P.J., Belmar-Beiny, M.T., 1991. Fouling of heat exchangers in the food industry: a chemical engineering perspective. Trends Food Sci. Technol. 2 (1), 33–37.

Huang, L., Liu, L.-S., 2009. Simultaneous determination of thermal conductivity and thermal diffusivity of food and agricultural materials using a transient plane-source method. J. Food Eng. 95 (1), 179–185.

Icier, F., Izzetoglu, G.T., Bozkurt, H., Ober, A., 2010. Effects of ohmic thawing on histological and textural properties of beef cuts. J. Food Eng. 99 (3), 360–365.

Knirsch, M.C., dos Santos, C.A., de Oliveira Soares Vicente, A.A.M., Vessoni Penna, C.T., 2010. Ohmic heating, a review. Trends Food Sci. Techn. 21 (9), 436–441.

Kreith, F., 1965. Principles of Heat Transfer, second ed. International Textbook Company, Scranton, PA.

Kreith, F., 1973. Principles of Heat Transfer. Dun-Donnelley Publisher, New York, NY.

Larkin, J.W., Spinak, S.H., 1996. Safety considerations for ohmically heated, aseptically processed, multiphase low-acid foods. Food Technol. 50 (5), 242–245.

Leizerson, S., Shimoni, E., 2005. Effect of ultrahigh-temperature continuous ohmic heating treatment on fresh orange juice. J. Agric. Food Chem. 53 (9), 3519–3524.

Lewis, W.K., Whitman, W.G., 1924. Principles of gas absorption. Indust. Eng. Chem. 16 (12), 1215–1224.

McAdams, W.H., 1954. Heat Transmission, third ed. McGraw-Hill, New York, NY.

Meredith, R., 1998. Engineering Handbook of Industrial Microwave Heating. The Institution of Electrical Engineers, London, UK.

Metaxas, A.C., Meredith, R.J., 1983. Industrial Microwave Heating. Peter Peregrinus Ltd, London, UK.

Nitin, N., Gadiraju, R.P., Karwe, M.V., 2006. Conjugate heat transfer associated with a turbulent hot air jet impinging on a cylindrical object. J. Food Proc. Eng. 29 (4), 386–399.

Orfeuil, M., 1987. Electrical Process Heating. Battelle Press, Columbus, OH.

Piyasena, P., Dussault, C., Koutchma, T., Ramaswamy, H.S., Awuah, G.B., 2003. Radio frequency heating of foods: principles, applications and related properties – a review. Crit. Rev. Fod Sci. Nutr. 43 (6), 587–606.

Reznick, D., 1996. Ohmic heating of fluid foods. Food Technol. 50 (5), 250–251.

Sakin, M., Kaymak-Ertekin, F., Ilicali, C., 2009. Convection and radiation combined surface heat transfer coefficient in baking ovens. J. Food Eng. 94 (3–4), 344–349.

Sarang, S., Sastry, S.K., Gaines, J., Yang, T.C.S., Dunne, P., 2007. Product formulation for ohmic heating: blanching as a pretreatment method to improve uniformity of solid–liquid food mixtures. J. Food Sci. 72 (5), 227–234.

Sarang, S., Heskitt, B., Tulsiyan, P., Sastry, S.K., 2009. Residence time distribution (RTD) of particulate foods in a continuous flow pilot-scale ohmic heater. J. Food Sci. 74 (6), E322–E327.

Sastry, S.K., Li, Q., 1996. Modeling the ohmic heating of foods. Food Technol. 50 (5), 246–248.

Sastry, S.K., Palaniappan, S., 1992. Ohmic heating of liquid–particle mixtures. Food Technol. 46 (12), 64–67.

Schiffmann, R.F., 2001. Microwave processes for the food industry. In: Datta, A., Anantheswaran, R.C. (Eds.), Handbook of Microwave Technology for Food Applications. Marcel Dekker, New York, NY.

Shynkaryk, M.V., Ji, T., Alvarez, V.B., Sastry, S.K., 2010. Ohmic heating of peaches in the wide range of frequencies (50 Hz to 1 MHz). J. Food Sci. 75 (7), E493–E500.

Singh, P.R., Heldman, D.R., 2003. Introduction to Food Engineering, third ed. Academic Press, Amsterdam, The Netherlands.

Stancl, J., Zitny, R., 2010. Milk fouling at direct ohm heating. J. Food Eng. 99 (4), 437–444.

Sweat, V.E., 1986. Thermal properties of foods. In: Rao, M.A., Rizvi, S.S.H. (Eds.), Engineering Properties of Foods. Marcel Dekker, New York, NY.

Szczesniak, A.S., 1983. Physical properties of foods: what they are and their relation to other food properties. In: Peleg, M., Bagley, E.B. (Eds.), Physical Properties of Foods. Avi Publishing Co. Inc., Westport, CT.

Tiwari, G., Wang, S., Tang, J., Birla, S.R., 2011. Computer simulation model development and validation for radio frequency (RF) heating of dry food materials. J. Food Eng. 105 (1), 48–55.

Wang, S., Tiwari, G., Jiao, S., Johnson, J.A., Tang, J., 2010. Developing postharvest disinfestation treatments for legumes using radio frequency energy. Biosystems Eng. 105 (3), 341–349.

CHAPTER

Reaction Kinetics

4

4.1 Introduction

Foods are highly reactive systems. Chemical reactions occur constantly between the component substances of foods, and between foods and their surroundings (air, packaging materials, equipment surfaces, etc.). Fresh fruits and vegetables undergo rapid post-harvest biochemical changes. Post-mortem reactions deeply affect the characteristics of meat and fish. Numerous types of chemical and biochemical reactions occur during food processing and storage. The systematic study of reactions in foods is the subject matter of food chemistry and food biochemistry. However, the *rate* at which these reactions take place is of utmost interest to the food process engineer (Earle and Earle, 2003; Toledo, 2007; Villota and Hawkes, 1992). The following are some of the most important applications of reaction kinetics in food engineering:

- Calculation of thermal processing for the destruction of microorganisms and for the inactivation of enzymes.
- Optimization of thermal processes with respect to quality.
- Optimization of processes with respect to cost.
- Prediction of the shelf life of foods as a function of storage conditions.
- Calculation of refrigeration load in the storage of respiring agricultural produce
- Development of time-temperature integrators.

Reactions in food processing may be classified into two groups (Toledo, 2007):

1. *Desirable or induced reactions*. These are reactions that are intentionally induced by the process in order to produce a desirable transformation in the food. The pyrolysis of carbohydrates during coffee roasting, the hydrolysis of collagen when meat is cooked, the conversion of lactose to lactic acid by fermentation in the production of yogurt, and the hydrogenation of oils to produce solid fats are examples of intentionally induced reactions.
2. *Undesirable reactions*. These are chemical or biochemical reactions that occur during processing or storage and result in undesirable effects on food quality. Maillard-type browning in lemon juice, onset of rancidity as a result of lipid

Food Process Engineering and Technology. DOI: http://dx.doi.org/10.1016/B978-0-12-415923-5.00004-6
© 2013 Elsevier Inc. All rights reserved.

oxidation in nuts and crackers, and spoilage reactions induced by microorganisms are examples of this group of reactions.

This classification is, obviously, imperfect. The same reaction may be desirable in one case and undesirable in another. In the production of wine, alcoholic fermentation is induced; in tomato juice, the same reaction is considered spoilage. Enzymatic browning is essential for the development of color in tea; in potatoes and apples, it is a defect. Very often, the characterization of a reaction as desirable or undesirable is a matter of degree. Proteolysis in some cheeses is essential for the development of flavor. Too much of it is spoilage.

Another way to classify reactions in food is based on the distinction between enzyme-catalyzed (enzymatic) and non-enzymatic reactions. Very often, non-enzymatic reactions are called *chemical reactions* while reactions where enzymes or cells are involved are distinguished as *biochemical reactions*.

4.2 Basic concepts

4.2.1 Elementary and non-elementary reactions

Elementary reactions are well-defined reactions resulting from a single collision between two (and rarely three) molecules or ions. The neutralization of OH^- with H^+ is an elementary reaction. *Non-elementary reactions* consist of a series of elementary reactions. Although the kinetics of each elementary reaction in the series affects the rate of the total change, non-elementary reactions are often treated as a "black box", where only the rate of disappearance of the reactants entering the "box" or the rate of formation of the final products is considered. We shall refer to non-elementary reactions as *overall reactions*. The thermal inactivation of microorganisms or the formation of dark pigments in the Maillard reaction are non-elementary reactions. The rate of Maillard browning is usually expressed as the rate of colorimetric darkening or the rate of formation of an intermediate molecule such as hydroxymethyl furfural (HMF). Most reactions of interest in food processing are of the overall type.

4.2.2 Reaction order

Consider an elementary chemical reaction between the molecules A and B, resulting in the formation of the molecules E and F, according to Eq. (4.1):

$$aA + bB \xrightarrow{k} eE + fF \tag{4.1}$$

The rate of a reaction is defined as the rate at which the number of molecules of the reacting species increases or decreases with time. In constant-volume processes, the number of molecules may be replaced by concentration. According to the law of mass action, the rate of any reaction at a given time is proportional to the concentrations of the reacting substances, raised to a power equal to the number of molecules participating in the reaction. The proportionality constant is called the *rate constant* and is designated by the symbol k. Thus, the rate of the reaction shown in Eq. (4.1), in terms of the disappearance of the species A, is:

$$-\frac{dC_A}{dt} = k(C_A)^a(C_B)^b \tag{4.2}$$

where C represents the concentration of the different chemical species.

Theoretically, reverse reactions occur simultaneously with each reaction. Assume that the rate constant of the reverse reaction shown in Eq. (4.3) is k′:

$$eE + fF \xrightarrow{k'} aA + bB \tag{4.3}$$

The rate of disappearance of A for the complete reversible reaction is then:

$$-\frac{dC_A}{dt} = k(C_A)^a(C_B)^b - k'(C_E)^e - (C_F)^f \tag{4.4}$$

In the case of the reactions of interest in food processing, the rate constant of the reverse reaction is very often negligible. Furthermore, the concentration of one of the reactants is usually much higher than that of the other and therefore not affected considerably by the reaction – for example, the concentration of water in hydrolysis, which is often (but not always) carried out with water in large excess. Then, the reaction may be written as a pseudo-monomolecular non-reversible process and Eq. (4.4) becomes:

$$-\frac{dC_A}{dt} = k(C_A)^n \tag{4.5}$$

The exponent n in Eq. (4.5) is called the *reaction order*. In elementary reactions it is usually equal to the number of molecules of the reactant participating in the reaction (molecularity, e.g., $n = a$ in the example given, with respect to the disappearance of A). In non-elementary reactions, such as the thermal inactivation of microorganisms, the reaction order is simply an experimental value with no meaning whatsoever as to the molecularity or mechanism of the reaction.

Zero-order kinetics

If $n = 0$, the rate of the reaction becomes equal to the rate constant k, independent of the concentration of the reactants:

$$-\frac{dC_A}{dt} = k \tag{4.6}$$

The SI units of k, for zero-order reactions, are $mol \cdot s^{-1}$. Assuming constant k (constant temperature, pH, etc.), integration between $t = 0$ and $t = t$ gives:

$$C_A = C_{A0} - kt \tag{4.7}$$

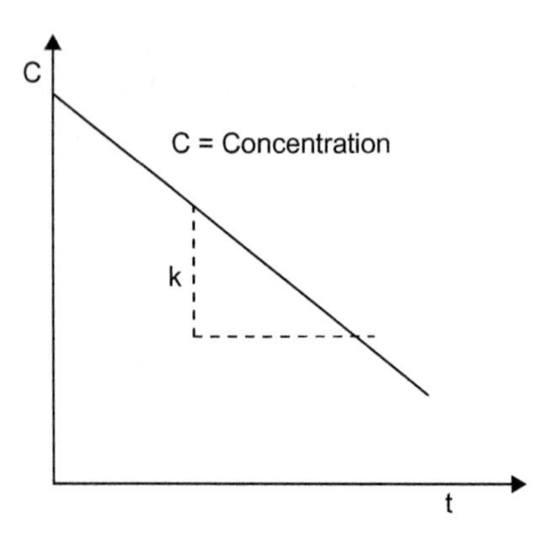

FIGURE 4.1

Graphical representation of zero-order kinetics.

The concentration of A decreases linearly with time (Figure 4.1). The slope of the C vs t line is $-$k. Zero-order food reactions are not very common. Zero-order kinetics has been reported in some cases of non-enzymatic browning, caramelization and lipid oxidation (Bimbenet *et al.*, 2002). Zero-order behavior for the accumulation of HMF (hydroxymethyl furfural, a key intermediate in non-enzymatic browning) was reported by Burdurlu *et al.* (2005).

First-order kinetics

With n = 1, Eq. (4.5) becomes:

$$-\frac{dC_A}{dt} = k\, C_A \tag{4.8}$$

The SI units of k for first-order reactions are s^{-1}. Assuming constant k as before, integration gives:

$$\ln\frac{C_A}{C_{A0}} = -\, kt \tag{4.9}$$

The logarithm of C_A decreases linearly with time (Figure 4.2). According to Eq. (4.9) reactant A cannot be totally depleted. A useful notion in first-order kinetics is that of *half life*, $t_{1/2}$. The half life of A in this reaction is the time required for the quantity of A to be reduced to half its original value. It follows that:

$$t_{1/2} = \frac{\ln 2}{k} \tag{4.10}$$

First-order kinetics is approximated by many phenomena of interest in food processing. Thermal destruction of microorganisms is treated as a first-order

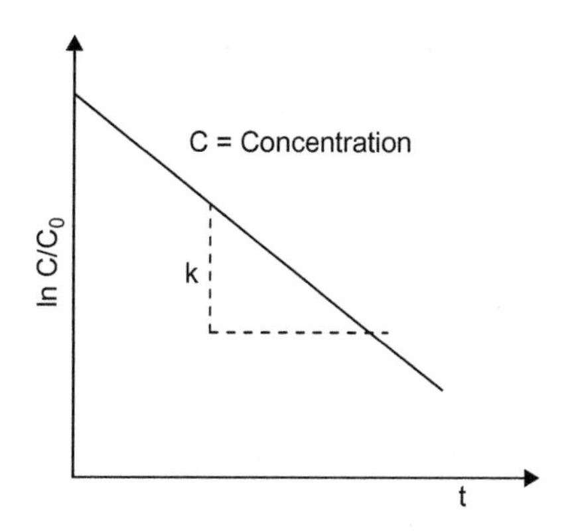

FIGURE 4.2

Graphical representation of first-order kinetics.

reaction. Non-enzymatic browning of citrus juice concentrates has been reported to follow first-order kinetics (Berk and Mannheim, 1986). First-order kinetics was confirmed for Vitamin C loss in orange juice concentrate during storage (Burdurlu *et al.*, 2005). Mathematical methods for the determination of the rate constants of consecutive first-order reactions were developed and tested by Erdoğdu and Şahmurat (2007).

In many cases, the concentration of the reactant studied (e.g., reactant A) is high and the reaction rate is relatively slow. In this case, after a short time, the concentration of A is still very high and not very different from its original value. The initial rate of reaction is nearly constant, as in a zero-order reaction. This may be the reason for the contradictory publications in the literature, where the same reaction is reported as zero order in some and first order in others.

EXAMPLE 4.1

The average retention of ascorbic acid in concentrated orange juice, after 8 weeks of storage at 28°C, was reported to be 68% (Burdurlu *et al.*, 2005). Assuming first-order kinetics, calculate the rate constant of ascorbic acid loss at 28°C.

Solution

For first-order kinetics, Eq. (4.9) applies:

$$\ln\frac{C_A}{C_{A0}} = -kt$$

$$k = -\frac{1}{t}\ln\frac{C}{C_0} = -\frac{1}{8\times 7}\ln\frac{100-68}{100} = 0.02\ \text{day}^{-1}$$

4.2.3 Effect of temperature on reaction kinetics

All chemical reactions are accelerated when the temperature is increased. The relationship between the reaction rate constant k and the temperature is described in Eq. (4.11):

$$k = A \exp\frac{-E}{RT} \tag{4.11}$$

where:

A = a constant, named the "pre-exponential factor"; its units are the same as those of the rate constant k, which in turn depend on the order of the reaction
R = universal gas constant = $8.314\ \mathrm{kJ \cdot K^{-1} \cdot kmol^{-1}}$
T = absolute temperature, K
E = activation energy, $\mathrm{kJ \cdot kmol^{-1}}$.

Equation (4.11) is known as the *Arrhenius equation* after the Swedish chemist Svante August Arrhenius (1889–1927, Nobel Prize in Chemistry, 1903). A graphical representation of the Arrhenius equation is shown in Figure 4.3.

The *activation energy* actually represents the sensitivity of the reaction rate to changes in temperature. If k_1 and k_2 are the rate constants at temperatures T_1 and T_2, respectively, then:

$$\ln\frac{k_2}{k_1} = \frac{E(T_2 - T_1)}{RT_1T_2} \tag{4.12}$$

Arrhenius' Law is extensively used for the prediction of quality loss as a function of temperature (e.g., Labuza and Riboh, 1982).

Another way of representing the sensitivity of a reaction to a change in temperature is the factor known as Q_{10} or the *temperature quotient*. Q_{10} is the ratio of the rate constant of a reaction to that of the same reaction at a temperature

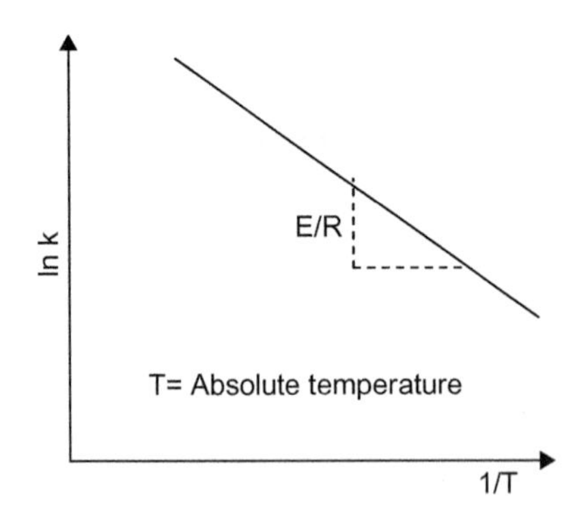

FIGURE 4.3

Graphical representation of the Arrhenius Law.

lower by 10°C. Applying this definition to Eq. (4.12), we obtain the relationship between the activation energy and the temperature quotient:

$$\ln Q_{10} = \frac{10\,E}{RT_1(T_1+10)} \cong \frac{10E}{RT_1^2} \tag{4.13}$$

Typical ranges of E and Q_{10} for different types of reactions are given in Chapter 17, Table 17.1.

In elementary reactions, the energy of activation has a physical meaning and is related to the reaction mechanism at the molecular level. In overall reactions of interest in food processing, the Arrhenius equation is simply an empirical approximation and E is an experimental parameter of that equation with no theoretical significance.

One of the applications of the Arrhenius model in food process engineering is the *accelerated storage test*. Investigation of the changes in food during normal storage may require a long time. The changes can be accelerated, however, by using a higher temperature for test storage. If the system is known to obey the Arrhenius Law and if the activation energy is known, the rate of change at the normal storage temperature can be calculated from the accelerated rate (Mizrahi *et al*., 1970; Labuza and Riboh, 1982).

EXAMPLE 4.2

It has been reported that the rate of an enzymatic reaction is increased by a factor of 3.2 if the reaction is carried out at 45°C (318 K) instead of 37°C (310 K). Calculate the energy of activation and the Q_{10} value.

Solution

Apply Eq. (4.12):

$$\ln\frac{k_2}{k_1} = \frac{E(T_2 - T_1)}{RT_1T_2} \quad \Rightarrow \quad \ln 3.2 = 1.163 = \frac{E(318-310)}{8.314\times 318\times 310} = \frac{E}{102\,450}$$

$$E = 102\,450 \times 1.163 = 119\,150 \ \text{kJ}\cdot\text{kg}^{-1}\cdot\text{mol}^{-1}$$

The Q_{10} value is calculated using Eq. (4.13):

$$\ln Q_{10} = \frac{10\,E}{RT_1(T_1+10)} = \frac{10\times 119\,150}{8.314\times 310\times 320} = 1.4446 \quad \Rightarrow \quad Q_{10} = 4.24$$

EXAMPLE 4.3

Pasteurized grapefruit juice is aseptically packaged in multi-layer containers and stored under refrigeration. The ascorbic content of the juice at time of packaging is 55 mg per 100 g. The nutritional information on the label specifies an ascorbic acid content of 40 mg per 100 g. What should be the maximum temperature of storage (assumed constant) be, if the product has to comply with the specification on the label after 180 days of storage?

Ascorbic acid loss in grapefruit juice follows first-order kinetics with a rate constant of 0.006 day^{-1} at 20°C. The activation energy of the reaction is 70 000 kJ/kmol.

Solution

First we calculate the rate constant at the storage temperature, using Eq. (4.9):

$$\ln\frac{C}{C_0} = -kt \quad \Rightarrow \quad \ln\frac{40}{55} = -k \times 180 \quad \Rightarrow \quad k = 0.00177\ \text{day}^{-1}$$

Next, we calculate the storage temperature T, using the Arrhenius law, substituting in the equation the values of k at 20°C (293 K), k at the (unknown) storage temperature, and the energy of activation.

$$\ln\frac{k_2}{k_1} = \frac{E(T_2 - T_1)}{RT_1T_2} \qquad \ln\frac{0.00177}{0.006} = \frac{70000(T-293)}{8.314 \times 293 \times T} = -1.221$$

$$T = 281\ \text{K} = 8°\text{C}$$

4.3 Kinetics of biological processes

4.3.1 Enzyme-catalyzed reactions

The rate of enzymatic reactions depends on a number of conditions, such as the concentration of the enzyme and the substrate, the temperature, pH, ionic strength, etc.

If the substrate is present in large excess, the rate of an enzymatic reaction is proportional to the concentration of the enzyme. This is the basis for the quantitative definition and determination of *enzyme activity*.

The effect of substrate concentration is more complex. The velocity of an enzymatic reaction, in terms of the rate of product formation is given by the well-known *Michaelis–Menten equation*:

$$v = v_{\max}\frac{s}{K_m + s} \tag{4.14}$$

where:

v = velocity of the reaction (rate of product formation)

v_{max} = a maximum value to which v tends asymptotically as the substrate concentration increases

K_m = a parameter of the reaction, known as the *Michaelis–Menten Constant*.

The effect of substrate concentration on the rate of a hypothetical enzymatic reaction, according to the Michaelis–Menten model is shown in Figure 4.4.

The rate of enzymatic reactions is strongly affected by the temperature. As the temperature is increased, the reaction rate increases up to a maximum value (at

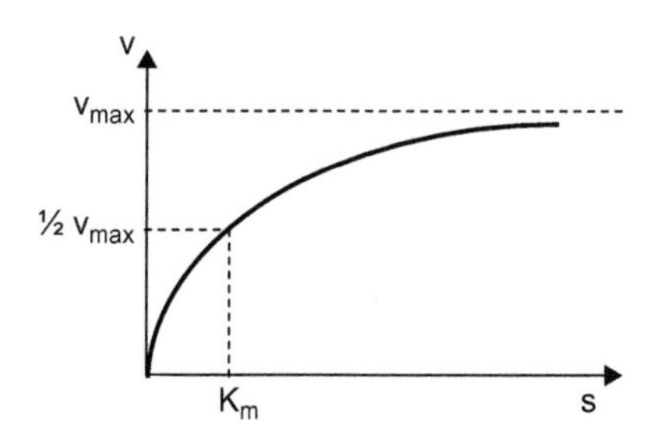

FIGURE 4.4

The Michaelis–Menten plot.

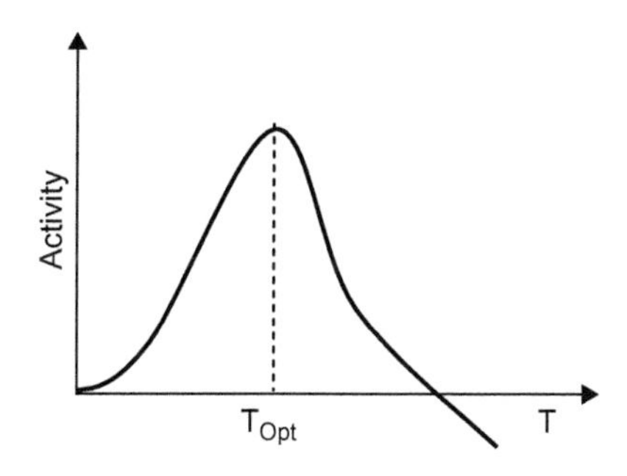

FIGURE 4.5

Effect of temperature on enzyme activity.

the optimal temperature) and then decreases (Figure 4.5). This behavior is not in contradiction with the Arrhenius model. The bell-shaped rate–temperature curve is the consequence of the simultaneous occurrence of two contradicting processes, namely the enzymatic reaction itself and the thermal inactivation of the enzyme, both enhanced by higher temperature.

The effect of pH on the rate of enzymatic reactions also follows a bell-shaped curve, with maximum activity at the optimal pH. Control of enzymatic activity by adjusting the pH is a frequently applied practice in food processing.

4.3.2 Growth of microorganisms

The term "growth of microorganisms" may be interpreted either as the increase in the number of living cells or as the increase in biomass. In this discussion, we shall study the kinetics of microbial growth with reference to the number of cells and not to their mass.

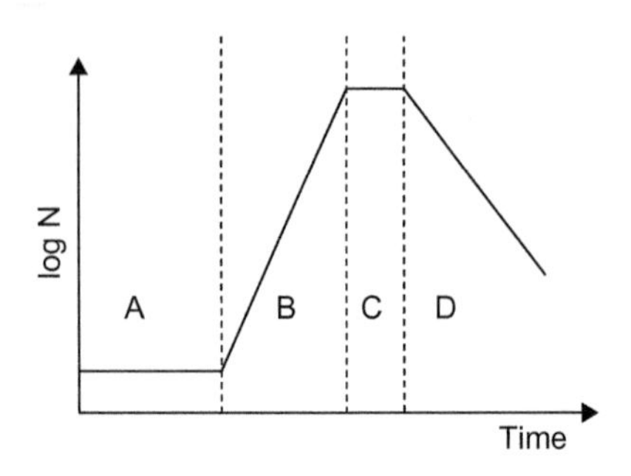

FIGURE 4.6

Schematic representation of microbial growth curve: A, lag phase; B, log phase; C, stationary phase; D, decline phase.

The classical curve of microbial growth (number of living cells N versus time, Figure 4.6) shows four phases (Loncin and Merson, 1979):

1. *Lag phase*. Initially, there is no growth. The cells may use food to increase in mass but not in number.
2. *Log (logarithmic, exponential) phase*. The number of living cells increases with time exponentially.
3. *Stationary phase*. The number of living cells remains nearly constant in time. This phase is usually explained as one during which the rate of new cell generation is equal to the rate of death.
4. *Decline*. The number of living cells decline with time, usually in an exponential manner.

The rate of microbial growth follows a pattern known as the "*Monod kinetics*", after the French biologist Jacques Lucien Monod (1910–1976, Nobel Prize in Physiology and Medicine, 1965). Since growth occurs as the result of division of a living cell into two, it may be assumed that the rate of growth at any moment will be proportional to the number of living cells at that moment:

$$\frac{dN}{dt} = \mu N \tag{4.15}$$

Equation (4.15) actually represents first-order kinetics. The rate constant μ is called the *specific growth rate*. One of the interesting features of μ is its dependence on substrate concentration. The relationship between μ and substrate concentration is very similar to the Michaelis–Menten equation:

$$\mu = \mu_{\max} \frac{S}{K_S + S} \tag{4.16}$$

where μ_{max} is a maximum value to which the specific growth rate tends asymptotically as the availability of substrate increases.

Cell death occurs simultaneously with cell generation. Assuming that cell death also follows first-order kinetics with a rate constant k_d, the actual increase in the number of living cells can be written as follows:

$$\frac{dN}{dt} = \text{rate of generation} - \text{rate of death} = \mu N - k_d N \qquad (4.17)$$

During the log phase, the substrate is abundant, μ is almost at its maximum and the rate of death is relatively negligible (assuming that the temperature, pH, water activity and other conditions are favorable to growth).

The growth rate of microorganisms is strongly temperature-dependent. Just like the rate constant of enzymatic reactions, it increases with increasing temperature up to a maximum, then declines rapidly. At a certain temperature it becomes negative – i.e., the number of living cells decreases with time. The minimum, optimum and maximum temperatures for microbial growth depend on the microorganism species (see Chapters 17 and 18).

4.4 Residence time and residence time distribution

4.4.1 Reactors in food processing

Residence time (RT) and *residence time distribution* (RTD) are notions that serve to describe, primarily, the length of time during which the food or portions of it have been subjected to a certain treatment (Bimbenet *et al.*, 2002). For convenience, this is defined as the length of time during which the food has "resided" in a reactor. In chemical process engineering, the term "reactor" often means a specialized device (generally a vessel with accessories) used to carry out a controlled reaction. In this section, however, any portion of the physical process system where reactions occur will be considered a reactor. By this definition, a fermentor, an oven, an extruder, a drying tunnel, an oak barrel used for aging wine, or a box of cookies are reactors.

In a batch reactor, the notion of "residence time" is trivial. It is simply the duration of the batch cycle, and it is the same for every portion of material in the batch. In contrast, residence time analysis in continuous operations must take into consideration the physical characteristics of the reactor and the operation conditions. Most reactors are too complex for accurate analysis. Therefore, reactors are typified according to a number of idealized models (Figure 4.7). The following are some of these models.

1. The *plug flow reactor (PFR)*. In this type of reactor, the material flows as a block (plug). Each part of the fluid has the same velocity. There is no mixing within the fluid. Consequently, the residence time is equal for every portion of the fluid. The residence time distribution (see below) is flat.
2. The *laminar flow reactor (LFR)*. The fluid moves through the reactor (usually a tubular reactor) in laminar flow, i.e., in parallel layers. The layers do not

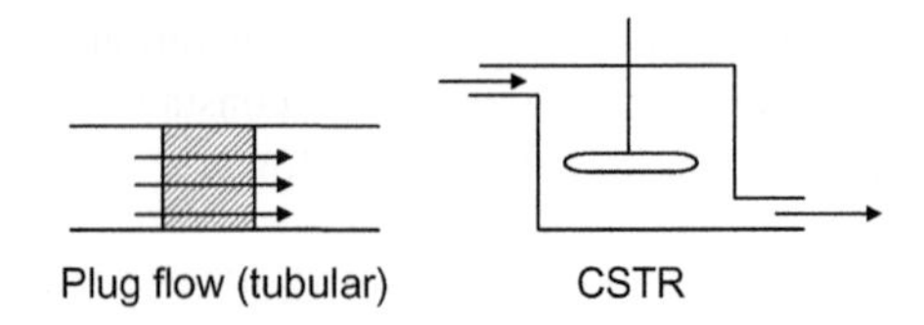

FIGURE 4.7

Two ideal reactor models.

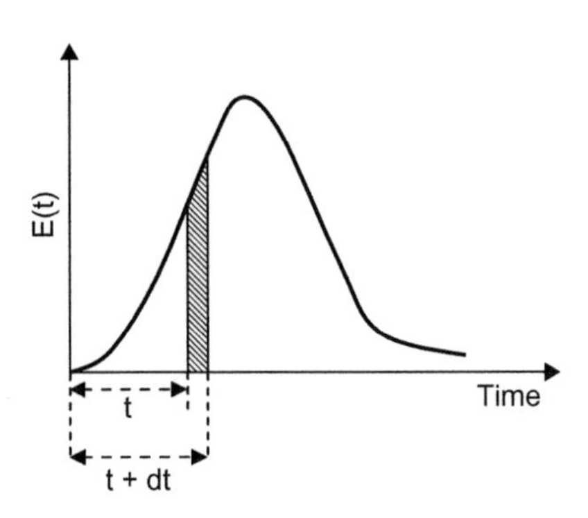

FIGURE 4.8

Residence time distribution, E(t) function.

move at the same velocity, but there is no mixing between the layers (see Chapter 2, Section 2.2.2). Residence time is not uniform.

3. The *continuous stirred tank reactor (CSTR)*. Physically, this type of reactor corresponds to a perfectly agitated vessel with continuous feeding and discharge. As a result of perfect mixing, the composition and all other conditions at a given moment are uniform at all points within the reactor. The composition of the fluid discharged is identical to that of the fluid bulk in the reactor at the same moment.

4.4.2 Residence time distribution

Residence time distribution (RTD) is treated with the help of classical statistical functions and parameters.

The RDT function E(t), known as the *frequency density function*, describes the probability of a given particle or portion spending a time t in the reactor. A hypothetical curve representing the E(t) function versus t is shown in Figure 4.8. In this figure, the area of the shaded strip gives the mass fraction of the fluid that has spent a time between t and $t + \Delta t$ in the reactor.

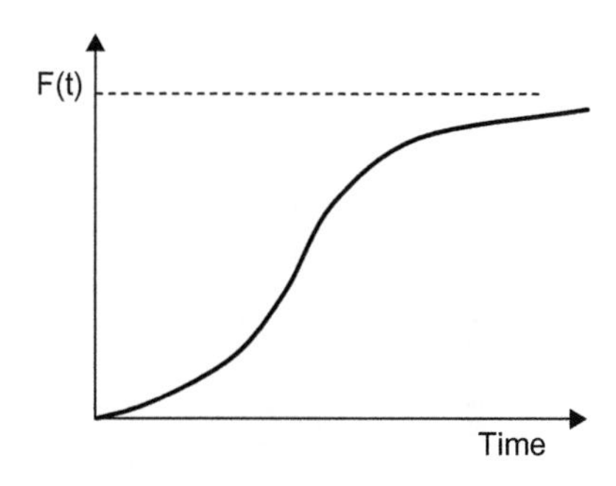

FIGURE 4.9

Residence time distribution, F(t) function.

Another useful RDT function is F(t), known as the *cumulative distribution function*. It represents the mass fraction of the fluid that has spent a time t or less in the reactor. A hypothetical curve corresponding to F(t) is shown in Figure 4.9.

The mean residence time t_m is:

$$t_m = \int_0^\infty t.E(t).dt \tag{4.18}$$

Ideally (i.e., with no diffusion, dispersion and no volume change), the mean residence time is equal to the mean travel time through the reactor (space time), τ:

$$t_m \approx \tau = \frac{V}{Q} \tag{4.19}$$

where V = active volume (capacity) of the reactor, m^3, and Q = volumetric rate of throughput, $m^3 \cdot s^{-1}$.

One of the procedures for the experimental determination of the RDT functions is the method of "pulse injection", described schematically in Figure 4.10.

At time t = 0, a small quantity of a tracer is rapidly injected into the reactor with the feed. The concentration of the tracer is measured at the reactor exit as a function of time. Denoting the exit concentration of the tracer as C, we obtain:

$$E(t) = \frac{C_t}{\int_0^\infty C_t dt} \tag{4.20}$$

E(t) and F(t) curves for a plug-flow reactor and a CSTR are shown in Figures 4.11 and 4.12, respectively.

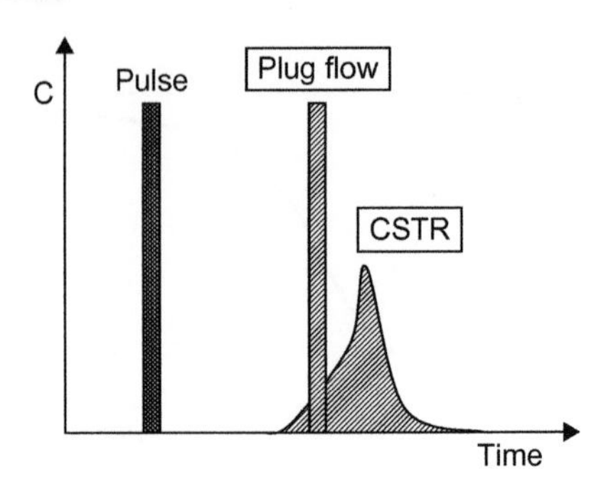

FIGURE 4.10

Reaction of a plug-flow reactor and a CSTR to a pulse injection.

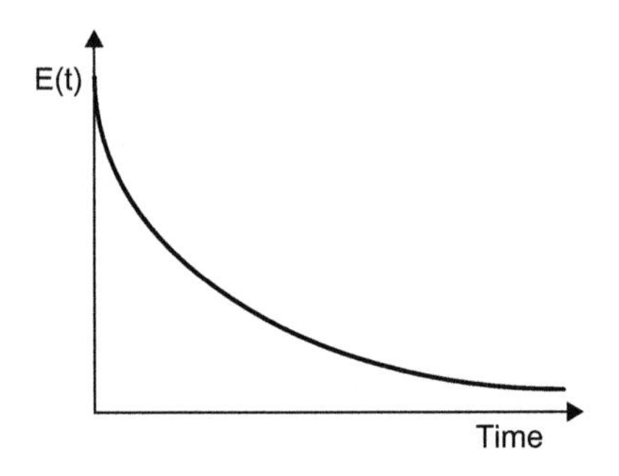

FIGURE 4.11

E(t) curve for CSTR.

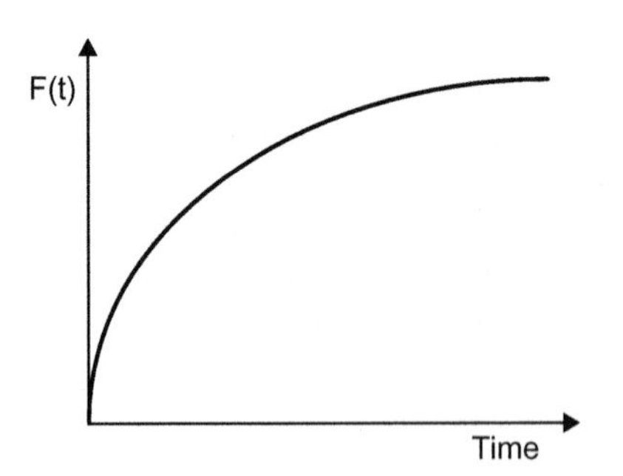

FIGURE 4.12

F{t} curve for CSTR.

References

Berk, Z., Mannheim, C.H., 1986. The effect of storage temperature on the quality of citrus products aseptically packed into steel drums. J. Food Process. Preservat. 10 (4), 281–292.

Bimbenet, J.J., Duquenoy, A., Trystram, G., 2002. Génie des Procédés Alimentaires. Dunod, Paris, France.

Burdurlu, H.S., Koca, N., Karadeniz, F., 2005. Degradation of vitamin C in citrus juice concentrates during storage. J. Food Eng. 74 (2), 211–216.

Earle, R.L., Earle, M.B., 2003. Fundamentals of Food Reaction Technology. Royal Society of Chemistry, Cambridge, UK.

Erdoğdu, F., Şahmurat, F., 2007. Mathematical fundamentals to determine the kinetic constants of first-order consecutive reactions. J. Food Proc. Eng. 30 (4), 407–420.

Labuza, T.P., Riboh, D., 1982. Theory and application of Arrhenius kinetics to the prediction of nutrient losses in foods. Food Technol. 36 (10), 66–74.

Loncin, M., Merson, R.L., 1979. Food Engineering, Principles and Selected Applications. Academic Press, New York, NY.

Mizrahi, S., Labuza, T.P., Karel, M., 1970. Feasibility of accelerated tests for browning in dehydrated cabbage. J. Food Sci. 35 (6), 804–807.

Toledo, R.T., 2007. Kinetics of chemical reactions in foods. *Fundamentals of Food Process Engineering*. Springer, New York, NY.

Villota, R., Hawkes, J.G., 1992. Reaction kinetics in food systems. In: Heldman, D.R., Lund, D.B. (Eds.), Handbook of Food Engineering. Marcel Dekker, New York, NY.

CHAPTER

Elements of Process Control

5

5.1 Introduction

Process control is the organization of activities with the purpose of:

1. Maintaining certain variables of a process (such as temperatures, pressures, concentrations, flow rates, etc.) within specified limits; or
2. Changing said variables according to a preset program.

The activities of type 1 are called *regulation* activities. Those of type 2 are *servo* activities.

The objectives of process control are, among others:

- To enhance the reliability of the process
- To improve process safety
- To increase production
- To reduce the proportion of production units which do not meet the specifications
- To improve process economics, by reducing production cost.

The use of control systems to replace some of the human work and at the same time improve process reliability and safety is called *automation*. Well established in non-food industries, automation is becoming an objective of increasing importance in food processing (Kondo, 2010; Mahalik and Nambiar, 2010).

The concept of "control" is not limited to industrial processes. Any process, be it biological, social, economical, political or military, usually features some sort of control mechanism. Control may be manual or automatic. This chapter, however, deals exclusively with *automatic process control*.

5.2 Basic concepts

Consider the simple case of a thermostated, electrically heated laboratory water bath. (Figure 5.1).

Assume that the *control objective* is to maintain the temperature of the bath at 37°C. We set the thermostat knob of the bath to 37°C and turn the heater on. A thermometer measures the temperature of the water. When the temperature of the water exceeds 37°C, the thermostat turns off the heater automatically. The

Food Process Engineering and Technology. DOI: http://dx.doi.org/10.1016/B978-0-12-415923-5.00005-8
© 2013 Elsevier Inc. All rights reserved.

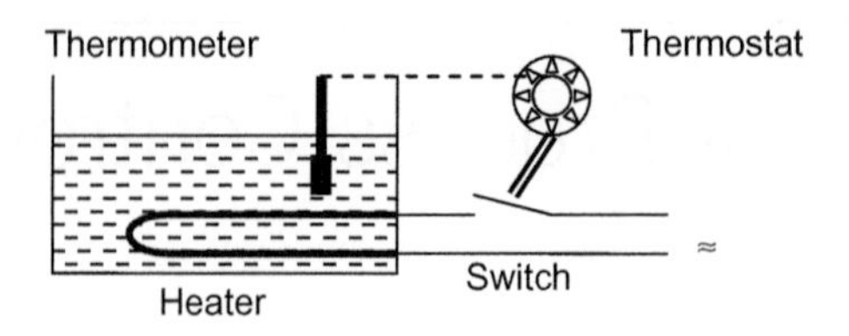

FIGURE 5.1

Thermostated water bath.

temperature of the water drops. When the temperature drops below 37°C, the thermostat turns on the heater automatically, and so on. We define a number of basic elements in this simple case:

- The water bath is the process.
- The water temperature is the controlled variable.
- The electric current is the manipulated variable. In our case, it can have only two values: on or off.
- 37°C is the set point. It may be stored in the thermostat under various forms (digital data, position of a cursor, tension of a spring, etc.).
- The thermometer is the sensor or the measuring element. It measures the temperature, and sends the thermostat a signal carrying information about the temperature (measurement signal).
- The difference between the set point and the actual temperature at any moment is the deviation or error.
- The thermostat is the controller. It compares the measured value (signal from the sensor) with the set point stored in its memory, and calculates the error. If the error is outside a preset range (known as the differential band), it issues a signal (correction signal) to the electric switch. The signal, which can be an electric current, a mechanical force or a digital signal, causes the switch to close or to open.
- The switch is the actuator, also known as the final control element.
- Any factor capable of making the system deviate from the set point is a disturbance.
- The path of the signals forms a loop known as the control loop (Figure 5.2). A control loop may be open or closed. In our case, the control loop is a closed loop.

5.3 Basic control structures

In classical process control, there are two basic control structures: *feedback control* and *feed-forward control* (Stephanopoulos, 1983).

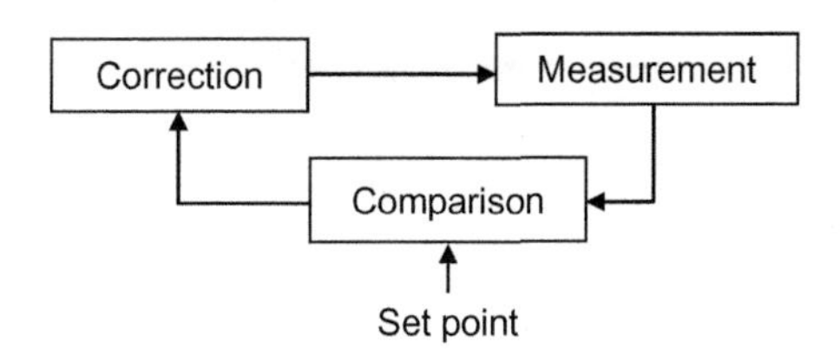

FIGURE 5.2

Basic control flow diagram.

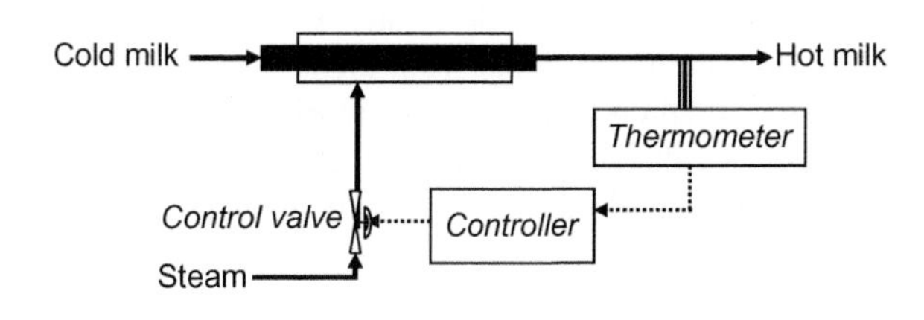

FIGURE 5.3

Feedback control of milk heat exchanger.

5.3.1 Feedback control

Consider the simplified control system of a milk pasteurization line (Figure 5.3).

The control objective in this case is to secure that the milk is heated to the correct pasteurization temperature. Automatic control is achieved in the following way:

1. A thermometer measures the temperature of the milk coming out of the heat exchanger, and sends the controller a measurement signal.
2. The controller compares the measurement value with the desired value (set point) and determines the error. A correction signal is sent to the steam valve.
3. The correction signal causes the valve to open or to close. As a result, the steam flow-rate to the heat exchanger is increased or reduced. Consequently, the temperature of the milk exiting the heat exchanger changes in the appropriate direction.
4. The thermometer sends to the controller a measurement signal which constitutes the feedback from the process to the controller.

This type of closed-loop control structure is called *feedback control*. This is the most common structure in food process industry.

5.3.2 Feed-forward control

Consider the same pasteurizer with a different control system, as shown in Figure 5.4.

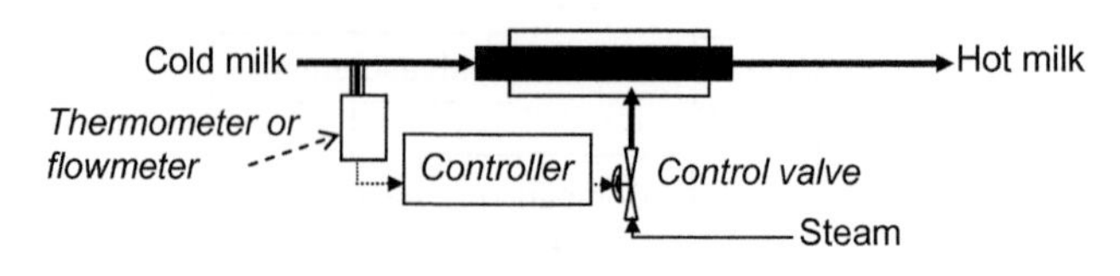

FIGURE 5.4

Feed-forward control of milk heat exchanger.

1. The thermometer measures the temperature of the milk entering the heat exchanger, and sends a measurement signal to the controller. (The measured variable can be the temperature, the flow-rate, or both.)
2. In the light of the measurement signal from the sensor, the controller *calculates* (or, rather, predicts) the temperature of the milk at the outlet of the pasteurizer. In this case, the calculation involves material and energy balances and knowledge of parameters such as the coefficient of heat transfer, the specific heat of the milk, and the area of the heat exchange surface. The difference between the predicted temperature and the set point is the error.
3. A correction signal is sent to the steam valve and operates the valve, according to the error. There is no feedback to the controller.

This type of control is known as *feed-forward* or *predictive* control. Its use in the food industry is less widespread.

5.3.3 Comparative merits of control strategies

Feedback control is simpler and less expensive than feed-forward control. It relies on measured parameters and not on predictions. Most importantly, it does not require knowledge of the process. This is a significant advantage over feed-forward control, particularly in the case of complex processes. On the negative side, feedback control corrects the process only after the error has occurred.

Feed-forward control predicts the error and reacts so as to prevent it before it occurs. It requires knowledge of the process, so that the reaction of the controlled variables to corrective actions may be calculated or estimated. It is therefore easier to apply to simple processes.

5.4 The block diagram

The hardware elements and the signals of a control system are graphically represented as block diagrams (Figure 5.5). Conventionally, the set point is denoted by the letter r and assumed to be positive. The measurement signal is represented by the letter c and assumed to be negative. The correction signal is shown as m and the error as e. By definition, $e = r - c$. However it is customary to present the

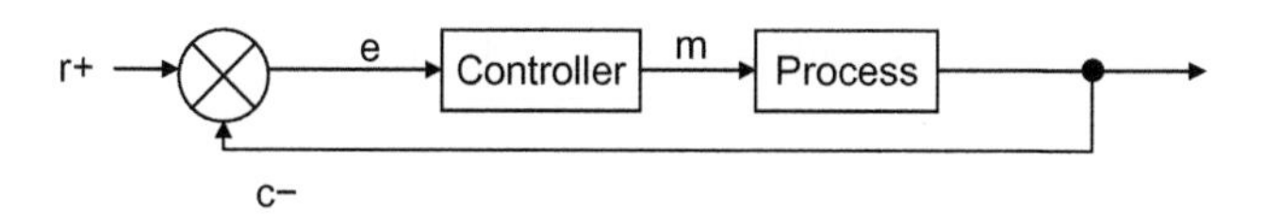

FIGURE 5.5

Basic control block diagram.

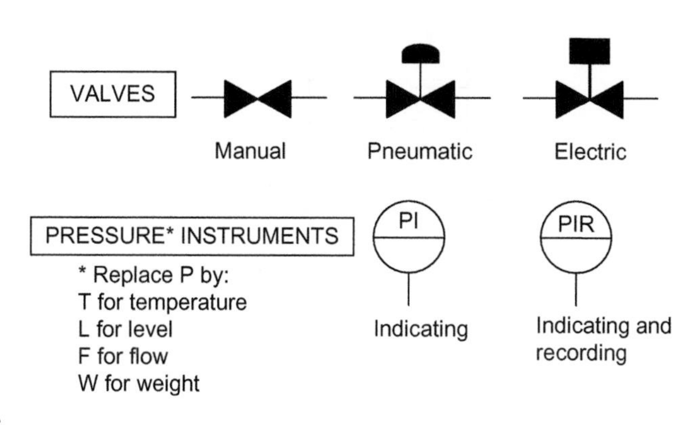

FIGURE 5.6

Standard symbols for some control elements.

error as a percentage of the measurement range. If the range of the measuring instrument is Δ, then the error is usually written as $e = 100\ (r - c)/\Delta\%$.

The hardware elements always found in a loop are the controller, the measuring instrument and the actuator. The real physical loop contains additional hardware such as signal transmission lines (pneumatic, electric, hydraulic, optic, etc.), transducers for the conversion of one type of signal to another (e.g., analog to digital, pneumatic to electric, etc.), amplifiers used to enhance weak signals, recorders, safety switches, alarms, etc. All these elements are shown in detailed diagrams, using conventional standard symbols (Figure 5.6).

5.5 Input, output and process dynamics

Measurement instruments, controllers, actuators, recorders, transducers and many other elements of the control system receive signals (inputs) and send signals (output). The output S_o is a function of the input S_i (Figure 5.7). The characteristics of this function constitute what is called "system response". In the following discussion we shall consider only single-input, single-output (SISO) systems.

FIGURE 5.7

Schematic representation of a SISO system.

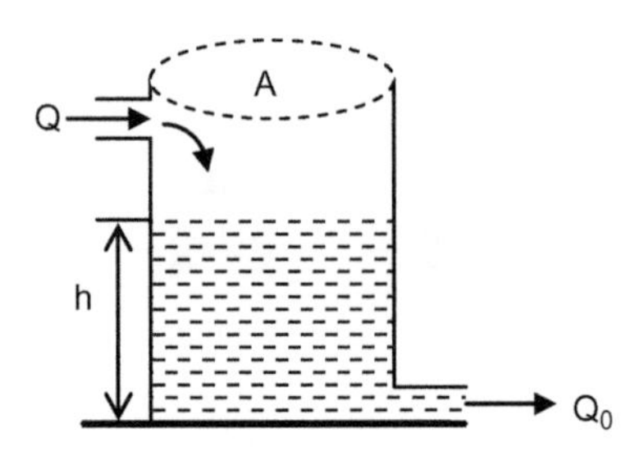

FIGURE 5.8

Water tank as a first-order system.

5.5.1 First-order response

Consider a cylindrical tank containing a liquid, shown in Figure 5.8. Liquid is continuously fed into the tank through a feed pipe at a volumetric flow-rate Q_i, and discharged continuously through a short pipe connected to the bottom of the pipe. Let Q_o be the discharge flow-rate and A the cross-section area of the tank.

In the system described, the input is the feed flow-rate. The output (i.e., the change in the system caused by the input) is the height h of the liquid in the tank. Let us analyze the relationship between Q_i and h. A material balance can be written as follows:

$$A\frac{dh}{dt} = Q_i - Q_o \tag{5.1}$$

Assume laminar flow in the discharge pipe. Then, the Hagen–Poiseuille rule, given in Chapter 2, Eq. (2.12), applies. The discharge flow-rate at any moment is proportional to the height of the liquid in the tank:

$$Q_o = \frac{h}{R} \tag{5.2}$$

R is the resistance to flow at the discharge pipe, assumed constant.

Substitution in Eq. (5.1) gives:

$$RA\frac{dh}{dt} + h = RQ_i \tag{5.3}$$

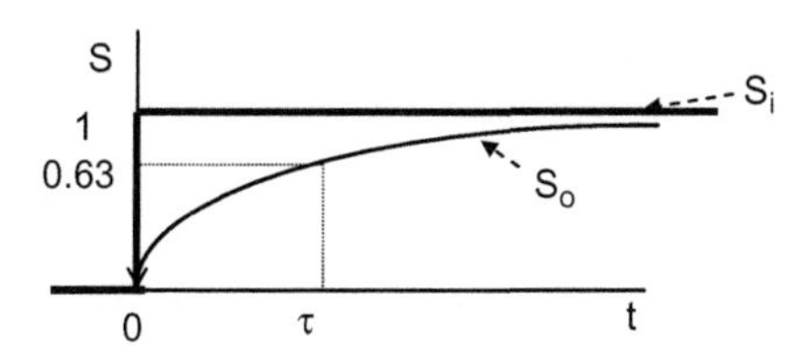

FIGURE 5.9

Response of a first-order system.

Recalling that Q_i is the input S_i and h is the output S_o, Eq. (5.3) can be written in the following general form:

$$KS_i = \tau \frac{dS_o}{dt} + S_o \quad (5.4)$$

Eq. (5.4) represents the response of systems known as *first-order systems*. The constant K is called the system *gain* and τ is the *time constant* of the system. In the tank example, R is the gain and RA is the time constant.

Now, let us examine the response of a first-order system to a step-function (jump) change in input. We indicate the jump in input as boundary conditions:

$$S_i = 0 \quad for \quad t < 0$$
$$S_i = 1 \quad for \quad t \geq 0$$

With these boundary conditions, the solution of Eq. (5.4) is:

$$S_o = 1 - e^{-t/\tau} \quad (5.5)$$

It can be seen (Figure 5.9) that the output lags after the input. In system dynamics, this is known as a *first-order lag*. Many real systems fit the first-order model well.

EXAMPLE 5.1

Show that a steam-jacketed stirred tank used to heat a batch of water behaves like a first-order system. Assume that the temperature of the steam and the overall heat transfer coefficient are constant. Neglect heat losses. Find the gain and the time constant of the system.

Solution

Let:

m = mass of the water in the tank
T = instantaneous temperature of the water, assumed uniform
C_p = specific heat of the water, assumed constant
T_s = temperature of the steam, assumed constant
U = overall heat transfer coefficient, assumed constant
A = heat transfer area, assumed constant.

$$m \cdot C_p \frac{dT}{dt} = UA(T_s - T) \quad \Rightarrow \quad \left[\frac{mC_p}{UA}\right] \frac{dT}{dt} + T = T_s$$

Remembering that the temperature of the steam is the input and the temperature of the water is the consequence (output), the result fits the first-order model, with:

$$\tau = \frac{m\,C_p}{UA} \quad \text{and} \quad K = 1$$

5.5.2 Second-order systems

The response of many real systems approximates a behavior represented by a second-order differential equation like Eq. (5.6):

$$\tau^2 \frac{d^2 S_o}{dt^2} + 2\zeta\,\tau \frac{dS_o}{dt} + S_o = KS_i \tag{5.6}$$

The additional constant ζ is called the *damping factor* because of its effect on the oscillatory character of the response of the system to step changes in input. At low ζ, the response is highly oscillatory (Figure 5.10). At high ζ ($\zeta > 1$), the response is non-oscillatory or over-damped.

Second-order response characterizes systems consisting of two first-order lags connected in series, mechanical devices containing a spring-loaded element together with a viscous damping component, thermometers in protective wells, etc.

5.6 Control modes (control algorithms)

Control modes (control algorithms) define the relationship between the error e and the correction signal m issued by the controller. The following are some of the frequently applied control modes:

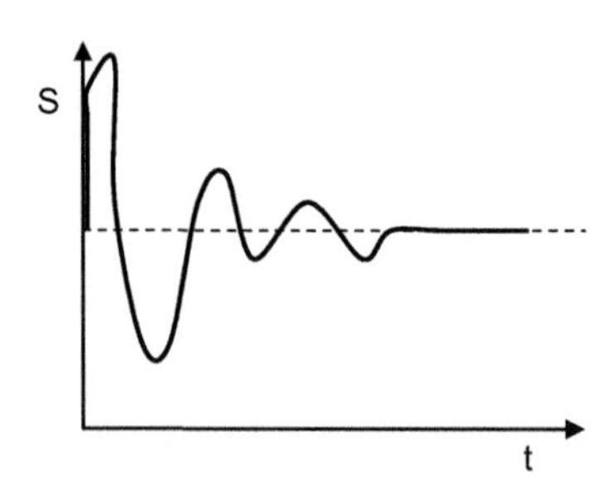

FIGURE 5.10

Response of a second-order system.

- On–off control
- Proportional (P) control
- Integral (I) control
- Proportional–integral (PI) control
- Proportional–integral–differential (PID) control.

5.6.1 On–off (binary) control

In this mode of control, the actuator can have only one of two positions: either on or off (open or closed, all or nothing, 1 or 0). There are no intermediate positions such as partially open. This is the simplest and the least costly of the controllers. On–off control is very common in household devices and laboratory instruments, and in some cases of industrial control systems (air compressors, refrigeration engines, limit switches and alarms, etc.).

In on–off control, the controlled variable must be allowed to fluctuate about the set point, within a range known as the *differential band* (or *dead zone*). For example, in a thermostat set to 25°C, the switch turns the heater on at 24°C and off at 26°C. Between these two values, the heater would be either on or off, depending on the direction of change. Without a differential band, the controller would cycle between two contradictory positions at very high frequency (chatter), causing damage to contactors and valves or causing the actuator to "freeze" altogether. The differential band or "hysteresis" is either stored in the controller as an analog or digital datum, or built-in in the hardware as the physical lag between the controller and the actuator. In practice, the width of the differential band is usually set between 0.5% and 2% of the control range. A narrow band results in more precise control but high frequency of change, and *vice versa*.

EXAMPLE 5.2

An electrically heated fryer is on–off controlled. The set point is 170°C. When the heater is on, the temperature of the oil rises at the rate of 4°C per minute. When the heater is off, the temperature drops at the rate of 4°C per minute. Calculate the duration of one cycle of the heater switch if the differential band on each side of the set point is:

a. 10°C
b. 1°C.

Solution

a. The switch is turned-off when the oil temperature reaches 180°C and on again when the temperature drops to 160°C. The gap is 20°C. The duration of the cycle is: $t = 20/4 = 5$ minutes.
b. The gap is now 2°C. The duration of the cycle is: $t = 2/4 = 0.5$ min.

The response of a first-order system to on-off control with hysteresis is shown in Figure 5.11.

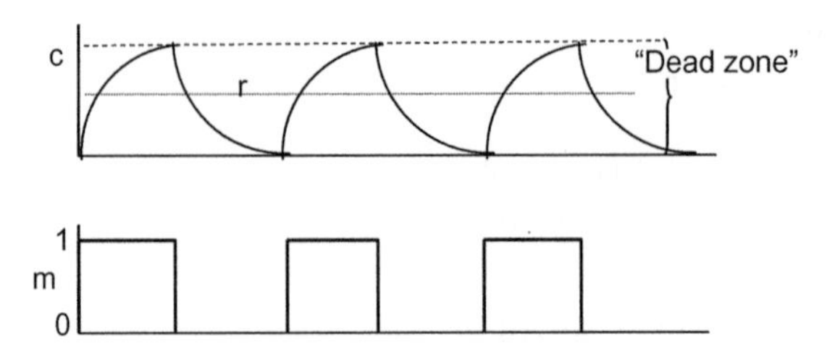

FIGURE 5.11

On–off control with differential band (dead zone).

5.6.2 Proportional (P) control

In proportional (P) control, the magnitude of the correction signal m is proportional to the error e.

$$m = Ke + M$$

The proportionality factor K is called *proportional gain*. The constant M is known as the controller *bias*, because it represents the magnitude of the correction signal when no correction is needed ($e = 0$). K is dimensionless. M, e and m are usually expressed as percentages.

Proportional control serves to eliminate the oscillation associated with on–off controllers. The magnitude of the corrective action is reduced as the controlled variable approaches the set point. In control systems of the proportional type, the actuator can assume intermediate positions between two extremes, depending on the amplitude of the correction signal from the controller. In some simpler systems, the actuator works in on–off fashion and the proportionality is achieved by regulating the ratio of on-time/off-time. At the set point, that ratio is equal to 1.

The proportional gain K is usually a fixed property of the controller, but in some proportional controllers K is manually adjustable. If K is increased, the sensitivity of the controller to error is increased but the stability is impaired. The system approaches the behavior of on–off controlled systems and its response becomes oscillatory. The bias M is usually adjustable. As a rule, it is customary to adjust M so as to stabilize the system at a state slightly different from the set point. The difference between the measurement c and the set point r at steady (stable) state is called the *offset*. It is possible to eliminate the offset by adjusting the bias. This action is called *reset*. However, adjusting the controller to zero offset under a given set of process conditions would require re-adjusting every time the load or any other process condition changes. In automatic control, this would be highly problematic.

EXAMPLE 5.3

A steam heated plate heat exchanger serves to pasteurize orange juice at the nominal rate of 2000 $kg \cdot h^{-1}$. A proportional controller regulates the flow rate, according to the exit temperature of the juice. The juice enters the pasteurizer at 20°C. The set point is 90°C. The steam valve is linear and normally closed. When the valve is fully open, the steam flow-rate is 400 $kg \cdot h^{-1}$.

The controller has been tuned according to the following equation:

$$m\,(\%) = 0.8\,e\,(\%) + 50$$

The temperature measurement range is 50–150°C.

Assume that the heat transferred to the juice is exactly equal to the heat of condensation of the steam, which is 2200 $kJ \cdot kg^{-1}$. The specific heat of the juice is 3.8 $kJ \cdot kg^{-1} \cdot K^{-1}$.

a. What is the exit temperature of the juice and the offset at steady-state?

b. It is desired to reduce the offset to half its value, by changing the bias, without changing the gain. What should be the bias?

Solution

Let J be the mass flow-rate of the juice and S the mass flow-rate of the steam. Let C_p be the specific heat of the juice and h_{fg} the latent heat of condensation of the steam. The notations of Section 5.4 will be used for all other parameters.

a. The system equation is:

$$J \cdot C_p(T_{out} - T_{in}) = S \cdot h_{fg} \quad \Rightarrow \quad 2000 \times 3.8(T_{out} - 20) = S \times 2200$$

The controller equation is:

$$m = 0.8e + 50 = 0.8\frac{(r - T_{out}) \times 100}{\Delta} + 50 \quad m = 0.8 \times \frac{(90 - T_{out}) \times 100}{100} + 50$$

The steam valve equation is:

$$S = S_{max} \times m/100 \quad \Rightarrow \quad S = 400 \times m/100$$

Solving the three equations for T_{out} we find $T_{out} = 83.7$°C.

The offset is $90 - 83.7 = 6.3$°C.

b. The gain m is:

$$m = 0.8 \times \frac{(90 - 83.7) \times 100}{100} + 50 = 55$$

Let the new bias be M. Now the offset is $6.3 / 2 = 3.15$°C.

$$55 = 0.8 \times \frac{3.15 \times 100}{100} + M \quad \Rightarrow \quad M = 52.48$$

In order to reduce the offset to half its value, the bias has to be changed to 52.48%.

5.6.3 Integral (I) control

In integral control, the rate of change of the correction signal (and not the actual value of that signal) is proportional to the error.

$$\frac{dm}{dt} = R\,e \tag{5.7}$$

Integration gives:

$$m = R\int e\,dt + M_i \tag{5.8}$$

The corrective signal m depends not on the actual value of the error but on its time interval – i.e., on the past history of the system. By virtue of Eq. (5.7), the corrective signal to the actuator continues to grow as long as the error persists. This results in the elimination of the offset.

Removal of the offset is the main advantage of I control. However the response of I-controlled systems is slow and may be highly oscillatory. The response of such systems to a step change in the controlled variable is a ramp, with a slope proportional to the proportionality constant R.

5.6.4 Proportional–integral (PI) control

Because of its shortcomings, integral control is usually applied in combination with other modes. In the PI mode, proportional control is combined with integral control. The combined algorithm is represented in Eq. (5.9):

$$m = K(e + \int e\,dt) + M \tag{5.9}$$

The integral term contributes the feature of automatic reset (removal of the offset). The proportional term increases the stability.

5.6.5 Proportional–integral–differential (PID) control

A third mode is incorporated in this popular type of control, known also as the "three-term control". The third term is that of differential control, defined in Eq. (5.10):

$$m = T\frac{de}{dt} \tag{5.10}$$

In the differential mode, the correction signal is proportional not to the instantaneous error but to the rate of change of the error. If the error increases rapidly, the correction signal is larger. It can be said that the differential action is predictive. The response of differential-controlled systems is fast, but highly unstable. The differential mode is therefore not applied alone but in combination with other control types.

The PID mode provides proportional control with the automatic reset feature of integral control and the rapidity and predictive action of differential control. The behavior of the PID control is represented by Eq. (5.11), containing the three terms, P, I and D:

$$m = K\left(e + R\int e\,dt + T\frac{de}{dt}\right) + M \tag{5.11}$$

The PID algorithm is extensively used in industrial process control. For optimal performance, the controller must be "tuned" by adjusting the parameters of the three terms K, R and T. The notion of "optimal performance" will be defined in the following paragraph. Several methods exist for the manual tuning of PID controllers (O'Dwyer, 2006). Automatic tuning (self-tuning) systems are also available (Bobàl *et al.*, 2005).

Figure 5.12 shows the response of three types of control to a disturbance. Note the large offset in the case of a P controller. In the PI controller, the offset has been removed but the response is oscillatory. In the PID system, the offset has been eliminated and the system is more stable.

5.6.6 Optimization of control

The definition of "quality of control" is not unequivocal. Consider the response of a controlled system to a disturbance. There are several ways to define the quality of the control and the objective of optimization (Ogunnaike and Ray, 1994):

- Since the fundamental objective of process control is to eliminate error as much as possible, the objective of optimization might be to minimize the maximum deviation emax. We have seen that this can be done by increasing the controller gain K. However, if K is increased beyond a critical value, stability is impaired and the response shows weakly damped oscillations. The time to reach stability (steady state) may be too long.

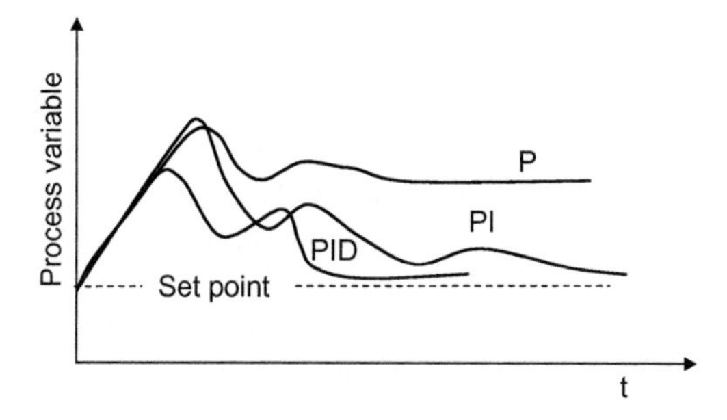

FIGURE 5.12

Response of P, PI and PID controlled systems to disturbance.

- The time to reach steady state, known as "recovery time", is an important factor. The objective of optimization could then be to shorten recovery time as much as possible. However, this would require decreasing the gain K, which would result in increased emax , i.e. in loss of accuracy.
- The previous criteria may be both incorporated in a parameter known as the *integral absolute error* (IAE), defined in Eq. (5.12):

$$IAE = \int_0^{\infty} |e| \, dt \tag{5.12}$$

  In this case, optimization would mean minimizing the IAE. Obviously, this criterion can be applied only to controllers comprising an integral (I) component, because otherwise the IAE would always be infinite because of the offset.
- Another measure, sometimes used as a criterion for control quality, is the *integral squared error* (ISE), defined in Eq. (5.13):

$$ISE = \int_0^{\infty} e^2 \, dt \tag{5.13}$$

  ISE is particularly sensitive to large deviations, and therefore optimization with respect to ISE enhances the protective action of the control against damage to the product or to the equipment. Like IAE, ISE can be applied only to controllers comprising an integral (I) feature.

5.7 Physical elements of the control system

5.7.1 The sensors (measuring elements)

The number of controlled variables in food processing is quite large, and often a considerable number of different sensors are available for measuring each parameter (Bimbenet *et al.*, 1994; Webster, 1999). Some of the most common types of measuring instruments will be described in this section. For a detailed survey of this highly specialized topic, please consult Kress-Rogers and Brimelow (2001).

A number of general principles apply to all the sensors used in food process control, regardless of the measured variable:

1. The output signal must be of a kind that can be transmitted to and read by the controller. Very often, a "converter" has to be used to transform the measurement signal to the desired type without distortion. In modern control systems, the desired format is a digital signal. In transmission, the signal must be protected against electrical and electromagnetic disturbances.
2. The range of measurement (scale) and the sensitivity must meet the process requirements. The measurement has significance only if the measured parameter falls within the range of the sensor. Thus, if the measurement range

of a thermometer is 20–100°C, a temperature of 120°C will be read as 100°C and a temperature of 10°C as 20°C.

3. Sensors can be on-line, at-line or off-line. On-line measurements, whenever feasible, are preferable. At-line measurement refers to rapid tests that can be performed on samples, near the production line.
4. Sensors can be contacting or non-contacting (remote). Remote sensing, whenever possible, is the generally preferred type.
5. On-line sensors are physically inserted in the process line and often come in contact with the food. In this case, the sensor and its position in the line must comply with the strict rules of food safety. Remote sensors eliminate this problem, but the number of commercially available remote sensing elements is still limited.

5.7.1.1 Temperature

The most common temperature-sensing devices are filled thermometers, bimetals, thermocouples, resistance thermometers, thermistors and infrared thermometers.

Filled thermometers measure temperatures either through the thermal expansion of a liquid or through changes in the vapor pressure of a relatively volatile substance. Thermal expansion thermometers are the most common type. The fluid in the thermometer is usually mercury or colored alcohol. Although the sturdy construction of industrial filled thermometers protects the product from contamination with glass, mercury or spirit in the case of breakage, filled thermometers are being replaced with other types that do not present that kind of risk. For traditional reasons, however, the use of mercury-in-glass thermometers as a temperature reference in food canning is still mandatory.

Bimetal thermometers consist of strips of two different metals, joined together. Due to the difference in the thermal expansion coefficient of the metals, a change in the temperature causes the strip to bend or twist. The displacement is usually read on a dial. Bimetal thermometers can serve as on–off actuators in simple thermostats in ovens, frying pans, etc. Bimetal thermometers are not accurate, and they lack stability.

Thermocouples (Reed, 1999) are among the most common industrial temperature measuring devices. They are based on a phenomenon discovered by the German physicist Thomas Johann Seebeck in 1821. Seebeck discovered that a voltage is generated in a conductor subjected to a temperature difference between its extremities. The value of the voltage generated varies from one metal to another. Consequently, an electric current flows in a closed circuit made of two different metals when their two junctions are held at different temperatures. The EMF created is a measure of the temperature difference between the junctions. Thus, a thermocouple measures a temperature difference, and hence the temperature of one junction if the temperature of the other (reference) junction is known.

The voltage V generated as a result of the thermoelectric effect is given approximately by:

$$V = S\Delta T \tag{5.14}$$

where S is the Seebeck coefficient of the material. The coefficient S is temperature-dependent. If two electrothermally dissimilar conductors A and B are joined at points 1 and 2, then the voltage generated between 1 and 2 is approximately:

$$V_{1-2} = (S_A - S_B)(T_1 - T_2) \tag{5.15}$$

If the Seebeck coefficients do not change much within the temperature range in question, then:

$$V_{1-2} = k \cdot \Delta T \tag{5.16}$$

Indeed, within a known range of temperatures, the response of a thermocouple is fairly linear – i.e., the EMF generated is proportional to the temperature difference. This EMF is generally in the order of a few mV per 100°C. The most common pairs used are copper/constantan and iron/constantan (constantan is a copper–nickel alloy). The Seebeck coefficients of copper, iron and constantan are $+6.5$, $+19$ and -35 $\mu V \cdot K^{-1}$, respectively. The measuring junction of a thermocouple may be very small, thus permitting measurement of the temperature in a precise location, and a rapid response time. Thermocouples with different kinds of tips (measuring junctions) are available for different applications (e.g., thermocouples for measuring the temperature inside a can). Thermocouples are usually inserted in a protective ceramic or stainless steel sheath, which necessarily results in slower response. In certain cases, such as use in ohmic heating, the sheath has to be fully insulated (Zell *et al.*, 2009).

Resistance thermometers (Burns, 1999) or "Resistance Temperature Detectors (RTD)" are based on the effect of temperature on the electric resistance of metals. Due to their accuracy and robustness, they are extensively used as in-line thermometers in the food industry. Within a wide range of temperatures, the resistance of metals increases linearly with temperature. The measuring element is usually made of platinum. The resistance of platinum changes by approximately 0.4% per K. Since electrical current flows through the measuring element during the measurement, there is some degree of self-heating of the thermometer, causing a slight error in the readings.

Thermistors (Sapoff, 1999) are also resistance thermometers, but the resistance of the measuring element, a ceramic semiconductor, decreases with the temperature. Thermistors are very accurate but highly non-linear. They are used where very high accuracy is a requisite.

Infrared thermometry (Fraden, 1999) measures temperatures by measuring the infrared emission of the object. Infrared thermometers may be remote (non-contact) or on-line (contact). In the contact type, a small black-body chamber is in contact with the object. A small infrared sensor installed inside the chamber measures the emission of the black-body walls. In the non-contact type, the lens is directed to the object. In infrared thermometry it is important to consider the emissivity of the object. Furthermore, in non-contact applications the instrument reads the average temperature of what it sees – i.e., the object and its surroundings. To overcome this problem, instruments can crop the image so as to consider only the part where the object is present. Remote infrared thermometry is extremely useful in measuring objects that cannot be accessed for contact (e.g., in microwave heating) and moving objects.

5.7.1.2 Pressure

Pressure is measured either as a variable by itself or as an indicator of another variable. Examples of the second case are the use of pressure measurement as an indicator of level (hydrostatic level measurement), flow-rate (orifice or *Venturi flowmeter*), temperature (autoclave), shear (homogenizer), etc. The measured value is expressed as absolute pressure, gauge pressure (absolute pressure minus atmospheric pressure) or differential pressure (the difference between the pressures at two points of the system).

Pressure-measuring instruments may be manometric, mechanical or electrical/electronic. Manometers measure pressure or pressure difference through the level of a fluid in a tube. These types of instruments are used in the laboratory but not in industry. The *MacLeod gauge*, used for the measurement of high vacuum (e.g., in a freeze dryer) is a special type of mercury manometer.

In mechanical devices, the pressure signal is converted to displacement. The process pressure is applied to a flexible surface (a Bourdon tube, a membrane or bellows), causing the surface to move. The movement may be transmitted to a pointer for direct reading, or converted to an analog or digital signal sent to a distant data acquisition element. A *Bourdon gauge* is shown schematically in Figure 5.13.

Electrical/electronic pressure transducers are devices that emit an analog or digital electrical signal as response to changes in applied pressure. There are many types of pressure transducers, each based on a different physical effect. Strain gauges generate an electric signal (change in resistance) as a response to deformation (strain). The most common configuration is a metallic foil pattern bound to an elastic backing. If the gauge is subjected to deformation under the effect of applied force, the resistance of the metal pattern changes according to the following expression:

$$\Delta R/R = (GF)\varepsilon \quad \varepsilon = \text{strain}\left(\text{e.g.}\frac{\Delta L}{L}\right) \tag{5.17}$$

where:

R = resistance of the unstrained gauge
ΔR = change in the resistance due to strain
L = length of the unstrained metal pattern

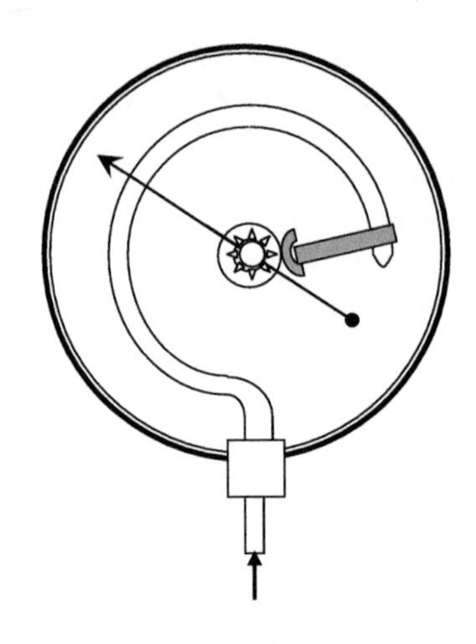

FIGURE 5.13

Structure of the Bourdon gauge.

ΔL = elongation.

In *capacitance pressure transducers*, the displacement of a membrane causes a change in the capacitance of a capacitor. The change in capacitance is converted to a voltage or current signal.

Piezoelectric transducers are based on the generation of an electric current across a quartz crystal submitted to pressure. Piezoelectric transducers respond to changes in pressure; they cannot be used to measure static pressures.

In general, electrical/electronic transducers are reliable measuring devices of great flexibility. They can be made quite small so as to measure pressure punctually in locations that are difficult to access (e.g., inside a cooking extruder).

5.7.1.3 Flow-rate

Control of flow-rate is important, particularly in continuous formulation processes involving fluids. The parameter measured is, in fact, fluid velocity, from which fluid flow-rate is inferred.

Differential pressure flowmeters measure a pressure drop between two points of the flow path and calculate the velocity from it. The relationship between pressure drop and velocity follows the Bernoulli Law, simplified as follows:

$$v = a\sqrt{\Delta P} \tag{5.18}$$

where a is a constant depending on the geometry. The pressure drop may be measured across an orifice (Figure 5.14), a nozzle (Figure 5.15), a Venturi tube (Figure 5.16), etc.

Variable area flowmeters, also known as *rotameters* (Figure 5.17), consist of a vertical tapered tube and a float. According to the velocity of the fluid, fed from

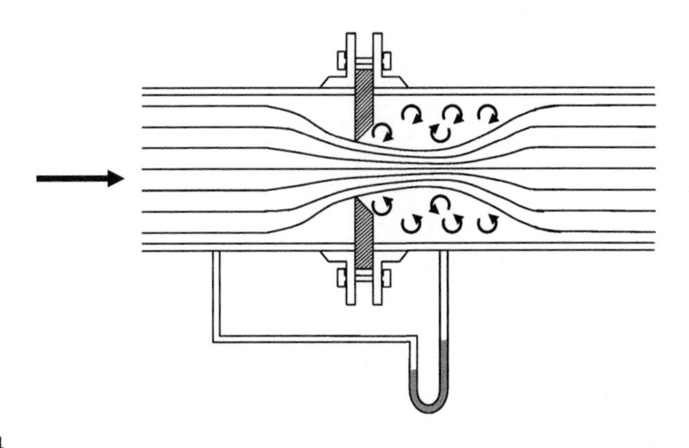

FIGURE 5.14

Orifice flowmeter.

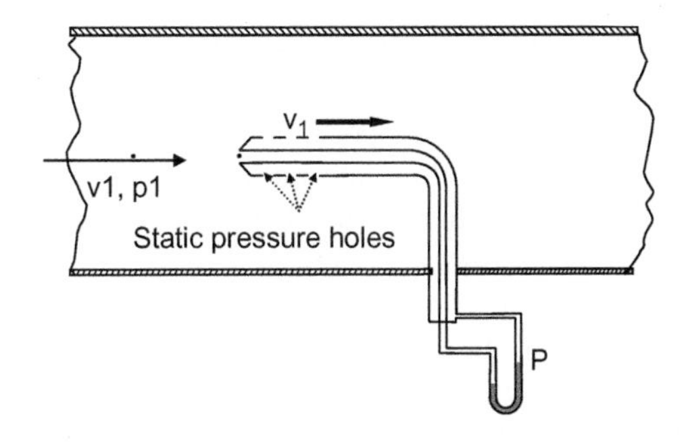

FIGURE 5.15

Pitot tube flowmeter.

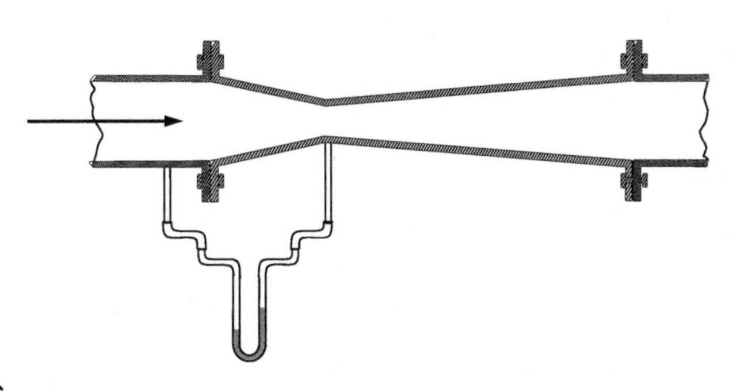

FIGURE 5.16

Venturi flowmeter.

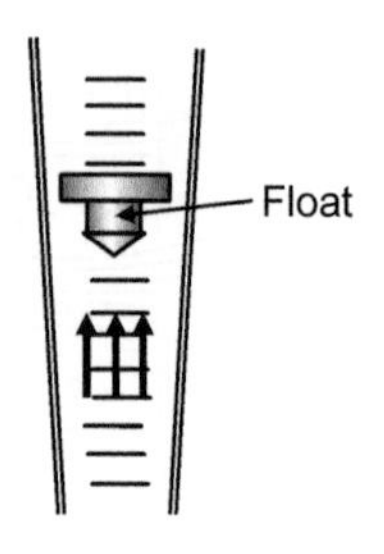

FIGURE 5.17

Rotameter.

below, the float occupies a position where the flow cross-section area corresponds to a pressure drop equal to its apparent weight.

Electromagnetic flowmeters operate on the principle of electromagnetic induction whereby a voltage is induced whenever a conductor moves through a magnetic field. A magnetic field is applied across the flow channel. The liquid, serving as the conductor, cuts the lines of the magnetic field at a rate proportional to its velocity. The induced voltage, proportional to the velocity, is read.

In mass flowmeters, the mass flow-rate rather than the velocity is measured. The principal types of mass flowmeters are thermal devices and Coriolis meters. In thermal flowmeters, mass flow-rate is inferred from the rate of dissipation of heat from a heated sensor by the flowing stream. The *hot wire anemometer*, used for measuring the flow-rate of gases, also belongs to this class. A wire heated by a known input of electric power is exposed to the stream of gas, and the flow-rate is inferred from the equilibrium temperature of the wire. The Coriolis meter is based on the *Coriolis effect*, whereby any object moving on the surface of the Earth is deflected from its direction of movement under the effect of the rotation of the Earth (Figura and Texeira, 2007). Similarly, a fluid subjected simultaneously to translational and rotational movement experiences the same type of "virtual" force, known as the Coriolis force. In the Coriolis meter, the fluid is passed through a flexible U-shaped tube that is vibrated at its natural frequency (instead of being rotated). The mass flow-rate is computed from the changes in frequency and amplitude of the tube vibration.

5.7.1.4 Level

Level measurement devices serve to measure and control the level of liquids or solids in vessels. They are of two kinds: limit switches and continuous meters.

Limit switches are on–off devices that react when the solid or the liquid reaches a certain level in the vessel. The most common type, used for liquids, is the familiar *float switch* found in many automotive and household applications. *Paddle switches* serve to control the level of solids. They consist of a mechanically rotated paddle. When the solid level reaches the instrument, the rotation is stopped or slowed down due to friction. Essentially the same kind of interaction occurs in vibrating instruments, used with both liquids and solids. *Conductive switches* may simply consist of a pair of electrodes. When a conducting fluid reaches the instrument, electrical contact is established between the electrodes.

Continuous sensors measure the level of solids or liquids, as well as changes in level, continuously. According to the physical principles on which they are based, we distinguish the following types, among others:

- Hydrostatic pressure instruments (for liquids)
- Differential pressure instruments (for liquids)
- Capacitance instruments (for liquids)
- Acoustic (ultrasonic) instruments (for liquids and solids)
- Microwave (radar) instruments (for liquids and solids).

5.7.1.5 Color, shape and size

Color is an important quality parameter. Automatic measurement of color is essential in many process control applications, such as sorting of fruits and vegetables in packing houses, and in control of roasting of coffee and nuts, frying of potato chips, oven toasting of breakfast cereals, browning of baked goods, etc.

Any color within the visible range can be represented with the help of three-dimensional coordinates (or three-dimensional color space) L, a, b. (Figura and Teixeira, 2007). The axis L represents "luminosity", with 0 = black and 100 = white. The "a" axis gives the position of the measured color between the two opponent colors red and green, with red at the positive end and green at the negative end. The "b" axis reflects the position of the color in the yellow (positive)–blue (negative) channel. With the help of the L*a*b space system, any color is represented by a simple equation containing the three parameters. Color measurement instruments (colorimeters) are photo-electric cell based devices, capable of reading the L*a*b values and "calculating" the color perceived. It is also possible to store predetermined standard colors in the memory of the system, as a reference.

Some simple devices for the evaluation of size or shape in special cases are available. One example is the method of measuring the height (thickness) of products moving on a conveyor belt by illuminating the items at an angle and measuring the length of their shadow by photo-electric means. Another example is the "dud detector", a device used to detect cans and jars with faulty sealing. Jars that have the proper vacuum inside will have their lid depressed. The dud detector identifies and rejects, often without contacting the product, the jars that do not show the proper depression of the lid. However, most modern devices for the acquisition of data concerning size and shape are based on computerized image analysis.

5.7.1.6 Composition

The development of advanced sensors for the on-line or rapid at-line measurement of composition is one of the most exciting areas of food engineering research (Figura and Teixeira, 2007). The techniques used include refractive index (for concentration), infrared (for moisture content), conductometry (for salt content), potentiometry (for pH and redox potential), viscosity (e.g., for blending), *Fourier Transform Infrared Spectroscopy* (FTRI, for composition in general), *near infrared reflectance* (NIR) *spectroscopy* (for composition, in general), etc. (Bhuyan, 2007).

5.7.2 The controllers

The most significant trend in process control technology is the rapid growth in the application of computers for control (Mittal, 1997). The computer-controller accepts and emits signals in digital form. In addition to the usual controller functions, the computer also stores the information in a convenient form, produces reports and graphs, processes the data, and performs calculations in the course of the process or after it. For example, it can calculate the F_0 value of a thermal process, at any point in a can (see Chapter 17, Section 17.3). The computer is usually dedicated, and known as a "*Programmable Logic Controller*" PLC.

Recent developments in controller technology include the introduction of *adaptive* and so-called *intelligent* controllers. Adaptive controllers (e.g., self-tuning controllers) are systems that can automatically adjust their settings according to changes in process conditions. Intelligent controllers are well adapted to deal with complex processes and with cases where the inputs are not precisely defined parameters. Intelligent controllers are those that use the approaches of *fuzzy logic* and *neural networks*. Many of the quality attributes of foods cannot be expressed in precisely quantitative terms: the operator or the taster uses heuristic terms such as "not bad" or "somewhat good". Fuzzy controllers accept data in heuristic terms and can process them. At the end of data processing, the result is "defuzzified" and produced as a precise or "crisp" output. Davidson *et al.* (1999) have used the fuzzy logic approach for the automatic control of peanut roasting, Perrot *et al.* (2000) have applied fuzzy sets for the control of quality in baking, and Chung *et al.* (2010) have used fuzzy control for processes requiring pH adjustment. Neural networks (Ungar *et al.* 1996) are particularly adapted for controlling complex processes. They can be visualized as a network of data processing steps linked by complex connections. Neural networks have been used for the optimization of olive oil production by the decanter process (Jiménez Marquez *et al.* 2009). An interesting trend in food process control is the use of fuzzy logic and trained neural networks in attempts to introduce *sensory attributes* as control set points (Kuponsak and Tan, 2006).

5.7.3 The actuators

The actuator is the control element that performs the correcting action, in response to a signal from the controller. The actuator may be a valve, a switch, a servo-motor, or a mechanical device capable of causing displacement. One of the

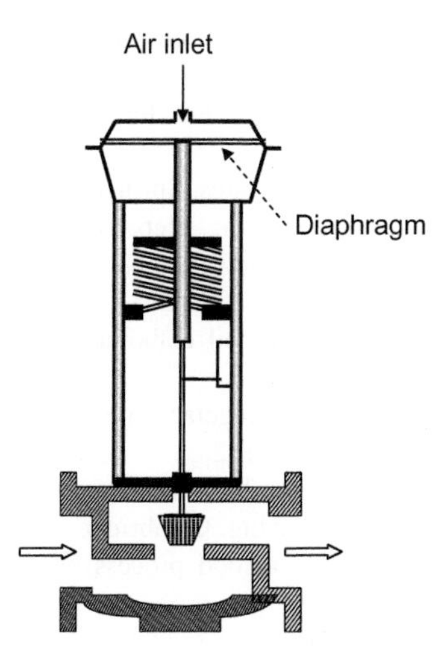

FIGURE 5.18

Pneumatic control valve, air-to-close type.

most common actuators, the control valve, will be discussed in some detail as a representative example.

A control valve regulates the flow-rate of a fluid (steam, cooling water, compressed air, liquid product, etc.) according to the correction signal, which can be electrical or pneumatic. The relationship between the amplitude of the correction signal and the response of the valve is known as the "valve equation". The valve is said to be linear if the flow-rate is proportional to the correction signal. *Normally closed* and *normally open* valves are, respectively, totally closed or totally open in the absence of a correction signal. In the case of pneumatic valves, the usual terminology is "air to open" or "air to close" (Figure 5.18).

If the controller signal is digital or electrical and the valve is pneumatic, a converter is used. If operating the valve requires considerable force, amplifiers are needed. In some designs, the converter, amplifier and valve form one unit.

References

Bhuyan, M., 2007. Measurement and Control in Food Processing. Taylor & Francis Group, Boca Raton, FL.

Bimbenet, J.J., Dumoulin, E., Trystram, G. (Eds.), 1994. Automatic Control of Food and Biological Processes. Elsevier, Amsterdam.

Bobàl, J., Böhm, J., Fessl, J., Machàcek, J., 2005. Digital Self-tuning Controllers. Springer, New York, NY.

Burns, J., 1999. Resistive thermometers. In: Webster, J.G. (Ed.), The Measurement, Instrumentation and Sensor Handbook. Springer, New York, NY.

Chung, C.-C., Chen, H.-H., Ting, C.-H., 2010. Grey prediction fuzzy control for pH processes in the food industry. J. Food Eng. 96 (4), 575–582.

Davidson, V.J., Brown, R.B., Landman, J.J., 1999. Fuzzy control system for peanut roasting. J. Food Eng. 41 (2–3), 141–146.

Figura, L.O., Teixeira, A.A., 2007. Food Physics. Springer-Verlag, Berlin, Germany.

Fraden, J., 1999. Infrared thermometers. In: Webster, J.G. (Ed.), The Measurement, Instrumentation and Sensor Handbook. Springer, New York, NY.

Jiménez Marquez, A., Aguilera Herrera, M.P., Uceda Ojeda, M., Beltrán Maza, G., 2009. Neural network as tool for virgin olive oil elaboration process optimization. J. Food Eng. 95 (5), 135–141.

Kondo, N., 2010. Automation on fruit and vegetable grading system and food traceability. Trends Food Sci. Technol. 21 (3), 145–152.

Kress-Rogers, E., Brimelow, C.J.B. (Eds.), 2001. Instrumentaton and Sensors for the Food Industry. second ed. Woodhead Publishing, Cambridge, UK.

Kuponsak, S., Tan, J., 2006. Control of a food process based on sensory evaluations. J. Food Proc. Eng. 29 (6), 675–688.

Mahalik, N.P., Nambiar, A.N., 2010. Trends in food packaging and manufacturing systems and technology. Trends Food Sci. Technol. 21 (3), 117–128.

Mittal, G.S., 1997. Computerized Control Systems in the Food Industry. CRC Press, New York, NY.

O'Dwyer, A., 2006. Handbook of Pi and Pid Controller Tuning Rules, second ed. Imperial College Press, London, UK.

Ogunnaike, B.A., Ray, W.H., 1994. Process Dynamics, Modeling and Control. Oxford University Press, New York, NY.

Perrot, N., Agioux, L., Ionnou, I., 2000. Feedback quality control in the baking industry using fuzzy sets. J. Food Proc. Eng. 23 (4), 249–280.

Reed, R.P., 1999. Thermocouple thermometers. In: Webster, J.G. (Ed.), The Measurement, Instrumentation and Sensor Handbook. Springer, New York, NY.

Sapoff, M., 1999. Thermistor thermometers. In: Webster, J.G. (Ed.), The Measurement, Instrumentation and Sensor Handbook. Springer, New York, NY.

Stephanopoulos, G., 1983. Chemical Process Control: An Introduction to Theory and Practice. Prentice Hall, New York, NY.

Ungar, L.H., Hartman, E.J., Keeler, J.D., Martin, G.D., 1996. Process modeling and control using neural networks. AIChE Proc. 312 (92), 57–67.

Webster, J.G. (Ed.), 1999. Instrumentation and Sensor Handbook. Springer, New York, NY.

Zell, M., Lyng, J.G., Morgan, D.J., Cronin, D.A., 2009. Development of rapid response thermocouple probes for use in a batch ohmic heating system. J. Food Eng. 93 (3), 344–347.

CHAPTER

Size Reduction

6

6.1 Introduction

In the food industry, raw materials and intermediate products must often be subjected to size reduction operations such as *cutting*, *chopping*, *grinding*, *milling*, and so on. In the case of liquids and semi-solids, size reduction operations include *mashing*, *atomizing*, *homogenizing*, etc. *Foaming* may also be seen as a size reduction operation, since it is based on the disintegration of large gas bubbles into smaller ones. The following are some important applications of size reduction in the food industry:

- Milling of cereal grains to obtain flour
- Wet grinding of corn in the manufacture of starch
- Wet grinding of soybeans in water, in the production of soy milk
- Fine grinding (refining) of chocolate mass
- Flaking of soybeans prior to solvent extraction
- Cutting of vegetables and fruits to desired shapes (cubes, strips, slices ...)
- Fine mashing of baby food
- Homogenization of milk and cream
- Foaming in the production of ice cream, batters, coffee drinks, etc.
- Emulsification.

Size reduction is a widespread, multi-purpose operation. It may serve a number of different objectives, such as:

- Accelerating heat and mass transfer (flaking of soybeans or grinding coffee in preparation to extraction, atomization of milk as a fine spray into hot air in spray-drying)
- Facilitating separation of different parts of a material (milling wheat to obtain flour and bran separately, filleting of fish)
- Obtaining a desirable product texture (refining of chocolate mass, meat grinding)
- Facilitating mixing and dispersion (milling or crushing ingredients for dry-mixing, homogenization of liquids to obtain stable emulsions)
- Portion control (slicing cold-cuts, bread, cakes)
- Obtaining pieces and particles of defined shapes (cubing meat for stew, cutting pineapple to obtain the familiar wheel-shaped slices, cutting dough to make cookies).

Food Process Engineering and Technology. DOI: http://dx.doi.org/10.1016/B978-0-12-415923-5.00006-X
© 2013 Elsevier Inc. All rights reserved.

In addition, size reduction of food at the moment of consumption (mastication) has a decisive effect on the perception of food quality (Jalabert-Malbos *et al.*, 2007).

Only size reduction of solids will be treated in this chapter. Due to its close relation with fluid flow and mixing, size reduction in liquids and gases are discussed in the next chapter.

6.2 Particle size and particle size distribution

Before discussing size reduction, we shall define some fundamental notions pertaining to particle size, size distribution and particle shape.

6.2.1 Defining the size of a single particle

The size of a regular-shaped particle can be defined with the help of a small number of dimensions: one in the case of spheres or cubes, two in the case of cylinders and ellipsoids, three in the case of prisms, etc. Our raw materials and products, however, are seldom made of regular particles. How, then, can we define the "size" of an irregular shaped particle, using a small number of dimensions – preferably one? Several approaches are possible. The most common is to assign to the particle an *equivalent diameter*. Most frequently, this is the diameter of a sphere that would behave like the particle in question when subjected to a certain measurement or test. Some of the best-known equivalent diameters are listed in Table 6.1.

Table 6.1 "Equivalent Diameter" Types Commonly Used to Define Particle Size.

Equivalent Diameter Type	Symbol	Equivalent Behavior
Sieve diameter	d_A	passing through the same sieve opening
Surface diameter	d_S	having the same surface area
Volume diameter	d_V	having the same volume
Surface–volume diameter (Sauter diameter)	d_{SV}	having the same surface/volume ratio
Laser diameter	d_L	having the same interaction (diffraction pattern) with a laser beam
Stokes diameter	d_{ST}	of equal density, falling at the same Stokes terminal velocity in a given fluid

EXAMPLE 6.1

A sugar crystal has the form of a rectangular prism, with dimensions $1 \times 2 \times 0.5$ mm. Calculate the diameter of a sphere having:

a. The same volume
b. The same surface area
c. The same surface/volume ratio (Sauter diameter).

What is the sphericity of the crystal?

Solution

Volume of the crystal $V = 1 \times 0.5 \times 2 = 1$ mm^3
Surface area of the crystal $S = 2(1 \times 0.5 + 1 \times 2 + 0.5 \times 2) = 7$ mm^2
Therefore $S/V = 7$ mm^{-1}.

a. Diameter of sphere with the same volume:

$$d_V = \sqrt[3]{6V/\pi} = \sqrt[3]{6 \times 1/\pi} = 1.24 \text{ mm}$$

b. Diameter of sphere with the same surface area:

$$d_S = \sqrt{S/\pi} = \sqrt{7/\pi} = 1.49 \text{ mm}$$

c. Sauter diameter:

$$\pi d_{SV}{}^2/(\pi d_{SV}{}^3/6) = 6/d_{SV} = 7 \qquad d_{SV} = 6/7 = 1.17 \text{ mm}$$

Sphericity of the crystal (see Section 6.2.4):

$$\text{Sphericity} = \frac{\pi\, d_V{}^2}{S} = \frac{\pi \times (1.24)^2}{7} = 0.6897$$

6.2.2 Particle size distribution in a population of particles; defining a "mean particle size"

Food materials often consist of particles of different size. To characterize such particulate materials, it is necessary to determine their particle size distribution (PSD) and to define a *mean particle size*. In the following discussion we shall refer mainly to powders, but the concepts developed are equally applicable to droplets in an emulsion or in a spray, to a shipment of potatoes arriving from the field, or to a batch of green peas leaving a viner.

Particle size distribution refers to the proportion of particles within a certain size range in a population of particles. The importance of particle size distribution in connection with food quality and processing is obvious (Servais *et al.*, 2002). Methods for the determination of particle size distribution include sifting, microscopic examination (usually coupled with automatic image analysis), laser diffraction techniques and others. Sieve analysis is a simple technique that is commonly used for the determination of PSD and for the quantitative evaluation of the "fineness" of powders. A weighed amount of material is placed on top of a stack of nested mesh sieves, with opening sizes decreasing from top to bottom. The assembly of sieves is mechanically shaken or vibrated for a period of time, and the amount of powder retained on each sieve is then weighed and recorded. Today, most standard test sieves are designated by the size of their openings in micrometers, but the traditional US standard and Tyler "mesh number" designations are still widely used. The mesh number is the number of openings per linear inch of mesh. The opening sizes corresponding to mesh numbers are given in Table A.15 (see Appendix).

Various types of particle size distribution can be defined, depending on the parameter by which the individual diameters are measured. If the particles are counted, the result is the *arithmetic* or *number* PSD. If the particles are weighed (as they are in sieving analysis), the result is the *mass* PSD. Assuming that the true density of the particles is uniform, this is also the *volume* PSD. Similarly *surface* and *surface/volume* PSDs can also be defined. Obviously, each type of PSD provides a different type of mean particle diameter.

EXAMPLE 6.2

The size distribution shown in Table 6.2 was determined by screening analysis of a beverage powder.

a. Calculate the mass average diameter.

b. Convert the table to a PSD table by number and calculate the number average diameter. The particle density is 1280 $kg \cdot m^{-3}$.

Solution

The screen analysis table is expanded as shown in Table 6.3.

Table 6.2 Sieving Test Results.

Sieve	x (μm)	mg
1	1200	0
2	1000	3
3	800	7
4	600	46
5	400	145
6	200	178
7	100	45
8	50	11
9	30	4
10	20	1.9
Pan		0
Σ		440.9

Table 6.3 Expansion of the Sieving Results.

Sieve	x (μm)	x mean	mg	m%	Δx	m %/Δx	Cum +	Cum −
1	1200		0	0.0		0	0	100
2	1000	1100	3	0.7	200	0.00342	0.6	99.4
3	800	900	7	1.6	200	0.00798	2.2	97.8
4	600	700	46	10.4	200	0.05216	12.6	87.4
5	400	500	145	32.9	200	0.16446	45.5	54.5
6	200	300	178	40.4	200	0.20186	85.9	14.1
7	100	150	45	10.2	100	0.10204	96.1	3.9
8	50	75	11	2.5	50	0.04988	98.6	1.4
9	30	40	4	0.9	20	0.04532	99.5	0.5
10	20	25	1.9	0.4	10	0.04304	99.9	0.1
Pan			0	0.0				
Σ			440.9	100.0				

a. Mass average diameter:

$$d_m = d_V = \frac{\sum x_i m_i}{\sum m_i} = \frac{175\,482}{440.9} = 398\ \mu\text{m}$$

b. Converting the PSD to number percentage, the number of particles in each size interval is:

$$n_i = \frac{6m_i}{\rho \pi x_i^3}$$

The PSD is shown in Table 6.4.
For comparison, Figure 6.1 shows the size distribution by mass and by number.

6.2.3 Mathematical models of PSD

A number of statistical functions have been proposed for representing, explaining and predicting size distribution in a population of particles. All the models described below predict size distribution with the help of only two parameters.

The *Gaussian* or *normal distribution* (after Carl Friedrich Gauss, German scientist, 1777–1855) is described by Eq. (6.1):

$$f(x) = \frac{d\varphi}{dx} = \frac{1}{\sigma\sqrt{2\pi}} \exp\left[-\frac{(x-\bar{x})^2}{2\sigma^2}\right] \tag{6.1}$$

where:

x = particle size (e.g., particle diameter)
$\bar{x}$ = mean particle size

Table 6.4 PSD by Number.

Sieve No.	d	d mean	m	n	n %
1	1200		0	0	0
2	1000	1100	3	5513	0.000974
3	800	900	7	23 486	0.00415
4	600	700	46	328 016	0.057955
5	400	500	145	2 837 197	0.501284
6	200	300	178	16 124 558	2.848929
7	100	150	45	32 611 465	5.76188
8	50	75	11	63 773 531	11.26768
9	30	40	4	152 866 242	27.00881
10	20	25	1.9	297 416 561	52.54834
Pan			0		
				565 986 569	

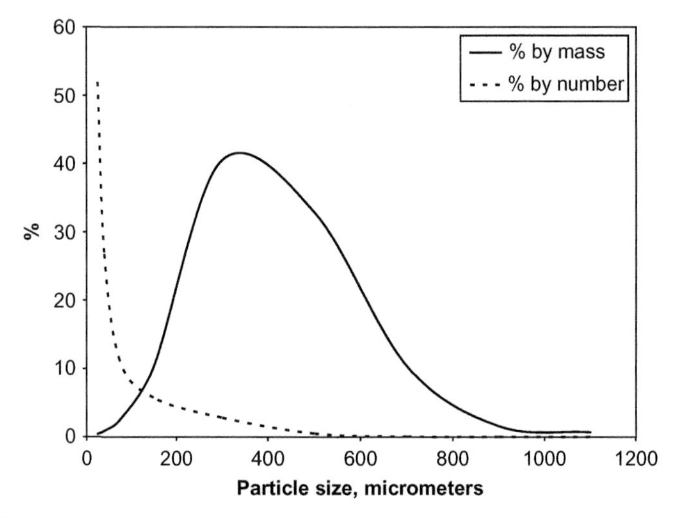

FIGURE 6.1

Particle size distribution of the powder.

f(x) = the probability density function
ϕ = frequency of particle size x in the population
σ = standard deviation.

The well-known bell-shaped curve representing the normal distribution model is shown in Figure 6.2.

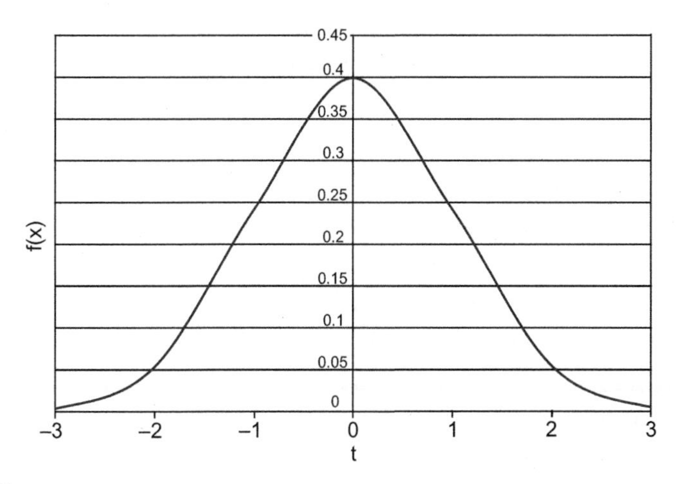

FIGURE 6.2

Normal (Gauss) distribution.

This famous statistical model may well fit size distribution in unprocessed agricultural produce consisting of discrete units (e.g., fruit), but it is not very useful for representing PSD in food powders, emulsions or sprays. In normal distribution:

a. The mean, the mode and the median are identical
b. 50% of the particles are ± 0.6745 standard deviations apart from the mean
c. About 20% of the particles are ± 0.25 standard deviations apart from the mean
d. The range between one standard deviation and two standard deviations, on both sides of the mean, contains about 27% of the population.

- The *log-normal distribution* is described by Eq. (6.2),

$$f(\ln x) = \frac{d\varphi}{dz} = \frac{1}{\sigma_z\sqrt{2\pi}}\exp\left[-\frac{(z-z^2)}{2\sigma_z^2}\right] \quad \textit{with} \; z \equiv \ln x \tag{6.2}$$

The log-normal distribution fits fairly well the PSD of liquid sprays and powders produced by spray-drying,

- The *Gaudin–Schuhmann* function, shown in Eq. (6.3), is an empirical model.

$$F(x) = \left(\frac{x}{x'}\right)^n \tag{6.3}$$

where:
F(x) = cumulative mass fraction of particles smaller than x
x′ = a parameter of the model, often called “size parameter”
n = another parameter of the model, often called “distribution parameter”.

- The *Rosin–Rammler* function (also known as the *Weibull* function), shown in Eq. (6.4), is also an empirical distribution model.

$$R(x) = 1 - e^{-\left(\frac{x}{x'}\right)^n} \tag{6.4}$$

where R(x) = cumulative mass fraction of particles smaller than x. The Rosin–Rammler model is fairly accurate in representing PSD of materials obtained by size reduction.

The characterization of selected food powders by five different PSD functions has been studied by Yan and Barbose-Cánovas (1997).

EXAMPLE 6.3

A sample of powdered spice was sifted on two screens. It was found that 15% (by weight) of the material consisted of particles larger than 500 μm and 54% of the powder consisted of particles larger than 200 μm. Predict the PSD of the spice:

a. According to the Rosin–Rammler model
b. According to the Gaudin–Schuhmann model.

Solution

a. The Rosin–Rammler model:

$$F(x) = 1 - e^{-(x/x')^n}$$

The parameters n and x′ are calculated from the data (two unknowns, two equations). We find: n = 1.227 and x′ = 297 μm.
The values found are substituted in the Rosin–Rammler equation and F(x) is calculated for different values of x. The results are shown in Table 6.5 and Figure 6.3.

b. The Gaudin–Schuhmann model:

$$F(x) = (x/x')^n$$

The parameters n and x′ are calculated from the data. We find: n = 0.67 and x′ = 637 μm.
The parameters are substituted in the equation and F(x) is calculated for various values of x. The maximum value of x is, of course, x′. The results are shown in Table 6.6 and Figure 6.4.

6.2.4 A note on particle shape

Particle shape is, of course, an important quality parameter. In powders, the shape of the particles has a marked effect on the bulk density and flowability of the material. In chocolate, the shape of sugar crystals may affect the mouth-feel of the product. The regularity and uniformity of the particles in a cut fruit product or in short pasta is an important quality factor. Size reduction operations modify the shape distribution of the particles.

Table 6.5 PDS According to Rosin–Rammler.

x	1 − F %	F %	f(x) %
20	96.4	3.6	7.1
50	89.4	10.6	12.5
100	76.9	23.1	12.0
150	64.9	35.1	10.9
200	54.0	46.0	9.5
250	44.5	55.5	8.2
300	36.3	63.7	6.9
350	29.4	70.6	5.8
400	23.6	76.4	4.7
450	18.9	81.1	3.9
500	15.0	85.0	3.1
550	11.9	88.1	2.5
600	9.3	90.7	2.0
650	7.3	92.7	1.6
700	5.7	94.3	1.3
750	4.4	95.6	1.0
800	3.4	96.6	0.8
850	2.6	97.4	0.6
900	2.0	98.0	0.5
950	1.5	98.5	0.4
1000	1.2	98.8	

There exist a number of methods for describing particle shape quantitatively. Image analysis coupled with appropriate algorithms is extensively used in research and in quality assessment. A simple (and necessarily imperfect) quantitative expression of particle shape is the concept of *sphericity*. The sphericity of a particle is the ratio of the surface area of a sphere having the same volume as the particle to the surface area of the particle. The sphericity of a sphere is, by definition, equal to 1; the sphericity of any other shape is, of course, less than unity. If sphericity is adopted as a quantitative expression of shape, the shape distribution of an assembly of particles may be described as the distribution of sphericity, just as PSD.

6.3 Size reduction of solids: basic principles

6.3.1 Mechanism of size reduction in solids

Compression and *shear* are the two types of force involved in size reduction of solids. *Impact*, sometimes given as a third type, is in fact strong compression

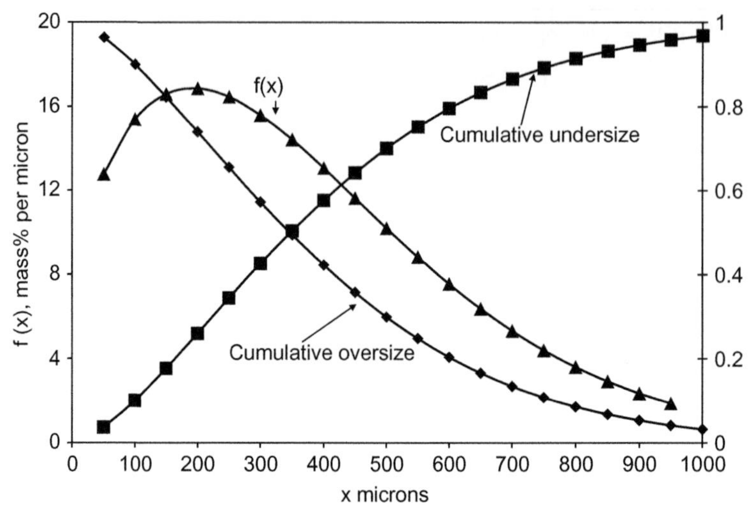

FIGURE 6.3

Rosin–Rammler plot of the data.

Table 6.6 Prediction of PSD using the Gaudin–Schuhman Model.

X	F(x)	1 – F(X)	f(x) %
20	9.838223	90.16178	8.339374
50	18.1776	81.8224	10.74429
100	28.92188	71.07812	9.027762
150	37.94965	62.05035	8.067182
200	46.01683	53.98317	7.420703
250	53.43753	46.56247	6.943124
300	60.38066	39.61934	6.56959
350	66.95024	33.04976	6.265886
400	73.21613	26.78387	6.011919
450	79.22805	20.77195	5.794962
500	85.02301	14.97699	5.606488
550	90.6295	9.3705	5.440529
600	96.07003	3.929972	3.929972
637	100	0	

applied for a very short time. When such forces act on a solid particle, elastic or plastic deformation occurs, depending on the material. As the deformation reaches a certain limit value determined by the nature and structure of the solid

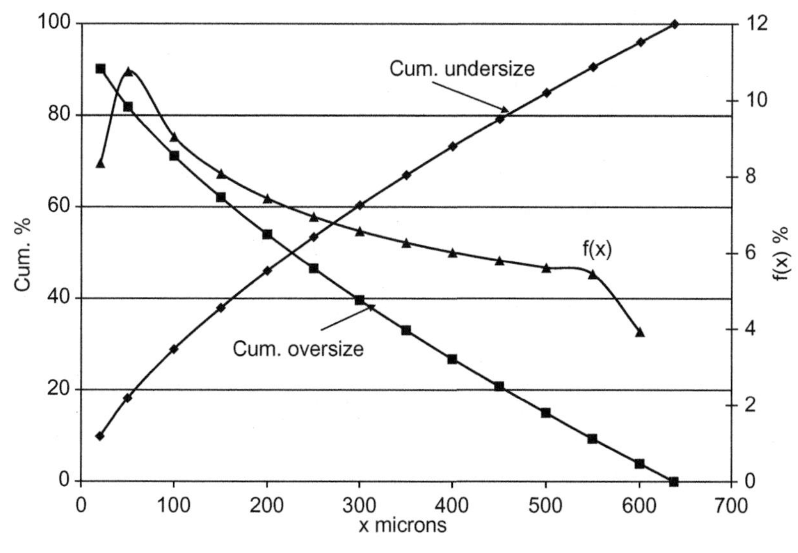

FIGURE 6.4

Gaudin–Schuhman plot of the data.

(*mechanical strength*), the particle breaks along certain planes (*planes of failure*). In any size reduction operation, all the types of force are usually involved, to different extents.

In evaluating the performance of a size reduction operation, two aspects have to be considered: the PSD of the size-reduced material, and the energy expenditure.

6.3.2 Particle size distribution after size reduction

Even if the feed consists of uniformly sized particles, the material discharged from the size reduction device will present a more or less wide PSD. Where size uniformity of the product is important, a step of classification (sorting) by size follows the size reduction operation. The degree of size uniformity in the product is an important issue. If product specifications require particle sizes within a given range, both the oversize and the undersize fractions represent either a loss of material or extra processing cost, or both.

Experimental data on the change in PSD during size reduction of food materials are scarce. Some information is available from milling experiments with minerals. When a batch of fairly uniform particles is milled and PSD is determined periodically during the operation, it is observed that:

- In the first stages of milling, PSD becomes bi-modal.
- As milling progresses PSD gradually becomes mono-modal.
- Gradually, the large particles disappear and the frequency of a certain size increases.

- After prolonged milling, a certain "final" PSD is reached. The final PSD depends on the mill characteristics, operation conditions, and the properties of the material milled, but is almost independent of the size of the particles in the feed (Lee *et al.*, 2003).

6.3.3 Energy consumption

As a rule, size reduction operations are heavy in energy consumption (Hassanpour *et al.*, 2004; Loncin and Merson, 1979). As an example, the cost of energy is the single largest item in the total cost of wheat milling. Milling of one bushel (approximately 27 kg) of wheat requires 1.74 kWh of electric energy (Ryan and Tiffany, 1998).

The total energy consumption of a mill consists of two parts: the energy imparted to the milled material, and that needed to overcome friction in bearings and other moving parts of the mill. The energy transferred to the material corresponds to the work of deformation and is stored in the particle as internal stress. When the particle fractures, the stored energy is released. Part of it provides the increment in surface energy resulting from increased surface area (see Chapter 1, Section 1.10), but most of it is released as heat. Eventually, friction losses also generate heat. Consequently, size reduction may result in considerable increase in the temperature of the treated material. Temperature rise as a result of size reduction may be an important technological issue, particularly with heat-sensitive products, thermoplastic substances, and materials with high fat content. When necessary, this problem is addressed by air- or water-cooling of the machine or by using cryogenics such as liquid nitrogen or dry ice (*cryo-milling*).

Crushing efficiency η_c is defined as the ratio of the increment in surface energy to the total energy imparted to the material:

$$\eta_c = \frac{E_s}{E_a} = \frac{\sigma(A - A_0)}{E_a} \tag{6.5}$$

where:

E_S = Surface energy increment per unit mass of material ($J \cdot kg^{-1}$)
E_a = Total energy transferred to the material per unit mass ($J \cdot kg^{-1}$)
σ = surface tension of the solid ($J \cdot m^{-2}$ or $N \cdot m^{-1}$)
A, A_0 = surface area per unit mass (specific surface area) of the milled material and of the feed, respectively ($m^2 \cdot kg^{-1}$).

Crushing efficiency is very low, usually a few percent or less.

Mechanical efficiency η_m of a size reduction device is defined as the ratio of the energy transferred to the material to the total energy consumption W of the device, per unit mass of material treated.

$$\eta_m = \frac{E_a}{W} = \frac{\sigma(A - A_0)}{\eta_c W} \tag{6.6}$$

Noting that the specific surface area of a (spherical) particle is inversely proportional to its diameter, one can write:

$$W = K\left(\frac{1}{x} - \frac{1}{x_0}\right) \tag{6.7}$$

where x, x_o = mean Sauter diameter of the milled material and the feed, respectively, and K comprises the surface tension and the two efficiencies. Assuming that these values are constant, K is a constant.

Equation (6.7) is known as *Rittinger's Law*. Due to the inaccuracy of the assumptions on which it is based and the difficulty in determining its parameters, Rittinger's Law is only approximate.

A different expression for energy requirement of size reduction has been proposed by Kick. Kick assumes that the energy needed to reduce the size of the material by a certain proportion (say, by half or by one order of magnitude) is constant (first-order relationship). *Kick's Law* is written as:

$$W = K' \log\left(\frac{x_0}{x}\right)$$

Rittinger's equation is said to fit fine milling better, while Kick's expression describes coarse grinding better (Earle, 1983). Both Kick's and Rittinger's Laws have been found to be valid in the initial stages of wet grinding of soybeans (Vishwanathan *et al.*, 2011).

EXAMPLE 6.4

Sugar crystals were ground from an average Sauter diameter of 500 μm to a powder with an average Sauter diameter of 100 μm. The net energy consumption was 0.5 kWh per ton. What would be the net energy consumption for grinding the crystals to 50-μm powder?

a. According to Rittinger's Law
b. According to Kick's Law.

Solution

a. Rittinger's Law:

$$\overline{E} = K\left(\frac{1}{x_2} - \frac{1}{x_1}\right)$$

K is calculated from the first milling data and applied to the second milling:
$K = 0.5/(1/100 - 1/500) = 62.5$ kWh · μm · ton^{-1}
$E = 62.5\ (1/50 - 1/500) = 1.125$ kWh · ton − 1.

b. Kick's Law:
$\overline{E} = K\log(x_1/x_2)$
$K = 0.5/\log(100/500) = -0.715$
$E = -0.715^*\ \log(50/500) = 0.715$ kWh · ton^{-1}.

6.4 Size reduction of solids: equipment and methods

The variety of size reduction equipment available to the food industry is extensive. For the selection of the appropriate machine, the following factors in particular must be considered:

- Structure, composition and mechanical properties of the material to be processed
- Desired PSD and particle shape of the product to be obtained
- Desired rate of throughput
- Control of product overheating
- Inertness of the surfaces in contact with the food
- Sanitary design, ease of cleaning
- Ease of maintenance
- Environmental factors (noise, vibration, dust, explosion hazard)
- Capital and operating costs (e.g., energy consumption, wear resistance, etc.).

Size reduction equipment types may be classified according to the main kind of action exerted on the processed material, as follows:

- Main action is impact
- Main action is pressure
- Main action is attrition
- Main action is shearing.

6.4.1 Impact mills

The principal representative of this class is the *hammer mill* (Figure 6.5).

The crushing elements are hammers fitted on a high-speed rotor inside a cylindrical chamber. The chamber walls may be smooth, or lined with corrugated breaker plates. The hammers may be fixed or swinging. Swinging hammers are used when it is necessary to reduce the risk of damage to the mill in the case of encounter between the hammer and large, hard chunks. The principal crushing action takes place as a result of the collision between the particles and the hammers and chamber walls. The leading face of the hammers may be blunt or sharp. Knife-like sharp hammers are useful in the case of fibrous materials where some shearing action is necessary. Normally, the chamber exit is fitted with interchangeable screens to allow continuous removal of the sufficiently small particles, while the oversize material is retained for further size reduction. In the case of heat-sensitive materials, the increased residence time of part of the feed in the mill chamber may be unacceptable. In this case, the mill exit is unrestricted. An external pneumatic loop is provided for screening the milled material and recycling the coarse fraction through the mill (Figure 6.6).

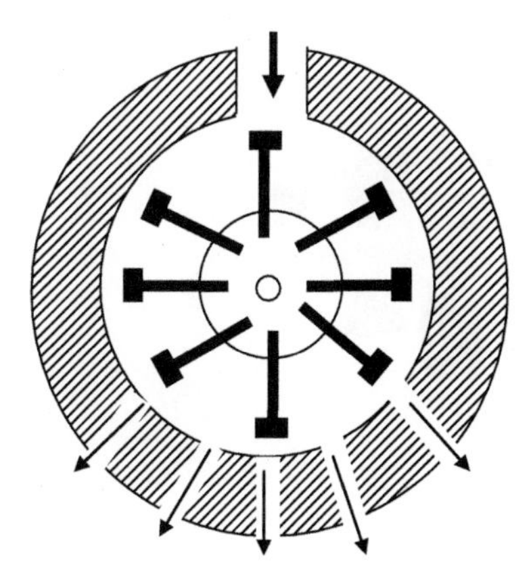

FIGURE 6.5

Basic structure of a hammer mill.

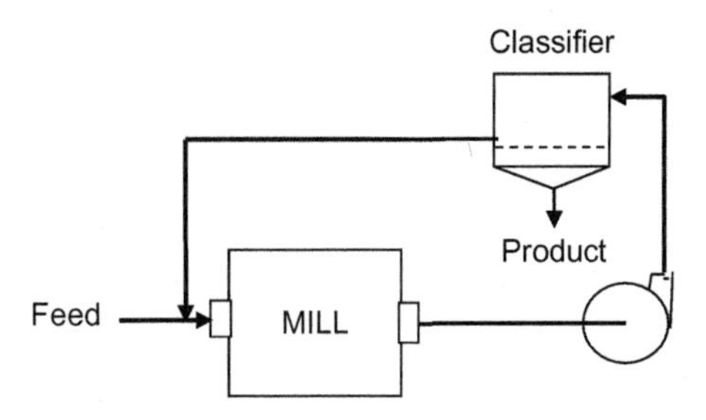

FIGURE 6.6

Recycling of coarse material in milling.

6.4.2 Pressure mills

One of the most widespread types of this family is the *roller mill* (Figure 6.7).

The material is compressed between a pair of counter-rotating heavy rollers with hardened surfaces, to the point of fracture (Gutsche and Fuerstenau, 2004). The rollers may be smooth, or corrugated to reduce slippage. The mill may consist of several rollers with gradually decreasing clearance between the rollers. Roller mills are extensively used in the grain milling industry (Figure 6.8). Mills equipped with corrugated rollers are used in the initial stages of wheat milling. They are called "breaks" because they serve to "break" the grain and to separate the feed into the main fractions. Breakers are also used in the dry milling of corn and for fracturing soybeans prior to flaking for the production of oil. In wheat

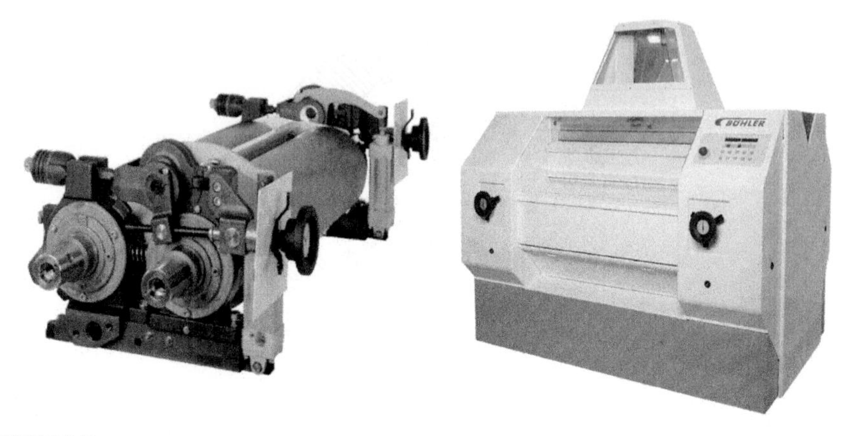

FIGURE 6.7

A four-roller mill and a pair of rolls.

Photograph courtesy of Bühler A.G.

FIGURE 6.8

Battery of "breaks" in a flour mill.

Photograph courtesy of Bühler A.G.

flour mills, units with smooth rollers are used as the further step of size reduction; these are called "refiners", as they serve to further refine the fractions to produce different flours and other products. A more detailed description of the wheat milling process is included at the end of this chapter. Mills consisting of a series of smooth rollers are standard equipment for refining chocolate mass (Figure 6.9). Smooth roller mills are also used as flaking machines – for example, in the production of cornflakes, oatmeal, etc. For flaking, the particles are first conditioned (plasticized) by humidification and heating in order to avoid disintegration.

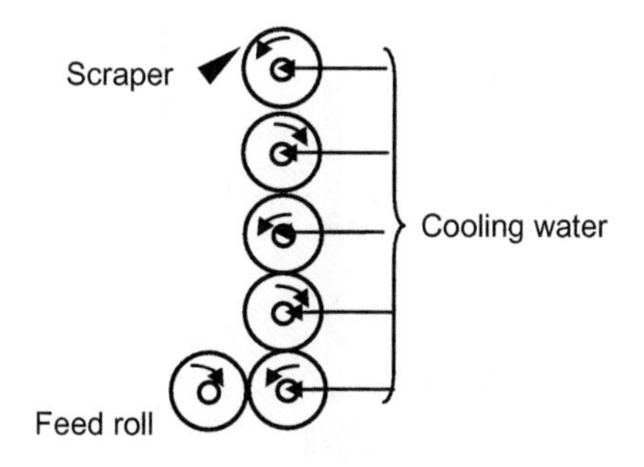

FIGURE 6.9

Schema of a five-roller refiner.

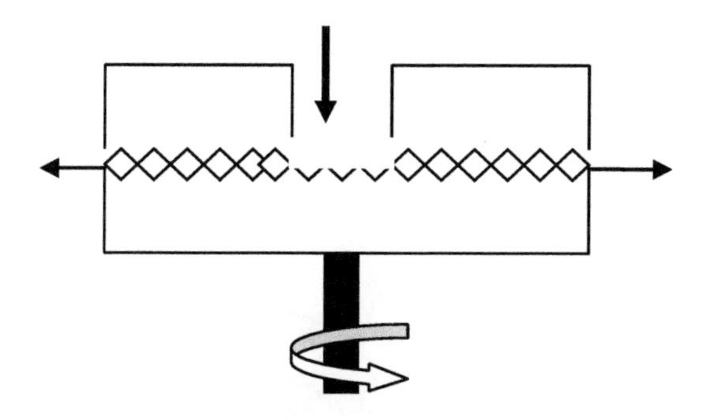

FIGURE 6.10

Attrition mill with flat grinding surface.

6.4.3 Attrition mills

This is a very large group of mills. In many of the types belonging to this class, the material is ground between two corrugated surfaces moving with respect to each other. The prehistoric stone mill and the familiar household coffee grinder are examples of this kind. The surfaces may be flat (Figure 6.10) or conical (Figure 6.11), vertical or horizontal. Usually, one of the surfaces rotates and the other is stationary.

Mills with circular grinding surfaces are also known as *disc mills* (Figure 6.12). The fineness of grinding is often controllable by adjusting the gap between the surfaces.

The grinding surfaces may be made of hardened metal or of coarse corundum. Corundum mills are used for reducing the size of particles in suspension down to colloidal range (micrometer range), and are therefore named "colloid mills" (Figure 6.13). They are used in the preparation of fine suspensions such as homogenized strained infant food, comminuted fruit pastes, and mustard.

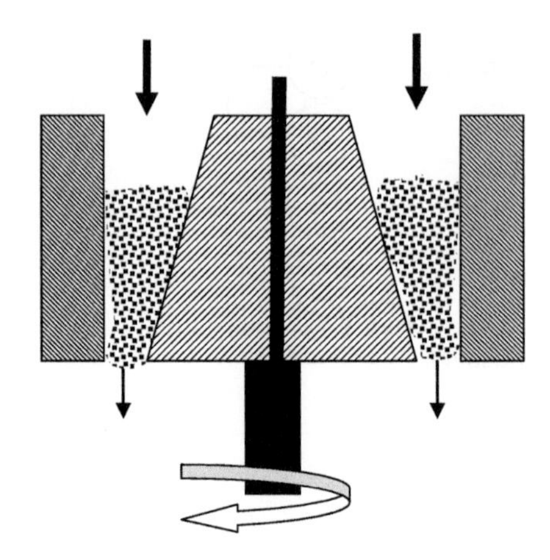

FIGURE 6.11

Attrition mill with conical grinding surface.

FIGURE 6.12

Disc mill.

Photograph courtesy of Retsch GmbH.

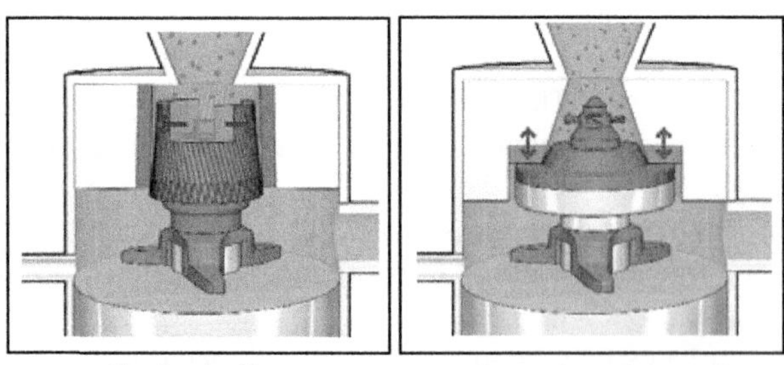

FIGURE 6.13

Colloid mills.

Photograph courtesy of FrymaKoruma GmbH.

The main mechanism of action in attrition mills is shear, due to friction between the particle and the pair of grinding surfaces. Unless the mill is cooled the temperature rise may be considerable, particularly in the case of dry milling. A different kind of attrition mill makes use of free "grinding media" instead of fixed grinding surfaces. One of the representatives of this kind is the *ball mill*; here, the feed is vigorously mixed in a vessel containing free spherical bodies (balls) made of materials with different degrees of hardness, from plastic to metal, according to the hardness of the particles to be disintegrated. Dry ball-milling (Figure 6.14) is used for fine grinding of pigments. Wet ball-milling (Woodrow and Quirk, 1982) is one of the methods used for the disintegration of cells in suspension (Figure 6.15).

6.4.4 Cutters and choppers

Cutting and *chopping* are size reduction operations based on shearing through the use of sharp-edged moving elements (knives, blades). The term "cutting" is

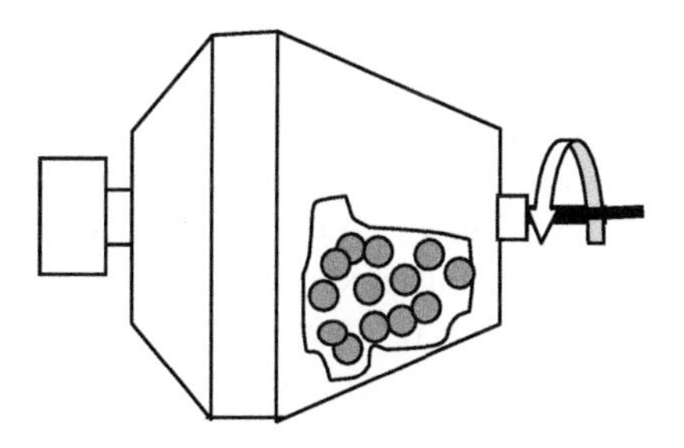

FIGURE 6.14

Dry ball mill.

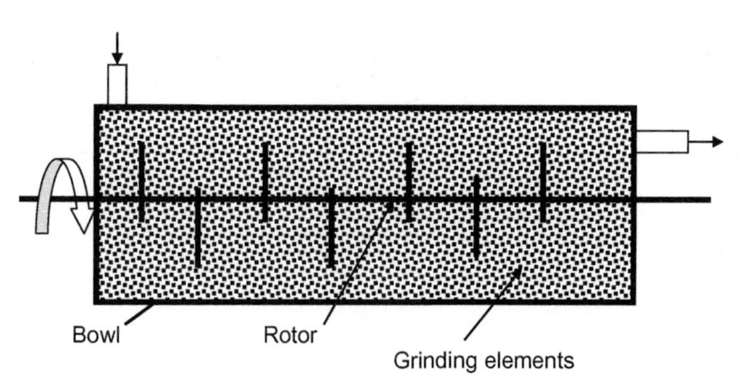

FIGURE 6.15

Wet ball mill.

usually reserved for operations resulting in particles with fairly regular geometric forms (cubes, juliennes, slices), while the term "chopping" is applied mainly to random cutting. The variety of cutting machines used in the food industry is vast. In the majority of cases, cutting is done by revolving knives or saws. The following are some examples of cutting and chopping machines.

- The *cube cutter* cuts along three mutually perpendicular planes. In the machine shown in Figure 6.16, the material is first cut into slices. In the second stage, the slices are cut longitudinally to produce strips. In the third and final stage, the strips are shortened to produce cubes.
- An interesting system is used for cutting potatoes in the French-fry industry. Peeled potatoes are hydraulically conveyed at high velocity through a tube and thrown against a stationary set of knives in quadratic array. The system has a number of advantages over other methods of cutting:
  a. There are no moving parts
  b. Hydraulic conveying orients the potatoes along their long axis, and the long strips thus obtained are preferred by the market
  c. Hydraulic conveying provides cutting and washing (removal of released starch granules) in one step.
- The *silent cutter* (Figure 6.17) is widely used in the meat industry for simultaneous chopping and mixing. A batch of the material to be processed is placed in a horizontal revolving dish. The dish circulates the material through a set of horizontal revolving knives. In the meat industry, the machine is often operated under vacuum to prevent lipid oxidation. Another related type of machine is the *bowl mixer-cutter* (Figure 6.18), similar in action to the kitchen blender or food processor. It is extensively used in the meat industry, but also

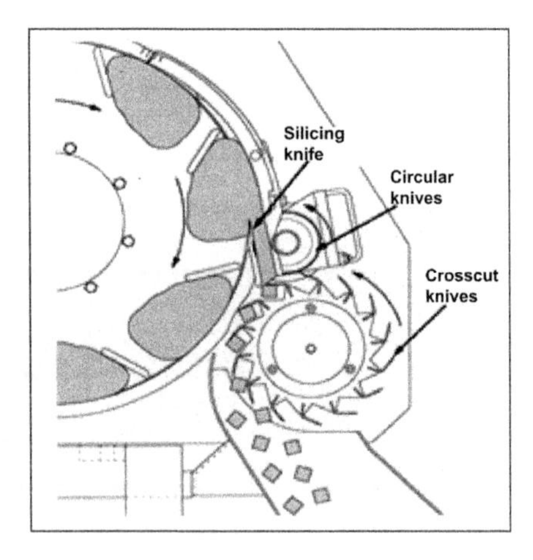

FIGURE 6.16

Cube cutter.

Figure courtesy of Urshel Laboratories.

FIGURE 6.17

Silent cutter.

Photograph courtesy of the Department of Biotechnology and Food Engineering, Technion.

FIGURE 6.18

Bowl mixer-cutter.

as a high-energy blender in the production of salads, and even as a dough kneader.

- The *meat grinder/chopper* is a familiar machine that is available in a vast range of capacities and variations. Basically, a worm (screw) conveyor forces the meat against one or more revolving knives and perforated plates (Figure 6.19).

FIGURE 6.19

Meat grinder.

Photograph courtesy of Hobart.

The need to use knives and blades for cutting presents two problems:

- The efficiency of the cutting machine strongly depends on the sharpness of the knives. Although the blades are made of special metals, loss of sharpness is always a problem, requiring costly maintenance (Marsot *et al.*, 2007).
- The knives are often a source of contamination that transfers microorganisms to the freshly produced new surfaces of the cut food. The steady increase of microbial count in the cut material in the course of a cutting operation is a frequently observed phenomenon. The problem is particularly serious in the meat and cold-cuts industry. Compliance with the need for frequent cleaning and sanitation operations is a factor to consider in the selection of cutting equipment.

This last point is elegantly addressed by a "knifeless" cutting method, making use of a narrow jet of water at very high velocity. In this method, water is pressurized up to many hundreds of MPa and released as a narrow high-velocity jet through nozzles made of very hard materials that are resistant to erosion. The jet is directed to the surface of the material to be cut. By virtue of its very high kinetic energy, the jet penetrates and cuts through like a sharp blade. The desired shape of the cut can be obtained by moving either the nozzle or the material to be cut. The movement is often computer-controlled (Figure 6.20). Originally applied mainly to the mining industry, *water-jet cutting* has been finding uses in some food industries (Becker and Gray, 1992; Heiland *et al.*, 1990).

Cutting assisted by ultrasonic vibrations is a new technique said to overcome some of the shortcomings of conventional cutting (Arnold *et al.*, 2011). Ultrasonic excitation was found to significantly reduce cutting force.

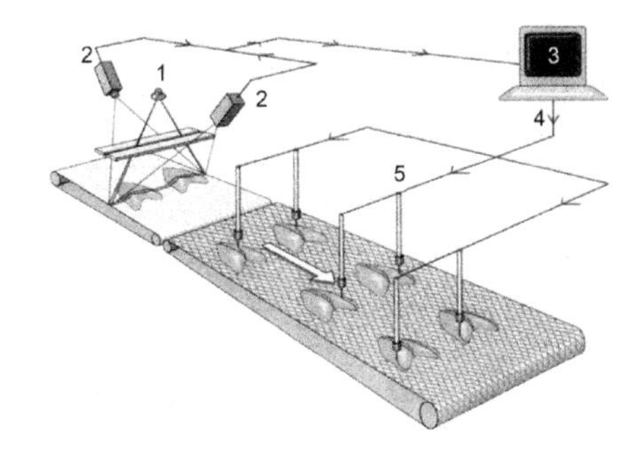

FIGURE 6.20

Computer-aided water jet cutting.

Figure courtesy of Stein.DSI/FMC FoodTech.

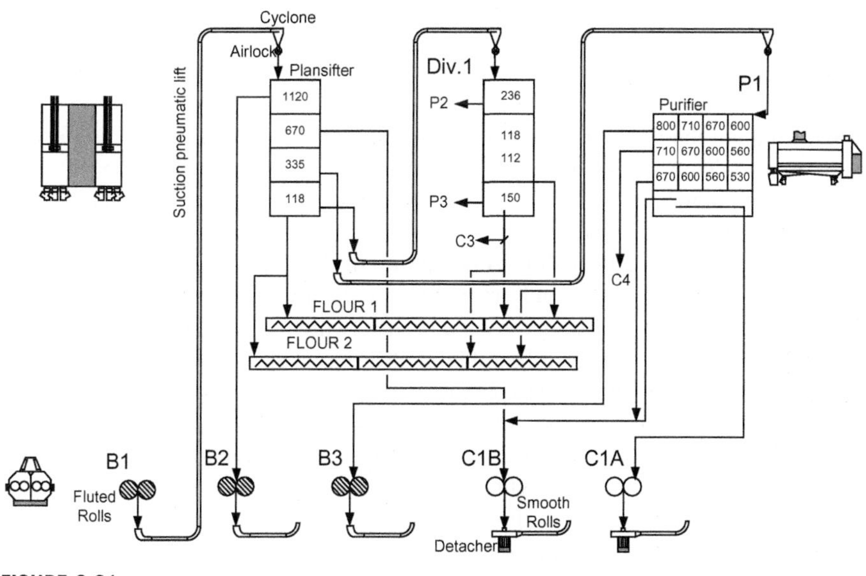

FIGURE 6.21

Simplified flow diagram of the break system in wheat milling.

Figure courtesy of Bühler A.G.

6.4.5 The wheat milling process

The process of milling wheat combines simultaneous size reduction and mechanical separation operations (Delcour and Hoseney, 2010). The wheat kernel consists essentially of three distinct parts: a tough seed coat (pericarp), a friable

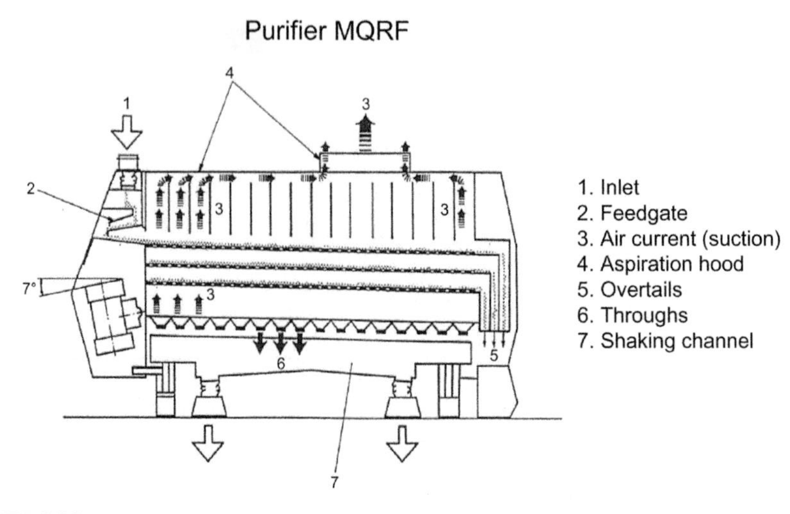

FIGURE 6.22

Schema of a purifier used in wheat milling.

Figure courtesy of Bühler A.G.

endosperm, and the germ. The principal objective of the milling process is to separate, as completely as possible, these three fractions. At the end of the process, the seed coat comes out as *bran*. The endocarp produces *flours* and *semolinas*. The *germ* may be separated as a product by itself, or added to another fraction. The milling process begins with the opening up of the grain to release its contents and, subsequently, separation and gradual size reduction of the released fractions. The opening-up step is performed by the *breaks*, which are roller mills consisting of pairs of corrugated (fluted) rolls rotating at different speeds to induce shearing. The gradual size reduction of the contents is carried out by a series of roller mills consisting of smooth rolls (*reduction rolls*), turning at nearly equal speed, where pressure and shear do the crushing. The separation of the fractions is achieved by sieving on plansifters, and by sieving and air entrainment in purifiers. A flow diagram of the first stages of the milling process is shown in Figure 6.21.

Cleaned, washed and tempered wheat is fed to the first break (B1), where mainly the coat is torn open with little size reduction. The resulting material is sent to a plansifter composed of a number of superimposed sieves. Most of the material is retained by the coarser (upper) sieve, and consists of the bran with coarse particles of adhering endosperm and germ. A small amount of size-reduced free endosperm (middlings, semolina and flour) is separated on the finer (lower) sieves. The coarse fraction is fed to the second break (B2), where more endosperm is released from the bran. This breaking–sifting sequence is repeated several (usually four to five) times, until practically all the seed contents is released from the bran. The endosperm particles are milled in *reduction rolls* (e.g., C1A, C1B, ... and so on) and separated in plansifters into products of different PSD, such as flours and semolinas. Small bran particles adhering to milled endosperm may be separated by air suction and vibrating screens in units called *purifiers* (Figure 6.22).

It should be pointed out that the various layers of the endosperm differ in composition and physical properties. The outer layers are harder, darker, and richer in protein. The inner layers are softer, whiter, and richer in starch (*starchy endosperm*). Consequently, the sequence of gradual milling and sifting may be used to produce flour fractions of different characteristics.

References

Arnold, G., Zahn, S., Legler, A., Rohm, H., 2011. Ultrasonic cutting of foods with inclined moving blades. J. Food Eng. 103 (4), 394–400.

Becker, R., Gray, G.M., 1992. Evaluation of water jet cutting system for slicing potatoes. J. Food Sci. 57 (1), 132–137.

Delcour, J.A., Hoseney, R.C., 2010. Principles of Cereal Science and Technology, third ed. American Association of Cereal Chemists, St Paul, MN.

Earle, R.L., 1983. Unit Operations in Food Processing, second ed. Pergamon Press, Oxford, UK.

Gutsche, O., Fuerstenau, D.W., 2004. Influence of particle size and shape on the comminution of single particles in a rigidly mounted roll mill. Powder Technol. 143–144, 186–195.

Hassanpour, A., Ghadiri, M., Bentham, A.C., Papadopoulos, D.G., 2004. Effect of temperature on the energy utilization in quasi-static crushing of α-lactose monohydrate. Powder Technol. 141 (3), 239–243.

Heiland, W.K., Konstance, R.P., Craig, J.C., 1990. Robotic high pressure water jet cutting of chuck slices. J. Food Proc. Eng. 12 (2), 131–136.

Jalabert-Malbos, M.L., Mishellany-Dutour, A., Woda, A., Peyron, M.A., 2007. Particle size distribution in the food bolus after mastication of natural foods. Food Qual. Preference 18 (5), 803–812.

Lee, C.C., Chan, L.W., Heng, P.W., 2003. Use of fluidized bed hammer mill for size reduction and classification. Pharm. Dev. Technol. 8 (4), 431–442.

Loncin, M., Merson, R.L., 1979. Food Engineering, Principles and Selected Applications. Academic Press, New York, NY.

Marsot, J., Claudon, L., Jacqmin, M., 2007. Assessment of knife sharpness by means of a cutting force measuring system. Appl. Ergon. 38 (1), 83–89.

Ryan, B., Tyffany, D.G., 1998. Energy use in Minnesota Agriculture. <www.ensave.com/MN_energy_page_pdf>.

Servais, C., Jones, R., Roberts, I., 2002. The influence of particle size distribution on the processing of food. J. Food Eng. 51 (3), 201–208.

Vishwanathan, K.H., Singh, V., Subramanian, R., 2011. Wet grinding characteristics of soybean for soymilk extraction. J. Food Eng. 106 (1), 28–34.

Woodrow, J.R., Quirk, A.V., 1982. Evaluation of the potential of a bead mill for the release of intracellular bacterial enzymes. Enzyme. Microb. Technol. 4, 385–389.

Yan, H., Barbosa-Cánovas, G., 1997. Size characterization of selected food powders by five particle size distribution functions. Food Sci. Technol. Intl. 3 (5), 361–369.

CHAPTER

Mixing

7

7.1 Introduction

The fundamental objective of mixing is to increase the homogeneity of material in bulk (Uhl and Gray, 1966; Paul *et al.*, 2004). In food process technology, however, mixing or agitation may be used to achieve additional effects, such as enhancing heat and mass transfer, accelerating reactions, creating a structure (e.g., dough), changing the texture, etc. Very often, mixing occurs simultaneously with size reduction, as is the case in foaming, solid–liquid mixing, homogenizing and emulsification.

The basic mechanism in mixing consists in moving parts of the material in relation to each other. Because of fundamental differences between the two operations, it is useful to discuss separately the mixing (blending) of liquids and the mixing of particulate solids. In liquid–liquid, liquid–solid and liquid–gas mixing, the collision between the moving domains results in interchange of momentum or kinetic energy. The scientific discipline most relevant to liquid mixing is therefore fluid mechanics. The energy input per unit volume is closely related to the quality of mixing. In contrast, the mixing of particulate solids such as powders is governed primarily by the laws of solid physics and statistics.

When mixing two-phase mixtures, more homogeneity means disruption of the dispersed phase. In the case of emulsification, such disruption is, of course, desirable. When blending a mixture of fragile solid particles in a liquid, mechanical damage to the solid particles may be detrimental to quality. The mechanism of damage to suspended particles as a result of agitation has been investigated by Wang *et al.* (2002) and Bouvier *et al.* (2011).

7.2 Mixing of fluids (blending)

7.2.1 Types of blenders

The simplest mixer for fluids is the *paddle mixer* (Figure 7.1), consisting of one or sometimes two pairs of flat blades mounted on a shaft. For mixing liquids in hemispherical kettles and vessels with a bowl-shaped bottom, *anchor mixers* (Figure 7.2) that conform to the shape of the vessel walls are used. Paddle and anchor mixers are usually operated at low speed (tens of revolutions per minute).

Food Process Engineering and Technology. DOI: http://dx.doi.org/10.1016/B978-0-12-415923-5.00007-1
© 2013 Elsevier Inc. All rights reserved.

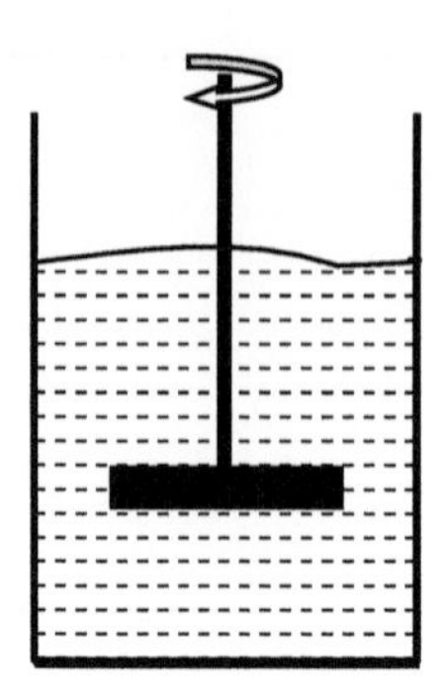

FIGURE 7.1

Paddle mixer.

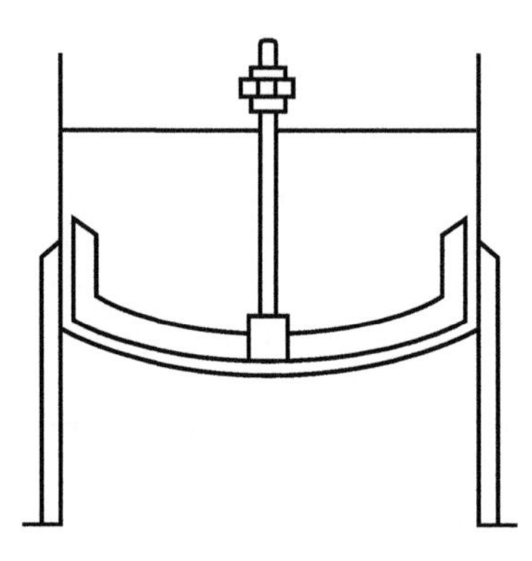

FIGURE 7.2

Anchor mixer.

Anchor mixers are frequently used in jacketed cooking kettles, and are often equipped with wipers that scrape the product off the heated surface to prevent scorching.

In *turbine mixers* (Figure 7.3), the impeller consists of a larger number (four or more) of flat or curved blades, mounted on a (usually vertical) shaft. Similar to the impeller of a centrifugal pump (Chapter 2), the turbine may be open, semi-closed (closed by a plate on one side) or shrouded (closed on both sides with a central opening or "eye" for circulation). Turbine mixers are usually operated at high speed (hundreds of revolutions per minute). They exert considerable shear on the fluid, and are therefore suitable in applications involving mass transfer (e.g., oxygen transfer in fermentors) or phase dispersion (e.g., emulsification and homogenization). The diameter of the impeller is, typically, one-third to one-half of the diameter of the vessel.

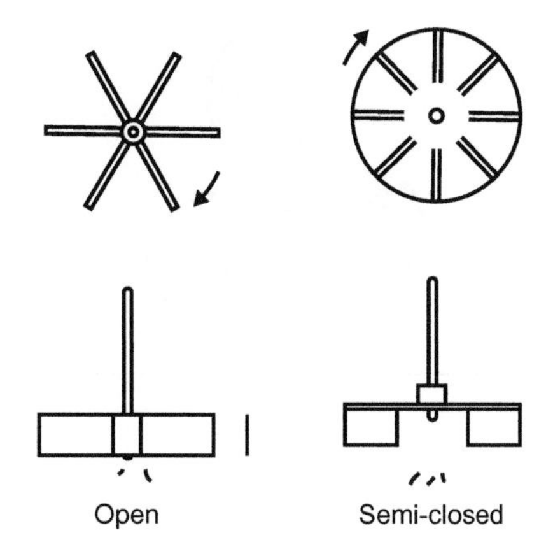

FIGURE 7.3

Turbine mixers.

Propeller mixers (Figure 7.4) are primarily used to blend low-viscosity liquids. The shaft is usually coupled directly to the motor. Rotation speed is high (hundreds to thousands of revolutions per minute). Impeller diameter is much smaller than that of turbine mixers. In many cases, these are portable mixers that can be moved from one vessel to another. The mixer shaft is usually positioned on an angle and off-center.

Helical mixers are equipped with a vertical rotating impeller with one or more helical ribbons. They are mainly used for mixing highly viscous fluids (Bouvier *et al.*, 2011).

7.2.2 Flow patterns in fluid mixing

Consider a vertical cylindrical vessel equipped with a centrally mounted mixer of any of the types described above, in rotation. The movement imparted to a volume element of the fluid can be characterized by three velocity components (Figure 7.5):

1. A radial velocity v_r pushing the liquid towards the walls of the vessel
2. An axial velocity v_a acting in the vertical direction, up or down
3. A tangential velocity v_t, tending to move the liquid in a circular motion on a horizontal plane.

The axial and radial velocities move fluid portions one with respect to another and thus cause mixing. The tangential velocity simply tends to rotate the fluid, creates vortexes, and has no significant mixing effect. The following actions may

FIGURE 7.4

Propeller mixer.

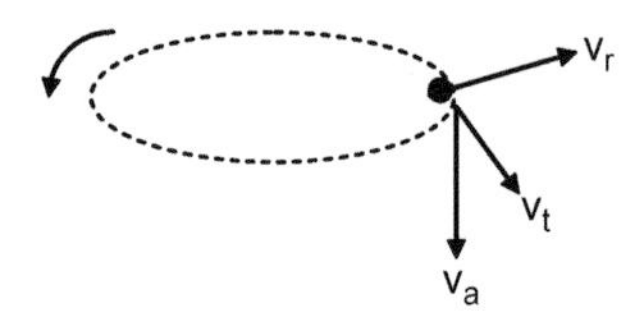

FIGURE 7.5

Velocity components in liquid mixing.

be taken in order to improve the mixing by reducing the rotational movement of the fluid:

- Attachment of baffles on the walls of the mixing vessel
- Positioning of the mixer shaft off-center
- Installation of two sets of impellers, turning in opposite directions.

Paddle mixers induce primarily radial flow with little movement in the axial direction. The mixing action is therefore concentrated near the horizontal plane of rotation and is not propagated to the rest of the liquid bulk in the vessel. In tall vessels, it may be necessary to mount several impellers at different height on the same shaft.

Turbine mixers act, to a certain degree, as centrifugal pumps. They induce radial and axial flow. If desired, the proportion of axial movement can be increased by using pitched blades.

In the case of propeller mixers the predominant flow pattern is axial flow.

7.2.3 Energy input in fluid mixing

The quality of mixing depends on the effective energy input by unit mass or unit volume of fluid. It has been long known, for example, that the rate of oxygen transfer in aerated fermentors equipped with turbine mixers is nearly proportional to the net mixing power input per unit volume of broth (Hixson and Gaden, 1950). The relationship between mixing power and the type, dimensions, and operation conditions of the mixer is expressed with the help of several correlations (McCabe and Smith, 1956). Most commonly, the mixing power correlations comprise the following dimensionless groups:

$$P_0 = \text{Power number} = \frac{P}{d^5 N^3 \rho}$$

$$\text{Re} = \text{Reynolds number} = \frac{d^2 N \rho}{\mu}$$

$$\text{Fr} = \text{Froude number} = \frac{dN^2}{g}$$

where:

P = mechanical power input to the mixer, w
N = speed of rotation, s^{-1}
d = impeller diameter, m
ρ = density of the fluid, $kg \cdot m^{-3}$
μ = viscosity of the fluid, Pa · s
g = gravitational acceleration, ms^{-2}.

The correlations are of the following general form:

$$P_0 = K(\text{Re})^n(\text{Fr})^m \tag{7.1}$$

The coefficient K and the exponents m, n are determined experimentally and depend on the type of mixer. The correlations are usually presented as log/log plots of $\Phi = (P_0)/(\text{Fr})^m$ versus (Re). The general form of these curves is shown in Figure 7.6.

Mixing power charts for turbine and propeller mixers are shown in Figures A.8 and A.9 (see Appendix).

The Froude number represents the action of gravitational forces. Its effect is significant when vortexes are formed. Therefore, in the absence of vortex formation, like in baffled vessels or at Reynolds number lower than 300, the Froude number is not an important factor. If the effect of the Froude number is neglected, simpler expressions can be obtained in the following cases:

1. The Reynolds number is 10 or lower (clearly laminar regime). In this region, the curve representing Φ versus Re is a straight line with a slope of −1 on log−log coordinates.

$$P_0 \propto \frac{1}{\text{Re}} \Rightarrow (P_0)(\text{Re}) = \frac{P}{N^2 d^3 \mu} = \text{Constant} \tag{7.2}$$

It follows that, under these conditions, the power input is proportional to the square of the rotational speed and to the cube of impeller diameter. Viscosity has an effect on mixing power, density has none.

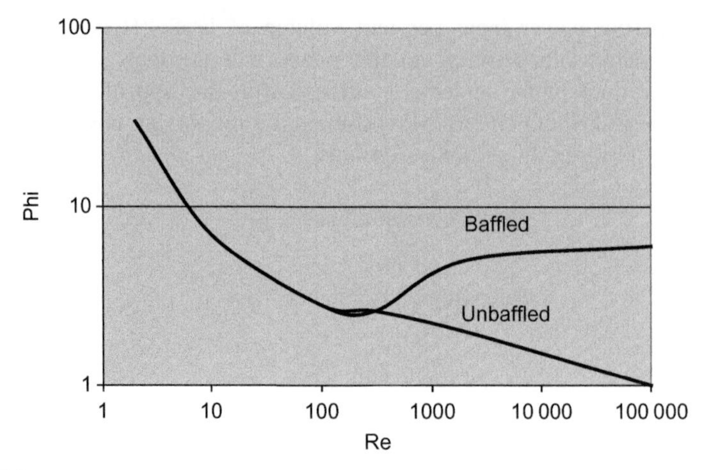

FIGURE 7.6

Example of a Power Function vs Reynolds number plot.

Figure adapted from McCabe and Smith (1956).

2. The Reynolds number is 10 000 or higher (clearly turbulent regime). In this region, Φ does not change with Re. The power number P_0 is therefore constant:

$$P_0 = \text{Constant} = \frac{P}{d^5 N^3 \rho} \tag{7.3}$$

In this region, mixing power is proportional to the cube of the rotation speed and to the fifth power of impeller diameter. It is proportional to the fluid density but insensitive to viscosity.

In high-power industrial mixers, more than one type of impeller may be used in the same unit. For example, sharp-edged impellers may be added for cutting through highly viscous or fibrous media (Figure 7.7).

EXAMPLE 7.1

A laboratory fermentor is equipped with a disc flat-blade turbine mixer with a diameter of 0.1 m. The fermentor is a cylindrical vessel with an inside diameter of 0.3 m, filled to the height of 0.3 m. Normally, the mixer is rotated at a speed of 600 rpm. The process is to be scaled-up to a fermentor with a capacity of 1 m^3, observing geometrical similitude and maintaining the same net agitation power per unit volume.

a. What should be the speed of revolution?
b. Will a 5 HP motor be sufficiently strong to turn the mixer?

Data: For the fluid in the fermentor: $\rho = 1000\ \text{kg} \cdot \text{m}^{-3}$; $\mu = 0.02\ \text{Pa} \cdot \text{s}$.

Solution

a. We first calculate the mixing power per unit volume in the laboratory unit.

$$\text{Re} = \frac{d^2 N \rho}{\mu} = \frac{(0.1)^2 \times 600/60 \times 1000}{0.02} = 5000$$

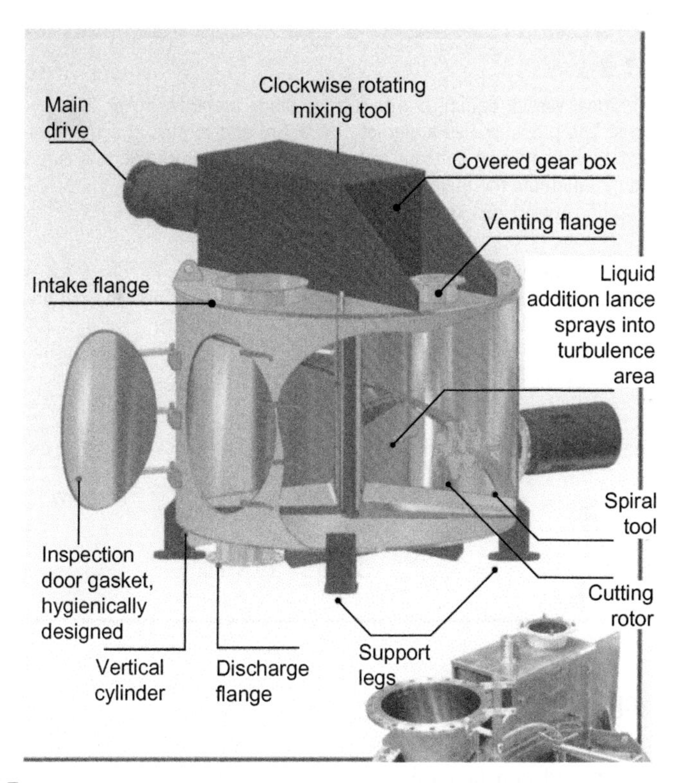

FIGURE 7.7

Combination vertical mixer.

Figure courtesy of Amixon Inc.

The regime is clearly turbulent, therefore P_O is constant. The mixing chart (Figure A.8) shows $P_O = 5$.

$$P_O = \frac{P}{\rho N^3 d^5} = 5 \quad \Rightarrow \quad P = 5 \times 1000 \times (10)^3 \times (0.1)^5 = 50\ \text{W}$$

The volume of the liquid in the laboratory fermentor is 0.021 m^3. Hence, the power per m^3 is 50/0.021 = 2381 $W \cdot m^{-3}$. The volume of the large fermentor is 1 m^3; hence, the mixing power in the large fermentor is 2381 W.

In order to satisfy geometric similitude, the vessel diameter will be 1.084 m and the impeller diameter 0.361 m. We assume that the conditions in the large fermentor will also be turbulent. (The validity of this assumption is tested at the end of the example.)

$$P = 2381 = 5 \times 1000 \times N^3 \times (0.361)^5 \quad \Rightarrow \quad N = 4.2\ \text{s}^{-1} = 256\ \text{rpm}$$

b. The power required in HP: P = 2381/750 = 3.2 HP. The 5-HP motor will, therefore, suffice.

(Note : In the large fermentor Re will be 27 400. The regime will be fully turbulent.)

EXAMPLE 7.2

A vertical cylindrical vessel, equipped with a three-blade propeller mixer, is to be used for blending syrups. The propeller has a diameter of 0.1 m and rotates at a speed of 60 rpm. Experimental data indicate that a mixing energy input of 2 J per kg of liquid is required for adequate mixing. Calculate the mixing time for a batch of 100 kg.

Data: Syrup density = 1200 $kg \cdot m^{-3}$; syrup viscosity = 4 $Pa \cdot s$.

Solution

We first determine the power of the mixer, using the Re vs P_O chart for propellers (Figure A.4). The mixing time is then calculated from the power and the required energy input.

$$Re = \frac{d^2 N \rho}{\mu} = \frac{(0.1)^2 \times 1 \times 1200}{4} = 3$$

The regime is laminar. The plot gives $P_O = 14$.

$$P_O = \frac{P}{\rho N^3 d^5} = 14 \quad \Rightarrow \quad P = 14 \times 1200 \times (1)^3 \times (0.1)^5 = 0.168 \text{ W}$$

$$t = \frac{2 \times 100}{0.168} = 1190 \text{ s} = 19.8 \text{ min}$$

7.2.4 Mixing time

Mixing time is defined as the time necessary to attain a certain degree of homogeneity under given and constant mixing conditions. It is conveniently determined by adding a tracer and measuring its concentration in samples taken periodically during the mixing.

Mixing time depends on the geometry of the impeller and the vessel, the speed of rotation and the physical properties of the fluids being mixed. Viscosity affects mixing time markedly, higher viscosity resulting in longer mixing time. When mixing two fluids of very different viscosities, mixing time is shorter if the fluid of higher viscosity is added, while mixing, to the fluid of lower viscosity than the other way around.

7.3 Kneading

Under this heading, we group intensive mixing operations applied to dough and paste-like products. Mixers used for this type of application are characterized by their ability to deliver high rotational momentum. Despite the relatively low rotational speed, the energy input per unit mass of product is considerable. Since the mechanical energy transferred to the material is finally converted to heat, it is

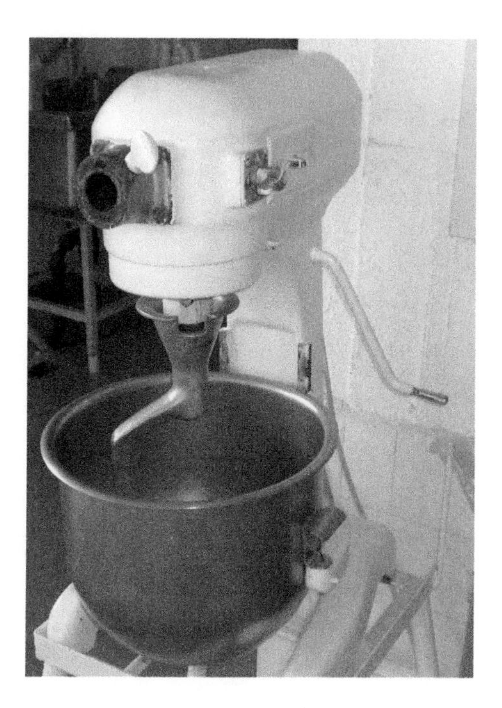

FIGURE 7.8

Planetary mixer.

Photograph courtesy of the Department of Biotechnology and Food Engineering, Technion.

often necessary to provide kneading machines with cooling jackets to prevent overheating.

Kneading machines include *planetary mixers* (Figure 7.8), *horizontal dough mixers* (Figure 7.9), *sigma-blade mixers* (Figure 7.10) and *cutter mixers* (discussed in Chapter 6). The planetary mixer is very popular for home use and small-scale industrial applications. It can be equipped with different types of mixing elements for whipping, blending or kneading. It has been shown (Chesterton *et al.*, 2011) that in a planetary mixer used for making cake batter, the maximum shear rate occurs at the bowl wall and is nearly uniform over most of the height of the mixer. Mixing and kneading are also among the effects performed by extruders. Extrusion is treated in a separate chapter (Chapter 15).

Because of the considerable shear induced by the operation, kneading often results in extensive changes in structure at molecular level (Hayta and Alpaslan, 2001). Kneading action is necessary for the development of the gluten proteins in wheat dough (Haegens, 2006). Plasticization of butter and margarine (Wright *et al.*, 2001) is, essentially, a kneading operation.

FIGURE 7.9

Horizontal dough mixer.

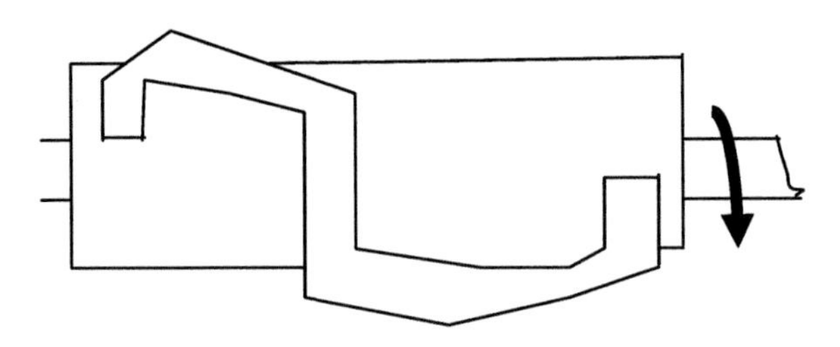

FIGURE 7.10

Inside a double sigma blade mixer.

EXAMPLE 7.3

A dough mixer is used to knead bread dough in batches of 100 kg each. The net mixing power input is 250 W per kg of dough. The machine is equipped with a water jacket for cooling. The objective is to maintain the dough temperature constant during the mixing cycle.

a. The increase in the temperature of the cooling water is to be maintained below 15°C. What is the minimum flow-rate of cooling water?

b. Due to malfunction of the control system, the flow of cooling water was interrupted for 8 minutes. Calculate the increase in dough temperature. For the dough: $C_p = 2424\ J \cdot kg^{-1} \cdot K^{-1}$.

Solution

a. Let G be the mass flow-rate of the water ($C_p = 4180\ J \cdot kg^{-1} \cdot K^{-1}$and ΔT the temperature increase of the water. Then:

$$G \cdot C_p.\Delta T = qG \times 4180 \times 15 = 250 \times 100 \quad \Rightarrow G = 1435\ \ kg \cdot h^{-1} = 0.4\ \ kg \cdot s^{-1}$$

b. If the heat generated by mixing is not removed, then:

$$(m \cdot C_p \cdot \Delta T)_{dough} = q \quad \Delta T_{dough} = \frac{q}{(m \cdot C_p)_{dough}} = \frac{250 \times 100 \times 8 \times 60}{100 \times 2424} = 49.5^{\circ}C$$

7.4 In-flow mixing

Liquid, gases or particulate solids flowing through piping or equipment may be mixed without mechanical mixers, simply by using turbulence. By inserting into the flow channel specially shaped elements, regular flow patterns may be distorted. This kind of mixing is also known as "passive mixing". Devices causing acceleration of the fluid, such as Venturi tubes and other kinds of restrictions, may be used for in-flow mixing. In-flow passive mixers are particularly useful in continuous processes. Their performance depends on turbulence. They are therefore not very efficient for heavy mixing duty, such as mixing very viscous liquids. The *Rotated Arc Mixer* (RAM), developed at CSIRO, Australia (Metcalfe and Lester, 2009), tries to solve this problem. The RAM consists of two snugly fitting concentric tubes: the inner tube is static, the outer tube is rotated. Short openings or slits are cut on the wall of the inner tube. The viscous fluid to be mixed is moved axially through the inner tube. As the outer tube rotates, viscous drag is transmitted to the fluid through the slits, which causes some transverse, chaotic flow, cutting through the laminar flow pattern and resulting in mixing. For optimal performance, the location and position of the slits are determined according to the properties of the fluid and the operation conditions.

7.5 Mixing of particulate solids

7.5.1 Mixing and segregation

As a rule, mixing powders is more difficult than blending liquids, mainly because of the tendency of powders to segregate (Kaye, 1997). Powder mixtures consist of particles of different size, shape and density. Different particles move differently under the action of mixing forces. Thus, movement in a powder bulk may promote, at the same time, mixing and segregation (demixing). Segregation is more serious in free-flowing powders than in cohesive ones. Sometimes, a powder in movement may go through successive cycles of mixing and demixing. The phenomenon of over-mixing – i.e., total segregation of the mixture as a result of continuing mixing – is known.

7.5.2 Quality of mixing: the concept of "mixedness"

The importance of good mixing of powder mixtures in food processing cannot be over-emphasized. Uneven distribution of a vitamin in a powder product intended for infant feeding can have disastrous consequences. Imperfect mixing of salt and leavening agent in self-rising flour would cause serious customer dissatisfaction. The definition of quantitative criteria for the quality of mixing in solid particulate products is, therefore, of considerable importance (Lacey, 1954; Harnby, 1997a; Harnby *et al*., 1997).

Consider a binary mixture of powders consisting of, say, 10 kg of salt and 90 kg of sugar. Assume that the mixture is well mixed. A sample of the mixture is taken to the laboratory for analysis. If the sample consists of a single particle, the result of the analysis will be either 100% salt or 100% sugar. As the sample size is increased, the result of the analysis approaches the true value of 10% salt, 90% sugar. It is clear that the deviation from the true composition depends on the size of the sample.

If n samples are analyzed, the root mean square S (or RMS) of the deviation of the measurements x from the true mean composition is:

$$S = \sqrt{\frac{1}{n}\sum_{i=1}^{i=n}(x_i - \bar{x})^2} \tag{7.4}$$

Note that the standard deviation is given by the same expression, with n − 1 instead of n in the denominator. In statistical quality assurance, S^2, known as the "variance", is an important parameter.

Consider two particulate components P and Q placed in a mixer. Let the proportion of each component, in mass fraction, be p and q respectively. In the beginning, when the components are totally separated, a sample would consist of either pure P or pure Q. The variance of the totally unmixed system would be:

$$S_0^2 = q(1 - q) = qp \tag{7.5}$$

Now, let us start the mixer. Assume that perfect mixing has been reached – i.e., the mixture has been totally randomized. As explained above, even in the case of a perfectly randomized mixture, the results obtained by analyzing a sample deviate from the true composition and the deviation depends on the number of particles in the sample. Statistics teaches that the variance of samples taken from a perfectly randomized mixture is:

$$S_R^2 = \frac{qp}{N} \tag{7.6}$$

where N is the number of particles in the samples. Obviously, the variance approaches zero as the sample size is increased.

The quality of mixing of a real mixture would fall between that of the totally unmixed (segregated) and that of the perfectly mixed (random) mixtures. A *mixing index* or *index of mixedness* M is defined as follows (Lacey, 1994):

$$M = \frac{S_0^2 - S^2}{S_0^2 - S_R^2} \tag{7.7}$$

where S^2 is the variance of the real mixture. If the samples are fairly large, (Eq. 7.7) becomes:

$$M = 1 - \frac{S^2}{S_0^2} \tag{7.8}$$

The mixing index M of a real mixture lies between 0 (totally segregated) and 1 (totally randomized). The concepts of mixedness and mixing quality are equally applicable to the blending of fluids. It would be logical to assume that complete mixedness, i.e., M = 1, is approached asymptotically as mixing is continued, and to seek a quantitative relationship between mixing time and mixing quality (Kuakpetoon *et al.*, 2001). Unfortunately, this assumption is not always confirmed by reality, mainly because it does not account for the phenomenon of de-mixing in free-flowing powders.

EXAMPLE 7.4

A pharmaceutical firm prepares a powder mixture for subsequent compaction into tablets. 22 kg of the active principle and 78 kg of an inert excipient are put into a powder mixer and mixed for 7 minutes at constant speed. Ten samples of the mixture, each weighing 20 g, were analyzed for the active principle content. The results, in weight percent, were: 21.8, 21.8, 23.0, 21.4, 22.3, 22.0, 22.7, 20.9, 22.0 and 21.7%.

a. Calculate the RMS of the deviation and the variance.
b. Calculate the mixing index, assuming that the particles are small and of equal size.

Solution

a. The true concentration of the active principle in the powder mixture is 22%. Let p be the individual result found for one sample. Calculate the deviation (p − 22) for this sample. Calculate $(p - 22)^2$ for each sample. Sum the results, then divide the sum by the number of samples (10). The result is the variance. Take the square root of the variance; this will be the RMS.

The calculations and the results are shown in Table 7.1.

b. Considering that the number of particles in the samples is large, the mixing index M is given by (Eq. 7.8):

$$M = 1 - \frac{S^2}{S_0^2}$$

$$S_0^2 = q(1 - q) = qp = 0.22 \times 0.78 = 0.1716$$

$$M = 1 - \frac{0.0000332}{0.1716} = 0.9998$$

The mixing quality is very good.

Table 7.1 Mixing Quality of Powders

Sample No.	p	$(p-0.22)^2$
1	0.218	0.00000400
2	0.218	0.00000400
3	0.23	0.00010000
4	0.214	0.00003600
5	0.223	0.00000900
6	0.22	0.00000000
7	0.227	0.00004900
8	0.209	0.00012100
9	0.22	0.00000000
10	0.217	0.00000900
Σ		0.00033200
$\Sigma/n = S^2$		**0.0000332**
Rms = S		**0.0057619**

7.5.3 Equipment for mixing particulate solids

Mixers for powders fall into two categories:

1. *Diffusive (passive) mixers*. In these mixers, homogeneity is achieved as a result of the random motion of the particles when the particle is in flow under the effect of gravity or vibration, without mechanical agitation. The mechanism is assimilated to "diffusion" (Hwang and Hogg, 1980), in reference to the random movement of molecules. There is almost no shear. Consequently, these mixers are particularly suitable for particulates that require gentle mixing, such as fragile agglomerates, but they do not perform well with cohesive powders.
2. *Convective (active) mixers*. In these mixers, mixing is achieved by mechanical agitation. Parts of the bulk material are conveyed with respect to each other by the action of impellers or turbulent gas flow. Shearing occurs, and may be considerable.

In practice, most powder mixers function by a combination of both mechanisms (Harnby, 1997b, 1997c). In addition to mixing, powder mixers may perform other functions in powder processing technology. Some of these additional functions are:

- Particle coating
- Agglomeration
- Admixture of liquids (such as fat into a dry soup mixture)
- Drying
- Size reduction, change of particle shape.

The *drum blender* (Figure 7.11) consists of a horizontal cylinder rotating about its axis. Its mixing action is essentially diffusive. The powder to be mixed is placed inside the drum. As the drum rotates, the powder is lifted up until the angle of repose (see Chapter 2) is exceeded. At this point, the powder falls back on the rest of the bulk and enters a new cycle of lifting and falling. Diffusive mixing takes place during the residence of the powder in air, while falling. Continuous operation can be made possible by tilting the drum.

Tumbler mixers also belong to the category of diffusive mixers. Two types of tumbler mixers are shown here. In the double-cone tumbler (Figure 7.12), the powder undergoes cycles of expansion and compaction as the vessel rotates. In the *V-shaped tumbler* (Figure 7.13), the powder is subjected to cycles of division and reassembly. Convectional elements such as rotating or stationary flow distortion bars (intensifier bars) are sometimes installed in both types of mixers.

Numerous types of convective mixers for particulate solids are available, two of which are described here. Figure 7.14 shows a powerful *paddle mixer* for solids. The rotating elements mix the powder both by moving the bed and by fluidizing. A liquid component may be sprayed on the powder while mixing. Mixers of this type are available both for batch and for continuous operation. In a recently developed version of the machine, the entire mixing chamber can be rotated upside down for rapid discharge of the product, and then rotated back for charging a new batch.

Trough mixers are so called because they consist of a conduct with a U-shaped cross-section and a longitudinal rotating shaft (or shafts) carrying various types of mixing elements. The agitating elements may be a series of paddles or a screw (like a screw conveyor). One of the best known trough mixers is the *ribbon mixer* (Figure 7.15).

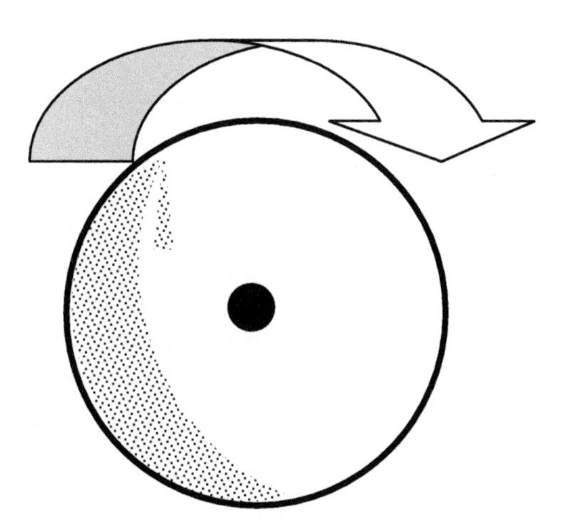

FIGURE 7.11

Cross-section of drum blender.

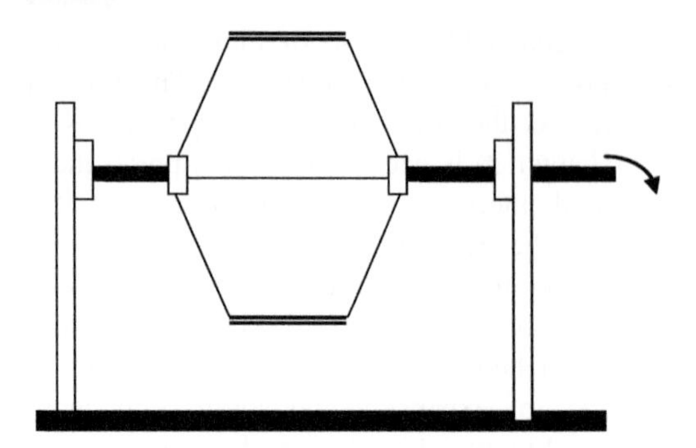

FIGURE 7.12

Double-cone tumbler mixer.

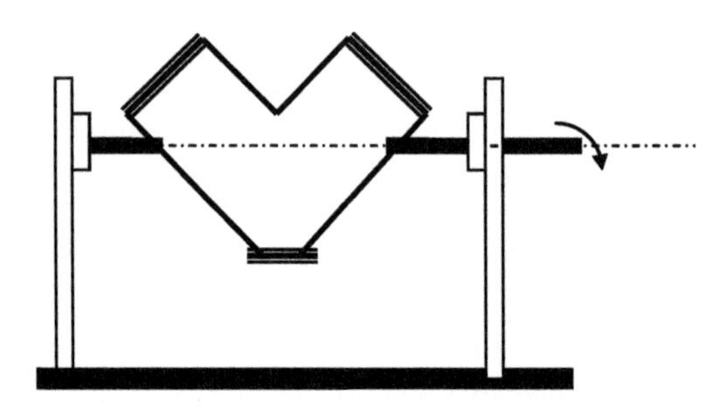

FIGURE 7.13

V-shaped tumbler mixer.

7.6 Homogenization

7.6.1 Basic principles

Homogenization of dispersed systems (suspensions and emulsions) involves reduction of the size of the dispersed particles. Such size reduction is achieved by the action of shearing forces (Peters, 1997). Shear can be applied to the fluid by mechanical agitation, by forcing the fluid to flow at very high velocity through a narrow passage, by shearing the fluid between two surfaces moving one with respect to the other, or by ultrasonic vibrations.

Homogenization is applied very frequently in food processing, the best-known application of this operation being in the processing of fluid milk, with the objective of preventing the separation of fat-rich cream from the bulk under the effect of gravity (Walstra *et al.*, 2005).Other applications include emulsification of salad dressings and sauces, fine mashing of strained infant food, stabilization of tomato

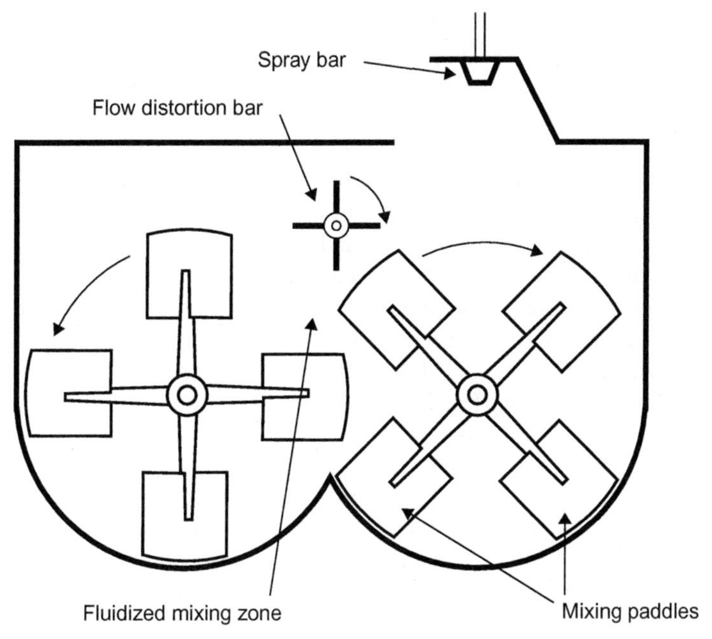

FIGURE 7.14

The "Forberg" mixer.

Figure courtesy of Forberg International AS.

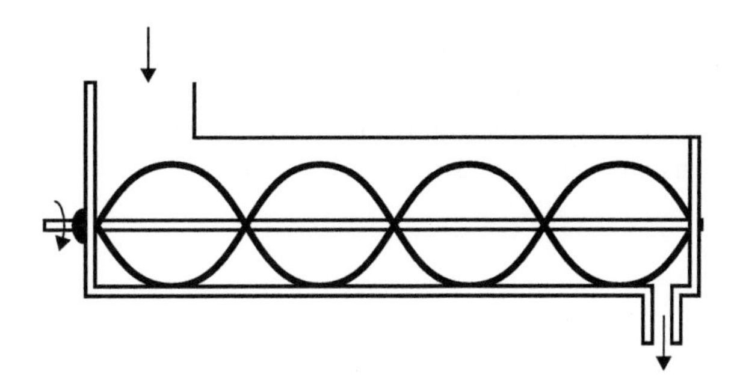

FIGURE 7.15

Ribbon mixer.

concentrates, etc. In biotechnology, high-pressure homogenization is used for cell rupture and release of intracellular material (Hetherington *et al.*, 1971; Kleinig and Middelberg, 1996; Floury *et al.*, 2004).

The subject of homogenization is closely connected to the analysis of the requirements for breaking apart dispersed particles under the effect of shear.

Since the ultimate objective of homogenization is to achieve a certain degree of reduction in the mean size of the particles (e.g., fat droplets in an oil/water emulsion), it is useful to study the effect of the relevant factors on particle size.

Consider a spherical droplet of fat of diameter d in an emulsion, being subjected to shear forces (Figure 7.16). Each force acting on the droplet may be decomposed into a component normal to the surface (compression or tension, see Chapter 1) and a component tangential to the surface (shear).

The shear tends to deform the droplet and eventually break it. The drop resists distortion and rupture under the effect of the surface (wall) tension σ. For disruption to occur, the shear must overcome the internal pressure created by the surface tension and governed by Laplace's Law (Pierre-Simon Laplace, French mathematician and physicist, 1749–1827). For a spherical drop or bubble, the expression of Laplace's Law is given in (Eq. 7.9):

$$\Delta P = \frac{4\sigma}{d} \tag{7.9}$$

The extent of particle disruption in homogenization depends on the balance between the disrupting forces (shear, turbulence of the continuous medium) and the forces resisting deformation and disruption (surface tension). Formally, one can write:

$$\frac{d}{d_0} = f(d, v, \rho, \mu, \sigma, \ldots)$$

where:

- v = local velocity, $m \cdot s^{-1}$
- d_0, d = initial and final mean droplet diameter, m
- ρ = density of the continuous medium, $kg \cdot m^{-3}$
- μ = (apparent) viscosity of the continuous medium, $Pa \cdot s$
- σ = surface tension, $N \cdot m^{-1}$ or $J \cdot m^{-2}$.

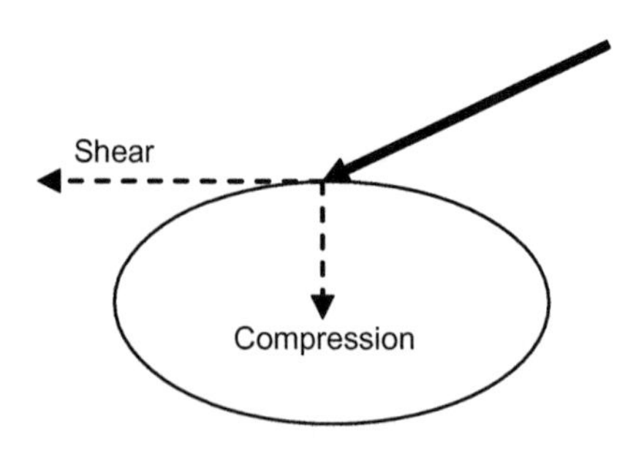

FIGURE 7.16

Decomposition of a force acting on a droplet.

These variables are incorporated in a dimensionless group known as the Weber number, We. The Weber number may be written in different forms, depending on the variable selected to represent the shear forces (local velocity v, shear stress τ, pressure drop due to friction ΔP_f):

$$We_v = \frac{dv^2\rho}{\sigma} \quad \text{or} \quad We_\tau = \frac{\tau\, d}{\sigma} = \frac{\gamma\mu d}{\sigma} \quad \text{or} \quad We_{\Delta P} = \frac{\Delta P_f d}{\sigma} \tag{7.10}$$

The relationship between We and the degree of droplet size reduction is given in (Eq. 7.11), where K and m are experimentally determined numericals (Bimbenet *et al.*, 2002):

$$We = K\left(\frac{d_0}{d}\right)^m \tag{7.11}$$

The dimensionless Weber number is also applicable to the disruption of gas bubbles in the process of foam generation (Narchi *et al.*, 2011).

7.6.2 Homogenizers

The shear forces required for homogenization can be generated in different ways. Accordingly, different types of equipment are used for homogenization. These are:

1. *High shear mixers*. These have been discussed in Section 7.2.
2. *Colloid mills*. These have been described in Chapter 6, Section 6.4.
3. *High-pressure homogenizers* (HPH, also known as *Gaulin homogenizers* after the inventor). In this type of equipment, homogenization is achieved by forcing the mixture to flow at high velocity through a narrow gap. The homogenizer consists of a high-pressure pump and a homogenizing head (Figure 7.17).

   The pump is usually a positive displacement reciprocating (piston) pump. The high pressure, in the range of 20–70 MPa, is necessary to overcome the friction in the homogenizing head. The homogenizing head houses the narrow gap assembly (homogenizing valve) which comes in different forms, according to design. The head may contain one valve (single-stage homogenization) or two valves in series (two-stage homogenization).

   The mechanisms responsible for particle disruption in high-pressure homogenization are:

   - Shear within the fluid, due to velocity gradients.
   - Impact of the particles with hard surfaces of the valve.
   - Cavitation. In its passage through the narrow gap, the liquid is accelerated to very high velocity, resulting in a decrease in pressure. The pressure may drop locally below the water vapor pressure, causing some evaporation to occur. The vapor bubbles thus formed expand and then collapse, creating shock waves that contribute to the breaking of particles. The occurrence of

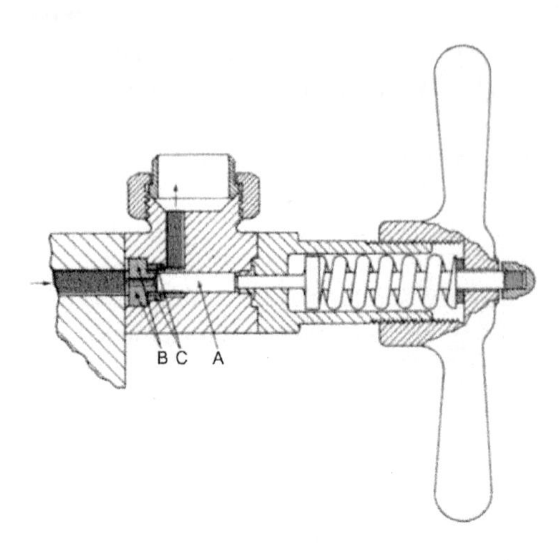

FIGURE 7.17

High-pressure homogenizing head. A: pressure rod; B: valve seat; C: fluid passage

cavitation in high-pressure homogenization has been validated indirectly by the detection of free-radical related oxidation as a result of the process (Lander *et al.*, 2000).

Recently, cavitation has been visually observed and acoustically measured in a model high-pressure homogenizer (Håkansson *et al.*, 2010). The droplet size reduction is related to the homogenization pressure drop, as shown in (Eq. 7.12):

$$\frac{d}{d_0} = (\Delta P)^n \tag{7.12}$$

As mentioned previously, high-pressure homogenization is also used in biotechnology for breaking microorganism cells. The extent of cell disruption B (proportion of broken cells after one pass through the valve, measured as the proportion of cytoplasmatic protein released) depends on the homogenization pressure as follows:

$$\ln(1 - B) = -k(\Delta P)^n \tag{7.13}$$

with n = 1.5–3.

4. *Ultrasonic homogenizers*. Acoustic waves in the frequency range of 20–30 kHz (ultrasonic waves) are capable of breaking droplets in an emulsion, disintegrating soft solid particles in suspension, and even depolymerizing polymers in solution (Berk, 1964; Berk and Mizrahi, 1965). The mechanisms of action include compression–expansion cycles and cavitation. The vibrations are applied to the liquid by immersed probes of different shapes (Figure 7.18). Ultrasonic homogenization, at net power levels

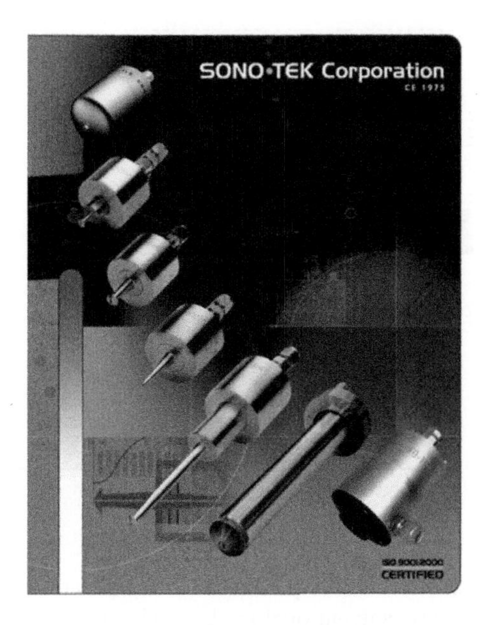

FIGURE 7.18

Different shapes of ultrasonic probes.

Photograph courtesy of SonoTek Corporation.

of up to 1 kW, is extensively used in the laboratory for emulsification and cell disruption (see, for example, Jena and Das, 2006). The use of ultrasound for homogenization and other applications in the food industry has been reviewed by Soria and Villamiel (2010).

5. *Magnetohydrodynamic effect*. Kerkhofs *et al.* (2011) found that when an orthogonal magnetic field is applied on a laminar flow of electrically conducting fluid containing dispersed particles, agglomeration or, on the contrary, disruption of the particles occurs, depending on whether the flow is laminar or turbulent. The authors have used this effect to disrupt oil droplets in a mayonnaise-type emulsion. The mixture is passed through a Venturi tube while a magnetic field, orthogonal to the flow, is applied by a pair of permanent block magnets placed on both sides of the conduct. The authors report that the technique resulted in the continuous production of mayonnaise of adequate viscosity and droplet size.
6. *Membrane emulsification*. Emulsions can be produced by forcing the dispersed phase into the continuous phase through a membrane (Figure 7.19). In this technique the droplet size is controlled by the size of membrane pores, and therefore is more uniform than in emulsions resulting from mechanical homogenization (Charcosset, 2008). The technique can be used for producing oil-in-water (o/w) or water-in-oil (w/o) emulsions. Membrane emulsification (ME) can be performed in two different ways, namely direct ME and premix

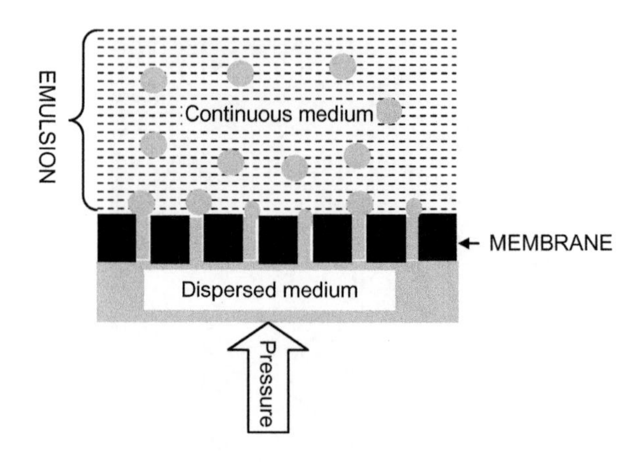

FIGURE 7.19

Principle of membrane emulsification.

ME (Trentin *et al.*, 2011). In the premix method, a coarse emulsion (premix) is first prepared by any conventional method. The premix is then forced through a membrane into a stream of continuous medium.

7.7 Foaming

The production of food foams involves two actions, namely, incorporation of air into a liquid or semi-liquid food, and uniform dispersion of the air in the liquid as small bubbles. Whipping with the help of planetary mixers equipped with wire whisks turning at high speed is the preferred method for the small-scale batch production of foams. Borcherding *et al.* (2008), in their study on the effect of homogenization and temperature on milk foam, produced foam by sparging air directly into the milk through fritted glass. Continuous rotor–stator mixers (Indrawati *et al.*, 2008; Narchi *et al.*, 2011) equipped with nozzles for gas injection are used for large-scale operations. Both the method of foam generation (type of equipment, operating conditions) and the composition of the material affect the *foamability*, stability, and organoleptic quality. Foamability or foaming capacity is often expressed in terms of *overrun*. Overrun is percent volume increase achieved by foaming. The term is extensively applied to ice-cream. Ideally, a good foam should consist of the smallest possible bubbles of uniform size (Müller-Fischer and Windhab, 2005).

References

Berk, Z., 1964. Viscosity of orange juice concentrates; effect of ultrasonic treatment and concentration. Food Tech. 18 (11), 153–154.

Berk, Z., Mizrahi, S., 1965. A new method for the preparation of low viscosity orange juice concentrates by ultrasonic irradiation. Fruchtsaftindustrie 10, 71–73.

Bimbenet, J.J., Duquenoy, A., Trystram, G., 2002. Génie des Procédés Alimentaires. Dunod, Paris.

Borcherding, K., Hoffmann, W., Lorenzen Chr., P., Schrader, K., 2008. Effect of milk homogenisation and foaming temperature on properties and microstructure of foams from pasteurised whole milk. LWT 41 (10), 2036–2043.

Bouvier, L., Moreau, A., Line, A., Fatah, N., Delaplace, G., 2011. Damage in agitated vessels of large visco-elastic particles dispersed in a highly viscous fluid. J. Food Sci. 76 (5), E384–E391.

Charcosset, C., 2008. Preparation of emulsions and particles by membrane emulsification for the food processing industry. J. Food Eng. 92 (3), 241–243.

Chesterton, A.K.S., Moggridge, G.D., Sadd, P.A., Wilson, D.I., 2011. Modelling of shear rate distribution in two planetary mixtures for studying development of cake batter structure. J. Food Eng. 105 (2), 343–350.

Floury, J., Bellettre, J., Legrand, J., Desrumaux, A., 2004. Analysis of a new type of high pressure homogenizer. A study of the flow pattern. Chem. Eng. Sci. 59, 843–853.

Haegens, N., 2006. Mixing, dough making and dough makeup. In: Hui, Y.H. (Ed.), Baking Products: Science and Technology. Blackwell Publishing, Ames, IA.

Håkansson, A., Fuchs, L., Innings, F., Revstedt, J., Bergenståhl, B., Trägårdh, C., 2010. Visual observations and acoustic measurements of cavitation in an experimental model of a high-pressure homogenizer. J. Food Eng. 100 (3), 504–513.

Harnby, N., 1997a. Characterisation of powder mixtures. In: Harnby, N., Edwards, M., Nienow, A.W. (Eds.), Mixing in the Process Industry, second ed. Butterworth and Co., London, UK.

Harnby, N., 1997b. The mixing of cohesive powders. In: Harnby, N., Edwards, M., Nienow, A.W. (Eds.), Mixing in the Process Industry, second ed. Butterworth and Co., London, UK.

Harnby, N., 1997c. Selection of powder mixers. In: Harnby, N., Edwards, M., Nienow, A.W. (Eds.), In *Mixing in The Process Industry*, second ed. Butterworth and Co., London, UK.

Harnby, N., Edwards, M.F., Nienow, A.W. (Eds.), 1997. second ed. Butterworth and Co., London, UK.

Hayta, M., Alpaslan, M., 2001. Effect of processing on biochemical and rheological properties of wheat gluten proteins. Nahrung/Food 45 (5), 304–308.

Hetherington, P.J., Follows, M., Dunnill, P., Lilly, M.D., 1971. Release of protein from baker's yeast (Saccharomyces cerevisiae) by disruption in an industrial homogenizer. Trans. Inst. Chem. Eng. 49, 142–148.

Hixson, A.W., Gaden Jr., E.L., 1950. Oxygen transfer in submerged fermentation. Ind. Eng. Chem. 42, 1792–1801.

Hwang, C.L., Hogg, R., 1980. Diffusion mixing of flowing powders. Powder Technol. 26 (1), 93–101.

Indrawati, L., Wang, Z., Narsimhan, G., Gonzalez, J., 2008. Effect of processing parameters on foam formation using a continuous system with a mechanical whipper. J. Food Eng. 88 (1), 65–74.

Jena, S., Das, H., 2006. Modeling of particle size distribution of sonicated coconut milk emulsions: effect of emulsifiers and sonication time. Food Res. Int. 39 (5), 606–611.

Kaye, B.H., 1997. Mixing of powders. In: Fayed, M.E., Otten, L. (Eds.), Handbook of Powder Science and Technology. Chapman & Hall, New York, NY.

Kerkhofs, S., Lipkens, H., Velghe, F., Verlooy, P., Martens, J.A., 2011. Mayonnaise production in batch and continuous process exploiting magnetohydrodynamic force. J. Food Eng. 106 (1), 35–39.

Kleinig, A.R., Middelberg, A.P.J., 1996. On the mechanism of microbial cell disruption in high-pressure homogenization. Chem. Eng. Sci. 5, 891–898.

Kuakpetoon, D., Flores, R.A., Milliken, G.A., 2001. Dry mixing of wheat flours: effect of particle properties and blending ratio. LWT 34 (3), 183–193.

Lacey, P.M.C., 1954. Developments in the theory of particle mixing. J. Appl. Chem. 4, 257–268.

Lander, R., Manger, W., Scouloudis, M., Ku, A., Davis, C., Lee, A., 2000. Gaulin homogenization, a mechanical study. Biotecnol. Prog. 1, 80–85.

McCabe, W., Smith, J.C., 1956. Unit Operations of Chemical Engineering. McGraw-Hill, New York, NY.

Metcalfe, G., Lester, D., 2009. Mixing and heat transfer of highly viscous food products with a continuous chaotic duct flow. J. Food Eng. 95 (1), 21–29.

Müller-Fischer, N., Windhab, E.J., 2005. Influence of process parameters on microstructure of foam whipped in a rotor-stator device within a wide static pressure range. Colloids Surf. A Physicochem. Eng. Asp. 263 (1–3), 353–362.

Narchi, I., Vial, C., Labbafi, M., Djelveh, G., 2011. Comparative study of the design of continuous aeration equipment for the production of food foams. J. Food Eng. 102 (2), 105–114.

Paul, E.L., Atiemo-Obeng, V.A., Kresta, S.M. (Eds.), 2004. Science and Practice. Wiley, Hoboken, NJ.

Peters, D.C., 1997. Dynamics of emulsification. In: Harnby, N., Edwards, M., Nienow, A.W. (Eds.), In *Mixing in The Process Industry*, second ed. Butterworth and Co., London, UK.

Soria, A.C., Villamiel, M., 2010. Effect of ultrasound on the technological properties and bioactivity of food: a review. Trends Food Sci. Tech. 21 (7), 323–331.

Trentin, A., De Lano, S., Güell, C., López, M., Ferrando, M., 2011. Protein-stabilized emulsions containing beta-carotene produced by premix membrane emulsification. J. Food Sci. 106 (4), 267–274.

Uhl, V.W., Gray, J.B., 1966. Mixing – Theory and Practice, vol. I. Academic Press, New York, NY.

Walstra, P., Wouters, J.T.M., Geurts, T.J., 2005. Dairy Science and Technology. CRC Press, Taylor & Francis Group, Boca Raton, FL.

Wang, Y.Y., Russel, A.B., Stanley, R.A., 2002. Mechanical damage to food particulates during processing in a scraped surface heat exchanger. Food Bioprod. Process. 80 (1), 3–11.

Wright, A.J., Scanlon, M.G., Hartel, R.W., Maragnoni, A.G., 2001. Rheological properties of milkfat and butter. J. Food Sci. 66 (8), 1056–1071.

CHAPTER

Filtration and Expression

8

8.1 Introduction

Filtration is an operation by which solid particles are separated from a liquid or gas by forcing the mixture through a porous medium that retains the particles. We distinguish between two modes of filtration: surface filtration and depth filtration.

In surface filtration, the fluid is forced through a porous medium (filter medium) that retains the particles on its surface. Particles larger than the openings of the porous surface are retained. The familiar laboratory operation of filtration through filter paper is an example of surface filtration. Filter media include paper, woven and non-woven fabrics, porous ceramic, sand, sintered glass or metal, membranes, etc. (Sutherland, 2011).

In the case of depth filtration, the mixture flows through a thick layer of porous material of fibrous (glass wool, rock wool) or particulate (sand) structure. The solid particles of the mixture are retained as a result of collision with the fibers or particles of the filter medium, or by adsorption. The passages available to flow are much larger than the retained particles. Retention is probabilistic and not absolute. This kind of filtration occurs in air filters, sand-bed filters or the familiar oil filters in cars.

The purpose of filtration may be the removal of undesirable solid particles from a liquid product (e.g., clarification of wine), or, alternatively, recovery of a solid product from a solid/liquid mixture (separation of sugar crystals from mother liquor).

The following are some examples of application of filtration in food processing:

- Clarification of fruit juices. In the production of clear apple juice, the raw juice is treated with pectolytic enzymes that cause flocculation of the colloidal particles in suspension. These are then removed by filtration in one or several stages.
- Clarification of wine. Wine also may contain colloidal particles in suspension (cloud). In the process of "fining", a small quantity of a protein solution is added to the cloudy wine. The protein combines with the tannins of the wine to give an insoluble complex. The complex precipitates, carrying the colloidal particles to the bottom. The precipitate is usually separated by filtration.

Food Process Engineering and Technology. DOI: http://dx.doi.org/10.1016/B978-0-12-415923-5.00008-3
© 2013 Elsevier Inc. All rights reserved.

- Bleaching of edible oils. Many unrefined edible oils contain undesirable pigments. In the process of bleaching, the oil is mixed with an adsorbent solid (bleaching earth) that removes the pigment by adsorption. The bleached oil is then separated from the solids by filtration.
- Filtration is an operation of major importance in the manufacture of sugar. The process of sugar refining, from the treatment of raw juice to the recovery of sugar crystals, comprises many steps of filtration, utilizing various types of equipment.
- The preparation of brines and syrups for use as liquid media for canned foods is a common operation in the food industry. Good manufacturing practice requires that these liquids are filtered before their incorporation in the product, in order to avoid introduction of foreign bodies (sand, hair, etc.). Because the quantity of solids to be removed from such liquids is usually very small, on-line filtration by strainers is possible.
- Filtration is an essential step in the treatment of drinking water, and of water used in the preparation of soft beverages and beer.
- Air filtration is of major importance in processes requiring an ultra-clean or aseptic environment. The application of *ultra-clean technologies* in the food industry is expanding.

The relationship between filtration and the quality and safety of foods and beverages is obvious (Heusslein and Brendel-Thimmel, 2010). For a detailed analysis of the filtration applications in the food and beverage industries, see Sutherland (2010).

8.2 Depth filtration

The principle of depth filtration is illustrated in Figure 8.1.

In the case of depth filtration, the solid particles are entrapped over the entire depth of the filter medium. The size range of the particles retained is very wide, from 0.1 μm (microorganisms) to 0.1 mm (dust and fine powder). As explained in the previous section, the removal of a particle in depth filtration depends on the probability of the particle coming within the effective range of action of the retaining surface (e.g., a fiber in the case of a filter medium consisting of rock wool) (Aiba *et al.*, 1965). The extent of removal is, therefore, well represented by a first-order type expression:

$$\frac{\partial C}{\partial z} = -kC \tag{8.1}$$

where:

C = concentration of solid particles in the stream (e.g., particles per m^3).
z = depth of filter, m.
k = collection efficiency.

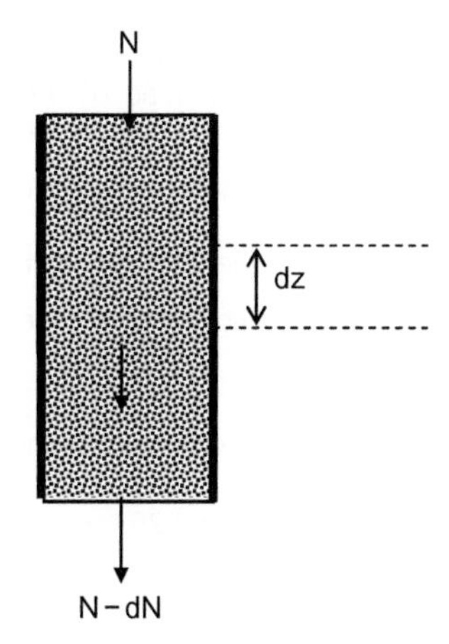

FIGURE 8.1

Depth filtration.

Integration of Eq. (8.1) gives:

$$\ln \frac{C}{C_0} = -kZ \tag{8.2}$$

where C, C_0 = particle concentration at the exit and the entrance to the filter, respectively, and Z = total depth of the filter medium.

The collection efficiency k depends on the size of the retained particles, the thickness and nature (e.g., electrostatic charge) of the filter fibers, the porosity of the filter, and the flow-rate (Gaden and Humphrey, 1956; Maxon and Gaden, 1956). The collection efficiency is higher if:

- The particles are large (however, extremely small particles are also well retained, due to Brownian movement)
- The fibers are fine
- The filter bed is less porous (more dense).

The effect of flow-rate is more complex (Payatakes, 1977). On one hand, the probability of a particle being removed by adsorption is higher if the retention time is longer – i.e., if flow-rate is lower. On the other hand, the probability of a particle being retained as a result of impact is higher if the particle velocity is higher.

EXAMPLE 8.1

A depth filter filled with a fibrous mat is used as a test unit for filtering air. The filter is cylindrical, 5 cm in diameter. The length of the filtering section is 25 cm. The filter is found to retain 99% of the particles in the air, at a flow-rate of $100\,\text{cm}^3 \cdot \text{s}^{-1}$. It is proposed that a larger and more efficient filter be built, using the same filling material. The new filter will have a particle retention ratio of 99.9% at an air flow-rate of 10 000 $cm^3 \cdot s^{-1}$. Calculate the dimensions of the filter in the following two cases:

a. Regarding pressure drop (energy cost), the air velocity will be the same as in the test filter.
b. Due to space limitation, the diameter of the filter will be limited to 20 cm.

It is assumed that the collection efficiency factor k increases with air velocity to the power 1/6.

Solution

a. The air velocity, hence the collection efficiency k will be the same as in the test run:

$$\ln\frac{C}{C_0} = -kZ \quad \ln 0.01 = -k \times 25 \quad \Rightarrow \quad k = 0.184$$

In the large filter:

$$\ln\frac{C}{C_0} = \ln 0.001 = -0.184Z \quad \Rightarrow \quad Z = 37.5 \text{ cm}$$

The diameter will be:

$$d = 5\sqrt{\frac{10\,000}{100}} = 50 \text{ cm}$$

b. The velocity of the air in the test unit:

$$v_1 = \frac{dV}{dt}\frac{1}{A} = 100\frac{1}{\pi \times 5^2/4} = 5.1 \text{ cm} \cdot \text{s}^{-1}$$

The velocity of the air in the projected unit with a diameter of 0.2 m:

$$v_2 = 10\,000\frac{1}{\pi \times 20^2/4} = 21.2 \text{ cm} \cdot \text{s}^{-1}$$

The collection efficiency of the large filter will be calculated from:

$$\frac{k_2}{k_1} = \frac{k_2}{0.184} = \left(\frac{21.2}{5.1}\right)^{1/6} = 1.27 \quad \Rightarrow \quad k_2 = 0.234$$

$$\ln\frac{C}{C_0} = \ln 0.001 = -0.234Z \quad \Rightarrow \quad Z = 29.5 \text{ cm}$$

8.3 Surface (barrier) filtration

8.3.1 Mechanisms

Surface filtration (also known as barrier filtration) is the most common type of filtration in the food industry. In surface filtration, a porous surface retains the particles solely on grounds of particle size. In contrast with depth filtration, there is a definite cut-out size (maximum size of particles passing through), and retention of particles above the cut-out size is absolute.

Surface filtration processes fall into two categories:

1. *Dead-end filtration* (also known as frontal filtration or cake filtration). Here, the direction of suspension flow is normal to the filter surface. The particles are stopped (come to a dead-end) on the filter surface and accumulate as a *cake*. The familiar laboratory filtration operation (Figure 8.2) is an example of dead-end filtration.
2. *Cross-flow filtration* (also known as tangential filtration, Murkes, 1988). Here, the direction of suspension flow is parallel (tangential) to the filter surface (Figure 8.3). The retained particles are carried forwards by the flowing suspension.

Since the main kind of filtration media used in tangential filtration are membranes, tangential filtration will be discussed in more detail in Chapter 10 (Membrane Processes).

8.3.2 Rate of filtration

The rate of filtration, defined as the volume of filtrate obtained per unit time, characterizes the capacity of a filter. Physically, dead-end filtration is a particular case of flow through porous media. The basic law of fluid flow through porous media is known as the *Darcy Law* (Henry P.G. Darcy, French engineer, 1803–1858):

$$Q = \frac{dV}{dt} = \frac{A\Delta P}{\mu R} \tag{8.3}$$

where:

V = filtrate volume, m^3
Q = filtrate flow-rate, $m^3 \cdot vs^{-1}$
A = filter area (normal to the direction of flow), m^2
ΔP = pressure drop across the porous bed, Pa
R = resistance of the bed to flow, m^{-1}
μ = viscosity of the fluid, $Pa \cdot s$.

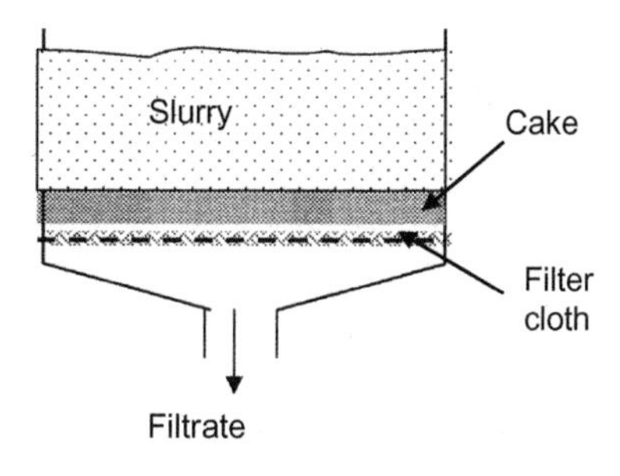

FIGURE 8.2

Dead-end (cake) filtration.

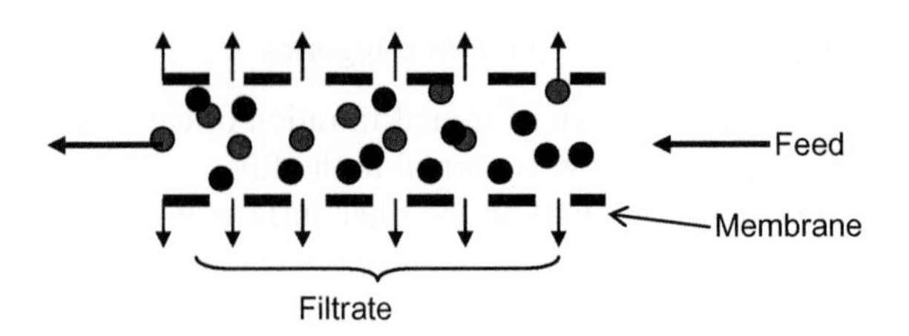

FIGURE 8.3

Tangential filtration.

Darcy's Law is, essentially, an application of the basic law of transport (Chapter 3, Eq. (3.1)), with pressure drop as the driving force (Darcy, translated by Bobeck, 2004).

In the case of cake filtration, the total resistance R is the sum of two resistances in series, namely, the resistance of the filter medium R_f and that of the cake R_c:

$$R = R_f + R_c \tag{8.4}$$

It is logical to assume that the resistance of the cake is proportional to its thickness L. The proportionality constant r (m^{-2}) is called "specific cake resistance". The resistance of the filter medium may be seen as the resistance of an additional fictitious cake thickness, L_f. Substituting in Eq. (8.3), we obtain:

$$Q = \frac{dV}{dt} = \frac{A\Delta P}{\mu\, r(L + L_f)} \tag{8.5}$$

In batch filtration, cake thickness L increases with increasing filtrate volume. If the volume of cake deposited per unit volume of filtrate obtained is v, one can write:

$$L = \frac{v\,V}{A} \tag{8.6}$$

In a similar manner, the fictitious cake thickness L_f can be related to a fictitious filtrate volume V_f:

$$L_f = \frac{v\,V_f}{A} \tag{8.7}$$

The value of v can be calculated by material balance from the concentration of the feed suspension and the density of the cake (see Example 8.2).

Substitution of Eqs (8.6) and (8.7) in Eq. (8.5) yields:

$$\frac{dV}{dt} = \frac{A^2 \Delta P}{r\,\mu\,v(V + V_f)} \tag{8.8}$$

Equation (8.8) is the fundamental differential equation of cake filtration. Its solution depends on the boundary conditions characterizing the process. Two cases will be considered (Cheremisinoff and Azbel, 1983):

1. *Constant rate filtration*: Q = dV/dt is kept constant. The pressure drop has to be increased in order to overcome the increasing resistance due to cake build-up.
2. *Constant pressure drop filtration*: ΔP is kept constant. The rate of filtration, Q = dV/dt, decreases as a result of increasing cake resistance.

It should be noted that both of the above cases are theoretical models. Real filtration processes behave differently, but may be closer to one of the models than to the other.

The condition of constant rate is almost met in batch filtration in a filter press. This type of filter is usually fed by a positive displacement pump, at nearly constant flow-rate. The batch is stopped when the pressure reaches the limit specified for the system.

Batch vacuum filtration in the laboratory is an example of constant pressure filtration. In this case, the pressure drop is the difference between atmospheric pressure and the vacuum created by a pump, both nearly constant.

1. *Solution of Eq. (8.8) for constant rate filtration*: the solution of the differential Equation (8.8) gives a linear relationship between the pressure drop ΔP and the cumulative filtrate volume V, as shown in Eq. (8.9) and Figure 8.4:

$$\Delta P = \frac{r\mu vQ}{A^2} V + \frac{r\mu vQ}{A^2} V_f \tag{8.9}$$

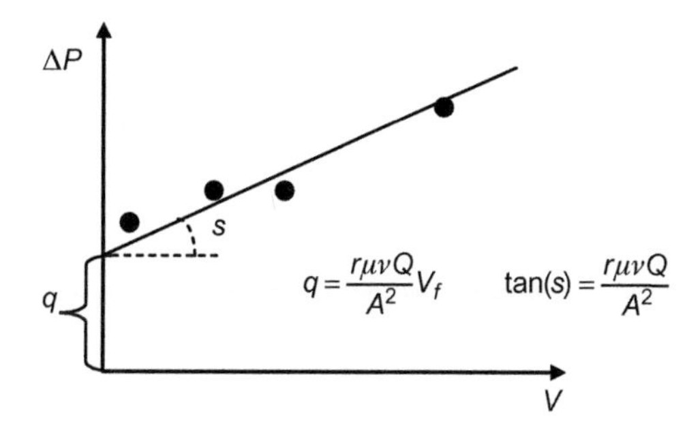

FIGURE 8.4

Treatment of experimental results of constant rate filtration.

The relationship between the pressure drop and filtration time, obtained by substituting $V = Qt$, is also linear.

2. *Solution of Eq. (8.8) for constant pressure filtration*: for constant pressure filtration, the solution of Eq. (8.8) gives a quadratic equation in V, shown in Eq. (8.10):

$$t = V^2\left(\frac{r\mu v}{2A^2\Delta P}\right) + V\left(\frac{r\mu v V_f}{A^2\Delta P}\right) \tag{8.10}$$

In both cases, the values of r and V_f are obtained experimentally, in the following manner:

1. In the case of constant rate filtration, the pressure is recorded as a function of cumulative filtrate volume. The plot ΔP versus V is linearized (Figure 8.4). From the slope ($rvQ\mu/A^2$) and the intercept ($rvQ\mu V_f$), r and V_f are calculated.
2. In the case of constant pressure filtration, Eq. (8.8) is written in the following form:

$$\frac{dt}{dV} \cong \frac{\Delta t}{\Delta V} = V\left(\frac{r\mu v}{A^2\Delta P}\right) + \left(\frac{r\mu v}{A^2\Delta P}\right)V_f \tag{8.11}$$

The constants are grouped in an expression known as Ruth's Coefficient, C, as follows:

$$C = \frac{2A^2\Delta P}{r\mu v} \quad (\text{m}^6\cdot\text{s}^{-1}) \tag{8.12}$$

Substitution in Eq. (8.11) gives:

$$\frac{\Delta t}{\Delta V} = V\frac{2}{C} + V_f\frac{2}{C} \tag{8.13}$$

Filtrate volume increments ΔV corresponding to filtration time intervals Δt are measured. The ratios $\Delta t/\Delta V$ are calculated and plotted against the cumulative filtrate volume V (Figure 8.5). From the slope (2/C) and the intercept ($2V_f/C$), r and V_f are calculated.

EXAMPLE 8.2

An aqueous slurry contains w kg solids per m^3 of water. The true density of the solid particles is ρ. The slurry is filtered. The porosity of the filter cake is ε. Develop an expression for v (volume of cake deposited for each m^3 of filtrate collected) in the following two cases:

a. The cake is drained free of liquid
b. The cake is saturated with liquid.

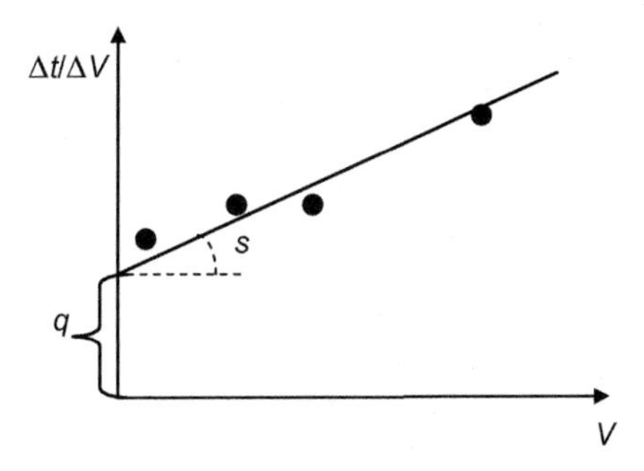

FIGURE 8.5

Treatment of experimental results of constant pressure filtration.

Solution

a. Assume that a slurry containing 1 m^3 of water and w kg of solid is filtrated. Since the cake is free of water, the volume of filtrate will be 1 m^3. The volume of cake will be:

$$\frac{w}{\rho(1-\varepsilon)}, \text{ hence: } v = \frac{w}{\rho(1-\varepsilon)}$$

b. If the cake is saturated with liquid (i.e., all the free volume is full of liquid), the volume of the cake will still be the same, but the volume of filtrate will be less than 1 m^3 by the amount held in the pores. Hence:

$$v = \frac{w}{\rho(1-\varepsilon)} \times \frac{1}{1 - \frac{w\varepsilon}{\rho(1-\varepsilon)}}$$

EXAMPLE 8.3

The results given in Table 8.1 were obtained in a constant pressure filtration test through cloth.

The height of cake formed, after collection of 0.01 $m^3$of filtrate, was 4.4 cm. The filter surface area was 450 cm^2. The pressure drop was 150 kPa.

Calculate the specific resistance of the cake and the equivalent resistance of the filter cloth.

Solution

The data are tabulated and then plotted as $\Delta t/\Delta V$ versus V (Table 8.2, Figure 8.6). The slope and the intercept are read:

$$\text{Slope} = 2/C = 6 \times 10^{-5}\ \text{s} \cdot \text{cm}^{-6} \quad \text{hence} \quad C = 0.167 \times 10^5\ \text{cm}^6 \cdot \text{s}^{-1}$$

By definition:

$$C = \frac{2A^2\Delta P}{r\mu v} = \frac{2 \times (450)^2 \times 150\,000}{r \times 0.001 \times 4.4 \times 450/10\,000} = 0.167 \times 10^5$$

$$r = 18.3 \times 10^6\ \text{cm}^{-2}$$

The intercept is practically nil. The resistance of the filter medium is therefore negligible.

Table 8.1 Experimental Filtration Data

Time, s	0	60	180	360	600	900	1260	1680	2160
Cumulative V, cm^3	0	1000	2000	3000	4000	5000	6000	7000	8000

Table 8.2 Transformation of the Filtration Data

V	t	dt/dv
0	0	
1000	60	0.06
2000	180	0.12
3000	360	0.18
4000	600	0.24
5000	900	0.30
6000	1260	0.36
7000	1680	0.42
8000	2160	0.48

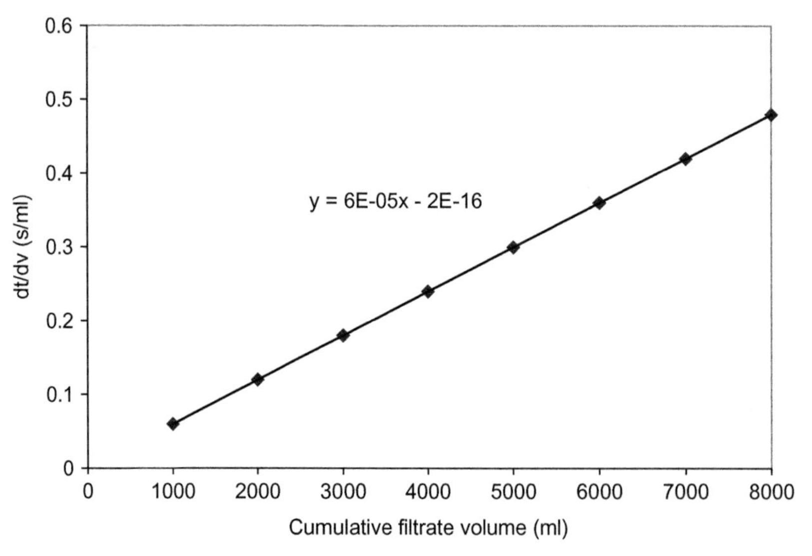

FIGURE 8.6

dt/dV versus V (Example 8.2).

8.3.3 Optimization of the filtration cycle

Many of the filtration processes in the food industry are batch operations. A batch is stopped either when the pressure becomes excessive (in the case of constant rate filtration) or when the filtration rate drops to an unsatisfactory level. In both cases, the feed is interrupted, the cake is removed, the filter is cleaned and a new batch is started. Optimization of the process aims at maximizing the cumulative volume of filtrate per working day.

Assume that the filtration cycle consists of filtration time θ_f and cleaning time θ_c, both in seconds. The number of cycles per day of 24 hours is $24 \times 3600/(\theta_f + \theta_c)$. Assuming that the volume of filtrate per cycle is V, the cumulative daily volume V_d can be calculated using Eq. (8.14):

$$V_d = \frac{24 \times 3600}{\theta_f + \theta_c} \cdot V \qquad (8.14)$$

By substituting V in Eq. (8.14), depending on the mode of filtration, the filtration time per cycle θ_f or the filtrate volume per cycle V that would result in maximum daily cumulative filtrate volume can be calculated. In the case of constant pressure filtration, optimization yields:

$$V_{opt} = \sqrt{\frac{2A^2 \Delta P \theta_c}{r \mu v}} \qquad (8.15)$$

8.3.4 Characteristics of filtration cakes

A filtration cake consists of a porous bed of solid particles and inter-particle space, totally or partially filled with filtrate. The specific resistance of the cake depends on the size and shape of the particles and the porosity of the bed.

One way of predicting cake resistance to flow is to assume that the cake behaves like an assembly of parallel, straight capillary passages through which the filtrate flows. Applying the Hagen-Poiseuille equation (see Chapter 2), the cake resistance corresponding to this model is found to be:

$$r = \frac{32}{\varepsilon\, d_C^2} \qquad (8.16)$$

where ε = cake porosity (dimensionless) and d_C = mean diameter of the fictitious capillary passages.

The above model has obvious shortcomings. The mean equivalent diameter is difficult to define. Furthermore, in a real cake, the filtrate flows through tortuous paths of unequal cross-section and unequal length.

A more refined model that takes into consideration the particulate structure of the porous bed is the Karman-Kozeny equation. According to this model, the flow-rate through a porous bed is given by Eq. (8.17):

$$Q = \frac{dV}{dt} = A\frac{1}{K} \cdot \frac{\varepsilon^3}{(1-\varepsilon)^2} \cdot \frac{1}{\mu S^2} \cdot \frac{\Delta P}{L} \tag{8.17}$$

where:

ε = cake porosity (void volume fraction, dimensionless)
S = specific area (total surface area of the particles per unit volume of cake, m^{-1})
K = an empirical coefficient, depending of particle shape distribution, dimensionless.

Comparing the Karman-Kozeny model with Darcy's Law, an expression for the specific resistance of the cake is obtained, as shown in Eq. (8.18):

$$r = \frac{K(1-\varepsilon)^2 S^2}{\varepsilon^3} \tag{8.18}$$

In all the filtration equations developed in the previous sections, the specific resistance of the cake was assumed to be independent of the pressure and constant in time. This assumption may be applicable to cakes formed from suspensions of rigid particles with regular shapes. Such cakes are characterized as "incompressible". Most filtration cakes encountered in the food industry consist of soft particles with irregular shapes and wide size distribution. Such cakes are compressible. Compaction under the effect of pressure results in reduced cake porosity and increased specific surface, hence in increased specific resistance. The effect of pressure on the specific resistance of compressible cakes is assumed to follow the exponential expression given in Eq. (8.19):

$$r = r_0(\Delta P)^s \tag{8.19}$$

The coefficient r_0 and the exponent s are empirical values.

8.3.5 The role of cakes in filtration

Cloudy liquids containing minute quantities of very small particles in suspension are difficult to filtrate because the particles rapidly clog the filter. In contrast, fluids containing a considerable quantity of large particles are easy to filtrate, because they form a thick but permeable cake. Any small particles are retained or retarded by the cake, and obstruction of the filter openings is prevented (Figure 8.7). The formation of a porous, incompressible cake from the start of the operation as a "pre-filter zone" results in high filtration rates for a considerable duration. If the feed liquid does not contain the type of solids necessary to form such a cake, solids known as "filter aids" may be added to the suspension.

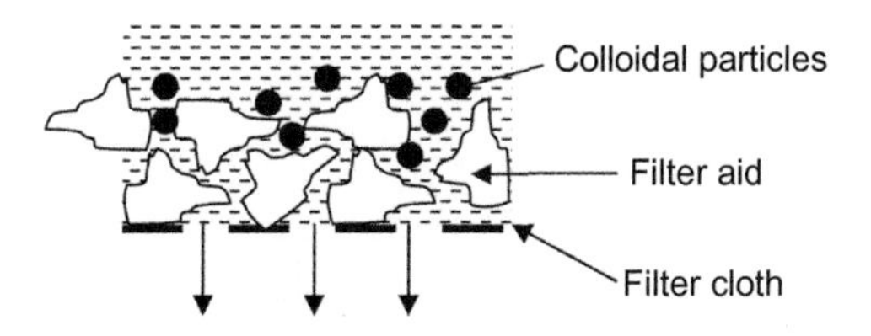

FIGURE 8.7

Function of filter aid.

Diatomaceous earth (kieselguhr), perlite (a natural siliceous rock) and cellulose powder, and mixtures thereof, are the most common types of filter aids. They consist of practically insoluble, rigid particles. Food-grade filter-aid materials are available. The use of filter aids is widespread in the production of clear juices, wine and edible oils. Obviously, filter aid can only be used if the filtrate is the desirable product and the cake is the waste.

It is sometime possible to improve the filterability of colloidal suspensions by certain kinds of pre-treatment, with the purpose of inducing coagulation or flocculation of the suspended particles. Depending on the nature of the colloid, pre-treatment may consist of heating, enzymatic reactions, change of pH, or the addition of certain electrolytes such as aluminum salts and synthetic or natural poly-electrolytes.

8.4 Filtration equipment

8.4.1 Depth filters

The most common depth filter is the sand filter, used for water purification (Stevenson, 1993). Sand filters range in size from very large tanks or boxes used for municipal water treatment to small portable vessels used in swimming pools. In open sand filters, water flows by gravity through a thick bed of sand or other particulate material (Figure 8.8).

In the so-called "fast sand filters", high filtration rate is achieved due to the moderate depth of the sand bed. Flocculants are usually added to the water before filtration. When the bed becomes too contaminated, it is cleaned by reversing the flow (back-wash). A rapid upward flux of water lifts the bed and removes the flocks entrapped between the sand particles. In slow sand filters, used mainly in municipal water treatment, the sand bed may be 1.5–2 meters deep. Due to the slow flow-rate, a thin, slimy layer of biomass forms on top of the sand. This layer contributes to the purification of the water, acting as a bio-filter, but when it becomes too thick it is mechanically removed to restore flow-rate. Additional purification effects can be achieved by incorporating adsorbents in the filtering bed. Thus, coating the sand particles with graphite oxide provides a filter capable of removing mercury traces from water efficiently (Anon, 2011).

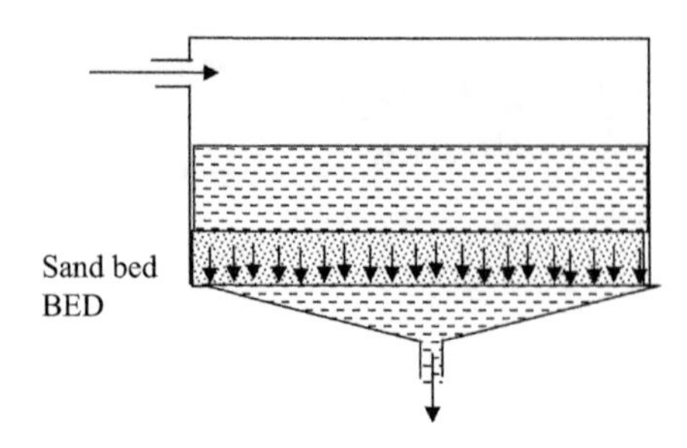

FIGURE 8.8

Schema of a sand bed filter.

8.4.2 Barrier (surface) filters

The pressure drop required to move the filtrate through the barrier may be created by applying a positive pressure at the feed side or a vacuum at the filtrate side, or centrifugal force. Accordingly, barrier filters fall into three categories: pressure filters, vacuum filters and filter centrifuges (Dickenson, 1992). The most common types of pressure filters are filter presses and cartridge filters. The typical vacuum filter is the continuous rotary filter. Centrifugal filters are also called basket centrifuges.

1. *Filter presses*. A filter press consists of a set of plates, pressed together in horizontal or vertical stacks, so as to form a series of compartments. The plates are covered with the filter medium (mesh, cloth, canvas, etc.). The solids accumulate in the compartments while the filtrate is collected and discharged from the space between the compartments (Figure 8.9). Filter presses are fed using positive displacement pumps capable of developing high pressures. The plates are pressed together by mechanical or hydraulic means. Most filter presses operate in batches. At the end of a batch, which may include a phase of cake washing, the filter is open, the cake drops from the compartments, the filtering surfaces are washed and the filter is closed again. Filter presses are, therefore, relatively demanding in labor cost. Filter presses wherein the entire filtration cycle is fully automated have been developed (Kurita *et al.*, 2010).

   In filter presses known as frame-and-plate filters, the compartments are formed by plates alternating with open frames.
2. *Cartridge filters*. Cartridge filters are used to continuously separate small quantities of suspended solids from a fluid in flow. Some of their most common uses are:
   - Filtration of compressed air used in control and instrumentation
   - Filtration of steam (steam strainer)
   - Filtration of clean water before spray nozzles, to prevent clogging of the nozzle by occasional dirt.

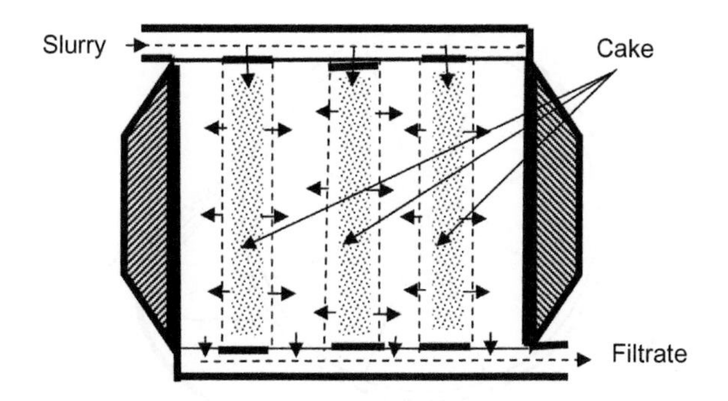

FIGURE 8.9

Schema of a filter press.

The filtering element of a cartridge filter may be a perforated metal surface, cloth, canvas, paper, mesh, porous ceramic, or a series of discs with a very narrow gap between them. The filtering elements are placed in housings with flow-diverting elements required to force the fluid through the filtering surfaces. Most cartridge filters can be cleaned in-place. In some cases, the filtering elements are replaceable or even disposable.

3. *Continuous rotary vacuum filters*. In continuous vacuum filters, the filtrate is sucked through a moving porous filter medium. The solids remaining on the filter medium are removed, usually by scraping off with a blade. The moving filter medium may take the form of an endless belt (belt filters) or may cover the face of a rotating cylinder (drum filter). The continuous vacuum drum filter (Figure 8.10) is the most common type. In one of the models, the drum consists of two concentric horizontal cylinders. The outer cylinder is slotted and covered by the filter cloth, while the inner cylinder has a solid wall. Radial partitions separate the space between the two cylinders into sections. Any liquor sucked through the filter cloth is admitted into the sections and discharged through pipes passing through the hollow shaft of the drum. Part of the drum is submerged in a trough containing the slurry to be filtered. In the submerged section, filtrate is sucked in and the cake forms on the surface of the drum. As the drum rotates, the cake emerges from the slurry and moves towards the scraper, known as the doctor blade. In its circular travel, the cake may be washed with sprays and dried. The section immediately underneath a certain portion of the cake is sequentially connected to vacuum (to collect the wash water), to hot air (to dry the cake) and to pressurized air (to lift the cake slightly off the cloth before scraping).

   In some models the filter cloth does not wrap around a single drum but passes over a second, much smaller drum, from which the cake drops without scraping, because of the strong curvature (Figure 8.11).

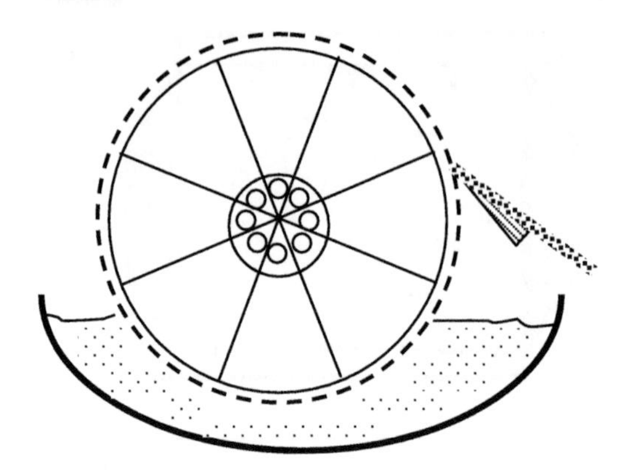

FIGURE 8.10

Schema of a rotary vacuum filter.

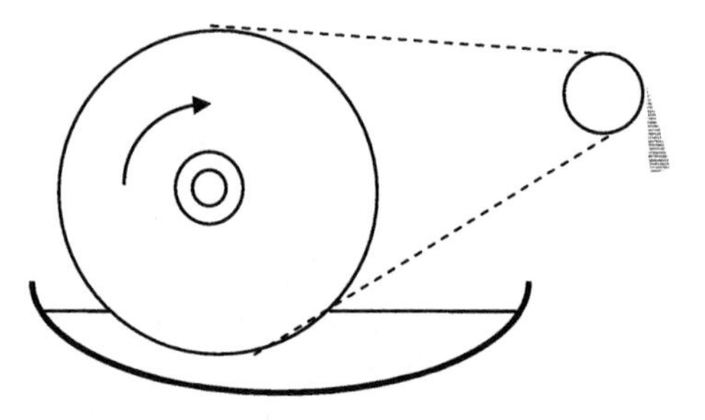

FIGURE 8.11

Rotary vacuum filter with auxiliary drum.

In vacuum filters, the driving pressure drop is necessarily limited to less than 100 kPa. For this reason, vacuum filtration is not recommended if:

- The viscosity of the slurry is high
- The slurry contains very small particles
- Rapid filtration is required.

If the cake is the waste and not the product, slurries containing colloidal particles may still be filtered using a continuous vacuum drum filter, provided that pre-coating with a filter aid is practiced. First, an adequate layer of filter-aid cake is formed by filtering a slurry of the filter aid in water. This layer is called the pre-coat. The doctor blade is adjusted so as to avoid excessive removal of the pre-coat. Nevertheless, some of the filter aid is removed and the pre-coat must be renewed periodically.

4. *Filter centrifuges.* Filter centrifuges consist of cylindrical baskets with perforated walls, spinning rapidly on a vertical or horizontal axis

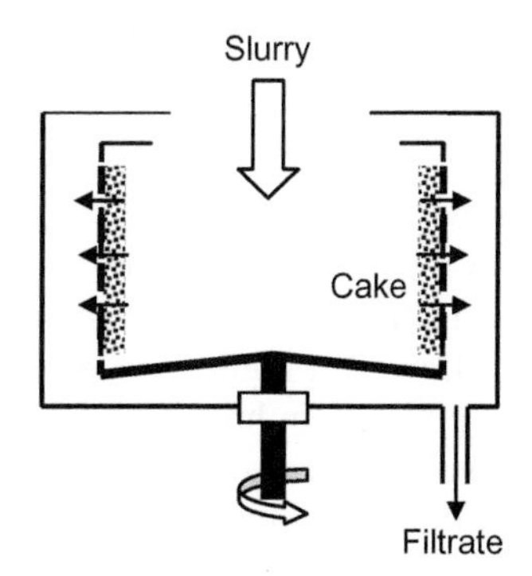

FIGURE 8.12

Horizontal basket filter centrifuge.

FIGURE 8.13

Vertical basket centrifuge showing scoop and scraper assembly.

Photograph courtesy of the Department of Biotechnology and Food Engineering, Technion, I.I.T.

(Figures 8.12 and 8.13). Their main use in the food industry is for the separation of sugar crystals from the mother liquor (molasses) in the final stage of sugar manufacture. In the sugar industry, these machines are called "centrifugals". The slurry is fed into the spinning basket, and the liquor is forced out through the perforations by the action of centrifugal force. The solids form a cake on the basket wall, and can be removed intermittently with a moving "plough" without stopping the machine. Usually, the speed of rotation is step-changed at the various phases of the cycle (charging, filtration, cake washing, cake drying, discharge). Models with devices for the continuous discharge of solids are available. Basket centrifuges are also used for the separation of salt crystals from mother liquor.

8.5 Expression

8.5.1 Introduction

Expression is a separation process whereby liquid is expelled from a wet material by applying pressure (Fellows, 2000). It is therefore often called pressing or squeezing. Expression is akin to filtration, at least in two respects: both are pressure driven, and both involve flow of a liquid through a porous bed. Incidentally, the porous bed is often called a "cake" in both cases.

Expression is used extensively in the food industry. The following are some of the applications of expression in food processing:

- Extraction (pressing) of juice and essential oil from citrus fruit
- Extraction of juice from crushed fruit (apples, pears, cranberries, pineapple)
- Separating tomato juice from crushed tomatoes
- Extraction of grape juice in wineries
- Recovery of crude edible oil from oilseeds, copra and nuts, using expellers
- Pressing cooked fish or slaughterhouse offal in the production of fish meal and meat meal, in order to remove some of the water and oil
- Obtaining emulsions by pressing crushed olives in the traditional method for making olive oil
- Expelling whey from curd in the manufacture of hard and semi-hard cheese
- Recovery of sugar juice from milled sugar cane
- Dewatering of various waste materials, such as citrus peels, spent sugar beet cossettes and spent coffee from soluble coffee plants.

The pressure applied in various applications range from less than 20 kPa up to 100 MPa (Table 8.3).

Expression equipment may carry different names: presses, extractors, expellers, etc. A wide variety of equipment types are available for batch or continuous expression.

Table 8.3 Pressures used in Various Applications of Expression

Pressure Range MPa	Application
0.015–0.08	Cheese curd pressing
1–3	Fruit juice extraction
7–15	Spent coffee dewatering
Up to 100	Expelling oil from oilseeds

Table adapted from Schwartzberg (1983).

8.5.2 Mechanisms

The mechanisms involved in expression processes and their kinetics have been reviewed by Schwartzberg (1983). Despite the wide diversity of the materials treated and the equipment used, certain common elements are found in all expression operations. The material may be considered as consisting of a porous solid matrix containing a fluid. The fluid may be practically free, retained by capillary forces, physically entrapped inside cells, or adsorbed on the surface of the solid matrix. Pressure is required to overcome the resistance of the solid matrix to deformation and compaction, to disrupt the cells, and to compensate for the pressure drop arising from the flow of the fluid through the solid. In continuous operations, pressure is also needed to move the pressed solid cake through the equipment and out. The efficiency of expression depends on the following factors, among others:

1. *Compressibility of the solid.* The response of the solid to applied pressure depends on its mechanical properties, such as rigidity, pliability, hardness. For a pliable cake, empty of fluid, the relationship between the applied pressure and compaction (reduction in volume) may be modeled as follows:

$$\log\frac{V - V_\infty}{V_0 - V_\infty} = -k_k P \qquad (8.20)$$

   where:

   V = volume of compressed cake in equilibrium with pressure P
   V_0 = volume of cake before compression
   V_∞ = volume of cake compressed by an infinitely high pressure (minimal volume)
   P = pressure, Pa
   k_k = compressibility constant, Pa^{-1}.

   Eq. (8.20) does not take into account the portion of the pressure drop used to break cells and to support fluid flow. It may be applicable to very wet materials consisting of a liquid of low viscosity and a totally inert solid matrix. With this assumption, and neglecting the volume of air entrapped in the slurry, Eq. (8.20) may be used to evaluate approximately the maximum yield of "juice" that can be expressed by applying a pressure P. Experimentation is needed to validate the model and to determine the compressibility constant.

2. *Viscosity of the liquid.* Both the rate of liquid release and the final yield are negatively affected by high liquid viscosity. In many cases, the viscosity of the liquid (juice) itself depends on the pressure applied and the yield of juice drawn. The viscosity of apple juice was found to be affected by the yield drawn: the higher the yield, the higher the juice viscosity (Körmendy, 1979). The yield-versus-quality function is particularly significant in expression processes because the composition of the expressed liquid changes with the

pressure applied and the duration of the process. Thus, the juice of white grapes expressed by applying moderate pressure is considered of higher quality for the production of white wine. Very often, the term "first press" is used to indicate premium quality. Multi-stage expression is sometimes applied for obtaining products with different quality grades.

3. *Pre-treatment of the material.* A variety of mechanical, thermal and chemical treatments are usually applied to materials before pressing. The most common treatment is comminution or crushing, applied to fruits, vegetables, olives, sugar cane, etc., and grinding, applied to oilseeds. Comminution disrupts cellular tissue, cuts fibers, releases liquid and increases the compressibility of the solid. On the other hand, extensive comminution may increase the amount of fines that can block the capillary passages of the cake and thus impair the rate of juice draining. Furthermore, fines are obviously not desirable if the product is not to be cloudy. Extensive comminution may also cause extraction of undesirable, less soluble substances such as bitter principles into the liquid. It is therefore necessary to optimize the extent of comminution, depending on the characteristics of the product.

   Thermal pre-treatment may be applied for a variety of reasons. Ground oilseeds are subjected to an essentially thermal operation known as conditioning, before entering the expeller. The objectives of conditioning are:

   - To plasticize the mass
   - To denature proteins and thus free the oil from emulsions
   - To cause coalescence of oil droplets
   - To reduce the viscosity of the oil
   - To remove moisture if necessary.

   In some cases enzymatic or chemical treatments are required, mainly to free the water that is bound to the solid matrix. In the production of juice from apples and pears, the crushed fruit is treated with enzymes to increase juice release as a result of controlled pectolysis. Commercial pectolytic enzyme preparations specifically formulated for this application are available. Regarding non-enzymatic chemical treatments, a typical example of this case is the pressing of citrus peels. The operation of citrus juice extraction (squeezing) leaves a voluminous waste consisting mainly of peels (about 50% of the mass of the fruit processed). The principal use of this waste material is as cattle-feed. Unless dispensed as fresh fodder, the peels are dehydrated and made into pellets. Pressing the peels before dehydration to remove part of the water helps reduce the cost of dehydration. However, practically no liquid can be pressed out of the peels, regardless of the pressure applied, because of the strong water binding power of the pectin. Therefore, the peels are chopped and then mixed with lime. After a short reaction time during which the pectin is transformed to calcium pectate, the material can be successfully pressed to remove a substantial quantity of press liquor.

   Other types of pre-treatment operations for increasing expression yields include freezing–thawing and pulsed electric discharges (Grimi *et al.*, 2010).

4. *Applied pressure*. Depending on the type of equipment, the element used for applying pressure to the slurry may be a piston, an inflatable tube or "bladder" (pneumatic press), a screw conveyor with progressively decreasing pitch, mechanical paddles pressing the slurry against a perforated surface, etc. As a rule, it is preferable to increase the pressure gradually. The flow of juice becomes increasingly difficult as the cake is compacted. A good balance between flow-rate and rate of compaction can be maintained by gradual increase of the pressure. Maximization of the press yield requires optimization of the time–pressure profile (Pérez-Gálvez *et al*., 2010). In some presses the pressure is regulated automatically, according to an optimal time–pressure profile.

8.5.3 Applications and equipment

1. *Pulpers and finishers*. These solid–liquid separators operate at a very low pressure created by moderate centrifugal force. Typically, they consist of a perforated horizontal or slightly inclined cylindrical shell, and a set of paddles attached to the rotating axis (Figure 8.14). The material is fed into the cylinder and thrown to the perforated wall by the centrifugal action of the paddles. The liquid (juice) flows out through the perforations and is collected in a trough below. The pulp is pushed towards the exit end. The yield of juice recovery depends on the diameter of the perforations, the speed of rotation of the paddles, the residence time inside the machine, and the gap between the paddles and the shell. The residence time can be adjusted by changing the lead angle of the paddles. If the material contains hard and abrasive particles, the stainless steel paddles are replaced by rubber strips or brushes. For tomato and citrus juice, the diameter of the perforations is 0.6–0.8 mm. Soft or cooked fruit (e.g., tomatoes) can be crushed and pressed in one step in a

FIGURE 8.14

Pulper finisher

Photograph courtesy of Rossi-Catelli.

pulper finisher. Hard fruit is first crushed in a mill (e.g., a hammer mill) before entering the pulper finisher (Downes, 1990).

2. *Helicoidal extractors and screw presses*. More pressure can be applied to the material by replacing the paddles with an auger (worm, screw) and by restricting the pulp exit with an adjustable gate. Compression is achieved by decreasing the cross-sectional area available for the forward flow of the material, and by adjusting the gate (end-plate) opening. The gate may be hydraulic, spring-loaded, or positioned by a screw. Helicoidal extractors for the extraction and finishing of juices (e.g., citrus juice finishers) are built for moderate pressures. On the other hand, screw presses (expellers) for expelling oil from oilseeds are sturdy machines, built for pressures in excess of 50 000 kPa (Williams, 2005).
3. *Hydraulic presses*. Hydraulic presses of one sort or another are traditionally used for the extraction of juice from apples and grapes, as well as oil from olives. One of the most common types of hydraulic presses is the vertical "pack press" (Downes, 1990). Batches of the crushed material are wrapped in pressing cloth and piled, one on top of another, with "lathes" between them. When the pile has been completed, the hydraulic piston is operated and the material is gradually compressed. The press cloth serves both as a package for the mash and as a filter for the juice. Pack presses are labor intensive but inexpensive, and they give a high juice yield. Another type of hydraulic press is the horizontal rotary press in which the mash is pressed in bulk. Both types of hydraulic presses operate batchwise.

   The effect of pressure, temperature and moisture content on the yield of oil obtained by hydraulic pressing of some oilseeds was studied by Willems *et al.* (2008).
4. *Inflated pouch presses*. Inflated or pneumatic presses are extensively used for extracting grape juice in the wine industry. The mashed grapes are introduced into a horizontal cylinder with perforated walls. A diaphragm or "bladder" located inside the cylinder is then inflated, pressing the mash gradually and evenly.
5. *Citrus juice extractors*. These are machines designed and built specifically, taking in consideration of the structure of citrus fruit and the requirement concerning the quality of the juice. Accordingly, non-specific expelling devices that obtain juice by pressing the entire fruit (apple, berries, grapes, etc.) cannot be used for citrus fruit. Today, there are essentially two types of industrial juice extractors for citrus: the Brown extractor and the FMC extractor (Berk, 1969; Rebeck, 1990). In the Brown extractor (Brown Citrus Machinery), the fruit is halved and reamed. In the FMC (Food Machinery Corporation) extractor, the fruit is not halved but pressed whole between the intermeshing fingers of two cups (Figure 8.15). As the fruit is pressed, a circular "plug" is cut at the bottom of the fruit by a sharp-edged tube. The whole "inside" of the fruit (i.e., juice, pulp, seeds) is forced into the tube with perforated walls. At the end of the pressing cycle, a second "plug" is cut from

FIGURE 8.15

Cups of the FMC Inline Juice Extractor.

Photograph courtesy of the Department of Biotechnology and Food Engineering, Technion, I.I.T.

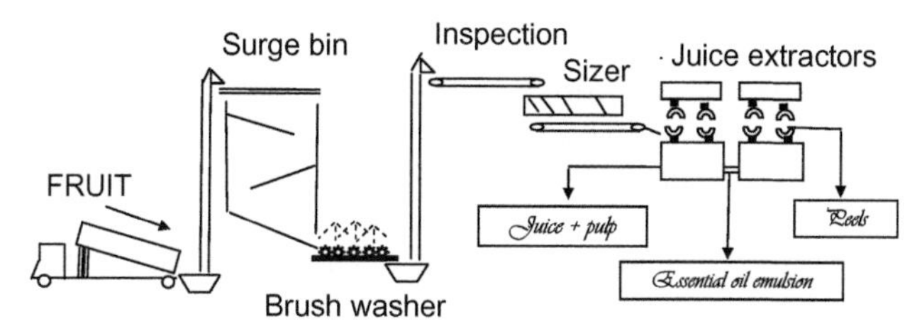

FIGURE 8.16

Flow diagram of citrus fruit pressing line.

the upper end of the fruit. A piston now compresses the content of the tube, forcing the juice out of the perforated walls, then discarding the rag, seeds, and two "plugs" through a special opening. In the course of the pressing action the essential oil in the peel is washed away by a small quantity of water, to be recovered later by centrifugation. A flow diagram of the citrus fruit juice extraction process is shown in Figure 8.16.

References

Aiba, S., Humphrey, A.E., Millis, N.F., 1965. Biochemical Engineering. Academic Press, New York, NY.

Anon, 2011. "Super sand" filters five times better. Filtr. Separat. 48 (4), 11.

Berk, Z., 1969. Industrial Processing of Citrus Fruit. United Nations, New York, NY.

Cheremisinoff, N.P., Azbel, D.S., 1983. Liquid Filtration. Ann Arbor Science Publishers, Woburn, MI.

Darcy, H., translated from French by Bobeck, P. 2004. The Public Fountains of the City of Dijon. Kendall/Hunt, Dubuque, France.

Dickenson, C., 1992. Filters and Filtration Handbook, third ed. Elsevier Advanced Technology, Oxford.

Downes, J.W., 1990. Equipment for extraction of soft and pome fruit juices, Production and Packaging of Non-carbonated Fruit Juices and Fruit Beverages, second ed. Blackie Academic and Professional, London, UK.

Fellows, P.J., 2000. Food Processing Technology: Principles and Practice. Woodhead Publishing Limited, Cambridge, UK.

Gaden, E.L., Humphrey, A.E., 1956. Fibrous filters for air sterilization. Ind. Eng. Chem. 48, 2172–2176.

Grimi, N., Vorobiev, E., Lebovka, N., Vaxelaire, J., 2010. Solid–liquid expression from denaturated plant tissue: filtration–consolidation behaviour. J. Food Eng. 96 (1), 29–36.

Heusslein, R., Brendel-Thimmel, U., 2010. Food and beverage: linking filter performance and product safety. Filtr. Separat. 47 (6), 26–31.

Körmendy, I., 1979. Experiments for the determination of the specific resistance of comminuted and pressed apple against its own juice. Acta Aliment. Acad. Sci. Hung. 4, 321–342.

Kurita, Y., Suwa, S., Murata, S., 2010. Filter presses: a review of developments in automatic filter presses. Filtr. Separat. 47 (3), 32–35.

Maxon, W.D., Gaden, E.L., 1956. Fibrous filters for air sterilization. Ind. Eng. Chem. 48 (12), 2177–2179.

Murkes, J., 1988. Crosslow Filtration: Theory and Practice. Wiley, Chichester, UK.

Payatakes, A.C., 1977. Deep Bed Filtration Theory and Practice. Summer School for Chemical Engineering Faculty, Snowmass, CO.

Pérez-Gálvez, R., Chopin, C., Mastail, M., Ragon, J.-Y., Guadix, A., Bergé, J.-P., 2010. Optimisation of liquor yield during the hydraulic pressing of sardine (*Sardina pilchardus*) discards. J. Food Eng. 93 (1), 66–71.

Rebeck, H.M., 1990. Processing of citrus fruit, Production and Packaging of Non-carbonated Fruit Juices and Fruit Beverages, second ed. Blackie Academic and Professional, London, UK.

Schwartzberg, H.G., 1983. Expression-related properties. In: Peleg, M., Baglet, E.B. (Eds.), Physical Properties of Foods. Avi Publishing Company, Westport, CT.

Stevenson, D.G., 1993. Granular Media Water Filtration: Theory, Design and Operation. Ellis Horwood, Chichester, UK.

Sutherland, K., 2010. Food and beverage: filtration in the food and beverage industries. Filtr. Separat. 47 (2), 28–31.

Sutherland, K., 2011. Filter media guidelines: selecting the right filter media. Filtr. Separat. 48 (4), 21–22.

Willems, P., Kuipers, N.J.M., De Haan, A.B., 2008. Hydraulic pressing of oilseeds: experimental determination and modeling of yield and pressing rates. J. Food Eng. 89 (1), 8–16.

Williams, M.A., 2005. Recovery of oils and fats from oilseeds and fatty materials, sixth ed. In: Shahidi, F. (Ed.), Bailey's Industrial Oil and Fat Products, vol. 5. Wiley-Interscience, New York, NY.

CHAPTER

9 Centrifugation

9.1 Introduction

Centrifugation and *decantation* (*sedimentation*, *settling*, *flotation*) are processes for the separation of heterogeneous mixtures of phases that differ from each other in their density. The physical principles governing these processes are the same. Decantation occurs under the effect of the Earth's gravity. In the case of centrifugation, accelerated separation is made possible by the effect of centrifugal forces which may be many times stronger than the Earth's gravity.

Centrifugation and decantation may be used to separate solid particles from a liquid, or two immiscible liquids of different densities from each other, or both.

Centrifugal separation is achieved not only with centrifuges but also in any system where a rotational movement is imparted to a mixture. One of the devices for centrifugal separation without a mechanical centrifuge is the cyclone, discussed briefly at the end of this chapter. However, the major part of the present chapter will deal only with mechanical centrifugation and centrifuges.

Centrifuges are relatively expensive machines, both in capital expenditure and in the cost of operation (energy consumption, wear of rapidly moving parts, need for sturdy construction capable of resisting the very high forces and pressures). Notwithstanding their cost, centrifuges are used extensively in industry. The following are a few of the many applications of mechanical centrifugation in the food industry.

- *Milk separation*. One of the oldest and most widespread uses of the centrifuge in the food industry is the separation of whole milk to skimmed milk and cream. The centrifuges used for this purpose are known as "separators" (Walstra et al., 2005).
- *Cheese production*. In modern dairy plants, centrifuges are used for the rapid separation of curd from whey.
- *Pulp control in fruit and vegetable juices*. Centrifugation is used for the reduction of pulp content in fruit juices, and for the production of clear juices by total removal of the pulp.
- *Edible oil processing*. Several operations in the production and refining of edible oils involve the separation of the oil from an aqueous phase. Centrifugation is the preferred method of separation. The modern methods for the production of olive oil are based on a special type of centrifuge, called a

Food Process Engineering and Technology. DOI: http://dx.doi.org/10.1016/B978-0-12-415923-5.00009-5
© 2013 Elsevier Inc. All rights reserved.

"decanter". A system for the production of olive oil is shown later in Figure 9.13, below.

- *Essential oil recovery*. Essential oils of citrus fruit are recovered by centrifugal separation of aqueous mixtures formed in the course of juice extraction.
- *Production of starch*. One of the methods for the separation of starch from suspensions is mechanical centrifugation.
- *Production of yeast*. Centrifuges are used for the separation of commercial yeast from the liquid growth medium.

9.2 Basic principles

As stated above, centrifugation is a process of *accelerated decantation*. It is therefore advantageous to analyze first the process of natural decantation in the field of Earth's gravity. The following discussion deals with the critical conditions for the separation of *solid particles* from a liquid. The same reasoning and the same results are also applicable to the separation of liquid droplets dispersed in a fluid. Specific engineering issues concerning liquid–liquid separation will be discussed separately.

9.2.1 The continuous settling tank

Consider the continuous settling tank shown in Figure 9.1.

A suspension of solid particles is continuously fed in at one end of the tank. The density of the particles being higher than that of the liquid medium, the particles sink to the bottom of the tank. The clear liquid flows out of the tank over a weir. Consider a solid particle, at the surface of the liquid at time $t = 0$. What are the conditions for the particle to be separated (i.e., to be retained in the tank and not to flow out over the weir with the liquid)?

Being heavier than the liquid, the particle sinks at a velocity u $(m \cdot s^{-1})$, while at the same time traveling with the bulk of the liquid towards the outlet at a velocity v $(m \cdot s^{-1})$. If the distance between the liquid surface and the tip of the

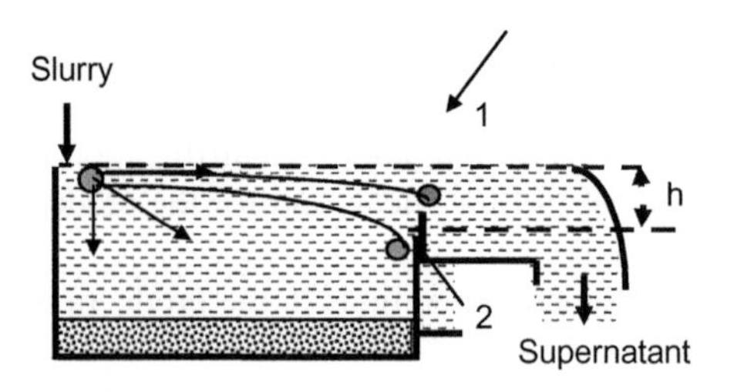

FIGURE 9.1

Continuous settling tank.

weir is h (m), the particle will be kept in the tank if its residence time t in the liquid above the weir satisfies the condition set in Eq. (9.1).

$$t \geq \frac{h}{u} \tag{9.1}$$

Let V (m^3) be the volume of the liquid mass above the weir, A(m^2) its horizontal cross-sectional area and Q ($m^3 \cdot s^{-1}$) the volumetric flow-rate of the suspension. Then:

$$t = \frac{V}{Q} = \frac{Ah}{Q} \tag{9.2}$$

It follows that the maximum flow-rate that can satisfy the condition in Eq. (9.1) is:

$$Q_{\max} = uA \tag{9.3}$$

Assuming that the particles are spherical and that Stoke's Law applies, the terminal sinking velocity u is given by:

$$u = \frac{d_p^2(\rho_s - \rho_l)g}{18\mu}$$

(see Eq. (2.22)). Substitution in Eq. (9.3) gives:

$$Q_{\max} = \frac{Ad_p^2(\rho_s - \rho_l)g}{18\mu} \tag{9.4}$$

where:

d_p = particle diameter, m
ρ_L, ρ_S = density of the liquid and the solid, respectively, $kg \cdot m^{-3}$
μ = viscosity of the liquid, $Pa \cdot s$
g = acceleration due to gravity, $m \cdot s^{-2}$.

Eq. (9.4) indicates that the maximum permissible feed rate of a settling tank for the separation of all the particles of diameter d or larger is Q_{max}. Inversely, Eq. (9.4) can be used to calculate the minimum size of particles that can be totally retained in a settling tank of cross-sectional area A, fed at a volumetric rate of Q.

EXAMPLE 9.1

The *tabling process* is a traditional rural method for the separation of starch from its suspensions in water. The slurry is fed to a long horizontal settling tank (a table). Starch granules settle to the bottom, and the clear supernatant flows out over a weir at the opposite end of the table. If the table is 1.6 m wide and starch granules are assumed to be spheres with a diameter of 30 μm and a density of 1220 $kg \cdot m^{-3}$, what is the minimal length of a table required for treating 4300 liters of slurry per hour?

Solution

The average settling velocity of the starch is:

$$u = \frac{d_p^2(\rho_s - \rho_l)g}{18\mu} = \frac{(30\times 10^{-6})^2 \times (1220 - 1000)\times 9.8}{18\times 0.001} = 107.8\times 10^{-6}\mathrm{m\cdot s^{-1}}$$

The minimal cross-sectional area of the table is found from Eq. (9.3):

$$Q_{max} = uA \Rightarrow A_{min} = \frac{Q}{u} = \frac{4.3}{3600\times 107.8\times 10^{-6}} = 11.08\ \mathrm{m^2}$$

The minimal length is then:

$$L_{min} = \frac{11.08}{1.6} = 6.92\ \mathrm{m}$$

9.2.2 From settling tank to tubular centrifuge

Imagine that the settling tank in Figure 9.1 is rotated around a horizontal axis. Now flip the entire system 90° so that the axis of rotation becomes vertical (Figure 9.2). The configuration obtained is that of the simplest type of centrifuge; namely, the tubular centrifuge. The driving force for sedimentation is now the centrifugal force, acting in radial direction. (Note: It will be assumed that vertical movement due to gravity is negligible in comparison with the radial movement under the effect of centrifugal forces.)

Consider a particle at distance r_1 from the rotation axis at time $t = 0$ (Figure 9.3). The tip of the weir is at a distance r_2. The suspension flows in vertical direction

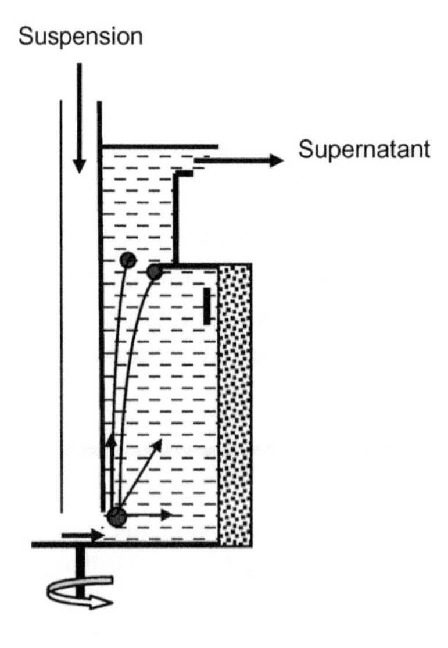

FIGURE 9.2

Basic structure of tubular centrifuge.

upwards. The particle will be separated (kept in the centrifuge) if it can travel in the radial direction a distance at least equal to the distance from r_1 to r_2, during its vertical travel with the fluid. What is the maximum flow-rate for such separation to occur?

The relations developed for the settling tank are valid with the following changes:

1. The gravity acceleration g is replaced by the centrifugal acceleration $\omega^2 r$ (ω being the angular velocity of rotation).
2. Since the axial acceleration is not constant, the velocity of radial displacement varies with the distance r from the axis of rotation.

$$u = \frac{dr}{dt} = \frac{d_p^2(\rho_s - \rho_l)\omega^2 r}{18\mu} \tag{9.5}$$

u (i.e., the velocity of centrifugal sedimentation) is not constant but increases with the distance r from the axis of rotation, according to Eq. (9.5).

The minimum residence time for retention t is obtained by integration of Eq. (9.5).

$$t = \frac{18\mu}{d_p^2(\rho_s - \rho_l)\omega^2} \ln \frac{r_2}{r_1} \tag{9.6}$$

The particle is retained if its residence time in the centrifuge is at least equal to t. Denoting by V the volume of the liquid mass between r_1 and r_2 (active volume), the maximum flow-rate for the separation to occur can be calculated from Eq. (9.7):

$$\frac{V}{Q_{max}} = \frac{\pi(r_2^2 - r_1^2)L}{Q_{max}} = \frac{18\mu}{d_p^2(\rho_s - \rho_l)\omega^2} \ln \frac{r_2}{r_1} \tag{9.7}$$

where L = length (height) of the active volume, m.

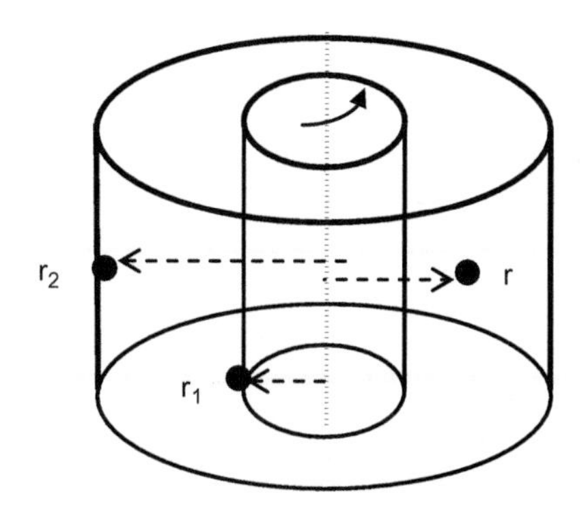

FIGURE 9.3

Position of a particle in a tubular centrifuge.

The "*theoretical equivalent area*" (denoted as Σ) of a centrifuge is defined as the surface area of an imaginary settling tank that would have the same separation capability as the centrifuge at the same maximum flow-rate. Comparing the limit separation expression for a tubular centrifuge (Eq. (9.4)) with that of a settling tank (Eq. (9.7)) we find:

$$\Sigma \text{ (for a tubular centrifuge)} = \frac{\pi L(r_2^2 - r_1^2)\omega^2}{g \ln\left(\frac{r_2}{r_1}\right)} \tag{9.8}$$

The theoretical equivalent area Σ is a useful concept for the characterization of centrifuges because it depends only on the geometry of the machine (L, r_1, r_2) and the speed of rotation (ω) and not on the properties of the material treated (particle size, density of the solid and the fluid, viscosity). Eq. (9.8) is valid for tubular centrifuges only. Other types of centrifuges have other expressions for Σ.

Eq. (9.7) illustrates the importance of density difference and particle size in centrifugal separation. Obviously, centrifugal or gravitational separation is possible only if the phases differ in their density. Only if a density difference exists can particles be separated according to their particle size, but, when this occurs, particle size has a strong influence on separation capability. In certain cases it is possible to improve centrifugal separation by size augmentation treatments before centrifugation (e.g., flocculation of solid particles, coalescence of droplet size in emulsions, etc.). In some cases, centrifugal separation can be improved by increasing the density difference (e.g., increasing the density of the aqueous medium by adding salt to an oil-in-water emulsion).

EXAMPLE 9.2

A laboratory procedure specifies centrifugation of a liquid "at 10 000 g". A centrifuge with an effective bowl radius of 12 cm is available. What should be the speed of rotation of a centrifuge to meet the specification?

Solution

The centrifugal acceleration a is to be equal to 10 000 times g.

$$\alpha = \omega^2 r = 4\pi^2 N^2 r = 4 \times \pi^2 \times 0.12 \times N^2 = 10\,000 \times 9.8$$
$$N = 143.9 \text{ rps} = 8634 \text{ rpm}$$

EXAMPLE 9.3

In a tubular centrifuge, the feed is introduced at a distance of 0.03 m from the rotation axis. The centrifuge is used to clarify a very dilute suspension of solid particles in water. The particles are assumed to behave as spheres with a diameter of 5 μm and a density of 1125 $kg \cdot m^{-3}$. To be retained totally, the particles must travel a radial distance of 0.035 m. The effective length of the tubular bowl is 0.3 m. The speed of rotation is 15 000 rpm.

What is the maximum capacity of the machine for total clarification? What happens if the centrifuge is fed at a rate higher than that capacity?

Solution

Equation 9.7 will be applied. The data are:

$r_1 = 0.03$ m, $r_2 = 0.03 + 0.035 = 0.065$ m, $N = 1500$ rpm $= 250\ s^{-1}$
$d_p = 5 \times 10^{-6}$ m, $\rho_s = 1125\ kg \cdot m^{-3}$, $\rho_1 = 1000\ kg \cdot m^{-3}$, $\mu = 0.001$ Pa·s.

Equation 9.7 is re-written as follows:

$$Q_{max} = \frac{\pi(r_2^2 - r_1^2)L\, d_p^2(\rho_s - \rho_l)(2\pi N)^2}{18\mu \ln\frac{r_2}{r_1}}$$

Substituting the data we find:

$$Q_{max} = 0.00173\ m^3 \cdot s^{-1} = 6.24\ m^3 \cdot h^{-1}$$

If the feed rate is increased above Q_{max}, the supernatant will not be clear.

EXAMPLE 9.4

Given a tubular centrifuge with a theoretical equivalent area $\Sigma = 1000\ m^2$. The centrifuge is fed a suspension of particles, assumed spherical, of different sizes, at the rate of 5000 liters per hour. Particle density is 1180 $kg \cdot m^{-1}$. The liquid has a density of 1050 $kg \cdot m^{-3}$ and a viscosity of 0.01 Pa·s. What is the diameter of the largest particle in the supernatant?

Solution

Comparing Eq. (9.6) with Eq. (9.7), one can write:

$$\Sigma = Q_{max} \frac{18\mu}{d_p^2(\rho_s - \rho_l)g}$$

Substituting the data, we obtain:

$$1000 = \frac{5}{3600} \times \frac{18 \times 0.01}{d_p^2 \times (1180 - 1050) \times 9.8} \Rightarrow d_p = 14\ \mu m$$

The diameter of the largest particle in the supernatant will be 14 μm.

9.2.3 The baffled settling tank and the disc-bowl centrifuge

The separation capacity of the simple settling tank of Figure 9.1 can be improved by introducing inclined baffles into the tank (Figure 9.4). The addition of baffles can be seen either as increasing the sedimentation area or as decreasing the sedimentation distance. Baffled tanks for gravitational separation have some applications in the food industry (e.g., flotation separation of green peas according to density, hence maturity).

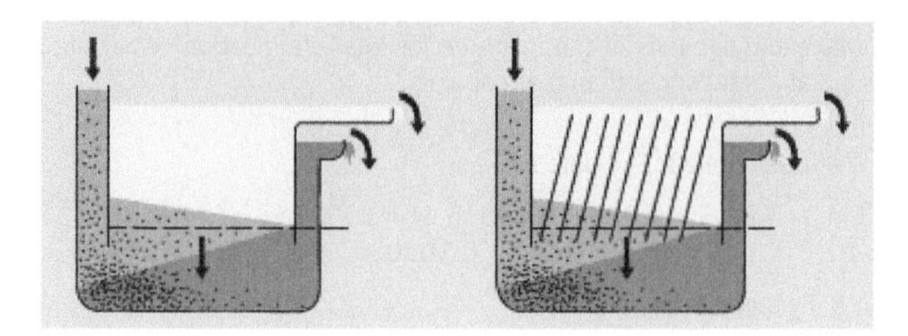

FIGURE 9.4

Simple and baffled setting tanks.

Figure courtesy of Alfa-Laval.

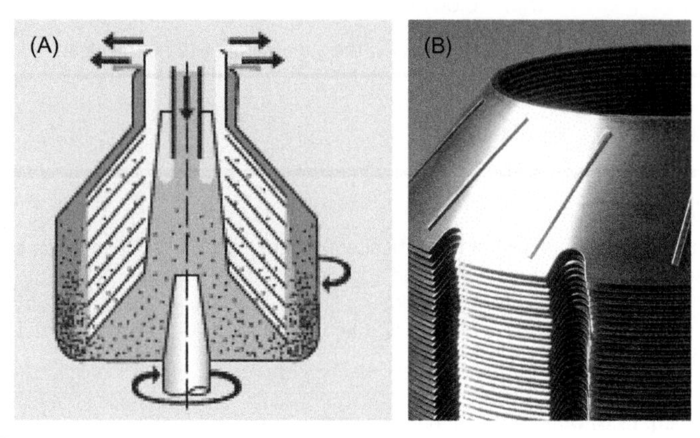

FIGURE 9.5

A. Basic structure of a disc-bowl centrifuge. B. Stack of conical dishes.

Figures courtesy of Alfa-Laval.

Now imagine the same transformation of rotation and flipping that was operated in the settling tank. The resulting configuration is that of a rotating bowl with conical dishes (discs) in it (Figure 9.5). This is, in principle, the structure of the disc-bowl centrifuge.

Applying the same reasoning as for the tubular centrifuge, an expression can be derived for the maximum flow-rate permissible, if all particles with diameter d or larger have to be separated. This expression is given in Eq. (9.9):

$$Q_{\max} = \left[\frac{d_p^2(\rho_s - \rho_l)}{18\mu}\right] \cdot \left[\frac{2\pi}{3}\omega^2 N \, tg\alpha(r_2^3 - r_1^3)\right] \tag{9.9}$$

where α = basis angle of the conical dish and N = number of dishes.

It follows that the theoretical equivalent area Σ of the disc-bowl centrifuge is:

$$\Sigma_{disc\text{-}bowl} = \left[\frac{2\pi}{3}\omega^2 N \, tg\alpha(r_2^3 - r_1^3)\right] \tag{9.10}$$

9.2.4 Liquid–liquid separation

The separation of two immiscible liquids that differ in their density is governed by the same principles as the density-driven solid–liquid separation. The continuous discharge of the two phases from the settling tank or the centrifuge, however, requires special consideration. Ideally, separation is complete when a sharp front of separation, or interface, forms between the two liquid phases. The separation front will be in stable equilibrium if the hydrostatic pressures of the two phases at the interface are equal. This requirement determines the location of the interface and hence that of the outlets for the heavy and light phases in the settling tank or the centrifuge.

Consider liquid–liquid separation in a gravity settling tank, as shown in Figure 9.6.

Separation of the phases is complete and a sharp interface forms at a distance from the inlet. The condition for hydrostatic equilibrium is given in Eq. (9.10):

$$(h_i - h_H)\rho_H = (h_i - h_L)\rho_L \tag{9.10}$$

where:

h = distance from an arbitrary line of reference
ρ = density
i, H, L = indices for interface, heavy phase, light phase respectively.

Now imagine that the line of reference becomes an axis of rotation and the entire system is flipped 90°, so that the axis of rotation becomes vertical. We now have a tubular centrifuge for liquid–liquid separation, but before we can apply the equilibrium condition to the centrifuge we need an expression for the pressure exerted by a rotating mass of fluid on its envelope, taking into consideration that this pressure is now due to centrifugal forces and not to gravity.

Consider a hollow cylinder of liquid and, in it, an annular element of thickness dr (Figure 9.7).

The incremental contribution of this element to the pressure is:

$$dP = \frac{2\pi\, r\, dr L \rho \omega^2 r}{2\pi\, r L} = r\rho\omega^2 dr \tag{9.11}$$

Integration between r_1 and r_2 gives the total pressure exerted by the liquid on the envelope at r_2:

$$P = \frac{\rho\omega^2}{2}(r_2^2 - r_1^2) \tag{9.12}$$

Applying Eq. (9.12) to Eq. (9.10) and replacing the distances h by radii r, we obtain:

$$\frac{\rho_L}{\rho_H} = \frac{r_i^2 - r_H^2}{r_i^2 - r_L^2} \tag{9.13}$$

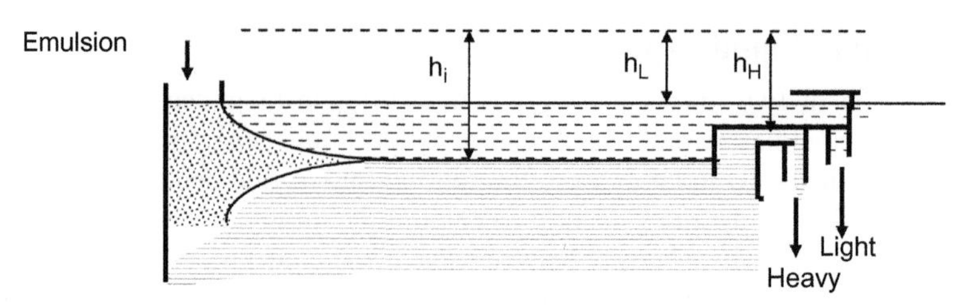

FIGURE 9.6

Liquid–liquid separation in settling tank.

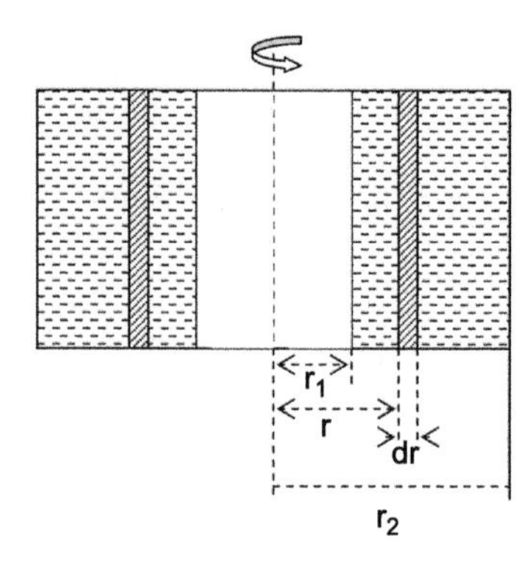

FIGURE 9.7

A rotating annular mass of liquid.

In a given centrifuge, r_H and r_L are fixed by the geometry of the machine. The location of the interface r_i is then determined by the density of the light and heavy phases. If the density of one of the components is changed, the location of the interface also changes and may fall outside the separation baffle. Some of the heavy component may then be discharged with the light phase or *vice versa.* To overcome this problem, centrifuges are often equipped with easily interchangeable *density rings* of different dimensions. The proper ring is selected according to the densities of the two phases.

EXAMPLE 9.5

A basket centrifuge with solid walls, 0.6 m in diameter and 0.4 m high, contains 50 kg of water. Calculate the pressure at the wall as the centrifuge rotates at 6000 rpm.

Solution

The pressure on the wall is given by Eq. (9.12)

$$P = \frac{\rho \omega^2}{2}(r_2^2 - r_1^2)$$

The inner radius of the annular water mass is calculated from its volume:

$$V = \frac{m}{\rho} = \pi(r_2^2 - r_1^2)h \Rightarrow r_2^2 - r_1^2 = \frac{m}{\rho \pi h} = \frac{50}{1000 \times \pi \times 0.4} = 0.0398 \text{ m}^2$$

$$P = \frac{1000 \times 4 \times \pi^2 \times (6000/60)^2}{2} \times 0.0398 = 7848\,241 \text{ Pa} = 7850 \text{ kPa}$$

This is a very high pressure indeed, justifying the strong construction required.

9.3 Centrifuges

Centrifuges may be classified by function or by structure (Leung, 1998, 2007). By function, we distinguish between centrifuges for solid–liquid separation and those for liquid–liquid separation. The former are designated "clarifiers" and the latter are called "separators". "Purifiers" are separator centrifuges used to free the light phase from the heavy phase as completely as possible. "Concentrators" are separator centrifuges used to produce a heavy phase as free as possible from traces of the light phase. Clarifiers used for treating suspensions with a high content of solids are called "desludgers" or "decanters"; those designed to remove small amounts of solids (or a minor liquid phase from a cloudy liquid) are known as "polishers". The main types of centrifuges according to structure are *tubular*, *disc-bowl* and *basket* centrifuges.

9.3.1 Tubular centrifuges

Tubular centrifuges are the simplest type of mechanical centrifuge (Figure 9.8). The tubular centrifuge consists of a vertical tubular rotor with an inlet for feed and outlets for the light and heavy phases, enclosed in a stationary housing. Tubular centrifuges serve mainly as separators. The two liquid phases are discharged continuously, while small amounts of solid impurities (sludge) are retained in the bowl and removed when the machine is stopped for cleaning. Tubular centrifuges rotate at high speed (typically 15 000 rpm), developing centrifugal accelerations in excess of 10 000 g. They can, therefore, separate liquids of slightly different densities and retain very fine solid particles. On the other hand, they have limited capacity due to their geometry (see Section 9.2).

9.3.2 Disc-bowl centrifuges

As explained in Section 9.2.3, centrifugal separation can be improved by distributing the flow to a number of parallel narrow channels between conical dishes. The disc-bowl centrifuges, also known as disc stack centrifuges, can serve both as clarifiers and as separators. The rotor (bowl) containing a stack of discs is placed on a vertical spindle and rotates inside a stationary housing. The capacity range

FIGURE 9.8

Tubular centrifuge.

Photograph courtesy of Tomoe Engineering Co. Tokyo.

of disc-bowl centrifuges extends to over 100 m^3 per hour. The types of disc-bowl clarifiers described below differ in their mode of discharging the solids.

In centrifuges with a *solid wall bowl* (Figure 9.9), the liquid phases are discharged continuously but, as in the tubular centrifuge, there is no outlet for the sludge. The accumulated solids are removed manually when the machine is stopped and the bowl is taken apart for cleaning. Such machines are therefore suitable mainly as separators (e.g., for separating cream from skim milk), but cannot handle suspensions with a significant concentration of solids.

Nozzle centrifuges (Figure 9.10) are equipped with nozzles for the continuous discharge of solids while the machine is turning. They serve as clarifiers for suspensions containing a moderate concentration of solids (up to about 10% by volume). The bowl is of a double-cone shape. The solids accumulate at the zone of maximum diameter, where the centrifugal force is the highest. Narrow nozzles located on the periphery of the bowl at that zone serve as outlets for the solids. The pressure drop through the nozzles must be sufficiently high to prevent outburst of liquid from the bowl. Therefore, the sludge must be sufficiently thick to provide controlled flow through the nozzle without completely plugging it. Interchangeable nozzles with different hole diameters serve to adjust the centrifuge according to the solids content of the feed.

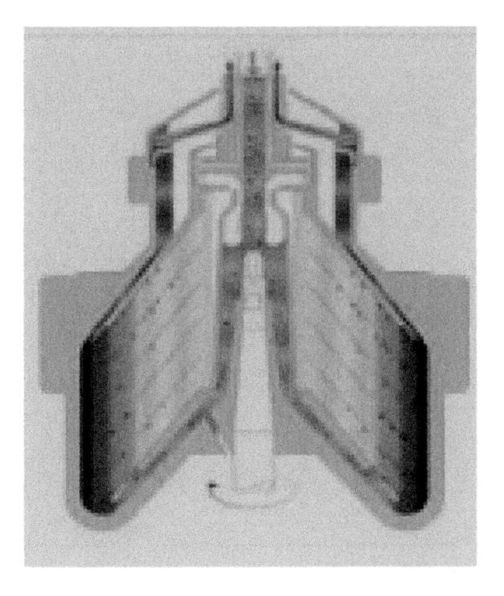

FIGURE 9.9

Solid wall bowl centrifuge.

Figure courtesy of Alfa-Laval.

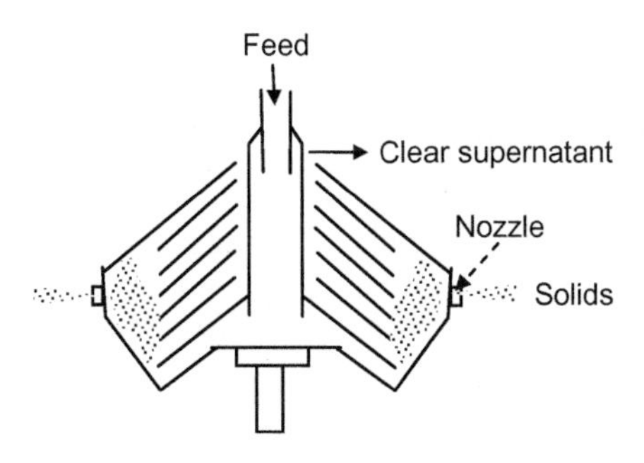

FIGURE 9.10

Schema of continuous solid discharge (nozzle) centrifuge.

Self cleaning desludger centrifuges (Figure 9.11) are used as clarifiers with suspensions containing a high proportion of solids, typically 30–40% by volume. In this type, the accumulated sludge is discharged intermittently. The bowl wall is not of one piece but consists of two separate conical parts pressed together by hydraulic force. In operation, the solids accumulate in the zone of maximum diameter, about the plane separating the two cones. When the zone of accumulation is full with solids the hydraulic system releases the bottom cone, which drops slightly to leave an opening between the two halves. The solids are rapidly

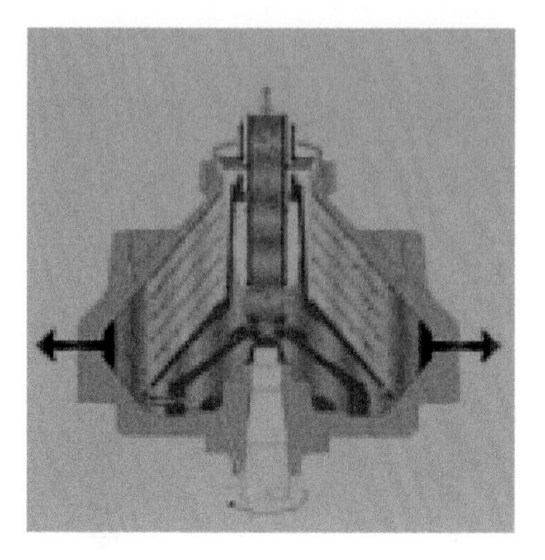

FIGURE 9.11

Intermittent solid discharge centrifuge.

Figure courtesy of Alva-Laval.

ejected through this opening, the bowl is closed again and a new cycle of accumulation begins. The entire operation, which may include a phase of rinsing of the accumulation chamber, is usually automated.

9.3.3 Decanter centrifuges

Decanter centrifuges have a variety of uses. They are primarily clarifiers, suitable for treating suspensions with very high content of solids (40–60%). Basically, decanter centrifuges consist of a solid-wall horizontal bowl with a cylindrical zone and a conical zone (Figure 9.12). The solids move to the wall and the clear liquid to the center. A screw conveyor or scroll scraps the solids off the wall and moves them mechanically towards the solids outlet. If equipped with the necessary baffles and outlets, decanters can also separate two immiscible liquids while continuously discharging a voluminous sludge. When operating in this mode, decanters are used in the production of olive oil. Decanter centrifuges can be used also for continuous solid–liquid extraction.

The flow diagram of an olive oil production process is shown in Figure 9.13.

9.3.4 Basket centrifuges

The main part of a basket centrifuge consists of a cylindrical chamber spinning rapidly about a vertical, horizontal or inclined axis. If the cylinder has a perforated wall the centrifuge acts as a filter, with the centrifugal force pushing the liquor through the perforations (see Chapter 8, Section 8.4.2). With a solid wall, the basket centrifuge is used for both liquid–liquid and liquid–solid separation.

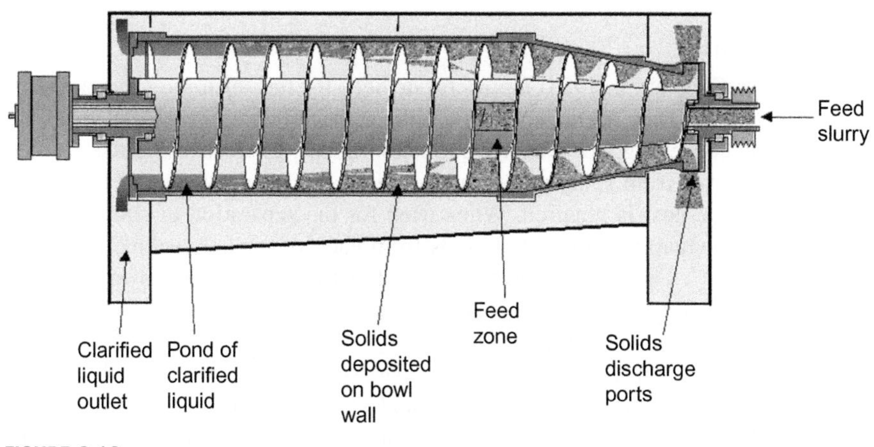

FIGURE 9.12

Decanter centrifuge.

Figure courtesy of Alfa-Laval.

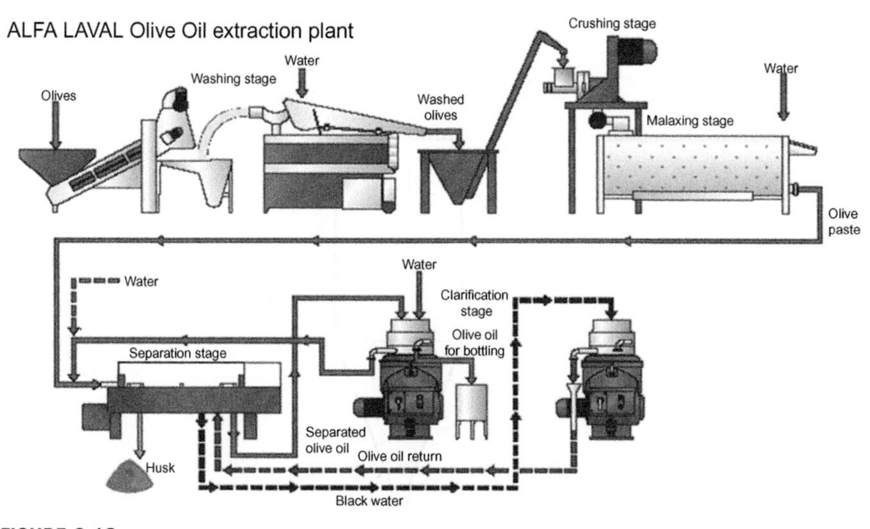

FIGURE 9.13

The Alfa-Laval system of olive oil production.

Figure courtesy of Alfa-Laval.

Basket centrifuges operate mostly in batches. After complete separation of the phases, liquids are discharged by introducing tubes to the appropriate distance from the wall, while the centrifuge is turning. The solid cake is removed much as in a filter centrifuge.

9.4 Cyclones

Cyclones (McCabe et al., 1985) are useful devices for the separation of solid particles from gases. They are extensively used for the separation of solids from air in pneumatic conveying (see Chapter 2, Section 2.5.4), for recovering the powder product at the exit from spray dryers, and in many situations where freeing an air stream from fine dust is required. When used for the separation of solids from a liquid or two mutually immiscible liquids from each other, the cyclone is named a "hydrocyclone". One of the notable applications of hydrocyclones in the food industry is their use for the concentration of starch slurries in the production of starch from corn or potato (Bradley, 1965).

Essentially, cyclones are centrifugal separators without mechanical moving parts. The field of centrifugal forces is created by the flow pattern of the suspension itself. Figure 9.14 shows the basic structure of a cyclone. The device consists of a vertical cylinder with a conical bottom. The feed inlet, usually of rectangular cross-section, connects to the cylinder tangentially. An outlet for the solids is provided at the bottom of the conical section. The outlet pipe for the solid-free fluid extends into the cylinder, below the feed entrance.

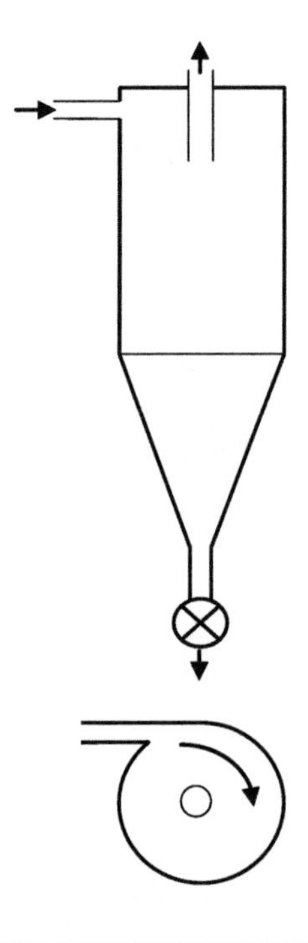

FIGURE 9.14

Basic structure of a cyclone separator.

Due to the tangential direction of the entrance, a rotational movement is imparted to the feed, thus giving rise to centrifugal forces. Under the effect of these forces the solid particles are accelerated radially in the direction of the wall and they quickly reach their settling velocity. Under the combined effect of the rotational movement and gravity, the particles spiral down to the bottom of the cone and are usually discharged through a rotary valve. The solid-free gas spirals upward and is discharged through the exit pipe.

Cyclone efficiency is defined as the mass fraction of solid particles of a given size retained in the cyclone. Obviously, cyclone efficiency is higher for larger particles (Coker, 1993).

References

Bradley, D., 1965. The Hydrocyclone. Pergamon Press, London, UK.

Coker, A.K., 1993. Understanding cyclone design. Chem. Eng. Prog. 89, 5155.

Leung, W.W.-F., 1998. Industrial Centrifugation Technology. McGraw-Hill Professional, New York, NY.

Leung, W.W.-F., 2007. Centrifugal Separations in Biotechnology. Academic Press, New York, NY.

McCabe, W., Smith, J., Harriot, P., 1985. Unit Operations in Chemical Engineering, fourth ed. McGraw-Hill International, New York, NY.

Walstra, P., Wouters, J.T.M., Geurts, T.J., 2005. Dairy Science and Technology. CRC Press, Taylor & Francis Group, Boca Raton, FL.

CHAPTER

Membrane Processes 10

10.1 Introduction

In this chapter, the term *membrane* indicates thin films of material (most commonly, but not always, a synthetic polymer) with selective permeability. Membrane-based separation processes make use of this selective permeability. A considerable number of different membrane processes have found industrial applications in the food sector (Mohr *et al.*, 1989). In this chapter, we shall deal mainly with four: *microfiltration* (MF), *ultrafiltration* (UF), *nanofiltration* (NF) and *reverse osmosis* (RO). The driving force for material transport through the membrane in those four processes is a pressure difference. These processes are therefore called *pressure-driven membrane processes*. *Electrodialysis*, in which the driving force is an electric field, has a number of interesting applications in food processing, and will be briefly discussed at the end of the chapter. *Pervaporation* (Huang, 1991), a separation technique based on vaporization through perm-selective membranes, is described in Chapter 13 (Distillation).

The importance of biological membranes as selective barriers in cells and tissues is well known. The property of selective permeability (perm-selectivity) of natural membranes was discovered by the French physicist Abbé Nollet as early as 1748. The industrial application of membranes, however, is a fairly recent field, rapidly expanding thanks to the continuing development of new man-made membranes with improved properties.

Microfiltration (MF) and ultrafiltration (UF) are genuine filtration processes where particle size is practically the sole criterion for permeation or rejection. In contrast, reverse osmosis (RO) membranes separate particles at molecular level, and their selectivity is based on the chemical nature of the particles. Nanofiltration (NF) is, in essence, a membrane process similar to reverse osmosis.

The approximate ranges of separation and typical operation pressures for the four pressure-driven membrane processes are given in Table 10.1 and Figure 10.1.

10.2 Tangential filtration

The vast majority of industrial membrane processes are tangential (cross-flow) filtration operations (see Chapter 8, Section 8.3.1). Consider a tubular membrane module used for the microfiltration of an aqueous suspension of solid particles (Figure 10.2).

Food Process Engineering and Technology. DOI: http://dx.doi.org/10.1016/B978-0-12-415923-5.00010-1
© 2013 Elsevier Inc. All rights reserved.

Table 10.1 Typical Range of Application of Pressure-Driven Membrane Separation Processes

Process	Typical operating pressure range: MPa	Limit particle size range: nm or (molecular weight)
MF	0.1– 0.3	100–10 000
UF	0.2–1.0	1–100 (10^2–10^6 Da)
NF	1–4	0.5–5.0 (10^2–10^3 Da)
RO	3–10	(10^1–10^2 Da)

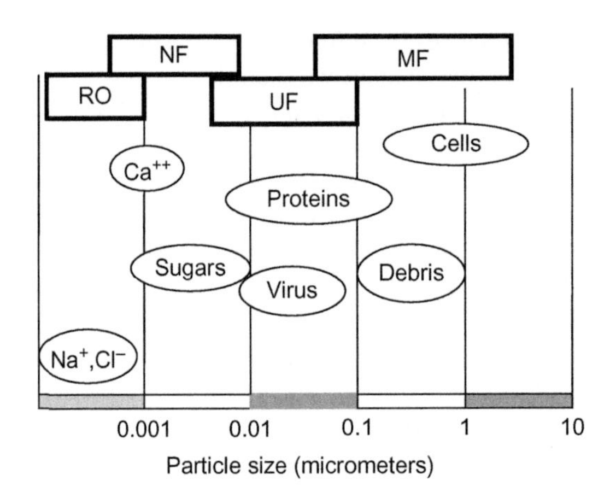

FIGURE 10.1

Separation range of pressure-driven membrane processes.

The suspension flows inside the tube. Assume that the particles are larger than the pores of the membrane and, therefore, only the continuous medium (water) passes through the membrane. The fraction that passes through the membrane is called the *permeate*. The fluid that is retained by the membrane is the *retentate*. As the suspension advances in the tube, water is gradually removed and the suspension becomes progressively more concentrated. As long as the axial flow-rate inside the tube is sufficiently high, the solid particles are carried along with the suspension and do not accumulate on the membrane surface as a cake.

Just as in ordinary filtration, the driving force for the transport of the permeate through the membrane is the pressure drop across the membrane, or the *trans-membrane pressure difference* (TMPD or ΔP_{TM}). The pressure at the permeate side is practically uniform. The pressure at the side of the retentate decreases in

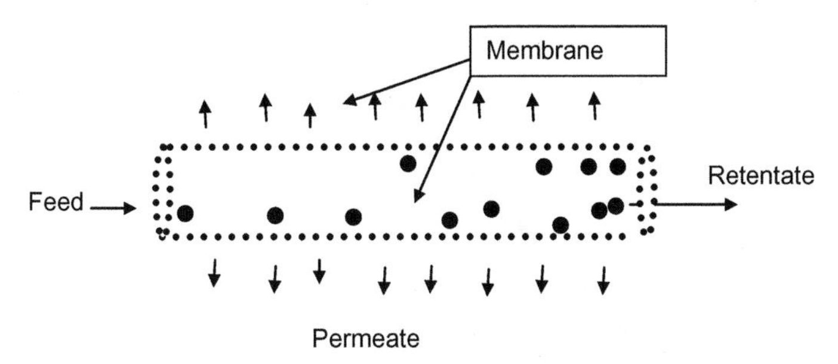

FIGURE 10.2

Separation in a tubular membrane.

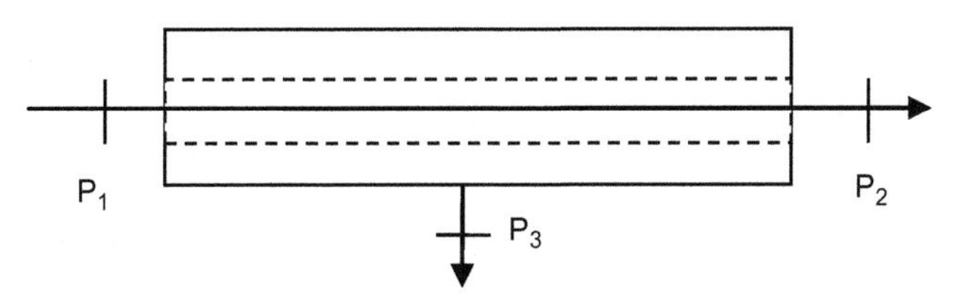

FIGURE 10.3

Definition of the transmembrane pressure drop (TMP).

the direction of the flow (Figure 10.3). The transmembrane pressure difference is defined as:

$$TMPD = \frac{P_1 + P_2}{2} - P_3 \tag{10.1}$$

where P_1, P_2 = retentate side pressure at the module inlet and outlet, respectively, and P_3 = permeate side pressure, assumed uniform.

Note: The notions developed above are valid for all the types of membrane processes. Microfiltration was used only as an example.

10.3 Mass transfer through MF and UF membranes

10.3.1 Solvent transport

As explained above, microfiltration and ultrafiltration are genuine filtration processes. The membranes used in these processes are porous. Therefore, the transport of permeate through the membrane follows the basic principles of flow

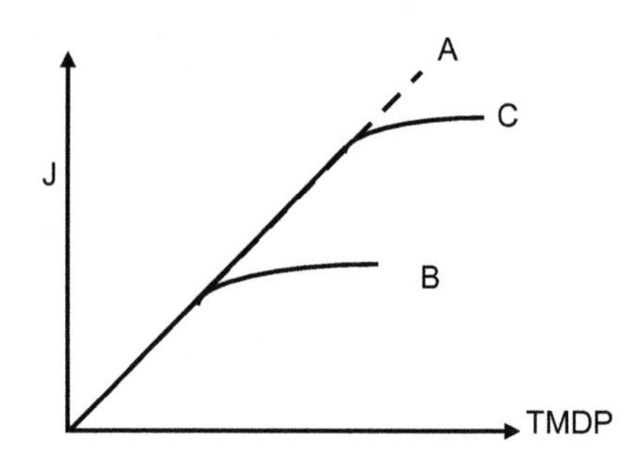

FIGURE 10.4

Effect of TMDP on flux.

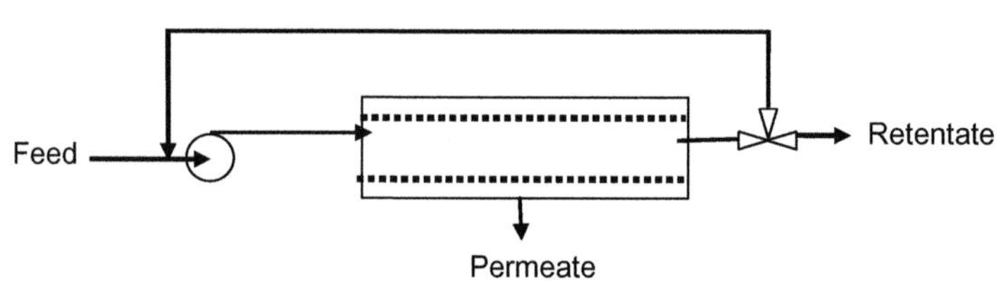

FIGURE 10.5

Retentate recycling to maintain axial flow-rate.

through porous media (see Section 8.3.2). For membrane filtration, Darcy's Law is written as follows:

$$J = L_p \, \Delta P_{TM} \qquad (10.2)$$

where J = permeate flux, $m \cdot s^{-1}$, and L_p = hydraulic permeability, $m\ s^{-1}\ Pa^{-1}$.

The hydraulic permeability is an important characteristic of the membrane, because it affects strongly the filtration capacity of the system.

The effect of transmembrane pressure difference on permeate flux is shown in Figure 10.4.

The straight line A represents the theoretical behavior according to Eq. (10.2). The curve B depicts the typical behavior observed in reality. The decline in flux may be explained in the light of a number of factors, such as build-up of a layer of solids on the membrane surface (fouling), concentration polarization (to be explained later) and membrane compaction. Increasing the tangential flow-rate results in increased flux (curve C). One of the ways to increase the tangential flow-rate is to recycle part of the retentate (Figure 10.5).

What are the physical properties that determine the hydraulic permeability of membranes? If the membrane is considered as a medium perforated by straight, parallel capillaries of radius r, then the hydraulic permeability, based on Poiseuille's Law, is given by:

$$L_p = \frac{\varepsilon\, r^2}{8\mu\, z} \qquad (10.3)$$

where:

ε = membrane porosity, dimensionless
z = membrane thickness, m.
μ = viscosity of the permeate, Pa s.

Application of the Poiseuille model to membranes has the same shortcomings as its use for the prediction of the specific resistance of filter cakes (see Section 8.3.4). Nevertheless, with certain corrections, the model is useful for the approximate prediction of membrane permeability. Both the porosity (number of openings per unit area) and the average radius of the pores can be determined by microscopy.

EXAMPLE 10.1

A membrane for MF was examined microscopically and found to have about 120 000 pores with an average diameter of 0.8 μm, per mm. square of membrane surface. It is desired to estimate the hydraulic permeability of the membrane to water, using the Hagen-Poiseuille capillary model. The thickness of the membrane is 160 μm.

Solution

Equation 10.3 is applied.

$$L_p = \frac{\varepsilon\, r^2}{8\mu\, z}$$

The porosity is equal to the proportion of the surface area occupied by holes:

$$\varepsilon = 120\,000 \times \frac{\pi(0.8\times10^{-3})^2}{4} = 0.06$$

$$L_p = \frac{0.06\times(0.4\times10^{-6})^2}{8\times0.001\times160\times10^{-6}} = 7.5\times10^{-9}\ \text{m}\cdot\text{Pa}\cdot\text{s}^{-1}$$

10.3.2 Solute transport; sieving coefficient and rejection

MF or UF membranes with a certain pore size will let through or retain particles according to their size. MF membranes have relatively large pore diameters (e.g., 0.1–1 μm). They retain suspended solid particles such as microorganisms and broken cells, but allow protein molecules to permeate. UF membranes have pores smaller by one or two orders of magnitude, and they retain solute particles such as protein molecules or even peptides but are permeable to sugars. However, the size limit for rejection or permeation is not exact. Particles with diameters close to the pore size (say one-half even one-quarter of pore diameter) are retarded, due to the effect of pore walls. Furthermore, the pore size distribution of membranes is seldom homogeneous.

The *sieving coefficient*, S, of a membrane with respect to a given solute is defined as follows:

$$S = \frac{C_{perm}}{C_{retn}} \tag{10.4}$$

where C_{perm} and C_{retn} are the concentration of the solute in the permeate and in the retentate (measured at the membrane interface) respectively.

For particles considerably larger than the widest pore, rejection is total, i.e., $S = 0$. Particles considerably smaller than the smallest pore are not retained at all ($S = 1$).

For solutes with particle size close to the pore size: $0 < S < 1$.

Rejection R is defined as follows:

$$R\,\% = (1 - S)^{*}100 \tag{10.5}$$

MF membranes are specified by their average pore diameter (e.g., 0.5 μm), while UF membranes are characterized by their *cutout molecular weight* (COMW). COMW is the molecular weight of the smallest molecule retained by the membrane. For the purpose of this definition, $R = 95\%$ is usually accepted as total rejection. It will be said, for example, that the cut-out molecular weight of a certain membrane is 100 000 daltons (Da).

10.3.3 Concentration polarization and gel polarization

Consider ultrafiltration of a solution of protein (Figure 10.6).

Assume total rejection of the protein by the membrane and unrestricted permeation of the solvent. At the membrane surface, the protein is separated from the solvent. A protein concentration gradient, normal to the membrane surface, is created. Protein concentration near the membrane is higher than in the bulk of the solution, a distance away from the membrane. This situation is called "concentration polarization".

Accumulation of a concentrated layer of protein near the membrane surface reduces the rate of filtration, for two reasons:

1. The high concentration at the upstream face of the membrane causes osmotic back-pressure, resulting in back-flow of permeate (solvent) towards the retentate. This effect is not particularly significant in UF and MF, where the retentates are suspensions of solid particles or solutions of substances with relatively high molecular weight, hence low osmotic pressure.
2. The concentrated and hence highly viscous layer at the membrane interface constitutes an additional resistance to flow towards the membrane.

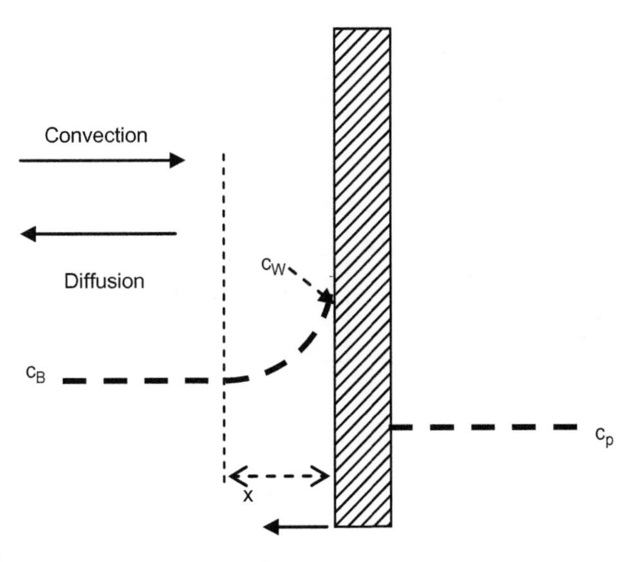

FIGURE 10.6

Concentration polarization.

The transport of solute at the upstream side of the membrane is composed of two components in countercurrent to each other:

1. Solute transport from the fluid bulk towards the membrane, by virtue of the flow of solvent under the effect of the TMPD.
2. Solute transport from the membrane interface towards the bulk under the effect of the concentration gradient (back-diffusion).

At steady state, the local concentration does not change with time, hence the two effects must be in equilibrium. Assuming Fick's Law for the back-diffusion, the steady-state condition can be written as follows:

$$J \cdot C = -D\frac{dC}{dx} \quad \Rightarrow \quad \frac{J}{D}\int_0^{\delta} dx = \int_{C_W}^{C_B} \frac{dC}{C} \tag{10.6}$$

where:

C = concentration of the protein (C_W at the membrane interface, C_B in the fluid bulk), kg · protein per m^{-3} of solvent.

J = solvent flux, $m^{-1}\,s^{-1}$

D = diffusivity of the protein in the solvent, $m^2 \cdot s^{-1}$

x = distance from the membrane

δ = thickness of the boundary layer for diffusion.

Integration of Eq. (16.6) gives:

$$J = \frac{D}{\delta}\ln\frac{C_W}{C_B} = k_L \ln\frac{C_W}{C_B} \tag{10.7}$$

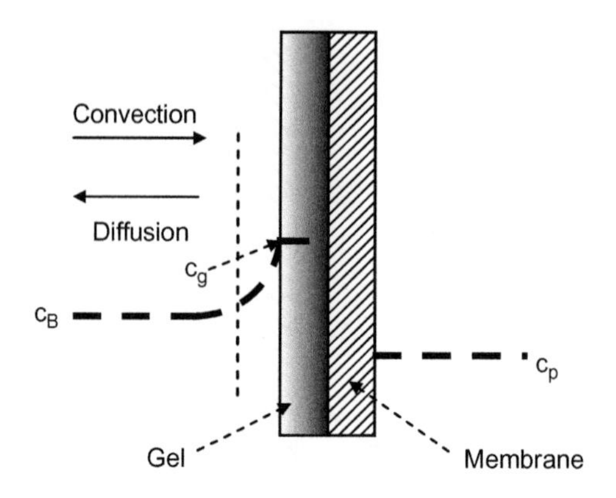

FIGURE 10.7

Gel polarization.

where k_L stands for the convective mass transfer coefficient in liquid phase.

Equation (10.7) establishes the relationship between the solvent flux J and the concentration of the protein near the membrane C_W. If the flux is increased (e.g., by raising the transmembrane pressure difference), C_W increases accordingly, resulting in increased resistance to solvent flow towards the membrane. This explains, at least partially, the deviation of membrane filtration rate from linearity.

The concentration of the protein in the liquid layer adjacent to the membrane cannot grow above a certain limit C_G, at which the layer becomes a gel. From this point on, the flux J cannot grow further and remains constant, independently of an increase in pressure. This phenomenon is called *gel polarization* (Figure 10.7). The maximum and constant value of the flux in a situation of gel polarization is given in Eq. (10.8):

$$J_{\max} = k_L \ln \frac{C_G}{C_B} \tag{10.8}$$

The gelation concentration C_G depends on the protein and on the operation conditions (ionic strength, temperature etc.)

The shape of curve describing the variation of the flux as a function of TMPD can be explained in the light of the three models described above, namely, Darcy's Law, concentration polarization and gel polarization (Figure 10.8):

1. At low flux (Segment I), Darcy (linear) behavior
2. At intermediate flux (Segment II), deviation from linearity due to gradual build-up of resistance as a result of concentration polarization
3. At high flux (Segment III), saturation (gelation, gel polarization), practically constant (TMPD-independent) flux.

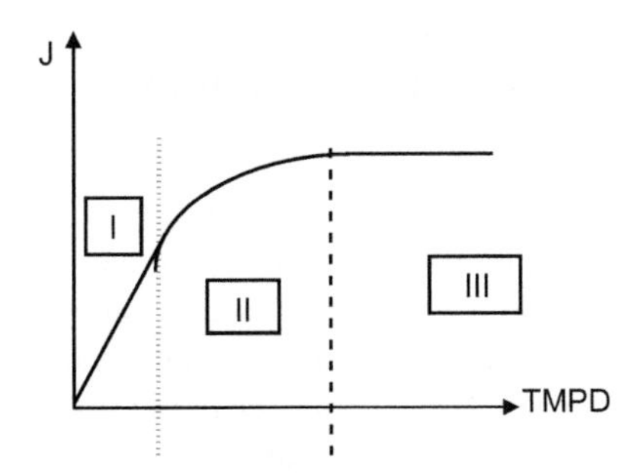

FIGURE 10.8

Phases in the change of flux with TMPD: (I) linear (Darcy); (II) concentration polarization; (III) gel polarization.

The effect of the mass transfer coefficient k_L on the flux is considerable. For a given set of material properties, k_L depends of the tangential velocity, hence on the overall flow-rate through the module (Figure 10.4, curve C). The effect of flow conditions (tangential velocity, turbulence), system geometry (shape and dimensions of the flow channel) and material properties (viscosity, density) on the coefficient k_L can be evaluated with the help of the correlations mentioned in Chapter 3, Section 3.4.

EXAMPLE 10.2

A UF membrane module is used to separate proteins from low molecular weight solutes. The mean bulk concentration of protein in the retentate is 3% w/v.

a. What is the maximal permissible permeate flux, if a protein concentration above 12% w/v is to be avoided at the membrane surface? It is assumed that the convective mass transfer coefficient at the membrane–fluid interface, under the conditions of operation, is $6 \times 10^{-6}\ \text{m} \cdot \text{s}^{-1}$; 100% retention of the protein is also assumed.

b. What is the maximal permeate flux attainable, if the gelation concentration of the protein is 22% w/v.

Solution

a. What is the permeate flux J that will cause the concentration of the protein at the wall to be 12%? Eq. (10.8) is applied:

$$J = k_L \ln \frac{C_W}{C_B} = 6 \times 10^{-6} \ln \frac{12}{3} = 8.31 \times 10^{-6} \text{m} \cdot \text{s}^{-1}$$

b. The flux attains its maximum value when the concentration at the wall reaches the gelation concentration.

$$J_{max} = k_L \ln \frac{C_G}{C_B} = 6 \times 10^{-6} \ln \frac{22}{3} = 11.95 \times 10^{-6}\ \text{m} \cdot \text{s}^{-1}$$

10.4 Mass transfer in reverse osmosis

10.4.1 Basic concepts

The term "osmosis" signifies the spontaneous transfer of water from a more dilute into a more concentrated solution through a membrane. In order to stop osmotic transfer of water into a solution, a certain pressure, called osmotic pressure, must be exerted against the direction of the transfer. Application of a pressure stronger than the osmotic pressure causes water transfer in the opposite direction, from the concentrated solution to the less concentrated medium (Figure 10.9). This is the principle of reverse osmosis (Berk, 2003).

As stated in the Introduction to this chapter, reverse osmosis is not a true filtration process because the separation of components is not based solely on the size of their particles. RO membranes are essentially homogeneous, non-porous, gel-like materials. Therefore, theories of flow through porous media (e.g., Darcy's Law) are not satisfactorily applicable to reverse osmosis. Different models have been proposed for selective mass transfer through RO membranes (Jonsson and Macedonio, 2010; Soltanieh and Gill, 1981). One of these models (Lonsdale *et al.*, 1965) assumes that both the solute and the solvent dissolve in the upstream face of the membrane, then cross through the membrane by molecular diffusion and are released into the permeate bulk in contact with the downstream face. This model, known as the *homogeneous solution–diffusion model*, explains the separation of the components as a consequence of differences in the solubility and diffusivity of the chemical species. A different model, called the *preferential sorption–capillary flow theory* (Sourirajan, 1970), assumes that the surface of the membrane acts as a microporous medium that preferentially adsorbs water and rejects the solute. The adsorbed water then penetrates the pores and is transported by capillary flow towards the downstream face of the membrane. Another frequently used model is the Kedem–Katchalsky formalism, based on the linear thermodynamics of irreversible processes (Kedem and Katchalsky, 1958) and its modifications.

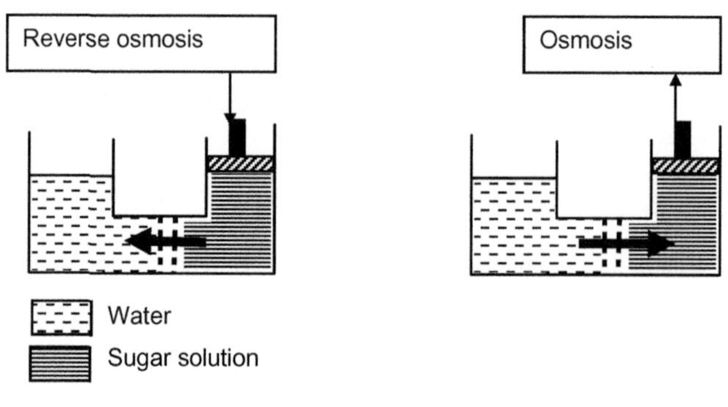

FIGURE 10.9

Osmosis and reverse osmosis.

10.4.2 Solvent transport in reverse osmosis

Like ultrafiltration and microfiltration, reverse osmosis is a pressure-driven process where the driving force for transport is the transmembrane pressure difference. In reverse osmosis, however, the pressure difference must overcome the difference in osmotic pressure $\Delta\pi$ between the retentate and the permeate, in addition to the resistance of the membrane to the transfer. The net available driving force is the "net applied pressure" (NAP), defined as follows:

$$NAP = TMPD - \Delta\pi \qquad (10.8)$$

Solvent flux is given by the following expression (Baker, 2004):

$$J_W = K_W(TMPD - \Delta\pi) \qquad (10.9)$$

The osmotic pressure π (Pa) of ideal solutions is calculated with the help of van't Hoff's equation (Jacobus Henricus van't Hoff, 1852–1911, Dutch chemist, 1901 Nobel Prize in Chemistry):

$$\pi = \varphi\, C_M RT \qquad (10.10)$$

where:

C_M = molar concentration of the solution, $\text{kmol} \cdot \text{m}^{-3}$
R = gas constant = 8314 $\text{Pa} \cdot (\text{kmol} \cdot \text{m}^{-3})^{-1} \cdot \text{K}^{-1}$
T = absolute temperature, K
ϕ = a dimensionless constant, depending on the dissociation of the solute. For non-ionic solutes such as neutral sugars, $\phi = 1$.

Assuming total rejection of the solute, the concentration ratio achieved by reverse osmosis is then written as follows:

$$\frac{C_{retn}}{C_{feed}} = \frac{Q_{feed}}{Q_{feed} - Q_W} = \frac{Q_{feed}}{Q_{feed} - AJ_W} \qquad (10.11)$$

where:

C_{feed}, C_{retn} = concentration of the solute in the feed and in the retentate, respectively
Q_{feed}, Q_W = volumetric flow-rate of the feed and of the the permeate (water), respectively.
A = area of the membrane.

The coefficient K_W is similar to the hydraulic permeability, defined in Eq. (10.2).

The relationship between solvent flux and TMPD is shown in Figure 10.10. The straight line refers to theoretical behavior according to Eq. (10.9). The curve represents typical behavior in practice. Just as in the case of UF and MF, the deviation from linearity in RO is due to fouling, concentration polarization, gel polarization and membrane compaction. The increase in osmotic pressure (i.e., decrease in NAP) as the retentate becomes more concentrated also contributes to the reduction

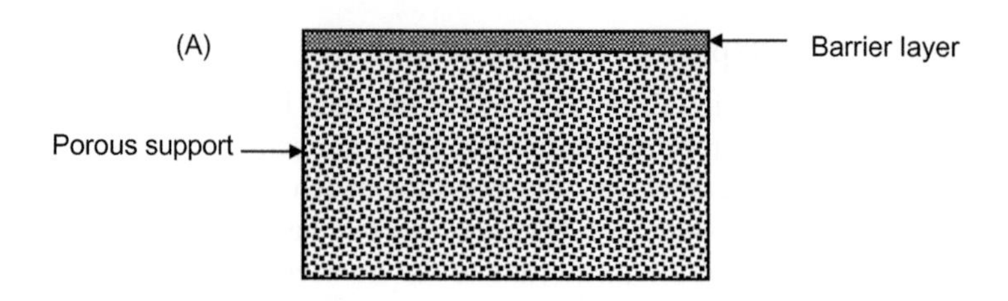

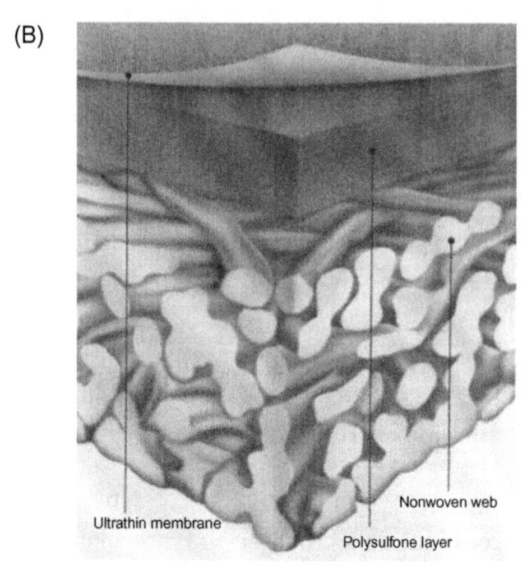

FT30 Membrane Composite

FIGURE 10.10

A. Structure of asymmetric membrane. B. Asymmetric membrane with three layers.

Figure courtesy of Filmtec – Dow Chemical Company.

of the flux below the theoretical value. As in UF and MF, fouling, concentration polarization and osmotic back-pressure can be counteracted by increasing the tangential flow-rate. This measure, however, results in lower concentration ratios, as predicted by Eq. (10.11), and creates the need for extensive retentate recycling.

The strong negative effect of high osmotic pressure on transmembrane flux puts an upper limit on the concentration ratio that can be achieved without unacceptable reduction in plant capacity.

EXAMPLE 10.3

The hydraulic permeability of a RO membrane is $8\ \text{l}\cdot\text{m}^{-2}\cdot\text{h}^{-1}\ \text{atm}^{-1}$ at 20°C.

a. Convert the hydraulic permeability to SI units.

b. Calculate the permeate flux at 20°C, if a TMPD of 5000 kPa is applied and the retentate is a 1.5% w/v sucrose solution, in the following cases:

(i) Rejection for sucrose is 100%

(ii) Rejection for sucrose is 90%.

Solution

a.

$$\frac{1\ \text{liter}}{m^2\cdot h\cdot atm}=\frac{0.001\ m^3}{m^2\times 3600s\times 101\,325\ Pa}=2.78\times10^{-12}\frac{m}{Pa\cdot s}$$

The hydraulic permeability of the membrane is:

$$k_p=8\times2.78\times10^{-12}=22.24\times10^{-12}\ m\cdot Pa\cdot s^{-1}=22.24\times10^{-9}\ m\cdot kPa\cdot s^{-1}$$

b. Permeate flux is given by:

$$J_W=K_W(TMPD-\Delta\pi)=L_p(TMPD-\Delta\pi)$$

We need the osmotic pressure difference $\Delta\pi$.

(i) If the rejection of sucrose is 100%, the osmotic pressure difference is the osmotic pressure of the 1.5% sucrose solution. The molecular weight of sucrose is 342. By virtue of Eq. (10.10);

$$\Delta\pi=\frac{CRT}{M}=\frac{15\times8.314\times293}{342}=106.8\ kPa$$

$$J_W=22.24\times10^{-9}(5000-106.8)=108824\times10^{-9}=0.1088\times10^{-3}\ m\cdot s^{-1}$$

(ii) If the rejection is 90% and the concentration of the retentate is 1.5%, then the concentration of the permeate will be calculated from:

$$R=1-\frac{C_{perm}}{C_{retent}}=1-\frac{C_{perm}}{1.5}=0.9\ \Rightarrow\ C_{perm}=0.15\%=1.5\ kg\cdot m^{-3}$$

$$\Delta\pi=\frac{(15-1.5)\times8.314\times293}{342}=92.16\ kPa$$

$$J_W=22.24\times10^{-9}(5000-92.16)=109\,150\times10^{-9}=0.1091\times10^{-3}m\cdot s^{-1}$$

EXAMPLE 10.4

What is the maximum theoretical concentration of orange juice, attainable by reverse osmosis concentration at a TMPD of 5000 kPa? Assume that the juice behaves osmotically as a solution of glucose (MW = 180) and that solute rejection is total.

Solution

The maximum retentate concentration is that corresponding to an osmotic pressure equal to the TMPD. The concentration of orange juice corresponding to an osmotic pressure of 5000 kPa is calculated from Eq. (10.10):

$$\pi=\frac{CRT}{M}=\frac{C\times8.314\times293}{180}=5000\ \Rightarrow\ C=369.4\ kg\cdot m^{-3}$$

The density of sugar solutions at that concentration is about 1175 $kg\cdot m^{-3}$. Accordingly, the concentration found above can be converted to w/w percentage (Bx):

$$C=369.4\ kg\cdot m^{-3}=\frac{369.4}{1175}=0.3144\ kg\cdot kg=31.44\ \%w/w=31.44\ Bx$$

10.5 Membrane systems

10.5.1 Membrane materials

The majority of commercial membranes are made from a wide variety of organic polymers: cellulose and its derivatives (mainly cellulose acetate), polyolefines, polysulfones, polyamides, chlorine and fluorine substituted hydrocarbons, etc. Inorganic (ceramic) membranes based on oxides of zirconium, titanium, silicium and aluminum have been developed. They are produced by precipitation of the oxides from salts onto a macroporous ceramic support, followed by sintering of the colloidal solid particles at high temperature to form a microporous inorganic film.

The suitability of membranes for use in the process industry in general, and in food processing in particular, depends on a number of characteristics. The principal requisites are:

1. High permeate flux at moderate TMPD
2. Good retention capability according to the specific application
3. Good mechanical strength
4. Chemical stability and inertness
5. Bio-inertness (if used with enzymes and other biologically active materials)
6. Thermal stability
7. Resistance to cleaning and sanitizing agents
8. Resistance against microbial action
9. Smooth, fouling resistant surface
10. Compliance with all other food safety requirements
11. Availability in module types compatible with the specific application
12. Long service life
13. Affordable cost.

Some of these characteristics may be mutually contradictory. For example, high retention often leads to low permeate flux; high mechanical strength may be achieved by increasing membrane thickness, at the expense of flux, etc. This problem is adequately addressed by anisotropic (asymmetric) membranes, also known as thin film membranes (TFMs), first developed for water desalination by RO. These membranes (Figure 10.10) are composed of two or more layers, each of different composition or structure, with each layer contributing a different characteristic to the assembly. A typical asymmetric membrane consists of a very thin film of dense material (skin) on top of a thick, macroporous support. The thin dense film, typically a fraction of a micrometer in thickness, contributes the required retention, while the thick, open support provides the adequate mechanical strength without adding much resistance to permeate flux. Additional mechanical strength and ease of handling are usually provided by a backing made of fabric or mesh material or a non-woven highly porous medium. Multi-layer membranes, consisting of microporous, mesoporous and macroporous layers, one on top of another, are also available. In some membranes, both the skin and the underlying porous layer are made of the same polymer material (e.g., cellulose acetate). In others, more appropriately termed "composite membranes", the different layers consist of different polymers (e.g., a dense thin film of polyamide on a more open layer made of polysulfone, the whole being cast on a highly porous backing).

The membrane material should be chemically inert to the material being processed. Adsorption by the membrane is of concern because it results in an undesirable change in feed composition (e.g., adsorption of pigments, phenolic compounds, etc., from wine) and shortens the service life of the membrane.

Lack of resistance to the common detergents and disinfectants used in the food industry (e.g., strong alkali, oxidizing agents) has long been a limiting factor to the application of membranes in food processing, but novel membranes with improved chemical resistance are now available.

Tolerance to high temperature is important, mainly for the following reasons:

- The filtration rate is strongly enhanced by a high temperature. Therefore, operating at a relatively high temperature results in increased plant capacity.
- Microbiological activity is efficiently controlled by operating at a high temperature. This is particularly important in dairy applications.
- In-place cleaning is more efficient at a high temperature.

Ultrasonic irradiation has been suggested as an option for cleaning membranes. The use of this option requires membranes that are also resistant to ultrasound (Kallioinen and Mänttäri, 2011).

Cost is, of course, an important factor, but it must be considered in all its aspects. Thus, for example, ceramic membranes are considerably more expensive than polymers, but their extended service life partly compensates for the higher initial capital cost.

10.5.2 Membrane configurations

"Membrane configuration" refers to the geometry of the membrane and its position in space in relation to the flow of the feed fluid and of the permeate. As most industrial membrane installations are of a modular design, membrane configuration also determines the manner in which the membrane is packed inside the modules.

Four main types of membrane configurations are used in the food industry. These are: plate-and-frame, spiral-wound, tubular, and hollow-fiber configurations. The membrane geometry is planar in the former two and cylindrical in the latter two. The desirable characteristics of a membrane configuration are:

- Compactness, i.e., the ability to pack as much membrane surface as possible into a module of limited volume
- Low resistance to tangential flow (less friction, less energy expenditure, less pressure drop along the retentate flow channel)
- No "dead" regions, uniform velocity distribution

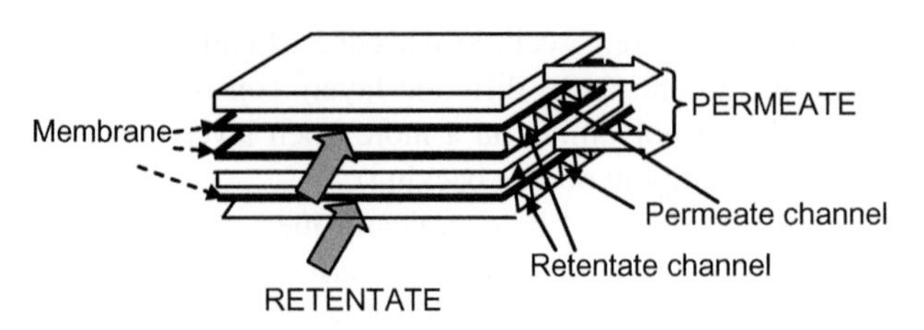

FIGURE 10.11

Plate-and-frame (stack) configuration.

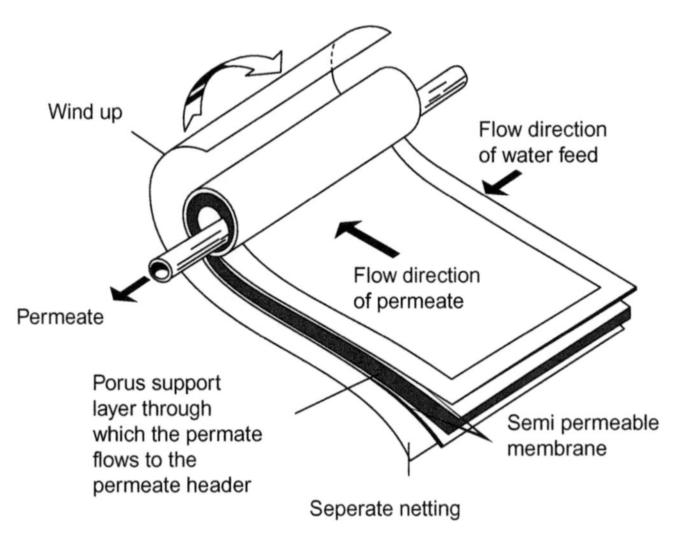

FIGURE 10.12

Spiral-wound configuration.

- A high degree of turbulence at the retentate side, in order to minimize fouling and promote mass transfer
- Easy cleaning and maintenance
- Low cost per unit membrane area.

The *plate-and-frame* configuration (Figure 10.11) resembles the plate-and-frame filter press (Section 8.4.2), with the filter media replaced by membranes. The membranes may be square or circular, arranged in vertical or horizontal stacks. Plate-and-frame modules cannot withstand very high pressure, and are therefore limited to MF and UF duty. The surface-area to volume ratio of plate-and-frame modules is not high.

In the *spiral-wound* configuration (Figure 10.12), two large sheets of membrane are heat-sealed on three sides, forming a bag. A flexible spacer mesh or a porous support layer is inserted into the bag, creating a free space between the

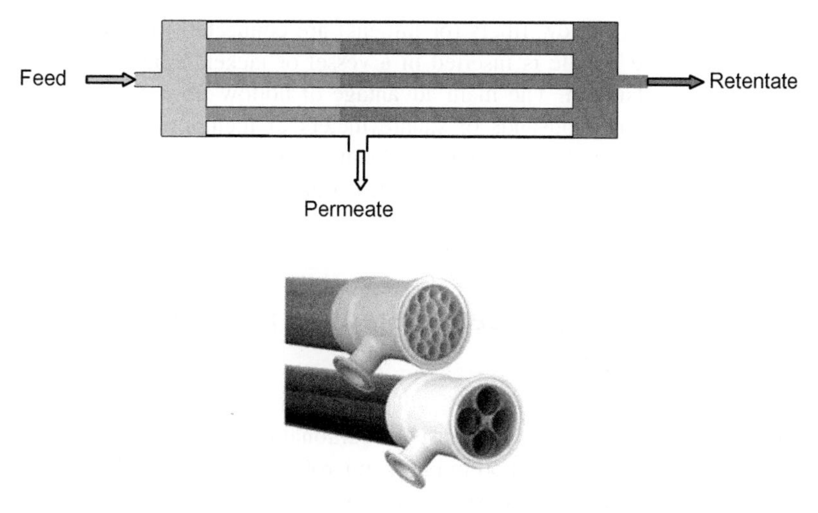

FIGURE 10.13

Tubular membrane module.

two membrane layers for permeate flow. The sandwich assembly thus formed is wound spirally, forming a cylindrical module. The open side of the bag is connected to a central perforated tube serving as a collector for the permeate. The rolled-up membrane sheets are separated by a mesh spacer, thus providing a flow channel for the retentate. Spirally-wound membranes are sold as cylindrical assemblies or cartridges, complete with central tube, spacers and connections. Their surface-area to volume ratio is high.

The *tubular* membrane configuration resembles a shell-and-tube heat exchanger. The membrane is cast on the inner wall of rigid porous tubes, made of polymer or ceramic. The tubes are connected to end-plates and installed as parallel bundles inside a shell (Figure 10.13). The tubes may have diameters in the range of 10–25 mm. Flow direction is usually inside-out, i.e., the retentate flows inside the tubes and the permeate is collected at the shell-side. It is often possible to reverse the flow (outside-in) for cleaning and unclogging of the membrane. Tubular configurations provide the possibility of maintaining high tangential velocity in the feed stream, and are therefore particularly suitable for applications where the feed contains a high proportion of suspended solids or must be strongly concentrated. Because of their relatively large diameter, tubular membranes are easy to clean and inspect. The surface-area to volume ratio of tubular modules, however, is not high.

Hollow fiber configurations are, in principle, similar to the tubular set-up. The tubes however, are much thinner, with diameters from 1 mm down to capillary size – hence the name "hollow fibers". The small diameter imparts sufficient mechanical strength to the tubes so an external rigid support is not necessary. A very large number of hollow fibers (or lumens) are connected to perforated end-plates and the entire bundle is inserted in a vessel or jacket. Flow direction may be inside-out or outside-in. The main advantage of hollow-fiber modules is their compactness, attaining thousands of square meters of membrane area per cubic meter of module bulk volume. Their disadvantage is their high susceptibility to fouling and clogging, limiting their use to clear fluids of relatively low viscosity.

10.6 Membrane processes in the food industry

10.6.1 Microfiltration

As a solid–liquid separation process, microfiltration is increasingly becoming a preferred alternative to the more expensive traditional processes of dead-end filtration and centrifugation. In the food industry, microfiltration is extensively used for the clarification of cloudy fluids. Membranes with nominal pore sizes in the order of 0.1–0.5 μm or less produce permeates that are practically free of microorganism cells. Consequently, microfiltration is being increasingly used for the purification of drinking water and of water for the production of soft beverages. Food fluids clarified by MF include clear fruit juices, wine, vinegar and sugar syrups.

Microfiltration frequently serves for the pre-treatment of fluids, before ultrafiltration or reverse osmosis. Removal of suspended particles and colloidal material by MF is essential for reducing the rate of fouling in subsequent UF or RO steps. Oil droplets and particles of fat (e.g., in whey) are also removed by MF. Membranes made of hydrophobic polymers, such as polyvinylidene fluoride, are particularly suitable for that application.

Brines used in the manufacture of cheese and in fish processing contain suspended solids and fats, which must be removed before recycling or disposal of the fluid. Other wastewater streams generated in food processing often require pre-filtration before further treatment. MF is used with advantage in such applications.

10.6.2 Ultrafiltration

10.6.2.1 Dairy applications

With the development of improved membranes, ultrafiltration has become a major separation process in the dairy industry. The two largest areas of application are the pre-concentration of milk for cheese manufacture, and the production of protein concentrates from whey. Both applications are related to cheese-making.

Whole cow's milk contains typically 3.5% protein, 4% fat, 5% lactose and 0.7% inorganic salts (ash). The balance (about 87%) is water. In the traditional cheese-making process, milk, usually after adjustment of the fat content and pasteurization, is coagulated through a combination of lactic acid fermentation and enzymatic reactions. After cutting and temperature adjustment, the coagulated mass separates into solid particles (curd) suspended in liquid (whey). The curd

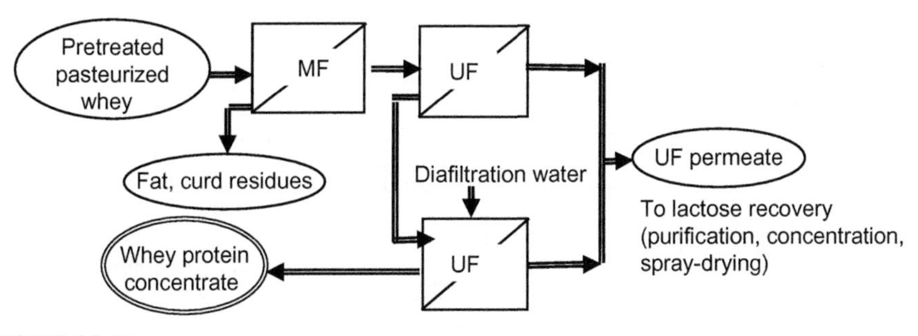

FIGURE 10.14

Simplified flow diagram of whey protein ultrafiltration.

contains the major milk protein casein, and most of the fat. The whey is essentially a dilute aqueous solution of lactose, mineral salts and non-casein milk proteins (e.g., lactalbumins and lactoglobulins). The curd is subjected to various methods of treatment and curing, resulting in the vast variety of different cheeses. The whey is usually a problematic waste, due to its high volume (often 90% of the mass of milk used) and high cost of disposal due to its extremely high BOD.

Ultrafiltration of milk retains the fat globules and the proteins. The inorganic salts and the lactose, along with some of the water, are partially removed as permeate. The resulting retentate is a partially concentrated milk with reduced lactose and mineral content. The concentration ratio is often three- to five-fold. If further reduction of lactose and mineral content is desired, the retentate is diluted with water and ultrafiltered once again. This kind of ultrafiltration is sometimes called *diafiltration*. The pre-concentrated milk is mainly used in the manufacture of many types of cheeses (most commonly soft cheeses). The advantages of using UF pre-concentrated milk for making cheese are as follows (Cheryan, 1986):

- Increased yield, probably because of the inclusion of some of the non-casein proteins in the curd
- Lower energy consuption
- Reduced volume of whey.

The use of UF in cheese-making was developed by Maubois, Macquot and Vassal in 1969. The method is therefore known as the MMV Process (Zeman and Zydney, 1996). Plants for the UF pre-concentration of milk operate at a TMPD range of 200–500 kPa.

The second major application of UF in the dairy industry is the fractionation of whey (Figure 10.14). As stated earlier, whey contains the major part of the non-casein proteins, together with low-molecular weight solutes (lactose, mineral salts). Ultrafiltration of whey produces a valuable whey protein concentrate as the retentate, and a protein-free permeate containing mainly lactose and minerals. The retentate is usually concentrated further by evaporation, and spray-dried.

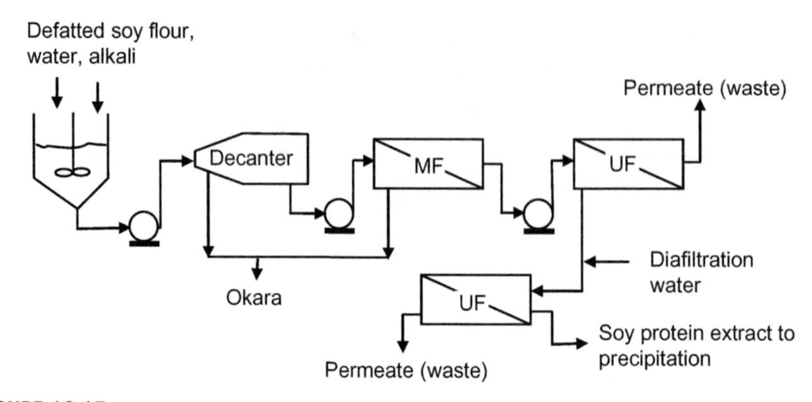

FIGURE 10.15

Simplified flow diagram of soy protein purification by diafiltration.

Whey proteins find extensive use in the manufacture of cheese and a considerable array of food products and health food specialities.

10.6.2.2 Other applications

The use of UF for the concentration and fractionation of plant protein extracts has been suggested (Hojilla-Evangelista *et al.*, 2004). Ultrafiltration is potentially useful in the process of isolated soybean protein production (Krishna Kumar *et al.*, 2003; Lawhon *et al.*, 1975, 1979; Omosaiye and Cheryan, 1979). In this process, defatted soy flour is extracted with water at high pH. The aqueous extract is then concentrated by UF. The sugars and other low molecular weight solutes may be partially removed by repeated steps of dilution and ultrafiltration (diafiltration), producing a purified and concentrated soy protein extract for further processing. (Figure 10.15).

Ultrafiltration is used commercially as an alternative to evaporative concentration in the production of gelatin.

10.6.3 Nanofiltration and reverse osmosis

Water desalination, the first industrial application of RO, is still the major use of this process. Some water-related RO and NF processes, such as wastewater treatment and water softening, find some use in the food industry. The main use of RO and NF in the food industry is, however, for concentrating liquid foods by membrane removal of water (Berk, 2003; Köseoğlu and Guzman, 1998).

The advantages of RO concentration over evaporative concentration are:

1. Tremendous saving in energy expenditure
2. Elimination of the risk of thermal damage to the product
3. Better retention of volatile aromas
4. Lower capital investment.

On the other hand, the general problems connected with the use of membranes in food processing (safety, sanitation, cleaning) are also relevant to RO and NF.

As stated earlier in this chapter, concentration by RO is limited to fluids of low osmotic pressure and relatively low viscosity. Therefore, RO is mostly used as a process of pre-concentration, before final concentration by evaporation. Not surprisingly, pre-concentration of maple sap was one of the first specific applications of RO in food processing (Willits *et al.*, 1967). Maple sap (or maple water) is an extremely diluted solution, containing on average 2.5% soluble solids (2.5 degrees Brix). Its osmotic pressure is therefore very low. To make maple syrup, the sap must be concentrated to 66–68 Bx by removing about 96 kg of water from every 100 kg of sap. Typically, 75–80% of this quantity of water can be removed by RO, without increasing the osmotic pressure and viscosity to unacceptable levels. Thus, to bring this intermediate product to the standard concentration of commercial maple syrup, only 20–25% of the water has to be removed by subsequent evaporation. The savings in energy and capital are, obviously, considerable. Maple sap has a relatively high tolerance to high temperature. Therefore RO is carried out at 80°C and above, resulting in higher capacity and improved microbiological safety. The concentration of birch sap by reverse osmosis has also been reported (Kallio *et al.*, 1985).

Other cases of RO concentration of very dilute solutions concern aromas (Braddock *et al.*, 1991; Matsuura *et al.*, 1975; Nuss *et al.*, 1997) and aqueous extracts of herbs and teas (Schreier and Mick, 1984; Zhang and Matsuura, 1991). In these cases, the non-thermal nature of RO is a distinct advantage.

Concentration of fruit and vegetable juices by RO has been investigated extensively (Morgan *et al.*, 1965; Matsuura *et al.*, 1974; Anon., 1989; Gostoli *et al.*, 1995, 1996; Köseoğlu *et al.*, 1991). The products thus treated include apple (Moresi, 1988; Sheu and Wiley, 1983), citrus fruits (Medina and Garcia, 1988; Anon.,1989; Jesus, 2007), grape (Gurak, 2010), tomato (Merlo *et al.* 1986a, 1986b) and pineapple (Bowden and Isaacs, 1989). Pre-concentration of tomato juice is particularly interesting because of its relatively low initial concentration (4–5 Bx). RO pre-concentration of fruit juices is usually carried to an upper concentration limit of 20–30 Bx. Jesus *et al.* (2007) concentrated orange juice to 36 Bx by RO, at a TMPD of 6 MPa, but the concentrate was devoid of the characteristic aroma of fresh juice. Garcia-Castello *et al.* (2011) tried to concentrate, by RO, a model solution simulating the waste fluid obtained by pressing orange peels, but encountered serious flux decline due to fouling.

Most types of fruit juices are processed at moderate temperature to prevent browning and thermal damage to the flavor. Tomato juice tolerates higher temperatures well, and is usually RO-treated at temperatures above 60°C. Juices containing suspended particles are first clarified by MF, and only the clear filtrate (serum) is concentrated by RO. If desired, the removed pulp is added back to the RO concentrate, for cloudiness and aroma.

As with ultrafiltration, reverse osmosis finds its most widespread application in the dairy industry. Concentration by RO is applied to whole milk, skim milk

(Fenton-May *et al.*, 1972; Schmidt, 1987) and whey (Nielsen *et al.*, 1972). RO-concentrated whole milk is used in the manufacture of cheese (Barbano and Bynum, 1984), yogurt (Guirguis *et al.*, 1987; Jepsen, 1979) and ice cream (Bundgaard, 1974). Pre-concentration of milk and other dairy fluids prior to conventional evaporative concentration and drying results in important savings in energy cost.

Concentration of liquid egg by RO has been reported (Conrad *et al.*, 1993).

The sector of beer, wine and other alcoholic beverages features interesting applications of reverse osmosis. Because of its low molecular weight and hydrophilic nature, ethanol is only partially rejected by RO membranes. Thus, when an alcoholic beverage such as wine or beer is subjected to RO, some of the alcohol is transported to the permeate while practically all the other solutes, including flavor and aroma compounds, are retained. The alcohol and the water in the permeate may be separated and recovered by distillation. By diluting the retentate with the recovered water, a beverage with full aroma but reduced alcohol content is obtained (Bui *et al.*, 1986; Nielsen, 1982), while admixture of the recovered alcohol to the original beverage results in a product with increased alcohol content (Bui *et al.*, 1988). Since both the water and the alcohol were recovered from the original beverage, their addition to the product is not considered an adulteration. Another application of RO in the wine industry is the concentration of grape must (Duitschaever *et al.*, 1991).

RO and NF are being increasingly used for the pre-concentration of thin juices in the production of sugar from beet and cane (Madsen, 1973; Tragardh and Gekas, 1988).

As explained in Chapter 16, Section 16.5, the condensates of evaporators are not pure water but contain varying amounts of organic matter. Treatment of these condensates by RO or NF helps recover important quantities of high-quality water and reduces the cost of waste disposal considerably (Guengerich, 1996; Morris, 1986). This operation is sometimes called "condensate polishing".

An interesting application of reverse osmosis deals with the removal of organic solvents from edible oil miscella. (Miscella is the name given to the solution resulting from the extraction of oil by organic solvents.) The advantages of solvent removal by membrane over the traditional process of desolventizing by evaporation are safety, and energy economy (Köseoglu and Engelgau, 1990; Köseoglu *et al.*, 1990). The membranes used are nanofiltration membranes resistant to oil and organic solvents (Darvishmanesh *et al.*, 2011; Kwiatkowski and Cheryan, 2005; Pagliero *et al.*, 2011). Despite the claimed potential advantages, the use of membranes for desolventizing solvent–oil mixtures has not been applied in industry, to date.

Nanofiltration is, in essence, similar to RO. The key difference between the two processes is in the rejection of monovalent anions and cations. While the rejection of monovalent ions by RO membranes is practically absolute, such ions are only partially rejected by NF membranes, the extent of penetration depending on the type of membrane, the concentration of the feed, and the TMPD. Due to this difference, NF is sometimes called "leaky RO". Polyvalent ions are

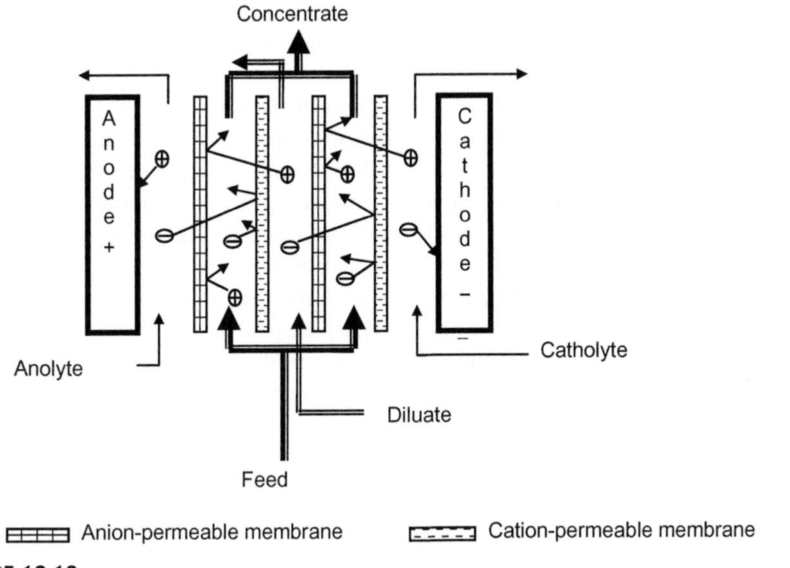

FIGURE 10.16

Principle of operation of electrodialysis.

preferentially rejected by NF. For non-charged molecules, the cut-off molecular weight is in the order of 100–300 Da. Being less dense than the RO membranes, the same permeate flux can be achieved by NF at lower TMPD. Furthermore, since monovalent ions pass through the membrane, the osmotic pressure difference $\Delta\pi$ is lower. This is particularly important in fluids containing mineral salts. For the same reason, NF is advantageously used for the demineralization and deacidification of UF permeate of whey. Another important application of NF is water-softening, where the preferential retention of the bivalent ions of calcium and magnesium produces a hardness-free permeate.

10.7 Electrodialysis

Electrodialysis (ED) is an electrochemical membrane process by which ions are transported through perm-selective membranes under the effect of an electrical field. (Baker, 2004; Strathmann, 1995). ED membranes are sheets of ion exchange materials. (Ion exchange and ion exchangers are discussed in Chapter 12.)

Ions are separated by ED membranes according to their charge. ED is most commonly used for the transfer of ions into or from pumpable fluids. One of the first uses of ED was the desalination of brackish water. The principal applications of ED in food processing are the operations of desalting (Shi *et al.*, 2010), demineralization, deacidification or acidification of liquid foods.

Figure 10.16 shows schematically the mode of operation of a multi-membrane electrodialysis system. The assembly consists of alternating anionic and cationic membranes arranged in a stack, similar to a plate-and-frame configuration. Anionic (anion exchanger) membranes are permeable to anions and reject cations. Cationic (cation exchanger) membranes do the opposite.

More discriminating selectivity among ions of the same charge depends on the pore size of the membrane. The pore size of commercial ED membranes varies between 10 and 100 Å (Bazinet *et al.*, 1998). Spacers placed between the membranes (not shown in the Figure) serve to provide mechanical support to the membranes and to promote turbulence. Electrodes of opposite charge are placed in the end compartments of the stack and connected to a source of direct current voltage. An electrolyte solution (the electrode stream) is fed to the end compartments. The product to be demineralized and a carrier electrolyte solution are allowed to flow through alternate compartments, as shown in the Figure. The purpose of the ED system shown is to deplete the product stream (diluate) from its ions by transporting them to the carrier stream (concentrate). Under the effect of the electric field, negatively charged anions tend to migrate towards the positive electrode (anode) but are stopped by cationic membranes. Cations tend to move in the opposite direction and are stopped by anionic membranes. The product does not touch the electrodes. Electrolysis (oxidation or reduction) occurs in the electrode stream and not in the product or carrier.

One of the first applications of ED in food processing was the partial deacidification of fruit juices that are too sour to be acceptable. Although the acidity can be reduced by precipitation of the acid (commonly citric or malic) as a calcium salt, or by adsorption on an anion exchanger in OH form (see Chapter 12), DE deacidification has the great advantage of avoiding addition of chemicals. Figure 10.16 shows the principle of removal of citric acid in a DE cell. Deacidification of fruit juices and grape must has been reported (Guérif, 1993; Vera *et al.*, 2007, 2009).

The high mineral content of whey and whey products is sometimes undesirable, particularly when the material is intended for infant feeding. Electrodialysis provides a technologically sound method for demineralizing whey products (Perez *et al.*, 1994) and desalting products such as soy sauce (Fidaleo *et al.*, 2012).

References

Anon, 1989. RO membrane system maintains fruit juice taste and quality. Food Eng. Intl. 14 (3), 54.

Baker, R.W., 2004. Ion exchange membrane processes – electrodialysis. In: Baker, R.W. (Ed.), Membrane Technology. John Wiley, New York, NY.

Barbano, D.M., Bynum, D.G., 1984. Whole milk reverse osmosis retentates for cheddar cheese manufacture: cheese composition and yield. J. Dairy Sci. 67, 2839–2849.

Bazinet, L., Lamarche, F., Ippersiel, D., 1998. Bipolar membrane electrodialysis: applications of electrodialysis in the food industry. Trends Food Sci. Technol. 9 (3), 107–113.

Berk, Z., 2003. Reverse osmosis. In: Heldman, D.R. (Ed.), Encyclopedia of Agriculture, Food and Biological Engineering. Taylor & Francis, London, UK.

Bowden, R.P., Isaacs, A.R., 1989. Concentration of pineapple juice by reverse osmosis. Food Aust. 41 (7), 850–851.

Braddock, R.J., Sadler, G.D., Chen, C.S., 1991. Reverse osmosis concentration of aqueous-phase citrus juice essence. J. Food Sci. 56 (4), 1027–1029.

Bui, K., Dick, R., Moulin, G., Gadzy, P., 1986. Reverse osmosis for the production of low ethanol content wine. Am. J. Enol. Viticult. 37 (4), 297–300.

Bundgaard, A.G., 1974. Hyperfiltration of skim milk for ice cream manufacture. Dairy Industr. 39 (4), 119–122.

Cheryan, M., 1986. Ultrafiltration Handbook. Technomic Publishing Co., Lancaster, PA.

Conrad, K.M., Mast, M.G., Ball, H.R., Froning, G., MacNeil, J.H., 1993. Concentration of liquid egg white by vacuum evaporation and reverse osmosis. J. Food Sci. 58 (5), 1017–1020.

Darvishmanesh, S., Robberecht, S., Luis, P., Degrève, J., 2011. Performance of nanofiltration membranes for solvent purification in the oil industry. *J. Am. Oil Chem. Soc.* 88 (8), 1255–1261.

Duitschaever, C.L., Alba, J., Buteau, C., Allen, B., 1991. Riesling wines made from must concentrated by reverse osmosis. Amer. J. Enol. Viticult. 42 (1), 19–25.

Fenton-May, R.I., Hill Jr., C.G., Amundson, C.H., Lopez, M.H., Auchair, P.D., 1972. Concentration and fractionation of skim-milk by reverse osmosis and ultrafiltration. J. Dairy Sci. 55 (11), 1561–1566.

Fidaleo, M., Moresi, M., Cammaroto, A., Ladrange, N., Nerdi, R., 2012. Soy sauce desalting by electrodialysis. J. Food Eng. 110 (2), 175–181.

Garcia-Castello, E.M., Mayor, L., Chorques, S., Argüelles, A., Vidal-Brotóns, D., Gras, M. L., 2011. Reverse osmosis concentration of press liquid from orange juice solid wastes: flux decline mechanisms. J. Food Eng. 106 (3), 199–205.

Gostoli, C., Bandini, S., di Francesca, R., Zardi, G., 1995. Concentrating fruit juices by reverse osmosis – low retention–high retention method. Fruit Process. 5 (6), 183–187.

Gostoli, C., Bandini, S., di Francesca, R., Zardi, G., 1996. Analysis of a reverse osmosis process for concentrating solutions of high osmotic pressure: the low retention method. Food Bioprod. Process. 74 (C2), 101–109.

Guengerich, C., 1996. Evaporator condensate processing saves money and water. Bull. Int. Dairy Fed. 311, 15–16.

Guerif, G., 1993. Electrodialysis applied to tartaric stabilization of wines. Rev. Oenol. Tech. Viticoles Oenol. 69, 39–42.

Guirguis, N., Versteeg, K., Hickey, M.W., 1987. The manufacture of yoghurt using reverse osmosis concentrated skim milk. Aust. J. Dairy Technol. 42 (1–2), 7–10.

Gurak, P.D., Cabral, L.M.C., Rocha-Leaõ, M.H., Matta, V.M., Freitas, S.P., 2010. Quality evaluation of grape juice concentrated by reverse osmosis. J. Food Eng. 96 (2), 421–426.

Hojilla-Evangelista, M.P., Sessa, D.J., Mohamed, A., 2004. Functional properties of soybean and lupin protein concentrates produced by ultrafiltration – diafiltration. J. Amer. Oil Chem. Soc. 81 (12), 1153–1157.

Huang, R.Y.M., 1991. Pervaporation Membrane Separation Processes. Elsevier, Amsterdam, The Netherlands.

Jepsen, E., 1979. Membrane filtration in the manufacture of cultured milk products – yoghurt, cottage cheese. Cult. Dairy Prod. J. 14 (1), 5–8.

Jesus, D.F., Leite, M.F., Silva, L.F.M., Modesta, R.D., Matta, V.M., Cabral, L.M.C., 2007. Orange (*Citrus sinensis*) juice concentration by reverse osmosis. J. Food Sci. 81 (2), 287–291.

Jonsson, G., Macedonio, F., 2010. Fundamentals in reverse osmosis. Comp. Membr. Sci. Eng. 2, 1–22.

Kallio, H., Karppinen, T., Holmbom, B., 1985. Concentration of birch sap by reverse osmosis. J. Food Sci. 50 (5), 1330–1332.

Kallioinen, M., Mänttäri, M., 2011. Influence of ultrasonic treatment on various membrane materials: a review. Separ. Sci. Technol. 46 (9), 1388–1395.

Kedem, O., Katchalsky, A., 1958. Thermodynamic analysis of the permeability of biological membranes to non-electrolytes. Biochim. Biophys. Acta 27, 229–246.

Köseoğlu, S.S., Guzman, G.J., 1998. Application of reverse osmosis technology in the food industry. In: Amjad, Z. (Ed.), Reverse Osmosis. Chapman and Hall, New York, NY.

Köseoğlu, S.S., Engelgau, D.E., 1990. Membrane applications and research in edible oil industry: assessment. J. Am. Oil Chem. Soc. 67 (4), 239–245.

Köseoğlu, S.S., Lawhon, J.H., Lusas, E.W., 1990. Membrane processing of crude vegetable oils. II. Pilot scale solvent removal from oil miscellas. J. Am. Oil Chem. Soc. 67 (5), 281–287.

Köseoğlu, S.S., Lawhon, J.H., Lusas, E.W., 1991. Vegetable juices produced with membrane technology. Food Technol. 45 (1), 124, 126–128.

Krishna Kumar, N.S., Yea, M.K., Cheryan, M., 2003. Soy protein concentrates by ultrafiltration. J. Food Sci. 68 (7), 2278–2283.

Kwiatkowski, J.R., Cheryan, M., 2005. Recovery of corn oil from ethanol extracts of ground corn using membrane technology. J. Am. Oil Chem. Soc. 82 (3), 221–227.

Lawhon, J.T., Lin, S.H.C., Cater, C.M., Mattil, K.F., 1975. Fractionation and recovery of cottonseed whey constituents by ultrafiltration and reverse osmosis. Cereal Chem. 52 (1), 34–43.

Lawhon, J.T., Manak, L.J., Lusas, E.W., 1979. Using industrial membrane systems to isolate oilseed protein without an effluent waste stream. Abstract of Papers. Am. Chem. Soc. 178 (1), 133 (Coll).

Lonsdale, H.K., Merten, U., Riley, R.L., 1965. Transport properties of cellulose acetate osmotic membrane. J. Appl. Polym. Sci. 9, 1341.

Madsen, R.F., 1973. Application of ultrafiltration and reverse osmosis to cane juice. Intl. Sugar J. 75, 163–167.

Matsuura, T., Baxter, A.G., Sourirajan, S., 1974. Studies on reverse osmosis for concentration of fruit juices. J. Food Sci. 39 (4), 704–711.

Matsuura, T., Baxter, A.G., Sourirajan, S., 1975. Reverse osmosis recovery of flavor components from apple juice waters. J. Food Sci. 40 (5), 1039–1046.

Medina, B.G., Garcia, A., 1988. Concentration of orange juice by reverse osmosis. J. Food Process Eng. 10, 217–230.

Merlo, C.A., Rose, W.W., Pederson, L.D., White, E.M., 1986a. Hyperfiltration of tomato juice during long term high temperature testing. J. Food Sci. 51 (2), 395–398.

Merlo, C.A., Rose, W.W., Pederson, L.D., White, E.M., Nicholson, J.A., 1986b. Hyperfiltration of tomato juice: pilot plant scale high temperature testing. J. Food Sci. 51 (2), 403–407.

Mohr, C., Engelgau, D., Leeper, S., Charboneau, B., 1989. Membrane Applications and Research in Food Processing. Noyes Data Corp., Park Ridge, NJ.

Moresi, M., 1988. Apple juice concentration by reverse osmosis and falling film evaporation. In: Bruin, S. (Ed.), Preconcentration and Drying of Food Materials. Elsevier Science Publishers, Amsterdam, The Netherlands.

Morgan Jr., A.I., Lowe, E., Merson, R.L., Durkee, E.L., 1965. Reverse osmosis. Food Technol. 19, 1790.

Morris, C.W., 1986. Plant of the year: golden cheese company of california. Food Eng. 58 (3), 79–90.

Nielsen, C.E., 1982. Low alcohol beer by hyperfiltration route. Brewing Distilling Intl. 12 (8), 39–41.

Nielsen, I.K., Bundgaard, A.G., Olsen, O.J., Madsen, R.F., 1972. Reverse osmosis for milk and whey. Process Biochem. 7 (9), 17–20.

Nuss, J., Guyer, D.E., Gage, D.E., 1997. Concentration of onion juice volatiles by reverse osmosis and its effects on supercritical CO_2 extraction. J. Food Proc. Eng. 20 (2), 125–139.

Omasaiye, O., Cheryan, M., 1979. Low-phytate, full-fat soy protein product by ultrafiltration aqueous extracts of whole soybeans. Cereal Chem. 56, 58–62.

Pagliero, C., Ochoa, N.A., Martino, P., Marchese, J., 2011. Separation of sunflower oil from hexane by use of composite membranes. J. Am. Oil Chem. Soc. 88 (11), 1813–1819.

Perez, L.J., Andres, L.J., Alvarez, R., Coca, J., Hill, C.G., 1994. Electrodialysis of whey permeates and retentates obtained by ultrafiltration. J. Food Proc. Eng. 17, 177–190.

Schmidt, D., 1987. Milk concentration by reverse osmosis. Food Technol. Aust. 39 (1), 24–26.

Schreier, P., Mick, W., 1984. Ueber das Aroma von schwartzem Tee: herstellung eines Teekoncentrates mittels Umgekehrosmose und dessen analytische Characterisierung. Zeit. Fuer Lebensmittel-Untersuch. Forsch 179, 113–118.

Sheu, M.J., Wiley, R.C., 1983. Preconcentration of apple juice by reverse osmosis. J. Food Sci. 48, 422–429.

Shi, S., Lee, Y.H., Yun, S.H., Hung, P.V.X., Moon, S.H., 2010. Comparisons of fish meat extract desalination by electrodialysis using different configurations of membrane stack. J. Food Eng. 101 (4), 417–423.

Soltanieh, M., Gill, W., 1981. Review of reverse osmosis membranes and transport models. Chem. Eng. Commun. 12 (1–3), 279–363.

Sourirajan, S., 1970. Reverse Osmosis. Logos Press Ltd., London, UK.

Strathmann, H., 1995. Electrodialysis and Related Processes. In: Noble, R.D., Stern, S.A. (Eds.), Membrane Separation Technology. Elsevier, Amsterdam, The Netherlands.

Tragardh, G., Gekas, V., 1988. Membrane technology in the sugar industry. Desalination 69 (1), 9–17.

Vera, E., Sandeaux, J., Persin, F., Pourcelly, G., Dornier, M., Ruales, J., 2009. Deacidification of passion fruit juices by electrodialysis with bipolar membrane after different pretreatments. J. Food Eng. 90, 68–73.

Vera, E., Sandeaux, J., Persin, F., Pourcelly, G., Dornier, M., Piombo, G., et al., 2007. Deacidification of tropical fruit juices by electrodialysis. Part II. J. Food Eng. 78, 1439–1445.

Willits, C.O., Underwood, J.C., Merten, U., 1967. Concentration by reverse osmosis of maple sap. Food Technol. 21 (1), 24–26.

Zeman, L.J., Zydney, A.L., 1996. Microfiltration and Ultrafiltration. Marcel Dekker Inc., New York, NY.

Zhang, S.Q., Matsuura, T., 1991. Reverse osmosis concentration of green tea. J. Food Process Eng 14, 85–105.

CHAPTER

Extraction 11

11.1 Introduction

Literally, the term "extraction" conveys the idea of pulling something out of something else (*ex*, out; *traction*, the action of pulling). It is used to indicate a wide variety of actions, from the surgical removal of a tooth to the retrieval of an item from a database. A device for pressing oranges is known as a "juice extractor". In this chapter, however, extraction will be defined as a separation process, based on differences in solubility. A solvent is used to solubilize and separate a solute from other materials with lower solubility in the said solvent.

It is customary to distinguish between two classes of extraction processes:

1. *Solid–liquid extraction.* This is where a solute is extracted from a solid phase with the help of a solvent. Examples include extraction of salt from rock using water as the solvent, extraction of coffee solubles from roasted and ground coffee in the production of "soluble coffee", extraction of edible oils from oilseeds with organic solvents, extraction of protein from soybeans in the production of isolated soybean protein, etc. Solid–liquid extraction is also termed "leaching" or "elution" (when applied to the removal of adsorbed solute from an adsorbent; see Chapter 12). The mechanism of solid–liquid extraction involves wetting the solid surface with solvent, penetration of the solvent into the solid, dissolution of the extractables, transport of the solutes from the interior of the solid particles to their surface, and dispersion of the solutes within the bulk of the solvent surrounding the solid particles by diffusion and agitation. In some cases, the solubilization step may include chemical changes promoted by the solvent, such as hydrolysis of insoluble biopolymers to produce soluble molecules. The process of extraction with a super-critical fluid (SCF) will be included in this category, it being mainly (but not exclusively) applied to solids.
2. *Liquid–liquid extraction.* This is a method for extracting a solute from a solution in a certain solvent, by using another solvent. Examples include extraction of penicillin from aqueous fermentation broth by butanol, extraction of oxygenated terpenoids from citrus essential oils using ethanol as a solvent, etc. Liquid–liquid extraction, also known as partitioning, is common in the chemical and pharmaceutical industries and in biotechnology, but much less so in food processing.

Food Process Engineering and Technology. DOI: http://dx.doi.org/10.1016/B978-0-12-415923-5.00011-3
© 2013 Elsevier Inc. All rights reserved.

Like adsorption (Chapter 12), distillation (Chapter 13) and crystallization (Chapter 14), extraction is a separation process based on molecular transport, in which molecules pass from one phase to another under the effect of a difference in chemical potential. The analysis and design of separation operations based on molecular transport require knowledge in three main areas:

1. *Equilibrium*. Net molecular transport stops when equilibrium is reached between the phases – i.e., when the chemical potential of the substance in question is the same in all phases in contact with each other. Although true equilibrium can never be reached within a finite duration, the concept of equilibrium is basic to the calculations. In practice, many of such separation processes consist of a number of consecutive stages through which the phases move, following various flow patterns. Countercurrent movement is the most frequent. The common approach is to simulate the real process as a series of theoretical *equilibrium stages* and to correct the deviation from theoretical behavior with the help of empirical or semi-empirical *efficiency factors*. Equilibrium data may be given in the form of equations, tables or charts. Each stage of the extraction process comprises two operations. First, the two phases are brought into contact for a certain length of time, during which mass transfer between the phases occurs. Subsequently, the two phases are separated from each other by a number of possible methods, such as filtration, decantation, centrifugation, squeezing, draining, etc. Physically, the two operations may take place in the same piece of equipment (e.g., an extraction column) or may require two separate devices. The conditions for optimizing the two operations may be mutually contradictory. Thus, division of one of the phases into fine particles improves the rate of inter-phase mass transfer (see Section 11.2.7) but makes subsequent separation more difficult.
2. *Material and energy balance*. Each stage of the process must satisfy the laws of conservation of matter and energy, expressed as: *In* = *Out* + *Accumulation*. In steady-state continuous processes there is no accumulation. In the graphical representation of the processes, the equations resulting from material balance are represented as *lines of operation*.
3. *Kinetics*. The rate of inter-phase molecular transport depends on diffusion coefficients and turbulence. Transport kinetics determines the rate at which equilibrium is approached. Kinetic effects are often accounted for with the help of the efficiency factors mentioned above.

11.2 Solid–liquid extraction (leaching)

Solid–liquid extraction is a separation process based on the preferential dissolution of one or more of the components of a solid mixture in a liquid solvent. In this context, the term "solid mixture" is used in its practical meaning. In reality, the physical state of the component to be extracted from the raw material is not always solid. In the case of solvent extraction of oil from oilseeds, for example, the oil is already in the form of liquid droplets in the raw material. In the case of extraction of sugar from beet, the sugar is in solution in the cell juices before contacting the solvent water.

Although the mechanisms involved and the mode of operation are the same in both cases, it is customary to distinguish between "extraction" and "washing". In the case of extraction, the useful product is the extractable solute. In the case of washing, the purpose of the process is to remove undesired solutes from the desired insoluble product. Extraction of coffee solubles in the manufacture of instant coffee is an example of the first case. Removal of lactose and other solutes from cheese curd by washing with water, and the removal of low molecular weight components from defatted soybean flour in the production of soybean protein concentrate, are examples of the second.

11.2.1 Definitions

Consider a stage (numbered as stage n) in a multi-stage countercurrent solid–liquid extraction process, each stage comprising the operations of contacting (mixing) and separation. Stage n is represented schematically in Figure 11.1. Assume that any of the streams of the system consists of some or all of the following three components: the extractable solute (C), the insoluble matrix (B) and the liquid solvent (A).

For the solid streams (slurries), we define:

$E = A + C$ (kg)
$y = C/(A + C)$ (dimensionless)
$N = B/(A + C)$ (dimensionless).

Note that N is simply the inverse of the solution-holding capacity of the insoluble matrix. It depends on the porosity of the matrix, and the density, viscosity

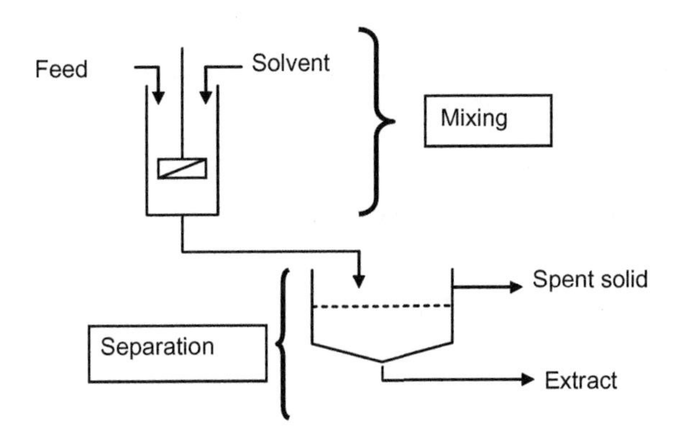

FIGURE 11.1

A stage in a solid–liquid extraction process.

and surface tension (wetting power) of the solution, as well as on the method of separation.

For the liquid streams (extracts), we define:

R = A + C (kg)
x = C/(A + C) (dimensionless)
N = B/(A + C) (dimensionless).

Note: In the case of clear extracts (complete solid–liquid separation after mixing), B = 0, hence N = 0. On the contrary, in turbid extracts, N > 0.

The above symbols will be indexed according to the number of the stage from which the represented stream *exits*.

11.2.2 Material balance

Refer to Chapter 12, Figure 12.1. Material balance for Stage n gives:

$$E_{n-1} + R_{n+1} = E_n + R_n \quad \Rightarrow \quad E_{n-1} - R_n = E_n - R_{n+1} \tag{11.1}$$

Since n can be any number, Eq. (11.1) is valid for any of the stages of the system. If the process comprises p stages in total, one can write:

$$E_0 - R_1 = E_1 - R_2 = E_2 - R_3 = \ldots = E_n - R_{n+1} = \ldots = E_p - R_{p+1} = \text{Constant} = \Delta \tag{11.2}$$

11.2.3 Equilibrium

As explained above, it is usually assumed that the streams leaving the stage are in equilibrium with each other. The inaccuracies introduced by this assumption are then corrected using efficiency factors. In the case of solid–liquid extraction, the extract from stage n is assumed to be in equilibrium with the solution imbibed in the slurry leaving the same stage. If the quantity of solvent is sufficient to dissolve all the extractable solute and if the solid matrix is truly inert, then the concentration of the solute in the extract must be equal to that of the imbibed solution. Expressed formally, the equilibrium condition under these assumptions is:

$$y_n^* = x_n \tag{11.3}$$

The asterisk indicates that y is the equilibrium concentration.

11.2.4 Multi-stage extraction

The objective of solid–liquid extraction is to extract as much as possible of the solute, using a limited quantity of solvent, so as to obtain a concentrated extract. Obviously, all these conditions cannot be met by single-stage extraction, hence the need for multi-stage processes. Multi-stage extraction can be continuous or semi-continuous. A number of options exist regarding the movement of the solid and the liquid streams with respect to each other (Figure 11.2). Countercurrent operation, in which the most dilute extract contacts the slurry with the least amount of residual extractables, is the preferred set-up.

Equation (11.2) in conjunction with an equilibrium function such as Eq. (11.3) may be used for the calculation of the composition of the extracts and the slurries from each stage, or for the determination of the number of contact stages needed to meet the process specifications. These problems can be solved by means of iterative arithmetic calculations, mathematical modeling (Veloso *et al.*, 2005), or graphically. Graphical methods, despite their lack of accuracy, are extremely useful for the analysis of different separation processes based on multiple contact stages. Later, we shall see their application in adsorption and distillation. Here, one of the many graphical methods for the treatment of countercurrent multi-stage extraction processes, known as the Ponchon–Savarit method, will be described.

Consider a continuous, countercurrent multi-stage solid–liquid extraction process consisting of p stages (Figure 11.3). The raw material to be extracted is fed to Stage 1. Fresh solvent is fed to Stage p. Spent solids are discharged from Stage p. Final extract is collected from Stage 1.

Assume that the process data and specifications are as follows:

1. The relationship $N = f(y)$ is known.
2. Equilibrium is assumed. The equilibrium function $y^* = f(x)$ is known (e.g., $y^* = x$).
3. The extracts are free of suspended solids.
4. The compositions of the starting solid and of the fresh solvent are known.

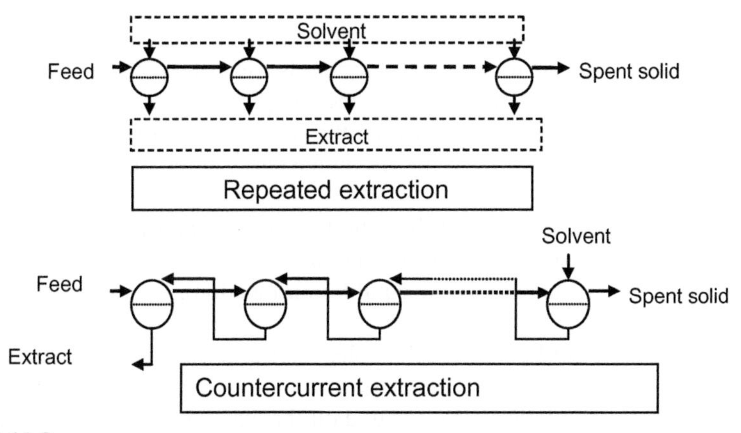

FIGURE 11.2

Repeated (cross-flow) and countercurrent extraction processes.

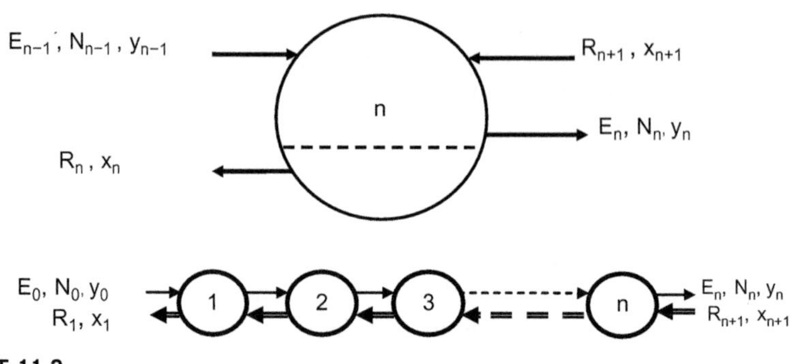

FIGURE 11.3

Multi-stage countercurrent solid–liquid extraction.

5. The feed/solvent mass ratio is specified. The expected yield of recovery for the extractable is also stated. From these data, the composition of the final spent solids and the concentration of the final extract are calculated.

The requirement is to calculate the number p of theoretical contact stages needed to achieve the process objectives stated above.

The problem is solved graphically on the Ponchon–Savarit diagram, (McCabe *et al.*, 2000; Treybal, 1980), which is a graph of N versus x or y (Figure 11.4).

- First, the line representing $N = f(y)$ is drawn. This is known as the "slurry line" or the "overflow line".
- The points representing the four streams entering or leaving the system are located. These are the two solid streams E_0 (N_0, y_0) and E_p (N_p, y_p), and the two liquid streams R_1 (R_1, x_1) and R_{p+1} (R_{p+1},x_{p+1}).
- Refer to Eq. (11.2). Analytical geometry teaches that the point representing the quantity $E_0 - R_1 = \Delta$ is located on the straight line connecting E_0 with R_1. By the same reasoning, Δ is located on the straight line passing through E_p and R_{p+1}. The point Δ is then marked on the chart, at the intersection of these two straight lines.
- From R_1, the point representing the solid E_1 is found. This point is on the slurry line, with y_1 being related to x_1 by the equilibrium conditions. In our case, it has been assumed that the equilibrium function is as formulated in Eq. (11.3).
- Connecting the point E_1 with Δ, the point R_2 is found on the x, y axis, since $N = 0$ for clear extracts.
- This two-step construction is repeated, until the extreme point R_{p+1} is reached or passed. The number of two-step constructions performed is the number of theoretical contact stages.

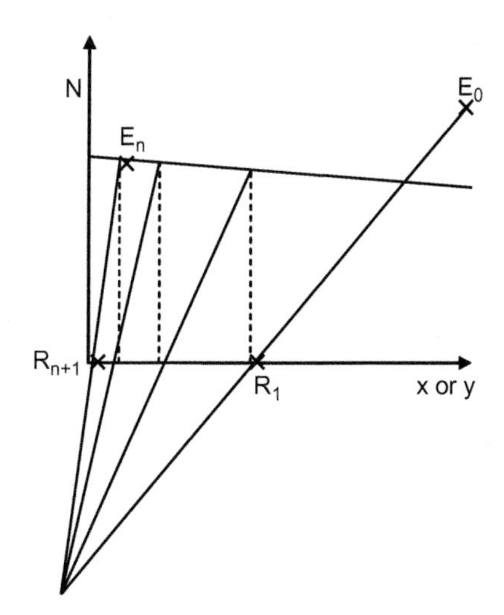

FIGURE 11.4

The Ponchon–Savarit graphical representation of multi-stage extraction.

EXAMPLE 11.1

Biologists have developed a variety of fungus that produces the carotenoid pigment lycopene in commercial quantity. Each gram of dry fungus contains 0.15 grams of lycopene. A mixture of hexane and methanol is to be used for extracting the pigment from the fungus. The pigment is very soluble in that mixture. The requirement is that 90% of the pigment be recovered in a countercurrent multi-stage process. Economic considerations dictate a solvent to feed ratio of 1. Laboratory tests have indicated that each gram of lycopene-free fungus tissue retains 0.6 gram of liquid, after draining, regardless of the concentration of lycopene in the extract. The extracts are free of insoluble solids.

What is the minimum number of contact stages required?

Solution

The Ponchon–Savarit diagram is constructed (Figure 11.5), as follows:

- The slurry line is N = 1/0.6 = 1.667 = Constant (for all x, y).
- The extract line is N = 0 (clear extract).
- Each gram of feed solid has 0.15 g solute, 0.85 g inert matrix and no solvent. Hence it is represented by the point:

$$N = \frac{0.85}{0.15} = 5.67 \quad \text{and} \quad y = 1$$

  (point E0).
- Pure solvent is fed to Stage 1. It is represented by the point: N = 0, x = 0 (point R_{p+1}).
- Points E_P (spent solids) and R_1 (rich extract) are calculated by a total material balance, as follows:

  In:

  1 kg dry fungus = 0.15 kg lycopene + 0.85 kg inert material 1 kg pure solvent.

  Out:

  Spent solids = 0.85 kg inert material + 0.85 × 0.6 kg liquid = 1.360 kg total Rich extract = 2 − 1.36 = 0.64 kg.

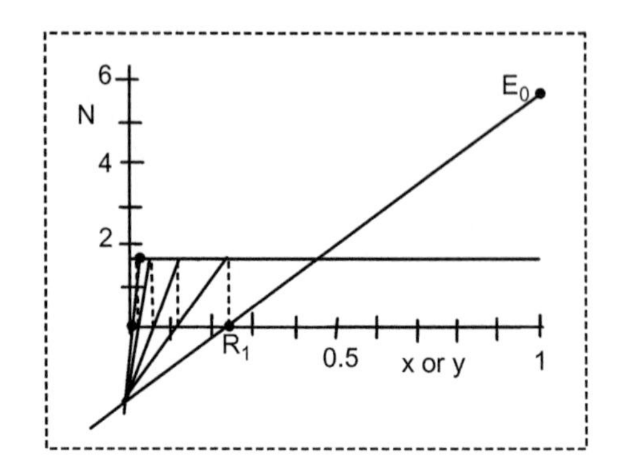

FIGURE 11.5

Lycopene extraction example.

Of the 0.15 kg of lycopene, 90% will be in the extract and 10% in the spent solids. Hence:

$$y_p = \frac{0.15 \times 0.1}{0.85 \times 0.6} = 0.0294 \quad \text{and} \quad x_1 = \frac{0.15 \times 0.9}{0.64} = 0.211$$

- The four points E_0, E_p, R_1, R_{p+1}, are marked on the graph. The difference point Δ is found by joining E_0 with R_1, and E_p with R_{p+1}.
- The equilibrium stages are drawn, assuming $y^* = x$.

The graph shows between four and five stages (rounded up to five stages).

11.2.5 Stage efficiency

The efficiency of a real contact stage is a measure of the deviation from theoretical (equilibrium) conditions. Stage efficiency in all stage contact processes is generally expressed as "Murphee efficiency", the exact definition of which varies with the process in question (McCabe *et al.*, 2000). In the case of leaching, Murphee efficiency η_M is defined as follows:

$$\eta_M = \frac{X_0 - X}{X_0 - X^*} \tag{11.4}$$

where:

X_0 = concentration of the extract entering the stage
X = actual concentration of the extract leaving the stage
X^* = concentration of the extract that would be reached at equilibrium.

The efficiency of a contact stage depends on the time of contact between the phases, the contact surface area, and agitation. Theoretically, it can be calculated based on mass transfer fundamentals. However, exact knowledge of the turbulence conditions, diffusion coefficient and residence time is usually not available. Stage efficiencies are therefore either determined experimentally or assumed in the light of previous experience. The average Murphee efficiency of a multi-stage leaching process is also approximately equal to overall efficiency, which is defined as the ratio of the number of theoretical equilibrium stages to the number of actual stages needed to obtain the same end results.

$$\eta_M \approx \eta_{overall} = \frac{N_{theoretical}}{N_{actual}} \tag{11.5}$$

EXAMPLE 11.2

Shelled and dried oil palm kernels contain 47% oil. The kernels are to be extracted with hexane in a countercurrent, multi-stage extractor. The feed to solvent ratio is 1 (i.e., 1 ton of solvent per ton of kernels fed to the extractor). It is assumed that the weight of miscela retained in the kernel is equal to the weight of oil removed. (Miscela is the solution of oil in the solvent). The planned recovery of oil is 99%. The average Murphee efficiency of the contact stages is expected to be 92%. How many contact stages should the extractor have?

Solution

Since the miscela replaces the oil on one-to-one basis, the weight of the spend slurry remains constant and equal to that of the feed.

For all the solids, including the feed, N = 53/47 = 1.13.

The material balance is as follows.

In:

1 kg kernels = 0.47 kg oil + 0.53 kg inert material 1 kg pure hexane.

Out:

1 kg spent kernels, containing 1% of the oil 1 kg extract, containing 99% of the oil.

The four points are:

E_0: y = 1, N = 1.13
E_p: y = 0.01, N = 1.13
R_{p+1}: x = 0, N = 0
R_1: x = 0.465, N = 0.

The Ponchon–Savarit construction shows five ideal stages (Figure 11.6). The number of real stages will be: 5/0.92 = 5.43. This will be rounded up to six real stages.

11.2.6 Solid–liquid extraction systems

Most large-scale solid–liquid extraction processes in the food industry are continuous or quasi-continuous. Batch extraction is used in certain cases, such as the extraction of pigments from plants, isolation of protein from oilseeds, or the production of meat and yeast extracts. In its simplest form, a batch extraction system consists of an agitated mixing vessel where the solids are mixed with the solvent,

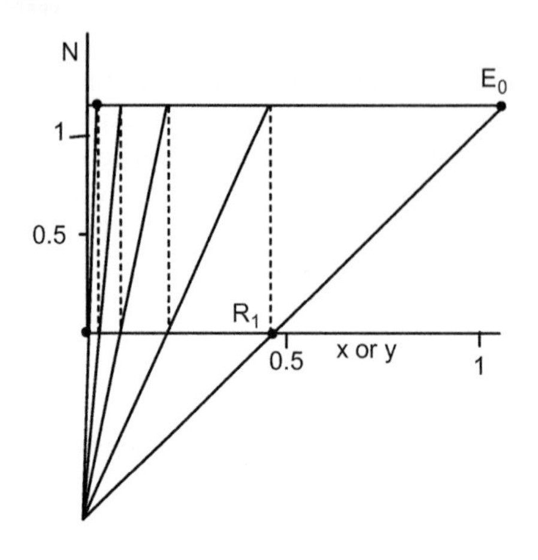

FIGURE 11.6

Palm kernels extraction example.

followed by a solid–liquid separation device. Decanter centrifuges are advantageously used as separators.

In continuous multi-stage extraction, the main technical problem is that of moving the solids from one stage to another continuously, particularly if the process is carried out under pressure. The different solid–liquid extraction systems in use differ mainly in the method applied for conveying the solid stream from stage to stage.

1. *Fixed bed extractors*. In fixed bed extractors, the solid bed is stationary. The countercurrent effect is achieved by moving the position where the fresh solvent is introduced. An example of fixed bed extractor is the quasi-continuous, countercurrent, high-pressure process used for the extraction of coffee in the manufacture of instant (soluble) coffee. The system consists of a battery of extraction columns or "percolators" (Figure 11.7), conventionally six in number.

   Suppose that at the start of the process five of the percolators are full with fresh roasted and ground coffee. Percolator No. 6, also full with fresh coffee, is at standby. Very hot water (over 150°C) is introduced into percolator No. 5 and trickles down through the bed of coffee, extracting the soluble substances. The entire battery is maintained under high pressure, corresponding to the high temperature of extraction. The extract from percolator No. 5 is pumped to percolator No. 4, to be further enriched in solubles, and so on following the sequence (5–4–3–2–1) through the battery of extractors. The most concentrated extract exits at percolator No. 1. The coffee in No. 5, having been extracted with pure water, is the one with the lowest concentration of extractables. When its contents are thoroughly exhausted, No. 5 is disconnected,

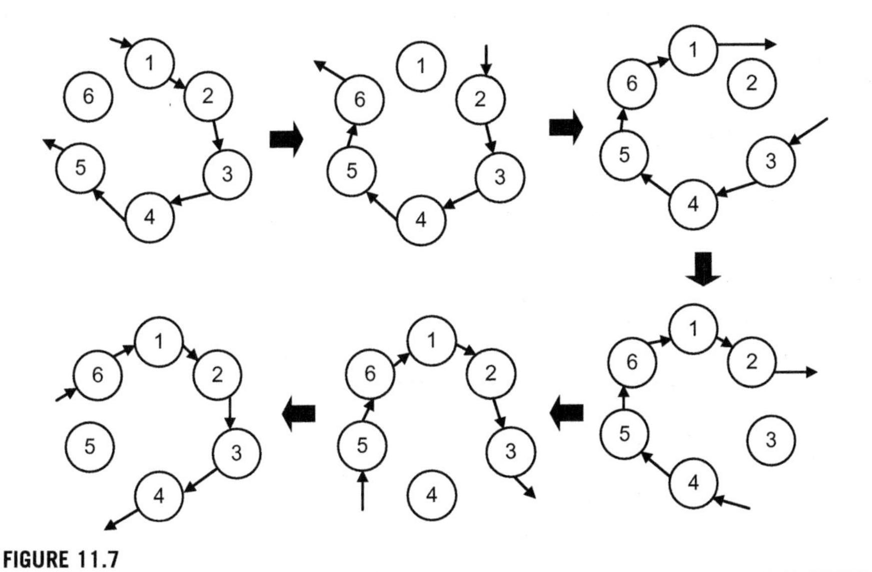

FIGURE 11.7

Sequence of stages in semi-continuous countercurrent extraction of coffee.

emptied by blowing-out the spent coffee solids, cleaned, and filled with fresh coffee, while at the same time percolator No. 6, full with fresh coffee, is connected to the tail-end of the battery. Hot water is now admitted to No. 4 and the liquid flow sequence becomes 4–3–2–1–6. At this stage, No. 5 becomes the standby unit. At the next stage, No. 4 is emptied, No. 5 is connected, and the sequence of liquid flow becomes 3–2–1–6–5, and so on. Thus, the countercurrent effect is achieved without having to move the solids from one stage to another.

2. *Belt extractors.* Belt extractors are used extensively for extracting edible oil from oilseeds and sugar from crushed sugar cane. The material to be extracted is continuously fed by means of a feeding hopper, so as to form a thick mat on a slowly moving perforated belt (Figure 11.8). The bed height is kept constant by regulating the feed rate. The mat is continuous, but distinct extraction stages are delimited by the way in which the liquid stream is introduced. Fresh solvent is sprayed on the solid at the tail "section" of the extractor, i.e., at the section nearest to the discharge outlet for spent solids. The first extract is collected at the bottom of that section and pumped over the section preceding the last. This process of spraying the liquid over the solids, percolation of the liquid through the bed thickness, collection of the liquid below the perforated belt, and pumping to the next section is repeated in the direction opposite to the movement of the belt. The most concentrated extract, known as "full micella", is collected at the bottom of the first (head) section. In their passage from one section to another, the

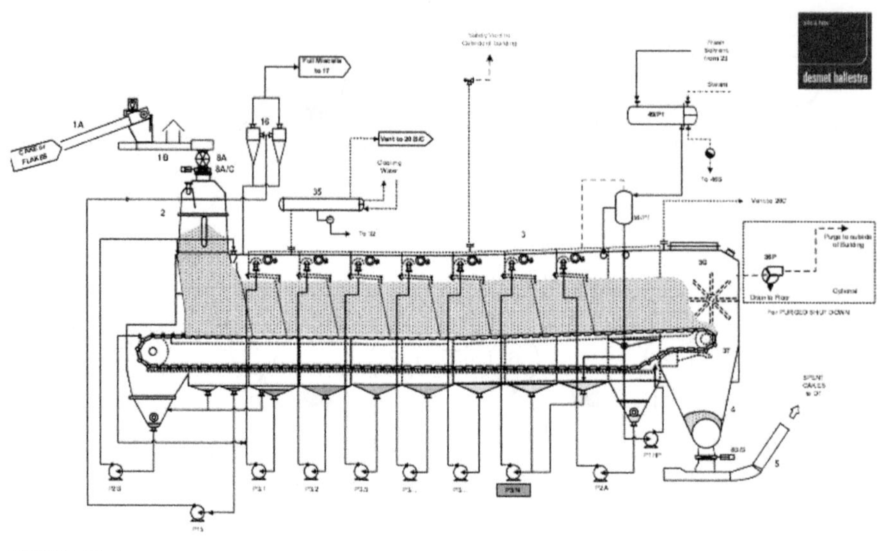

FIGURE 11.8

Belt extractor.

Figure courtesy of Desmet Ballestra A.G.

extracts may be reheated with the help of heat exchangers. In the case of extraction with volatile solvents, the entire system is enclosed in an airtight case where a slight negative pressure is maintained to prevent leakage of solvent vapors. In the so-called "two-stage" extractor, two belts in series are used. The transfer of the solids from one belt to the next causes mixing and resettling of the bed. Belt extractors are high-capacity units (e.g., 2000–3000 tons per day in the case of solvent extraction of flaked soybeans).

3. *Carousel extractors.* Carousel extractors are also most commonly used for the extraction of edible oil from oilseeds (Weber, 1970). These extractors consist of a vertical cylindrical vessel inside which a slowly revolving concentric rotor is installed (Figure 11.9). The rotor is divided into segments by radial partition walls, and rotates over a slotted bottom. The segments contain the solid to be extracted. The liquid extractant is introduced at the top and percolates through the solid bed. The extract exits through the slotted bottom and is collected in chambers separated by weirs, to be pumped back onto the solid bed at the next section. The sequence of liquid collection and pumping is in the direction opposite to that of rotation. At the end of one rotation, the segment passes over a hole in the bottom plate through which the spent solids are discharged. At the next station, the chamber is filled with fresh material and the cycle continues. Carousel extractors are designed for capacities comparable to those of belt extractors.

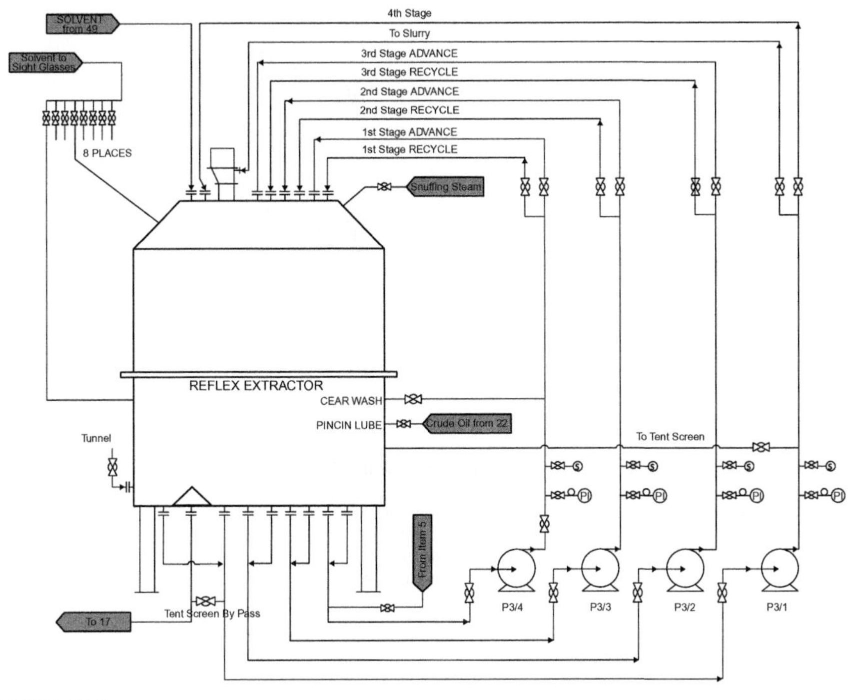

FIGURE 11.9

Carousel extractor.

Figure courtesy of Desmet Ballestra A.G.

4. *Auger extractors*. In auger extractors (Figure 11.10) the solids are conveyed vertically by a large screw conveyor rotating inside a cylindrical enclosure, against a descending stream of extractant liquid. Inclined versions of this class of extractors are also available. Variants of the auger extractor (often under the name of "diffusers") are extensively used for the extraction of sugar from cut sugar-beet chips (cossettes) with hot water.
5. *Basket extractors*. In basket extractors, the material to be extracted is filled into baskets with perforated bottoms. The baskets are moved vertically or horizontally. In the case of vertical basket extractors, also known as bucket elevator extractors, the solvent flows down by gravity through the buckets and is collected at the bottom of the chain. Vertical basket extractors are among the first large-scale continuous solid–liquid extractors (Berk, 1992).

11.2.7 Effect of processing conditions on extraction performance

Processing conditions strongly affect the rate and yield of extraction, as well as the quality of the extracted product. Understanding the influence of processing

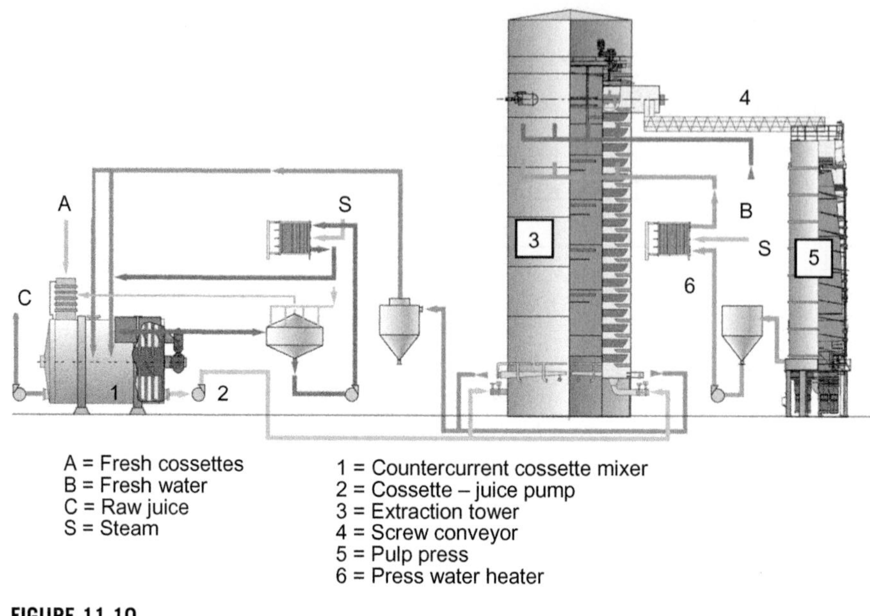

FIGURE 11.10

Sugar beet cossette extractor.

Figure courtesy of BMA Braunschweigische Maschinenbauanstalt A.G.

parameters on performance is essential for the optimal design and operation of extraction processes and systems.

1. *Temperature*. Where thermal damage is not an issue, high temperatures are preferred for their positive effect on yields and rate. At high temperature the solubility of the extractables in the solvent is higher and solvent viscosity is lower, resulting in enhanced wetting and penetration capability and higher diffusion coefficients. In the case of volatile, inflammable solvents such as hexane, ethanol or acetone, safety considerations determine the highest applicable temperature. In certain cases, lower temperatures may be preferred if the selectivity of the solvent towards the desired extractable is improved by lowering the temperature.
2. *Pressure*. Solid–liquid extraction at very high temperature implies pressurization to maintain the solvent in liquid state. As mentioned in Section 11.2.6, water at high temperature–high pressure is utilized for the extraction of coffee. The polarity of water decreases as its temperature is raised. Cacace and Mazza (2007) applied pressurized liquid water at very high temperature (pressurized low-polarity water) for the extraction of biologically active substances from plant tissue. Corrales *et al.* (2009) extracted anthocyanin pigments from grape skins, using high (600-MPa) hydrostatic

pressure. Other applications of pressurized solvents in extraction processes have been reviewed by Pronyk and Mazza (2009).

3. *Particle size*. The rate of extraction is improved by reducing the size of the solid particles. For this reason, sugar beet is sliced into thin strips (cossettes) and soybeans are ground and flaked prior to extraction. Size reduction facilitates both the internal transport of solute (by reducing the distance to the surface) and the external transport (by increasing the contact area with the solvent). Extraction of proteins from soybean flour is also improved by size reduction (Vishwanathan *et al.*, 2011).
4. *Agitation*. Agitation accelerates external transport to and from the particle surface, but does not affect extraction rate if the rate-controlling factor is internal diffusion (Cogan *et al.*, 1967).
5. *Ultrasound-assisted extraction*. Intense sonication of the fluid facilitates release of intercellular materials from suspended solids by disrupting cell walls. Ultrasonication is a widely applied laboratory technique for the extraction of enzymes from biomass. Systems for ultrasound-assisted solvent extraction at pilot-plant scale are commercially available.
6. *Application of pulsed electric fields (PEFs) to extraction*. Pulsed electric fields are known to open pores in cell membranes (electroporation). Originally investigated as a method for the preservation of foods by inactivating microorganisms, PEF treatment has been applied lately as a means to improve both solid–liquid extraction yield and extract purity (Loginova *et al.* 2010, 2011a,b).
7. *Surfactand-assisted extraction*. Do and Sabatini (2011) extracted oil from peanut and canola, using water as a solvent in the presence of surfactants. Oil-recovery yield, in a two-stage semi-continuous process at pilot plant scale, was over 90%.

11.3 Supercritical fluid extraction

11.3.1 Basic principles

A supercritical fluid (SCF) is a substance at a temperature and pressure above those of the critical point. The supercritical region for a pure substance (in this case, carbon dioxide) is shown on the phase diagram in Figure 11.11.

The critical temperature T_C is the temperature above which a gas cannot be liquefied at any pressure. Hence, the critical point C represents the end of the gas–liquid equilibrium curve on the temperature–pressure plane. The density of supercritical fluids is close to that of the liquid, while their viscosity is low and comparable to that of a gas. These two properties are the key to the functionality of SCFs as extractants. The relatively high density imparts to SCFs good solubilization power, while the low viscosity results in particularly rapid permeation of the solvent into the solid matrix.

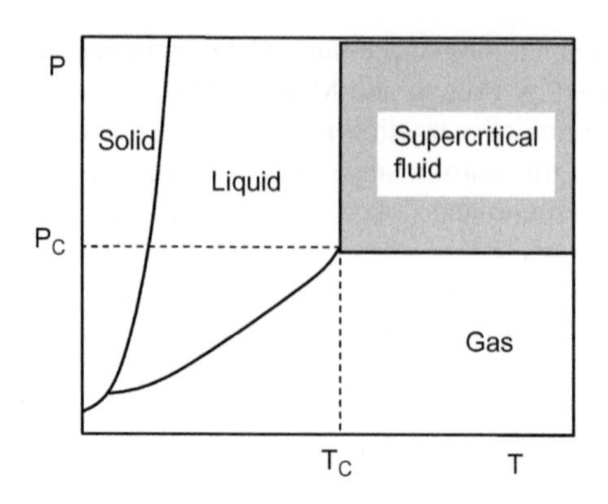

FIGURE 11.11

Phase diagram showing supercritical region.

Supercritical fluid extraction (SCFE or SFE) is an extraction process carried out using a supercritical fluid as a solvent (Brunner, 2005; King, 2000). Although a number of substances could serve as solvents in SFE, carbon dioxide is by far the most commonly used extraction medium. Carbon dioxide near its critical point is a fairly good solvent for low molecular weight non-polar to slightly polar solutes. However, the solubility of oils in supercritical CO_2 is considerably lower than in conventional hydrocarbon solvents. Nevertheless, in view of its other advantageous characteristics, the use of SFE for the extraction of lipids and fatty acids is not to be dismissed (Döker *et al.*, 2010; Pradhan *et al.*, 2010; Sahena *et al.*, 2009).

Carbon dioxide is non-toxic, non-flammable and relatively inexpensive. Its critical temperature is 31.1°C (304.1 K), which makes it particularly suitable for use with heat-sensitive materials. Its critical pressure is relatively high (7.4 MPa), hence the high capital cost of the equipment.

11.3.2 Supercritical fluids as solvents

The solvation capability of SCFs is of considerable economic importance as it determines the extraction yields and the mass ratio of solvent to feed, hence the physical size of the system and the operating cost.

The capability of a supercritical fluid to dissolve a certain substance is approximately represented by a value called the *solubility parameter* (Rizvi *et al.*, 1994). The solubility parameter δ is related to the critical pressure, the density of the gas and that of the liquid, as follows:

$$\delta = 1.25\, P_C^{0.5} \left(\frac{\rho_g}{\rho_l} \right) \tag{11.6}$$

At low pressure, the density of a gas is low while the density of the liquid is almost independent of the pressure. Hence, at low pressure the solubility parameter is low. The density of the gas increases with the pressure and at the critical point it reaches its maximum value, which is the density of the liquid. Near the critical point, the performance of a SCF as a solvent is strongly affected by the pressure. The relationship between the solubility parameter and the pressure follows an S-shaped curve (Rizvi *et al.*, 1994), as shown in Figure 11.12. The pressure-dependence of the SCF is the phenomenon on which supercritical extraction is based. The feed material is extracted with the SCF at a pressure and temperature corresponding to the maximum solubility. The charged solvent (gaseous solution) is separated and the solute is then precipitated from that solution, simply by lowering the pressure.

The effect of temperature on solubility in SCFs is more complex, because the temperature affects both the density of the solvent and the molecular mobility in general. For example, it was found that the solubility of evening primrose oil in supercritical carbon dioxide decreases with temperature at a pressure of 10 MPa, but increases with temperature if the extraction is carried out at 30 MPa (Lee *et al.*, 1994). At intermediate pressure, the solubility is almost independent of the temperature.

The dissolution power of supercritical carbon dioxide can be increased considerably by adding a small amount of another fluid, known as the co-solvent, modifier, enhancer or entrainer. The principal function of the entrainer is often to modify the polarity of SCF CO_2 so as to increase the solubility of more polar substrates. Thus, for example, the solubility of caffeine in supercritical carbon dioxide is greatly enhanced by adding a small amount (e.g., 5%) of ethanol to the solvent (Kopcak and Mohamed, 2005). The extraction of lycopene from tomato skins with supercritical carbon dioxide is facilitated by the addition of water, ethanol or canola oil as modifiers (Shi *et al.*, 2009). The co-solvent, being of lower volatility, remains in the product when the CO_2 is vaporized, and can be removed by a subsequent step of desolventizing or by another separation process.

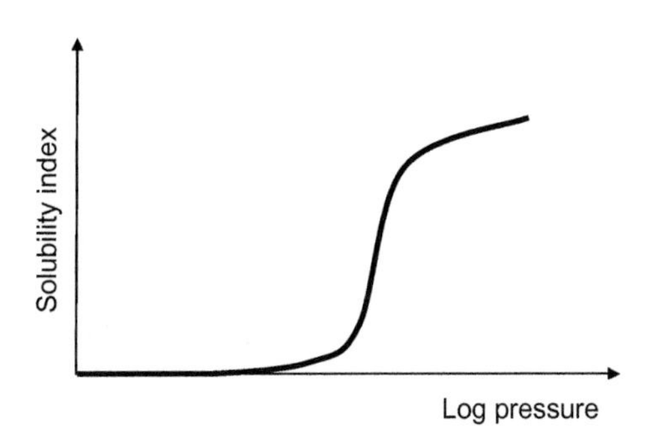

FIGURE 11.12

Effect of pressure on the solvation power of SCF.

11.3.3 Supercritical extraction systems

Most industrial- and laboratory-scale SCF extractions are batch processes. Because of the limited solubilization, it is necessary to apply relatively high solvent to feed ratios. In order to supply a high flow-rate of fresh solvent, the SCF must be continuously recovered and recycled. The basic components of a carbon dioxide SCE system are as follows (Figure 11.13):

1. An extraction vessel where a batch of feed material is treated with the SCF solvent.
2. An expansion vessel or evaporator, where the pressure is reduced, causing the solute to precipitate from the extract-laden CO_2.
3. A separator that serves to separate the product from the CO_2 gas.
4. A heat exchanger (condenser) for cooling and condensing the CO_2.
5. A reservoir or tank for storing the liquid CO_2 and adding make-up CO_2 as necessary.
6. A CO_2 pump (compressor) for pressurizing the liquid CO_2 above the critical pressure.
7. A heater for bringing the temperature to the desired level (above the critical temperature) before introducing the SCF into the extraction vessel.

The system can be divided into a high-pressure region (from the compressor to the expansion valve) and a low-pressure region (from the expansion valve to the compressor). Because of the high sensitivity of the solubility to pressure change, the pressure drop between the two regions is not very large. The extracted material is often precipitated completely by a slight pressure release. Consequently, the compression ratio and energy consumption of the compressor

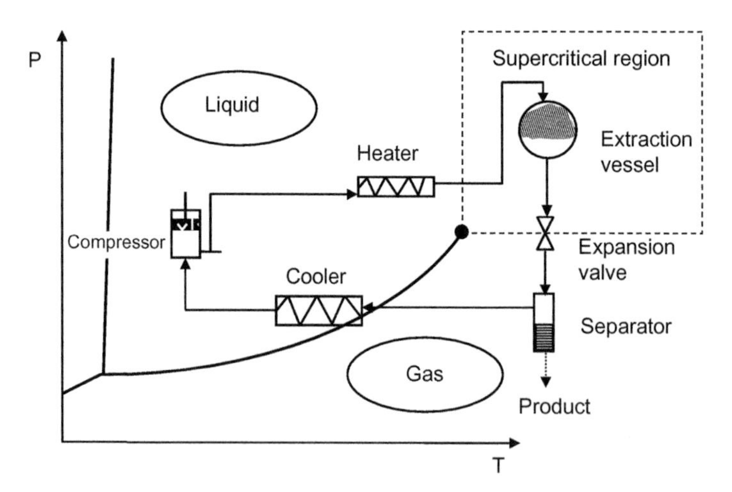

FIGURE 11.13

Supercritical extraction system, shown on the CO_2 phase diagram.

are usually moderate. In some cases, removal of the extracted material from the charged supercritical CO_2 is carried out by washing the CO_2 with another, better solvent (e.g., removal of caffeine from supercritical CO_2 by washing with water).

11.3.4 Applications

The advantages of SCF extraction over conventional solvent extraction include:

- Moderate temperature of operation
- Non-toxic, non-flammable, nature-friendly solvents
- Very volatile solvents that leave no solvent residues in the product
- Good mass transfer, due to the low viscosity of the solvent
- Selective dissolution (e.g., extraction of caffeine from green coffee without removal of flavor precursors)
- Low overall energy expenditure, with no energy consumption for desolventizing.

The disadvantages include:

- Limited solvation power, requiring high solvent to feed ratio, partially overcome by the use of entrainers
- High operation pressure, thus expensive equipment
- Difficulty in running it as a continuous process.

Based on the above, it may be concluded that SCF extraction should be applicable particularly in:

- Low-to-moderate volume, high added-value processes
- Where heat-sensitive biomaterials are involved
- Where selection of a "green technology" is a predominant condition.

The main commercial-scale applications include:

- *Extraction of hops*. Modern breweries make extensive use of hop extracts. Industrial hop extraction is one area where SCF extraction has largely replaced older technologies (Gardner, 1993). SCF extraction has the advantage of extracting mainly the desirable flavor components without the undesirable heavy resins (gums). Furthermore, by varying the extraction conditions hop extracts richer in aroma or richer in bitter principles can be produced, according to the requirements of the brewing industry. Pelleted hops are extracted with supercritical CO_2 in extraction vessels with capacities ranging from a few hundreds to a few thousands of liters.
- *Decaffeination of coffee and tea*. This is another area where SCF extraction has been successful in replacing other extraction methods, mainly because of the absence of objectionable solvent residues. In the case of coffee, the extraction is carried out on humidified whole green coffee beans. Both the decaffeinated coffee or tea and the extracted caffeine (after further purification

and crystallization) are valuable products (Lack and Seidlitz, 1993). Decaffeination of green tea by supercritical extraction has also been investigated (Kim *et al.*, 2008).

- *Other applications*. Other food-related applications of SCF extraction include extraction of aromas (Sankar and Manohar, 1994), pigments and physiologically active substances (Higuera-Ciapara *et al.*, 2005; Rossi *et al.*, 1990; Shi *et al.*, 2009; Zeidler *et al.*, 1996), purification of cooking oils (Hong *et al.*, 2010), etc,. With the increasing interest in natural nutraceuticals, the use of SCF extraction for the production of substances such as plant antioxidants (Nguyen *et al.*, 1994), phytosterols, omega fatty acids from fish oil (Rubio-Rodríguez *et al.*, 2012), etc., can be expected to grow.

In addition to its industrial applications, SCF extraction is a useful laboratory technique for extraction and isolation. Small-scale complete systems are available for laboratory use.

11.4 Liquid–liquid extraction

11.4.1 Principles

Liquid–liquid extraction, also known as partitioning, is a separation process consisting of the transfer of a solute from one solvent to another, the two solvents being immiscible or partially miscible with each other. Frequently, one of the solvents is water or an aqueous mixture and the other is a non-polar organic liquid. As in all extraction processes, liquid–liquid extraction comprises a step of mixing (contacting) followed by a step of phase separation. It is important to consider both steps in the selection of solvents and modes of operation. Thus, while vigorous mixing is favorable to the transfer of the extractable from one solvent to the other, it may also impair the ease of phase separation by forming emulsions.

Equilibrium is reached when the chemical potential of the extractable solute is the same in the two phases. Practically, this rule leads to the definition of a "distribution coefficient", K, as follows:

$$K = \frac{C_1}{C_2} \tag{11.7}$$

where C_1 and C_2 are the equilibrium concentrations of the solute in the two phases, respectively. The distribution coefficient is an expression of the relative preference of the solute for the solvents. In ideal solutions (i.e., where the chemical potential may be assumed to be proportional to the concentration) the distribution coefficient at a given temperature is practically constant, i.e., independent of the concentration.

In some cases, the efficiency of a liquid–liquid extraction process can be strongly improved by modifying the distribution coefficient. Thus, an organic acid would prefer the non-polar solvent when not dissociated (i.e., at low pH) and the aqueous solvent when dissociated (i.e., at high pH).

11.4.2 Applications

Liquid–liquid extraction is an important separation method in research and chemical analysis. As a commercial process it is frequently used in the chemical and mining industries and in the downstream recovery of fermentation products (antibiotics, amino acids, steroids). Its applications to food are restricted to isolated cases, such as the transfer of carotenoid pigments from organic solvents to edible oils, or the production of "terpeneless" essential citrus oil by extracting the oxygenated compounds of the essential oil with aqueous ethanol.

References

Berk, Z., 1992. Technology of Production of Edible Flours and Protein Products From Soybeans. FAO, Rome.

Brunner, G., 2005. Supercritical fluids: technology and application to food processing. J. Food Eng. 67 (1–2), 21–33.

Cacace, J.E., Mazza, G., 2007. In: Shi, J. (Ed.), Pressurized Low Polarity Water Extraction of Biologically Active Compounds From Plant Products. Taylor & Francis Group, Boca Raton, FL.

Cogan, U., Yaron, A., Berk, Z., Mizrahi, S., 1967. Isolation of soybean protein: effect of processing conditions on yield and purity. J. Am. Oil Chem. Soc. 44 (5), 321–324.

Corrales, M., Garcia, A.F., Butz, P., Tauscher, B., 2009. Extraction of anthocyanins from grape skins assisted by high hydrostatic pressure. J. Food Eng. 90 (4), 415–421.

Do, L.D., Sabatini, D.A., 2011. Pilot scale study of vegetable oil extraction by surfactant-assisted aqueous extraction process. Sep. Sci. Technol. 46 (6), 978–985.

Döker, O., Salgın, U., Yıldız, N., Aydoğmuş, M., Çalımlı, A., 2010. Extraction of sesame seed oil using supercritical CO_2 and mathematical modeling. J. Food Eng. 97 (3), 360–366.

Gardner, D.S., 1993. Commercial scale extraction of alpha acids and hop oils with compressed CO_2. In: King, M.B., Bott, T.R. (Eds.), Extraction of Natural Products Using Near-Critical Solvents. Blackie Academic & Professional, Glasgow, UK.

Higuera-Ciapara, I., Toledo-Guillen, A.R., Noriega-Orozco, L., Martinez-Robinson, K.G., Esqueda-Valle, M.C., 2005. Production of a low-cholesterol shrimp using supercritical extraction. J. Food Proc. Eng. 28 (5), 526–538.

Hong, S.A., Kim, J., Kim, J.-D., Kang, J.W., Lee, Y.-W., 2010. Purification of waste cooking oils vie supercritical carbon dioxide extraction. Sep. Sci. Technol. 45 (8), 1139–1146.

Kim, W.-J., Kim, J.-D., Kim, J., Oh, S.-G., Lee, Y.-W., 2008. Selective caffeine removal fro green tea using supercritical carbon dioxide extraction. J. Food Eng. 89 (3), 303–309.

King, J.W., 2000. Advances in critical fluid technology for food processing. *Food Sci. Technol. Today* 14, 186–191.

Kopcak, U., Mohamed, R.S., 2005. Caffeine solubility in supercritical carbon dioxide/co-solvent mixtures. J. Supercrit. Fluids 34 (2), 209–214.

Lack, E., Seidlitz, H., 1993. Commercial scale decaffeination of coffee and tea using supercritical CO_2. In: King, M.B., Bott, T.R. (Eds.), Extraction of Natural Products Using Near-Critical Solvents. Blackie Academic & Professional, Glasgow, UK.

Lee, B.C., Kim, J.D., Hwang, K.Y., Lee, Y.Y., 1994. Extraction of oil from evening primrose seed with supercritical carbon dioxide. In: Rizvi, S.S.H. (Ed.), Supercritical Fluid Processing of Food and Biomaterials. Blackie Academic & Professional, London, UK.

Loginova, K.V., Shynkaryk, M.V., Lebovka, N.I., Vorobiev, E., 2010. Acceleration of soluble matter extraction from chicory with pulsed electric fields. J. Food Eng. 96 (3), 374–379.

Loginova, K., Loginov, M., Vorobiev, E., Lebovka, N.I., 2011a. Quality and filtration characteristics of sugar beet juice obtained by "cold" extraction assisted by pulsed electric field. J. Food Eng. 106 (2), 144–151.

Loginova, K.V., Vorobiev, E., Bals, O., Lebovka, N.I., 2011b. Pilot study of countercurrent cold and mild heat extraction of sugar from sugar beets, assisted by pulsed electric fields. J. Food Eng. 102 (4), 340–347.

McCabe, W., Smith, J., Harriot, P., 2000. Unit Operations of Chemical Engineering, sixth ed. McGraw-Hill Science, New York, NY.

Nguyen, U., Evans, D.A., Frakman, G., 1994. Natural antioxidants produced by supercritical extraction. In: Rizvi, S.S.H. (Ed.), Supercritical Fluid Processing of Food and Biomaterials. Blackie Academic & Professional, London, UK.

Pradhan, R.Ch., Meda, V., Rout, P.K., Naik, S., Dalai, A.K., 2010. Supercritical CO_2 extraction of fatty oil from flaxseed and comparison with screw press expression and solvent extraction processes. J. Food Eng. 98 (4), 393–397.

Pronyk, C., Mazza, G., 2009. Design and scale-up of pressurized fluid extractors for food and bioproducts. J. Food Eng. 95 (2), 215–226.

Rizvi, S.S.H., Yu, Z.R., Bhaskar, A.R., Chidambara Raj, C.B., 1994. Fundamentals of processing with supercritical fluids. In: Rizvi, S.S.H. (Ed.), Supercritical Fluid Processing of Food and Biomaterials. Blackie Academic & Professional, London, UK.

Rossi, M., Spedicato, E., Schiraldi, A., 1990. Improvement of supercritical CO_2 extraction of egglipids by means of ethanolic entrainer. Ital. J. Food Sci. 4, 249.

Rubio-Rodríguez, N., de Diego, S.M., Beltrán, S., Jaime, I., Sanz, M.T., Rovira, J., 2012. Supercritical fluid extraction of fish oil from fish by-products: a comparison with other extraction methods. J. *Food Eng*. 109 (2), 238–248.

Sahena, F., Zaidul, I.S.M., Jinap, S., Karim, A.A., Abbas, K.A., Norulaini, N.A.N., et al., 2009. Application of supercritical CO_2 in lipid extraction – a review. J. Food Eng. 95 (2), 240–253.

Sankar, U.K., Manohar, B., 1994. Mass transfer phenomena in supercritical carbon dioxide extraction for production of spice essential oils. In: Rizvi, S.S.H. (Ed.), Supercritical Fluid Processing of Food and Biomaterials. Blackie Academic & Professional, London, UK.

Shi, J., Yi, Ch., Xue, S.J., Jiang, Y., Ma, Y., Li, D., 2009. Effects of modifiers on the profile of lycopene extracted from tomato skins by supercritical CO_2. J. Food Eng. 93 (4), 431–436.

Treybal, R.E., 1980. Mass Transfer Operations, third ed. McGraw-Hill, New York, NY.

Veloso, G.O., Krioukov, V.G., Vielmo, H.A., 2005. Mathematical modeling of vegetable oil extraction in a counter-current cross flow horizontal extractor. J. Food Eng. 66 (4), 477–486.

Vishwanathan, K.H., Singh, V., Subramanian, R., 2011. Influence of particle size on protein extractability from soybean and okara. J. Food Eng. 102 (3), 240–246.

Weber, K., 1970. Solid/liquid extraction in the Carrousel extractor. Chem Ztg/Chem. Apparatur Verfahrenstechnik 94, 56–62.

Zeidler, G., Pasin, G., King, A.J., 1996. Supercritical fluid extraction of cholesterol from liquid egg yolk. J. Clean. Prod. 4, 143.

CHAPTER

Adsorption and Ion Exchange

12

12.1 Introduction

Adsorption is the phenomenon by which molecules from a gas or from a liquid mixture are fixed preferentially on a solid surface. The forces responsible for adsorption may be London–van der Waals forces of intermolecular attraction (physical adsorption), electrostatic attraction (e.g., ion exchange) or chemical affinity (chemical adsorption or "chemisorption"). Physical adsorption is reversible and of low selectivity. It involves relatively low levels of adsorption energy. Chemical adsorption is not reversible and may be highly selective. Fixation of free molecules on a solid surface liberates heat, as a result of the reduction in the kinetic energy of the adsorbed particles. In the case of physical adsorption, the energy of adsorption is in the order of 10^0 to $10^1\ kJ \cdot mol^{-1}$. The higher the energy of adsorption, the stronger is the bond between the adsorbed molecules (adsorbate) and the adsorbent.

In food technology, the importance of adsorption as a physical phenomenon is considerable. Adsorption of water vapor and its relation with water activity has been discussed in Chapter 2. Food quality may be affected by the adsorption and desorption of odorous volatiles. In this chapter, however, adsorption will be discussed only as a separation process in food process engineering.

Most adsorbents utilized in technological applications are porous solids with very large specific surfaces (in the order of thousands of m^2 per gram), resulting in very high adsorption capacities. These include activated silica, activated clay and activated charcoal (carbon) (Bansal and Goyal, 2005). Recently, microporous edible supports with specific surfaces of about 270 m^2 per gram have been produced and used as adsorbents for food flavors (Zeller and Saleeb, 1996; Zeller *et al.*, 1998). These materials are produced from solutions of sugars or sugar salts by rapid freezing of fine droplets of the solution, followed by freeze-drying.

The following are some of the engineering applications of adsorption as a separation process in the food industry:

- Decolorization of edible oils with "bleaching earths" (activated clays)
- Decolorization of sugar syrup with activated carbon in the manufacture of sugar
- Removal of bitter substances from fruit juices by adsorption on polymers (Fayoux *et al.*, 2007)

Food Process Engineering and Technology. DOI: http://dx.doi.org/10.1016/B978-0-12-415923-5.00012-5
© 2013 Elsevier Inc. All rights reserved.

- Recovery and purification of phenolic substances (Soto *et al.*, 2010)
- Odor abatement by passing gaseous emanations through activated carbon
- Removal of chlorine from drinking water by adsorption on carbon
- Various applications of ion exchange.

In practice, adsorption may be carried out as a batch, semi-continuous or continuous process. In a typical batch process with a liquid solution, a batch of the solution is mixed with a portion of the adsorbent for a determined length of time, after which the two phases are separated. In a semi-continuous process, the gas or the solution is continuously passed through a static bed of adsorbent in a column. In a continuous process, both the adsorbent and the gas (or liquid solution) are continuously fed into a system (usually in countercurrent fashion) where contact between the two phases occurs. Truly continuous adsorption processes are not commonly applied in the food industry.

12.2 Equilibrium conditions

Consider a liquid solution or a gas mixture in equilibrium with a solid adsorbent. Let x be the concentration of the adsorbate in the adsorbent (usually expressed as weight by weight or as mol fraction) and y^* the equilibrium concentration of the adsorbate in the solution or gas (usually expressed as weight per volume, or in gases as partial pressure). Various theoretical and semi-empirical models have been proposed to explain and predict the equilibrium conditions $y^* = f(x)$. The graph representing $y^* = f(x)$ at constant temperature is termed a *sorption isotherm* (see also Chapter 1, Section 1.7.4, on water vapor sorption isotherms).

Langmuir's model of adsorption (Irving Langmuir, 1881–1957, American scientist, Nobel Prize in Chemistry 1932) assumes that the solid surface is uniform, i.e., the availability of every site of adsorption to the adsorbate is equal. The adsorbate is bound to the solid as a mono-molecular layer. No molecules are adsorbed after the mono-molecular layer is filled. The rate of adsorption at any moment is proportional to the concentration of the adsorbate molecules in the solution or in the gas, and to the fraction of free sites on the solid surface. Simultaneously, molecules are desorbed from the solid and the rate of desorption is proportional to the fraction of occupied sites. At equilibrium, the rate of adsorption is equal to the rate of desorption. Based on these assumptions, the following expression is developed:

$$x = \frac{x_m K y^*}{1 + K y^*} \tag{12.1}$$

The value of x at saturation is x_m. K is a constant depending on the adsorbent and the adsorbate. The general shape of an adsorption isotherm corresponding to the Langmuir model is shown in Figure 12.1.

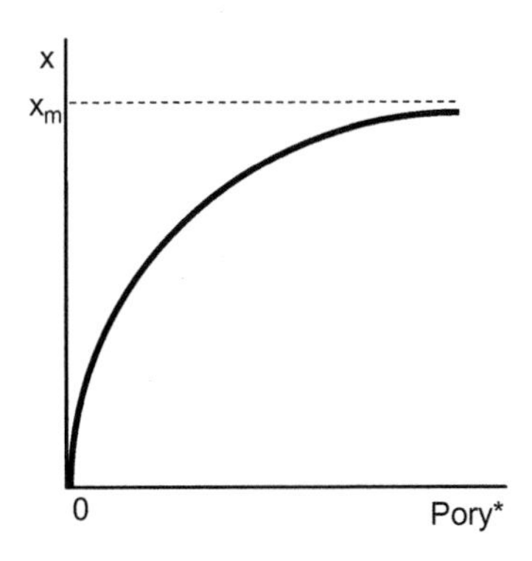

FIGURE 12.1

The Langmuir model of adsorption.

One of the shortcomings of Langmuir's theory is the assumption that no further net adsorption occurs once the mono-molecular layer is full. In reality, additional molecules do adsorb on the monolayer, although with a considerably weaker bond. To address this difficulty, models of multi-layer adsorption have been proposed. One of these models is the *BET isotherm*, developed by Brunauer, Emmett and Teller. The BET theory admits multilayer adsorption with increasingly weaker bonds (smaller energy of adsorption) for each subsequent layer. The equilibrium expression corresponding to the BET theory is given in Eq. (12.2) in terms of partial pressures.

$$x = \frac{x_m K\left(P/P_0\right)}{\left[1 + (K-1)\left(P/P_0\right)\right]\left[1 - \left(P/P_0\right)\right]} \tag{12.2}$$

where:

P = partial pressure of the adsorbate in the gas
P_0 = vapor pressure of the pure adsorbate
x_m = monolayer value of x
K = a constant.

Note that the ratio P/P_0 is, in fact, the *activity* of the adsorbate.

Yet another useful adsorption model is the empirical *Freundlich model*, formulated as in Eq. (12.3):

$$y^* = mx^n \tag{12.3}$$

The coefficient m and the exponent n are numerical constants, determined experimentally. In dilute solutions, the linear form of the equation ($n = 1$) provides a fair approximation. Note that in the case of adsorption from a gas, the linear form of the Freundlich rule is identical with Henry's Law.

EXAMPLE 12.1

One of the methods for the determination of the specific surface of powders is based on adsorption. A sample of the powder is brought into contact with nitrogen vapors at the temperature of liquid nitrogen. The quantity of nitrogen adsorbed by the powder as a function of the partial pressure of nitrogen is determined. The BET monolayer value is calculated from the data. From that, the specific surface of the powder is calculated, assuming that 1 g of nitrogen covers, as a monomolecular layer, a solid surface of 3485 m^2.

An absorbent powder was subjected to the nitrogen adsorption experiment described above. The quantities of nitrogen adsorbed per 1 g of powder at different pressures of nitrogen are given in Table 12.1. Calculate the specific surface area of the powder.

Solution

The BET equation (Eq. 12.2) can be re-written as follows:

$$\left(\frac{1}{x}\right)\left(\frac{\pi}{1-\pi}\right)=f=\frac{(K-1)\pi}{Kx_m}+\frac{1}{Kx_m} \qquad \pi=P/P_0$$

If the left-hand term f of the equation is plotted against the activity $\pi = P/P_0$, a straight line is obtained. From the slope and the intersect of the line, both x_m and K can be calculated.

Since the measurements are carried out at the temperature of liquid nitrogen, P_0 is vapor pressure of boiling liquid nitrogen, i.e., 1 atm or 100 kPa. From the data, we calculate π and f (Table 12.2).

The plot of f versus π gives a straight line (Figure 12.2). The intercept is 0.1393 and the slope is 6.83.

Solving for K and x_m, we obtain:

$$K = 50.03 \quad x_m = 0.1436\ \text{g}\cdot\text{g}^{-1}$$

Hence, the specific surface area S is:

$$S = (3485)x_m = 500\ \text{m}^2\cdot\text{g}^{-1}$$

Table 12.1 Nitrogen Adsorption Data

N_2 pressure P (kPa)	5	10	15	20
N_2 adsorbed per gram of powder (grams)	0.1095	0.1351	0.1516	0.1661

Table 12.2 Results of Nitrogen Adsorption Calculations

π	0.05	0.1	0.15	0.2
x	0.1095	0.1351	0.1516	0.1661
f	0.480654	0.822436	1.164054	1.505117

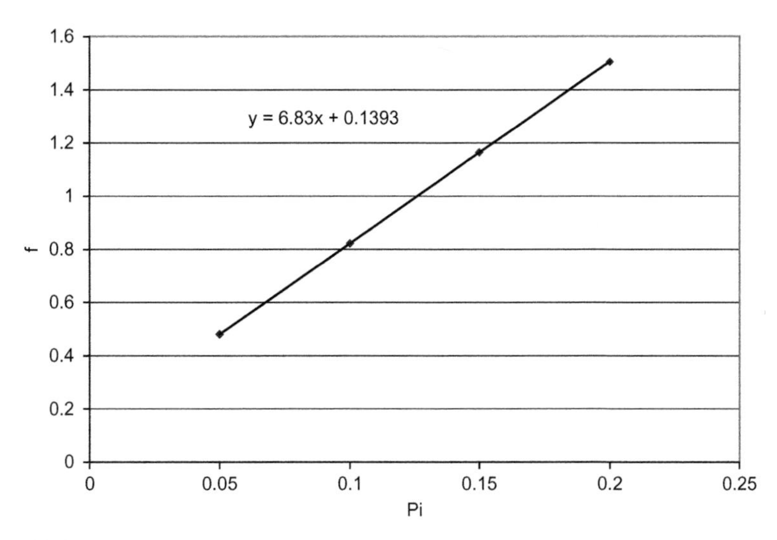

FIGURE 12.2

Nitrogen adsorption test – f versus π plot.

12.3 Batch adsorption

Consider a single-step adsorption operation, wherein a liquid solution is treated with a solid adsorbent, as shown in Figure 12.3.

For convenience, the quantity of the solution is given in terms of adsorbate-free solvent (G, kg) and the quantity of the solid in terms of adsorbate-free adsorbent (L, kg). Assume that equilibrium has been reached, no solvent is adsorbed, and no adsorbent is lost to the liquid. Frequently, the question to be answered is: Given the quantities and the concentrations of the solution and the solid entering the system, what will be the concentration of adsorbate in the solution leaving the operation?

A material balance can be established (Eq. 12.4):

$$G(y_0 - y^*) = L(x_1 - x_2) \Rightarrow y^* - y_0 = -\frac{L}{G}(x_1 - x_2) \tag{12.4}$$

If the equilibrium function $y^* = f(x)$ is given as an algebraic expression, Eq. (12.4) can be solved analytically for y^*. For example, if the Freundlich model of equilibrium is applicable and the incoming adsorbent may be assumed not to contain any absorbate (as is generally the case in practice), we obtain:

$$\frac{L}{G} = \frac{y_0 - y^*}{(y^*/m)^{1/n}} \tag{12.5}$$

If, on the other hand, the equilibrium conditions are given as a graph (y^* vs x), the problem is solved graphically (Figure 12.4).

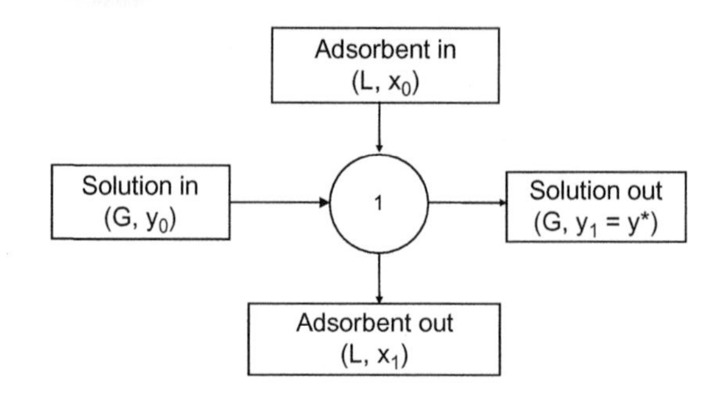

FIGURE 12.3

Schematic representation of a batch adsorption process.

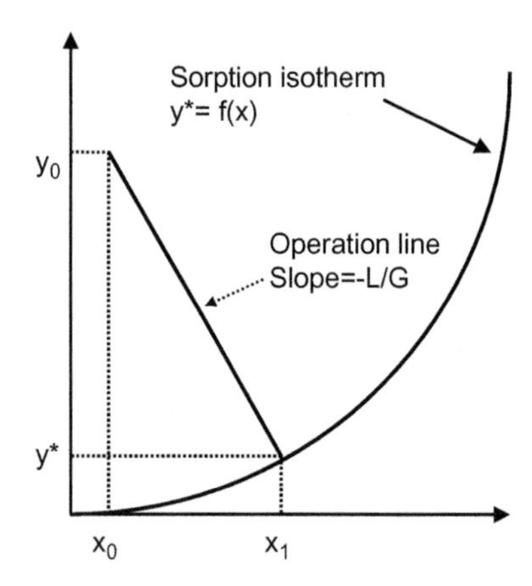

FIGURE 12.4

Graphical representation of an equilibrium adsorption process.

On the x, y plane, Eq. (12.4) is represented by a straight line with a negative slope equal to $-L/G$ and passing through the point (x_0,y_0). The intersection between this line (known as operation line) and the equilibrium curve (sorption isotherm) gives the values of x_1 and y^*.

As stated before, truly continuous adsorption processes are difficult to implement. On the other hand, "repeated batch adsorption", also known as "multistage crosscurrent adsorption" (Treybal, 1980), is frequently practiced. Repeated adsorption of a batch of solution with small fractions of adsorbent is more efficient than treatment with the total quantity of adsorbent in one stage (see Example 12.3).

EXAMPLE 12.2

The clear juice of a tropical fruit contains a bitter terpenoid. It has been found that the bitter principle can be removed by adsorption on a polyamide. Laboratory tests have shown that the sorption isotherm of the terpenoid on polyamide powder follows the Freundlich equilibrium model. Assuming equilibrium, calculate the quantity of fresh adsorbent to be applied to 1 kg of juice, in order to reduce the concentration of the bitter substance from 20 mg/kg to 0.1 vmg/kg, in the following cases:

- Case 1: The equilibrium function is linear $y = 0.00012\ x$
- Case 2: The equilibrium function is exponential $y = 0.00003\ x^{1.4}$

x and y represent the concentration (mg/kg) of the terpenoid in the absorbent and in the juice, respectively.

Solution

The values of y_0, y^*, n and m are substituted in Eq. (12.5), which is solved for L/G

For case 1: $y_0 = 20$ mg/kg , $y^* = 0.1\ \text{mg} \cdot \text{kg}^{-1}$, $n = 1$, $m = 0.00012$.
The result is: L/G = 0.0239 kg adsorbent per kg juice.
For case 2: $y_0 = 20$ mg/kg , $y^* = 0.1\ \text{mg} \cdot \text{kg}^{-1}$, $n = 1.4$, $m = 0.00003$.
The result is: L/G = 0.0606 kg adsorbent per kg juice.

EXAMPLE 12.3

A raw sugar syrup contains a dark pigment. The concentration of the pigment is measured by colorimetry and expressed as "color units", CU. Assume that the CU value is proportional to the pigment concentration (mg per kg). It is desired to reduce the CU by treatment with active carbon. In laboratory tests, the colored syrup with a CU value of 120 was treated with varying proportions of carbon. After equilibration and filtration, the CU of the liquid was measured. The results are given in Table 12.3.

a. How well do the data fit the Freundlich model? What are the parameters of the model?
b. How much carbon is needed in order to decolorate 1 kg of syrup from CU = 120 to CU = 6 in one step?
c. What would be the CU if the final syrup if the treatment were carried out in two steps, each step with one-half of the quantity of adsorbent found in part (b) above?

Solution

a. The Freundlich equation is linearized as follows:

$$\log y^* = \log m + n \log x$$

If the system follows the Freundlich model, a plot of log y^* versus x should be linear. The parameters m and n are then found from the intercept and slope of the straight

Table 12.3 Data of Bleaching with Active Carbon

g carbon per kg syrup	0	2	4	6	8
CU after treatment	120	63.6	36.0	20.4	12.0

line. In our case, y is defined as CUs in the syrup and x as CUs absorbed per kg of carbon. The material balance (Eq. 12.4) is written as follows:

$$x = \frac{y_0 - y^*}{L/G} = \frac{120 - y^*}{L/G}$$

where L/G = kg of carbon added to 1 kg of pigment-free syrup. (Note: It will be assumed that the mass concentration of the pigment is extremely low. Consequently, L/G is very nearly equal to kg of carbon to kg of syrup as is. Furthermore, the material balance, as written, assumes that the fresh carbon does not contain any pigment.) The value of x is calculated from the experimental data for each value of L/G. The results are shown in Table 12.4.

The plot of log y* versus log x is found to be very nearly linear (linear regression coefficient $R^2 = 0.9954$) with a slope of n = 2.266 (Figure 12.5). The intercept is log m = −7.3617.

Table 12.4 Syrup Decoloration Example, Equilibrium Values

y^*	x
63.6	28200
36.0	21000
20.4	16600
12.0	13500

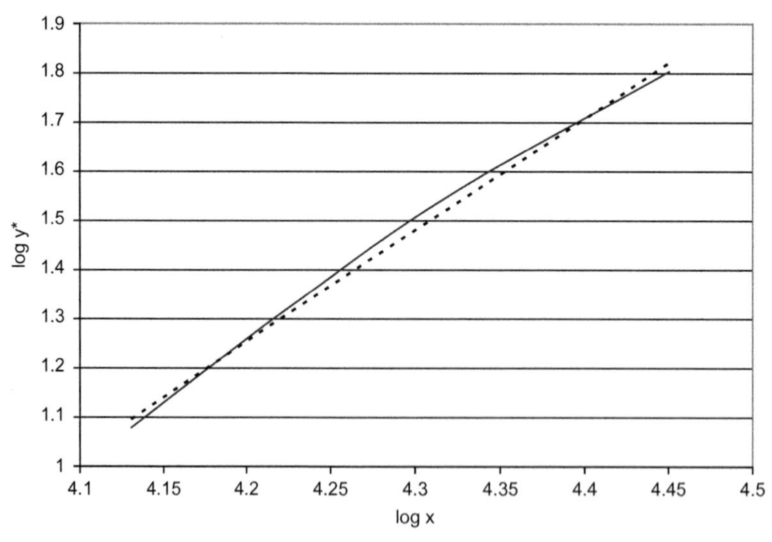

FIGURE 12.5

Conformity with the Freundlich model.

Therefore, the adsorption of the pigment on carbon seems to follow the Freundlich equation with $n = 2.266$ and $m = \text{antilog } -7.3617 = 5.45 \times 10^{-9}$.

b. $n = .66$, $m = 5.45 \times 10^{-9}$, $y_0 = 120$CU and $y^* = 6$ CU are substituted in Eq. (12.5), which is solved for L/G. The result is L/G = 0.0117 kg carbon per kg of syrup.

c. Two step adsorption:

- First step: L/G = 0.0117/2 = 0.0058 kg carbon per kg syrup
  $y_0 = 120$ CU
  $n = 2.66$
  $m = 5.45 \times 10^{-9}$.
  Substituting the data in Eq. (12.5) and solving graphically for y^*, we obtain:

$$y^* = 22 \text{ CU}$$

- Second step: L/G = 0.0117/2 = 0.0058 kg carbon per kg syrup
  $y_0 = 22$ CU (color of the treated syrup from the first step)
  $n = 2.66$
  $m = 5.45 \times 1^{-9}$.
  The color of the treated syrup from the second step $(y^*)_2$ is found by substituting the data and solving Eq. (12.5) for y^*. The result is:

$$(y^*)_2 = 0.65 \text{ CU}$$

Conclusion: The two-step decoloration is much more efficient than the one-step process using the same total quantity of adsorbent.

12.4 Adsorption in columns

Adsorption processes often use columns. The gas or liquid containing the adsorbate is passed through a porous bed of adsorbent contained in a column. The process may be seen as continuous with respect to the mobile phase, but discontinuous with respect to the stationary bed. Adsorption in columns can be treated as a multi-stage separation process. The efficiency of separation is better than that of batch operation. Furthermore, both the contacting and separation steps as well as the regeneration of the absorbent are carried out in the same equipment. Adsorption columns are widely used in adsorption from gas and clear solutions, and in ion exchange.

Consider a vertical column filled with fresh adsorbent (Figure 12.6).

As a mixture containing the adsorbate is slowly introduced from the top, the upper layers of the adsorbent become saturated with adsorbate. As feeding is continued, additional layers and finally the entire column become saturated. Let y_0 and y be the concentrations of adsorbate in the solution entering and leaving the column, respectively. A graph representing y as a function of the quantity of solution fed to the column is called a "breakthrough curve". A typical breakthrough plot is shown in Figure 12.7. The quantity of solution that passes through the column can be expressed as volume (m^3), as "number of

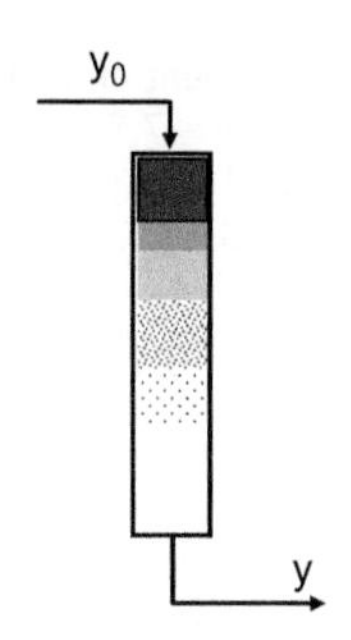

FIGURE 12.6

Adsorption in a column.

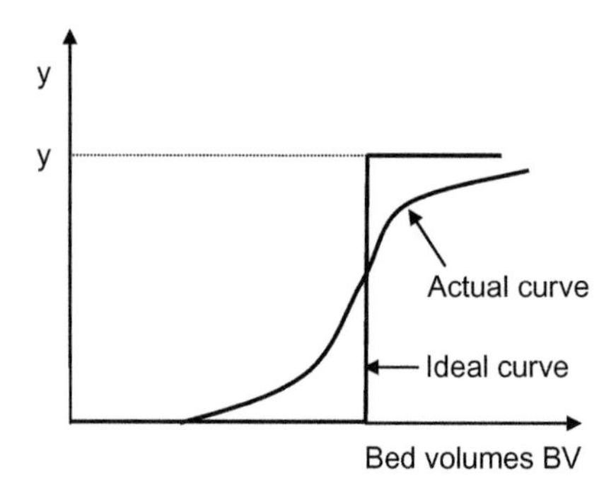

FIGURE 12.7

Typical breakthrough curve.

bed volumes" (dimensionless) or as time (assuming that the volumetric flow-rate is known and constant).

12.5 Ion exchange

12.5.1 Basic principles

Ion exchange is a type of reversible adsorption of ions on a special type of adsorbents called "ion exchangers". Ion exchangers are *charged* insoluble mineral or synthetic polymers (poly-ions) (Helferich, 1962; Dyer *et al.*, 1999; Zagorodni, 2007). Charged functional groups such as $-SO_3^-$, $-COO^-$ or $-NH_3^+$ are covalently bound to the insoluble polymer matrix. These fixed charges (charged functional groups or counter-ions) are neutralized by mobile, interchangeable ions of the opposite sign. Eq. (12.6) shows, as an example, the interchange between sodium and potassium cations when a resin loaded with mobile sodium cations is brought to contact with a solution containing a potassium salt. In this equation, R represents the insoluble polymeric matrix with fixed negative charges:

$$RNa + K^{+}(\text{in solution}) \leftrightarrow RK + Na^{+}(\text{in solution}) \qquad (12.6)$$

While the mineral ion exchangers of the soil are of paramount importance in nature, the industrial exchangers are almost exclusively synthetic organic polymers called *ion exchange resins*. For the exchange of small ions, the most commonly used polymer matrix is polystyrene, cross-linked with divinylbenzene. For larger charged particles (e.g., citrate anion), more porous matrices, known as macro-reticular resins, are used. For the separation of proteins in biotechnology, special hydrophilic polymer matrices such as dextran, agarose or cellulose are preferred. The matrix can be a poly-anion (negatively charged) or a poly-cation (positively charged). The first is neutralized by interchangeable cations and therefore functions as a *cation exchanger*. The second is an *anion exchanger*.

The most widespread industrial application of ion exchange is in the area of water treatment, described in some detail at the end of this section. Other applications in food processing include:

- Removal of calcium from solutions (beet or cane juices) in the manufacture of sugar
- Decolorization of sugar-beet juice (Coca *et al.*, 2008)
- Preparative separation of organic acids (citric, lactic, fumaric, etc.)
- Removal of excess acidity from fruit juices (Johnson and Chandler, 1985; Vera *et al.*, 2003)
- Demineralization of syrups in the manufacture of glucose
- Demineralization of slaughterhouse blood plasma (Moure *et al.*, 2008)
- pH control in must and wine (Walker *et al.*, 2004)
- Partial removal of potassium and/or tartrate in wine to reduce potassium tartrate ("tartar" or "Weinstein") precipitation (Mira *et al.*, 2006)
- Removal of metal ions (e.g., iron and copper) from wine (Palacios *et al.*, 2001)
- Demineralization of whey (e.g., Noel, 2002).

In addition, certain ion exchangers can also adsorb uncharged molecules and are used, for example, for the removal of unwanted color from clear solutions and for the chromatographic separation of sugars (Al Eid, 2006; Barker *et al.*, 1984; Coca *et al.*, 2008; Wilhelm *et al.*, 1989).

12.5.2 Properties of ion exchangers

The performance of ion exchangers depends on certain characteristics, some of which are discussed below:

1. *Exchange capacity*. The total capacity of an ion exchanger reflects the total amount of exchange sites (counter-ions) per unit mass or volume of the

exchanger. It is usually expressed as milli-equivalents per gram or per milliliter. In the case of the adsorption of large ions, the actual capacity may be smaller than the total capacity due to steric difficulties in accommodating large ions within the polymer matrix. The capacity of ion exchangers is of obvious economic significance.

EXAMPLE 12.4

It is desired to evaluate the total exchange capacity (meq per gram dry resin) of a cation exchanger consisting of sulfonated polystyrene in sodium (Na) form, cross-linked with divinylbenzene. It will be assumed that each benzene ring carries one sulfonic acid residue. The effect of cross-linking with divinylbenzene on the capacity will be neglected.

Solution

The "building block" of sulfonated polystyrene in sodium form is $-[C_8H_7SO_3^-\ Na^+]-$ with a formula weight of 206. Each unit provides one exchange site. The exchange capacity is therefore $1/206 = 0.00485$ equivalents per gram, or 4.85 meq per gram.

Note: The result is within the range of the actual exchange capacity of commercial resins of this type. It should be noted, however, that total exchange capacity is often expressed as meq per ml of wet (fully swollen) resin, which is about 40% of the capacity expressed as meq per gram of dry resin.

2. *Equilibrium characteristics, selectivity*. Consider ion exchange between two ions A and B of the same sign. Let z_A and z_B represent the valences of A and B, respectively. Denote the ion A in the solution and in the resin as A_S and A_R, respectively, and use the same distinction for the ion B. The exchange reaction is:

$$z_B A_S + z_A B_R \leftrightarrow z_B A_R + z_A B_S \tag{12.7}$$

where z_A and z_B are the valences of A and B, respectively.
To characterize the equilibrium conditions by a formal equilibrium coefficient, the activities of the different species in the solution must be known. Since this information is usually not available, a "pseudo-coefficient of equilibrium", K′, expressed in terms of concentrations, is defined as follows:

$$K' = \frac{[A_R]^{z_B}}{[B_R]^{z_A}} \times \frac{[B_S]^{z_A}}{[A_S]^{z_B}} \tag{12.8}$$

The brackets represent concentrations of A and B in terms of equivalents per liter (normality) in the solution or in the resin. We now define, for each one of the ions A and B, equivalent fractions *u* in the solution and *v* in the resin, as follows:

$$u_A = \frac{[A_S]}{N} \quad u_B = \frac{[B_S]}{N} \quad and \quad v_A = \frac{[A_R]}{Q} \quad v_B = \frac{[B_R]}{Q} \tag{12.9}$$

where N = total equivalents per liter of solution and Q = total equivalents per liter of resin (total exchange capacity of the resin)

Substitution in Eq. (12.8) gives:

$$K' = \left(\frac{v_A}{u_A}\right)^{z_B} \left(\frac{u_B}{v_B}\right)^{z_A} \left(\frac{N}{Q}\right)^{z_A - z_B} \quad (12.10)$$

Eq. (12.10) is further simplified in the case of two ions of the same valence, $z_A = z_B = z$. In this case:

$$K' = \left(\frac{v_A}{u_A} \times \frac{u_B}{v_B}\right)^{z} \quad (12.11)$$

The expression inside the brackets in Eq. (12.11) is called the *selectivity* of the resin for ion A against ion B. In the case of exchange between ions of the same charge, ion exchange resins are selective towards the ion of smaller solvated volume. Thus, in the series of monovalent cations, the order of selectivity is:

$$Cs^+ > Rb^+ > K^+ > Na^+ > Li^+ > H^+$$

Note: Ions with higher atomic number (more protons in the nucleus) have smaller volumes.

In the case of exchange between a monovalent and a bivalent ion, the resin is selective towards the bivalent ion if the total concentration of the solution in ions is low and towards the monovalent ion if the total concentration is high. This behavior is particularly important in water-softening.

Different types of selectivity curves are shown in Figure 12.8.

3. *Electrolytic properties*. The ion-counter ion system behaves as an electrolyte with given dissociation characteristics. Cation exchangers with anions of strong acids (such as sulfonic acid) attached as functional groups are qualified as "strongly acidic", while those carrying carboxyl ions as fixed functional groups are weak resins. Anion exchangers are classified in the same way as strong bases or weak bases, depending on the dissociation constant of their functional group. The full exchange capacity of a resin can be utilized only if the functional groups are fully dissociated.
4. *Cross-linkage*. The polymer chains of ion exchange resins are cross-linked to varying degrees. Polystyrene-based resins are usually cross-linked with divinylbenzene in the course of polymerization. The degree of cross-linkage affects some of the properties of the resin.
   a. Swelling. Dry resins swell considerably when they are wetted. Resins with a higher degree of cross-linkage swell less and are more rigid. Resins with low cross-linkage are softer, gel-like, and swell more. Higher swelling results in accelerated mass transfer and faster equilibration rate.

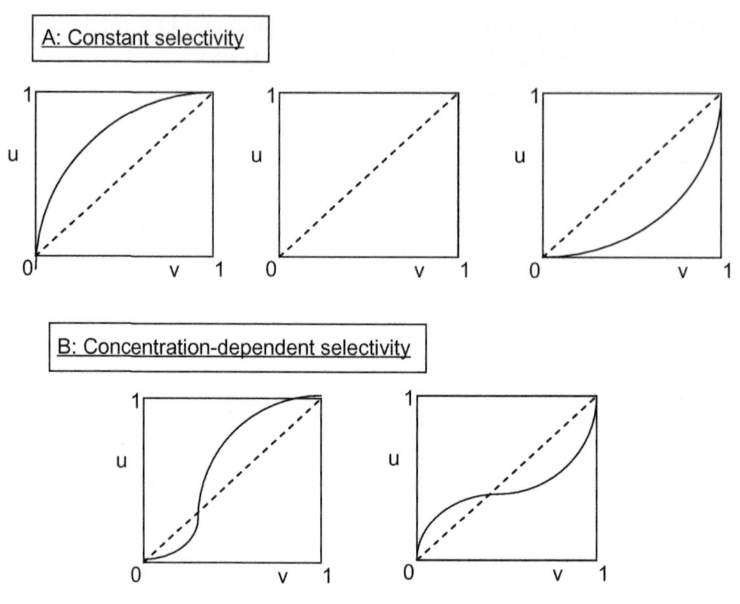

FIGURE 12.8

Different selectivity curves.

  b. Cross-linkage reduces the flexibility of the polymer chains and therefore enhances the selectivity for smaller ions that are easier to accommodate within the polymer network.

5. *Particle characteristics*. Ion exchange resins are usually produced as small spheres (beads), but other shapes (such as short cylinders) are available. Obviously, the size, shape and internal porosity of the particles affect mass transfer kinetics and column performance.

12.5.3 Water-softening using ion exchange

Water-softening is an important industrial operation, often necessary in food processing. Some of the processes requiring softened water are:

- Preparation of feed water for steam boilers, with the purpose of preventing scaling
- Treatment of water used for cooling containers after thermal processing, with the purpose of preventing unsightly "spots" left after the water drops dry out
- Preparation of softened process water, for the production of beverages, for cooking legumes, etc.

The "hardness" of water is due to the presence of calcium and magnesium cations. We distinguish between two kinds of "hardness":

1. Hardness due to calcium and magnesium salts in all forms. This kind of hardness affects the reaction of hard water with proteins (in legumes), certain anions and, particularly, fatty acid anions (soaps).
2. Hardness due to calcium and magnesium salts in bicarbonate form. This kind of hardness causes precipitation of insoluble carbonates (scaling) when the water is heated, according to the following equation:

$$Ca(HCO_3)_2(\text{soluble}) \rightarrow CaCO_3(\text{insoluble}) + CO_2 + H_2O$$

In water-softening by ion exchange, calcium and magnesium cations are exchanged with Na^+ or H^+ cations. In certain applications the hardness cations are exchanged with Na^+ and the resin is regenerated with a concentrated solution of NaCl according to the following equation:

$$Ca^{++}(\text{soln}) + 2NaR \leftrightarrow CaR_2 + 2Na^+$$

In the stage of softening, the medium (hard water) is a dilute solution and therefore the resin is highly selective for the bivalent calcium and magnesium ions. At the stage of regeneration, the medium (concentrated brine) is a concentrated solution and therefore the resin is selective for the monovalent sodium ion. Practical water-softening processes by ion exchange are based on this shift in selectivity.

Exchanging hardness ions with sodium converts water from hard to salty. This may prevent scaling but cause other problems such as increased corrosiveness. An alternate process is the total demineralization of water, using a double anion-and-cation exchange process. The cation exchanger is in the H^+ form and the anion exchanger is in the OH^- form. The cation exchanger adsorbs the cations in the feed water and releases H^+ ions. The anion exchanger exchanges the anions in the water with OH^- anions that are neutralized by the H^+ ions:

$$Ca^{++} + 2HR \rightarrow CaR_2 + 2H^+$$

$$2Cl^- + 2R'OH \rightarrow 2R'Cl + 2OH^-$$

$$2H^+ + 2OH^- \rightarrow 2H_2O$$

The cation exchanger is regenerated with HCl and the anion exchanger with NaOH.

12.5.4 Reduction of acidity in fruit juices using ion exchange

Ion exchangers can be used for the reduction of excess acidity in fruit juices. When applied to citrus juices, this treatment has been found to remove, simultaneously, some of the bitterness of the product.

The following equation illustrates the reduction of acidity due to citric acid, using an anion exchanger in OH form. The three-basic citric acid is the main source of acidity in citrus fruit juices:

$$H_3Citrate + 3ROH \rightarrow R_3Citrate + 3H_2O$$

The citrate ion is relatively large. The resin used for this application is therefore a macro-reticular polymer, providing the internal porosity required for the accommodation of large counter-ions.

Other carboxylic acids (malic, fumaric, lactic, tartaric) are adsorbed in the same way.

References

Al Eid, S., 2006. Chromatographic separation of fructose from date syrup. J. Food Sci. Nutr. 57 (1–2), 83–96.

Bansal, R.C., Goyal, M., 2005. Activated Carbon Adsoption. CRC Press, Taylor & Francis Group, Boca Raton, FL.

Barker, P.E., Irlam, G.A., Abusabah, E.K.E., 1984. Continuous chromatographic separation og glucose-frucrose mixtures using ion exchange resins. Chromatographia 18 (10), 567–574.

Coca, M., García, M.T., Mato, S., Cartón, A., González, G., 2008. Evolution of colorants in sugarbeet juices during decolorization using styrenic resins. J. Food Eng. 89 (4), 72–77.

Dyer, A., Hudson, M.J., Williams, P.A. (Eds.), 1999. Progress in Ion Exchange: Advances and Applications. Woodhead Publishing Ltd, Cambridge, UK.

Fayoux, S.C., Hernandez, R.J., Holland, R.V., 2007. Debitering of navel orange juice using polymeric films. J.Food Sci. 72 (4), E143–E154.

Helferich, F., 1962. Ion Exchange. Mc-Graw-Hill, New York, NY.

Johnson, R.L., Chandler, B.V., 1985. Ion exchange and adsorbent resins for removal of acids and bitter principles from citrus juice. J. Sci. Food Agric. 36 (6), 480–484.

Mira, H., Leite, P., Ricardo da Silva, J.M., Curvelo Garcia, A.S., 2006. Use of ion exchange resins for tartrate wine stabilization. J. Int. Sci. Vigne. Vin. 40 (4), 223–246.

Moure, F., Del Hoyo, P., Rendueles, M., Diaz, M., 2008. Demineralization by ion exchange of slaughterhouse porcine blood plasma. J. Food Process. Eng. 31 (4), 517–532.

Noel, R.,2002. Method of processing whey for demineralization purposes. US Patent 6383540.

Palacios, V.M., Caro, I., Pérez, L., 2001. Application of ion exchange techniques to industrial processes of metal ion removal from wine. Adsorption 7 (2), 131–138.

Soto, M.L., Moure, A., Dominguez, H., Parajó, J.C., 2010. Recovery, concentration and purification of phenolic compounds by adsorption: a review. J. Food Eng. 105 (1), 1–27.

Treybal, R.E., 1980. Mass-Transfer Operations, third ed. McGraw-Hill, New York, NY.

Vera, E., Dornier, M., Ruales, J., Vaillant, F., Reynes, M., 2003. Comparison between different ion exchange resins for the deacidification of passion fruit juice. J. Food Eng. 57 (2), 199–207.

Walker, T., Morris, J., Threlfall, R., Main, G., 2004. Quality, sensory and cost comparison for pH reduction of Syrah wine using ion exchange or tartaric acid. J. Food Qual. 27 (6), 483–496.

Wilhelm, A.M., Casamatta, G., Carillon, T., Rigal, L., Gaset, L., 1989. Modelling of chromatographic separation of xylose-mannose in ion exchange resin columns. Bioprocess. Biosyst. Eng. 4 (4), 147–151.

Zagorodni, A.A., 2007. Ion Exchange Materials: Properties and Applications. Elsevier, Amsterdam, The Netherlands.

Zeller, B.L., Saleeb, F.Z., 1996. Production of microporous sugars for adsorption of volatile flavors. J. Food Sci. 61 (4), 749–752.

Zeller, B.L., Saleeb, F.Z., Ludescher, R.D., 1998. Trends in the development of porous carbohydrate food ingredients for use in flavor encapsulation. Trends Food Sci. Technol. 9 (11–12), 389–394.

CHAPTER

Distillation

13

13.1 Introduction

Distillation is a separation operation based on differences in volatility. If a mixture containing substances that differ in their volatility is brought to ebullition, the composition of the vapors released will be different from that of the boiling liquid. After condensation, the vapors constitute the "distillate". The remaining liquid is called "residue" or "bottoms".

One of the oldest separation processes, distillation is of central importance in the chemical process industry. In the food sector, its main application is in the production of ethanol and alcoholic beverages from fermented liquids. Other food-related applications include the recovery, fractionation and concentration of volatile aromas, recovery of organic solvents (desolventation) in the production of edible oils by solvent extraction and removal of undesirable odorous substances, e.g. deodorization of cream, deodorization of edible oils and fats (Calliauw et al., 2008). Distillation may be carried-out as a batch or as a continuous process.

13.2 Vapor–liquid equilibrium (VLA)

Consider a binary liquid solution consisting of substances A and B. If the mixture behaves as an ideal solution, the vapor pressure of substance A, p_A, is given by Raoult's Law (François-Marie Raoult, French chemist, 1830 – 1901), as formulated in Eq. (13.1):

$$p_A = x_a p_A^0 \tag{13.1}$$

where x_a = concentration of A in the solution, in mol fraction, and p_A^0 = vapor pressure of pure A, at the temperature of the solution.

If, furthermore, the vapor phase behaves as an ideal gas mixture, then Dalton's Law (John Dalton, English physicist and chemist, 1766–1844) applies. The partial pressure of A in the vapor is then:

$$\overline{p}_A = y_A P \tag{13.2}$$

where y_A = concentration of A in the vapor phase, in mol fraction, and P = total pressure.

Food Process Engineering and Technology. DOI: http://dx.doi.org/10.1016/B978-0-12-415923-5.00013-7
© 2013 Elsevier Inc. All rights reserved.

At equilibrium, the partial pressure of A in the gas must be equal to the vapor pressure of A over the solution. The equilibrium concentration of A in the vapor, y_A^* is then given by Eq. (13.3):

$$y_A^* = x_A \frac{p_A^0}{P} \tag{13.3}$$

Unless the total pressure is abnormally high, the assumption of ideal gas mixture for the gas phase is fairly safe. In contrast, few liquids of interest in food processing behave like ideal solutions. For example, the vapor pressure behavior of ethanol–water mixtures deviates considerably from Raoult's Law. The vapor pressure of component A over a non-ideal solution is given by Eq. (13.4):

$$p_A = \gamma_A x_A p_A^0 \tag{13.4}$$

where γ_A is the activity coefficient of A in the solution. The activity coefficient is not constant, but varies with the temperature and the composition of the mixture. The equilibrium concentration of A in the vapor phase is then:

$$y_A^* = x_A \gamma_A \frac{p_A^0}{P} \tag{13.5}$$

The use of Eq. (13.5) for the prediction of vapor–liquid equilibrium data is problematic, mainly because of the difficulty in finding activity coefficient values for mixtures as a function of temperature and composition. Another expression used for the prediction of VLE data is the "relative volatility", defined as follows:

$$\alpha_{A \to B} = \frac{y_A^* (1 - x_A)}{x_A (1 - y_A^*)} \tag{13.6}$$

where $\alpha_{A \to B}$ is the volatility of A relative to B.

In an ideal solution, the relative volatility is simply the ratio of the vapor pressures of the pure components A and B. Within a limited range of concentrations, the relative volatility may be assumed to be fairly constant.

Detailed experimental VLE data for many mixtures of industrial interest are available in the literature. Data for the ethanol–water system are given in table form in Table A.14 (see Appendix) and as a graph in Figure 13.1.

One of the peculiarities of the ethanol–water solution is the formation of an azeotrope at the ethanol concentration of 0.894 (mol fraction). An azeotrope is a homogeneous mixture that has the same composition in the vapor phase as in the liquid phase at boiling point. Therefore, an azeotrope behaves in distillation as a pure substance and not as a mixture. It is thus impossible to separate an azeotrope into its components by simple distillation (see Section 13.8 for separation of the azeotrope by pervaporation). The ethanol–water azeotrope has a boiling point of 78.15°C at atmospheric pressure, slightly lower than that of pure ethanol. Many mixtures of two or three substances are known to form azeotropes.

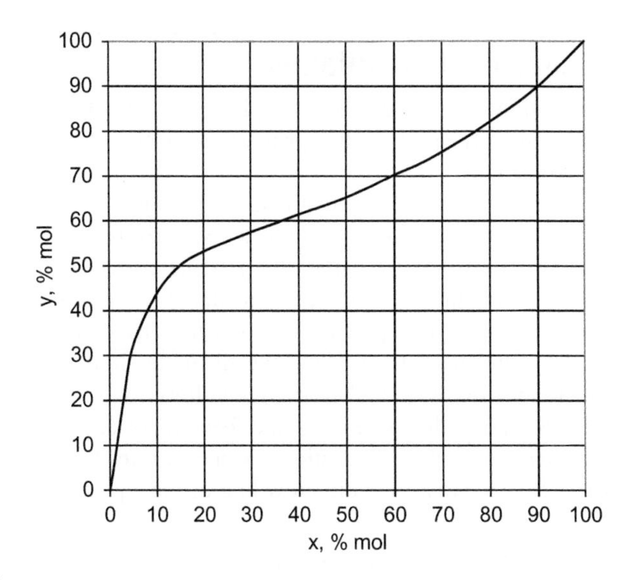

FIGURE 13.1

Ethanol–water vapor–liquid equilibrium curve, in mol%.

EXAMPLE 13.1

Using the VLE data for the ethanol–water system from Table A.14, calculate the relative volatility of ethanol vs water in liquid mixtures containing 10, 20 and 50% (mol) alcohol.

Solution

The relative volatility is defined as:

$$\alpha_{A \to B} = \frac{y_A^*(1 - x_A)}{x_A(1 - y_A^*)}$$

Table A.14 shows:

x = 0.1	y* = 0.437
x = 0.2	y* = 0.532
x = 0.5	y* = 0.622

Substitution in the equation gives:

For x = 0.1	α = 6.99
For x = 0.2	α = 4.55
For x = 0.5	α = 1.65

Clearly, for the ethanol–water system, the relative volatility is not constant in the range tested.

13.3 Continuous flash distillation

Continuous flash distillation is one of the simplest methods of distillation. In the food industry, it is mainly used in the primary recovery of aroma from fruit juices, or for deodorization.

Consider the process described schematically in Figure 13.2. The feed mixture is preheated and then introduced (flashed) into the vaporization chamber, where a pressure considerably lower than the saturation pressure corresponding to the temperature of the feed is maintained. The feed immediately comes to ebullition. Part of it evaporates adiabatically and part of it leaves the chamber as a liquid (bottoms). The vapor is condensed to produce the distillate. It is assumed that the vapor and the liquid leaving the chamber are in equilibrium.

Let F, V and B represent the quantity of the feed, vapor and bottoms, in mols, respectively. (The symbols for concentrations, x and y, refer to the more volatile component and are not indexed, to avoid crowding of the equations.)

The overall material balance is:

$$F = B + V \tag{13.7}$$

A material balance for the volatile component gives:

$$F \cdot x_F = V_y^* + Bx_B \tag{13.8}$$

Substitution and elimination of F gives:

$$y^* = -\frac{B}{V}(x_B - x_F) + x_F \tag{13.9}$$

Equation (13.9) is based solely on material balance and is linear on the x, y plane. The straight line corresponding to such material balance is called an "operating line". The operating line of Eq. (13.9) has a slope equal to $-B/V$ and passes through the point $x = y = x_F$. The slope $-B/V$ depends on the vaporization ratio –i.e., the quantity of vapor generated from F mols of feed. It is calculated by writing an energy balance for the vaporization process, based on the enthalpies of the feed, the bottoms and the vapor, and assuming adiabatic evaporation. The vaporization ratio is usually regulated by adjusting the feed temperature.

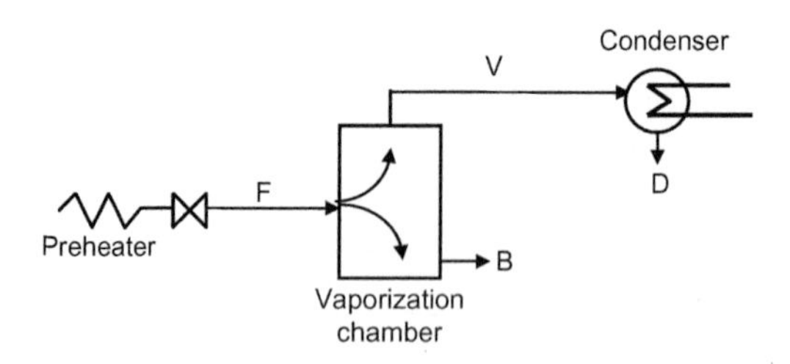

FIGURE 13.2

Schema of flash distillation.

The composition of the distillate, y*, must satisfy, on one hand, the operating equation Eq. (13.9) and, on the other hand, the equilibrium function $y^* = f(x_B)$. If the equilibrium function is available in graphical form (equilibrium curve), the composition of the distillate and that of the bottoms is obtained from the intersection of the operating line and the equilibrium curve (Figure 13.3).

Flash distillation is a relatively simple and inexpensive process. Its main disadvantage is in its being a single-step equilibrium process, resulting in limited enrichment and low yield. It is therefore mainly applied to mixtures containing components of widely different volatility.

EXAMPLE 13.2

Raw soy milk has a characteristic "beany" odor which is objectionable to some consumers. The main factor responsible for the beany flavor is heptanal (MW = 114). It has been found that the objectionable odor can be substantially reduced by flash distillation.

Raw soy milk, containing 100 mg heptanal per kg, is heated to 100°C and flashed into an evaporation vessel where the pressure is maintained at 20 kPa. Most of the heptanal is removed with the vapor. The remaining liquid (bottoms) is found to contain only 4 mg^{-1} heptanal.

Calculate the average activity coefficient of heptanal in water, assuming that it is not affected considerably by temperature or concentration within the conditions of the process. Assume that the thermal properties of the soy milk (which is a dilute aqueous mixture) are like those of water.

The vapor pressure of heptanal at 60°C (saturation temperature of water at 20 kPa) is 4 kPa.

Solution

The activity coefficient will be calculated from Eq. (13.5):

$$y^* = x\gamma\frac{p^0}{P}$$

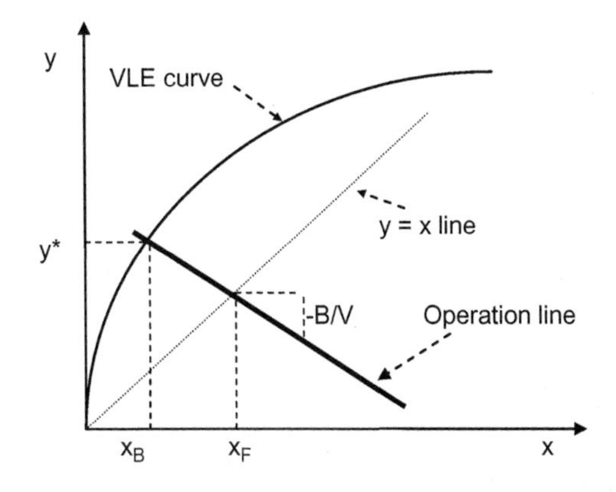

FIGURE 13.3

Graphical representation of flash distillation.

The total pressure P and the vapor pressure p^0 of heptanal at 60°C are known. We need x and y*. These can be calculated from the operating line equation:

$$y^* = -\frac{B}{V}(x - x_F) + x_F$$

The vaporization ratio will be calculated from an enthalpy balance:

$$Fh_F = Vh_v + Bh_B$$

For 1 kg of feed (F = 1):

$$h_F = Vh_v + (1 - V)h_B$$

All the solutions and the vapor are very diluted in hexanal. Their thermal properties will be assumed to be equal to those of water, and will be read from saturated steam tables:

$h_F = 419\ \text{kJ}\cdot\text{kg}^{-1}$, $h_V = 2610\ \text{kJ}\cdot\text{kg}^{-1}$, $h_B = 251\ \text{kJ}\cdot\text{kg}^{-1}$

Substitution gives: B/V = 0.07/0.93 = 0.075.
Substitution in the operation line equation gives:

$$y^* = -0.075(4 - 100) + 100 = 107.2\ \text{mg}\cdot\text{kg}^{-1}$$

x and y* have to be converted to mol fractions:

$$x = \frac{4/114}{10^6/18} = 0.63 \times 10^{-6} \quad y^* = \frac{107.2/114}{10^6/18} = 16.9 \times 10^{-6}$$

$$\gamma = \frac{y^* P}{x p^0} = \frac{16.9 \times 20}{0.63 \times 4} = 134$$

EXAMPLE 13.3

Recently, technologies have been developed for the production of aromas by microorganisms. A certain fermentation broth, produced to that purpose, contains an aroma substance at the concentration of 0.5% w/w. The aroma is to be recovered by flash distillation. The broth will be heated to temperature T and flashed into the evaporation vessel maintained at 20 kPa.

Find the minimal value of T for a recovery of 70% of the aroma in the distillate.

The liquid–vapor equilibrium function for the water-aroma system is: y* = 63x, where x and y are the concentrations of the aroma in the liquid and in the vapor, in weight percentage.

The thermal properties of the broth are like those of water.

Solution

Assume 100 kg of broth is flashed into the vaporization chamber. The quantity of aroma in the feed is 0.5 kg, of which 70% (0.35 kg) must be in the vapors and 30% (0.15 kg) in the bottoms. Hence, the concentration of the aroma in the vapor and in the bottoms is, respectively:

y* = 35/V %
x = 15/(100 − V) %.

We now use the equilibrium function given, to calculate V and B. y* = 63x, hence

$$\frac{35}{V} = 63 \times \frac{15}{100 - V} \Rightarrow V = 3.57 \text{ kg} \quad B = 100 - 3.57 = 96.43 \text{ kg}$$

Enthalpy balance:

$$Fh_F = Vh_v + Bh_B$$

The vapor and the bottoms approximate saturated water vapor and liquid at 20 kPa: $h_V = 2610\ \text{kJ} \cdot \text{kg}^{-1}$ and $h_B = 251\ \text{kJ} \cdot \text{kg}^{-1}$ (from Table A.10).

$$100h_f = 3.57 \times 2610 + 96.43 \times 251 = 33521.6 \Rightarrow h_f = 335.2\ \text{kJ} \cdot \text{kg}^{-1}$$

We find from the Table A.10 that the temperature of water for that enthalpy is 80°C.

13.4 Batch (differential) distillation

Batch distillation is, in essence, the simplest type of distillation, familiar to all from laboratory exercises (Figure 13.4). A batch of the mixture to be distilled is boiled in a closed vessel. The vapors are condensed by cooling, and the distillate is collected.

In contrast to continuous distillation, batch distillation does not take place at steady state. The composition of the boiling liquid, that of the vapors and that of the collected distillate changes continuously during the process.

Let L designate the quantity (mols) of the liquid mixture in the vessel (still) at a given time. Let x be the concentration (mol fraction) of the more volatile component in that mixture. An infinitesimal amount dL of the liquid is evaporated. Since the vapor emitted is assumed to be in equilibrium with the liquid, the concentration of the more volatile component in the vapor will be denoted as y*. A material balance can be written, as follows:

$$xL = (L - dL)(x - dx) + y^* dL \tag{13.10}$$

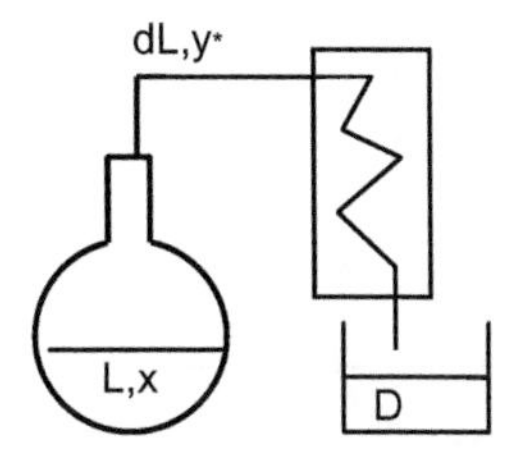

FIGURE 13.4

Batch distillation.

Separating the variables and integrating, we obtain:

$$\int_{L}^{L_0} \frac{dL}{L} = \ln\frac{L_0}{L} = \int_{x}^{x_0} \frac{dx}{y^* - x} \tag{13.11}$$

Equation (13.11) is known as Rayleigh's Law (John William Strutt, Lord Rayleigh, English physicist, 1842–1919, Nobel Prize in Physics, 1904), and differential distillation is often called "Rayleigh distillation". Note that the same formula can be used with concentrations in mass fraction and L in mass, if the VLE data are also in terms of mass.

The integral is solved analytically, if the equilibrium conditions are given as an algebraic expression. Otherwise the integral is calculated graphically. Computerized models for the calculation of batch distillation are available (Claus and Berglund, 2009).

In addition to being the standard method of distillation in the laboratory, differential batch distillation is the accepted process for small-scale distillation of brandy, perfumes and aromas. In these applications, the batch still may be equipped with a reflux condenser (Figure 13.5).

EXAMPLE 13.4

A manufacturer of brandy distils wine in batches. The wine contains 11% w/w ethanol. The distillate contains 40% w/w ethanol. (This distillate is later distilled again to a higher alcohol content.)

How much wine is needed to produce 100 kg of distillate?

Solution

The Rayleigh equation will be applied. Since the VLE function for ethanol–water is not given in algebraic form, it will be necessary to compute the integral by increments.

- Write a table with the values of x and corresponding y* from the VLE chart for mass fraction (Figure 13.6), starting with the initial value of x = 11% and decreasing x by a constant interval Δ (say, 1%).

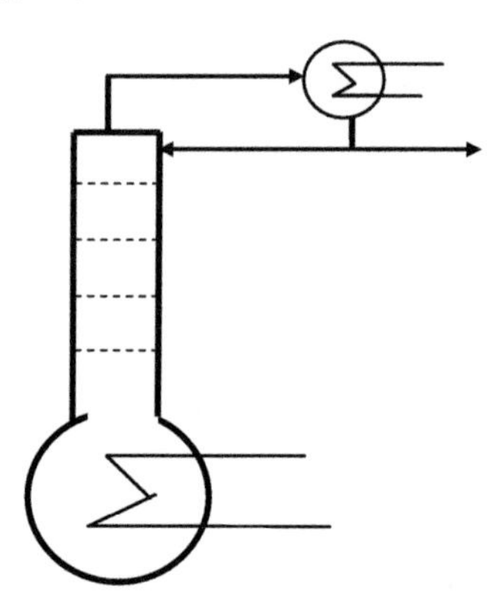

FIGURE 13.5

Batch distillation with reflux.

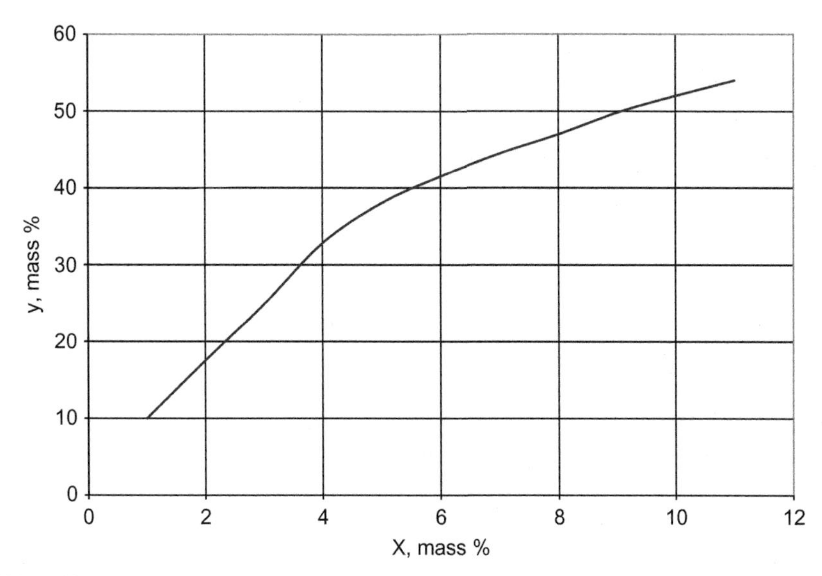

FIGURE 13.6

Batch distillation example.

- Calculate $\Delta Z = \Delta x/(y^* - x)$ for each increment. Note in a column.
- Sum the ΔZ values from the first row. Note that the sum is equal to $\ln(LO/L)$.
- Calculate L_O/L. Take $L_O = 100$ kg as a basis. Calculate L for each row. Mark in a column. Calculate $D = 100 - L$ for each row. Mark in another column.
- Calculate the quantity of alcohol in the bottoms L as Lx at each row.
- Calculate, by difference $11 - Lx$, the corresponding quantity of alcohol in the distillate.
- Calculate the alcohol content of the distillate. Find the row at which that content is 40%. This gives the quantity of 40% distillate obtainable from 100 kg of wine (Table 13.1).
- Convert to the quantity of wine needed to obtain 100 kg of distillate. The result is 435 kg of wine.

13.5 Fractional distillation

13.5.1 Basic concepts

Just as in the separation processes previously discussed, the recovery of a distillate rich in the more volatile component at relatively high yield requires the application of the "multistage countercurrent contact process" concept. In the case of distillation, this concept takes the form of *fractional distillation*, carried out in a *distillation column*. The distillation column contains the multiple contact stages,

Table 13.1 Calculations of Batch Distillation Example

x	y	ΔZ	ln L_0/L	L_0/L	L	D	Lx	11 − Lx	y_D
11	54								
10	52	0.02	0.02	1.02	97.67	2.33	9.77	1.23	53.00
9	49.8	0.02	0.05	1.05	95.34	4.66	8.58	2.42	51.95
8	47	0.03	0.07	1.08	92.98	7.02	7.44	3.56	50.76
7	44.5	0.03	0.10	1.10	90.58	9.42	6.34	4.66	49.48
6	41.5	0.03	0.13	1.13	88.14	11.86	5.29	5.71	48.14
5	38	0.03	0.16	1.17	85.60	14.40	4.28	6.72	46.67
4	32.8	0.03	0.19	1.21	82.87	17.13	3.31	7.69	44.87
3	24.8	0.04	0.23	1.26	79.66	20.34	2.39	8.61	42.34
2	17.5	0.05	0.28	1.32	75.50	24.50	1.51	9.49	38.74
1	10	0.08	0.36	1.44	69.58	30.42	0.70	10.30	33.88

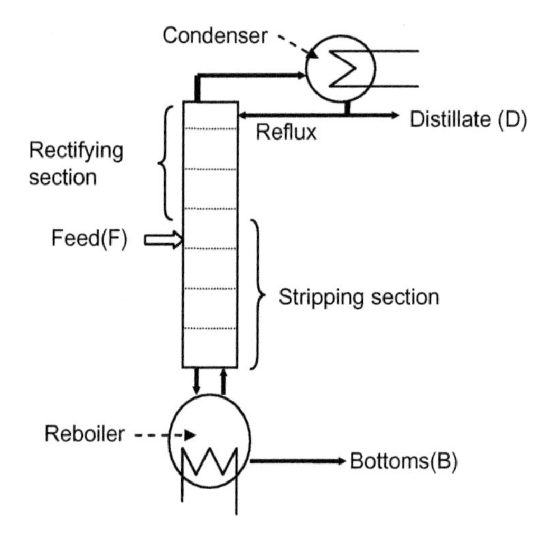

FIGURE 13.7

Schema of a fractional distillation column.

known as *plates* or *trays*, through which liquid and vapor move in opposite directions (Figure 13.7). The vapor moves up and the liquid moves down. The mixture to be distilled is introduced at a certain location. The portion of the column above that location is called the *rectification zone*, while the portion below the feed entrance is the *stripping zone*. The column is connected to a *reboiler*, where heat is supplied for vaporization (boiling), and a *condenser*, where the vapor is condensed by cooling.

In its travel up the column, the vapor is cooled by the descending liquid and the liquid is heated by the vapor. This exchange of heat results in the condensation of the less volatile component from the gas and the evaporation of the more volatile component from the liquid. The gas is thus enriched in volatile component as it moves up the column, and the liquid is stripped of its more volatile component as it flows down the column. The liquid obtained by condensation in the condenser is divided into two parts: one part is fed back into the column in order to provide sufficient liquid for contact with the gas. This part is called the *reflux*. The other part is the *distillate*. Vapor can be taken out and condensed at any position on the column, producing distillates or fractions of different compositions – hence the name "fractional distillation".

13.5.2 Analysis and design of the column

Designing a distillation column involves determination of the number of contact stages, selection of the type of "plates", calculation of the column dimensions, calculation of the heat exchange at the reboiler and the condenser, the design and selection of the auxiliary equipment (piping, pumps, measurement and control systems), etc. (Petlyuk, 2004; Stichlmair, 1998; Treybal, 1980.) The process specifications are the flow-rate, the composition and thermal properties of the feed, the desired composition of the distillate, and the expected yield. A basic requirement for the design is knowledge of the liquid–vapor equilibrium data for the mixtures involved. Software programs are available for the complete design of the process and the equipment. Here, we shall describe a simplified graphic method for the fractional distillation of a binary mixture.

The method, known as the McCabe–Thiele procedure, assumes that the molar heat of evaporation of most liquids of interest is about $40\ \mathrm{kJ \cdot mol^{-1}}$, and that there is little variance between them (e.g., $39.2\ \mathrm{kJ \cdot mol^{-1}}$ for ethanol, $40.6\ \mathrm{kJ \cdot mol^{-1}}$ for water).

The two sections of the column are analyzed separately.

1. *The rectifying section*. The stages (plates) will be numbered in descending order: 1, 2, 3, . . ., n. The streams entering and leaving stage n are shown in Figure 13.8. The concentrations in volatile component are denoted as x for the liquid and y for the vapor, both in mol fraction.

   Let us do a material balance for the section of the column from stage n to the condenser, inclusive (Figure 13.9).

   Overall mass balance:

$$V_{n+1} = L_n + D \tag{13.12}$$

   Material balance for the volatile component:

$$y_{n+1}V_{n+1} = x_nL_n + x_DD \tag{13.13}$$

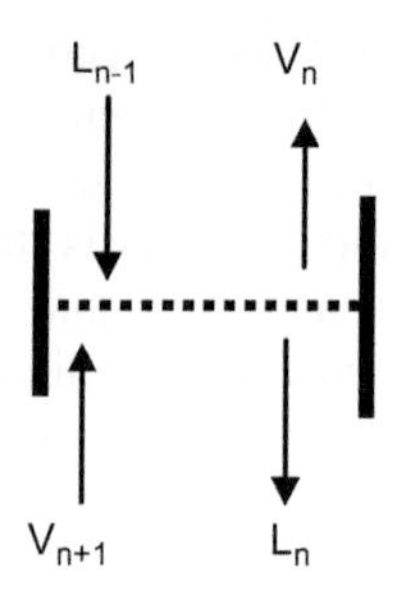

FIGURE 13.8

Streams entering and leaving plate n.

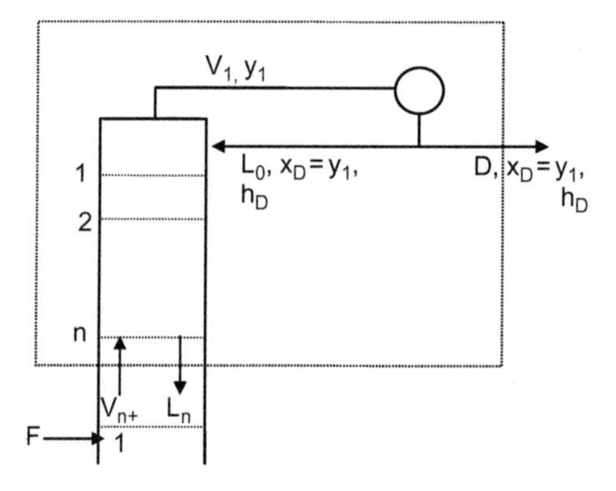

FIGURE 13.9

Material balance for the rectifying section.

By virtue of the equality of the heat of evaporation, the molar flow-rates of the liquid streams and the gas streams are equal, regardless of the composition, i.e., $V_1 = V_2 = V_3 = \ldots = V_n = V_{n+1}$ and $L_0 = L_1 = L_2 = \ldots L_n$.

We define the reflux ratio, R, as:

$$R = \frac{L_0}{D} \tag{13.14}$$

Substitution in Eqs (13.12) and (13.13) gives:

$$y_{n+1} = \frac{R}{R+1}(x_n - x_D) + x_D \tag{13.15}$$

Eq. (13.15), based solely on material balance, is linear. On the x, y plane it is represented by a straight line passing through the point $y = x = x_D$, with a slope equal to $R/(R+1)$. This line is called the "operation line of the

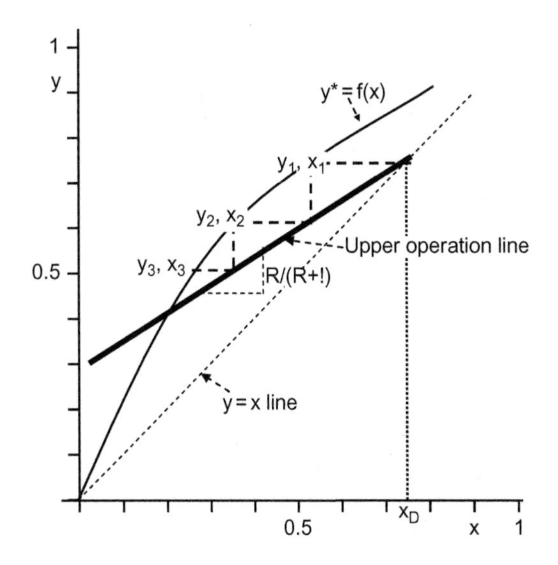

FIGURE 13.10

Construction of the upper operation line.

rectification section", or the "upper operation line". Figure 13.10 shows the construction of the upper operation line for $R = 2$ and $x_D = 0.75$. The same figure shows how we can find the composition of vapors and liquids leaving each stage in the rectification section. On the same figure, we draw the equilibrium curve for the mixture. We start with the point $y_1 = x_D$. We find the point y_1, x_1 on the equilibrium curve. From x_1 we find the point x_1, y_2 on the operation line, and so on. Each "step" in this graphical construction represents one plate in the rectification section.

2. *The stripping section*: the stages will be numbered from top to bottom, starting with the plate just below the feed inlet: 1, 2, 3 ... m. The molar flow-rates in this section will be marked as $\overline{L}$ and $\overline{V}$, to distinguish them from the rates L and V above the feed inlet. A material balance for the portion delimited in Figure 13.11 gives:

$$\overline{L}_m = B + \overline{V}_{m+1} \tag{13.16}$$

$$\overline{L}_m \cdot x_m = B \cdot x_B + \overline{V}_{m+1} \cdot y_{m+1} \tag{13.17}$$

$$y_{m+1} = \frac{\overline{L}_m}{\overline{V}_{m+1}}(x_m - x_B) + x_B \tag{13.18}$$

Equation (13.18) is the algebraic expression of the operation line below the feed inlet (lower operation line). It passes through the point $x = y = x_B$ and its slope is $\frac{\overline{L}_m}{\overline{V}_{m+1}}$. In contrast to the upper operation line, however, the

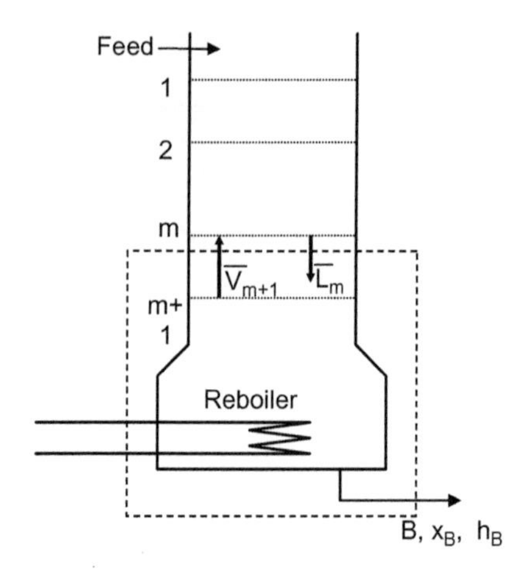

FIGURE 13.11

Material balance for the stripping section.

lower operation line cannot be built yet because the values in the slope are not known. Obviously, the flow-rates of liquid and vapor below the feed depend on the thermal conditions of the feed. If the feed is, for example, a cold liquid, then the quantity of liquid in the stripping section will be higher. It is customary to characterize the thermal condition of the feed with the help of the "thermal quotient θ": defined as follows:

$$\theta = \frac{h_V - h_F}{h_V - h_L} = \frac{\overline{L} - L}{F} \tag{13.19}$$

where the symbol h represents molar enthalpies and F represents the feed. The quotient θ is, in fact, the ratio of the heat needed to evaporate 1 mol of the feed to the total molar latent heat of evaporation of the mixture. If follows that if the feed is a saturated liquid then $\theta = 1$, and if the feed is a saturated vapor then $\theta = 0$.

We now can define the point of intersection between the two lines of operation. Combining Eqs (13.15) and (13.18) and substituting $\theta = \frac{\overline{L} - L}{F}$, we obtain:

$$y = \frac{\theta}{\theta - 1} \cdot x - \frac{x_F}{\theta - 1} \tag{13.20}$$

Equation (13.20) is then the geometric locus of the points of intersection between the two operation lines. It is represented by a straight line, passing through the point $x = y = x_F$ and having a slope equal to $\frac{\theta}{\theta - 1}$.

Example 13.5 illustrates the calculation of the number of ideal stages needed to perform a given separation task by fractional distillation, using the McCabe-Thiele graphical method.

EXAMPLE 13.5

A liquid mixture of ethanol and water, containing 35% (mol) ethanol, is to be distilled. The distillate should contain 70% (molar) and the bottoms not more than 2% (molar) ethanol. The reflux ratio $R = 1$. The feed is saturated liquid. The ethanol–water liquid–vapor equilibrium curve is given. Find the number of ideal contact stages and the optimal location for the introduction of the feed.

Solution

Refer to Figure 13.12.

Step 1: The equilibrium curve $y^* = f(x)$ is plotted.

Step 2: The upper operation line is built on the same plot, using Eq. (13.15), with $R = 1$ and $x_D = 0.7$.

Step 3: The θ line is built, using Eq. (13.20).

Step 4: The intersection of the upper operation line and the θ line is joined to the point $x = y = x_B$ by a straight line. This is the "lower operation line".

Step 5: Starting with the point $y_1 = x_D$, we construct the "steps" representing the stages going from the operation line to the equilibrium curve and back. The optimal location for the feed inlet (that which gives the smallest number of plates) is at or about the intersection between the two operation lines (Stage 3 in this example).

Step 6: We count the number of "plates".

The number of ideal stages needed is between six and seven (rounded to seven). The reboiler is considered as one plate. Therefore, the column should have the equivalent of six ideal plates. The ideal location for the introduction of the feed is between the third and the fourth stages.

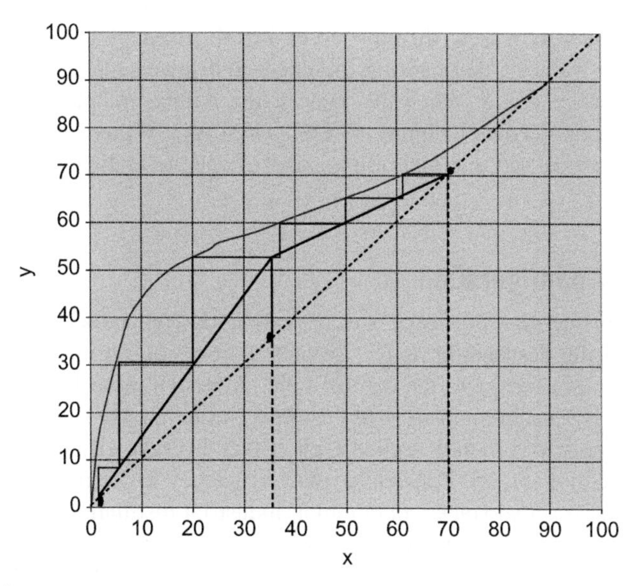

FIGURE 13.12

Fractional distillation example.

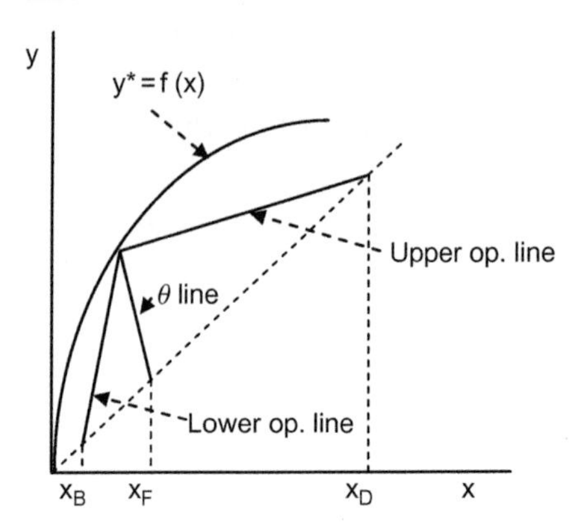

FIGURE 13.13

Minimal reflux ratio.

13.5.3 Effect of the reflux ratio

The reflux ratio may be modified by the operator, within certain limits. Its maximum value is ∞, meaning total reflux. Under total reflux conditions, the slope of the upper operation line is 1. The upper operation line, the 45° auxiliary line and the lower operation line merge together and the number of stages is minimal. This is, of course, a theoretical situation, because total reflux means no distillate. If the reflux ratio is now gradually reduced, more distillate is produced but the number of plates increases, thereby increasing the cost of the equipment. The minimal reflux value is that causing the upper and lower operation lines to meet on the equilibrium curve. When this occurs, the number of plates necessary for the separation is infinite (Figure 13.13). Thus, the reflux ratio is regulated between total reflux and minimal reflux, so as to operate under economically optimal conditions.

13.5.4 Tray configuration

The purpose of the plates or trays is to provide good contact between the ascending vapor and the descending liquid. Deviation from equilibrium depends on the contact surface provided and the contact time. Although these factors can be calculated by mass transfer analysis, it is customary to define some sort of "tray efficiency". The most commonly applied type is the "average efficiency" defined in Chapter 11, Section 11.2.5. Practically, plate efficiency is the ratio of ideal (equilibrium) stages to actual stages. In common industrial distillation plants, the practical efficiency varies between 50% and 90%.

There are many types of trays in fractional distillation columns. The simplest consist of a flat, perforated plate with a weir. The quantity of liquid on a perforated plate is determined by equilibrium between the pressure drop of the

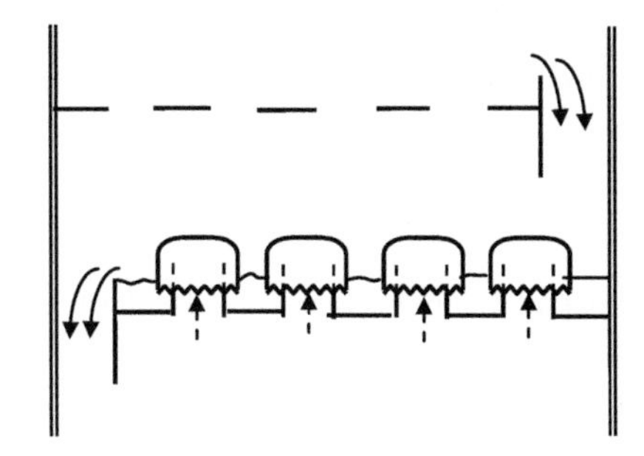

FIGURE 13.14

Schema of the bubble-cap tray.

ascending vapor and the hydrostatic pressure of the liquid. The traditional tray in distillation columns for food-grade alcohol is the more efficient but much more expensive "bubble-cap" tray, also known as the Barbet tray (Figure 13.14).

Packed columns do not contain trays. The contact between the liquid and the vapor is provided by the packing material, which consists of orderly or random particles (ceramic or stainless steel rings, saddles or irregular pieces). In the case of packed columns, it is customary to define a *height of theoretical equilibrium plate* (HTEP), which is the height of packing functionally equivalent to one ideal contact stage.

13.5.5 Column configuration

The distillation column is not always in the form of a single tower. The reboiler is usually a separate heat exchanger. Very often, the rectifying and stripping sections are built as interconnected but separate units. In the case of multi-component feeds, auxiliary columns can be used for the further purification of distillate fractions.

13.5.6 Heating with live steam

If one of the components of the mixture is water (e.g., ethanol−water mixtures), it is possible to replace the boiler by injection of live steam as a heating medium. The advantage of this method of heating is a reduction in both capital cost and energy expenditure. Steam supplies heat but also increases the amount quantity of water, and must be taken into account in the material balance. Practically, the upper operation line is not affected. The lower operation line passes through the point $y = 0$, $x = x_B$ instead of the point $y = x = x_B$.

13.5.7 Energy considerations

With respect to the operation cost of a fractional distillation system, two points have to be considered: the energy supplied to the reboiler, and the cooling water consumption of the condenser.

Refer to the scheme in Figure 13.15.

If the vapor is totally condensed, with no sub-cooling, the rate of heat removed in the condenser is:

$$q_c = V_1\lambda = D(R+1)\lambda \Rightarrow \frac{q_c}{D} = (R+1)\lambda \tag{13.21}$$

where λ is the molar heat of evaporation. The effect of the reflux ratio on the cooling load per mol of distillate is evident.

Energy balance for the entire system gives:

$$Fh_F + q_R = q_c + Bh_B + Dh_D \tag{13.22}$$

From overall material balance and material balance on the volatile, we obtain:

$$F = D\left(\frac{x_D - x_B}{x_F - x_B}\right) \tag{13.23}$$

The expression in brackets is composed of the composition of the feed, the distillate and the bottoms. We shall denote it as ξ for simplicity. The heating load at the reboiler, per mol of distillate, can now be expressed as:

$$\frac{q_R}{D} = (R+1)\lambda + h_D + (\xi - 1)h_D - \xi h_F \tag{13.24}$$

Equation (13.24) shows the strong effect of the reflux ratio and the thermal conditions of the feed on the heating load per unit product.

13.6 Steam distillation

Steam distillation is one of the principal methods for the manufacture of essential oils and fragrances (Ames and Matthews, 1968; Heath, 1981).

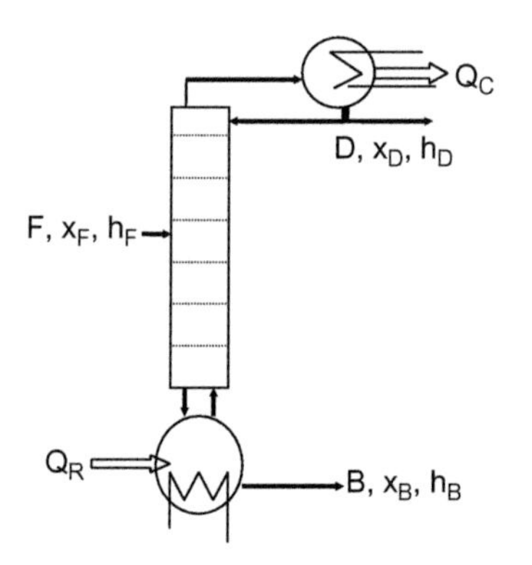

FIGURE 13.15

Energy balance in fractional distillation.

Essential oils are mostly mixtures of terpenoid substances. They have relatively high boiling temperatures (frequently above 200°C). Their recovery by ordinary distillation at atmospheric pressure is impractical because they undergo thermal decomposition at such high temperatures. They can be recovered by vacuum distillation, but a more economical approach is steam distillation.

Essential oils are practically insoluble in water. The total vapor pressure of a mixture of immiscible substances is equal to the sum of the vapor pressures of the pure components. Consequently, such a mixture will boil at a temperature lower than the boiling point of each of the components. A mixture of an essential oil and water will therefore boil at a temperature below 100°C at atmospheric pressure. This is the basic principle on which steam distillation is based. Saturated steam is bubbled through the material containing essential oils (juices, extracts, spices, herbs, etc.). The essential oils are volatilized into the steam and entrained towards the condenser. When the vapors are condensed, a liquid consisting of two immiscible layers is obtained, from which the essential oil is separated by centrifugation or by decantation. A variant of steam distillation is *hydrodistillation*, wherein a mixture of water and the aromatic material (e.g., a spice, a herb, a flower, etc.) is distilled, the vapor is condensed, and the essential oil is recovered from the condensate. Yet another variant is so-called "solvent-free extraction" (Lucchesi et al., 2007), which is not an extraction process at all but rather a laboratory distillation technique. The material is heated by microwave, whereby the essential oil is vaporized by the *in situ* water and recovered from the distillate.

EXAMPLE 13.6

The citrus terpene d-limonene (MW = 136, assumed to be totally insoluble in water) is to be steam distilled at atmospheric pressure. The vapor pressure of the pure limonene as a function of the temperature is given below:

Temperature, °C	Vapor Pressure, Pa
20	8
40	34
60	125
80	391
90	658
98	984
99	1033
99.5	1059
100	1084

a. Find the boiling temperature of the liquid mixture.
b. Calculate the percentage (w/w) of limonene in the vapor.

Table 13.2 Steam Distillation of Limonene

Temperature (°C)	Vapor Pressure of Limonene (Pa)	Vapor Pressure of Water (Pa)	Total Vapor Pressure (Pa)
20	8	2339	2347
40	34	7384	7418
60	125	19 940	20 065
80	391	47 390	47 781
90	658	70 140	70 798
98	984	94 500	95 484
99	1033	97 200	98 233
99.5	1059	99 900	100 959
100	1084	101 300	102 384

Solution

a. The data are reproduced in Table 13.2. A column with the vapor pressure of water at the indicated temperatures is added. The sum of the vapor pressure of the terpene and water is computed for each temperature. It is found by interpolation that at 99.5°C, the total vapor pressure is very nearly 1 atm (101 kPa). The mixture will therefore boil at 99.5°C.

b. At 99.5°C, the vapor pressure of limonene is 1.059 kPa. The total pressure is 101.3 kPa. The concentration of the terpene in the vapor is then (in mol fraction):

$$y = 1.059/101.3 = 0.01 \text{ (1\% molar)}$$

The molecular weight of limonene is 136. The concentration in mass fraction is, therefore:

$$C = \frac{136 \times 0.01}{136 \times 0.01 + 18 \times 0.99} = 0.071 \quad (7.1\%)$$

13.7 Distillation of wines and spirits

Brandies and other alcoholic spirits, such as whisky, rum and vodka, are made by distillation of fermented materials (wines, fruit juices, beers, etc.). The basic equation showing the production of ethyl alcohol from a hexose by fermentation is:

$$C_6H_{12}O_6 \rightarrow 2C_2H_5OH + 2CO_2$$

Obviously, ethyl alcohol is not the only volatile in the fermented material. While the presence of some of the additional volatiles may be desirable for their contribution to the flavor, many of the additional molecules (methyl alcohol, aldehydes, higher alcohols) are undesirable and must be removed in the course of the distillation (Faundez et al., 2006).

Cognac and Scotch whisky are traditionally batch-distilled in stills. Cognac is made from grape wine by a process of double-distillation in pot-stills. The first distillation produces a distillate with an alcohol content of about 25%, called the "brouilly"; this is then subjected to a second distillation, producing a distillate with about 70% ethanol. This is diluted and aged in oak barrels (Amerine et al., 1967).

Scotch malt whisky is also double-distilled. Large pot-stills made of copper and heated directly by coke or peat are used (Marrison, 1957).

Other brandies, whiskies, vodka, and food-grade alcohol are produced by fractional distillation in column. Bubble-cap columns are customary, but other types, such as sieve-plate columns, are also used. In the production of food-grade alcohol, it is often necessary to remove the fraction of higher alcohols. This fraction is called "fusel oil".

13.8 Pervaporation

13.8.1 Basic principles

Pervaporation is a separation process that combines membrane permeation with evaporation – hence its name (Néel, 1991).

Consider the pervaporation chamber of the set-up shown in Figure 13.16. The chamber is divided by a perm-selective membrane.

Assume that a liquid binary mixture of components A and B is fed to the chamber on one side (upstream side) of the membrane. Vacuum is maintained in the opposite downstream side. If the membrane is selectively permeable to B, then B will preferentially penetrate the membrane and be desorbed (evaporated) at the downstream face. Condensation of the permeate vapor will produce a distillate rich in B.

Vacuum is maintained at the downstream side for the efficient desorption of the permeate from the membrane surface. Alternatively, a *sweeping gas* can be

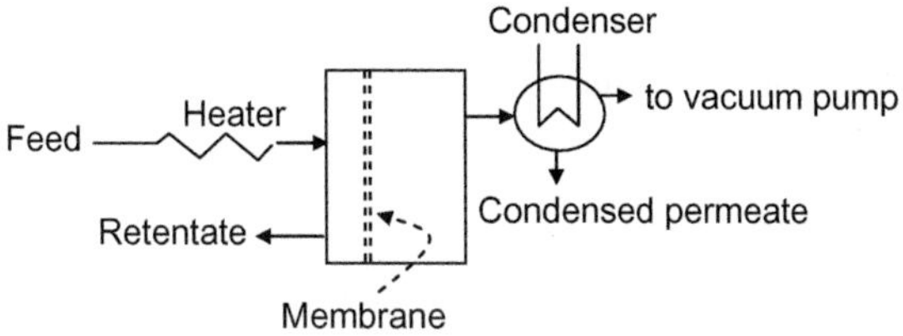

Condenser
to vacuum pump
Heater
Feed
Condensed permeate
Retentate
Membrane

FIGURE 13.16

Schema of pervaporation process.

used to carry away the B-rich vapor to a condenser and maintain the concentration gradient needed for transport through the membrane.

If, on the contrary, the membrane is selectively permeable to A, then A will be removed preferentially from the liquid mixture and the distillate will be rich in A. Thus, the direction of separation by pervaporation depends on the selectivity of the membrane and not on the relative volatility of the components. It follows that the component removed from the liquid mixture is not necessarily the more volatile one. Depending on the membrane installed, the more volatile or the less volatile component will be vaporized preferentially. This is a fundamental difference between pervaporation and simple distillation.

In the particular set-up described in Figure 13.16, the energy needed for vaporization is supplied by heating the feed liquid. In another variant of pervaporation, known as *vapor permeation*, the feed consists of a gaseous mixture of vapors. No phase transition is involved, and no energy is needed for evaporation.

13.8.2 Pervaporation membranes

Membranes for pervaporation are selected according to the type of separation to be performed (Garg et al., 2011; Koops and Smolders, 1991). As in all membrane separation processes, the primary membrane characteristics to be considered are permeate flux and selectivity. Unfortunately, these two parameters vary in opposite directions. A more permeable membrane provides higher permeate flux but poor selectivity, and *vice versa*. As explained in Chapter 3, Section 3.3.4.4, permeability is the product of solubility and diffusivity. The principle of multi-layer composite membranes (anisotropic membranes, see Chapter 10, Section 10.5.1) is used advantageously in pervaporation membranes as well.

The most common types of separation by pervaporation are: removal of small amounts of water from organic mixtures, removal of small amounts of organic substances from water, and separation of mixtures of organic substances (Garg et al., 2011). For the first type of pervaporation, membranes made of hydrophilic polymers, such as cellulose acetate or cross-linked polyvinyl alcohol, are selected. For the removal of organic substances from water and for the separation of organic mixtures, membranes made of hydrophobic elastomers are used. The mass transfer rate in hydrophobic membranes for different organic substances was studied by Overington et al. (2009). A hydrophobic membrane with a selective layer of polyoctylmethyl siloxane was used for the separation of crab-meat volatiles from dilute solutions (Martinez et al., 2011).

13.8.3 Applications

Processes related to ethanol separation are among the most widespread applications of pervaporation (Peng et al., 2010). Dehydration of the ethanol − water azeotrope for the production of quasi-anhydrous ethanol is carried out at industrial scale, using hydrophilic membranes (Dutta and Sridhar, 1991). Typically, the feed is the azeotrope from ordinary fractional distillation containing 96.4% (by volume) ethanol. The retentate is 99.3% ethanol. The permeate, containing 88.5% alcohol, is returned to the distillation column (Maeda and Kai, 1991). Another interesting application combines pervaporation with fermentation. In continuous alcoholic fermentation, the ethanol produced by the microorganisms must be simultaneously removed from the broth to keep the fermentation going. Pervaporation is one of the options for the *in situ* removal of ethanol from fermenting broths (Vane, 2005, 2008).

The use of pervaporation for the separation and concentration of fruit aromas has been investigated (Raisi et al., 2009; Raisi and Aroujalian, 2011; Rajagopalan and Cheryan, 1995).

References

Amerine, M.A., Berg, H.W., Cruess, W.V., 1967. The Technology of Wine Making, second ed. The Avi Publishing Company Inc., Westport, CT.

Ames, G.R., Matthews, W.S.A., 1968. The distillation of essential oils. Trop. Sci. 10, 136–148.

Calliauw, G., Ayala, J.V., Gibon, V., Wouters, J., De Greyt, W., Foubert, I., et al., 2008. Models for FFA-removal and changes in phase behavior of cocoa butter by packed column steam refining. J. Food. Eng. 89 (3), 274–284.

Claus, M.J., Berglund, K.A., 2009. Defining still parameters using Chemcad batch distillation model for modeling fruit spirits distillations. J. Food Proc. Eng. 32 (6), 881–892.

Dutta, B.K., Sridhar, S.K., 1991. Separation of azeotropic organic liquid mixtures by pervaporation. J. AIChE 37 (4), 581–588.

Faundez, C.A., Alvarez, V.H., Valderrama, J.O., 2006. Predictive models to describe VLE in ternary mixtures water + ethanol + congener for wine distillation. Thermochim. Acta 450 (1–2), 110–117.

Garg, P., Singh, R.P., Choudhary, V., 2011. Pervaporation separation of organic azeotropes using poly(dimethyl siloxane/clay) nanocomposite membranes. Separ. Purif. Tech. 80 (3), 435–444.

Heath, H.B., 1981. Source Book of Flavors. The Avi Publishing Company Inc., Westport, CT.

Koops, G.H., Smolders, 1991. C.A. Estimation and evaluation of polymeric materials for pervaporation membranes. In: Huang, R.Y.M. (Ed.), Pervaporation Membrane Separation Processes. Elsevier, Amsterdam, The Netherlands.

Lucchesi, M.e., Smadia, J., Bradshaw, S., Louw, W., Chemat, F., 2007. Solvent free oil microwave extraction of *Elletaria cardamomum* L.: A multivariate study of a new technique for the extraction of essential oil. J. Food Eng. 79 (3), 1079–1086.

Maeda, Y., Kai, M., 1991. Recent progress in pervaporation membranes for water/ethanol separation. In: Huang, R.Y.M. (Ed.), Pervaporation Membrane Separation Processes. Elsevier, Amsterdam, The Netherlands.

Marrison, L.W., 1957. Wines and Spririts. Penguin Books Ltd., Harmondsworth, UK.

Martinez, R., Sanz, M.T., Beltrán, S., 2011. Concentration by pervaporation of representative brown crab volatile compounds from dilute model solutions. J. Food Eng. 105 (1), 96–104.

Néel, J., 1991. Introduction to pervaporation. In: Huang, R.Y.M. (Ed.), Pervaporation Membrane Separation Processes. Elsevier, Amsterdam, The Netherlands.

Overington, A.R., Wong, M., Harrison, J.A., Ferreiea, L.B., 2009. Estimation of mass transfer rates through hydrophobic pervaporation membranes. Separ. Sci. Tech. 44 (4), 787–816.

Peng, P., Shi, B., Lan, Y., 2010. A review of membrane materials for ethanol recovery by pervaporation. Separ. Sci. Tech. 46 (2), 234–246.

Petlyuk, F.B., 2004. Distillation Theory and its Application to Optimal Design of Separation Units. Cambridge University Press, Cambridge, UK.

Raisi, A., Aroujalian, A., 2011. Aroma compound recovery by hydrophobic pervaporation: the effect of membrane thickness and coupling phenomena. Separ. Purif. Techn. 82, 53–62.

Raisi, A., Aroujalian, A., Kaghazchi, T., 2009. A predictive mass transfer model for aroma compounds recovery by pervaporation. J. Food Eng. 95 (2), 305–312.

Rajagopalan, N., Cheryan, M., 1995. Pervaporation of grape juice aroma. J Membrane Sci. 104 (3), 243–250.

Stichlmair, J.G., 1998. Distillation – Principes and Practices. Wiley-VCH, New York, NY.

Treybal, R.E., 1980. Mass Transfer Operations, third ed. McGraw-Hill, New York, NY.

Vane, L.M., 2005. A review of pervaporation for product recovery from biomass fermentation processes. J. Chem. Technol. Biotechnol. 80, 603–629.

Vane, L.M., 2008. Separation technologies for the recovery and dehydration of alcohols from fermentation broths. Biofuels, Bioproducts, Biorefining 2, 553–588.

CHAPTER

Crystallization and Dissolution

14

14.1 Introduction

Crystallization is the process by which solid crystals are formed from a liquid (Decloux, 2002; Hartel, 2001; Mersmann, 1995) or from an amorphous solid (Jouppila and Roos, 1994). In the case of crystallization from a liquid, the liquid medium is either a supersaturated solution or a supercooled melt. Examples of crystallization from a supersaturated solution are the crystallization of sugar and salt from their concentrated solutions. The formation of ice crystals in the process of food freezing and of fat crystals from the molten mass in the manufacture of chocolate are examples of crystallization from a supercooled "melt". An example of solid-phase crystallization from amorphous material is the crystallization of amorphous lactose during storage (Jouppila and Roos, 1994) or spray-drying (Das *et al.*, 2010).

In the food industry, crystallization may serve for the recovery of crystalline products (sugar, glucose, lactose, citric acid, salt), for the removal of certain undesirable components (winterizing of edible oils, a process consisting of chilling certain oils in order to solidify and remove waxes and other high melting-point components), or for the modification of certain food products in order to obtain a desirable structure (crystallization of sugar in the production of fondant and marzipan in confectionery, crystallization of fat in chocolate and margarine, candying of fruit). In foods, crystallization may also be an undesirable change that must be avoided. Typical examples are the crystallization of sugar in jams and preserves, the crystallization of lactose in ice cream, which is the cause of a defect known as "sandiness" (Livney *et al.*, 1995), and the re-crystallization of cocoa butter in chocolate in the form of white crystals (bloom) on the surface of the product.

The present chapter deals mainly with crystallization from solutions, and with the formation of crystalline polymorphs from melted lipids. Processes based on the crystallization of water to form ice crystals (food freezing, manufacture of ice cream, freeze-drying and freeze concentration) are discussed in Chapter 19, and solid-state crystallization in Chapter 22.

Crystallization from a solution can occur only if the solution is supersaturated – i.e., if the concentration of the solute exceeds its equilibrium value at the given temperature. A solution can be brought to *supersaturation* by a number of methods:

- Removal of the solvent by evaporation, membrane separation or freeze-concentration

Food Process Engineering and Technology. DOI: http://dx.doi.org/10.1016/B978-0-12-415923-5.00014-9
© 2013 Elsevier Inc. All rights reserved.

- Cooling the solution (assuming that the solubility increases with the temperature, which is the case with most but not all solutes)
- Changing the pH or the ionic strength
- Addition of a second solvent, miscible with the solution, so as to reduce the solubility of the solute
- Chemical complexation, chemical precipitation.

Solvent removal (concentration) and cooling are the main methods used for crystallization in the food process industry.

The process of crystallization occurs in two stages, namely *nucleation* and *crystal growth*, both requiring the condition of supersaturation.

In inorganic crystals, the forces of attraction between the molecules are strong (e.g., ionic bonds). On the contrary, in the crystals of organic compounds, weak forces, such as van der Waals forces and weak dipoles, bind the molecules together. Consequently, organic compounds may crystallize in a number of different forms. This is known as *polymorphism*. Due to the weakness of the inter-molecular forces in the crystal lattice, these forms may be quite unstable, giving rise to transitions from one form to another. In food processing, polymorphism is particularly important in the crystallization of fats (chocolate, stabilization of peanut butter).

In crystallization, adequate control of the size and shape of the crystals is essential. When crystallization serves to produce a crystalline product, the size and shape of the crystals may determine the ease of separation between the solids and the mother liquor (filtration, centrifugation, etc.) and the subsequent drying and handling of the product (flow properties, dustiness, solubility).

Dissolution, which is a transport phenomenon similar to crystallization but in reverse, is briefly discussed towards the end of this chapter.

14.2 Kinetics of crystallization from solutions

The process of crystallization involves two stages: nucleation and crystal growth.

14.2.1 Nucleation

Consider a solution consisting of one pure solute and one pure solvent. By definition, "saturation" defines the maximum concentration of the solute that can exist in solution in the given solvent, at a given temperature. Experience shows, however, that if a saturated solution is further concentrated or cooled slowly, the solute may still stay in solution. The solution is then said to be "supersaturated". Supersaturation is not a thermodynamically stable state. The degree of supersaturation or the "supersaturation ratio", β, is defined as follows:

$$\beta = \frac{C}{C_S} \tag{14.1}$$

where C and C_S are the actual concentration (kg solute per kg solvent) and the saturation concentration at the same temperature, respectively.

As the degree of supersaturation is increased, the solutes start to aggregate in the solution in orderly "clusters". At low supersaturation, the clusters are small and unstable. However, when a certain degree of supersaturation is reached, the clusters become stable and sufficiently large to possess the elements of the crystal lattice. At this point they become "nuclei", i.e., solid particles that can "fix" additional molecules of the solute and grow to become crystals. The nucleation by the process described here is called *homogeneous nucleation*, because the nuclei are formed from the solute originally present in the solution.

The time required for the appearance of the first detectable nuclei at a given condition of supersaturation is termed the *induction time*. The rate of homogeneous nucleation (e.g., number of nuclei formed per unit time in unit volume of solution), J, is given by the following theoretical equation (Hartel, 2001):

$$J = a\exp\left[-\frac{16\pi}{3} \times \frac{V^2\sigma^3}{(kT)^3(\ln\beta)^2}\right] \qquad (14.2)$$

where:

a = a constant
k = the Boltzmann constant
T = temperature, K
V = molar volume of the solute
σ = nucleus–solution surface tension
β = supersaturation ratio.

Equation (14.2) is obtained on thermodynamic grounds, with certain simplifying assumptions, such as sphericity of the nuclei. Nevertheless, the equation is of value in showing the effect of the different variables and, particularly, the influence of supersaturation. At low supersaturation, the rate of nucleation is practically negligible. A certain degree of supersaturation β_n must be reached and exceeded for homogeneous nucleation to be significant. The supersaturation range between $\beta = 1$ and $\beta = \beta_n$ is called the *metastable zone* (Figure 14.1).

In practice (e.g., in the process of crystallization of sugar in the manufacture of sugar), homogeneous nucleation is seldom the main mechanism of nuclei formation and the initiation of crystallization. More commonly, crystallization is induced by the presence (intentional or accidental) of solid particles of an origin external to the solution. This is called *heterogeneous nucleation*. Heterogeneous nucleation may also be considerable at supersaturation ratios below β_n, where the rate of homogeneous nucleation is negligible (Figure 14.2). The rate of heterogeneous nucleation is given by an expression similar to Eq. (14.2), to which a factor f, representing the influence of the surface tension (wetting angle) between the foreign particles and the solute clusters, has been added:

$$J = a\exp\left[-f\frac{16\pi}{3}\frac{V^2\sigma^3}{(kT)^3(\ln\beta)^2}\right] \qquad (14.3)$$

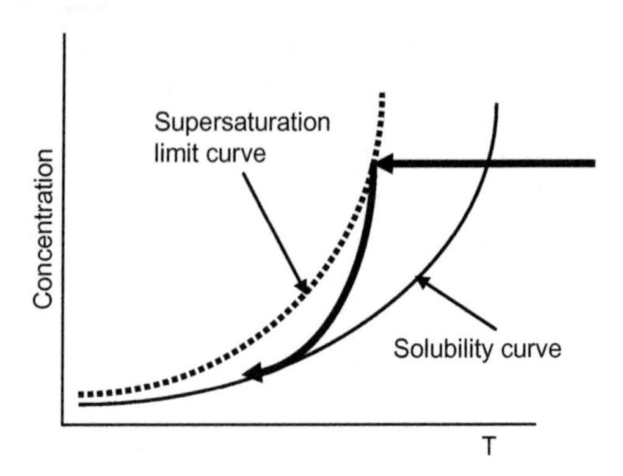

FIGURE 14.1

Solubility and supersaturation.

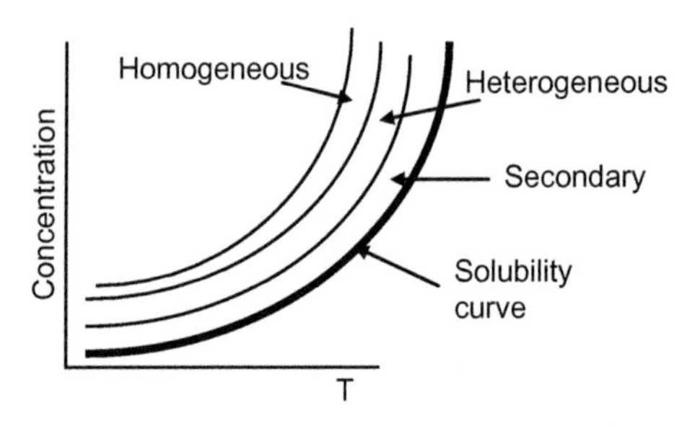

FIGURE 14.2

Types of nucleation most likely to occur as function of supersaturation.

Most often, the "foreign" particles are crystals of the solute in the solution. In many crystallization processes, these are small crystals added intentionally to the supersaturated solution. This operation is called *seeding*. However, the foreign nuclei can also be solid particles of impurities such as dust, imperfections on the surface of the vessel, or even tiny gas bubbles.

A third type of nucleation, known as *secondary nucleation*, occurs mainly in the upper portion of the metastable zone. Small pieces, filaments and dendrites, detached from already existing crystals as a result of agitation or friction, may act as nuclei for further crystallization. In general, secondary nucleation is undesirable in industrial crystallization because it results in less uniform crystal size distribution.

Figure 14.2 shows the zones of supersaturation where the different kinds of nucleation are most likely to occur.

14.2.2 Crystal growth

Crystals grow around the nuclei as a result of the deposition of solute molecules from the supersaturated solution. The rate of crystal growth G is defined as the incremental mass of solute, dm, deposited per unit surface area, per unit time:

$$G = \frac{dm}{Adt} \tag{14.4}$$

Generally, the rate of crystal growth is expressed as the following empirical equation (Cheng *et al.*, 2006):

$$G = k\sigma^g \tag{14.5}$$

where:

σ = driving force for deposition
k = an empirical coefficient depending on the temperature and agitation
g = a numerical known as the "growth order".

If the driving force for deposition is considered to be the difference between the chemical potential of the solute at the crystal/solution interface and the chemical potential at a given temperature, and if pressure is assumed to be approximately proportional to the concentration, then σ can be replaced by the difference between the concentration of the (supersaturated) solution and the equilibrium concentration at the solid surface (i.e., saturation concentration or solubility):

$$G = k'(C - C_s)^g \tag{14.6}$$

where:

C = concentration of the supersaturated solution
C_S = concentration at saturation
k′ = a combined empirical rate constant, depending on the solute, the solvent, the presence of "impurities", the temperature, agitation, etc.
g = a numerical exponent, representing the "order" of crystal growth.

Crystal growth involves two consecutive and independent processes: transport of the solute molecules from the bulk of the solution to the surface of the crystal, and "accommodation" of the incoming molecules in the orderly pattern of the crystal lattice (Figure 14.3). The rate-controlling factor may be one or the other of these processes, depending on the conditions. Solute transport occurs by diffusion, often assisted by agitation. It is highly dependent on the diffusion coefficient, and hence on the temperature and viscosity. As explained in Chapter 1, Section 1.8.1, crystallization is practically impossible in supersaturated solutions of very high viscosity.

Just like adsorption and condensation, crystal growth is an exothermic process. Heat is liberated as molecules of the solute are "immobilized" within the crystal lattice. At steady state, the liberated heat has to be dissipated and

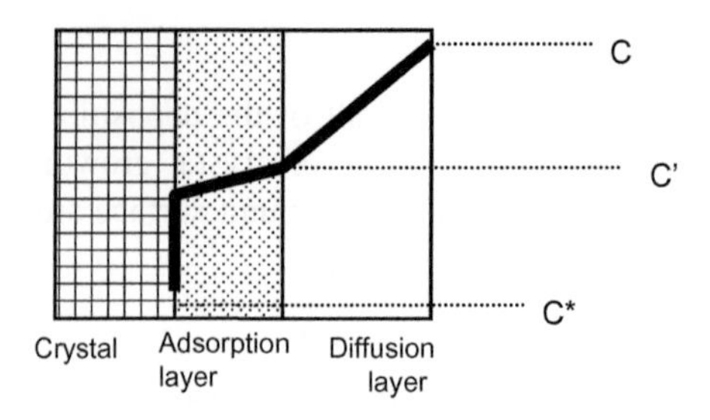

FIGURE 14.3

Mass transfer in crystal growth.

removed from the system. Although the thermal aspects of crystallization are of considerable theoretical relevance, their practical importance is often negligible.

The concentration of the liquid phase (mother liquor) and hence the rate of crystal growth decreases as more and more solute molecules pass to the solid phase in the course of crystal growth. Equation (14.6) shows the importance of maintaining a sufficiently high degree of supersaturation for sustained crystal growth. In practice, this is achieved by continuously evaporating some of the solvent throughout the process of crystallization. This mode of operation, known as *evaporative crystallization*, is standard procedure in the production of sugar and salt.

The effect of temperature on the rate of crystal growth is more complex. On one hand, transport of the solute to the crystal surface is faster at high temperature. On the other hand, increasing the temperature results in an increase of solubility and hence in the tendency of the molecules to leave the solid rather than attaching to it. Generally, the dependence of crystal growth on temperature follows a bell-shaped curve, with a maximum (Umemeto and Okui, 2005). However, other types of behavior may be observed in specific cases (Cheng *et al.*, 2006). Within the range of moderate temperatures, crystal growth rate increases with temperature in most cases.

The mixture of crystals and mother liquor is called "magma". A typical magma usually contains crystals of different sizes. The small crystals possess considerably larger specific surfaces, and hence higher surface energy per unit mass, than larger crystals. As a result, very small crystals are more "soluble", i.e., can be in equilibrium with supersaturated mother liquor. Such equilibrium is unstable (McCabe *et al.*, 2001), and the system will move in the direction of dissolution of the small crystals and growth of the large ones. This process is known as *Ostwald ripening*, and is also responsible for the disappearance of small bubbles and simultaneous growth of large bubbles in a foam. Ostwald ripening is one of the reasons for the undesirable change in the texture of ice cream (recrystallization of ice) during storage. Fluctuations in the storage temperature of ice cream may cause preferential melting of the small ice crystals, followed by further growth of the larger crystals, thus destroying the creamy texture of the product.

The solubility of a small crystal is related to its size through the Kelvin equation (McCabe, 2001):

$$\ln \beta = \frac{4V\sigma}{nRTL} \tag{14.7}$$

where L is the size of the crystal and n is the number of ions per molecule (n = 1 in the cases of undissociated substances such as sugar).

EXAMPLE 14.1

What is the size of a nucleus of a sugar crystal in equilibrium at 20°C with mother liquor at 115% supersaturation? The surface tension at the crystal–liquid interphase is estimated to be 0.003 $J \cdot m^{-2}$. The density of the sugar crystal is 1600 $kg \cdot m^{-3}$.

Solution

The molecular weight of sucrose is M = 342, hence the molecular volume of solid sugar is:

$$V = \frac{M}{\rho} = \frac{342}{1600} = 0.21375\ m^3 \cdot kmol^{-1}$$

For sugar, n = 1. Applying Eq. (14.7):

$$\ln 1.15 = 0.1398 = \frac{4 \times 0.21375 \times 0.003}{8.314 \times 293 \times L} \quad \Rightarrow \quad L = 7.5 \times 10^{-6} m = 7.5\ \mu m$$

The final average size, size distribution, shape and quantitative yield of the crystals are factors of considerable technological importance. The liquid–solid separation and drying characteristics of the product after crystallization are affected by these factors, as are the flow properties and bulk density of the final product. Usually, product specifications, dictated by market demand, include requirements regarding the size and shape of the crystals. It is possible to obtain a crystalline product with specified properties by controlling the temperature–supersaturation–purity profile of the material during crystal growth. The effect of impurities on the "habit" and size of the crystals is considerable (Martins *et al.*, 2006).

The rate of growth is not equal for all the faces of a crystal. Smaller faces grow faster than large ones. Consequently, the overall shape (habit) of the crystals may change as they grow, with a tendency to become more symmetrical. Impurities are known to modify crystal habit, in addition to their negative effect on crystal growth in general. The rate of crystal growth in organic substances is slower than that observed with inorganic materials.

Crystals do not grow indefinitely. There is a critical size or size range beyond which further growth does not occur. Inorganic crystals may attain very large dimensions, while the critical size of organic substances is much smaller. Rapid crystallization (e.g., by sudden cooling of highly supersaturated solutions) tends to produce small crystals, while larger, more symmetrical crystals are formed by slow crystallization. Seeding with a large number of nuclei per unit volume of solution results in the formation of a large number of small crystals.

The maximum amount of crystals obtainable from a given solution is known as the *crystalline yield* (Hartel, 2001). In a binary system, the crystalline yield is easily calculated from solubitity data and mass balance, assuming equilibrium. Let F kg be the initial mass of the solution (feed), C_0 the initial concentration (mass fraction), Y kg the mass of crystalline phase at the end of the process, and P and C_S the quantity (kg) and concentration (mass fraction) of the mother liquor, respectively. Overall mass balance gives:

$$F = Y + P$$

Mass balance for the solute gives:

$$FC_0 = Y + PC_S$$

Hence:

$$Y = F\frac{C_0 - C_s}{1 - C_S} \tag{14.8}$$

EXAMPLE 14.2

A mass of 100 kg of a saturated aqueous solution of pure sucrose at 90°C is cooled to 60°C. Calculate the crystalline yield. Repeat the calculation for final temperatures of 50, 30 and 20°C.

Solution

Equation (14.8) will be applied.

$$Y = F\frac{C_0 - C_s}{1 - C_S}$$

F = 100 kg. The feed solution is saturated at 90°C. The solubility of sucrose in water at 90°C (see Appendix, Table A.18) is 80.6%. Hence C_0 (in mass fraction) is 0.806.

$$Y = 100\frac{0.806 - C_S}{1 - C_S}$$

Values of C_S at the different final temperatures are also taken from Table A.18. The results are tabulated below:

Final Temperature (°C)	**60**	**50**	**30**	**20**
C_S	0.742	0.722	0.687	0.671
Yield Y (kg)	24.8	30.2	38.0	41.0

14.3 Polymorphism in lipid crystals

Crystallization of lipid molecules in fats, oils and related products is of utmost importance, as it determines to a large extent the appearance, texture, mouthfeel and stability of such products. One of the distinctive characteristics of lipids is their ability to solidify in a number of crystalline states, differing in shape, melting point, X-ray diffraction pattern, induction time and other properties (Garti and Sato, 2001). The different crystalline polymorphs have been classified in crystal groups and subgroups. The present system of nomenclature recognizes three main groups, α, β and β', and their subgroups. The sequence of formation of the different groups depends both on thermodynamic stability and on kinetics (Larson, 1994).

14.4 Crystallization in the food industry

14.4.1 Equipment

In industry, crystallization is carried out in *crystallizers*, commonly called *pans*. Pans can be batch or continuous, vacuum or atmospheric (Figures 14.4 and 14.5). In the pan, supersaturation is maintained by evaporation (evaporative crystallizers), by cooling (cooling crystallizers), or by both. Continuous pans are extensively used in modern, large-scale plants. Batch pans are used for the production of "seeds" (primary by homogeneous nucleation) or for small-scale crystallization processes.

Typically, a crystallization pan is a vessel equipped with heat exchange areas (for heating or cooling) and providing agitation. The heat exchange areas may be provided by a jacket or by a tubular heat exchange element. Agitation may be provided either by a mechanical stirrer or by recycling. Crystallization pans can operate in batch (Figure 14.6) or continuous mode. Recycling pans with external heat exchangers (Figure 14.7) are frequently used in the production of salt, citric acid, etc., but not in the sugar industry because of the high viscosity of the material. Agitation must be sufficiently vigorous to facilitate mass transfer and good mixing, but gentle enough to prevent crystal fracture, abrasion and secondary nucleation. Vacuum is applied to lower the boiling temperature and prevent thermal damage when necessary (e.g., inversion and caramelization/browning in sugar syrups).

14.4.2 Processes

14.4.2.1 Crystallization of sucrose

Crystallization is the final step in the recovery of sugar from sugar-cane or sugar-beet. Also called "sugar boiling", crystallization of sugar is a complex process requiring precise control, skill and experience. In modern plants, crystallization is a fully automated process. The controlled parameters include temperature, pressure, concentration (Brix), purity, crystal/liquid ratio, crystal mean size and shape, and flow-rates. The progress of crystal growth is often monitored by on-line automatic digital image processing techniques (Velazquez-Camilo *et al.*, 2010).

FIGURE 14.4

Large crystallizers in wastewater treatment.

Photograph courtesy of General Electric.

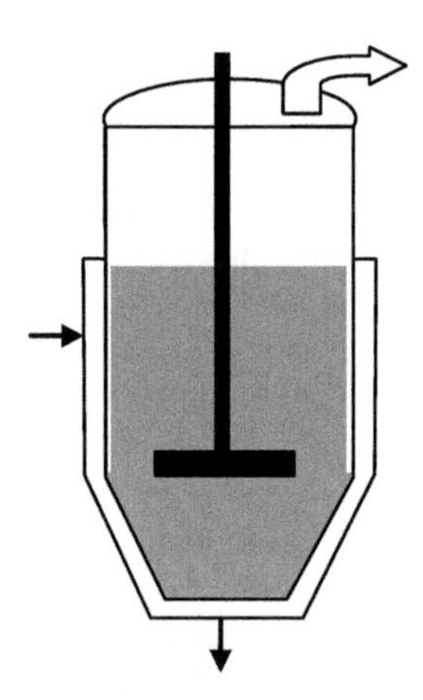

FIGURE 14.5

Batch crystallization pan with agitation.

A key notion in this process is the concept of "purity". Purity Q, a dimensionless value, is the ratio of sugar to total soluble substances in a mixture:

$$Q\% = \frac{C_{sucrose}}{C_{total\ solubles}} \times 100 \tag{14.9}$$

Crystallization of sugar may be carried out in batches or as a continuous process. In large-scale production, the modern practice is to use a mixed process

FIGURE 14.6

Batch crystallization pans in sugar plant.

Photograph Courtesy of BMA Braunschweigische Maschinenbauanstalt A.G.

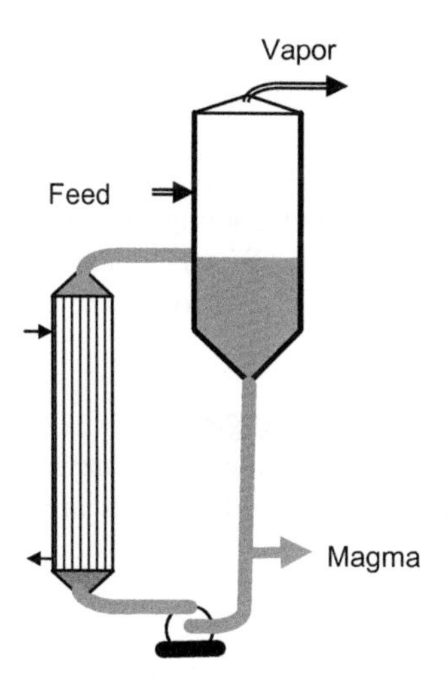

FIGURE 14.7

Recycling crystallization pan with external heat exchanger.

where some operations, such as the preparation of the "seed", are done in batch processes. In a batch process, two significant changes occur as more sugar is crystallized: namely, the consistency of the crystal–mother liquor mixture (called "massecuite" or magma) increases, and the purity of the mother liquor decreases. Both effects are detrimental to the proper growth of sugar crystals. As a result of the increase in the crystal content of the magma, mixing becomes more difficult, flow channels may be blocked, and crystal rupture due to friction may occur. The decrease in purity (or increase in the concentration of impurities) in the mother liquor interferes with crystal growth. The remedy is multi-stage crystallization, whereby the first crystals are separated and the mother liquor is sent to the next stage for further concentration/cooling and crystallization. A three-stage batch crystallization scheme is shown in Figure 14.8. The first "crop" of crystals is the purest (white sugar). Crystals formed in subsequent stages are progressively less pure (raw sugar or low-grade sugar). These are usually recycled (remelted or mingled) into the incoming syrup. The mother liquor from the last stage is called "molasses". It typically has a purity as high as 60%, but further crystallization of sugar from molasses is usually not economical. With the exception of a small quantity of high-quality molasses used for human consumption, the main uses of molasses are as animal feed and as a raw material for fermentation processes.

In a combined batch–continuous process, the starting syrup is brought to a supersaturation of about 1.1 by batch evaporation (Figure 14.9) and then seeded with a slurry of very fine crystals in high-purity syrup. The nuclei are allowed to grow to about 100 μm and to a crystal content of about 20%. The resulting massecuite, called "1st seed", is mixed with heavy syrup and fed to the continuous crystallizer system, consisting of a series of stages, each equipped with agitation and heat surface areas for heating or cooling. The stages may be vertically superimposed to form a "tower" (Figure 14.10).

In sugar crystallizing plants the desired degree of supersaturation is maintained primarily by evaporation, but cooling crystallization is often practiced in the latter stages of crystal growth. Water-cooled heat exchangers or vacuum (flash) evaporation, or both, are used for cooling the mixture.

14.4.2.2 Crystallization of other sugars

Commercial *dextrose* (glucose) is obtained by enzymatic or chemical hydrolysis of starch, mainly from corn. Starch hydrolyzates are purified, concentrated, and

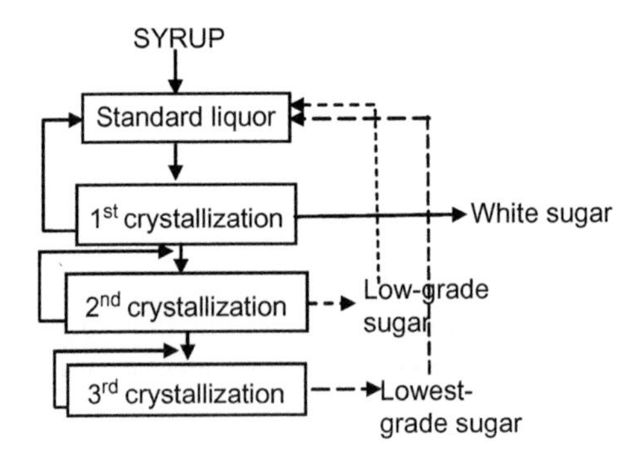

FIGURE 14.8

Three-stage crystallization of sugar.

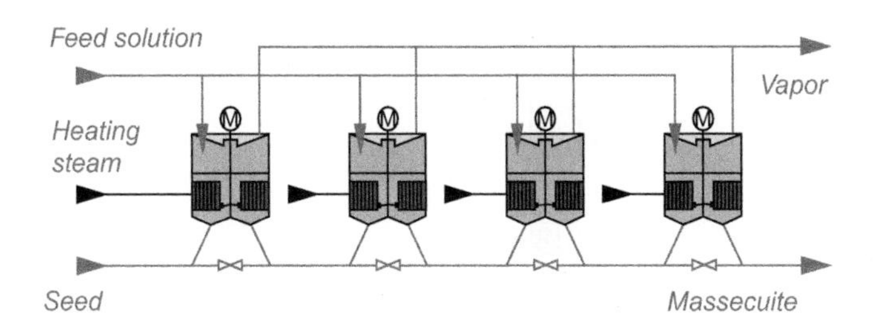

FIGURE 14.9

Vacuum pans in series for continuous crystallization of sugar.

Figure courtesy of BMA Braunschweigische Maschinenbauanstalt A.G.

sold mainly as thick syrups (corn syrup) or spray-dried amorphous powders (maltodextrins). The degree of hydrolysis is expressed as the DE (dextrose equivalent) of the product. Crystalline dextrose is obtained from totally hydrolyzed starch. The process of crystallization, like that of sucrose, is performed in stages.

Lactose is obtained from sweet whey, a by-product of cheese and casein manufacture. Crystallization is a key operation in the production of lactose from whey (Mimouni *et al.*, 2005). A very small proportion of the lactose available is purified and crystallized. Pure, solid lactose has limited use in foods and is used mainly as an excipient (filler) in pharmaceutical tablets. Due to its low solubility in water (19 g per 100 g of solution at 25°C, 37 g per 100 g of solution at 60°C), it is easily crystallized by seeding and cooling. The temperature must be kept low to prevent the Maillard browning reaction with nitrogen-containing impurities.

Most of the lactose in spray-dried milk powders is amorphous lactose. The amorphous (glassy) state is unstable and, given appropriate moisture and temperature conditions, tends to undergo crystallization during storage (Roos and Karel, 1992). Crystallization of lactose in milk powders impairs flowability and accelerates caking (Thomas *et al.*, 2004).

14.4.2.3 Crystallization of salt

Traditionally, food-grade salt has been produced by solar evaporation of sea water in shallow ponds (Walter, 2005). *Solar sea salt* is still produced in respectable quantities, and commands increasing demand at a premium price. The major part of industrial salt production, however, uses multiple effect evaporation (Figure 14.11; see also Chapter 19). Due to the fact that the solubility of salt does not change considerably with temperature, the process of production is based on evaporative crystallization and not on cooling. In the case of production from rock, water is forced into the salt bed in the rock. The brine thus produced by dissolution of the salt is subjected to a series of filtration and purification

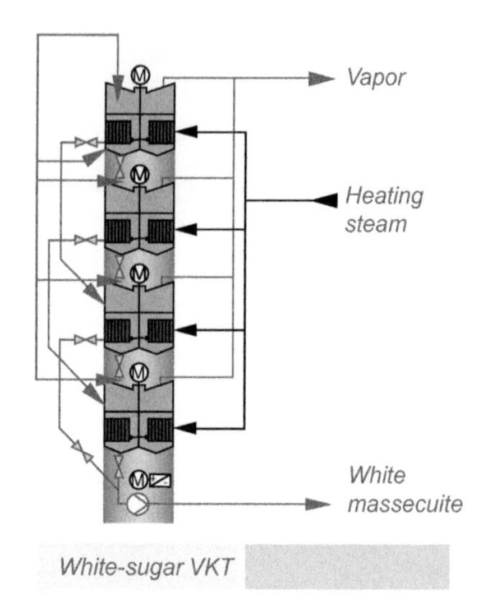

FIGURE 14.10

Crystallization pans arranged in tower.

Figure courtesy of Braunschweigische Maschinenbauanstalt A.G.

procedures. The purified brine is fed into multi-effect vacuum evaporators, and concentrated until nuclei start to form and then grow to the desired size in the mother liquor. Crystals are separated from the liquid by centrifugation or filtration, and then dried.

14.4.2.4 Crystallization of cocoa butter in the production of chocolate

Physically, chocolate is a dispersion of solid particles (sugar, milk solids, cocoa solids) in a continuous matrix of cocoa butter (Beckett, 2008). Cocoa butter is solid at room temperature, and typically melts at temperatures slightly below the body temperature of 37°C. "Melting in the mouth" is one of the principal quality properties of chocolate. At the final stages of the production process, chocolate is a liquid that is eventually cast into molds and solidified. Prior to molding, the liquid chocolate is subjected to a rigorously controlled cooling–heating–shearing process known as *tempering*, the purpose of which is to promote the desired crystallization pattern. The fat-crystallization behavior of chocolate strongly affects its mechanical properties, microstructure and appearance. Both under-tempering and over-tempering result in a product of inferior quality (Afoakwa *et al.*, 2008). Proper tempering is essential in order to obtain a final product with the following characteristics:

- A shiny surface (gloss)
- A brittle (not chewy) texture ("snap")
- A low tendency to produce "fat bloom" on the surface (see below).

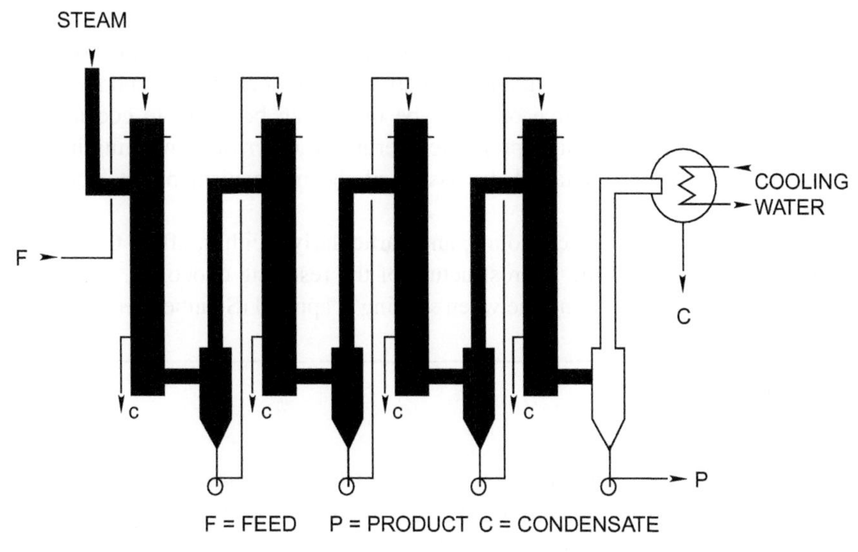

FIGURE 14.11

Schema of a multiple-effect evaporator.

Figure courtesy of APV Ltd.

In addition, proper tempering is needed to induce sufficiently low viscosity of the mass (pumpability into the molds) and adequate contraction upon cooling (release from the molds). The effect of temperature, contact time and relative humidity of the surrounding atmosphere on the adhesion of chocolate to the molds was studied by Keijbets *et al.* (2010).

The ideal time–temperature–shearing protocol of tempering depends on the raw materials (triglyceride composition of the cocoa butter) and on the type of chocolate produced. A typical tempering process (Hartel 2001) is described below.

- Stage 1: Cooling. the liquid chocolate is rapidly cooled to 26–27°C under agitation.
- Stage 2: Nucleation. The chocolate is held at that temperature to promote nucleation. A large number of small crystals of the α and β' types are formed. These polymorphs are not stable, but their nucleation rates are higher than that of the more stable β group. Due to the considerable mass of crystals, the viscosity at this stage is high.
- Stage 3: Melting. The chocolate is heated to 31–32°C.
- Stage 4: Transformation. the mass is held at that temperature to promote transformation to the stable β polymorph and to melt the unstable crystals that have lower melting points.
- Stage 5: Solidification. at this point, the viscosity is sufficiently low to permit molding or coating. Upon final cooling, a chocolate with a large number of small crystals of the stable β type, adequate gloss and good snap is obtained.

Fat bloom is a type of physical deterioration frequently observed in chocolate. It consists of the appearance of white, feather-like crystals on the surface of chocolate. It is caused by the melting and separation of low-melting fat, and its migration to the surface. The source of that type of fat can be unstable cocoa butter polymorphs left as a result of improper tempering, or other fat-containing ingredients in chocolate. For a detailed discussion of the mechanism of fat blooming, see Sato and Koyano (2001).

The usual ingredients of chocolate, and particularly lecithin, affect the kinetics of fat crystallization and the microstructure of the resulting chocolate. Lecithin is found to shorten the induction time when seeding is applied (Svanberg *et al.*, 2011).

14.5 Dissolution

14.5.1 Introduction

In its strict sense, the term "dissolution" applies to the passage of a *soluble* substance from a solid matrix to a liquid solution. In practice, however, the term is also used to describe the dispersion of an insoluble substance in a liquid (e.g., dispersion of cocoa powder or milk powder in water). Both true dissolution and dispersion are important in foods. The so-called *instant powders* are expected to be rapidly dispersible or totally soluble in a liquid (water or milk, hot or cold). In automatic dispensing machines, powders must undergo rapid and total dissolution or dispersion in water in order to deliver a beverage in a few seconds. In medicine, the rate of dissolution of a drug in the gastro-intestinal tract is of obvious importance.

14.5.2 Mechanism and kinetics

Dissolution kinetics has been studied extensively for over 100 years. One of the oldest works is the classical paper by Noyes and Whitney (1897), still widely quoted.

As explained above, the subject is of utmost interest not only in food engineering, but also in soil science, agriculture, chemical industry and pharmacy.

The dissolution of a solid in solvent comprises the following successive steps:

1. Wetting of the solid surface
2. Penetration of the solvent into solid
3. Release of solute molecules from the wetted surfaces of the solid
4. Migration of the solute into the liquid bulk by diffusion and convection.

Although each one of the steps has its own kinetic behavior, it is usually assumed that the rate-limiting step is the migration of the solute away from the solid matrix. This is the basic hypothesis of the so-called "diffusion controlled models" of dissolution kinetics. Other assumptions usually made in the development of such models are that:

- The solid particles are uniform, compact and spherical

- "Sink" conditions prevail (i.e., the solid particle is suspended in a very large volume of liquid) and therefore concentration of the bulk liquid does not change noticeably as a result of the dissolution.

Consider a spherical particle of a pure soluble solid, dissolving in a large volume of solution (Figure 14.12).

Let:

C_B = bulk concentration of the solution (assumed constant under "sink" conditions")
C_S = saturation concentration (constant, at constant temperature)
ρ = density of the solid
k_L = coefficient of mass transfer at the solid-liquid limit zone.

At steady state, the rate of dissolution must be equal to the rate of mass transfer from the interface into the solution:

$$-\frac{dr}{dt}(4\pi\, r^2 \rho) = 4\pi\, r^2 k_L (C_S - C_B) \tag{14.10}$$

The time necessary for reducing the radius of the particle form R_0 to R is found by integration of Eq. (14.10):

$$R_0 - R = k_L \frac{C_S - C_B}{\rho} t \tag{14.11}$$

The time for total dissolution of the particle (R = 0) is proportional to the starting radius:

$$t_{total} = \frac{\rho}{k_L(C_s - C_B)} R_0 \tag{14.12}$$

In terms of the mass of the particle:

$$m_0^{1/3} - m^{1/3} = Kt \quad \text{where:} \quad K = \left(\frac{4\pi\rho}{3}\right)^{1/3} \frac{k_L(C_S - C_B)}{\rho} \tag{14.13}$$

and the time for total dissolution:

$$t_{total} = \frac{m_0^{1/3}}{K} \tag{14.14}$$

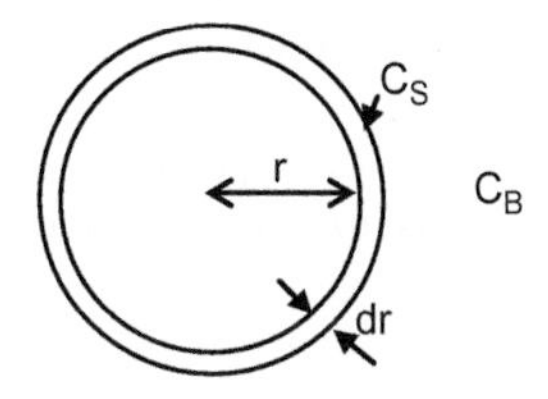

FIGURE 14.12

Dissolution of a spherical particle.

According to this model, the time necessary for the total dissolution of a particle of pure solute is proportional to the cubic root of the mass of that particle. Accordingly, the model is often called the "*cubic root model*" (Wang and Flanagan, 1999). Despite the simplifying assumptions on which it is based, the cubic root model predicts dissolution kinetics quite accurately, as long as the particles are not too small. The group K in Eq. (14.13) is not really constant. In a fully turbulent regime, the mass transfer coefficient k_L varies with the diameter of the particle. Furthermore, as the particle becomes smaller and smaller, the approximation used for the material balance introduces a significant error. Other diffusion-controlled models suggest proportionality to the square root or to the 2/3 root of the particle mass. Marabi *et al.* (2008) studied dissolution kinetics by observing the dissolution of a single powder particle and applying calorimetry.

Empirical logarithmic models have been suggested for the dissolution of high molecular substances. The most common, developed for hydrocolloids, is a first-order type model (Larsen *et al.*, 2003).

$$\log\left(\frac{m}{m_0}\right) = -kt \tag{14.15}$$

The obvious shortcoming of this model is that it does not allow for complete dissolution.

References

Afoakwa, E.O., Paterson, A., Fowler, M., Vieira, J., 2008. Effects of tempering and fat crystallisation behaviour on microstructure, mechanical properties and appearance in dark chocolate systems. J. Food Eng. 89 (2), 128–136.

Beckett, S.T., 2008. The Science of Chocolate, second ed. The Royal Society of Chemistry, Cambridge, UK.

Cheng, F., Bai, Y., Chang, L., Xiaohua, L., Chuang, D., 2006. Thermodynamic analysis of temperature dependence of the crystal growth of potassium sulfate. Ind. Eng. Chem. Res. 45, 6266–6271.

Das, D., Husni, H.A., Langrish, T.A.G., 2010. The effects of operating conditions on lactose crystallization in a pilot-scale spray dryer. J. Food Eng. 100 (3), 551–556.

Decloux, M., 2002. Crystallisation. In: Bimbenet, J.J., Duquenoy, A., Trystram, G. (Eds.), Génie des Procédés Alimentaires. Dunod, Paris, France.

Garti, N., Sato, K., 2001. Crystallization Processes in Fats and Lipid Systems. Marcel Dekker Inc., New York, NY.

Hartel, R.W., 2001. Crystallization in Foods. Aspen Publishers, Gaithersburg, MD.

Jouppila, K., Roos, Y.H., 1994. Glass transitions and crystallization in milk powders. J. Dairy Sci. 77 (10), 2907–2915.

Keijbets, E.L., Chen, J., Viera, J., 2010. Chocolate demoulding and effects of processing conditions. J. Food Eng. 98 (1), 133–140.

Larsen, C.K., Gåserød, O., Smidsrød, 2003. A novel method for measuring hydration and dissolution kinetics of alginate powders. Carbohyd. Polym. 51, 125–134.

Larson, K., 1994. Lipids. Molecular Organization, Physical Functions and Technical Applications. The Oily Press, Dundee, UK.

Livney, Y.D., Donhowe, D.P., Hartel, R.W., 1995. Influence of temperature on crystallization of lactose in ice-cream. Intl. J. Food Sci. Technol. 30 (3), 311–320.

Marabi, A., Mayor, G., Raemy, A., Burbidge, A., Wallach, R., Saguy, I.S., 2008. Assessing dissolution kinetics of powders by a single particle approach. Chem. Eng. J. 139, 118–127.

Martins, P.M., Rocha, F.A., Rein, P., 2006. The influence of impurities on the crystal growth kinetics according to a competitive adsorption model. Cryst. Growth Des. 6 (12), 2814–2821.

McCabe, W.L., Smith, J.C., Harriott, P., 2001. Unit Operations of Chemical Engineering. McGraw-Hill, New York, NY.

Mersmann, A., 1995. Crystallization Technology Handbook. Marcel Dekker, New York, NY.

Mimouni, A., Schuck, P., Bouhallab, S., 2005. Kinetics of lactose crystallization and crystal size as monitored by refractometry and laser light scattering: effect of protein. Lait 85 (4–5), 253–260.

Noyes, A., Whitney, W.R., 1897. The rate of solution of solid substances in their own solutions. J. Am. Chem. Soc. 19, 930–934.

Roos, Y., Karel, M., 1992. Crystallization of amorphous lactose. J. Food Sci. 57 (3), 775–777.

Sato, K., Koyano, T., 2001. Crystallization properties of cocoa butter. In: Garti, N., Sato, K. (Eds.), Crystallization Processes in Fats and Lipid Systems. Marcel Dekker, New York, NY, pp. 429–456.

Svanberg, L., Ahrné, L., Lorén, N., Windhab, E., 2011. Effect of sugar, cocoa particles and lecithin on cocoa butter crystallisation in seeded and non-seeded chocolate model systems. J. Food Eng. 104 (1), 70–80.

Thomas, M.E.C., Scher, J., Desroby-Banon, S., Desroby, S., 2004. Milk powders ageing: effect on physical and functional properties. Crit. Rev. Food Sci. Nutr. 44 (5), 297–322.

Umemeto, S., Okui, N., 2005. Power law and scaling for molecular weight dependence of crystal growth rate in polymeric materials. Polymer 46, 8790–8795.

Velazquez-Camilo, O., Bolaños-Reynoso, E., Rodriguez, E., Alvarez-Ramirez, J., 2010. Characterization of cane sugar crystallization using image fractal analysis. J. Food Eng. 100 (1), 77–84.

Walter, H.H., 2005. From a flat pan to a vacuum crystallizer, simmering salt production in the 19th and 20th century. Mitt. Geselschaft Deut. Chem. Fachgruppe Geschich. Chem. 18, 59–72.

Wang, J., Flanagan, D.R., 1999. General solution for diffusion-controlled dissolution of spherical particles. 1. Theory. J. Pharma. Sci. 88 (7), 731–738.

CHAPTER

Extrusion

15

15.1 Introduction

Extrusion is a process of central importance and widespread application in the food industry. The extensive development of extrusion technology represents one of the most significant achievements in food process engineering in the past 50 years. The expansion of food extrusion technology has been accompanied by considerable research activity that has generated an impressive volume of new knowledge regarding the physics and chemistry of the process.

Literally, extrusion (from the Latin word *extrudere*) means the action of pushing out. In engineering, it describes an operation of forcing a material out of a narrow gap. The familiar action of pressing toothpaste out of a tube corresponds well to this definition. Actually, the literal meaning describes only a small portion of the industrial applications of extrusion – namely, the cases in which the sole function of extrusion is to impart to the product a certain shape or form, without otherwise affecting the properties of the material. This is still the role of extrusion in pasta presses and feedstuff pelletizers. However, the most significant advance in food extrusion has been the development of extrusion cooking since the 1950s. Extrusion cooking may be defined as a thermo-mechanical process in which heat transfer, mass transfer, pressure changes and shear are combined to produce effects such as cooking, sterilization, drying, melting, cooling, texturizing, conveying, puffing, mixing, kneading, conching (chocolate), freezing, forming, etc. The extruder-cooker is a pump, a heat exchanger and a continuous high-pressure–high temperature reactor, all combined in one piece of equipment.

A short historical summary of the development of food extrusion is shown in Table 15.1.

Some of the specific advantages of extrusion cooking over other processes serving the same purpose are listed below.

- Extrusion cooking is a one-step process. A number of operations are simultaneously carried out in one piece of equipment. This does not exclude the need for additional operations upstream and downstream of extrusion. These operations include preparation (conditioning, formulation, modification,

Food Process Engineering and Technology. DOI: http://dx.doi.org/10.1016/B978-0-12-415923-5.00015-0
© 2013 Elsevier Inc. All rights reserved.

Table 15.1 Stages in the Development of Food Extrusion

Decade	Equipment	Commercial Uses
Prior to 1950	Forming, non-thermal extruder (e.g., pasta press)	Pasta
1950	Single-screw cooking extruder	Dry animal feed
1960	Single-screw cooking extruder	Texturized vegetable protein (TVP), ready-to-eat (RTE) cereals, puffed snacks, pellets, dry pet-food
1970	Twin-screw cooking extruder	Moist pet-food, upgrading of raw materials
1980	Twin-screw cooking extruder	Flat bread, croutons, confectionery, chocolate
1990	Twin-screw with long cooling dye	Moist texturized proteins
2000	Refrigerated (ultra-cold) extruder	Ice cream, frozen bars

cleaning, etc.) before extrusion, and various finishing operations applied to the extrudate after extrusion (drying, frying, addition of flavoring ingredients, etc.)

- Extrusion is a continuous process.
- For its output, the extruder is a relatively compact machine and requires little floor space.
- Extrusion requires little labor.
- The extruder is versatile. The same equipment with slight modifications may be used for achieving different objectives, or for processing many different products.
- Cooking extrusion is a genuine high temperature–short time process. Retention time in the extruder is relatively short. Consequently, extrusion cooking can be used for the disinfection and sterilization of the product and for the inactivation of heat-resistant toxins, such as aflatoxin (Saalia and Phillips, 2011).
- The energy expenditure of cooking extrusion is usually lower than that of alternative processes, because the major part of the energy (heat or mechanical work) is delivered to the product directly and not through an intermediary medium.

The first cooking extruders were single-screw machines. They are still in use in some processes, but most recent installations are now twin-screw extruders. Being simpler, the single-screw extruder will be described first.

15.2 The single-screw extruder

15.2.1 Structure

The basic structure of a single-screw extruder consists of the following elements (Figure 15.1):

- A hollow cylindrical enclosure, called the *barrel*. The barrel can be smooth or grooved. Conical barrels have been described (Meuser and Wiedmann, 1989), but they are not common in food extruders.
- A sturdy Archimedean *screw* or *worm*, with a thick root and shallow flights, turning inside the barrel. The flights of the rotating screw propel the material along a helicoidal channel (*flow channel*) formed between the screw root and the barrel. The width of the flow channel, resulting from the screw pitch, is considerably larger than its thickness. The gap between the screw tip and the barrel surface is made as narrow as possible.
- A restricted passage element, known as the *die*, at the exit end of the extruder. The functions of the die are to serve as a pressure-release valve and to impart to the extrudate the desired shape, determined by the cross-section of the aperture(s). The die is sometimes preceded by a perforated plate (breaker plate) which helps distribute the compressed material evenly across the die.
- A device for cutting the extrudate emerging from the die. In its simplest form, this consists of a rotating knife.
- Different kinds of devices for heating or cooling the barrel (steam or water jackets, electrical resistance heaters, induction heaters, etc.) These elements, external to the barrel, are usually divided into individual segments in order to impose different temperature conditions at different sections of the extruder.
- A hopper for gravity feeding or an auger for positive feeding.
- Ports for the injection of steam, water and other fluids as needed. Ports for pressure release.
- Measurement instruments (feed rate, temperature, pressure) and controls.
- A drive, usually with speed variation capability and torque control.

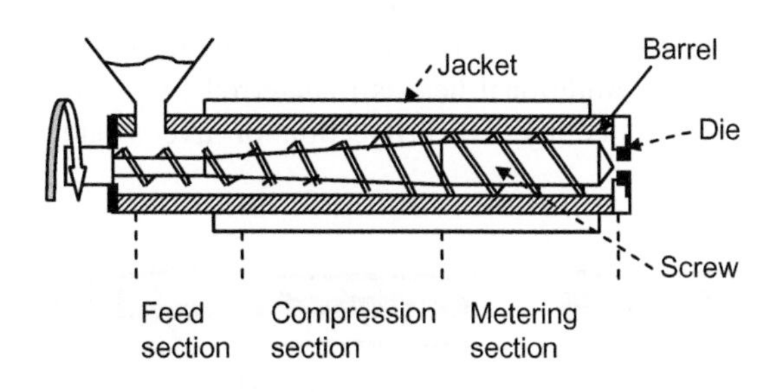

FIGURE 15.1

Basic structure of a single-screw extruder.

15.2.2 Operation

The materials processed in cooker-extruders are particulate moist solids or high-viscosity dough-like fluids. As the screw rotates, the flights drag the material towards the exit. The flow channel, described above, is delimited by two solid surfaces, namely, the screw and the barrel. Friction with the moving material occurs on both surfaces. Ideally, the friction at the barrel surface should be the stronger of the two, in order for internal shear to occur. Were the friction weak at the barrel surface (for example, as a result of lubrication) and strong at the screw surface, the material would stick to the screw and turn with it, without shear and without forward movement. Grooving helps reduce slippage at the barrel surface.

Screw configuration is such that the flow area along the flow channel is progressively reduced. Consequently, the material is progressively compressed as it moves down the barrel. The *compression ratio* is the ratio of the cross-sectional area of the flow channel at the feed end to that at the exit end. Reduction of the flow area can be achieved by several types of screw configurations. The most common are the progressively decreasing screw pitch and the progressively increasing root (core) diameter (Figure 15.2). Screw configurations corresponding to compression ratios between two and four are common. The pressure developed in a cooking extruder can be in the order of a few MPa.

Referring to Figure 15.1, it is customary to divide a single-screw extruder into three sections:

1. *The feed section*. The main function of this section is to act as a screw conveyor, transporting the material from the feed entrance to the subsequent sections. Almost no compression or modification of the mass occurs in this section.
2. *The transition section*. This is the section where the material is compressed and heated.
3. *The metering section*. This is the section where most of the objectives of the extrusion process (melting, texturization, kneading, chemical reactions, etc.) occur through shear and mixing.

Through the friction, most of the power used for turning the screw shaft is dissipated into the material as heat. Thus, a portion of the heat delivered to the product is generated *in situ*. Additional heat is transferred from the externally heated barrel surface and supplied by direct injection of live steam. In single-screw

FIGURE 15.2

Three methods to achieve compression.

extruders, the internally generated heat constitutes the major part of the energy input. Consequently, heating in a cooking extruder is extremely rapid.

Because of the high pressure in the extruder, the moist material can be heated to temperatures well above 100°C (sometimes up to 180–200°C). When the pressure is suddenly released at the exit from the die, some of the water in the product is flash evaporated and, as a result, the product is puffed. The degree of puffing can be controlled by releasing some of the pressure and cooling the mass at the last section of the extruder, before the die. Figure 15.3 shows an example of a pressure–temperature profile along the extruder, in a process where puffing has to be avoided (e.g., in the production of pellets).

As a result of compression, a pressure gradient is built in the opposite direction to the average movement of the mass. Therefore, the flow along the continuous flow channel contains two components: drag flow from the feed end to the die end, caused by the mechanical thrust of the screw flights, and pressure-driven back-flow in the opposite direction, caused by the pressure difference between the two extremities of the extruder. The net flow-rate is the difference between the flow-rates associated with the two velocity fields. Mixing in the single-screw extruder is in great part due to the existence of these two opposing flow patterns. The intensity of back-flow, and hence the "pumping efficiency" of the extruder, depends on the resistance of the die and other restrictions placed on the flow path (Figure 15.4).

15.2.3 Flow models, extruder throughput

Various models have been proposed for the analysis of the complex flow patterns in a cooking extruder (Bounié, 1988; Harper, 1980; Tadmor and Gogos, 1979; Tayeb *et al.,* 1992). The most common approach is mentally to "peel" the

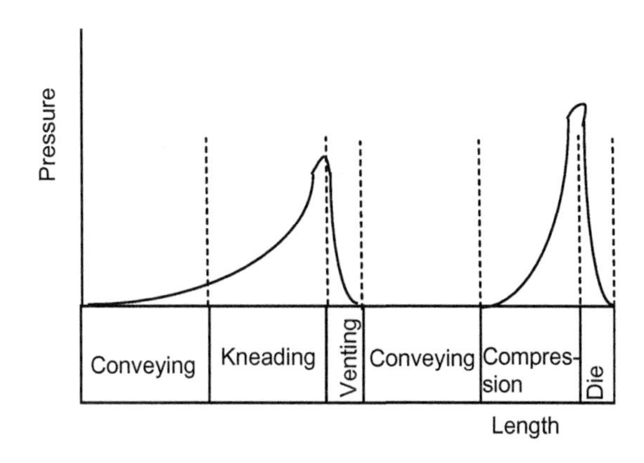

FIGURE 15.3

Pressure profile with venting.

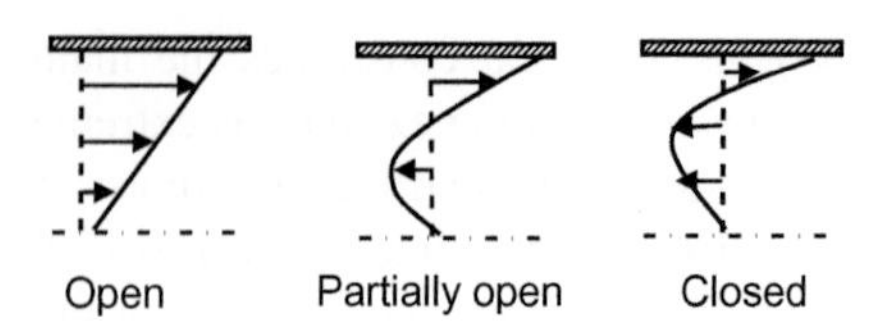

FIGURE 15.4

Back-flow as a function of back-pressure.

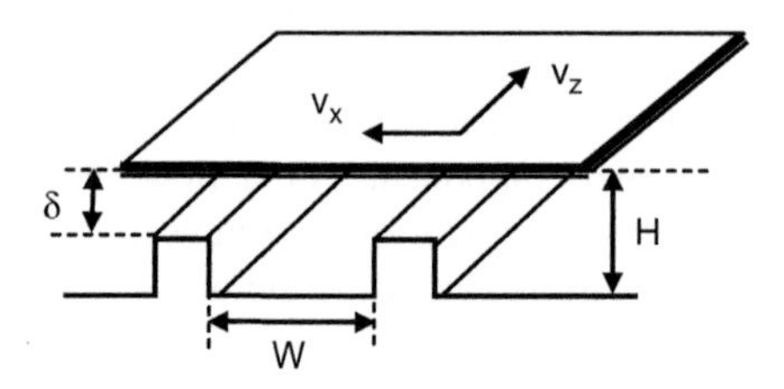

FIGURE 15.5

Flow channel model.

Figure adapted from Tadmor and Gogos (1979).

helicoidal flow channel off the screw, so as to have a flat strip of rectangular cross-section (Tadmor and Gogos, 1979). The movement of a fluid element within the flow channel is defined by the components of its velocity in the direction of the axes x, y and z, corresponding to the width, depth and length of the channel, respectively (Figure 15.5).

The longitudinal velocity v_z determines the throughput of the extruder. The cross-channel velocity v_x contributes to mixing. For ease of visualization, the screw is considered to be static and the barrel surface to be mobile. Referring to Figure 15.5, one can write:

$$v_B = \pi DN, \quad v_x = \pi DN \sin\theta, \quad v_z = \pi DN \cos\theta \tag{15.1}$$

where:

D = screw diameter, m
N = rotation speed, s^{-1}
θ = angle of the helix.

The volumetric throughput of the channel is approximated by Eq. (15.2), below:

$$Q = \left[\frac{v_z WH}{2}\right] + \left[\frac{WH^3}{12\mu}\left(\frac{dP}{dz}\right)\right] \tag{15.2}$$

where:

W, H = width and height of the flow channel, respectively
μ = apparent viscosity of the mass
dP/dz = pressure gradient along the extruder.

(The ratio W/H is termed the "channel aspect ratio".)

The first bracket in Eq. (15.2) represents the drag flow. The second bracket represents the back-flow due to the (negative) pressure gradient. The drag flow is the positive displacement element. The pressure flow represents the deviation of the single-screw extruder from true positive displacement pump performance.

In terms of the physical variables, Eq (15.2) can be written as follows:

$$Q = \left[\frac{(\pi DN \cos\theta)WH}{2}\right] - \left[\frac{WH^3}{12\mu}\left(\frac{P_2 - P_1}{L}\right)\right] \tag{15.3}$$

where P_1 and P_2 = pressure at the feed end and exit end, respectively, and L = length of the extruder.

Examination of Eq. (15.3) leads to the following conclusions:

- The throughput due to drag is proportional to the tip diameter of the screw, to the speed of revolution, and to the cross-sectional area of the flow channel.
- The back-flow is inversely proportional to the viscosity of the mass. In the case of high viscosity melts, frequently encountered in food extrusion, the influence of back-flow is greatly reduced.

Equation (15.3) is useful for directional evaluation of the effect of extruder geometry and operating conditions on extruder capacity, but not for design purposes because of the many simplifying assumptions made for its development. Most food materials of interest in food extrusion are non-Newtonian fluids, and the apparent viscosity of many of them cannot be represented by simple models such as power law. The assumption of constant cross-sectional area of the flow channel is another approximation.

The Tadmor–Gogos model has been tested at relatively high screw-channel aspect ratios (W much larger than H). A somewhat different flow model, more suitable for intermediate values of the channel aspect ratio, has been proposed by Alves *et al.* (2009).

EXAMPLE 15.1

Estimate the volumetric output of single-screw extruder operating with the following data:

Screw tip diameter, D = 0.1 m
Average screw height, H = 0.002 m
Average width of pitch, W = 0.05 m
Speed of rotation, N = 6 per second.
Angle of the helix, $\theta = 20°$
Length of the screw, L = 1.2 m

The average viscosity of the melt is estimated as 5 Pa · s.

The extruder is fed at atmospheric pressure. The pressure at the die is 1.3 MPa, gage.

Solution

Equation 15.3 is applied.

$$Q = \left[\frac{(\pi DN\cos\theta)WH}{2}\right] - \left[\frac{WH^3}{12\mu}\left(\frac{P_2 - P_1}{L}\right)\right]$$

$$Q = \frac{\pi \times 0.1 \times 6 \times 0.94 \times 0.05 \times 0.002}{2} - \frac{0.05 \times (0.002)^3 \times 1.3 \times 10^6}{12 \times 5 \times 1.2}$$

$$= 88.5 \times 10^{-6} - 7.2 \times 10^{-6} = 81.3 \times 10^{-6} m^3 s^{-1} = 0.293\ \mathrm{m^3 \cdot h^{-1}}$$

As stated above, the major part of the energy applied to the shaft is dissipated as heat. In single-screw extruders, the portion of the mechanical power used for building pressure and for pushing the mass through the die represents at most 28% of the total net energy input (Janssen, 1989).

The pressure at the die and the volumetric flow-rate are interrelated through a Darcy-type equation:

$$Q = \frac{\Delta P}{R} \tag{15.4}$$

where R is the total resistance of the die to flow. Separating the total resistance into the part of the resistance due only to die geometry and the part due to the properties of the fluid (viscosity), Eq. (15.4) can be rewritten as follows:

$$Q = \frac{\Delta P}{k_D \mu} \tag{15.5}$$

where k_D represents the resistance associated with die geometry only. Due to the irregularity of shape and the importance of end effects, it is difficult to calculate k_D analytically (e.g., with the help of laminar flow theory) and experimentation is required for its determination (Janssen, 1989).

15.2.4 Residence time distribution

Residence time distribution (RTD, see Chapter 4, Section 4.4) is particularly important in extrusion because of the utilization of the extruder as a reactor and, in some cases, as an autoclave. Because of the complexity of flow, residence time distribution cannot be deduced from knowledge of extruder geometry and the rheological properties of the processed material. The opposite approach is more common. RTD in extruders is determined experimentally, using pulse-injection techniques (Section 4.4). The experimental results are then utilized for the study of flow in extruders (Bounié, 1988; Levine, 1982). Residence time distribution in single-screw extruders is relatively wide (Harper, 1981) as a result of the deviation of the extruder from plug-flow behavior.

15.3 Twin-screw extruders

15.3.1 Structure

Twin-screw extruders comprise a pair of parallel screws, rotating inside a barrel with a figure-of-eight shaped cross-section (Figure 15.6). At the exit end, each half of the barrel converges into a short conical section, each with a die at the apex.

The screws can be co-rotating or counter-rotating (Figure 15.7). The type most commonly used in the food industry is the co-rotating, self-wiping twin-screw extruder (Harper, 1989). In self-wiping models, the two screws are closely intermeshed (Figure 15.8). In some models the barrel is split into two halves longitudinally, allowing opening for freeing the screws, inspection and cleaning. In other models, the barrel consists of short, separable modules (Figure 15.9). In most models, different screw configurations can be assembled by the user by sliding onto a shaft various screw elements of different compression ratios, mixing sections, reverse-pitch screw sections (to enhance mixing or to reduce pressure), restrictions (for building pressure), etc. This possibility imparts to the extruder a degree of versatility that is not found in older models with a fixed-screw configuration.

15.3.2 Operation

Most of the material transport occurs in C-shaped chambers formed between the screws (Figure 15.10). The proportion of mechanical power used for pumping is

FIGURE 15.6

Cross-section of the barrel of a twin-screw extruder.

Photograph courtesy of Clextral.

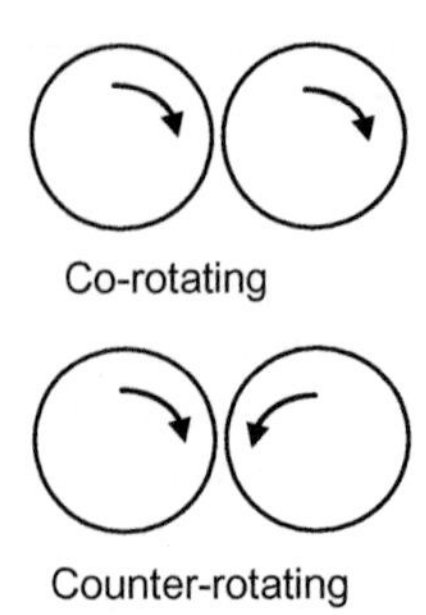

FIGURE 15.7

Directions of rotation of twin-screws.

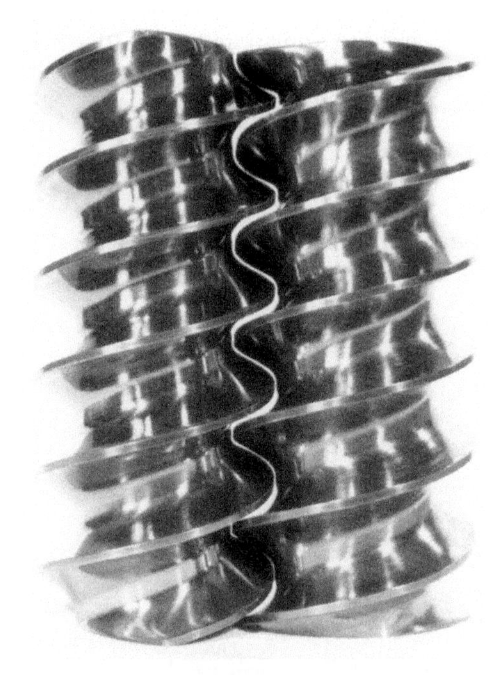

FIGURE 15.8

Intermeshing twin screws.

Photograph courtesy of Clextral.

higher than in single-screw extruders; therefore, a larger proportion of the energy for heating must be supplied from external sources. Slippage or friction on the barrel surface is less relevant. The first section of the extruder, the conveying section, extending over the major part of the screw length, acts as a heated screw

FIGURE 15.9

Twin-screw extruder with conditioner, showing the barrel sections.

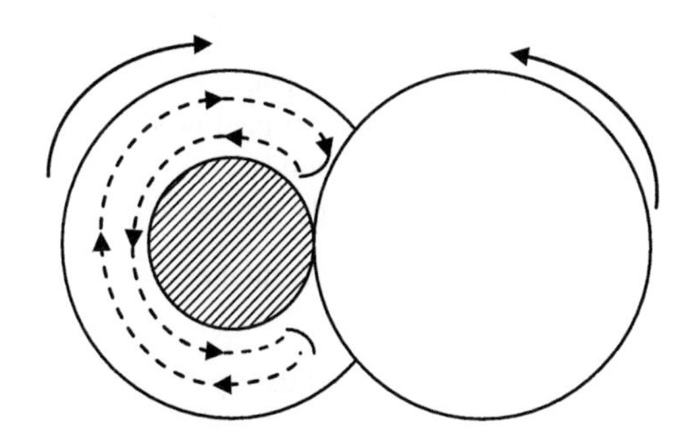

FIGURE 15.10

C-shaped flow channels in counter-rotating twin screws.

Figure adapted from Harper (1989).

conveyor. The particularly large screw angle in this section imparts to the material relatively high forward axial velocity. As a result, the filling ratio at this section is low. Heating occurs as a result of heat transfer from the barrel surface, itself heated externally. Because of the low hold-up, the material is rapidly heated and reaches temperatures close to that of the barrel. On the other hand, the conveying section does not contribute significantly to the pressure build-up. Pressurization is achieved at the next section, known as the melt pumping or metering section, with the help of flow restriction elements included in the configuration of the screw and, of course, the dies. The filling ratio in this second section increases gradually towards the die region. At a given time, different parts of the screw surface fulfill different functions (Sastrohartono *et al.*, 1992). The part of the screw surface facing the barrel performs the action of translation, while most of the mixing occurs in the intermeshing section between the screws.

15.3.3 Advantages and shortcomings

A detailed comparative study of twin-screw and single-screw extruders is given in Harper (1992). The main advantages of twin-screw over single-screw extruders are as follows:

- Pumping efficiency is better and less dependent on the flow properties of the material.
- Mixing is more efficient.
- The heat exchange rate from the barrel surface to the material is faster and more uniform.
- Residence time is more uniformly distributed.
- Thanks to the possibility of flexible modular configuration of the screw and the barrel, the twin-screw extruder is a more versatile machine.
- High moisture and sticky materials can be handled. This is probably the most important advantage, because it enables processing of materials that could not be handled hitherto by single-screw extruders. The introduction of twin-screw technology has resulted in a significant expansion of the application of extrusion to foods and the spectrum of extruded products.
- Self-wiping reduces the risk of residue build-up.
- Feeding problems with cohesive materials are less serious.

On the negative side, the relative disadvantages of the twin-screw extruder are:

- Higher cost: both capital expenditure and the operating costs are higher.
- Complexity: the twin-screw extruder is mechanically more complex and less robust; consequently, the machine is more sensitive to mechanical abuse (such as high torque).

15.4 Effect on foods

15.4.1 Physical effects

The feed entering the extruder is, most commonly, granular material that has been pre-treated (formulated, humidified or dried, heated, agglomerated etc.) In the conveying section of the extruder, the material keeps its particulate structure. As the temperature and the pressure increase and more shear is applied, the particulate structure disappears and a dough-like material is obtained. This mass is called *melt*, in analogy to the phenomenon of true melting in the extrusion of plastic polymers. As the melt continues its movement, a new internal structure and some degree of phase separation may occur, mainly as a result of shear.

Orientation of protein molecules is believed to be the basic mechanism in extrusion texturization of soybean flour (Stanley, 1989). Puffing (see Section 15.2.2) is essential for the creation of the characteristic porous structure of many extruded products. Evaporation of volatile substances, along with some water vapor, is sometimes considered to be one of the desirable physical effects of extrusion. Known in polymer processing as a method for the removal of volatile solvent and monomer residues (volatilization), evaporation of objectionable odorous components is the main objective of extrusion conching of chocolate (see below). In a different application, it has been found that dry extrusion cooking of soybeans before oil expression by mechanical expulsion results in considerably increased oil yields (Nelson *et al.,* 1987).

One of the primary objectives of extrusion is *forming*. In addition to the age-old shaping of long and short pasta by extrusion, specialized extruder types are now being used for forming dough products, candy bars, chocolate centers, ice-cream bars and more. Co-extrusion, which is the extrusion of two different masses through a common die, is used extensively for the production of composite items such as filled rolls and multi-phase snacks.

15.4.2 Chemical effect

15.4.2.1 Effects on starch

Historically, starchy foods and cereals were the first class of raw materials to be processed by extrusion, for human food and animal feed. The principal effect of extrusion cooking on starch is gelatinization. A theoretical model of starch gelatinization in a single-screw extruder was proposed by Gopalakrishna and Jaluria (1992), in the light of flow, heat transfer and moisture distribution. Gelatinization (also known as "pasting") is the process whereby the starch granules swell and eventually disappear, the crystalline regions are progressively "melted", and the starch molecules are unfolded and hydrated, resulting in the formation of a continuous viscous paste. Gelatinization occurs when a suspension of starch in water is heated. The temperature of gelatinization depends on the water content. In extrusion cooking, gelatinization can be achieved at relatively low moisture levels. The purpose of gelatinization in starchy foods and cereals is primarily to improve their digestibility, as gelatinized starch is more easily hydrolyzed by amylolytic enzymes. Another objective of starch gelatinization in extrusion cooking is to create the thermoplastic mass that can assume a stable porous structure upon puffing. The expansion of starchy melts was studied by Robin *et al.,* (2010). The effect of extrusion cooking on starch does not stop at gelatinization; it can also cause partial depolymerizing (dextrinization) of the starch molecule (Cheftel, 1986; Colonna *et al.,* 1989; Davidson *et al.,* 1984; Karathanos and Saravacos, 1992).

15.4.2.2 Effect on proteins

A vast variety of proteins and protein-rich foods have been processed by extrusion cooking. These include defatted soybean flour, soybean protein concentrate, other oilseed meals (peanut, sunflower, sesame), pulses (Chauhan *et al.*, 1988; Gahlawat and Sehgal, 1994; Korus *et al.*, 2006), algae, milk proteins and meat.

The effect of extrusion cooking on proteins has been studied extensively. A considerable proportion of the research in this field deals with the molecular background of texturization. The modification of proteins during extrusion cooking is mainly attributed to thermal effects and to shear. Protein denaturation is the primary thermal effect. Under the influence of high temperature and moisture, native proteins lose their structure (globular, miscellar, etc.), unfold, adsorb water and "melt". As in starch gelatinization, in extrusion cooking protein denaturation occurs at lower moisture content, resulting in a high-viscosity melt. In the case of soy protein, it has been shown that the protein in the extrudate is completely denatured if the extrusion temperature is above 130°C (Kitabatake and Doi, 1992). Texturization (i.e., restructuring of the melt into lamellar or fibrous forms) is believed to be largely the result of orientation of the protein chains of the melt, under the effect of unidirectional shear near to and at the die. The molecular processes associated with protein texturization have not been entirely elucidated. Protein–protein intermolecular cross-linking is thought to be responsible for the stabilization of the new texture (Burgess and Stanley, 1976). The formation of disulfide bonds at the expense of sulfhydryl groups is often proposed as the principal mechanism (see, for example, Hager, 1984). On the other hand, analysis of soy extrudates does not show any increase in the proportion of disulfide bonds, and the opposite result is often observed.

The principal thermal reaction involving proteins but not associated with restructuring is the Maillard reaction. Maillard-type reactions and the ensuing browning discoloration are significant only in the presence of reducing sugars.

15.4.2.3 Nutritional aspects

As with all high-temperature processes, extrusion cooking may be expected to affect the nutritional quality of foods adversely (e.g., by the destruction of vitamins) or favorably (e.g., by inactivation of anti-nutritional factors). A general review on the nutritional aspects of food extrusion is available (Singh *et al.*, 2007).

15.5 Food applications of extrusion

15.5.1 Forming extrusion of pasta

Pasta products are manufactured by mixing milled wheat with water and other ingredients into a homogeneous dough and extruding this dough through a variety of dies, depending on the shape of the pasta product (Brockway, 2001). Both kneading and forming take place in the pasta extruder, also known as pasta press (Figures 15.11 and 15.12). Air is removed from the mixture prior to extruding, to avoid air bubbles in the finished product. In contrast to extrusion cooking, pasta

FIGURE 15.11

Short pasta press.

Photograph courtesy of Isolteck Cusinato SRL.

FIGURE 15.12

Die for "rigatoni".

Photograph courtesy of Isolteck Cusinato SRL.

extrusion is *cold extrusion* (Le Roux *et al.,* 1995). The extruder is often equipped with a water jacket for cooling.

15.5.2 Expanded snacks

The familiar "corn curls" are made by extrusion cooking of corn grits to which some water has been added to bring the moisture content to about 15%. The extrusion temperature is typically 160–180°C and the compression ratio is 2:1 or

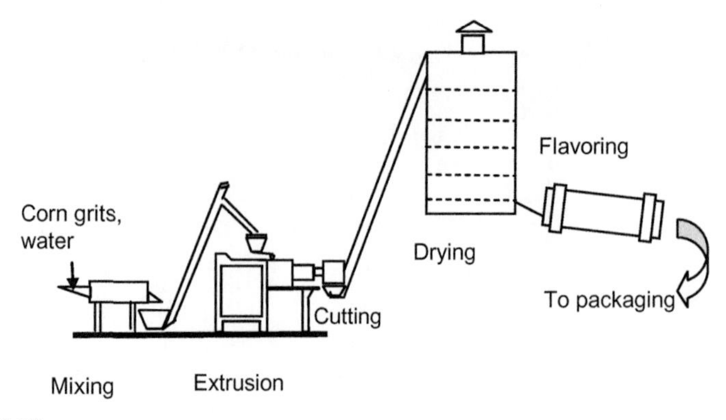

FIGURE 15.13

Flow diagram of expanded corn snack (corn curl) production process.

less. The melt is extruded through a circular die to produce a highly puffed rope, which is cut to the desired size by a revolving blade. At this stage, the curls still contain about 6% moisture and must be dried by baking or frying. The bland curls are flavored by spraying with oil and coating with various flavoring mixtures (cheese, hydrolyzed yeast, spices, peanut butter, etc.) and cooled to room temperature, at which the product becomes crisp and crunchy. Although the curved cylindrical form is still the most common shape of puffed corn snacks, other shapes such as spheres, tubes, cones, etc., are also produced. A flow diagram of the corn-curl production process is shown in Figure 15.13.

15.5.3 Ready-to-eat cereals

The process of manufacturing ready-to-eat breakfast cereals is essentially similar to that of puffed snacks; however, the formulation is different and more complex. The degree of puffing is controlled according to the desired product bulk density. Sugar, if needed, is added after extrusion, to prevent excessive browning. Heat-sensitive flavors and vitamins are also added after extrusion.

Extrusion cooking is also used for the preparation of raw materials for the production of ready-to-eat cereal products. For example, milled corn is cooked with other ingredients such as malt and extruded without puffing in the form of large granules. These granules are then flaked and toasted to produce corn flakes. The use of the reconstructed corn granules instead of the traditional "hominy grits" results in considerable improvement of raw material utilization efficiency.

15.5.4 Pellets

Expanded cereal products can also be produced by a two-step process. The first step is extrusion cooking and forming of particles, without puffing. These particles are known as pellets. At a later stage, and often in a different location, the pellets are puffed by heating or frying. Pellets are essentially non-porous particles of pre-gelatinized (cooked) starchy materials, having a particular shape, made by extrusion cooking of moist starchy mixtures. The main function of cooking extrusion in this case is to gelatinize the starch, and particularly the amylose fraction. Puffing is avoided by releasing the pressure and cooling before exit from the die. The shape and size of the pellets are defined by the die and the cutting operation. After extrusion, the pellets are stabilized by drying to a water content that warrants long-term preservation but leaves sufficient moisture for plasticizing and puffing the pellet when exposed to a very high temperature. The mechanism of puffing of pellets is quite similar to that of popping corn. The advantages of the two-step process are:

- The ability to store the intermediate high-density (low storage volume), shelf-stable product until needed for puffing.
- The separation between the high-capacity extrusion plant and the final production facility where the pellets are thermally puffed, baked, fried, flavored and packaged, according to market demand.

Single-screw cooking extruders are quite suitable for the production of puffed snacks, ready-to-eat breakfast cereals and pellets because of the low moisture content of the mass. Nevertheless, twin-screw extruders are often preferred for high-capacity production.

15.5.5 Other extruded starchy and cereal products

Bread-like structures can be produced by extrusion cooking of cereal doughs. Although sliced bread shapes can be obtained by using large dies, the organoleptic characteristics of the product are too different from those of conventional yeast-leavened, oven-baked bread to be commercially acceptable. On the other hand, flatbread (a product resembling the familiar crispbread or knäckebrot), bread sticks and croutons are commercially produced by extrusion. While extrusion provides the cooking and expansion, subsequent oven baking or toasting is still needed for the production of a crunchy crust. A detailed description of an industrial process for flatbread production by extrusion is given by Meuser and Wiedmann (1989).

Nixtamalization is a traditional process in Mexico and Central America whereby corn is treated with lime, cooked, and dried and ground to produce the flour used to make tortilla. In addition to the development of the characteristic flavor, the principal objective of nixtamalization is to improve the nutritional quality of maize by rendering its niacin nutritionally available. The traditional process is lengthy and tedious. Extrusion cooking provides a simplified and convenient alternative (Mensah-Agyapong and Horner, 1992).

Extrusion cooking of cornmeal produces a raw material that can be used directly in bakery products (Curic *et al.*, 2009).

15.5.6 Texturized protein products

The primary application of extrusion texturization is in the production of meat analogs from plant proteins. Extrusion cooking of soy flour produces a lamellar mass with the chewiness of meat, known as *texturized soy protein* (TSP) or *texturized vegetable protein* (TVP). TVP made from soy flour (44–50% protein) has some of the characteristic flavor of the starting material. A better product is made by extrusion texturization of soy protein concentrate (70% protein). Soy protein concentrate is made from soy flour by extracting the undesirable sugars and other low molecular weight components with aqueous alcohol. Texturized soy protein concentrate is free of undesirable flavor and has, of course, a higher protein content. Before extrusion, the starting material (soy flour or concentrate) is conditioned with steam at a moisture content of about 20%. In the extruder, the material reaches temperatures in the order of 160–180°C and becomes a thermoplastic "melt". The unidirectional shear causes some orientation of the protein chains. The melt is expelled through the die, cut, and dried (Berk, 1992).

Other protein-rich flours (peanut, sunflower, cottonseed, etc.) have been texturized by extrusion cooking, alone or with other ingredients. Texturized soy products are, however, practically the only texturized vegetable proteins on the market.

Texturized soy protein, made by ordinary extrusion cooking as described above, has the chewiness of meat but not its fibrous structure. Recently, new extrusion technologies have been developed for the production of fibrous structures from vegetable or animal protein sources. The fibrous structure is obtained by forcing the material through a long die, where orientation of the protein chains takes place, while the extrudate is slowly cooled. For this application, twin-screw extruders, capable of handling material at high moisture content, are used (Akdoğan, 1999; Chen *et al.*, 2010).

Another interesting application of extrusion cooking for texturization is the production of processed cheese from different sources of dairy proteins, usually with added fat.

15.5.7 Confectionery and chocolate

Extrusion is now used extensively in the confectionery industry for mixing, cooking, cooling and forming in one step. Sugar, corn syrup, flour, starch, fruit concentrates, fats and other ingredients of the recipe are directly fed into twin-screw co-rotating extruders, where the mixture is heated, mixed and homogenized at a high throughput rate. The product is shaped by using the proper die and cutting method. In addition, co-extrusion of different materials allows the production of multi-colored items such as fruit twists, as well as multi-phase filled items. Materials with different viscosities, from fondant to soft fruit jellies, can be handled. Volatile or heat-sensitive flavors are usually added at the end of the extrusion cooking process.

Licorice candy is one of the confectionery items frequently produced by extrusion (Gabriele *et al.,* 2001). Since the production of this candy requires cooking and mixing of wheat flour with the licorice extract, sugar, etc., to form a homogeneous mass, extrusion cooking is particularly suitable for the process.

Conching is one of the most important steps in the manufacture of chocolate. Before molding, the chocolate mass is subjected to a process of conching, which performs the functions of mixing, aerating, evaporation, homogenizing and size reduction. Conching is an intense mixing process in which the chocolate mass is kneaded by heavy reciprocating rollers or by rotary blades. The heat generated by friction liquefies the mass and causes the evaporation of water, acetic acid and other volatiles, while the shearing action helps reduce the size of the solid particles and coats them uniformly with cocoa fat. Traditional conching is a slow process, taking from a few hours to a few days. Extrusion performs the same functions in a much shorter time (Aguilar *et al.,* 1995). Considerable reduction in energy expenditure was reported when a special type of extruder (reciprocating multihole extruder) was used in addition to the chocolate mixer (Jolly *et al.,* 2003). The quality of chocolate produced by fast conching in extruders is usually considered less good than that of fine chocolate treated by conventional slow conching. On the other hand, extrusion conching provides efficient thermal treatment and mixing of the chocolate.

15.5.8 Pet foods

The production of pet-foods is one of the highest volume applications of extrusion cooking. The processes utilized for the production of dry pet-foods are essentially similar to those used for cereal products. By-products of animal origin are sometimes included in the formulation. Twin-screw extruders are used for the production of moist pet food.

References

Aguilar, C.A., Dimick, P.S., Hollender, R., Ziegler, G.R., 1995. Flavor modification of milk chocolate by conching in a twin-screw, co-rotating, continuous mixer. J. Sens. Stud. 10, 369–380.

Akdoğan, H., 1999. High moisture food extrusion. Intl. J. Food Sci. Technol. 34 (3), 195–207.

Alves, M.V.C., Barbosa Jr., J.R., Prata, A.T., 2009. Analytical solution of single screw extrusion applicable to intermediate values of screw channel aspect ratio. J. Food Eng. 92 (2), 152–156.

Berk, Z., 1992. Technology of Production of Edible Flours and Protein Products from Soybeans. FAO, Rome, Italy.

Bounié, D., 1988. Modelling of the flow pattern in a twin-screw extruder through residence time distribution experiments. J. Food Eng. 7, 223–246.

Brockway, B.E., 2001. Pasta. In: Dendy, D.A.V., Dobraszczyk, B.V. (Eds.), Cereal and Cereal Products Chemistry and Technology. Aspen Publishers, Gaithersburg, MD.

Burgess, G.D., Stanley, D.W., 1976. A possible mechanism for thermal texturization of soybean protein. Can. Inst. Food Sci. Technol. 9, 228–231.

Chauhan, G.S., Verma, N.S., Bains, G.S., 1988. Effect of extrusion cooking on the nutritional quality of protein in rice–legume blends. Die Nahrung 32 (1), 43–46.

Cheftel, J.C., 1986. Nutritional effects of extrusion cooking. Food Chem. 20, 263.

Chen, F.L., Wei, Y.M., Zhang, B., Ojokoh, A.O., 2010. System parameters and product properties response of soybean protein exruded at wide moisture range. J. Food Eng. 96 (2), 208–213.

Colonna, P., Tayeb, J., Mercier, C., 1989. Extrusion cooking of starch and starchy products. In: Mercier, C., Linko, P., Harper, J.M. (Eds.), Extrusion Cooking. American Association of Cereal Chemists, Inc., St Paul, MN.

Curic, D., Novotni, D., Bauman, I., Kricka, T., Dugum, J., 2009. Optimization of extrusion cooking of cornmeal as raw material for bakery products. J. Food Proc. Eng. 32 (2), 294–317.

Davidson, V.J., Paton, D., Diosady, L.L., Laroque, G., 1984. Degradation of wheat starch in a single-screw extruder: characteristics of extruded starch polymers. J. Food Sci. 49, 453–458.

Gabriele, D., Curcio, S., De Cindio, B., 2001. Optimal design of single screw extruder for licorice candy production: a rheology based approach. J. Food Eng. 48 (1), 33–44.

Gahlawat, P., Sehgal, S., 1994. Shelf life of weaning foods developed from locally available foodstuffs. Plant Foods Hum. Nutr. 45 (4), 349–355.

Gopalakrishna, S., Jaluria, Y., 1992. Modeling of starch gelatinization in a single-screw extruder. In: Kokini, J.L., Ho, C.-T., Karwe, M.V. (Eds.), Food Extrusion Science and Technology. Marcel Dekker, New York, NY.

Hager, D.F., 1984. Effect of extrusion upon soy concentrate solubility. J. Agr. Food Chem. 32, 293–296.

Harper, J.M., 1980. Food Extrusion. CRC Press, Boca Raton, FL.

Harper, J.M., 1981. Extrusion of Foods. CRC Press, Boca Raton, FL.

Harper, J.M., 1989. Food extruders and their applications. In: Mercier, C., Linko, P., Harper, J.M. (Eds.), Extrusion Cooking. American Association of Cereal Chemists, Inc., St Paul. MN.

Harper, J.M., 1992. A comparative study of single- and twin-screw extruders. In: Kokini, J.L., Ho, C.-T., Karwe, M.V. (Eds.), Food Extrusion Science and Technology. Marcel Dekker, New York, NY.

Janssen, L.P.B.M., 1989. Engineering aspects of food extrusion. In: Mercier, C., Linko, P., Harper, J.M. (Eds.), Extrusion Cooking. American Association of Cereal Chemists, Inc., St Paul, MN.

Jolly, M.S., Blackburn, S., Beckett, S.T., 2003. Energy reduction during chocolate conching using a reciprocating multihole extruder. J. Food Eng. 59 (2–3), 137–142.

Karathanos, V.T., Saravacos, G.D., 1992. Water diffusivity in the extrusion cooking of starch materials. In: Kokini, J.L., Ho, C.-T., Karwe, M.V. (Eds.), Food Extrusion Science and Technology. Marcel Dekker, New York, NY.

Kitabatake, N., Doi, E., 1992. Denaturation and texturization of food protein by extrusion cooking. In: Kokini, J.L., Ho, C.-T., Karwe, M.V. (Eds.), Food Extrusion Science and Technology. Marcel Dekker, New York, NY.

Korus, J., Gumul, D., Achremowicz, B., 2006. The influence of extrusion on chemical composition of dry seeds of bean (*Phaseolus vulgaris* L.). Electron. J. Polish Agricult. Univ. Food Sci. Technol. 9, 1.

Le Roux, D., Vergnes, B., Chaurand, M., Abécassis, J., 1995. A thermo-mechanical approach to pasta extrusion. J. Food Eng. 26 (3), 351–368.

Levine, L., 1982. Estimating output and power of food extruders. J. Food Process Eng. 6, 1–13.

Mensah-Agyapong, J., Horner, W.F.A., 1992. Nixtamalization of maize (*Zea mays* L.) using a single screw cook-extrusion process on lime-treated grits. J. Sci. Food Agric. 60 (4), 509–514.

Meuser, F., Wiedmann, W., 1989. Extrusion plant design. In: Mercier, C., Linko, P., Harper, J.M. (Eds.), Extrusion Cooking. American Association of Cereal Chemists, Inc., St Paul, MN.

Nelson, A.I., Wijeratne, S.W., Yeh, T.M., Wei, L.S., 1987. Dry extrusion as an aid to mechanical expelling of oil from soybeans. J. Am. Oil. Chem. Soc. 64 (9), 1341–1347.

Robin, F., Engmann, J., Pineau, N., Chanvrier, H., Bovet, N., Della Valle, G., 2010. Extrusion, structure and mechanical properties of complex starchy foams. J. Food Eng. 98 (1), 19–27.

Saalia, F.K., Phillips, R.D., 2011. Degradation of aflatoxins by extrusion cooking: effects on nutritional quality of extrudates. LWT 44 (6), 1496–1501.

Sastrohartono, T., Karwe, M.V., Jaluria, Y., Kwon, T.H., 1992. Numerical simulation of fluid flow and heat transfer in a twin-screw extruder. In: Kokini, J.L., Ho, C.-T., Karwe, M.V. (Eds.), Food Extrusion Science and Technology. Marcel Dekker, New York, NY.

Singh, S., Gamlath, S., Wakeling, L., 2007. Nutritional aspects of food extrusion: a review. Intl. J. Food Sci. Technol. 42 (8), 916–929.

Stanley, D.W., 1989. Protein reactions during extrusion processing. In: Mercier, C., Linko, P, Harper, J.M. (Eds.), Extrusion Cooking. American Association of Cereal Chemists, Inc., St Paul, MN.

Tadmor, Z., Gogos, C., 1979. Principles of Polymer Processing. Wiley, New York, NY.

Tayeb, J., Della Valle, G., Barès, C., Vergnes, B., 1992. Simulation of transport phenomena in twin-screw extruders. In: Kokini, J.L., Ho, C.-T., Karwe, M.V. (Eds.), Food Extrusion Science and Technology. Marcel Dekker, New York, NY.

CHAPTER

Spoilage and Preservation of Foods

16

The principal objective of food processing is to preserve the overall quality of the food over a certain duration known as *shelf life*. While operations such as heat processing, concentration, drying, irradiation, curing, smoking, etc., may serve other purposes, their main objective is, or was historically, to prevent or retard spoilage.

16.1 Mechanisms of food spoilage

For the purpose of our discussion, spoilage will be defined as any process leading to the deterioration of the safety, sensory quality (taste, flavor, texture, color and appearance) or nutritional value of food. The different types of food spoilage include:

1. *Microbial spoilage* – deterioration due to the activity and/or presence of microorganisms.
2. *Enzymatic spoilage* – undesirable changes due to enzyme catalyzed reactions.
3. *Chemical spoilage* – undesirable changes due to non-enzymatic chemical reactions between the components of food (e.g., Maillard browning), or between the food and its environment (e.g., lipid oxidation).
4. *Physical spoilage* – undesirable changes in the physical structure of the food (e.g., sugar crystallization in preserves, separation of emulsions, collapse of gels).

Obviously, the most important type of deterioration in foods is microbial spoilage, since it may affect both the quality and the safety of food.

16.2 Food preservation processes

The main preservation technologies (Desrosier, 1970) by which the various types of food spoilage can be controlled are as follows.

1. *Preservation by heat (thermal processes)*. Microorganisms and enzymes are destroyed at high temperatures. The extent of destruction depends on the temperature, the time of exposure and, of course, on the heat resistance of the microorganism or enzyme in question in the given medium. Exposure to high

Food Process Engineering and Technology. DOI: http://dx.doi.org/10.1016/B978-0-12-415923-5.00016-2
© 2013 Elsevier Inc. All rights reserved.

temperature not only destroys microorganisms and enzymes; it also accelerates a multitude of chemical reactions leading to changes in the texture, flavor, appearance, color, digestibility and nutritional value of the food. Some of these changes are desirable and constitute the complex process known as "cooking". Others are objectionable and are collectively named "thermal damage". Thorough knowledge of the kinetics of thermal preservation, cooking reactions and thermal damage mechanisms is essential for the optimization of thermal processes.

2. *Preservation by removal of heat (low-temperature processes).* The activity of microorganisms and enzymes as well as the velocity of chemical reactions are depressed at low temperatures. In contrast to heat, low temperature does not destroy enzymes and microorganisms to any significant extent, but merely depresses their activity. Such depression remains in place only as long as the low temperature is maintained. It ceases to function when the temperature is raised. It follows that, unlike thermal processing, the effect of preservation by cold is not permanent; hence the importance of maintaining the *cold chain* throughout the shelf life of the food product. Preservation by cold encompasses two distinct technological processes: chilling (maintaining the food at low temperature above the freezing point of the food) and freezing (below the freezing point). The difference between the two is not only in temperature. A substantial part of the preservation effect of freezing is due to the change of phase, from liquid to solid, with the corresponding reduction in molecular mobility. Transition of part of the water from liquid to ice results in drastic reduction of water activity, which in turn depresses microbial and enzyme activity, as we shall see. It should be kept in mind that the phase change may also produce undesirable and irreversible changes in texture. These will be discussed in some detail, later on.
3. *Preservation by reduction of water activity.* It is well known that microorganisms cannot develop at water activity levels below a limit value depending on the microorganism. Enzyme activity is also water-activity dependent. Drying, concentration, and addition of solutes (sugar, salt) are preservation techniques based on the reduction of water activity. As we shall see, the effect of water activity on chemical spoilage is more complex. The phenomenon of glass transition (see Chapter 1) and the depression of molecular mobility resulting from it are also consequences of the combined effect of low water activity and low temperature.
4. *Ionizing radiation.* Ionizing radiation has the capability of destroying microorganisms and inactivating enzymes. This powerful preservation technique has great potential as a solution to many problems in food production and distribution. Its widespread application is hindered by a number of issues, mainly in the area of consumer perception.
5. *Chemical preservation.* Two of the oldest food preservation techniques, namely salting and smoking, are based on the effects of salt and smoke chemicals on microorganisms. Many pathogens cannot develop at low pH,

hence the use of acids as food preservatives. The pH of foods can be lowered either by the addition of acids (acetic acid or vinegar, citric acid or lemon juice, lactic acid, etc.) or by the *in situ* production of acids (mainly lactic) through fermentation. Although defining wine as "preserved grape juice" would be absurd today, it may be assumed that alcoholic fermentation was originally a type of chemical preservation. Some spices and herbs are said to prevent or retard food spoilage. Many non-natural chemical substances (e.g., sulfur dioxide, benzoic acid, sorbic acid) are effective in preventing or retarding some kinds of food spoilage (food preservatives against microbial spoilage, antioxidants against oxidative deterioration, stabilizers against undesirable changes in texture and structure, etc.). The use of such chemicals is subject to food laws and regulations.

16.3 Combined processes (the "hurdle effect")

Many food products owe their safety and shelf-stability not to a single preservation technique but to the combined effect of a number of mechanisms acting simultaneously. While each mechanism alone would be insufficient to provide the desired protection, the combination does the job. As an example, one can assume that the shelf life and safety of dried sausage is provided by the combined effect of low water activity, relatively low pH, smoking, salt, low temperature and perhaps spices, each one of these "hurdles" being not "high" enough to achieve the expected preservation when acting alone. Additional examples of preservation by the combined effect of processes will be discussed in the following chapters dealing with food preservation. While the ability to preserve food by combining a number of preserving factors has been known for many years in different cultures, the concept of the "hurdle effect" is relatively new.

16.4 Packaging

The main objective of packaging is to provide a protective barrier between the food and the environment. The preservation functions of food packaging include prevention of microbial contamination, control of materials exchange (water vapor, oxygen, odorous substances) between the food and its environment, protection from light, etc. Recently, packaging materials containing preserving agents (e.g., antioxidants) have been developed. This new technology is known as *active packaging*.

Additional Reading

Blackburn, C. (Ed.), 2005. Food Spoilage Microorganisms. Woodhead Publishing Ltd., Cambridge, UK.

Coles, R., McDowell, D., Kirwan, M.J. (Eds.), 2003. Food Packaging Technology. Blackwell Publishing, Oxford, UK.

Deak, T., Beuchat, L.R., 1996. Handbook of Food Spoilage, Yeasts. CRC Press, Boca Raton, FL.

Desrosier, N.W., 1970. The Technology of Food Preservation, third ed. The Avi Publishing Company Inc., Westport, CT.

Harris, R.S., Karmas, E., 1975. Nutritional Evaluation of Food Processing. Avi Publishing Co., Westport, CT.

Karel, M., Fennema, O., Lund, D., 1975. Physical Principles of Food Preservation. Marcel Dekker, New York, NY.

Leistner, L., Gould, G.W., 2002. Hurdle Technologies, Combination of Treatments for Food Stability, Safety and Quality. Kluwer Academic/Plenum Publishers, New York, NY.

Ohlsson, T., Bengtsson, N. (Eds.), 2002. Minimal Processing Technologies in the Food Industry. Woodhead Publishing Ltd., Cambridge, UK.

Pitt, J.I., Hocking, A.D., 1997. Fungi and Food Spoilage, second ed. Blackie Academic & Professional, Glasgow, UK.

Rahman., M.S. (Ed.), 1999. Handbook of Food Preservation. Marcel Dekker, New York, NY.

CHAPTER

Thermal Processing

17

17.1 Introduction

The industrial application of food preservation by heat began with the work of the French inventor Nicolas Appert (1749–1841), who first demonstrated that long-term preservation of different kinds of foods can be achieved by heating the foods for a long time (many hours) in hermetically closed containers. The microbial origin of food spoilage and the relationship between thermal destruction of microorganism and food preservation were demonstrated only later by Louis Pasteur (French chemist and biologist, 1822–1895). The quantitative study of the kinetic aspects of thermal processing (thermo-microbiology) began in the early 20th century (see, for example, Olson and Stevens, 1939) and soon became an active area of research. Today, knowledge of the laws of heat transfer, combined with thermo-microbiology, constitutes the base for the rational design of thermal processes (Ball and Olson, 1957; Bimbenet *et al.*, 2002; Holdsworth, 1997; Leland and Robertson, 1985; Lewis and Heppel, 2000; Richardson, 2004).

Depending on their intensity, thermal preservation processes are classified into two categories:

1. *Pasteurization*: heat processing at relatively mild temperature (say, 70–100°C). Pasteurization destroys vegetative cells of microorganisms but has almost no effect on spores.
2. *Sterilization*: heat processing at high temperature (above 100°C) with the objective of destroying all forms of microorganisms, including spores.

Sterilization alone provides long-term preservation of foods, on the condition that recontamination is prevented by proper packaging. Pasteurization, on the other hand, provides only short-term stability or requires additional preserving factors (hurdles) such as refrigeration or low pH for long-term effectiveness. The following are some cases where pasteurization gives satisfactory results:

- The process objective is to destroy non-spore forming pathogens (e.g., *Mycobacterium tuberculosis*; Salmonella, Listeria, etc., in milk; Salmonella in liquid egg).
- The product is intended for consumption within a short time after production and is distributed under refrigeration (pasteurized dairy products, ready-to-eat meals prepared by cook–chill technologies).

Food Process Engineering and Technology. DOI: http://dx.doi.org/10.1016/B978-0-12-415923-5.00017-4
© 2013 Elsevier Inc. All rights reserved.

- The acidity of the product is high enough (pH < 4.6) to prevent growth of spore forming pathogens, particularly *Clostridium botulinum* (fruit juices, canned fruit, pickles).
- The process objective is to prevent "wild" fermentation and/or to stop fermentation (wine, beer).

In addition to pasteurization and sterilization, *blanching* may be considered a mild thermal treatment, the main purpose of which is to inactivate enzymes. It is mainly applied as a step in the preparation of vegetables prior to canning, freezing or dehydration. Blanching is carried out by immersing the vegetables in hot water or exposing them to open steam. Although its main objective is to inactivate certain enzymes, blanching has additional desirable effects such as enhancing the color, expelling air from the tissue and cleaning the surface. The blanching process has been reviewed in detail by Selman (1987).

The rational design of a thermal process requires data from two areas: the kinetics of thermal inactivation (thermal destruction, thermal death), and the distribution of the time–temperature function within the mass of the material (heat transfer, heat penetration). The first part of this chapter deals with the kinetics of thermal inactivation. The second part discusses heat penetration. The combination of the two sets of data for the design of thermal processes is the subject of the third part.

17.2 The kinetics of thermal inactivation of microorganisms and enzymes

17.2.1 The concept of decimal reduction time

If a suspension of cells (vegetative cells or spores) of a certain microorganism is heated above a certain temperature, death of the microorganism occurs – i.e., the number of living cells is gradually reduced. Temperatures at which such destruction occurs are called "lethal temperatures".

Experience shows that when a homogeneous suspension of cells is held at a constant lethal temperature, the decrease in the number of living cells with time is nearly logarithmic. This relationship was first demonstrated by Viljoen (1926) in a now classical experiment with bacterial spores. Such observations led to the development of a first-order kinetics model for the thermal destruction of microorganisms. If N is the number of surviving cells at time t, the first-order model is written as follows:

$$-\frac{dN}{dt} = kN \tag{17.1}$$

Integration gives:

$$\log\frac{N}{N_0} = -kt \tag{17.2}$$

The decimal reduction time D is defined as the duration (usually in minutes) of heating time at a constant lethal temperature required for the reduction of the number of living cells by a factor of 10 (i.e., by one log factor). It follows that:

$$\log\frac{N}{N_0} = -\frac{t}{D} \tag{17.3}$$

Figure 17.1 is a graphical representation of the logarithmic reduction model.

The decimal reduction time depends on the microorganism, the temperature and the medium (pH, redox potential, composition).

According to the first-order destruction model, complete sterility ($N = 0$) can never be achieved. A concept of "commercial sterility" is therefore conventionally defined as the objective of practical thermal sterilization (see Section 17.2.3).

The kinetic model of thermal death of microorganisms also applies to the thermal inactivation of enzymes.

It is important to emphasize that the first-order kinetics model of thermal destruction of microorganisms is not based on any known biological mechanism of the thermal death of cells. The model simply provides one of the many possible curve-fitting expressions for representing observed experimental results. For example, Peleg (2006) proposes the following expression as a possible (Weibullean) kinetics model, requiring two parameters, a and b, and eliminating the concept of decimal reduction time):

$$\log\frac{N}{N_0} = -\frac{at}{b-t} \tag{17.4}$$

The log-linear model is remarkably simple and fits experimental results fairly well, particularly as long as the number of survivors is still "countable". Yet deviations from the formal log-linear behavior are apparently observed so frequently that some researchers claim the model is "the exception rather than the rule" (Peleg, 2006). Conceptually, the log-linear model is unable to explain why some of the cells die, and others do not, when exposed to the same temperature for the same duration of time. Clearly, consideration of a *thermal resistance*

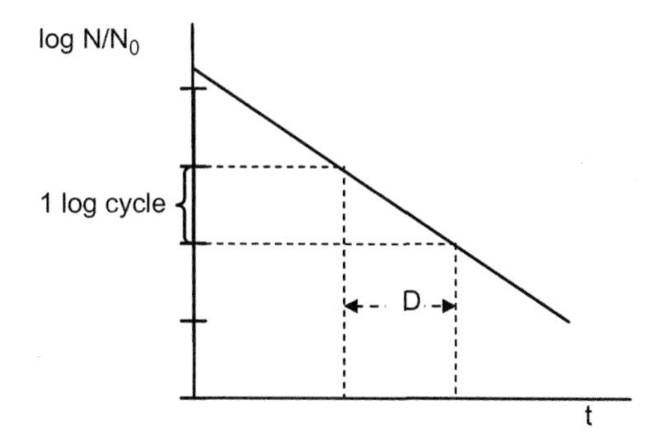

FIGURE 17.1

The log-linear model of thermal reduction of microorganisms.

distribution in the population should be brought into play in the construction of the thermal inactivation kinetics model. Despite the existence of more advanced models (Corradini and Peleg, 2007; Peleg, 2006; Peleg et al., 2008), the first-order approach is, for the time being, the principal route used in practice for the design and evaluation of thermal processes.

EXAMPLE 17.1

A suspension of bacterial spores containing 160 000 spores per ml is heated at 110°C. The number of survivors is determined in samples withdrawn every 10 minutes. The results are:

Heating Time	N (Survivors per ml)
0	160 000
10	25 000
20	8000
30	1600
40	200

Assuming first-order kinetics, calculate the decimal reduction time.

Solution

Log N is plotted against the heating time t. Good agreement with the first-order theory is observed, within the range of the data. The decimal reduction time is determined from the curve (Figure 17.2). The result is: D = 13.4 minutes.

(Note: The data were read from Viljoen's historical report).

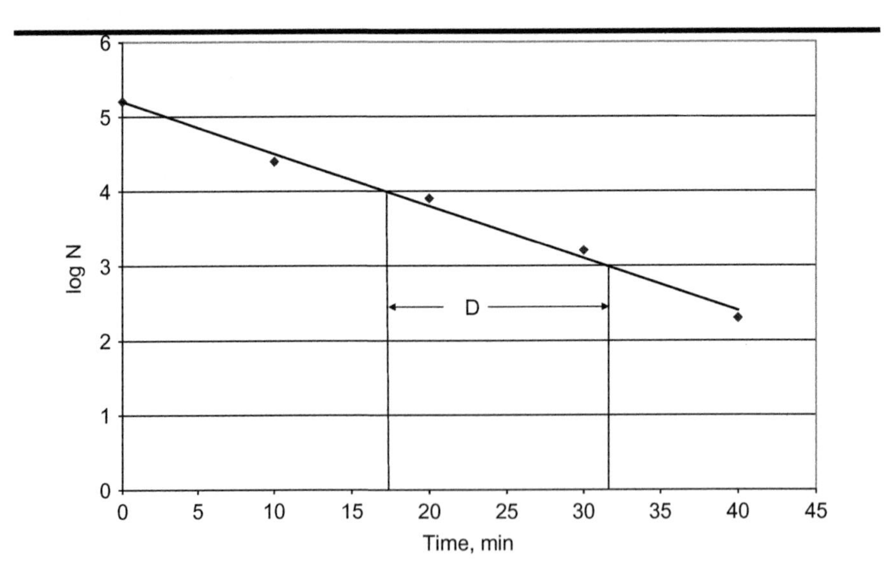

FIGURE 17.2

Conformity with the log-linear reduction model.

17.2.2 Effect of the temperature on the rate of thermal destruction/inactivation

Within the range of lethal temperatures, the rate of destruction is accelerated (D is shortened) by increasing the temperature. Experiments show a nearly linear relationship between the logarithm of D and the temperature (Figure 17.3)

$$\log\frac{D_1}{D_2} = \frac{T_2 - T_1}{z} \tag{17.5}$$

where D_1 and D_2 are the decimal reduction times at temperatures T_1 and T_2, respectively.

The "z value" is defined as the temperature increment needed for a 10-fold acceleration of the rate of thermal destruction (i.e., for shortening D by a factor of 10). For many spore-forming bacteria of interest in food processing, z = 8–12°C.

The logarithmic relationship between the rate of thermal inactivation and the temperature is in agreement with the Arrhenius model for the effect of temperature on the rate of chemical reactions, expressed as the following equation:

$$\log\frac{k_1}{k_2} = \frac{-E}{2.3R}\left(\frac{T_2 - T_1}{T_1 T_2}\right) \tag{17.6}$$

where:

k_1, k_2 = rate constants at absolute temperatures T_1 and T_2 respectively
E = energy of activation
R = universal gas constant = $8.31\ \text{J}\cdot\text{mol}^{-1}\cdot\text{K}^{-1}$.

The activation energy E (see Chapter 4) represents the sensitivity of the reaction rate to temperature, just as the z value indicates the sensitivity of the rate of thermal inactivation to temperature. The quantitative relationship between E and z is given by:

$$E = \frac{2.3RT_1T_2}{z} \tag{17.7}$$

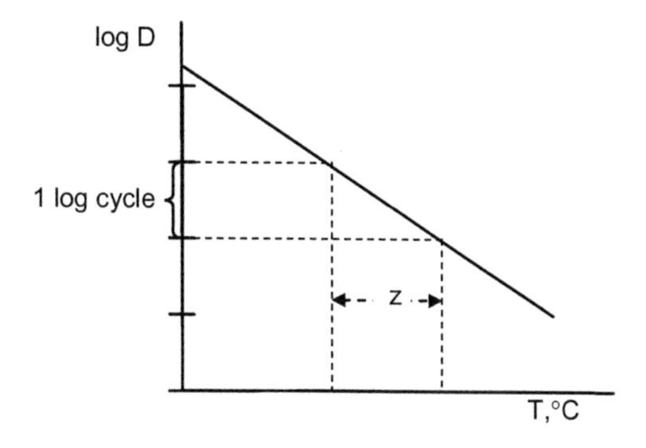

FIGURE 17.3

Effect of the temperature on D.

Table 17.1 Representative Approximate Values of z and E for Different Thermal Effects

Thermal Effect	z (°C)	E (10^3 kJ · $kmol^{-1}$ at T_m = 393 K (120°C))
Cell death (spores)	9–10	300–330
Enzyme inactivation	15–20	150–200
Chemical reactions (not enzyme-catalyzed)	30–40	75–100

If the absolute temperatures are not too far apart, one can write:

$$E \cong \frac{2.3RT_m^2}{z} \tag{17.8}$$

T_m is the average absolute temperature.

It is again important to emphasize that the energy of activation for thermal destruction of cells has no molecular significance. With this warning in mind, it is interesting to note the z value and the corresponding energy of activation of different thermal effects (Table 17.1).

EXAMPLE 17.2

In a laboratory experiment it was found that heating a suspension of spores at 120°C for 100 seconds results in a 9-log killing of the spores. To achieve the same reduction at 110°C, a time of 27.5 seconds is needed. Calculate the decimal reduction time at the two temperatures, the z value, the energy of activation, and the Q_{10} of the thermal inactivation process at these temperatures.

Solution

The decimal reduction times at the two temperatures:

$$D_{120} = \frac{100}{9} = 11.1 \text{ seconds} \quad D_{110} = \frac{27.7 \times 60}{9} = 183.3 \text{ seconds}$$

The z value is calculated from Eq. (17.5):

$$\log\frac{D_1}{D_2} = \frac{T_2 - T_1}{z} \quad \log\frac{183.3}{11.1} = \frac{120 - 110}{z} \Rightarrow z = 8.21°\text{C}$$

The activation energy E is calculated from Eq. (17.7):

$$E = \frac{2.3RT_1T_2}{z} \quad E = \frac{2.3 \times 8.314 \times 393 \times 383}{8.21} = 350.6 \times 10^3 \text{ kJ}\cdot\text{kmol}^{-1}\cdot\text{K}^{-1}$$

The Q_{10} value is found, applying Eq. (4.13) (Chapter 4):

$$Q_{10} = \frac{10E}{RT(T+10)} = \frac{10 \times 350.6 \times 10^3}{8.314 \times 393 \times 383} = 2.8$$

17.2.3 Lethality of thermal processes

From the foregoing discussion it can be concluded that the same lethality (i.e., the same reduction in the number of microorganisms) can be achieved under different time–temperature combinations. In order to compare different processes regarding their lethality, the concept of the "F value" used. The F value is defined as *the duration (in minutes) required to achieve a given reduction ratio in the number of microorganism at a given constant temperature*. It follows that:

$$F = D \log\left(\frac{N_0}{N}\right) \tag{17.9}$$

If, for example, log (N_0/N) = 12 is specified, then F = 12D. For commercial sterility in low-acid foods, 12-log reduction or 12D processes are usually specified (Pflug and Odlaugh, 1978). The practical consequence of a 12-log reduction process is that if the food originally contains 10^3 spores of the target microorganism per gram, it will contain 10^{-9} spores per gram after processing. If the food is packed in units of 500 g each, then one in two million packages will contain one viable spore. This "deterministic" calculation is obviously an over-simplification. The more sophisticated methods for the evaluation of the risk of spoilage in thermally processed foods take into consideration some of the factors neglected by the simplified method. At any rate, it should be remembered that the 12D practice is just a convention, considered to be safe. The validity of extrapolation of the decimal reduction concept over 12 log cycles is doubtful, and a process with the duration of 12D cannot be believed with certainty to cause a 12-log reduction of the living cells.

Consider a thermal process during which the temperature T of the product varies according to a known time–temperature profile, T = f(t). A certain log reduction ratio log(N_0/N) is achieved. We would like to calculate the duration in minutes (the F value) of an equivalent process at a given *constant temperature* (reference temperature R) that would result in the same reduction ratio. The calculation requires specification of the reference temperature R and knowledge of the z value of the microorganism considered. The F value of the equivalent process is:

$$F_R^z = \int_0^t 10^{\frac{T-R}{z}} dt \tag{17.10}$$

For thermal sterilization processes, the standard reference temperature R is 250°F or 121°C and the standard z value is 18°F or 10°C. The R and z indices are omitted for convenience and the standard F is written as F_0. Then:

$$F_{121}^{10} \equiv F_0 = \int_0^t 10^{\frac{T-121}{10}} dt \tag{17.11}$$

The F_0 value of any thermal sterilization process is the number of minutes of heating at 121°C required to achieve the same thermal destruction ratio of a specified target microorganism. With all the reserve expressed with respect to the formal models used as a basis for its definition, the F_0 concept is useful for comparing thermal processes with different time–temperature profiles and for specifying recommended thermal processing conditions for commercial products.

The calculation of F_0 involves calculating the integral in Eq. (17.11), based on the time–temperature profile of the process. This profile is either obtained experimentally or calculated according to heat transfer theory.

EXAMPLE 17.3

For the flash sterilization of milk, a thermal treatment of 2 seconds at 131°C is recommended. Calculate the F_0 value of the process.

Solution

Equation (17.11) is applied:

$$F_{121}^{10} \equiv F_0 = \int_0^t 10^{\frac{T-121}{10}} dt \quad F_0 = t \times 10^{\frac{T-121}{10}} = 2 \times 10^{\frac{131-121}{10}} = 20 \text{ seconds}$$

EXAMPLE 17.4

For the evaluation of pasteurization processes, it is recommended that an F value based on a reference temperature of 70°C and a z value of 7°C be utilized (Bimbenet *et al.*, 2002). For the evaluation of cooking processes and other chemical changes that occur during thermal processing (e.g., destruction of vitamins), the recommended reference temperature is 100°C and z = 30°C. Calculate the *pasteurization value* and the *cooking value* of the following constant temperature processes:

Process	Temp.(°C)	Time (s)
A	74	15
B	92	6
C	65	150
D	105	220

Solution

For processes at constant temperature, Eq. (17.10) reduces to:

$$F_R^z = t \times 10^{\frac{T-R}{z}}$$

Substitution of the data gives:

A.

$$F_{70}^7 = 15 \times 10^{\frac{74-70}{7}} = 55.9 \text{ s} \quad F_{100}^{30} = 15 \times 10^{\frac{74-100}{30}} = 2.04 \text{ s}$$

B.

$$F_{70}^7 = 6 \times 10^{\frac{92-70}{7}} = 8337 \text{ s} \quad F_{100}^{30} = 6 \times 10^{\frac{92-100}{30}} = 3.25 \text{ s}$$

C. $F_{70}^{7} = 150 \times 10^{\frac{65-70}{7}} = 29.0\ \text{s} \quad F_{100}^{30} = 150 \times 10^{\frac{65-100}{30}} = 10.2\ \text{s}$

D. $F_{70}^{7} = 220 \times 10^{\frac{105-70}{7}} = 22 \times 10^{6}\ \text{s} \quad F_{100}^{30} = 220 \times 10^{\frac{105-100}{30}} = 322.9\ \text{s}$

17.3 Optimization of thermal processes with respect to quality

The optimization of thermal processes with respect to the quality of food means the search for processing conditions that will provide the required preservation, with the least amount of damage to the organoleptic and nutritional qualities of the product (Awuah et al., 2007a; Van Loey et al., 1994). Destruction of microorganisms is not the only consequence of thermal processing. Other thermal effects include:

- Inactivation of enzymes: this is desirable and essential for long-term stability.
- *Cooking*: a large number of different chemical reactions that affect the quality of the product – changes in texture, flavor, color, appearance. Such changes are usually desirable up to a certain extent but objectionable beyond.
- Destruction of nutritionally significant components, such as heat-sensitive vitamins.

The kinetic parameters of these effects are different from those of the thermal destruction of microorganisms. As we have seen, the z value of ordinary chemical reactions is larger than that of thermal death of microorganisms. It follows that, *for an equal F_0, processing at higher temperature for a shorter time results in less thermal damage to quality*. This is the theoretical background of the *high temperature–short time (HTST)* concept (Jacobs et al., 1973).

The HTST approach is widely applied (e.g., in the production of ultra-high temperature UHT milk), but it has certain practical limitations:

- The z value for enzyme inactivation is also higher than that of sterilization (Awuah et al., 2007b). Therefore, for an equal F_0 value, enzyme inactivation will be less extensive if the process is carried out at higher temperature. The highest permissible process temperature of a HTST process is, therefore, that at which the residual enzyme activity of the product will not endanger long-range stability.

- If *cooking* is one of the objectives of thermal processing (e.g., in the production of canned white beans), then a HTST process may result in less than desirable cooking (undercooking).
- For safety reasons, thermal processes are designed to achieve the desired F_0 *at the coldest point of the product* (e.g., at the geometric center of a can heated by conduction, the central axis of a tube in continuous processing in a tubular heat exchanger). Consequently, a considerable portion of the food is over-processed. At higher processing temperatures, over-processing of the food outside the coldest region and therefore thermal damage to the average quality is more extensive, particularly if resistance to heat transfer is high (solid foods).

For all these reasons, the design of a thermal process for maximum security with minimum damage becomes a complex exercise in multi-variable optimization (Abakarov et al., 2009).

EXAMPLE 17.5

A liquid food is processed at a constant temperature of 110°C for 30 seconds. The process results in a 25% loss of a vitamin present in the food. The requirement is to change the constant temperature so that the same destruction of microorganisms is achieved with only 10% loss of the vitamin. Calculate the new constant temperature and the new processing time.

The z value of the thermal destruction of microorganisms is 10°C.

The Q_{10} value of the thermal destruction of the vitamin is 2.

Solution

For the inactivation of microorganisms:

$$\frac{t_1}{t_2} = 10^{\frac{T_2-T_1}{z}} \quad \frac{30}{t_2} = 10^{\frac{T_2-110}{10}} \qquad \text{(Equation A)}$$

For the thermal destruction of the vitamin (with C = the vitamin content):

$$\log\frac{C}{C_0} = -kt \quad \log(0.75) = -k_1 \times 30 \quad \log(0.90) = -k_2 t_2$$

$$\frac{k_2}{k} = \frac{\log(0.90)}{\frac{\log(0.75)}{30} \times t_2} = 2^{\frac{T_2-110}{10}} \qquad \text{(Equation B)}$$

Solving Equation A and Equation B for t_2 and T_2, we find:

$T_2 = 117°C$
$t_2 = 5.9$ s.

The result confirms the validity of the HTST approach.

17.4 Heat transfer considerations in thermal processing

Typical processing time for the sterilization of solid food in a 0.5-kg can in a retort at 120°C may be in excess of 1 hour, while the net heating time necessary to achieve the desired thermal death of the target microorganism at that temperature may be just a few minutes. The long duration of the process is due to the slowness of heat transfer (heat penetration) to the coldest point of the can. It can be stated that in thermal process design, the rate of heat transfer rather than the thermal resistance of the microorganism often determines the duration of the process.

17.4.1 In-package thermal processing

Heat transfer into the food in hermetically closed packages occurs in three consecutive steps:

1. Heat transfer from the heating medium to the package surface
2. Heat transfer through the packaging material
3. Internal heat transfer from the inner surface of the package to the coldest point of the product.

17.4.1.1 The heating medium

Saturated steam

In practice, the most efficient heating medium is saturated steam, for the following reasons:

- The heat transfer coefficient of a film of condensing vapor is very high.
- The temperature of saturated steam is easily controlled through its pressure.
- The "heat content" carried by unit mass of steam (heat of evaporation) is very high.
- The water content of most food products is very high. The pressure of the steam outside the package counteracts the pressure developed inside as the contents are heated. Deformation and rupture of the package as a result of excessive pressure differences are thus avoided.

In some cases the over-efficiency of steam as a heating medium is a disadvantage. In the case of packaging in jars and bottles, excessive temperature gradients are created in the glass as a result of too rapid heating of the surface and relatively low thermal conductivity of the glass. Such gradients cause large internal stress, resulting in shattering. Excessive temperature gradients are also detrimental to the quality of heat-sensitive solid and semi-solid products, particularly in large packages.

Hot water

Heat transfer from hot water to the package is less efficient. Hot water (heated by direct contact with steam, or indirectly in heat exchangers combined with recirculation) is the preferred medium for thermal processing of food in glass, trays or flexible packages, and heat-sensitive products. Good circulation of the heating water is essential for preventing "cold pockets". Air is often introduced in order to provide the needed over-pressure and some agitation. In some types of retorts, the product is not completely immersed in the heating medium. In the so-called *cascade retorts* (Van Loey et al., 1994), hot water (process water) is pumped out from the bottom of the retort, recirculated through a heat exchanger and reintroduced at the top of the retort, where a perforated plate maintains even distribution along the vessel. In this type of retort, the hot water cascades over containers. Cooling water is distributed in the same way. In *spray retorts*, the heating or cooling medium is introduced through multiple spray nozzles installed close to the retort wall. Spray retorts are especially suitable for processing food in flexible packages.

Steam–air mixture

A steam–air mixture is a fairly common heating medium. Heat transfer efficiency is comparable to that of water.

Hot gas (combustion gases)

In the so-called "flame sterilization" process, packages are heated directly by combustion gases and radiation. At present, flame sterilization of canned foods is not often practiced.

Microwave heating

In view of the fact that the long duration of processing cycles is mainly due to the slowness of heat penetration, the use of microwave energy for in-package thermal processing appears to be an attractive option. However, the industrial application of this option faces a number of problems. Capital and operating costs of microwave systems are much higher than for conventional thermal processing systems. Metal containers cannot be used. Sterilization at temperatures above 100°C requires external pressurization (Tang et al., 2008). On the other hand, microwave pasteurization at temperatures below 100°C has been applied successfully to acidified vegetables (Koskiniemi et al., 2011; Lau and Tang, 2002) and to ready-to-eat items for short storage (Burfoot et al., 1996, 1988). The quality of microwave treated products has been found to be high, but the higher cost is a problem.

17.4.1.2 The packaging material

The thermal conductivity of aluminum and tinplate is high, that of glass is relatively low (see Chapter 3, Table 3.1). In all cases, the resistance of the package to heat transfer is relatively insignificant because of its small thickness.

17.4.1.3 Internal heat transfer

Heat transfer through the product may be by convection or conduction, or by both. In solid foods (meat loaf, solid-pack tuna, etc.), conduction is the principal mode. In liquid foods, convection heat transfer predominates. In products consisting of solid particles in a liquid medium (e.g., fruit in syrup, vegetables in brine, meat in sauce, tomatoes in tomato juice), heat travels from the package wall to the liquid and from the liquid to the solid chunks. The location of the coldest point depends on the mode of heat transfer. In purely conductive transfer, the coldest point is the geometric center of the package. In convective transfer without agitation in a vertical can, the coldest point is at one-third the height from the bottom. In the case of large solid particles in a low-viscosity liquid medium, the coldest points may be distributed at the center of the solid particles.

Internal resistance to heat transfer is, obviously, the predominant factor. In products containing a liquid or semi-liquid medium, this resistance can be reduced considerably by agitation. Agitation can be achieved by several modes.

- In the end-over-end method, the container is tumbled at a given speed (Figure 17.4). Each time the container is inverted, the head-space tends to rise and agitates the contents. The speed of movement of the head-space from one end to the other depends on the viscosity of the product. If the speed of rotation is too fast, the head-space is immobilized and no agitation occurs. The optimal speed depends on the size of the container and the viscosity of its contents.
- In the spin method (Figure 17.5), the container is rotated about its axis. Agitation occurs as the result of the velocity gradient from the container wall to its center.
- In the reciprocating mode (see Chapter 18, Figure 18.8), the crates containing the product are gently rocked to and fro.

Mathematical models for the prediction of the heat penetration rate in containers are based on the theory of transient heat transfer, discussed in Chapter 3.

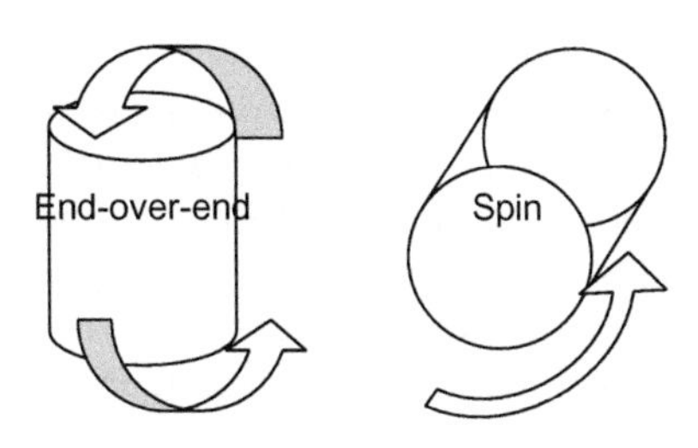

FIGURE 17.4

End-over-end and spin agitation of a can.

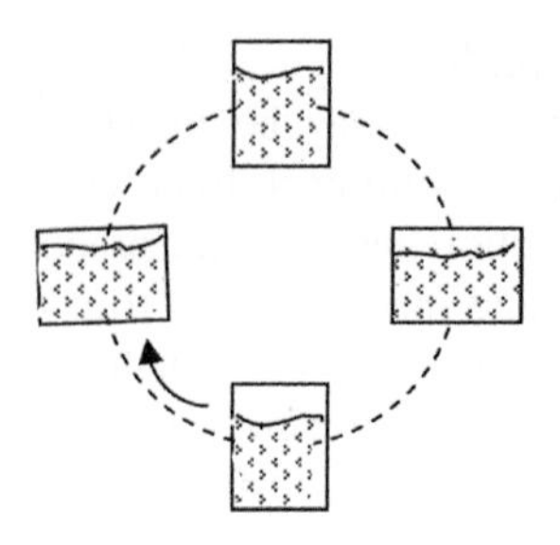

FIGURE 17.5

Position of the headspace in end-over-end rotation.

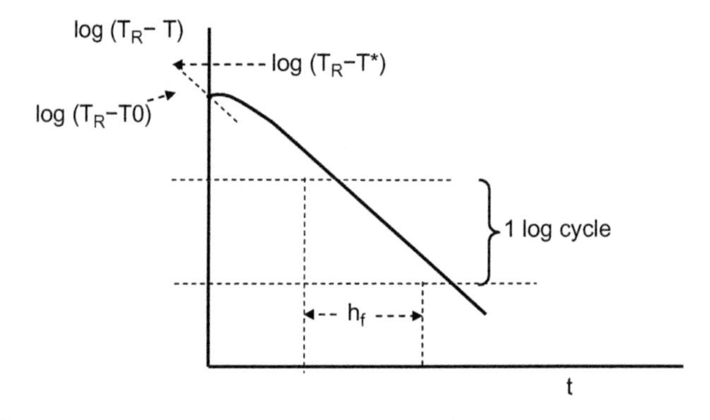

FIGURE 17.6

Theoretical heat penetration curve.

When a package of food is heated, the temperature T of every particle of the food tends towards the temperature of the heating medium T_R. The theory of unsteady-state heat transfer predicts that, after an initial period of adjustment, log (T_R−T) decreases linearly with time, as shown in Figure 17.6.

In the case of containers processed in a retort, T_R is the retort temperature. T^* is the imaginary initial temperature, obtained by extrapolation of the straight segment. The straight segment of the curve can be defined by its slope and intercept. The slope is conventionally expressed by the factor f_h (heating time required to reduce $T_R - T$ to one-tenth its value). The intercept T^* is usually transformed into the dimensionless expression j, known as the *heating lag factor*, as follows:

$$j = \frac{T_R - T^*}{T_R - T_0} \tag{7.12}$$

Table 17.2 Corn Thermal Process Data

t (min.)	0	2	4	8	11	14	20	40	45	47	49	51
T (°C)	27.8	102.8	110	111.7	108.9	111.1	115.6	120	120.5	106	84	68

Real heat penetration curves usually differ from the theoretical log-linear model, mainly due to changes in the properties (viscosity, thermal conductivity) of the material or even transition from one mechanism of heat transfer to another (e.g., from mainly conductive to mainly convective) during thermal processing. In such cases, the time–temperature relationship is sometimes approximated by a broken line, i.e., a sudden change of the slope f_h (Berry and Bush, 1987).

EXAMPLE 17.6

The temperature at the center of cans of corn in water was measured in the course of still retorting. The readings are shown in Table 17.2. The retort temperature was constant at 121°C. The first measurement was taken when the temperature of the retort stabilized. The steam was cut off at 45 minutes and the cooling water valve was opened at 47 minutes.

a. Calculate the F_0 of the process.
b. At what moment should the heating be stopped in order to have $F_0 = 8$ minutes?
c. From the data, plot the heat penetration curve (log $(T_R - T)$ versus t). Assuming that you believe the data, how would you explain the deviation from the theoretical straight line?

Solution

a. The F_0 of the process is given by Eq. (17.11):

$$F_0 = \int_0^t 10^{\frac{T-121}{10}} dt$$

Since the time–temperature history of the food is not given in analytical form, the integral will be calculated empirically. Equation (17.10) is written as follows:

$$F_0 = \sum_0^t 10^{\frac{T-121}{10}} \Delta t$$

Table 17.2 is expanded (Table 17.3) as follows:

- The time is divided into intervals. For each interval, the average (mid-point) temperature T is calculated.
- For each time interval Δt, the expression $10^{\frac{T-121}{10}} \Delta t$ is calculated. This is the contribution of the given interval to the F_0 of the process.
- The F_0 of the entire process is the sum of the contributions.

Figure 17.7 is a plot of the lethality F_0 versus t.

The result is: $F_0 = 17.3$ minutes.

b. The required process time for $F_0 = 8$ minutes.

The contribution of the cooling period to the total F_0 is 0.43 minutes. Therefore, the steam should be cut off when $F_0 = 8 - 0.43 = 7.57$ min. Going back to Table 17.2, we find that this occurs at t = 32 min. The steam should be cut off at t = 32 min.

Table 17.3 Corn Process Calculation

t (min.)	T (°C)	Average T (°C)	Interval F_0	Cumul. F_0
0	27.8		0	0
2	102.8	65.3	0.000	0.000
4	110	106.4	0.069	0.069
8	111.7	110.85	0.386	0.456
11	108.9	110.3	0.255	0.711
14	111.1	110	0.238	0.949
20	115.6	113.35	1.031	1.980
40	120	117.8	9.573	11.553
45	120.5	120.25	4.207	15.760
47	106	113.25	0.336	16.095
49	84	95	0.005	16.101
51	68	76	0.000	16.101

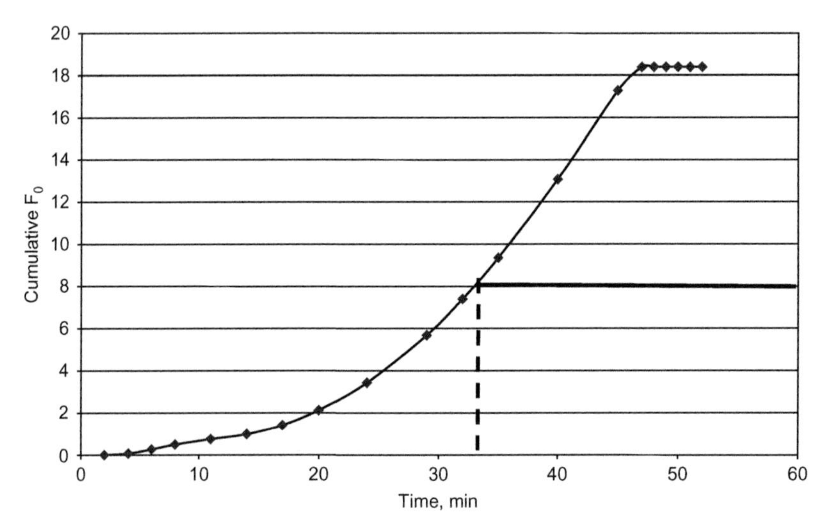

FIGURE 17.7

Thermal processing of corn: F_0 versus t plot.

c. Figure 17.8 is a plot of log (T_R−T) versus t. The plot is far from linear and can be represented, at best, as two straight lines with widely different slopes. This behavior may be due to the existence of two mechanisms of heat transfer (convection + conduction) and to internal changes in the properties of the contents (e.g., swelling and gelatinization of the free starch).

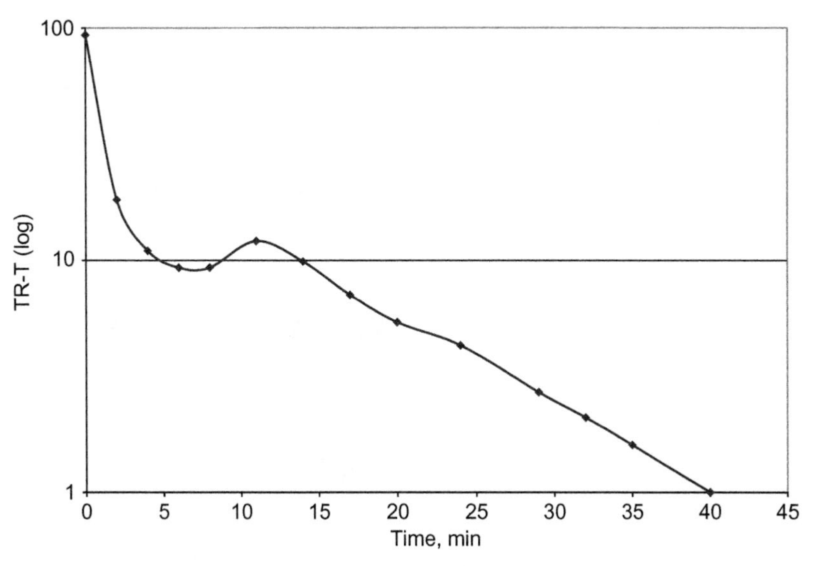

FIGURE 17.8

Heat penetration curve from data.

Example 7.6 illustrates a method for the calculation of thermal processes for food in hermetic containers, based on time–temperature data obtained experimentally. This method, known as *Bigelow's General Method* (Simpson et al., 2003), has been extensively used in practice since 1920. Computer programs for real-time monitoring of thermal processes in the light of measured time–temperature data are available (Tucker and Featherstone, 2011; Tung and Garland, 1978). An alternative method, known as *Ball's Formula Method*, makes use of theoretical thermal survival and heat penetration equations. The accurancy of several formula methods was examined by Smith and Tung (1982). Calculation methods based on finite elements mathematics have been proposed (Naveh et al., 1983).

17.4.2 In-flow thermal processing

Heat exchangers are used extensively for the pasteurization or sterilization of pumpable products. Let us calculate the time–temperature relationship in a pumpable food product passing through a tubular heat exchanger at a mass flow rate G (Figure 17.9).

The heat balance as the food passes over an element dA of heat transfer area gives:

$$U\frac{dA}{dt}(T_R - T) = GC_p\frac{dT}{dt} \tag{7.13}$$

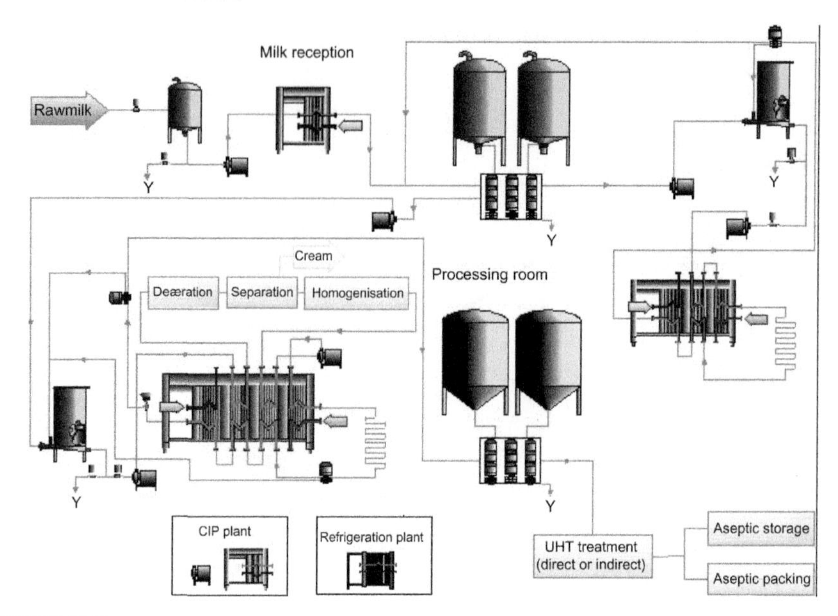

FIGURE 17.9

Flow diagram of in-flow thermal processing of milk (UHT) prior to aseptic packaging.

Figure courtesy of Alfa-Laval.

where U = overall coefficient of heat transfer and T_R = temperature of the heating medium.

dA/dt is related to the mass flow-rate as follows:

$$\frac{dA}{dt} = \frac{4G}{D\rho} \tag{7.14}$$

Substitution gives:

$$\left(\frac{4U}{C_p\, D\rho}\right)(T_R - T) = \frac{dT}{dt} \tag{7.15}$$

Integration gives:

$$\ln\frac{T_R - T}{T_R - T_0} = -\left(\frac{4U}{C_p\, D\rho}\right)t \tag{7.16}$$

We see that the time–temperature relationship is log-linear, just as in the case of in-package heat penetration. It should be noted that T is the average temperature of the product. The actual temperature distribution depends on the degree of turbulence and on the physical structure of the product, as in the case of products consisting of solid particles in a liquid medium.

In continuous in-flow heating, the temperature of the product usually rises very rapidly and residence time at the lethal temperature range is too short for complete sterilization or pasteurization (Trägårdh and Paulsson, 1985). It is therefore necessary to hold the heated product at high temperature for the required length of time, with no further heat transfer. This is usually done by installing a *holding tube* or *holding vessel* of appropriate dimensions after the heating section of the heat exchanger.

Another option for in-flow thermal processing is ohmic heating (see Chapter 3, Section 3.9). Ohmic heating is rapid – practically instantaneous – and does not require heat transfer surfaces. There are no temperature gradients and there is no fouling. Theoretically, these features would make ohmic heating the preferred method of in-flow thermal processing, particularly for fluids containing solid particles (Shim et al., 2010). However, ohmic heating, at its present state, does not provide the uniformity and reliability required for thermal processing of low-acid foods. In addition, even though heating is rapid, cooling still depends on heat transfer through surfaces. For all these reasons, the application of ohmic heating is limited at present to the pasteurization of fruit juices and, possibly, liquid egg.

EXAMPLE 17.7

A food liquid is given a thermal treatment consisting of three consecutive stages:

a. Heating in a heat exchanger. The temperature increases linearly from 30°C to 120°C in 90 seconds.
b. Holding at 120°C for 70 seconds.
c. Cooling in a heat exchanger. The temperature drops linearly from 120 to 10°C in 90 seconds.

Calculate the F_0 of each stage and of the entire process.

Solution

a. For the heating stage:

$$F_0 = \int_0^t 10^{\frac{T-121}{10}} dt \text{ with } T = 30 + \frac{120-30}{90}t = 30 + t$$

Hence:

$$F_0 = \int_0^{90} 10^{0.1t-9.1} dt = \left[\frac{10^{0.1t-9.1}}{0.1\ln 10}\right]_0^{90} = \frac{10^{-0.1}}{0.23} - \frac{10^{-9.1}}{0.23} \approx 5.47 \text{ s}$$

b. For the holding (constant temperature) stage:

$$F_0 = t \times 10^{\frac{T-121}{10}} = 70 \times 10^{\frac{120-121}{10}} = 55.6 \text{ s}$$

c. For the cooling stage:

$$F_0 = \int_0^t 10^{\frac{T-121}{10}} dt \text{ with } T = 120 - \frac{120-10}{90}t = 120 - 1.22t$$

Hence:

$$F_0 = \int_0^{90} 10^{\frac{120-1.22t-121}{10}} dt = \int_0^{90} 10^{-0.122t-0.1} = \left[-\frac{10^{-0.122t-0.1}}{-0.122\ln 10}\right]_0^{90} \approx 2.83 \text{ s}$$

The total lethality of the process is $5.47 + 55.6 + 2.83 = 63.9$ s. About 87% of the lethality is achieved during holding.

EXAMPLE 17.8

A liquid food is continuously heated in a heat exchanger, from 70°C to 130°C in 60 seconds. It is assumed that the temperature increase is linear with time. The purpose of the process is to inactivate a certain target microorganism. If the food originally contained 10^5 living cells of the target microorganism per gram, what will be the number of surviving cells per gram at the end of the process?

Data: the heating time at a constant temperature of 110°C for a 12-log reduction of the target microorganism is 21 minutes. The z value is 9°C.

Solution

The time–temperature profile of the process is: $T = 70 + t$ (t in seconds).

We shall calculate the lethality of the process in relation to a constant temperature process at 110°C with z = 9°C.

$$F_R^z = \int_0^t 10^{\frac{T-R}{z}} dt = \int_0^{60} 10^{\frac{70+t-110}{9}} = \int_0^{60} 10^{0.11t-4.44} = \left[\frac{10^{0.11t-4.44}}{0.11\ln 10}\right]_0^{60} = 571 \text{ s} = 9.52 \text{ min}$$

Let S be the survival ratio of our process. Comparing it with the reference process we find:

$$\frac{\log S}{-12} = \frac{9.52}{21} \Rightarrow \log S = -5.44$$

Hence:

$$\log\frac{N}{N_0} = \log\frac{N}{10^5} = -5.44 \quad \frac{N}{10^5} = 0.36 \times 10^{-5} \Rightarrow N = 0.36 \text{ cells per gram.}$$

EXAMPLE 17.9

A dairy desert is rapidly heated to 121°C in a heat exchanger and then pumped into an insulated holding tube, 0.1 m in diameter and 10 m long. What is the maximum volumetric flow-rate if the F_0 value of the holding must be at least 3 minutes? Assume perfect insulation.

Note: Considering the high viscosity of the product, it will be assumed that the flow regime will be laminar.

Solution

The *minimum* residence time in the holding tube should be 3 minutes. This means that the fastest moving particle of the fluid should reside in the tube at least 3 minutes.

If Q is the volumetric flow-rate, the average velocity of the fluid is:

$$v_{ave} = \frac{4Q}{\pi D^2} = 127.4Q \text{ m·s}^{-1}$$

In laminar flow in tubes, the maximum velocity is related to the average velocity as per Eq. (2.14):

$$v_{average} = \frac{v_{max}}{2}$$

Hence:

$$v_{max} = 2v_{ave} = 2 \times 127.4Q$$

On the other hand, the condition for minimum residence time implies:

$$t = \frac{L}{v_{max}} \geq 3\text{min} \Rightarrow v_{max} \leq \frac{10}{180}\text{m}\cdot\text{s}^{-1}$$

Hence:

$$Q \leq \frac{10}{180 \times 2 \times 127.4} \Rightarrow Q \leq 0.00022\ \text{m}^3\cdot\text{s}^{-1}$$

The maximum permissible flow-rate is 0.00022 $m^3 \cdot s^{-1}$, or 0.785 $m^3 \cdot h^{-}1$.

References

Abakarov, A., Sushkov, Y., Almonacid, S., Simpson, R., 2009. Multiobjective optimization approach: thermal food processing. J. Food Sci. 74 (9), E471–487.

Awuah, G.B., Ramaswamy, H.S., Economides, A., 2007a. Thermal processing and quality: principles and overview. Chem. Eng. Process 46, 584–602.

Awuah, G.B., Ramaswamy, H.S., Simpson, B.K., Smith, J.P., 2007b. Thermal inactivation kinetics of trypsin at aseptic processing temperatures. J. Food Proc. Eng. 16 (4), 315–328.

Ball, O.C., Olson, F.C.W., 1957. Sterilization in Food Technology. McGraw-Hill, New York, NY.

Berry Jr, M.R., Bush, R.C., 1987. Establishing thermal processes for products with broken heating curves from data taken at other retort and initial temperatures. J. Food Sci. 52 (4), 958–961.

Bimbenet, J.J., Duquenoy, A., Trystram, G., 2002. Génie des Procédés Alimentaires. Dunod, Paris, France.

Burfoot, D., Griffin, W.J., James, S.J., 1988. Microwave pasteurisation of prepared meals. J. Food Eng. 8 (3), 145–156.

Burfoot, D., Railton, C.J., Foster, A.M., Reavell, S.R., 1996. Modeling the pasteurization of prepared meals with microwaves at 896 MHz. J. Food Eng. 30 (1–2), 117–133.

Corradini, M.G., Peleg, M., 2007. In: Brul, S., Van Gerwen, S., Zwietering, M. (Eds.), Modelling Microorganisms in Food. Woodhead Publishing, Cambridge, UK.

Holdsworth, S.D., 1997. Thermal Processing of Packaged Foods. Chapman & Hall, London, UK.

Jacobs, R.A., Kempe, L.L., Milone, N.A., 1973. High-temperature–short-time (HTST) processing of suspensions containing bacterial spores. J. Food Sci. 38 (1), 168–172.

Koskiniemi, C.B., Truong, V-D., Simunovics, J, McFeeters, R.F., 2011. Improvement of heating uniformity in packaged acidified vegetables pasteurized with a 915 MHz continuous microwave system. J. Food Eng. 105 (1), 149–160.

Lau, M.H., Tang, J., 2002. Pasteurization of pickled asparagus using 915 MHz microwaves. J. Food Eng. 51 (4), 283–290.

Leland, A.C., Robertson, G.L., 1985. Determination of thermal processes to ensure commercial sterility of food in cans. In: Thorne, S. (Ed.), Developments in Food Preservation, vol. 3. Elsevier Applied Science Publishers, London, UK.

Lewis, M.S., Heppell, N.J., 2000. Continuous Thermal Processing of Foods. Springer, New York, NY.

Naveh, D., Kopelman, I.J., Pflug, I.J., 1983. The finite element method in thermal processing of foods. J. Food Sci. 48 (4), 1086–1093.

Olson, F.C.W., Stevens, H.P., 1939. Thermal processing of canned foods in tin containers. J. Food Sci. 4 (1), 1–20.

Peleg, M., 2006. Advanced Quantitative Microbiology for Foods and Biosystems. CRC, Taylor & Francis Group, London, UK.

Peleg, M., Normand, M.D., Corradini, M.G., van Asselt, A.J., de Jong, P., ter Steeg, P. F., 2008. Estimating the heat resistance parameters of bacterial spores from their survival ratios at the end of UHT and other heat treatments. Crit. Rev. Food Sci. Nutr. 48, 634–648.

Pflug, I.J., Odlaug, T.E., 1978. A review of z and f-values used to ensure the safety of low-acid canned foods. Food Technol. 32 (6), 63–70.

Richardson, P. (Ed.), 2004. Improving the Thermal Processing of Foods. Woodhead Publishing Ltd., Cambridge, UK.

Selman, J.D., 1987. The blanching process. In: Thorne, S. (Ed.), Developments in Food Preservation – Vol. 4. Elsevier Applied Science Publishers Ltd., London, UK.

Shim, J.Y., Lee, S.H., Jun, S., 2010. Modeling of ohmic heating patterns of multiphase food products using computational fluid dynamics codes. J. Food Eng. 99 (2), 136–141.

Simpson, R., Almonacid, S., Texeira, A., 2003. Bigelow's general method revisited: development of a new calculation technique. J. Food Sci. 68 (4), 1324–1333.

Smith, T., Tung, M.A., 1982. Comparison of formula methods for calculating thermal process lethality. J. Food Sci. 47 (2), 624–630.

Tang, Z., Mikhaylenko, G., Liu, F., Mah, J.H., Pandit, R., Younce, F., Tang, J., 2008. Microwave sterilization of sliced beef in gravy in 7-oz trays. J. Food Eng. 89 (4), 375–383.

Trägårdh, C., Paulsson, B-O., 1985. Heat transfer and sterilization in continuous flow heat exchangers. In: Thorne, S. (Ed.), Developments in Food Preservation, vol. 3. Elsevier Applied Science Publishers Ltd., London, UK.

Tucker, G., Featherstone, S., 2011. Thermal Processing. Wiley-Blackwell, Chichester, UK.

Tung, M.A., Garland, T.D., 1978. Computer calculation of thermal processes. J. Food Sci. 43 (2), 365–369.

Van Loey, A., Fransis, A., Hendrickx, M., Maesmans, G., de Noronha, J., Tobback, P., 1994. Optimizing thermal process for canned white beans in water cascading retorts. J. Food Sci. 59 (4), 828–832.

Viljoen, J.A., 1926. Heat resistance studies. 2. The protective effect of sodium chloride on bacterial spores in pea liquor. J. Infectious Diseases 39, 286–290.

CHAPTER

Thermal Processes, Methods and Equipment

18

18.1 Introduction

Thermal processes may be applied either to food in hermetic containers or to food in bulk before packaging. Thermal treatment before packaging is most commonly applied to pumpable products. The main types of thermal processes are summarized in Table 18.1.

18.2 Thermal processing in hermetically closed containers

This category of thermal processes includes the technology commonly known as "canning". In this method, the food is heated and cooled while contained in hermetically closed packages. The hermetic package protects the sterilized or pasteurized food from recontamination. The method is suitable for foods in all physical forms: solids, liquids, or liquids with solid particles. The packages can be cans, jars, bottles, trays, tubes, pouches, etc. The dimensions and shapes of the containers have commercial as well as technological significance. Suitability of the package for microwave heating is sometimes an issue. Recently, a can of highly unusual shape (toroid can) has been proposed and its advantages and shortcomings have been experimentally and computationally analyzed (Karaduman, 2012).

This section will deal mainly with processing in cylindrical metal cans. The specific requirements of processing in other packages will be outlined.

18.2.1 Filling the containers

There is a wide variety of methods and equipment for filling cans (Lopez, 1981). The choice of a method depends on the product, the size of cans, and the production rate.

Products can be filled into containers by volume (volumetric filling) or by weight (gravimetric filling). Volumetric filling is simpler and less expensive. Although package contents are usually specified by weight, volumetric filling is often the method of choice.

Food Process Engineering and Technology. DOI: http://dx.doi.org/10.1016/B978-0-12-415923-5.00018-6
© 2013 Elsevier Inc. All rights reserved.

Table 18.1 Classification of Thermal Processes

In-Package Processing	Bulk Processing	
	Hot Filling	Aseptic Filling
Food	Food	Food
Preheating	Heating (heat exchanger)	Heating (heat exchanger)
Filling	Holding	Holding
Exhausting	Hot filling	Cooling (heat exchanger)
Sealing	Sealing	Aseptic filling
In-package heating	In package cooling	Aseptic cooling
In-package cooling		

Cans are never filled completely. Some free space (*head-space*) is left above the product in order to form some vacuum in the sealed can at the end of the process.

Hand filling is practiced with fragile products such as grapefruit segments, with products such as sardines or stuffed vegetables requiring orderly arrangement, or when the production rate is too low to justify mechanical filling. In *hand-pack filling*, the product is filled manually into cylinders of known volume, then transferred mechanically into the cans (Figure 18.1).

Tumbler fillers (Figure 18.2) consist of a rotating drum fitted with baffles and, in it, a belt conveyor carrying the cans in the direction of the drum axis. The product is fed into the drum. As the drum rotates, portions of the product are lifted by the baffles, then fall into the cans when the baffle reaches a certain angle. The can conveyor is slightly tilted to prevent over-filling and leave a void volume (head-space) in the can.

Piston fillers (Figure 18.3) are suitable for filling pumpable products. The "filling heads" are actually piston pumps, transferring a fixed volume of product from a buffer reservoir to individual cans. Piston fillers are usually equipped with a "no-container–no-fill" control device, to prevent spillage when a container is accidentally not presented under the filling head.

In the case of products consisting of solid particles in a liquid medium (vegetables in brine, meat in sauce, fruit in syrup), the solid and the liquid are usually filled separately. Care must be taken to minimize the risk of leaving air pockets in the container.

18.2.2 Expelling air from the head-space

Some of the air in the head-space must be expelled, for several reasons.

FIGURE 18.1

Hand-pack filler.

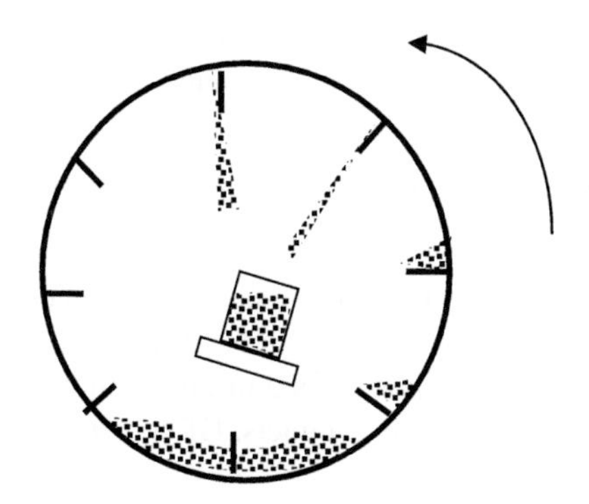

FIGURE 18.2

Operation of a tumbler filler.

- Air in the head-space expands and creates excessive internal pressure when the sealed can is heated. Damage to the seams and deformation of the can may result. This is particularly important in flexible packages (e.g., pouches).

FIGURE 18.3

Piston filler.

- Expulsion of most of the air (reduction of the partial pressure of oxygen) helps reduce the risk of oxidative damage to the product and internal corrosion of the metal can during storage.
- Vacuum in the container is particularly important in the case of products in jars. For most types of jar closures, internal vacuum is the main condition for keeping the lids tightly attached to the jar body.
- In flexible packages, good contact between the packaging material and the food is essential for efficient heat transfer during thermal processing. Vacuum causes the package to shrink as a "skin" around the food, and thus provides the required contact.

A number of methods are available for expelling air from packages before closure.

1. *Hot filling*. It is desirable to fill the product at the highest possible temperature, so as to minimize the quantity of dissolved oxygen and to create a head-space rich in water vapor instead of air. Continuous preheating is possible in the case of pumpable products. Of course, hot filling also shortens the processing time after closure.
2. *Thermal exhausting*. When filling at sufficiently high temperature is not a feasible option, the product is heated in the container, before closure. The filled open containers are conveyed through a bath of hot (near boiling) water, where the water level is kept about 1 cm below the container mouth (Figure 18.4). During this treatment, known as exhausting, air pockets are expelled, the quantity of dissolved oxygen is reduced, and the partial pressure of water vapor in the head-space is increased. A slight bumping movement or vibration imparted to the conveying belt helps expel the air bubbles.

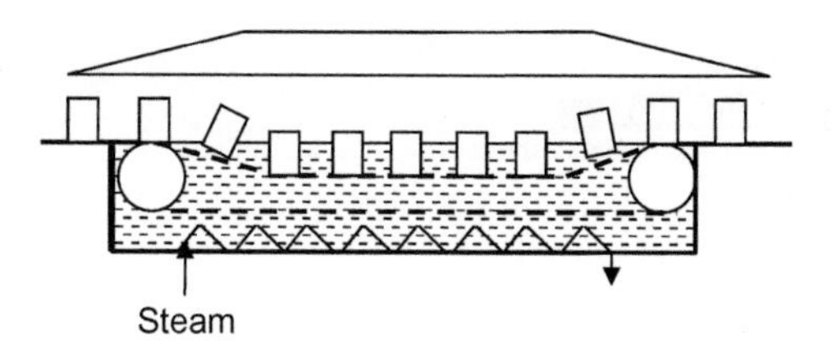

FIGURE 18.4

Thermal exhaust box.

3. *Steam injection*. Some closing machines are equipped with nozzles for injecting food-quality superheated steam into the head-space just before the lid is applied. To be effective, steam injection must be combined with hot filling.
4. *Mechanical vacuum*. For vacuum-closing, the seaming machines are equipped with mechanical vacuum pumps. Mechanical vacuum-closing is applied to products containing little or no liquid, and to flexible packages.

18.2.3 Sealing

When the canning industry began, metal cans were closed by soldering. At that time, one skilled worker would seal about 10 cans per hour (Ball and Olson, 1957). In addition to the extremely low rate of production, the use of lead-containing solder was a matter of serious health concern. The solderless "double seam" method of sealing, which is the only one used in industry today, was invented at the beginning of the 20th century. Modern can-seaming machines (Figure 18.5) can close up to 3000 cans per minute or more.

The double seam is formed by mechanically bending the edge of the can body and the lid and then pressing the two to create an interlocking seam (Figure 18.6). A thin ring of PVC or rubber, applied on the lid, acts as a gasket. The double seam is formed in two steps or "operations": a *first operation* of bending, and a *second operation* of tightening. The form and dimensions of the double seam must be in strict conformity with certain standards. In the food canning industry, routine inspection of the seams and adjustment of the machines are essential quality assurance operations (Lopez, 1981).

18.2.4 Heat processing

The purpose of in-package heat processing may be either sterilization or pasteurization.

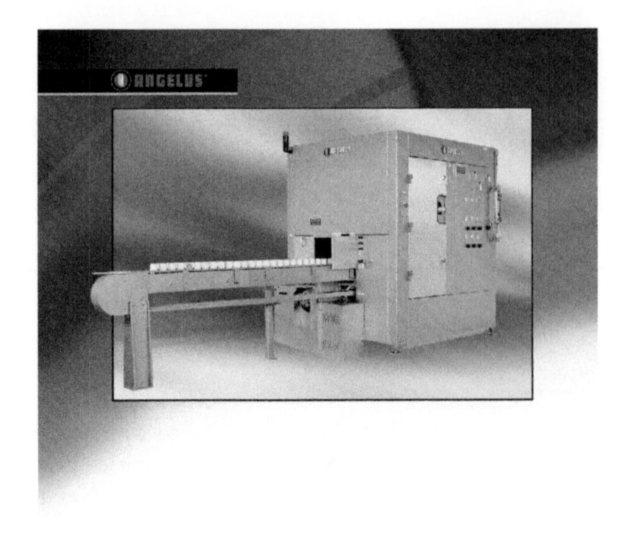

FIGURE 18.5

High speed can seamer.

Photograph courtesy of Pneumatic Scale – Angelus.

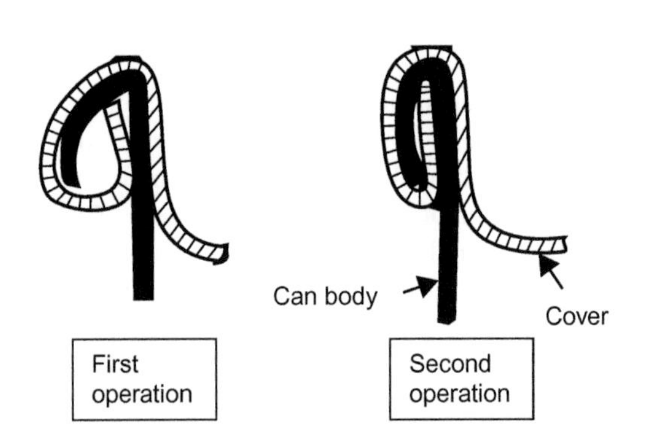

FIGURE 18.6

Formation of the double seam.

18.2.4.1 Sterilization in hermetic containers

Low-acid foods (pH > 4.5) are preserved by sterilization (Ball and Olson, 1957; Bimbenet *et al.*, 2002; Holdsworth, 1997; Lopez, 1981; Tucker and Featherstone, 2011). Practical sterilization requires processing the food at temperatures well above 100°C. Since most products contain water, pressures higher than

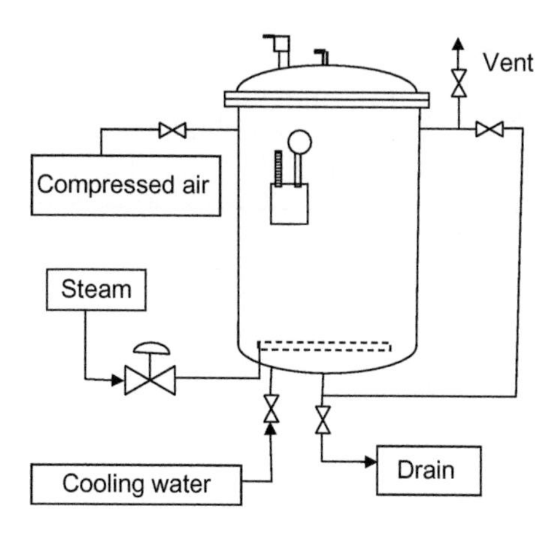

FIGURE 18.7

Simplified hook-up of a vertical retort.

atmospheric are developed inside the package. (The saturation pressure of water vapor at 120°C is approximately 200 kPa – about double the atmospheric pressure.) In order to prevent deformation and damage to the container and the seam, external pressure equal to or slightly higher than the internal pressure must be maintained. For this reason, sterilization is carried out in pressure vessels known as *autoclaves* or *retorts*. In the case of metal cans, the heating medium is usually saturated steam at the appropriate pressure. In the case of glass containers and flexible packages, the heating medium is hot water combined with compressed air to provide the necessary external pressure. Autoclaves are of two general types: batch and continuous.

Batch retorts

Batch retorts are pressure vessels fitted with inlets for steam, hot water, cooling water and pressurized air, and outlets for water recirculation, draining, pressure release and venting. They are equipped with a pressure gauge and thermometer, and usually connected to automatic control systems. They can be vertical (Figure 18.7) or horizontal. Horizontal retorts are easier to load and unload, and can be equipped with rotating baskets for end-over-end agitation or reciprocating movement (Figure 18.8), but they are less economical regarding floor-space. Very large (long) horizontal retorts (Figure 18.9) are usually fitted with internal fans to assure uniform temperature distribution. The cans are loaded into the retort in baskets (crates). Their arrangement in the crates may be either orderly or random.

"Crateless retorts" (Figure 18.10) are large vertical vessels. They are first partially filled with hot water. The cans are fed from the top, and the water in the vessel acts as a cushion for the incoming cans. When the retort is full, the vessel is closed and steam is admitted, pushing out the water through a drain-pipe and

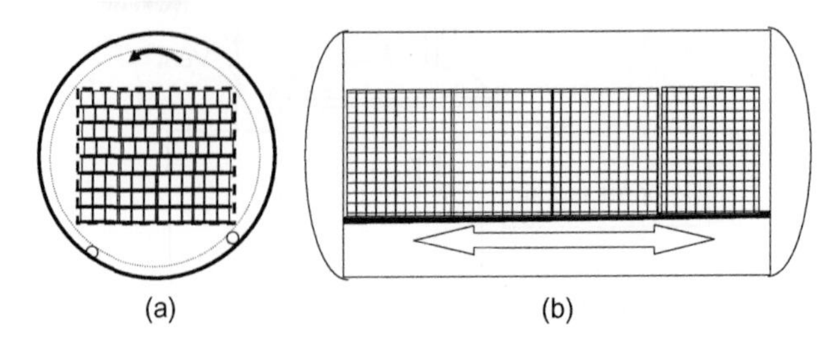

FIGURE 18.8

Horizontal retort with agitation: (a) rotating; (b) reciprocating.

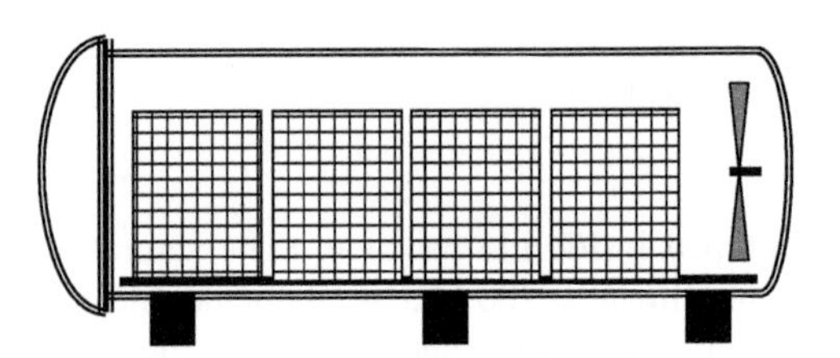

FIGURE 18.9

Arrangement of crates in long horizontal retort.

FIGURE 18.10

Crateless retort.

Photograph courtesy of Mălo.

starting the sterilization cycle. After the heating and cooling cycles are complete, the cans are carried out by water, through an opening at the bottom of the vessel.

The following is a typical sequence of operations during one cycle of batch sterilization in metal cans, using saturated steam for heating:

1. The full crates are loaded into the autoclave.
2. The autoclave is closed.
3. The air is expelled (purging): steam is introduced while a purging valve is kept wide open, and purging is complete when the temperature in the retort is equal to the saturation temperature of steam at the retort pressure. Nevertheless, a small venting cock is left throughout the process. Expelling air from the retort as completely as possible is important, as any remaining air impairs the rate of heat transfer to the can and produces a temperature lower than the saturation temperature at retort pressure.
4. Steam is introduced at a high flow-rate until the specified processing pressure and the corresponding temperature are reached (retort *come-up*). It is customary to start measuring the process time from that moment.
5. The steam flow-rate is controlled so as to maintain the specified retort pressure (temperature) during the specified process time.
6. At the end of the specified process time, the steam is shut off. Compressed air is introduced to replace the steam, while maintaining the pressure. At this stage, the temperature of the food inside the cans is still high, resulting in high internal pressure. External pressure must therefore be maintained to avoid deformation of the can and distortion of the seam.
7. Cooling water is admitted while overriding air pressure is maintained. Cooling can be achieved by spraying the water over the cans or by filling the retort with water. Excess water flows out through the overflow. Only clean water should be used for cooling. Chlorination of the cooling water is standard practice. Ozonation and UV treatment are possible alternatives. The compressed air is shut off when the temperature inside the cans is estimated to have dropped to below 100°C. It is preferable to stop the cooling while the cans are somewhat warm (40–50°C) in order to facilitate drying of the can surface and avoid corrosion. In the case of "white" (unprinted) cans, this also improves the application of labels with adhesives.
8. The retort is drained, opened and unloaded.

In practice, most of the operations of the process cycle described above are automated. A representative sample of each batch is placed in incubators and checked for sterility. Official regulations specify the size of the sample, the incubation time and temperature, and the tests to be performed on the incubated cans.

In the case of processing with hot water, the sequence of operations is slightly different. If vertical retorts are used, the retort is filled with water that is preheated before the crates with the product are introduced. If horizontal autoclaves are used, process water must be first preheated in a separate vessel and then introduced into the retort already loaded with product. At the end of the heating cycle, the hot water is pumped back to the preheating vessel, to be used in the next batch.

Continuous sterilization

Considerable savings in labor cost, energy expenditure and equipment downtime may be achieved by making in-package sterilization a continuous process. The advantages of continuous retorting are particularly significant in high-volume production lines, and where frequent changes in processing conditions or in container size are not required. Batch retorting is more suitable for smaller processing plants that produce a number of different products simultaneously. A technique for the optimal scheduling of small and medium-sized canneries producing different products simultaneously using batch retorts has been proposed (Simpson and Abakarov, 2009).

There are various types of continuous retorts, differing mainly in the method used for introducing and removing the cans continuously, without releasing the retort pressure.

- In continuous retorts with a hydrostatic lock (*hydrostatic sterilizers*), the containers enter the pressure chamber and leave it through two columns of water of sufficient height to counterbalance the chamber pressure (Figure 18.11). Cartridges riding on a chain conveyor (Figure 18.12) carry the cans through the system.
- In continuous retorts with *mechanical locks* (Figure 18.13), the cans enter the pressure chamber and leave it through rotary gates. (Figure 18.14). The pressure chamber consists of a horizontal cylindrical shell where the cans are moved forward along a spiral path while slowly revolving. For cooling under

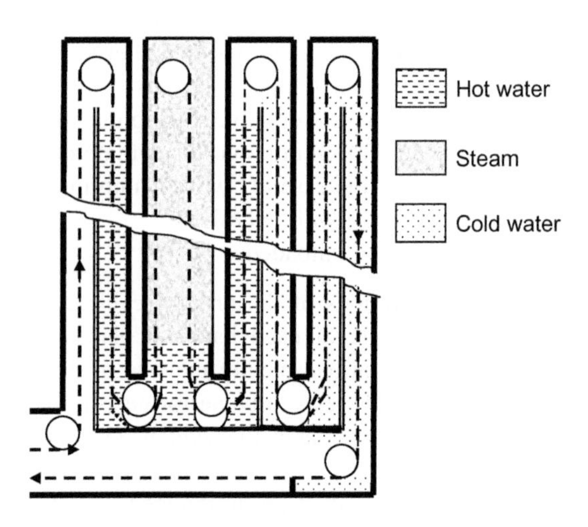

FIGURE 18.11

Hydrostatic sterilizer.

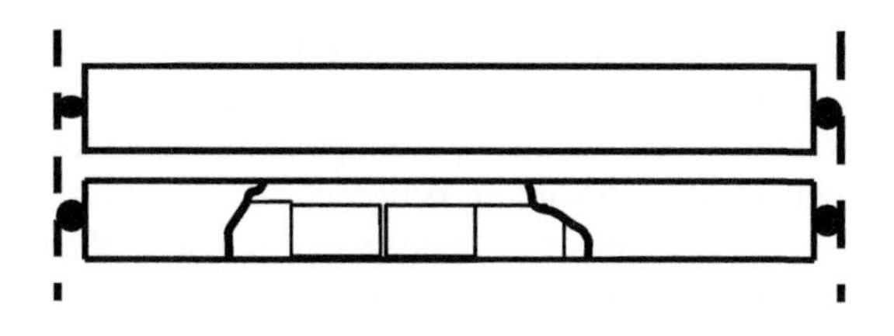

FIGURE 18.12

Arrangement of cans in the canisters of a hydrostatic sterilizer.

FIGURE 18.13

Continuous horizontal cooker–cooler.

Photograph courtesy of FMC FoodTech.

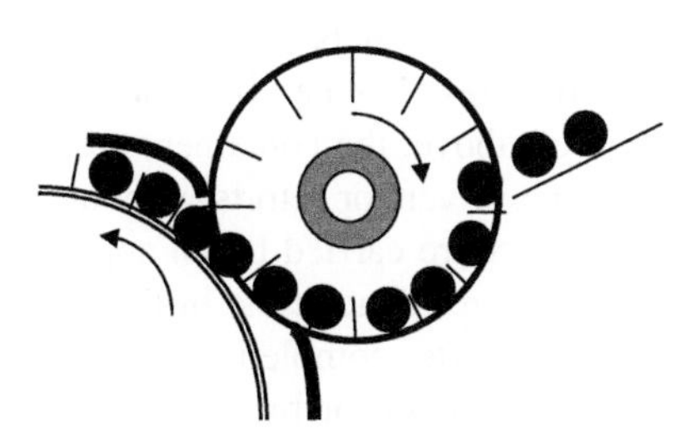

FIGURE 18.14

Schema of the feeding lock of a continuous horizontal sterilizer.

pressure, the cans are transferred to a second similar shell, also through a rotating gate.

Flame sterilization is a method of continuous sterilization without a retort (Heil, 1989). The heating medium is not steam or hot water but hot gas (1000°C or higher) from flames, at atmospheric pressure. After preheating in a tunnel with steam at atmospheric pressure, the rapidly spinning cans are heated using a direct flame, then maintained in a holding space before being cooled. Processing time is shortened considerably because of the high temperature of the heating medium. It is claimed that high quality is maintained because of the short processing time. Since the entire process takes place at atmospheric pressure, there is no need for special devices for feeding or discharging the containers; however, other problems limit the widespread applicability of flame sterilization:

- At sterilization temperatures (e.g., 120°C), the pressure inside the cans is considerably higher than the external atmospheric pressure. Since there is no steam pressure to counteract the internal pressure, only small and particularly rigid containers can withstand such a pressure difference.
- Due to direct contact with flames, the container surface reaches extremely high temperatures. In order to prevent scorching of the product inside, very efficient heat transfer must be provided between the can walls and the can contents. This is achieved by rapid spinning of the cans during their passage in front of the flames. Obviously, the method is not suitable for processing solid foods where heating is solely by conduction.

In addition to these shortcomings, it is difficult to assure complete sterilization of the product at every point, in every container. For all these reasons, the application of flame sterilization for the thermal processing of low-acid foods has been practically discontinued.

18.2.4.2 Pasteurization in hermetically closed containers

In products with a pH of below 4.5 (e.g., some tomato products, pickles, fruit in syrup and artificially acidified products), appropriate preservation may be achieved at temperatures below 100°C – i.e., by pasteurization. After removal of the air and sealing as described above, the containers are heated with near-boiling water. Batch processing is done in vats or retorts without pressurization. For continuous processing, the containers are carried by an appropriate conveyor through a bath of near-boiling water. The residence time in the bath depends on the product and the size of the container. After completion of the specified heat treatment, the containers are cooled by immersion or by spraying. Systems for rotating the cans during the heating and/or the cooling stages are available.

Recently, technologies for the production of pasteurized ready-to-eat meals packaged under vacuum in trays or flexible pouches ("*sous-vide*" products) have been developed, particularly in Europe (Baird, 1990; Church and Parsons, 1993; Creed and Reeve, 1998; Rybka, 2001; Schelleken, 1996; Shakila, 2009). These products have a specified shelf life of 10–30 days under refrigeration, from production to consumption, and pasteurization at temperatures below 100°C is usually sufficient for their preservation. In most cases, these products are batch-pasteurized in vats containing near-boiling water. Usually, the process of pasteurization also serves as in-package cooking. The microbial safety of *sous-vide* products has been studied extensively (Ghazala *et al.*, 1995; Gonzales-Fandos *et al.*, 2005; Rhodehamel, 1992; Shakila *et al.*, 2009; Simpson *et al.*, 1995).

18.3 Thermal processing in bulk, before packaging

In the case of pumpable products (liquids, semi-liquids, purees, suspensions of relatively small solid particles in liquids), it is possible to apply part or all the thermal treatment process continuously, in heat exchangers, before filling the product into containers. Two possibilities exist:

1. Only the heating is done in bulk, usually in a heat exchanger. After hot filling and sealing, the product is cooled in the hermetic container.
2. Both the heating and cooling stages are carried out in bulk, in heat exchangers. A holding tube or holding vessel is provided between the heating and the cooling sections. The treated product is filled and sealed.

18.3.1 Bulk heating–hot filling–sealing–cooling in container

This is the standard method for canning high-acid pumpable products such as fruit juices and purees. The material is heated to pasteurization in continuous heat exchangers and then hot-filled and sealed. The filled containers are cooled by water, usually by spraying. In the case of metal cans, cooling may be accelerated by spinning the container around its axis. The main disadvantage of the method is the relatively long cooling time, resulting in over-cooking of the product and requiring long cooling lines. Furthermore, hot-filling does not allow packaging in heat-sensitive plastics. Aseptic processing has largely replaced this way of processing in the case of delicate fruit juices (citrus, pineapple), but considerable quantities of tomato juice and other tomato products for retail marketing are still being processed by this method.

18.3.2 Bulk heating–holding–bulk cooling–cold filling–sealing

In this mode of operation, the entire process is carried in a system consisting of heat exchangers and piping. The product leaves the system after continuous sterilization or pasteurization (depending on the time–temperature profile), holding, and continuous cooling. If filling and sealing occurs in open space, recontamination of the cold product may occur. This may not be objectionable if the product is to be marketed under refrigeration and its planned marketing time is sufficiently short. This is the case with pasteurized milk and dairy products (Walstra *et al.*, 2005). A typical process for the continuous pasteurization of liquid milk is shown schematically in Figure 18.15.

The pasteurizer, in this case, is a plate heat-exchanger consisting of three sections. Raw milk is admitted to Section 1 (preheating, regeneration), where it is

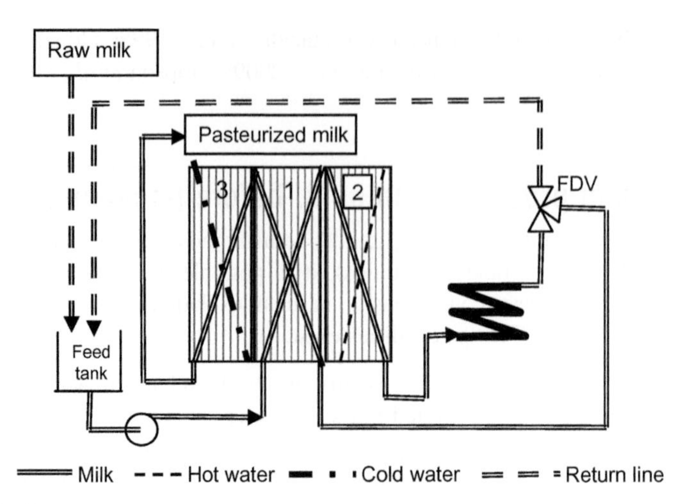

FIGURE 18.15

Schema of the continuous flash pasteurization of milk.

heated to about 55°C by hot pasteurized milk flowing on the other side of the plates. It then passes to Section 2 (heating), where it is heated to the specified process temperature, with hot water on the other side. The hot water for this purpose is heated by steam and circulated in closed circuit by appropriate pumps (not shown in this figure). The milk, thus pasteurized, passes through a holding tube of appropriate dimensions (according to the specified holding time). A three-way control valve (flow diversion valve, FDV), governed by the control system, diverts any under-processed portion back to the feed tank. Milk that has reached the specified process temperature (normally, minimum 72°C) is sent to Section 1, where it is partially cooled by the incoming raw milk, and then to Section 3 (cooling) for final cooling with cooling water and ice-water, and finally to packaging as pasteurized and cooled (normally 4°C) milk.

While considerable energy saving is achieved by including the "regeneration" section in the system, two-stage heating has additional technological benefits. The intermediate temperature of about 55°C is ideal for centrifugal separation of fat and for high-pressure homogenization. These processes are then performed on the preheated milk from the regeneration section.

18.3.3 Aseptic processing

Aseptic processing, also known as aseptic packaging or aseptic filling, is, without doubt, the most significant of recent developments in food technology (David *et al.*, 1996; Reuter, 1993; Sastry and Cornelius, 2002; Sastry, 1989; Willhoft, 1993). The basic principle of aseptic processing is shown in Figure 18.16.

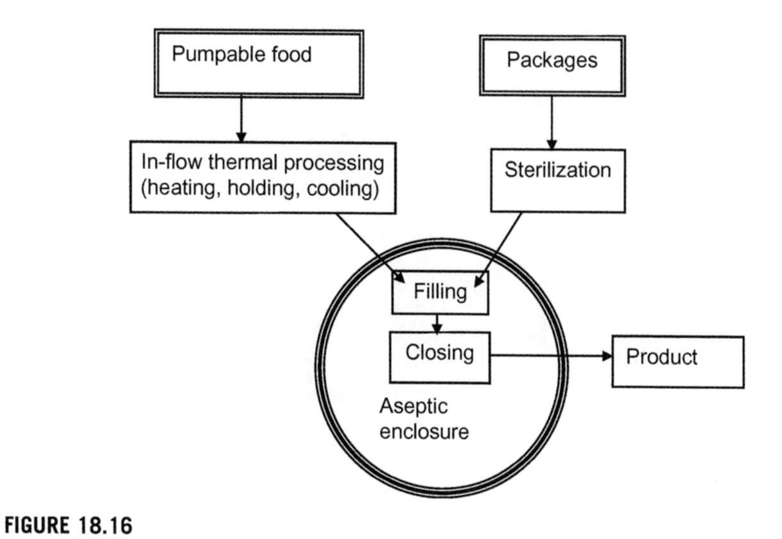

FIGURE 18.16

Schematic flow diagram of aseptic processing.

"Aseptic" processes had already been developed by World War II and were being applied commercially in the 1950s. The early applications were liquid and semi-liquid foods, such as cocoa drinks, custards and banana puree. The packages were, invariably, metal cans, and the process was therefore called "aseptic canning". The pumpable food was continuously heated to sterilizing temperature in heat exchangers, and, after holding, continuously cooled. The cans and lids were sterilized with steam or a mixture of superheated steam and air. The sterilized food and the sterile cans then met in an aseptic enclosure that contained the filling machine and the seamer, closely coupled. Aseptic conditions were maintained by a number of measures, such as disinfectants, a steady stream of superheated steam, UV radiation, etc. The aseptic enclosure was kept at a slight over-pressure to prevent penetration of air from outside. Compared with normally canned foods, the products were of superior quality, but the process was cumbersome, slow and expensive.

New technologies that allowed aseptic filling into flexible packages were developed in the 1960s. The output of the lines increased steadily. At first, the main product of commercial aseptic processing was UHT (ulta-high temperature) milk in cartons. Gradually, applications expanded to almost any pumpable low- or high-acid food, such as soups, gravy, dairy desserts, creams, soy milk, fruit juices, nectars, etc. The innovations included in-place formation of the package from laminated paper, plastic sheet or film, then sterilization using hydrogen peroxide, followed by hot air, filling and sealing in one machine. The packages suitable for aseptic processing now include carton boxes, pouches, trays, cups, large bags in boxes, metal barrels, etc. (Figure 18.17).

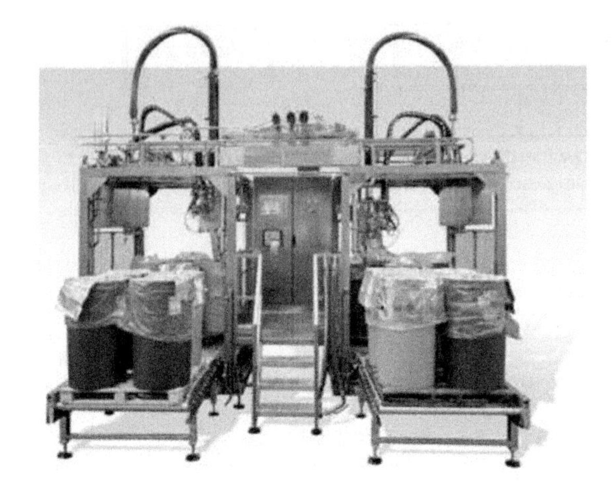

FIGURE 18.17

Aseptic filling of barrels.

Photograph courtesy of Rossi & Catelli.

FIGURE 18.18

Heat exchanger in aseptic processing.

Figure courtesy of Rossi & Catelli.

The adaptation of continuous heat exchangers (Figure 18.18) to the specific requirements of aseptic processing is one of the most significant factors for the success of this technology. These requirements relate to the sanitary design as well as to heat transfer and flow characteristics. Tubular, plate and swept-surface heat exchangers are used. Residence–time distribution in heat exchangers is not uniform, particularly in swept-surface exchangers (Trägårdh and Paulsson, 1985).

Consequently, and considering the very high contribution of the holding period to the lethality of the process (see Chapter 17, Example 17.7), holding vessels of adequate capacity must be provided. Although heating and cooling are done almost exclusively in a conventional heat exchanger, other heating methods have been investigated. These include ohmic heating and continuous microwave heating (Coronel *et al.*, 2005).

The issue of temperature distribution in foods containing solid particles in a liquid medium is particularly relevant in the case of continuous heating associated with aseptic processing, because of the different thermal properties and flow pattern of the particles. This subject has been researched extensively (see, for example, Jasrotia *et al.*, 2008; Palazoğlu and Sandeep, 2002; Yang *et al.*, 1992). Adequate holding capacity provides a practical solution to this problem as well.

References

Baird, B., 1990. *Sous vide*: what's all the excitement about? Food Technol. 44, 92–96.

Ball, O.C., Olson, F.C.W., 1957. Sterilization in Food Technology. McGraw-Hill, New York, NY.

Bimbenet, J.J., Duquenoy, A., Trystram, G., 2002. Génie des Procédés Alimentaires. Dunod, Paris, France.

Church, I.J., Parsons, A.L., 1993. Review: *sous vide* cook-chill technology. Intl. J. Food Sci. Technol. 35, 155–162.

Coronel, P., Truong, V.D., Simunovic, J., Sandeep, K.P., Cartwright, G.D., 2005. Aseptic processing of sweetpotato purees using a continuous flow microwave system. J. Food Sci. 70 (9), 531–536.

Creed, P.G., Reeve, W., 1998. Principles and application of *sous-vide* processed foods. In: Ghazala, S. (Ed.), Sous-Vide and Cook–Chill Processing for the Food Industry. Aspen Publishers Inc., Gaithersburg, MD.

David, J.R.D., Graves, R.H., Carlson, V.R., 1996. Aseptic Processing and Packaging of Foods. CRC Presss, Boca Raton, FL.

Ghazala, S., Ramaswamy, H.S., Smith, J.P., Simpson, M.V., 1995. Thermal processing simulations for *sous vide* processing of fish and meat foods. Food Res. Intl. 28, 117–122.

Gonzalez-Fandos, E., Villarino-Rodriguez, A., Garcia-Linaves, M.C., Garcia-Arias, M.T., Garcia-Fernandez, M.C., 2005. Microbiological safety and sensory characteristics of salmon slices processed by the *sous vide* method. Food Con. 16, 77–85.

Heil, J.R., 1989. Flame sterilization of canned foods. In: Thorne, S. (Ed.), Developments in Food Preservation – 5. Elsevier Science Publishers, London, UK.

Holdsworth, S.D., 1997. Thermal Processing of Packaged Foods. Chapman & Hall, London, UK.

Jasrotia, A.K.S., Simunovic, J., Sandeep, K.P., Palazoglu, T.K, Swartzel, K.R., 2008. Design of conservative simulated particles for validation of a multiphase aseptic process. J. Food Sci. 73 (5), E193–E201.

Karaduman, M., Uyar, R., Erdoğdu, F., 2012. Toroid cans – an experimental and computational study for process innovation. J. Food Eng. 111 (1), 6–13.

Lopez, A., 1981. A Complete Course in Canning, eleventh ed. The Canning Trade, Baltimore, MD, Book 1.

Palazoğlu, T.K., Sandeep, K.P., 2002. Assessment of the effect of fluid-to-particle heat transfer coefficient on microbial and nutrient destruction during aseptic processing of particulate food. J. Food Sci. 67 (9), 3359–3364.

Reuter, H., 1993. Aseptic Processing of Foods. B. Behr's Verlag, Hamburg, Germany.

Rhodehamel, E.J., 1992. FDA's concerns with *sous-vide* processing. Food Technol. 46, 73–76.

Rybka, S., 2001. Improvement of food safety design of cook-chill foods. Food Res. Intl. 34, 449–455.

Sastry, S.K., 1989. Process evaluation in aseptic processing. In: Thorne, S. (Ed.), Developments in Food Preservation – 5. Elsevier Science Publishers, London, UK.

Sastry, S.K., Cornelius, B.D., 2002. Aseptic Processing of Foods Containing Solid Particles. John Wiley & Sons, New York, NY.

Schellekens, M., 1996. New research issues in *sous-vide* cooking. Trends Food Sci. Technol. 7, 256–262.

Shakila, J.R., Jeyasekaran, G., Vijayakumar, A., Sukumar, D., 2009. Microbiological quality of *sous-vide* cook chill fish cakes during chilled storage (3°C). Intl. J. Food Sci. Technol. 44 (11), 2120–2126.

Simpson, M.V., Smith, J.P., Dodds, K.L., Ramaswamy, H., Blanchfield, B., Simpson, B.L., 1995. Challenge studiesClostridium botulinum in a *sous-vide* spaghetti with meat sauce product. J. Food Prot. 58, 229–234.

Simpson, R., Abakarov, A., 2009. Optimal scheduling of canned food plants including simultaneous sterilization. J. Food Eng. 90 (1), 53–59.

Trägårdh, C., Paulsson, B.-O., 1985. Heat transfer and sterilization in continuous flow heat exchangers. In: Thorne, S. (Ed.), Developments in Food Preservation – 3. Elsevier Applied Science Publishers Ltd., London, UK.

Tucker, G., Featherstone, S., 2011. Thermal Processing. Wiley-Blackwell, Chichester, UK.

Walstra, P., Wouters, J.T.M., Geurts, T.J., 2005. Dairy Science and Technology. CRC Press, Taylor & Francis Group, Boca Raton, FL.

Willhoft, E.M.A. (Ed.), 1993. Aseptic Processing and Packaging of Particulate Foods. Kluwer Academic Publishers Group, New York, NY.

Yang, B.B, Nuñez, R.V., Swartzel, K.R., 1992. Lethality distribution within particles in the holding section of aseptic processing system. J. Food Sci. 57 (5), 1258–1265.

CHAPTER

19 Refrigeration: Chilling and Freezing

19.1 Introduction

The preservation of food by refrigeration is based on a very general principle in physical chemistry: *molecular mobility is depressed and, consequently, chemical reactions and biological processes are slowed down at low temperature*. In contrast to heat treatment, low temperature does not destroy microorganisms or enzymes but merely depresses their activity. Therefore:

- Refrigeration retards spoilage but it cannot improve the initial quality of the product, hence the importance of assuring particularly high microbial quality in the starting material.
- Unlike thermal sterilization, refrigeration is not a method of "permanent preservation". Refrigerated and even frozen foods have a definite "shelf life", the length of which depends on the storage temperature.
- The preserving action of cold exists only as long as low temperature is maintained, hence the importance of maintaining a reliable "cold chain" all along the commercial life of the product.
- Refrigeration must often be combined with other preservation processes (the "hurdle" principle).

Natural ice, snow, cold nights and cool caves have been used for preserving food since pre-history. However, to become a large-scale industrial process, low-temperature preservation had to await the development of mechanical refrigeration in the late 19th century. Frozen food made its appearance shortly before World War II. The following events were milestones in the history of mechanical refrigeration:

1748: W. Cullen demonstrates refrigeration by vacuum evaporation of ether.
1805: O. Evans. First vapor compression system.
1834: J. Perkins. Improved vapor compression machine.
1842: J. Gorrie uses refrigeration to cool sick room.
1856: A. Twinning. First commercial application of refrigeration.
1859: F. Carré. First ammonia machine.
1868: P. Tellier attempts refrigerated transatlantic maritime transport of meat.
1873: C. von Linde. First industrial refrigeration systems in brewery.
1918: First household refrigerators.

Food Process Engineering and Technology. DOI: http://dx.doi.org/10.1016/B978-0-12-415923-5.00019-8
© 2013 Elsevier Inc. All rights reserved.

1920: W. Carrier. Start of commercial air conditioning.
1938: C. Birdseye. Start of the frozen food industry.
1974: S. Rowland and M. Molino. Refrigerant gases in the atmosphere suspected of destroying the ozone layer.

Food preservation at low temperature comprises two distinct processes: chilling and freezing. Chilling is the application of temperatures in the range of 0°C to 8°C (i.e., above the freezing point of the food), while freezing uses temperatures well below the freezing point, conventionally below −18°C. The difference between the two processes goes beyond the difference in temperature. The stronger preserving action of freezing is due not only to the lower temperature but also (and mainly) to the depression of water activity as a result of conversion of part of the water to ice. The so-called "superchilling process" makes uses of temperatures slightly below the freezing temperature of the food (typically −1°C to −4°) and can be seen as partial freezing (Kaale *et al.*, 2011).

The use of refrigeration in the food industry is not limited to preservation. Refrigeration is applied for a number of other purposes, such as hardening (butter, fats), freeze concentration, freeze-drying, air conditioning including air dehumidification, and cryomilling (strong cooling of plastic or fibrous materials to facilitate their disintegration by milling).

19.2 Effect of temperature on food spoilage

19.2.1 Temperature and chemical activity

The relationship between the temperature and the rate of chemical reactions is described by the Arrhenius equation, already discussed in Chapter 4:

$$k = Ae^{-\frac{E}{RT}} \quad \Rightarrow \quad \ln k = \ln A - \frac{E}{RT} \qquad (19.1)$$

where:

k = rate constant of the chemical reaction
E = energy of activation, $J \cdot mol^{-1}$
T = absolute temperature, K
R = gas constant
A = frequency factor, almost independent of temperature and therefore nearly constant for a given reaction.

The kinetics of chemical spoilage reactions such as non-enzymatic browning and loss of some vitamins during storage is found to fit the Arrhenius model quite closely within a fairly wide temperature range. The Arrhenius model is therefore widely used for the prediction of chemical spoilage during storage, following non-monotonic time–temperature profiles (Berk and Mannheim, 1986; Mizrahi *et al.*, 1970). Attention should be paid to the possibility of discontinuity in the Arrhenius model due to phase transition phenomena.

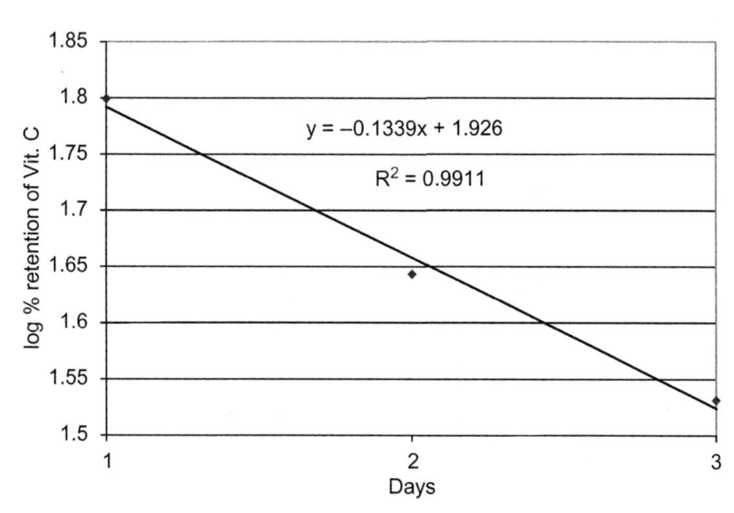

FIGURE 19.1

Loss of vitamin C in spinach during cold storage (Example 19.1).

EXAMPLE 19.1

It has been reported (Hill, 1987) that the percentage loss of vitamin C in spinach upon refrigerated storage at 2–3°C was 37% after 1 day, 56% after 2 days and 66% after 3 days. Do the data fit the "first-order" model of reaction kinetics?

Solution

If the "first-order" model is applicable, then log (C/C_0) should be linear with time ($C =$ vitamin concentration). The plot (Figure 19.1) is nearly linear ($y = -0.134x + 1.926$, with $R^2 = 0.991$).

EXAMPLE 19.2

The relationship between the rate of non-enzymatic browning and the temperature in a concentrated solution of carbohydrates is shown in Figure 19.2 in the form of an Arrhenius plot.

a. How would you explain the discontinuity in the curve?
b. Calculate the energy of activation at 70°C and at 40°C.
c. Assuming that the curve can be extrapolated, what is the rate of browning at 2°C?

Solution

a. The discontinuity occurs at $1/T = 0.00305\ K^{-1}$, corresponding to $T = 328\ K = 55°C$. This is in the range of the glass transition temperatures of concentrated solutions of carbohydrates (52°C for sucrose). The discontinuity is most probably due to phase transition.

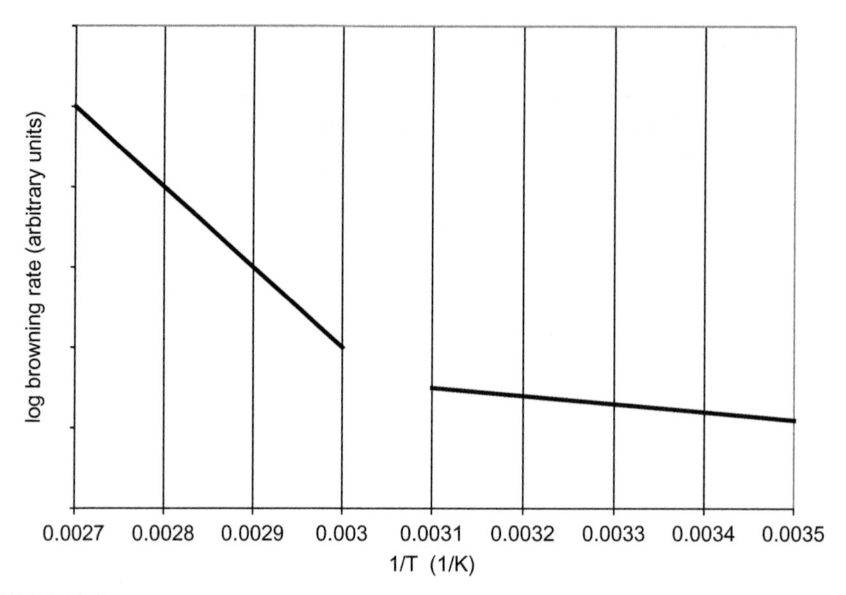

FIGURE 19.2

Rate of non-enzymatic browning vs temperature. (Example 19.2).

b. At 70°C ($1/T = 0.0029\ K^{-1}$), the slope of the curve is −213 63 K. Applying Eq. (19.1):

$$\frac{E}{R} = 21\,363 \quad \Rightarrow \quad E = 21\,363 \times 8.314 = 177\,600\ \text{kJ}\cdot\text{kmol}^{-1}\cdot\text{K}^{-1}$$

At 40°C ($1/T = 0.00319\ K^{-1}$), the slope of the curve is −5000 K.

$$\frac{E}{R} = 5000 \quad \Rightarrow \quad E = 5000 \times 8.314 = 41\,570\ \text{kJ}\cdot\text{kmol}^{-1}\cdot\text{K}^{-1}$$

c. At 2°C ($1/T = 0.00364\ K^{-1}$). Extrapolation would give $\ln(k) \approx -11$. Hence: $k = 16.7 \times 10^{-6}$, meaning that the rate of browning at 2°C would be very slow.

EXAMPLE 19.3

Orange juice is aseptically packed in multi-layer carton boxes. At the moment of packing, the juice contained a minimum of 50 mg vitamin C per 100 g. The label claims a vitamin C content of 40 mg per 100 g. What must be the maximal storage temperature, if the product has to comply with the claim on the label after 180 days of storage?

Assume that the loss of vitamin C follows first-order kinetics with a rate constant of $k = 0.00441\ \text{day}^{-1}$ and an energy of activation of $E = 700\ 00\ \text{kJ}\cdot\text{kmol}^{-1}$, at 27°C.

Solution

Let C (mg per 100 g) be the concentration of the vitamin in the juice. For a first-order reaction:

$$\ln\frac{C}{C_0} = -kt$$

The rate constant at 27°C is $k_1 = 0.00441$ day^{-1}.
Let the rate constant at the unknown storage temperature be k_2.

$$\ln\frac{C}{C_0} = \ln\frac{40}{50} = -k_2 t = -k_2 \times 180 \quad \Rightarrow \quad k_2 = 0.00124$$

The Arrhenius equation is written as follows:

$$\ln\frac{k_1}{k_2} = \frac{E}{R}\left(\frac{1}{T_2} - \frac{1}{T_1}\right)$$

$$\ln\frac{0.00441}{0.00124} = \frac{70000}{8.314}\left(\frac{1}{T_2} - \frac{1}{273+27}\right) \quad T_2 = 287 \text{ K} = 14°\text{C}$$

The maximum storage temperature is 14°C.

19.2.2 Effect of low temperature on enzymatic spoilage

The relationship between enzymatic activity and temperature follows the well-known bell-shaped curve, with maximum activity corresponding to the optimal temperature characteristic to each enzyme (Figure 19.3). At first sight, this behavior may seem to be in contradiction with the Arrhenius model. In reality however, the bell-shaped behavior is the result of two simultaneous and contrary processes, *both* depressed by low temperature. The first is the enzyme-catalyzed reaction itself; the second is the thermal inactivation of the enzyme (Figure 19.4).

Enzyme activity is strongly slowed down but not totally eliminated by refrigeration, even below freezing temperature. This is the reason for the need to inactivate the enzymes by blanching, particularly in frozen vegetables.

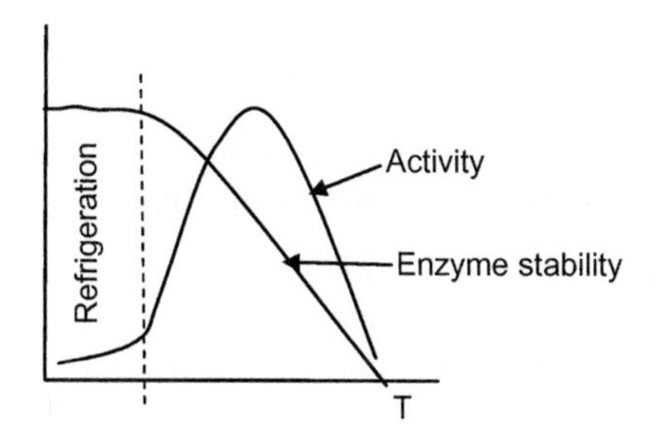

FIGURE 19.3

Effect of temperature on enzyme activity and stability.

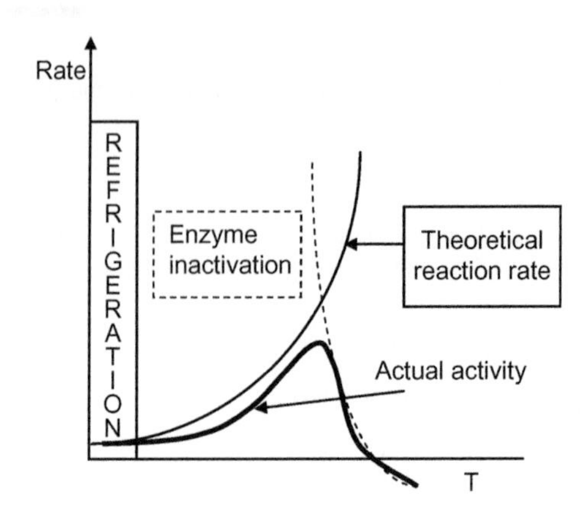

FIGURE 19.4

Explanation of the bell-shaped activity curve.

Enzymatic activity in chilled and frozen foods is of considerable technological significance. Such activity may be desirable, as in the case of the aging of chilled meat or in the development of flavor in many kinds of cheeses, but it can also be a source of deterioration, such as the activity of proteases in fish, peroxydases and lipoxydases in vegetables, or lipases in meat, etc.

19.2.3 Effect of low temperature on microorganisms

With respect to the effect of temperature on their activity, microorganisms are grouped in four categories: thermophiles, mesophiles, psychrotropes and psychrophiles (Hawthorn and Rolfe, 1968; Mocquot and Ducluzeau, 1968). Typical growth temperature ranges corresponding to the four groups are given in Table 19.1.

The four groups differ not only in their temperature requirement for growth but also in their rate of growth. Figure 19.5 shows qualitatively the interdependence of growth rate (generations per hour) and temperature for the four groups. Psychrophilic bacteria grow much more slowly, even at their optimal temperature (Stokes, 1968).

In refrigerated foods, psychrotropic and psycrophilic microorganisms are obviously the main reason for concern.

The relationship between temperature and shelf life is evident from the graph in Figure 19.6. It can be seen that, at lower temperature:

- The induction period (lag phase) is longer
- Growth rate in the logarithmic phase is slower
- As a result of the above, the bacterial count after a given storage time is considerably lower.

The microbiological quality of a refrigerated food depends therefore on the time–temperature profile during the life of the product. Obviously, additional

Table 19.1 Bacteria, Grouped by their Growth Temperature

	Growth temperature (°C)		
Group	**Minimum**	**Optimum**	**Maximum**
Thermophiles	34 to 45	55 to 75	60 to 90
Mesophiles	5 to 10	30 to 45	35 to 47
Psychrotropes	−5 to 5	20 to 30	30 to 35
Psychrophiles	−5 to 5	12 to 15	15 to 20

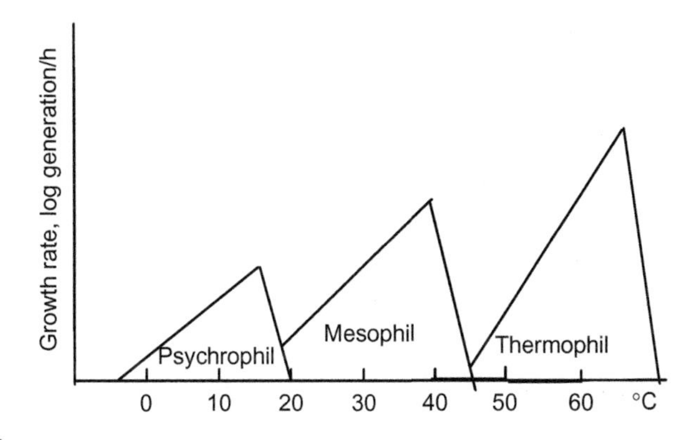

FIGURE 19.5

Classification of microorganisms by their response to temperature.

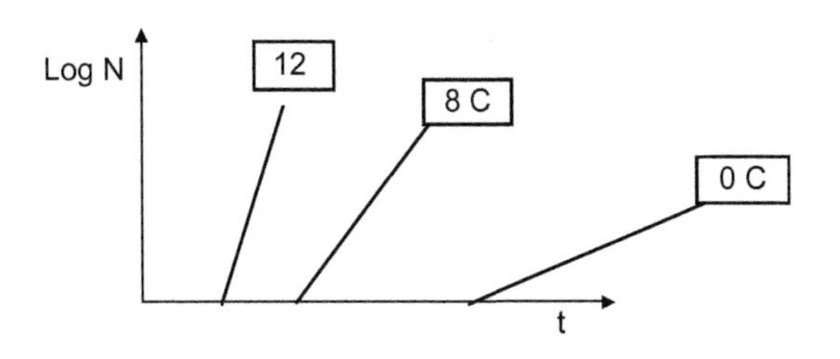

FIGURE 19.6

Schematic representation of the effect of storage temperature on the microbial load of a food.

factors such as pH and water activity play an important role in the microbial activity in foods. Predictive microbiology (McMeekin, 2003; McMeekin *et al.*, 1993) is a new science that involves trying to develop tools for the prediction of the microbiological quality of foods as a function of storage conditions.

19.2.4 Effect of low temperature on biologically active (respiring) tissue

By "active tissues" we mean foods such as fruits and vegetables after harvest, or meat after dressing, in view of their biochemical activity post-harvest or *post-mortem*.

The principal biochemical post-harvest process that occurs in fruits and vegetables is *respiration*, whereby sugars are "burnt", oxygen is consumed and carbon dioxide is produced. The rate of respiration is usually determined by measuring the rate of oxygen consumption or carbon dioxide release. Respiration is the most important (but not the only) cause of deterioration of fruits and vegetables during storage. It is usually said that the shelf life of fresh produce is inversely related to the rate of respiration. The rate of respiration is closely related to the temperature, increasing by two- to four-fold for every 10°C increase in the temperature within the range of usual storage temperatures. However, too low a storage temperature may cause a condition known as "chill injury" in certain commodities, and must be avoided (Fidler, 1968).

Fruits and vegetables differ in the intensity of their post-harvest respiration rate. The following is a rough classification of some fresh produce with respect to respiration rate during storage:

- High respiration rate – avocados, asparagus, cauliflower, berries
- Medium respiration rate – bananas, apricots, plums, carrots, cabbage, tomatoes
- Low respiration rate – citrus, apples, grapes, potatoes.

Many crops undergo a process of post-harvest ripening during storage. These are called "climacteric" species, because their rate of respiration increases to a maximum in the process. Ethylene is produced and acts as a ripening hormone. Apples, bananas and avocados are climacteric, while citrus fruits and grapes are non-climacteric (Figure 19.7).

It is possible to control the rate of respiration of fruits, vegetables and cut flowers with the help of refrigeration alone or through the combination of refrigeration and controlled atmosphere storage. "Controlled atmosphere", or CA, consists of long-term refrigerated storage in closed chambers where the atmosphere has been artificially modified so as to depress the rate of respiration to the desired level. The optimum composition of the atmosphere depends on the commodity, and often on the particular variety of the same commodity. Table 19.2 shows typical optimal storage conditions for selected fruits and vegetables.

Respiration is an exothermic process. The heat released by respiring commodities during refrigerated storage and transportation must be taken into account in calculating the refrigeration load required. Table 19.3 gives the approximate rate of heat evolution during storage of selected commodities.

Low temperature alone is not always sufficient for extending the shelf life of fruits and vegetables. Another important condition that must be controlled in post-

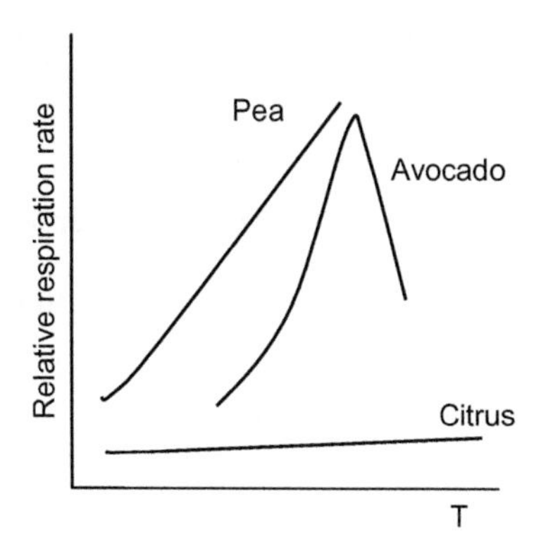

FIGURE 19.7

Effect of temperature on the respiration rate of some crops.

Table 19.2 Optimal Storage Conditions for Some Fruits and Vegetables

Fruits in controlled atmosphere:			
		Atmosphere	
Produce	**Temperature (°C)**	**% Oxygen**	**% CO_2**
Pineapple	10–15	5	10
Avocado	12–15	2–5	3–10
Pomelo	10–15	3–10	5–10
Lemon	10–15	5	0–10
Mango	10–15	5	5
Papaya	10–15	5	10
Melon (cantaloupe)	5–10	3–5	10–15

Vegetables in ordinary cold storage:		
Produce	**Temperature (°C)**	**Relative humidity (%)**
Artichoke	0–2	90–95
Asparagus	0–2	95–100
Broccoli	0–2	95–100
Carrot	0–2	98–100
Eggplant	10–14	90–95
Onion	0–2	65–75
Potato	8–12	90–95

Table 19.3 Approximate Rate of Heat Evolution during Refrigerated Storage of Selected Commodities, According to Storage Temperature (Singh and Heldman, 2003)

	Heat Evolution Rate, Watts per ton, when stored at:			
Commodity	**0°C**	**5°C**	**10°C**	**15°C**
Apple	10–12	15–21	41–61	41–92
Cabbage	12–40	28–63	36–86	66–169
Carrot	46	58	93	117
Sweet corn	125	230	331	482
Green peas	90–138	163–226	–	529–599
Orange	9	14–19	35–40	38–67
Strawberry	36–52	48–98	145–280	210–275

harvest storage is relative humidity. Loss of water is often the cause of texture deterioration, wilting, shriveling, etc. Water loss may be minimized by storing at high relative humidity, but humidity levels that can promote growth of mold must be avoided. To maintain high humidity, water may be sprayed as a fine "mist". Preserving chemicals (hydrogen peroxide, chlorine) are sometimes applied during refrigerated storage of fresh vegetables and fruit, as an additional hurdle.

19.2.5 The effect of low temperature on physical properties

Numerous changes in physical properties caused by exposure to low temperatures may have significant effects on food quality. These changes can include:

- An increase in viscosity
- A decrease in solubility, resulting in crystallization, precipitation, cloudiness (e.g., in beer)
- Hardening, transition to rubbery and glassy state in carbohydrate systems
- Hardening of fats
- Decomposition of colloidal systems such as emulsions and gels
- Freezing.

19.3 Freezing

Freezing is one of the most widespread industrial methods of food preservation (Tressler *et al.*, 1968). The transition from chilling to freezing is not merely a continuous change that can be explained on grounds of the lower temperature alone. On the contrary, freezing represents a point of sharp discontinuity in the relationship between temperature and the stability and sensory properties of foods.

- The exceptional efficiency of freezing as a method of food preservation is, to a large extent, due to the depression of water activity. Indeed, when food is frozen, water separates as ice crystals and the remaining non-frozen portion becomes more concentrated in solutes. This "freeze concentration" effect results in the depression of water activity. In this respect, freezing can be compared to concentration and drying.
- On the other hand, the same phenomenon of "freeze concentration" may accelerate reactions (Poulsen and Lindelov, 1981), inducing irreversible changes such as protein denaturation, accelerated oxidation of lipids and destruction of the colloidal structure (gels, emulsions) of the food.
- The *rate of freezing* has an important effect on the quality of frozen foods. Physical changes such as the formation of large ice crystals with sharp edges, expansion, and disruption of the osmotic equilibrium between the cells and their surroundings may cause irreversible damage to the texture of vegetables, fruits and muscle foods. It has been established that such damage is minimized in the case of rapid freezing.

19.3.1 Phase transition, freezing point

Figure 19.8 shows "cooling curves" describing the change in temperature as heat is removed from a sample. Figure 19.8(a) describes the cooling behavior of pure water. As the sample is cooled, the temperature drops linearly (constant specific heat) until the first crystal of ice is formed. The temperature at that moment is the *freezing temperature* of pure water, 0°C at atmospheric pressure, by definition. The freezing point is the temperature at which the vapor pressure of the liquid is equal to that of the solid crystal. Under certain conditions (absence of solid particles, slow undisturbed cooling) the sample may undergo a metastable state of supercooling. Some proteins, known as "antifreeze proteins", are capable of preventing crystallization of water ice at the freezing point.

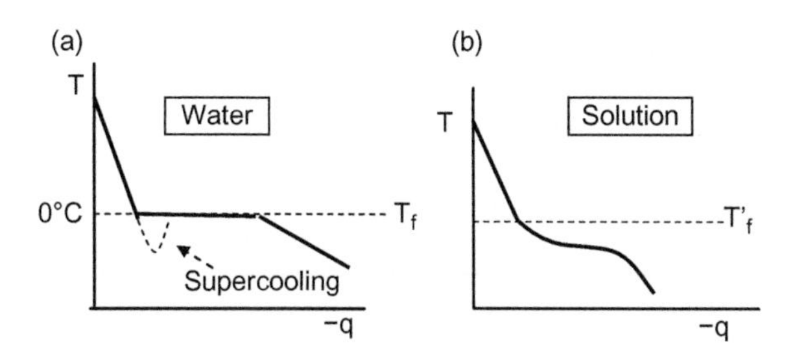

FIGURE 19.8

Cooling curves for (a) pure water and (b) aqueous solution.

Figure 19.8(b) represents the cooling curve of a solution or a real food material. As the sample is cooled, the temperature drops linearly. The first ice crystal appears at the temperature T'_f. This is the temperature at which the water vapor pressure of the solution is equal to that of pure water ice. Since the water vapor pressure of a solution is lower than that of water at the same temperature, T'_f is lower than the freezing temperature of pure water. The difference, called the *freezing point depression*, increases with the molar concentration of the solution. Transformation of some of the liquid water to ice results in higher concentration of the solutes, which in turn lowers the freezing point, and so on. There is no sharp phase transition at a constant temperature but rather a gradual *zone of freezing*, starting at the *temperature of initial freezing*, T'_f.

Experimental data pertaining to the initial freezing temperatures of foods are available in the literature (Earle, 1966). Assuming ideal behavior, the initial freezing temperature can be estimated from food composition data (Jie *et al.*, 2003; Miles *et al.*, 1997; Van der Sman and Boer, 2005). The initial freezing temperature of common fruits and vegetables falls between −0.8 and −2.8°C (Fennema, 1973).

EXAMPLE 19.4

Estimate the initial freezing temperature of 12-Bx orange juice and of 48-Bx orange juice concentrate. Assimilate the juices to 12% and 48% w/w ideal solutions of glucose (MW = 180).

Solution

The freezing point of a solution is the temperature at which its vapor pressure is equal to that of pure water ice. Assuming ideal behavior (Raoult's Law), the vapor pressure of a solution is $p = x_w p_0$, where x_w is the molar fraction of water in the solution and p_0 is the vapor pressure of pure water. The molar fraction of the two solutions is as follows.

For 12Bx:

$$x_w = \frac{(100-12)/!8}{(100-12)/18 + 12/180} = 0.987$$

For 48Bx:

$$x_w = \frac{(100-48)/18}{(100-48)/18 + 48/180} = 0.915$$

Values of vapor pressure of water and ice at different temperatures are given in Table A.13 (see Appendix). The vapor pressure of each solution at different temperatures is calculated from these values and the temperature at which they equal the vapor pressure of ice is searched. A graphical solution is shown in Figure 19.9.

Results:

The freezing point of 12-Bx juice is approximately −1.5°C.
The freezing point of 48-Bx juice concentrate is approximately −9.2°C.

It is not possible to freeze out all the water from a solution. When the concentration of the solute in the non-frozen portion reaches a certain level, that entire

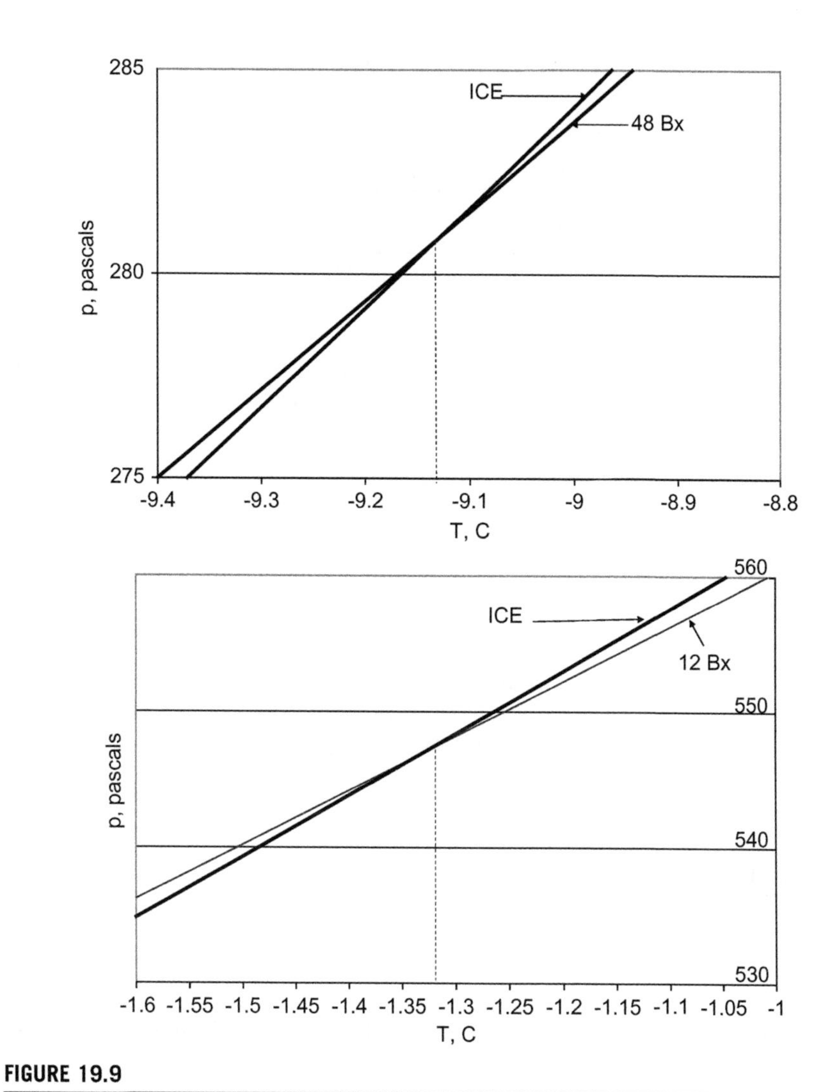

FIGURE 19.9

Freezing point of orange juice and concentrate (Example 19.4).

portion solidifies as though it were a pure substance. This new solid phase is called "eutecticum". A theoretical phase diagram for a salt solution is shown in Figure 19.10.

In the case of sugar solutions and most food materials it is practically impossible to reach the eutectic point, because glass transition of the non-frozen concentrated cold solution occurs before the eutectic point. Molecular motion in the

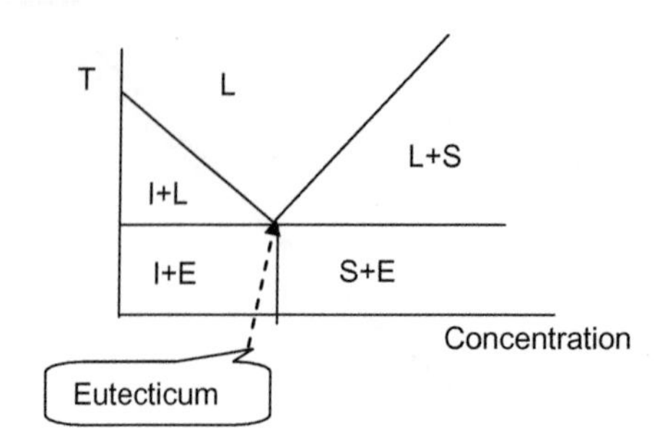

FIGURE 19.10

Phase diagram of a salt solution: S, salt; E, eutecticum; L, liquid; I, ice.

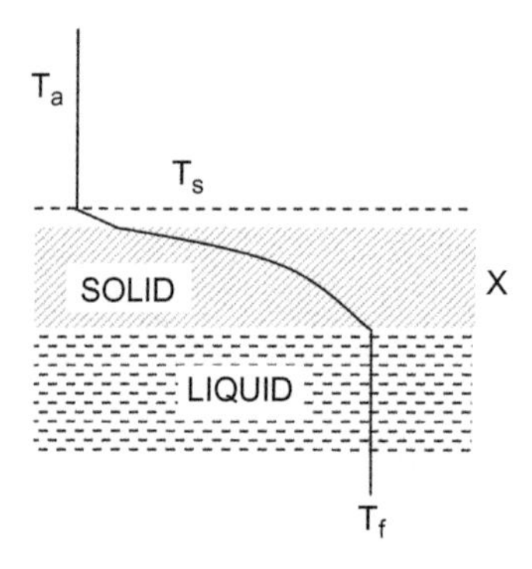

FIGURE 19.11

Temperature profile in freezing

glassy solid becomes extremely slow, and any further crystallization of water ice becomes practically impossible.

19.3.2 Freezing kinetics, freezing time

As stated in the introduction to this chapter, the quality of frozen foods is strongly affected by the speed of freezing. Furthermore, freezing time is of obvious economic importance, as it determines the product throughput of the freezing equipment. Analysis of the factors that affect freezing time is, therefore, of interest.

Let us consider a mass of liquid, cooled by cold air blowing over its surface (Figure 19.11). We shall assume that the liquid is initially at its freezing temperature, so that all the heat removed from the liquid comes from the latent heat of freezing, released when some of the liquid is transformed to ice.

The rate of heat transfer from the freezing front of the liquid to the cold air is therefore equal to the rate of heat release as a result of forming additional ice.

$$q = A\rho\lambda \frac{dz}{dt} = A\frac{1}{h + \frac{z}{k}}(T_f - T_a) \tag{19.2}$$

where:

q = rate of heat removal, W
A = area of heat transfer, m^2.
ρ = density of the liquid, $kg \cdot m^{-3}$
λ = latent heat of freezing of the liquid, $J \cdot kg^{-1}$
z, Z = thickness of the frozen phase
h = convective heat transfer coefficient at the air–ice interface, $W \cdot m^{-2} \cdot K^{-1}$
k = thermal conductivity of the frozen phase, $W \cdot m^{-1} \cdot K^{-1}$
T_a = temperature of the cooling medium (in this case, cold air)
T_f = freezing temperature.

Rearranging and integrating, we obtain the time necessary to freeze a mass of liquid of thickness Z:

$$t = \frac{\rho\lambda}{T_f - T_a}\left(\frac{Z}{h} + \frac{Z^2}{2k}\right) \tag{19.3}$$

Equation (19.3) is known as the *Plank equation*, originally proposed by R. Z. Plank in 1913 (López-Leiva and Hallström, 2003; Pham, 1986).

Equation (19.3) gives the freezing time for a semi-infinite slab. For other geometries, Plank's formula is written in the following general form:

$$t = \frac{\rho\lambda}{T_f - T_a}\left(\frac{d}{Qh} + \frac{d^2}{Pk}\right) \tag{19.4}$$

For a plate of thickness d, cooled from both sides	Q = 2	P = 8
For an infinite cylinder of diameter d:	Q = 4	P = 16
For a sphere of diameter d:	Q = 6	P = 24

Plank's formula is only approximate, due to the inaccuracy of some of the assumptions made:

- In reality, latent heat of freezing is not the only kind of energy exchanged. There are also some sensible heat effects, such as the further cooling of the ice formed, and lowering the temperature of the non-frozen material to the freezing point. In practice, the error resulting from this assumption is not large, since the latent heat effects far exceed sensitive heat effects.
- As explained above, there is no sharp freezing point in foods. T_f is therefore an average value.

- λ refers to the latent heat of freezing of the food and not of pure water. If the mass fraction of water in the food is w and the latent heat of freezing of pure water is λ_0, then λ can be calculated as $\lambda = w\lambda_0$. This is also an approximation, since not all the water is freezable. Furthermore, if the food contains fats that are solidified in the process, the enthalpy of solidification should be taken into account in the calculation of λ.

More precise methods for the calculation of freezing time have been proposed (Mannapperuma and Singh, 1989; Pham, 1986; Succar, 1989; Chevalier *et al.*, 2000). Some of these methods make use of numerical rather than analytical techniques. Despite its shortcomings, Plank's formula is valuable in process design and for visualizing the effect of process conditions on freezing time. The following are some of the practical consequences of Plank's equation:

- Freezing time is inversely proportional to the overall temperature difference $T_f - T_a$.
- Freezing time increases with increasing water content of the food.
- Freezing time is proportional to the sum of two terms: a convective term proportional to the size d and a conductive term proportional to d^2. When freezing larger items, such as beef carcasses, whole birds or cakes, the conductive term (the internal resistance to heat transfer) becomes the predominant factor and the convective term becomes less significant – i.e., increasing the velocity is practically ineffective. On the other hand, when freezing small particles the convective term is important, and improving convective heat transfer to the surface (e.g., increasing the turbulence) results in considerably faster freezing.

EXAMPLE 19.5

Blocks of filleted fish, 5 cm thick are to be frozen in a plate freezer (by contact with a cold surface on both sides).

a. Estimate the time for complete freezing of the blocks.

Data:

Temperature of the plates	$-28°C$, constant
Average freezing temperature	$-5°C$.
Density of the fish	$1100\ kg \cdot m^{-3}$
Water content of the fish	70% w/w.
Thermal conductivity of frozen fish	$1.7\ W \cdot m^{-1} \cdot K^{-1}$
Latent heat of freezing of water	$334\ kJ \cdot kg$

Sensible heat effects and heat loss will be neglected. Assume perfect contact between the fish blocks and the plate surface.

b. What will be the freezing time if the blocks are packaged in carton boxes? The thickness of the carton is 1.2 mm and its thermal conductivity is 0.08 $W \cdot m^{-1} \cdot K^{-1}$.

Solution

a. Equation (19.4) is applied, with Q = 2 and P = 8 (plate cooled from both sides).

$$t = \frac{\rho\lambda}{T_f - T_a}\left(\frac{d}{2h} + \frac{d^2}{8k}\right)$$

The surface contact is ideal, hence h is infinite.
Substituting the data:

$$t = \frac{1100 \times 334000 \times 0.7}{-5-(-28)}\left(\frac{(0.05)^2}{8 \times 1.7}\right) = 2055 \text{ s} = 0.57 \text{ h}$$

The freezing time for the unpackaged fish is 0.57 hours.

b. The surface resistance to heat transfer is now not zero but equal to the thermal resistance of the package of thickness z:

$$\frac{1}{h} = \frac{z}{k} = \frac{1.2 \times 10^{-3}}{0.08} = 0.015 \quad \Rightarrow \quad h = 66.7$$

$$t = \frac{1100 \times 334000 \times 0.7}{-5-(-28)}\left(\frac{0.05}{2 \times 66.7} + \frac{(0.05)^2}{8 \times 1.7}\right) = 6246 \text{ s} = 1.73 \text{ h}$$

The freezing time for the packaged fish is 1.73 hours.

EXAMPLE 19.6

a. Meatballs, 4 cm in diameter, are to be frozen in a blast of cold air, at −40°C and at "moderate" velocity. The convective heat transfer coefficient at the air–meatball interface is h = 10 $W \cdot m^{-2} \cdot K^{-1}$ (determined from previous experiments with copper spheres). All other data are as for fish (see Example 19.5). Calculate the time for complete freezing.

b. Repeat the calculations for a "very moderate" air current (h = 1 $W \cdot m^{-2} \cdot K^{-1}$) and a "very turbulent" blast (h = 100 $W \cdot m^{-2} \cdot K^{-1}$).

c. Repeat the calculations for the three cases, but with "mini" meatballs, 1 cm in diameter.

Solution

a. Plank's equation for spheres is:

$$t = \frac{\rho\lambda}{T_f - T_a}\left(\frac{d}{6h} + \frac{d^2}{24k}\right)$$

Substitution of the data yields:

$$t = \frac{1100 \times 334000 \times 0.7}{-5-(-40)}\left(\frac{0.04}{6 \times 10} + \frac{(0.04)^2}{24 \times 1.7}\right) = (7348000)(0.000706) = 5188 \text{ s}$$

b. For very slow air (h = 1 $W \cdot m^{-2} \cdot K^{-1}$):

$$t = 7348000\left(\frac{0.04}{6 \times 1} + \frac{(0.04)^2}{24 \times 1.7}\right) = (7348000)(0.00671) = 49300 \text{ s}$$

Table 19.4 Effect of Particle Size and Turbulence on Freezing Rate:Summary

Diameter	Freezing time (s)		
	Low h	Moderate h	High h
1 cm	12 271	1241	141
4 cm	49 300	5188	779

For very turbulent air ($h = 100\ W \cdot m^{-2} \cdot K^{-1}$):

$$t = 7348000\left(\frac{0.04}{6 \times 100} + \frac{(0.04)^2}{24 \times 1.7}\right) = (7348000)(0.000106) = 779\ \text{s}$$

c. For the mini-meatballs (d = 1 cm):

At moderate air velocity:

$$t = 7348000\left(\frac{0.01}{6 \times 10} + \frac{(0.01)^2}{24 \times 1.7}\right) = (7348000)(0.000169) = 1241\ \text{s}$$

At slow air velocity:

$$t = 7348000\left(\frac{0.01}{6 \times 1} + \frac{(0.01)^2}{24 \times 1.7}\right) = (7348000)(0.00167) = 12271\ \text{s}$$

At high air velocity:

$$t = 7348000\left(\frac{0.01}{6 \times 100} + \frac{(0.01)^2}{24 \times 1.7}\right) = (7348000)(0.0000192) = 141\ \text{s}$$

The results are summarized in Table 19.4.

Conclusion: The effect of heat transfer at the surface (e.g., air turbulence in blast-freezing) is much stronger in the case of small items where internal resistance to heat transfer is less significant.

19.3.3 Effect of freezing and frozen storage on product quality

Van Arsdel *et al.* (1969) provided one of the earliest reviews on the quality and stability of frozen foods. For a very large number of food products, freezing represents the best preservation method with respect to food quality. The nutritional value, flavor and color of foods are affected very slightly, if at all, by the process of freezing itself. The main quality factor that may be adversely affected by freezing is the texture. On the other hand, unless appropriate measures are taken, the deleterious effect of long-term frozen storage and of thawing on every aspect of product quality may be significant.

19.3.3.1 Effect of freezing on texture

In a vegetal or animal tissue, the cells are surrounded by a medium known as the extracellular fluid. The extracellular fluid is less concentrated than the protoplasm inside the cells. The concentration difference results in a difference in osmotic pressure which is compensated by the tension of the cell wall. This phenomenon, known as *turgor*, is the reason for the solid appearance of meat and the crispiness of fruits and vegetables. When heat is removed from the food in the course of freezing, the extracellular fluid, being less concentrated, starts to freeze first. Its concentration rises and the osmotic balance is disrupted. Fluid flows from the cell to the extracellular space. Turgor is lost and the tissue is softened. When the food is thawed, the liquid that was lost to the extracellular space is not reabsorbed into the cell but is released as free juice in the case of fruits or "drip" in the case of meat.

It is generally believed that freezing damage to the texture of cellular foods is greatly reduced by accelerating the rate of freezing. This is explained by the fact that the condition of osmotic imbalance, created at the onset of freezing, disappears when the entire mass is frozen. Furthermore, rapid freezing results in the formation of smaller ice crystals, presumably less harmful to the texture of cellular systems.

Another probable reason for the deterioration of the texture is the volumetric expansion caused by freezing. The specific volume of ice is 9% higher than that of pure water. Because cellular tissues are not homogeneous with respect to water content, parts of the tissue expand more than the others. This creates mechanical stress that may result in cracks. Obviously this effect is particularly strong in foods with high water content, such as cucumbers, lettuce and tomatoes. This kind of texture damage is partially prevented by adding solutes. Adding sugar to fruits and berries before freezing was a widespread practice before the development of ultra-rapid freezing methods.

The rate of freezing affects the size of ice crystals. Slow freezing produces large crystals. It has been suggested that large crystals with sharp edges may break cell walls and contribute to texture deterioration in cellular foods subjected to slow freezing.

There is no controversy regarding the fact that slow freezing results in a higher percentage of drip in meat and fish (Jul, 1984). It is also accepted that rapid freezing causes less damage to the texture of particularly fragile fruits. On the other hand, the general applicability of the theory stating the superiority of rapid freezing seems to be questionable (Jul, 1984). Nevertheless, quick freezing continues to be the practical objective of food freezing process design.

19.3.3.2 Effect of frozen storage on food quality

Frozen storage, even at a fairly low temperature, does not mean the absence of deteriorative processes. On the contrary, frozen foods may undergo profound quality changes during frozen storage. While the rate of reactions is generally (but not always) slower in frozen foods, the expected shelf life, and therefore the time available for the reactions to take place, is long. Some of the frequent types of deterioration in frozen foods are protein denaturation resulting in toughening of muscle foods, protein–lipid interaction, lipid oxidation, and oxidative changes in general (e.g., loss of some vitamins and pigments).

Extensive studies on the effect of frozen storage on product quality were undertaken in the 1960s, by researchers and by the Western Research Laboratories of the US Department of Agriculture (Canet, 1989). A large number of commodities were tested for changes in chemical composition and sensory characteristics. A concept known as "time–temperature tolerance" (TTT) was developed (Van Arsdel *et al.*, 1969). The studies showed a linear relationship between the storage temperature and the logarithm of storage time for equal reduction in quality (loss of a certain vitamin, loss of color or loss of organoleptic score). The logical conclusion of these studies could be that lower storage temperature *always* results in higher quality. This is not always the case, however. Reactions may be accelerated by the "freeze concentration" effect more than they are attenuated by the lower temperature. In this case, the rate of deterioration (e.g., lipid oxidation) may increase as the storage temperature is lowered, passing through a maximum and then diminishing at very low temperatures.

Mass transfer phenomena during frozen storage (oxygen transfer, loss of moisture) may be a major cause of quality loss. The quality of packaging is therefore particularly important in frozen foods. The PPP (product–process–package) approach consists of paying attention of all three factors in evaluating and predicting the effect of frozen storage on product quality (Jul, 1984).

Another type of change in frozen foods during storage is the process of recrystallization. As explained in Chapter 14, smaller crystals are more soluble than large ones. Equally, small ice crystals have a lower melting point than large ones. Consequently, if the storage temperature undergoes fluctuations, small ice crystals may melt and then solidify on the larger crystals. This may be the reason why foods frozen rapidly and those frozen slowly sometimes show similar ice crystal size distribution after storage. Recrystallization is particularly objectionable in ice cream, where conversion of small ice crystals to large ones results (Ostwald ripening) in loss of the smooth, creamy texture. Remedies are avoiding temperature fluctuation during storage as much as possible, or using ice-structuring proteins.

19.4 Superchilling

Superchilling is the process of cooling food to a temperature slightly below the initial freezing temperature. At such temperatures, only a small proportion of the water is converted to ice. Although the removal of some water results in a slight depression of water activity, the main preserving effect is due to the lower temperature. Since most of the water is in a liquid state, the damaging effects of complete freezing are avoided. This is an important advantage, particularly in fishing.

The fish are not frozen and at the same time freshness is maintained for a longer period, compared to storage and transportation in ice, by virtue of the lower temperature. This allows the fishing boats to continue fishing at sea for a longer period. Superchilling, sometimes combined with modified atmosphere packaging MAP (see Chapter 27), is a technology applicable as for short-term refrigerated storage of fish, seafood and meat (Duun *et al.*, 2008; Fatima *et al.*, 1988; Lauzon *et al.*, 2009; Liu *et al.*, 2010; Magnussen *et al.*, 2008; Sivertsvik *et al.*, 2003).

References

Berk, Z., Mannheim, C.H., 1986. The effect of storage temperature on the quality of citrus products aseptically packed into steel drums. J. Food Process. Preservat 10 (4), 281–292.

Canet, W., 1989. Quality and stability of frozen vegetables. In: Thorne, S. (Ed.), Developments in Food Preservation – 5. Elsevier Applied Science, London, UK.

Chevalier, D., Le Bail, A., Ghoul, M., 2000. Freezing and ice crystal formed in a cylindrical food model. Part I: freezing at atmospheric pressure. J. Food Eng. 46 (4), 277–285.

Duun, A.S., Hemmingsen, A.K.T., Haugland, A., Rustad, T., 2008. Quality changes during superchilled storage of pork roast. LWT 41 (10), 2136–2143.

Earle, R.L., 1966. Unit Operations in Food Processing. Pergamon Press, Oxford.

Fatima, R., Khan, M.A., Qadri, R.B., 1988. Shelf life of shrimp (Penaeus merguiensis) stored in ice (0°C) and partially frozen (−3°C). J. Sci. Food. Agric. 42 (3), 235–247.

Fennema, O., 1973. Solid–liquid equilibria. In: Fennema, O., Powrie, W.D., Marth, E.H. (Eds.), Low Temperature Preservation of Foods and Living Matter. Marcel Dekker Inc., New York, NY.

Fidler, J.C., 1968. Low temperature injury to fruits and vegetables. In: Hawthorne, J., Rolfe, E.J. (Eds.), Low Temperature Biology of Foodstuffs. Pergamon Press, Oxford, UK.

Hawthorn, J., Rolfe, E.J. (Eds.), 1968. Low Temperature Biology of Foodstuffs. Pergamon Press, Oxford, UK.

Hill, M.A., 1987. The effect of refrigeration on the quality of some prepared foods. In: Thorne, S. (Ed.), Developments in Food Preservation – 4. Elsevier Applied Science Publishers Ltd., London, UK.

Jie, W., Lite, L., Yang, D., 2003. The correlation between freezing point and soluble solids of fruits. J. Food Eng. 60 (4), 481–484.

Jul, M., 1984. The Quality of Frozen Foods. Academic Press, London.

Kaale, L.D., Eikevik, T.M., Rustad, T., Kolsaker, K., 2011. Superchilling of food: a review. J. Food Eng. 107 (2), 141–146.

Lauzon, H.L., Magnússon, H., Sveinsdóttir, K., Gudjónsdóttir, M., Martinsdóttir, E., 2009. Effect of brining, modified atmosphere packaging, and superchilling on the shelf life of cod (*Gadus morhua*) loins. J. Food Sci. 74 (6), M258–M267.

Liu, S.-L., Lu, F., Xu, X.-B., Ding, Y.-T., 2010. Original article: super-chilling maintains freshness of modified atmosphere-packaged *Lateolabrax japonicus*. Intl. J. Food Sci. Technol. 45 (9), 1932–1938.

López-Leiva, M., Hallström, B., 2003. The original Plank equation and its use in the development of food freezing rate prediction. J. Food Eng. 58 (3), 267–275.

Magnussen, O.M., Haugland, A., Hemmingsen, A.K.T., Johansen, S., Nordtvedt, T.S., 2008. Advances in superchilling of food – process characteristics and product quality. Trends Food Sci. Technol. 19 (8), 418–426.

Mannapperuma, J.D., Singh, R.P., 1989. A computer-aided method for the prediction of properties and freezing/thawing times of foods. J. Food Eng. 9, 275–304.

McMeekin, T.A., 2003. An essay on the unrealized potential of predictive microbiology. In: McKellar, R.C., Lu, X. (Eds.), Modelling Microbial Response in Foods. CRC Press, Boca Raton, FL.

McMeekin, T.A., Olley, N.J., Ross, T., Ratkowsky, D.A., 1993. Predictive Microbiology. Wiley, Chichester, UK.

Miles, C.A., Mayer, Z., Morley, M.J., Housaeka, M., 1997. Estimating the initial freezing point of foods from composition data. Intl. J. Food Sci. Technol. 22 (5), 389–400.

Mizrahi, S., Labuza, T.P., Karel, M., 1970. Feasibility of accelerated tests for browning in dehydrated cabbage. J. Food Sci. 35 (6), 804–807.

Mocquot, G., Ducluzeau, R., 1968. The influence of cold storage of milk on its microflora and its suitability for cheese-making. In: Hawthorne, J., Rolfe, E.J. (Eds.), Low Temperature Biology of Foodstuffs. Pergamon Press, Oxford, UK.

Pham, Q.T., 1986. Simplified equation for predicting the freezing time of foodstuffs. J. Food Technol. 21, 209–219.

Poulsen, K.P., Lindelov, F., 1981. Acceleration of chemical reactions due to freezing. In: Rockland, L.B., Stewart, G.F. (Eds.), Water Activity: Influence on Food Quality. Academic Press, New York, NY.

Sivertsvik, M., Rosnes, J.T., Kleiberg, G.H., 2003. Effect of modified atmosphere packaging and superchilled storage on the microbial and sensory quality of atlantic salmon (*Salmo salar*) fillets. J. Food Sci. 68 (4), 1467–1472.

Stokes, J.L., 1968. Nature of psychrophilic microorganisms. In: Hawthorne, J., Rolfe, E.J. (Eds.), Low Temperature Biology of Foodstuffs. Pergamon Press, Oxford, UK.

Succar, J., 1989. Heat transfer during freezing and thawing of foods. In: Thorne, S. (Ed.), Developments in Food Preservation – 5. Elsevier Applied Science, London, UK.

Tressler, D.K., Van Arsdel, W.B., Copley, M.J., Woolrich, W.R., 1968. The Freezing Preservation of Foods. The Avi Publishing Co., Westport, CT.

Van Arsdel, W.B., Copley, M.J., Olson, R.L., 1969. Quality and Stability of Frozen Foods. Wiley Interscience, New York, NY.

Van der Sman, R.G.M., Boer, E., 2005. Predicting the initial freezing point and water activity of meat products from composition data. J. Food Eng. 66 (4), 469–475.

CHAPTER

Refrigeration

20

Equipment and Methods

20.1 Sources of refrigeration

Low temperatures may be delivered by three types of sources:

1. Natural sources – ice, snow, climatic conditions
2. Mechanical refrigeration
3. Cryogenic agents.

Only the latter two are suitable for industrial implementation.

20.1.1 Mechanical refrigeration

A heat engine operating in reverse is capable of using work in order to transfer heat from a low-temperature to a high-temperature source (Figure 20.1). If the purpose of the process is to deliver heat to the high-temperature source, the device is called a "heat pump". If, on the contrary, the objective is to cool the low-temperature source, the same device is named a "refrigeration machine" (ASHRAE, 2006; Dosset and Horan, 2001).

Thermodynamically, refrigeration may be produced by a number of different processes. Three different thermodynamic effects are used commercially to produce refrigeration, namely the *vapor compression cycle*, the *absorption cycle* and the *Peltier effect*.

The basic principle of the most common type of mechanical refrigeration is a cyclic thermodynamic process known as the *Rankine cycle* (William John Macquorn Rankine, 1820–1872, Scottish engineer) or a *vapor compression cycle*. A theoretical reverse Rankine cycle is shown in Figure 20.2 as a T–S (temperature–enthropy) plot. The cycle consists of four sections:

Step 1 – *Compression*. Saturated vapor at pressure P_1 (Point 1) is compressed to pressure P_2 (Point 2). Ideally, isenthropic (adiabatic and reversible) compression is assumed. Mechanical work is supplied to the *compressor*.
Step 2 – *Condensation*. The compressed vapor is cooled until completely condensed as a saturated liquid (Point 3). Ideally, cooling is assumed to take place at constant pressure. The heat removed from the condensed vapor is

Food Process Engineering and Technology. DOI: http://dx.doi.org/10.1016/B978-0-12-415923-5.00020-4
© 2013 Elsevier Inc. All rights reserved.

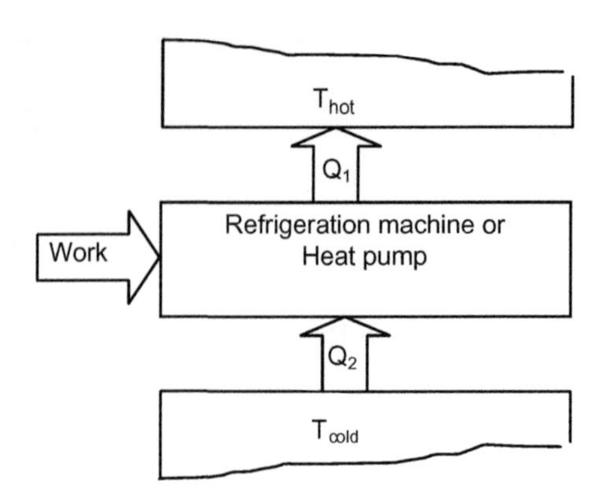

FIGURE 20.1

Principle of the refrigeration machine or heat pump.

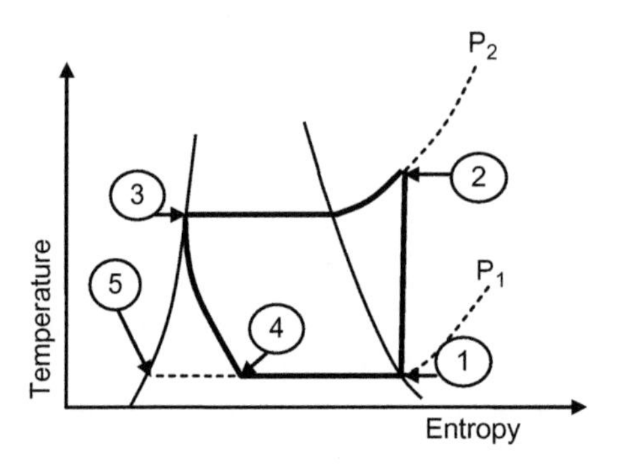

FIGURE 20.2

Vapor compression refrigeration cycle.

transferred to a cooling medium such as air or water. Physically, this step takes place in a heat exchanger serving as a *condenser*.

Step 3 – *Expansion*. The pressure of the liquid is released through a throttling element (e.g., an *expansion valve*) down to pressure P_1 (Point 4). The throttling process is supposed to be isenthalpic, not involving any exchange of energy. Point 4 represents a mixture of saturated vapor (Point 1) and saturated liquid (Point 5).

Step 4 – *Evaporation*. Heat is transferred to the liquid–vapor mixture until all the liquid is evaporated (back to Point 1). This is the step of the cycle where useful refrigeration is generated. Physically, this step occurs in a heat exchanger known as the *evaporator* or *diffuser*.

The fluid undergoing the cycle is called a *refrigerant*. Although, theoretically, the reverse Rankine cycle could be run with any fluid, only certain compounds and mixtures are suitable for use in practical refrigeration. The criteria for the selection of refrigerants will be discussed later.

The *refrigeration capacity*, q_e , of a mechanical refrigeration cycle is the rate at which heat is extracted from the surroundings at the step of evaporation. Energy balance for the evaporator gives:

$$q_e = m(h_1 - h_4) \tag{20.1}$$

where m = rate of circulation of the refrigerant, $kg \cdot s^{-1}$, and h_1, h_2, etc. = specific enthalpy of the refrigerant at Points 1, 2, etc., $J \cdot kg^{-1}$.

In the SI system, refrigeration capacity is expressed in watts. An old engineering unit, called the "commercial ton of refrigeration", is still in use. It is equal to the rate at which heat needs to be removed to produce 1 ton (2000 lb in the USA) of ice at 0°C in 24 hours from water at the same temperature. Its equivalent in SI units is 3517 watts.

The rate at which heat is removed to the surroundings, q_c, is obtained from the energy balance for the condenser:

$$q_c = m(h_3 - h_2) \tag{20.2}$$

The theoretical mechanical power w (watts) supplied to the compressor is:

$$w = m(h_2 - h_1) \tag{20.3}$$

By virtue of the first law of thermodynamics, for the entire cycle:

$$q_e + q_c - w = 0 \tag{20.4}$$

The useful output of the refrigeration machine is its refrigeration capacity. The cost factor is the power input to the compressor. The energy *efficiency* of a refrigeration machine is therefore expressed as its *coefficient of performance* (COP, dimensionless), defined as:

$$COP = \frac{q_e}{w} = \frac{h_1 - h_4}{h_2 - h_1} \tag{20.5}$$

Note*:* COP is a thermodynamic concept, different from the formal definition of "efficiency" of a heat engine. Therefore, COP can be, and usually is, considerably larger than 100%. Another practical expression of "energy efficiency" of a refrigeration machine is *horsepower per ton of refrigeration*.

A real refrigeration cycle deviates from the theoretical cycle described above. The main reasons for deviation are as follows:

- The compression is neither adiabatic nor reversible. The power requirement of the compressor is higher than the theoretical value due to heat losses as well as volumetric and mechanical inefficiencies.

- There is a pressure drop due to friction along the refrigerant flow channels and not only in the expansion valve.
- The low-pressure vapor entering the compressor (Point 1) is usually slightly superheated to avoid "wet compression" (presence of liquid droplets in the gas, which may cause accelerated wear in the compressor).
- Despite thermal insulation, there is some heat transfer (loss of refrigeration) to the refrigerant flow tubing between the expansion valve and the evaporator.

EXAMPLE 20.1

A refrigeration machine is to deliver a refrigeration capacity of 70 kW at −40°C. The refrigerant is R-134a. The condenser is air cooled, and the condensation temperature is assumed to be 35°C. Complete condensation with no supercooling is assumed. To avoid wet compression, 5°C superheat is allowed at the exit from the evaporator. Calculate the theoretical compressor power, the heat rejected at the condenser, and the coefficient of performance of the machine.

Solution

The thermodynamic properties of the refrigerant at the different "points" of the cycle will be first determined, with the help of the appropriate refrigerant table (see Appendix, Tables A.15 and A.16).

Point 1:

Pressure = p_1 = saturation pressure at −40°C = 51.8 kPa (Table A.15).

Temperature = T_1 = −35°C (−40°C + 5°C superheat)

The thermodynamic properties at Point 1 are found from Table A.16:

$h_1 = 377.3\ kJ \cdot kg^{-1}$; $s_1 = 1.778\ kJ \cdot kg^{-1} \cdot K^{-1}$; $v_1 = 0.371\ m^3 \cdot kg^{-1}$

Point 2:

Pressure = p_2 = saturation pressure at 35°C = 887.6 kPa

Entropy = $s_2 = s_1 = 1.778\ kJ \cdot kg^{-1} \cdot K^{-1}$

Other thermodynamic properties at Point 2 are found from Table A.16:

$T_2 = 49.3$ $h_2 = 438.1\ kJ \cdot kg^{-1}$

Point 3:

Saturated liquid at $p_3 = p_2 = 887.6$ kPa

$h_3 = 249.1$ kJ/kg (from Table A.15)

Point 4:

$h_4 = h_3 = 249.1\ kJ \cdot kg^{-1}$

The refrigerant mass flow-rate is calculated from Eq. (20.1):

$$q_e = m(h_1 - h_4)$$

$$m = \frac{q_e}{h_1 - h_4} = \frac{70}{377.3 - 249.1} = 0.546\ kg \cdot s^{-1}$$

The theoretical mechanical power w is calculated from Eq. (20.3):

$$w = m(h_2 - h_1)$$

$$w = 0.546\ (438.1 - 377.3) = 33.2\ kW$$

The coefficient of performance is:

$$COP = \frac{70}{33.2} = 2.11$$

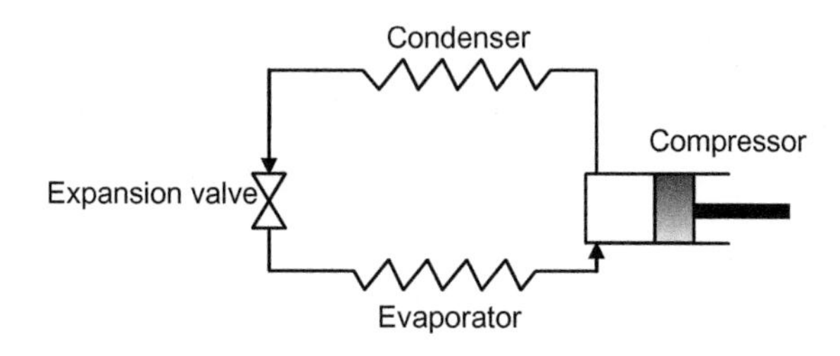

FIGURE 20.3

Vapor compression refrigeration system.

The following are the principal physical components of the vapor compression refrigeration machine (Figure 20.3):

- The *compressor*. The most common type is the air-cooled reciprocating piston compressor (see Chapter 2, Section 2.4.2). Depending on the capacity, the compressor may have one or a number of cylinders. Screw compressors are used for large capacities at constant load. Centrifugal (turbo) compressors are suitable for low compression ratios and are used mainly for air conditioning. The compressors of household refrigerators are usually hermetic, permanently lubricated units. For very low temperatures, two-stage compression may be necessary.
- The *condenser* may use air or water as the cooling medium. Air-cooled condensers are finned-tube radiators. Air flow is induced by fans. Water-cooled condensers are more compact, shell-and-tube type heat exchangers. The problems of fouling and scaling should be addressed by proper treatment of the cooling water in the closed circuit.
- The *expansion valve* or throttle often serves as the regulator of refrigerant flow-rate and is the main control element of the cycle. Different types (thermodynamic, floater-controlled, pressostatic, electronic) are available.
- The *evaporator* is the place where refrigeration is delivered to the system. Its geometry depends on the nature of the delivery. In a swept-surface freezer, it is the jacket of the heat exchanger. In a cold room, it is a finned-tube radiator-type heat exchanger with fans. In a milk cooler, it is a helical tube immersed in milk. In a plate freezer, it is the hollow plates of the unit. In order to utilize fully the refrigeration capacity of the machine, it is important to ensure that only gas (slightly superheated) leaves the evaporator.
- *Auxiliary hardware items* usually installed in the loop include buffer storage vessels for refrigerant, refrigerant pumps for large distribution systems, oil and water separators, filters, internal heat exchangers, valves, measurement and control instruments, sight glasses, etc.

As stated above, vapor compression systems are by far the most common type of refrigeration machines. Systems based on *absorption cycles* are used in specific

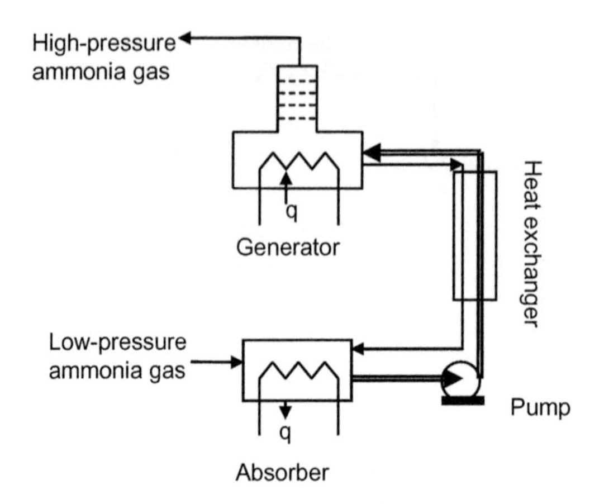

FIGURE 20.4

Absorption refrigeration aqua–ammonia system.

cases (experimental set-ups, refrigeration where electric power for running compressors is not available, etc.). Interest in absorption refrigeration may increase, however, in connection with the expansion of application of solar and geothermic energy. The principle of absorption refrigeration is shown schematically in Figure 20.4. Two fluids are involved: a refrigerant and an absorbent.

Typically, the refrigerant is ammonia and the absorbent is water. The thermodynamic process is still the reversed Rankine cycle, but the step of vapor compression is replaced by a sequence of operations consisting of the following:

- Absorption of the refrigerant at low pressure, coming from the evaporator, to produce a "strong" solution. This is an exothermic process and requires cooling (air or water). It takes place in the "absorber".
- Transfer (pumping) of the strong solution to the high-pressure region of the machine.
- Desorption of the refrigerant from the strong solution to deliver refrigerant vapor at high pressure. This is an endothermic process requiring a source of heat (fuel, solar or geothermic, etc.). It takes place in the *generator* or *boiler*.
- Return of the weak solution to the absorber.

The high-pressure refrigerant vapor emerging from the generator undergoes the same sequence of stages (condensation–expansion–evaporation) as in the vapor compression cycle.

The *Peltier effect* is the reverse of the Seebeck effect, discussed in connection with thermocouples (see Chapter 5, Section 5.7.1.1). A temperature difference is induced between the junctions of two different metals when a voltage is applied between the junctions. At present, Peltier refrigeration is used in devices requiring very small capacities, such as cooling and dehumidification of instrument housings, vending machines, etc.

20.1.2 Refrigerants

The refrigerants used in vapor compression cycles are volatile liquids. The desirable properties of a refrigerant are as follows:

- High latent heat of evaporation
- High vapor density
- Ability to be liquefied and evaporated at moderate positive pressure
- Non-toxicity, non-irritating, non-flammable, non-corrosive
- Miscibility with lubricating oils
- Environmental friendliness
- Inexpensive and easily available.

Ammonia was one of the first substances used as a commercial refrigerant. It satisfies most of the properties listed above, but it is toxic, irritant and may catch fire. It is used extensively in high-capacity installations.

Halogenated hydrocarbons are the preferred refrigerants for household refrigerators and air conditioning, and they are used extensively in industrial refrigeration as well. They consist of hydrocarbons substituted with chlorine and fluorine, and are numbered as Rxxx according to the following convention:

- The first digit from the right is the number of fluorine atoms in the molecule
- The second digit from the right is the number of hydrogen atoms plus one
- The third digit from the right is the number of carbon atoms minus one.

Thus, difluorochloromethane ($CHClF_2$) is R-22. The third digit, being zero, is not shown, and the number of chlorine atoms is found from the valence of carbon not occupied by hydrogen or fluorine. In the same manner, R-12 is difluorodichloromethane, ($CCl_2 F_2$), while tetrafluoroethane ($C_2 H_2 F_4$) is R-134.

We distinguish between CFCs (chlorofluorocarbons, such as R-12), HCFCs (hydrochlorofluorocarbons, such as R-22) and HFCs (hydrofluorocarbons, such as R-134).

In 1974 Rowland and Molina (1995 Nobel Prize in Chemistry) discovered that vapors of refrigerants may combine irreversibly with ozone and thus deplete the protective ozone layer of the stratosphere (Molina and Rowland, 1974). In 1987, governments of most countries of the world, under the auspices of the UN, signed the 1987 Montreal Protocol stipulating the gradual banning of ozone-depleting chemicals, including ozone-depleting refrigerants. The use of the most harmful group, CFCs, has been practically discontinued in developed countries. The less harmful HCFCs will gradually be phased out and, starting in 2020, manufacturers will no longer be allowed to produce HCFCs. These measures have prompted the chemical industry to develop new, more environmentally friendly refrigerants, such as HFCs.

It should be noted that vapors of refrigerants are also among the *greenhouse gases* responsible for global warming, but their ozone-depleting effect is much more harmful.

20.1.3 Distribution and delivery of refrigeration

The refrigeration capacity generated by the refrigeration machine may be delivered to the points of demand either directly or indirectly.

In direct delivery, called "direct expansion", the liquefied (condensed) refrigerant is conveyed directly to the point of demand, where it evaporates. The heat exchanged is the latent heat of evaporation. In indirect distribution, the evaporator is used to cool an intermediate liquid medium, which is pumped in closed circuit to the points of demand. The heat exchanged is sensible heat.

Direct expansion systems are less expensive, simpler, and more energy efficient. However they require long distribution lines carrying the volatile refrigerant, with the risk of leakage and contamination. The direct linkage between the refrigeration machine and the demand is problematic, as it impairs the flexibility of the system. Indirect distribution systems require higher capital investment and have higher operating costs; however, they are more suitable for multi-point distribution and fluctuating demand as they allow for the "storage" of refrigeration in the distribution medium. The desirable properties of the intermediate medium are:

- High specific heat
- Low freezing point
- Moderate and stable viscosity
- Low corrosivity
- Absence of toxicity
- Low cost.

The most commonly used intermediate media are solutions (brines) of calcium chloride and solutions of glycols.

20.2 Cold storage and refrigerated transport

Refrigeration installations may be divided into two groups:

1. Devices the main purpose of which is to maintain the low temperature of chilled or frozen goods. Examples include cold storage facilities, refrigerated vehicles.
2. Devices the main purpose of which is to lower the temperature of food rapidly. Examples include chillers, freezers.

Cold storage is distinguished from ambient storage by the presence of two features: thermal insulation and a source of cold. In a cold room of moderate size operating at 4°C, these two features may represent two-thirds of the erection cost.

Thermal insulation is provided by an appropriate thickness of porous material with a cellular (polymer foam) or fibrous (mineral wool) structure. The thermal conductivity of commercial insulation materials is about $0.05\ W\cdot m^{-1}\cdot K^{-1}$. The low thermal conductivity is due to the air entrapped in the porous structure (the thermal conductivity of stagnant air is $0.024\ W\cdot m^{-1}\cdot K^{-1}$). Moisture drastically reduces the efficiency of thermal insulation. It is therefore necessary to provide a moisture barrier, in the form of a plastic film or aluminum foil applied on the outer (warmer) face of the insulation.

Adequate thickness of the thermal insulation is calculated according to the design temperature of the cold room, the room geometry, the thermal conductivity of the insulation material, and the environmental conditions (temperature, winds, exposure to sun, etc.). Calculation of the economically optimal thickness weighs the cost of insulation against the cost of refrigeration. These factors vary with time and location. In the absence of more precise data, and as a rule of thumb, insulation thickness is designed for a refrigeration loss of $9\ W\cdot m^{-2}$.

EXAMPLE 20.2

A company builds cold storage facilities in different parts of the world. The specified storage temperature is 4°C. It is desired to develop a formula for the economically optimal thickness of the insulation, as a function of the following local variables:

Maximum outside temperature = T
Total cost of refrigeration = R currency units per kJ
Cost of insulation, including installation, I currency units per m^3.

The thermal conductivity of the insulation is $0.04\ W\cdot m^{-1}\cdot K^{-1}$. The useful life of the insulation is 7 years. The cold storage rooms are operative 24 hours a day, every day.

Solution

Let z be the insulation thickness, A its surface area, and k its thermal conductivity.

Let C_R and C_I be the hourly cost of refrigeration and insulation, respectively.

$$C_R = \frac{kA\Delta T}{z} \times \frac{3600R}{1000} = \frac{0.04 \times A \times (T-4) \times 3.6 \times R}{z} = \frac{0.144 \times R \times (T-4)}{z}$$

$$C_I = \frac{AzI}{7 \times 365 \times 24} = \frac{AzI}{61320}$$

The total hourly cost C_T is:

$$C_T = \frac{0.144 \times A \times (T-4) \times R}{z} + \frac{A \times z \times I}{61320}$$

To find z for minimum cost, we take the derivative of the total cost and set it to 0:

$$\frac{dC_T}{dz} = -0.144 \times A \times (T-4) \times R \times \frac{1}{z^2} + \frac{A \times I}{61320} = 0$$

$$z_{optimum} = 93.97\sqrt{\frac{(T-4)R}{I}}$$

The source of cold (the heat sink) is usually the evaporator of a vapor compression refrigeration machine. This is a heat exchanger, generally known as a *diffuser*. Typically, it consists of a finned coil with one or more fans. The temperature of the refrigerant in the diffuser is a few degrees centigrade lower than the temperature in the room. This temperature difference determines the required size of the diffuser (i.e., the size of the heat exchange area). A large temperature difference requires a smaller diffuser but results in higher cost of refrigeration. Furthermore, a colder heat exchange surface causes greater drying of the air. If the room contains unpackaged products, this may result in loss of product weight during storage, and excessive frost on the heat exchange surface of the diffuser. Thus, the selection of an evaporation temperature for the diffuser is a matter of cost optimization.

The refrigeration requirement (refrigeration load) of a cold storage room comprises the following elements:

- Heat transfer through the insulation
- Air changes, both intentional and unintentional
- Introduction of goods at temperatures higher than that of the room
- Heat generated by respiration (fruits and vegetables)
- Defrosting cycles
- Energy spent by fans, fork lifts, conveyors, lighting, etc.
- People working in the room.

EXAMPLE 20.3

A cold room is used to store frozen meat at −30°C. Calculate the refrigeration load of the facility.

Data:

Internal volume of the room = 1000 m^3
Surface area exposed to the exterior = 700 m^2
Loading program: 25 000 kg meat at −20°C per day is introduced to the room; an equal quantity of meat at −30°C is removed
The specific heat of meat at that range of temperatures is 1.8 $kJ \cdot kg^{-1} \cdot K^{-1}$
The exposed surface area is covered with polystyrene foam insulation, 25 cm thick; the thermal conductivity of the insulation is 0.04 $W \cdot m^{-1} \cdot K^{-1}$
Ambient temperature = 25°C.
Air changes: 2000 m^3 per day (assume dry air)
Allow 1.2 kW for lights, fans, defrost, people, conveyors etc.

Solution

Loss through insulation:

$$q_1 = \frac{kA\Delta T}{z} = \frac{0.04 \times 700 \times (25-(-30))}{0.25} = 6160 \text{ W}$$

Product load:

$$q_2 = \frac{25000 \times (-20-(-30)) \times 1800}{24 \times 3600} = 5208 \text{ W}$$

Air changes: the density of air at 25°C is 1.14 $kg \cdot m^{-3}$. Its specific heat is 1 $kJ \cdot kg^{-1} \cdot K^{-1}$.

$$q_3 = \frac{2000 \times 1.14 \times 1000 \times (25 - (-30))}{24 \times 3600} = 1450 \text{ W}$$

The total refrigeration load is:

$$Q_T = 6.16 + 5.21 + 1.45 + 1.2 = 14.02 \text{ kW}$$

The openings in cold storage rooms are often equipped with an *air curtain* to reduce air exchange when the door is open. A fan installed over the door blows a vertical jet of air across the opening as long as the door is open. Reducing unintentional air changes is important, not only as a way to reduce the refrigeration load but also because penetration of moist warm air from outside causes fogging in the room.

Refrigerated transportation is the most critical and often the weakest link of the cold chain.

For short-range transportation and distribution, vehicles with thermal insulation but without autonomous generation of refrigeration may be sufficient. In this case, cold air is blown into the vehicle chamber before loading with thoroughly refrigerated merchandise. Ice or dry ice (solid carbon dioxide) may be used, with obvious limitations and safety issues. The use of ice is particularly frequent in fishing boats. For long-range transportation, the vehicle must be equipped with its own source of cold. Mechanical refrigeration is the usual source, but the use of cryogenics (liquid nitrogen) is being increasingly applied.

20.3 Chillers and freezers

The engineering aspects of food freezing have been reviewed by Holdsworth (1987); see also Tressler *et al.* (1968).

Methods for chilling or freezing foods may be classified according to the heat transfer mechanism as follows:

- The use of a stream of cold air (blast chillers or freezers)
- Contact with cold surfaces (contact freezer)
- Immersion in cold liquid (immersion coolers or freezers)
- Evaporative cooling (cryogenic cooling or freezing)
- Pressure-shift freezing.

20.3.1 Blast cooling

The simplest embodiment of blast cooling is the batch cabinet chiller or freezer. A batch of food is placed in a cabinet, usually on trays or trolleys, or hung on hooks (meat carcasses). Cold air is blown over the food and circulated through a diffuser by means of fans (Figure 20.5).

A more advanced version is the continuous tunnel cooler or freezer (Figure 20.6), where the food is conveyed by trolleys or conveyor belts. One variation of the continuous belt freezer is the spiral freezer (Figure 20.7), in which the food is conveyed on a spiral made of a continuous flexible mesh belt. Spiral belt freezers occupy little floor space at high processing capacity.

In fluidized bed cooling, the cold air is used both to cool and to fluidize a bed of particulate foods. When used for freezing, this mechanism has the advantage of freezing the particles individually. The IQF (individual quick freezing) freezer is the most common type of this class (Figure 20.8).

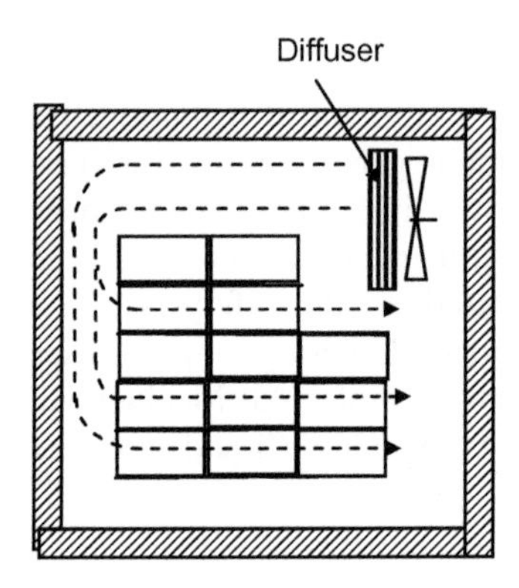

FIGURE 20.5

Cabinet freezer.

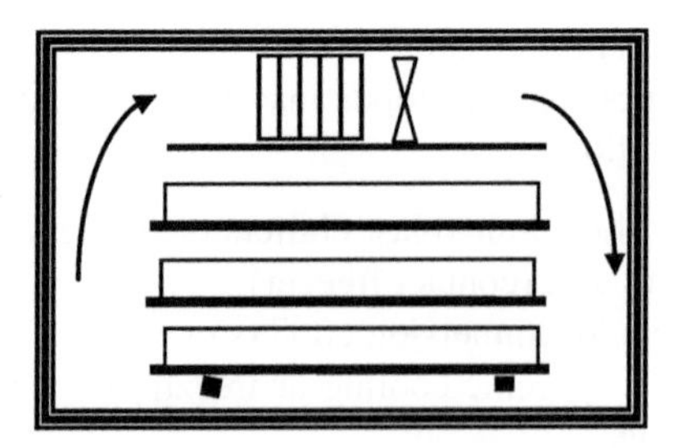

FIGURE 20.6

Tunnel freezer.

One variation of the IQF freezer features two sections. The first is a fluidization section where the air blast is sufficiently strong to lift the product. The purpose of this section is to rapidly freeze the surface of individual particles so as to form a crust and prevent adhesion. The rapid formation of a frozen crust also reduces moisture loss. The interior of the particle is still unfrozen. Freezing is completed on a belt at the second section, without fluidization (Figure 20.9).

The power required to circulate the blast of air constitutes a significant part of the energy expenditure of blast coolers in general. Moisture loss, which represents a loss in both product weight and quality, is also a common problem in blast cooling. The methods used to reduce moisture loss include:

- Proper packaging before chilling or freezing
- IQF
- Glazing: a small quantity of water is sprayed on the product so as to form a thin film of water on the surface. Upon freezing, this becomes a film of ice, impermeable to water vapor. Glazing is mainly applied to unpackaged fish and seafood.

FIGURE 20.7

Spiral belt freezer.

Photograph courtesy of FMC FoodTech.

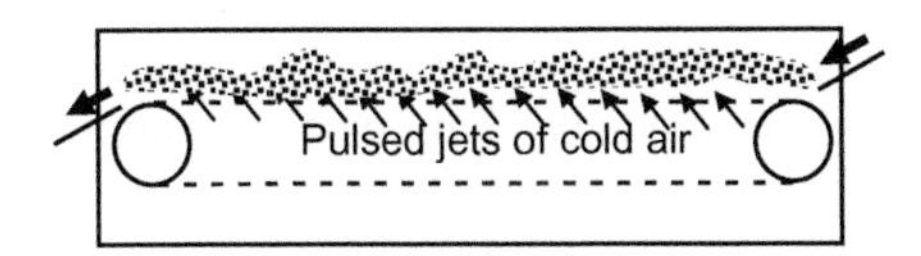

FIGURE 20.8

Principle of fluidized bed (IQF) freezing.

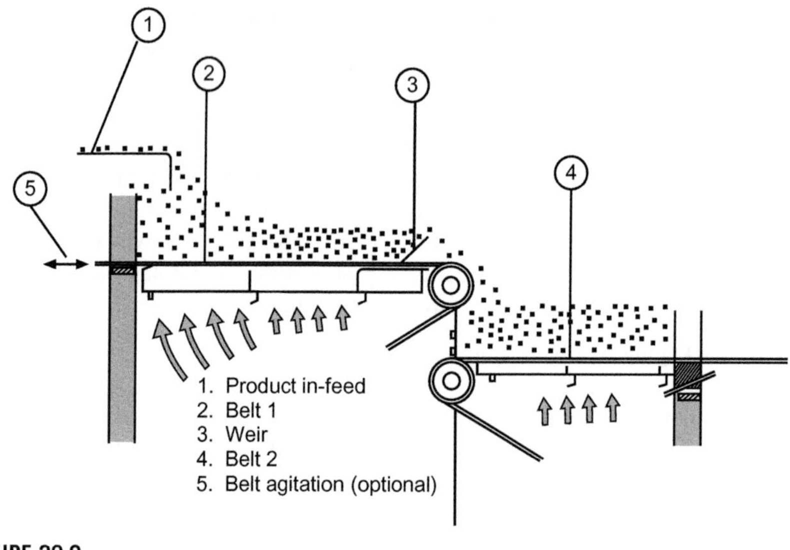

FIGURE 20.9

Two-stage fluidized bed (IQF) freezing.

Figure courtesy of FMC FoodTech.

20.3.2 Contact freezers

The most important types of contact freezers are the plate freezer, the swept-surface freezer and the extruder freezer.

The *plate freezer* (Figure 20.10) consists of hollow shelves (plates) cooled by refrigerant evaporating inside them. The food is pressed between the plates, ensuring good contact between the cold surface of the plate and the food. The freezing time depends on the thickness of the product between the shelves. Typically, it takes 2 hours to freeze a 6 cm thick block of fish to a core temperature of −18°C, with the plate temperature at −35°C. Plate freezers are energy efficient and compact. They are, however, appropriate for use only with product shapes delimited by two parallel planes, such food packed in rectangular boxes, or blocks of filleted fish. They are usually loaded and unloaded manually, and operate in batches. Systems with automated loading–unloading are available.

The scraped-surface (or swept-surface) freezer (Figure 20.11) is a continuous freezer, operating on the principle of the swept-surface heat exchanger (see Chapter 3, Section 3.7.4) with refrigerant in the jacket. It can be used only for partial freezing of pumpable products. This type of equipment is used for soft-freezing ice cream and for slush-freezing concentrated fruit juices.

In *extrusion freezing*, a liquid feed is cooled and frozen by contact with the wall of a cooled barrel while moderate shear is applied by slowly rotating screws. Extrusion freezing is mainly applied in the manufacture of ice cream, the main

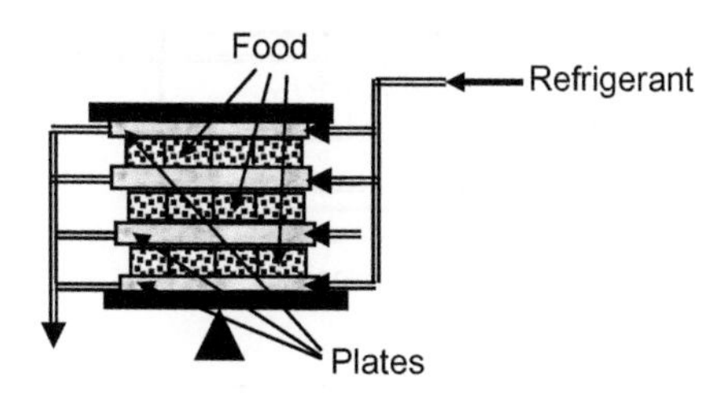

FIGURE 20.10

Schema of a plate freezer.

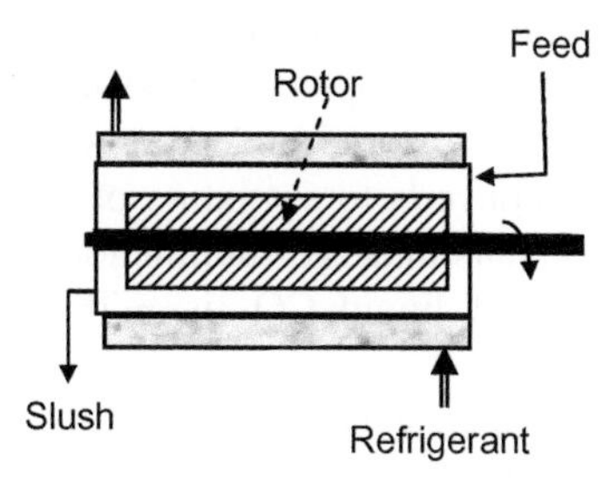

FIGURE 20.11

Swept-surface slush freezer.

advantage of the process being the formation of smaller ice crystals and air bubbles, evenly distributed in the mass (Bolliger *et al.*, 2000).

20.3.3 Immersion cooling

Immersion in cold water is a method used for the rapid cooling of fruits and vegetables to remove "field heat" after harvest. This operation is often called "hydro-cooling". In the poultry industry, dressed birds are cooled in a continuous bath of chilled water. Shrimp is often cooled by immersion in cold brine.

20.3.4 Evaporative cooling

Under this heading we include cooling methods based on direct contact with an evaporating substance.

In the case of "cryogenic cooling", the evaporating substance is liquid nitrogen or solid carbon dioxide (dry ice). At atmospheric pressure, liquid nitrogen boils at $-196°C$. Its latent heat of evaporation is $199\ kJ \cdot kg^{-1}$. For freezing, liquid nitrogen stored under pressure is sprayed over the food traveling on a conveyor belt inside a tunnel (Figure 20.12). Due to the very high temperature

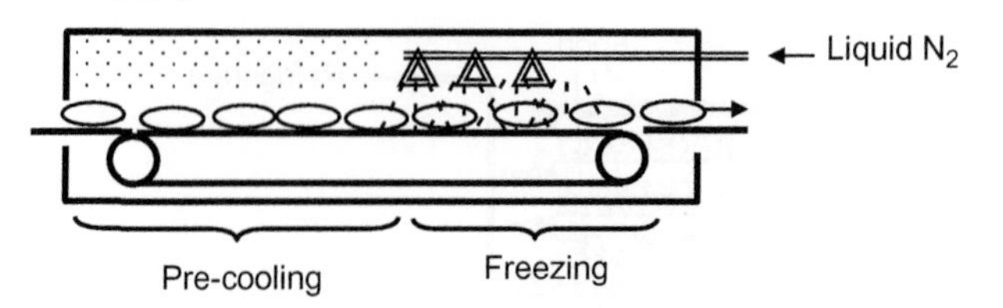

FIGURE 20.12

Liquid nitrogen freezer.

drop, freezing is extremely rapid. The cold vapors of nitrogen serve to pre-cool the product entering the tunnel, before it meets the liquid nitrogen spray. Practically, about 2.5 kg of liquid nitrogen are needed to freeze 1 kg of food.

Liquid carbon dioxide is also stored under pressure, but when it is released through the spray nozzles it converts to a mixture of gaseous CO_2 and a snow of solid CO_2 crystals. These crystals undergo sublimation on the surface of the food. The evaporation temperature of liquid CO_2 is $-78.5°C$ and its latent heat of evaporation is $571\ kJ \cdot kg^{-1}$.

Cryogenic freezing and cooling can be also applied to liquid materials by spraying the liquid into the refrigerant (Yu *et al.*, 2004). Cryogenic spray freezing is particularly useful for preparing small frozen droplets prior to freeze-drying (Wang *et al.*, 2006).

The main advantages of cryogenic freezing are:

- Compact, inexpensive equipment; low capital investment
- Extremely rapid freezing, very low moisture loss
- Versatile, quick to install anywhere.

The disadvantages are:

- High cost of the cryogens
- Dependence on a reliable supply of cryogens
- Not suitable for large items, as the very large temperature gradients may create large mechanical stress, cracks and breakage.

In total, cryogenic freezing is more expensive than mechanical refrigeration. Nevertheless, cryogenic freezing is valuable in the following situations:

- The freezing line is used occasionally
- Economy of floor space is critical (e.g., the need to accommodate the line in an existing, crowded plant layout, location with expensive land cost, etc.)
- Experimentation, product development at pilot scale.

A particular type of evaporative cooling uses water as a refrigerant. Water is sprayed over the food and allowed to evaporate under reduced pressure. "Vacuum" cooling is used for rapid cooling of vegetables (particularly lettuce) for the fresh market.

20.3.5 Pressure-shift freezing

The phase diagram of water indicates that increasing the pressure results in depressing the freezing temperature. If water is subjected to very high pressure and cooled, it may remain liquid at a temperature much lower than its freezing temperature at atmospheric pressure. If now the pressure is suddenly released, rapid freezing will occur. This method of freezing is called *pressure-shift freezing* (Kalichevsky *et al.*, 1995; Kalichevsky-Dong *et al.*, 2000). The rapid freezing is said to result in the formation of smaller and more uniform ice crystals (Kalichevsky-Dong *et al.*, 2000; Zhu *et al.*, 2004) and can be expected to cause potentially less damage to the texture. However, Buggenhout *et al.* (2005) reported that any slight advantage of pressure-shift freezing of carrots with respect to texture is lost during frozen storage.

References

ASHRAE, 2006. ASHRAE Refrigeration Handbook. ASHRAE, New York, NY.

Bolliger, S., Kornbrust, B., Goff, H.D., Tharp, B.W., Windhab, E.J., 2000. Influence of emulsifiers on ice cream produced by conventional freezing and low temperature extrusion processing. Intl. Dairy J. 10, 497–504.

Buggenhout, S.V., Messagie, I., Van Loey, A., Hendrickx, M., 2005. Influence of low-temperature blanching combined with high-pressure shift freezing on the texture of frozen carrots. J. Food Sci. 70 (4), S304–S308.

Dosset, R.J., Horan, T.J., 2001. Principles of Refrigeration, fifth ed. Prentice Hall, Englewood Cliffs, New Jersey.

Holdsworth, D.S., 1987. Physical and engineering aspects of food freezing. In: Thorne, S. (Ed.), Developments in Food Preservation – 4. Elsevier Applied Science, London, UK.

Kalichevsky, M.T., Knorr, D., Lillford, P.J., 1995. Potential food application of high pressure effects on ice–water transition. Trends Food Sci. Technol. 6 (8), 253–259.

Kalichevsky-Dong, M.T., Ablett, S., Lillford, P.J., Knorr, D., 2000. Effects of pressure-shift freezing and conventional freezing on model food gels. Int. J. Food Sci. Technol. 35 (2), 163–172.

Molina, M.J., Rowland, F.S., 1974. Stratospheric sink for chlorofluoromethanes: chlorine atom-catalysed destruction of ozone. Nature 249, 810–812.

Tressler, D.K., Van Arsdel, W.B., Copley, M.J., Woolrich, W.R., 1968. The Freezing Preservation of Foods. The Avi Publishing Co., Westport, CT.

Wang, Z.I., Finlay, W.H., Peppler, M.S., Sweeney, L.G., 2006. Powder formation by atmospheric spray-freeze-drying. Powder Technol. 170 (1), 45–62.

Yu, Z., Garcia, A.S., Johnston, K.P., Williams 3rd, R.O., 2004. Spray freezing into liquid nitrogen for highly stable protein nanostructured microparticles. Eur. J. Pharm. Biopharm. 58 (3), 529–537.

Zhu, S., le Bail, A., Ramaswamy, H.S., Chapleau, N., 2004. Characterization of ice crystals in pork muscle formed by pressure-shift freezing as compared with classical freezing methods. J. Food Sci. 69 (4), FEP190–FEP197.

CHAPTER

Evaporation

21

21.1 Introduction

Literally, the unit operation known as evaporation may be defined as the vaporization of a volatile solvent *by ebullition* to increase the concentration of *essentially non-volatile components* of a solution or suspension. Vaporization by boiling is emphasized in order to differentiate between *evaporation* and *dehydration*. The non-volatile nature of the components other than the solvent leads to the difference between *evaporation* and *distillation*. In the case of foods, the volatile solvent is almost always water.

The main objectives of evaporation in the food industry are the following:

- Mass and volume reduction, resulting in reduced cost of packaging, transportation and storage.
- Preservation, by virtue of the reduced water activity. It should be noted, however, that the rate of chemical deterioration increases with concentration. Thus, concentrated juices, although more resistant than their single-strength form to microbial spoilage, tend to undergo browning more readily. (Refer to the effect of water activity on the rate of various types of food deterioration, discussed in Chapter 1, Section 1.7.5.)
- Preparation for subsequent treatment such as crystallization (sugar, citric acid), precipitation (pectin, other gums), coagulation (cheese, yogurt), forming (candy), dehydration (milk, whey, coffee solubles), etc.
- Building a desired consistency (jams and jellies, tomato concentrates, ketchup).

Evaporation is one of the most "large-scale" operations in the food industry. Plants with evaporation capacities of up to hundreds of tons of water per hour are not uncommon. Made of special types of stainless steel, constructed according to high sanitary standards and equipped with sophisticated automatic control systems, evaporators for the food industry are relatively expensive. The most important *running cost* item in evaporation is energy. However, energy consumption per unit production can be reduced considerably through the use of multi-effect evaporation or vapor recompression, or both. Since such measures also result in increased capital investment while reducing energy cost, their application is subject to optimization *with respect to total cost.*

Food Process Engineering and Technology. DOI: http://dx.doi.org/10.1016/B978-0-12-415923-5.00021-6
© 2013 Elsevier Inc. All rights reserved.

In the food industry, the issue of thermal damage to product quality deserves particular attention. Evaporation under reduced pressure and correspondingly lower boiling temperature is widely practiced. However, lower temperature means higher viscosity of the boiling liquid, slower heat transfer, and therefore longer retention of the product in the evaporator. Thus, the selection of evaporation type and of operating conditions is also subject to optimization *with respect to product quality*.

One of the problems in the evaporative concentration of fruit juices is the loss of volatile aroma components. Evaporators may be equipped with essence recovery systems for entrapping and concentrating the aroma, which is then added back to the concentrate (Mannheim and Passy, 1974).

Evaporation is not the only way to concentrate liquid foods (Thijssen and Van Oyen, 1977). Alternative concentration processes are reverse osmosis, osmotic water transfer and freeze concentration, discussed elsewhere in this book.

21.2 Material and energy balance

Consider the continuous evaporation process described in Figure 21.1.

Overall material balance gives:

$$F = V + C \tag{21.1}$$

where:

F = mass flow-rate of feed, $kg \cdot s^{-1}$
V = mass flow-rate of vapor, $kg \cdot s^{-1}$
C = mass flow-rate of concentrate (product), $kg \cdot s^{-1}$.

Material balance for the solids is written as follows:

$$F \cdot x_F = C \cdot x_C \tag{21.2}$$

where x_F and x_C = mass fraction of solids in feed and concentrate, respectively

Combining Eqs (21.1) and (21.2) and defining "concentration ratio" as $R = x_C/x_F$:

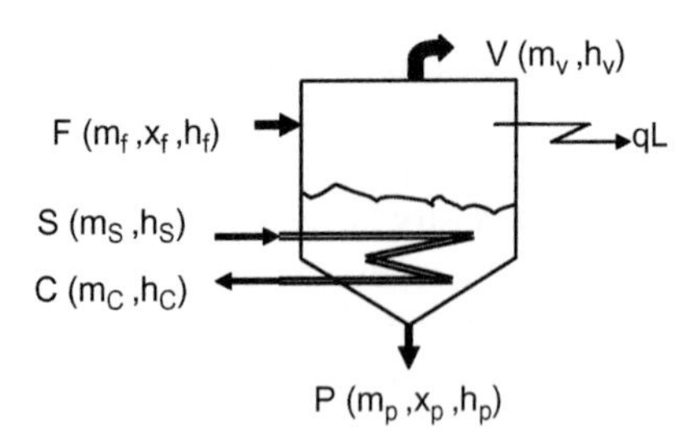

FIGURE 21.1

Material and energy balance in an evaporator unit.

$$V = F\left(1 - \frac{1}{R}\right) \quad (21.3)$$

Note: Equation (21.2) assumes that the vapor does not contain solids. In practice the vapor does contain small quantities of solids, carried over by droplets of the boiling liquid.

Heat balance gives:

$$F \cdot h_F + S(h_s - h_{SC}) = C \cdot h_C + V \cdot h_V + q_l \quad (21.4)$$

where h_F, h_S, h_{SC}, h_C, h_V = specific enthalpies ($J \cdot kg^{-1}$) of feed, steam, steam condensate, concentrate and vapor, respectively; and q_l = rate of heat loss, w.

EXAMPLE 21.1

A single-effect continuous evaporator is used to concentrate a fruit juice from 15 Bx to 40 Bx. The juice is fed at 25°C, at the rate of 5400 $kg \cdot h^{-1}$ (1.5 $kg \cdot s^{-1}$). The evaporator is operated at reduced pressure, corresponding to a boiling temperature of 65°C. Heating is by saturated steam at 128°C, totally condensing inside a heating coil. The condensate exits at 128°C. Heat losses are estimated to amount to 2% of the energy supplied by the steam. Calculate:

a. The concentration ratio R
b. The required evaporation capacity V ($kg \cdot s^{-1}$)
c. The required steam consumption S ($kg \cdot s^{-1}$).

Solution

a.

$$R = \frac{0.40}{0.15} = 2.667$$

b.

$$V = 5400 \times \left(1 - \frac{1}{2.667}\right) = 3375\ kg \cdot h^{-1} = 0.938\ kg \cdot s^{-1}$$

c. Boiling point elevation for 40 Bx will be assumed to be negligible (see Example 21.2). The vapor will be assumed to be saturated vapor at 65°C. Its enthalpy is read from steam tables as 2613 $kJ \cdot kg^{-1}$. The enthalpies of the steam and its saturated condensate are 2720.5 and 546.3 $kJ \cdot kg^{-1}$, respectively. The enthalpies of the feed and the product are calculated using the approximate formula for sugar solutions:

$$h = 4.187(1 - 0.7x)T \quad (21.5)$$

$$h_F = 4.187 \cdot 25 \cdot (1 - 0.7 \times 0.15) = 93.7\ kJ \cdot kg^{-1}$$

$$h_C = 4.187 \cdot 65 \cdot (1 - 0.7 \times 0.40) = 195.9\ kJ \cdot kg^{-1}$$

Substituting the data in Eq. (21.4), we find:

$$S = 1.1135\ kg \cdot s^{-1}$$

V/S = 0.83 kg water evaporated per kg of live steam consumed.

Notes on Example 21.1:

- Bx (degrees Brix) is an expression (already used in previous chapters) indicating concentration of total soluble solids in solutions, in which the solutes are sugar-like compounds (e.g., fruit juices, syrups, concentrates). One degree Brix is equivalent to 1 kg sugar-like soluble solids per 100 kg of solution. It is most commonly measured by refractometry.
- The quantity of water evaporated per unit time (evaporation capacity), V, is the nominal *size indicator* of a commercial evaporator.
- The ratio of evaporation capacity to steam consumption is a measure of the thermal efficiency of the operation. In single-effect evaporators receiving a cold feed, the thermal efficiency is, obviously, below 100% (83% in this example).
- The temperature of the feed may have a substantial influence on the energy balance. Let us assign, for convenience, the temperature of evaporation (i.e., T_C or T_v) as the reference temperature for enthalpies. The heat balance, Eq. (21.4) now becomes:

$$S(h_s - h_{SC}) = FC_{pF}(T_C - T_F) + V\lambda_v + q_l = [I] + [II] + [III] \tag{21.6}$$

where C_{pF} is the specific heat of the feed and λ_V is the latent heat of evaporation.

Equation (21.6) shows that the heat supplied by the steam is used to compensate three thermal loads, indicated in the equation as [I], [II], [III]:

[I] is the heat used to raise the temperature of the feed, up to the temperature of evaporation
[II] is the heat used to produce the vapor
[III] is the heat used to compensate heat losses

In the example, these three loads represent 10%, 88% and 2% of the total energy consumed, respectively.

21.3 Heat transfer

An evaporator is essentially a heat exchanger equipped with appropriate devices for the separation of the vapors from the boiling liquid. The evaporation capacity of the system is determined, first and foremost, by the rate of energy transfer from the heating medium to the boiling liquid. In the following discussion we shall analyze ways to maximize the rate of heat transfer in evaporators.

In an evaporator, heat is typically transferred from a film of condensing steam to a film of boiling liquid through a heat-conducting solid wall (Figure 21.2).

The rate of heat transfer per unit area (heat flux) is given by:

$$\frac{q}{A} = U(T_S - T_C) \tag{21.7}$$

where U is the overall coefficient of heat transfer ($W \cdot m^{-2} \cdot K^{-1}$).

Heat flux can be increased by increasing the coefficient U or the temperature difference $T_S - T_C$ or both.

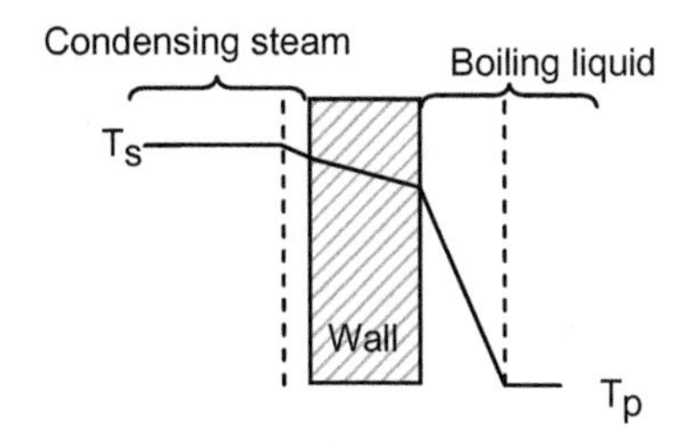

FIGURE 21.2

Path of heat transfer in an evaporator.

21.3.1 The overall coefficient of heat transfer U

The overall resistance to heat transfer is the sum of three resistances in series, namely that of the condensing steam film, that of the solid wall and that of the boiling liquid.

$$\frac{1}{U} = \frac{1}{\alpha_c} + \frac{\varepsilon}{k} + \frac{1}{\alpha_b} \tag{21.8}$$

where:

α_c and α_b = heat transfer coefficients of condensing steam and boiling liquid films, respectively. (Note: Elsewhere in the book, convective heat transfer coefficients are represented by the symbol h. In this chapter, the symbol h is used for specific enthalpy. To avoid confusion, the symbol α is used for heat transfer coefficients.)
ε = thickness of the solid wall, m.
k = thermal conductivity of the solid, $W \cdot m^{-1} \cdot K^{-1}$.

The three resistances are not of equal importance. Typical values of the individual coefficients are given in Table 21.1.

21.3.1.1 Heat transfer at the steam side

Semi-empirical methods for the prediction of heat transfer from condensing vapor films are available (McAdams, 1954; Perry and Green, 1997). The resistance of condensing films is usually low and its contribution to the total resistance is negligible. Care should be taken, however, to avoid superheat (apply de-superheating if necessary), to purge non-condensable gases (air) and to remove the condensates from the heat transfer surface as soon as they are formed.

21.3.1.2 Heat transfer through the wall and the problem of fouling

The thermal resistance of the solid wall is also relatively low. The wall (wall of tubes or heating jackets, etc.) is usually thin. The wall material is usually a metal with fairly high thermal conductivity, even in the case of stainless steel, which now replaces copper for obvious food safety reasons (typical k values are 17 $W \cdot m^{-1} \cdot K^{-1}$ for

Table 21.1 Typical Values of the Three Components of Eq. (21.8)

Coefficient	Characterization	$W \times m^{-2} \cdot K^{-1}$
α_c	Steam, condensing as film	10 000
	Steam, condensing as drops	50 000
k/ε	Stainless steel, 1 mm thick	15 000
α_b	Water, forced convection, no boiling	2000
	Water, forced convection, boiling	5000
	Tomato paste, boiling	300

stainless steel, $360\ W \cdot m^{-1} \cdot K^{-1}$ for copper). The thermal resistance of the "wall', however, may increase considerably and even become a serious rate-limiting factor as a result of *fouling* (Jorge *et al.*, 2010). Fouling is the deposition of certain solids on the heat transfer area, particularly at the product side. These solids may be coagulated proteins (milk), caramelized sugars (fruit juices, coffee solubles), scorched pulp (tomato paste), or solutes that reach their limit of solubility as the material is concentrated (calcium salts in milk, hesperidin in orange juice). Excessive browning is detrimental not only to the rate of heat transfer but also to product quality and to equipment lifetime. As fouling builds up on the heat transfer surface, the temperature of the wall increases and the sediments may "burn", imparting to the product a "burnt" taste and unsightly black specks. The rate of fouling often determines the frequency and duration of down periods of the evaporator (for cleaning), and can frequently be reduced by increasing the turbulence on the product side (flow, agitation, swept surface) or by pre-treatment of the feed. An example of a pre-treatment solution is found in a particular process for producing high-concentration tomato paste (Dale *et al.*, 1982). In this process, tomato juice is centrifuged to separate the pulp from the liquid (serum). The serum, which produces much less fouling, is concentrated by evaporation and then remixed with the pulp. A similar process has been proposed for the production of concentrated orange juice (Peleg and Mannheim, 1970).

21.3.1.3 Heat transfer at the product side

This is the most critical part of the overall heat transfer (Coulson and McNelly, 1956; Jebson *et al.*, 2003; Minton, 1986). It is also the most difficult to calculate or predict with confidence. Its analytical treatment is made complicated by the presence and action of a vapor phase mixed with the liquid, the discontinuity of the heat transfer surface, and the tremendous change in liquid properties as water is removed. The theoretical study of two-phase heat transfer, an extremely active area of research in recent decades, has not generated reliable and convenient methods for analytical calculation of the heat transfer coefficient for the boiling liquid side, for the geometries, physical conditions and complex material properties prevailing in most cases of evaporation in the food industry. Qualitatively, the heat transfer coefficient at the product side increases with the intensity of movement of the liquid in relation to the wall and decreases with liquid viscosity. Regarding viscosity, it should be noted that:

- It increases strongly with concentration.
- It is strongly affected by temperature.
- Many food liquids are non-Newtonian, shear-thinning fluids. Their "viscosity" is affected by flow and agitation. The beneficial effect of agitation on the rate of heat transfer is therefore more strongly accentuated in the case of shear-thinning foods such as fruit purees or ketchup.
- A partial solution to the problem of excessive viscosity during evaporation of pulpy juices consists in the separation of the pulp before concentration, evaporation of the pulpless juice, and remixing of the resulting concentrate with the pulp. The process is applied industrially in the production of tomato juice concentrate and has been tried on orange juice (Peleg and Mannheim, 1970).

The production of vapor has contradictory effects on heat transfer. On one hand, the bubbles formed as a result of boiling agitate the liquid and thus improve heat transfer. On the other hand, vapors have poor heat conductivity and therefore tend to impair heat transfer when part of the heat transfer area becomes covered with vapor. The relative importance of these two contradictory effects depends on the rate of vapor production, and hence on the temperature difference ΔT between the surface and the boiling liquid. In classical experiments with pool boiling of water, it has been observed that at low ΔT, vapor is formed as individual bubbles on the heating surface. When the bubbles become sufficiently large, buoyancy overcomes surface tension and the bubbles detach from the surface to be replaced by new ones. The coefficient of heat transfer increases, due to agitation by the rising bubbles. This is "*nucleate boiling*". Now, if ΔT is increased, vapor is generated at a higher rate. A point is reached where vapor is formed faster than it is removed and the surface becomes covered with an insulating blanket of vapor. The coefficient of heat transfer decreases. This is "*film boiling*". A value of ΔT exists for which the coefficient is at a maximum (Figure 21.3).

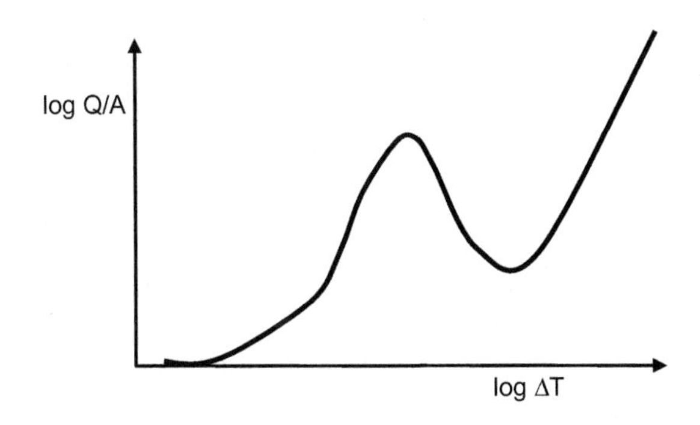

FIGURE 21.3

Effect of ΔT on heat flux.

Although these two stages with a clear transition between them are not observed in actual evaporation with food materials, dependence of coefficient of heat transfer on ΔT has been often reported to follow the same trend. Considering the two-phase nature of heat transfer in evaporation, it might be expected that factors such as surface tension, vapor specific gravity and vapor mass ratio should have an effect on the heat transfer coefficient. In fact, a number of semi-empirical correlations for α_b featuring dimensionless groups containing these parameters have been proposed (Malkki and Velstra, 1967; McNelly, 1953; Holland, 1975; Jebson *et al.*, 2003; Stephan and Abdelsalam, 1980). Many of these correlations have been developed for "pool boiling" (i.e., for stagnant boiling liquids with no flow or agitation). These correlations may well be useful in the analysis and design of steam boilers and evaporators in refrigeration systems, but they are not easily applicable to the rational design of evaporators for food liquids.

21.3.2 The temperature difference $T_S - T_C$ (ΔT)

The temperature difference is, of course, the driving force for heat transfer. It can be increased either by elevating the temperature of the steam or by lowering the temperature of evaporation. Each one of these two actions is applied in practice, but each has its practical limitations.

- *Increasing the steam temperature.* Higher steam temperature means higher wall temperature and increased risk of thermal damage to the product in direct contact with the wall, particularly if heat transfer at the product side is not sufficiently rapid (lack of turbulence, highly viscous products). Furthermore, this action implies higher steam pressure, therefore a mechanically stronger, more expensive structure.
- *Lowering the boiling temperature.* The boiling temperature of a solution is a function of concentration and pressure. The boiling temperature of an aqueous solution is higher than that of pure water at the same pressure. The difference is known as the "boiling point elevation", BPE. The boiling point elevation of a solution is a consequence of the water vapor pressure depression caused by the presence of solutes. BPE increases with the molar concentration of the solutes. In ideal solutions, the water vapor depression and hence the boiling point elevation can easily be calculated (see Example 21.2). Due to the relatively high molecular weight of the solutes, the boiling point elevation of food liquids is usually not large. In practice, the boiling temperature in an evaporator is lowered by reducing the pressure at the product side. Thus, evaporation under vacuum has the dual purpose of increasing the rate of heat transfer and avoiding excessive heat damage to the product. In an evaporator, pressure reduction is achieved by condensing the vapors in the condenser and expelling the non-condensable gases by means of a vacuum pump or ejector. The condenser is usually water-cooled. The temperature of condensation, and

hence the pressure in the evaporator, depends on the temperature and flow-rate of cooling water. Thus, the unavailability of sufficiently cold water in adequate quantity may limit the vacuum that can be achieved in the evaporator. Theoretically, mechanical refrigeration could be used to cool the condenser, but the cost would be prohibitive in most applications with food products. A second factor to consider is the effect of pressure on the specific volume of the vapors. Lowering evaporation pressure implies increasing considerably the volume of vapors to be handled, requiring oversize separators and ducts. A third factor, and probably the most important of all, is the tremendous increase in liquid viscosity as the temperature is lowered. This works against the reason for which low boiling temperature was sought; it considerably impairs the rate of heat transfer at the product side. In modern evaporators, moderate vacuum is preferred, resulting in somewhat higher evaporation temperature but allowing shorter residence time.

EXAMPLE 21.2

Calculate the boiling temperature and BPE of 45-Bx orange juice concentrate at the pressure of 20 kPa. Assimilate the concentrate to a 45% (w/w) solution of glucose (MW = 180) and assume ideal solution behavior.

Solution

The molar concentration (mol fraction) of the solution is:

$$x = \frac{45/180}{45/180 + (100-45)/18} = 0.076$$

The vapor pressure of pure water (p, in kPa) at the boiling temperature must be such that:

$$p(1-0.076) = 20 \text{ kPa} \quad \text{hence:} \quad p = 21.6 \text{ kPa}$$

From steam tables, we find that the temperature at which the vapor pressure of water is 21.6 kPa is T = 61.74°C. From the same tables, the boiling temperature of pure water at 20 kPa is 60.06°C.

The boiling temperature of the concentrate at 20 kPa = 61.74°C.

BPE = 61.74 − 60.06 = 1.68°C

EXAMPLE 21.3

A continuous evaporator is used to concentrate a liquid with strong tendency to cause fouling. The heat exchange area is 100 m^2. Boiling temperature is maintained at 40°C. The evaporator is heated by saturated steam condensing at 120°C. It has been determined that for every 1000 kg of water evaporated, a 1-μm thick uniform layer of sediment is formed on the heat exchange area. The thermal conductivity of the sediment is 0.2 $W \cdot m^{-1} \cdot K^{-1}$. The evaporation capacity of the evaporator, when clean, is 4 $kg \cdot s^{-1}$. The operation is stopped for cleaning when the evaporation capacity drops below 3 $kg \cdot s^{-1}$.

a. How much water is evaporated between cleaning periods? What is the thickness of the sediment layer when the operation is stopped?
b. What is the duration of the operation cycle between clean-ups?

Solution

a. Let:

m = mass of water evaporated
z = thickness of the sediment layer
U_0, U_t, U_f = overall coefficient of heat transfer at t = 0, at t = t and at the end of the cycle
λ = latent heat of evaporation of water at 40°C = 2407 $kJ \cdot kg^{-1}$.

$$dm/dt = \frac{UA\Delta T}{\lambda}$$

$$U_0 = \frac{4 \times 2407000}{100(120-40)} = 1203\ W \cdot m^{-2} \cdot K^{-1} \quad U_f = \frac{3 \times 2407000}{100(120-40)} = 903\ W \cdot m^{-2} \cdot K^{-1}$$

But

$$1/U_t = 1/U_0 + z_t/k \rightarrow z_t = 55.4 \times 10^{-6}\ m$$

$$m_f = 55\,400\ kg.$$

The final thickness of the sediment is, therefore, 55.4 μm.

b. Let us expand dm/dt as follows:

$$dm/dt = (dm/dz)(dz/dt) = \frac{UA\Delta T}{\lambda} \Rightarrow (dz/dt) \times \frac{1}{U} = (dz/dt)\left[\frac{1}{U_0} + \frac{z}{k}\right] = \frac{A\Delta T}{(dm/dz) \times \lambda}$$

Remembering that dm/dz is a constant numerical data (K = kg water evaporated per unit thickness of sediment formed), the following differential equation is obtained and solved as follows:

$$\left(\frac{1}{U_0} + \frac{z}{k}\right)dz = \frac{A\Delta T}{K\lambda}dt \quad \int_0^{z_t}\left(\frac{1}{U_0} + \frac{z}{k}\right)dz = \left[\frac{A\Delta T}{K\lambda}\right]t \Rightarrow \frac{z}{U_0} + \frac{z^2}{2k} = \left[\frac{A\Delta T}{K\lambda}\right]t$$

Setting $z = z_t$, substituting the data and solving for t, we find t = 16 265 s = 4.5 hours.

Notes on Example 21.3: A quick approximate calculation based on the arithmetic mean of the evaporation capacities would give t = 4.4 hours. The error would be larger in the case of a greater rate of fouling, or a deposit with a lower thermal conductivity.

EXAMPLE 21.4

The evaporator system shown in Figure 21.4 serves to concentrate a fruit juice from 12 Bx to 48 Bx. This is a forced circulation evaporator with an external shell-and-tube heat exchanger (calandria). Assume that the pressure maintained in the heat exchanger is such that no boiling occurs in the tubes, and all the evaporation takes place at the expansion chamber as a result of flash. The reduced pressure in the expansion chamber corresponds to a boiling temperature of 45°C. The feed is also at 45°C. The recirculated liquid reaches

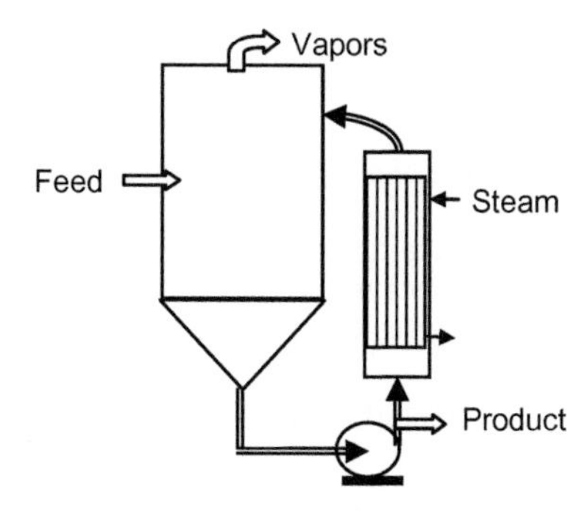

FIGURE 21.4

Forced-circulation evaporator example.

a temperature of 80°C before being released into the expansion chamber. The heat exchanger has 64 tubes with an internal diameter of 18 mm and an effective length of 2.40 m. The exchanger is heated by saturated steam, condensing in the shell at 130°C.

Evaluate:

a. The rate of recirculation (kg recirculated per kg product)
b. The maximum feed rate
c. The evaporation capacity
d. The overall coefficient of heat transfer.

Assimilate the thermodynamic properties of the juice to those of a sucrose solution at the same concentration. Heat losses, boiling point elevation, gravity effects, the resistance of the condensing steam and that of the tube wall are negligible.

Solution

Let:

G = mass flow-rate of recirculated juice, $kg \cdot s^{-1}$
C = mass flow-rate of product, $kg \cdot s^{-1}$
F = mass flow-rate of feed, $kg \cdot s^{-1}$
h_G = enthalpy of the recirculated juice at the heat exchanger exit
h_C = enthalpy of product = enthalpy of recirculated juice entering heat exchanger
q = rate of heat transfer at the heat exchanger, w
λ = latent heat of evaporation at 45°C = 2394 800 $J \cdot kg^{-1}$ (from steam tables).

a. Rate of recirculation.

Material balance: $V = F(1 - 12/48) = 0.75\ F$

Heat balance: $q = G\ (h_G - h_C) = V\lambda = 0.75\ F \cdot \lambda$

Hence: $G/F = 0.75\lambda/(h_G - h_C)$.

Using Eq. (21.5) to calculate the enthalpies h_R and h_C, according to the temperatures and concentrations of these solutions:

$h_G = 222\,400\ J \cdot kg^{-1} \cdot K^{-1}$

$h_p = 125\,100\ J \cdot kg^{-1} \cdot K^{-1}$

$G/F = 0.75 \times 2394\,800/(222\,400 - 125\,100) = 18.46$.

The rate of recirculation is: $G/C = 18.46 \times 48/12 = 73.84$.

b, c, d. Feed flow-rate m_f and overall heat transfer coefficient U:

$q = G(h_G - h_C) = UA\Delta T_{m\ log}$ (refer to Chapter 3, Section 3.7, on heat exchangers).

The heat transfer area $A = \pi NLd = 8.68\ m^2$ (where N, L, d = number, length, diameter of tubes).

Hence U/G = 169.8.

Assuming that the resistances of condensing steam and tube wall are negligible: $U \approx \alpha_b$.

Since there is no boiling in the tubes, α_b will be estimated using a correlation for single-phase heat transfer to liquids inside tubes. Assuming that the flow is turbulent (to be checked later), the Dittus–Boelter equation, Eq.(3.24), will be applied:

$$Nu = 0.023(Re)^{0.8}(Pr)^{0.4}\left(\frac{\mu}{\mu_w}\right)^{0.14}$$

The Reynolds number cannot be calculated at this stage, because the flow-rate in the tubes G is not yet known. The liquid properties (viscosity, specific gravity, thermal conductivity) will be obtained from appropriate tables (see Appendix). Substituting in the Dittus–Boelter equation, $U \approx \alpha_b$ is extracted as a function of G:

$U \approx \alpha_b = 280G^{0.8}$

Combining with U/G = 169.8 and solving for U and G:

$U = 2070\ W \cdot m^{-2} \cdot K^{-1}$ and $G = 12.19\ kg \cdot s^{-1}$, hence $F = 0.66\ kg \cdot s^{-1} = 2376\ kg \cdot h^{-1}$.

Notes on Example 21.4:

- The evaporation capacity of the system, compared to its heat exchange area, is rather small (about 200 kg per hour per square meter). This is a characteristic shortcoming of purely flash evaporators.
- The rate of recirculation is very high. This implies a large recirculation pump, considerable energy consumption for pumping, possible shear damage to the product and, worst of all, very long retention time.
- These disadvantages of the flash evaporator are among the reasons for the development and preferential use of "film" evaporators (see below).

Although conventional convective heat transfer is applied exclusively in industrial evaporation, microwave-heated evaporation has been investigated at laboratory scale (Assawarachan and Noomhorm, 2011).

21.4 Energy management

The energy economy of evaporation can be improved in a number of ways. In practice, the main types of energy-saving actions are:

1. Application of multiple effect evaporation
2. Vapor recompression
3. Utilization of waste heat (e.g. steam condensates).

21.4.1 Multiple-effect evaporation

Multiple-effect evaporation is evaporation in multiple stages, whereby the vapors generated in one stage serve as heating "steam" for the next stage. Thus, the first

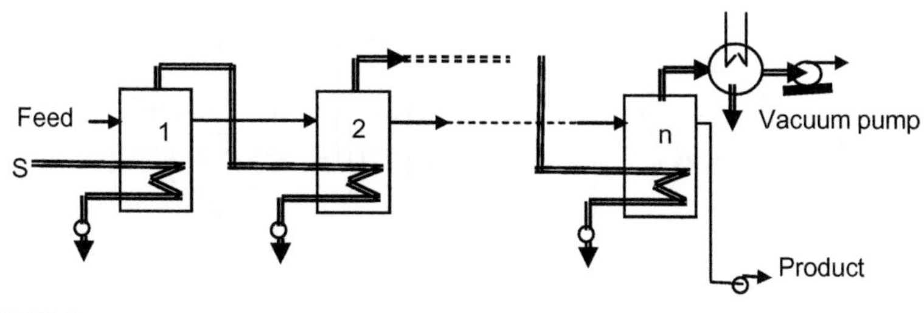

FIGURE 21.5

Multi-effect feed-forward evaporator.

stage acts as a "steam generator" for the second stage, which acts as a condenser to the first, and so on. The number of "effects" is the number of stages thus arranged. The first effect is heated with boiler steam. The vapors from the last effect are sent to the condenser (Figure 21.5).

For heat transfer to occur, a temperature drop must exist at each effect. In other words, the temperature of the vapors generated in a given effect must be superior to the boiling temperature in the following effect:

$T_1 > T_2 > T_3 > \therefore \therefore \therefore \therefore \therefore > T_n$, hence: $P_1 > P_2 > P_3 > \therefore \therefore \therefore \therefore \therefore > P_n$.

The boiling-point elevation (BPE) has a particular importance in multiple-effect evaporation. Because of the BPE, the vapors leaving a boiling solution are superheated vapors. If the BPE is Θ°C, then the superheat is exactly equal to Θ. When such vapors condense in the heating element of the next effect, they condense at their *saturation temperature*. The driving force for heat transfer is therefore reduced by Θ degrees at each effect. The total temperature drop in the system is:

$$\Delta T_{total} = T_S - T_W \quad (21.9)$$

where T_S and T_W are the saturation temperature of the steam fed to the first effect and that of the vapors from the last effect, respectively. The total driving temperature drop useful for heat transfer is that temperature difference, diminished by the sum of BPEs in each effect.

$$\Delta T_{total\ useful} = T_S - T_W - \sum BPE \quad (21.10)$$

In reality, the loss of thermal driving force is larger than the sum of BPEs. The pressure drop in the vapor ducts between the effects adds a further loss to the thermal driving force.

In the particular multi-effect setup shown in Figure 21.4, the liquid and the vapors travel in the same direction. This is called "*forward feeding*". Other feeding patterns exist, as shown in Figure 21.6. The choice of a flow sequence in a

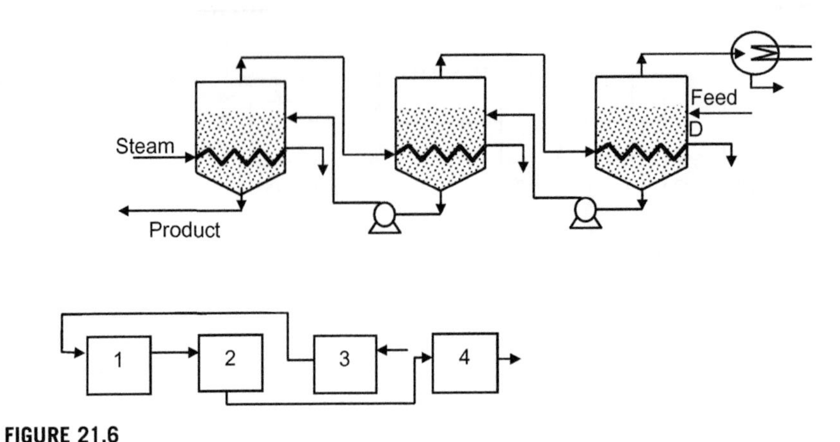

FIGURE 21.6

Triple-effect evaporator with backward feed and a quadruple-effect evaporator with mixed flow pattern.

multi-effect evaporator depends on a number of factors, such as the concentration ratios at each effect, the viscosity of the liquid and the heat-sensitivity of the product.

Detailed and accurate material and energy balances can be worked out for each one of the units or for the entire system. A rapid approximate calculation is possible if certain simplifying assumptions are made. First, let us assume that for each kg of vapor condensing on one side of the heat transfer surface, 1 kg of water is evaporated on the other side. This assumption implies that:

- The latent heat of evaporation does not change markedly with pressure
- The heat needed to heat the feed up to the boiling temperature is negligible in comparison with the heat consumed for the generation of vapors
- Heat losses and boiling point elevation are negligible.

With these assumptions one can write:

$$S = V_1 = V_2 = \ldots = V_n \Rightarrow V_{Total} = n \cdot S \Rightarrow \frac{V_{Total}}{S} = n \tag{21.11}$$

The economic and environmental consequences of this result are obvious. In a multi-effect evaporator, for each kg of water evaporated:

1. The quantity of steam consumed is inversely proportional to the number of effects. This conclusion has been tested experimentally in the case of tomato paste concentration in single, double and triple effect evaporators (Rumsey *et al.*, 1984).
2. The quantity of cooling water utilized in the condenser is inversely proportional to the number of effects.

Now, let us analyze heat transfer in a multi-effect system. The simplifying assumption made above leads to the conclusion that the same quantity of heat is exchanged in each one of the units:

$$U_1A_1(\Delta T)_1 = U_2A_2(\Delta T)_2 = \ldots = U_nA_n(\Delta T)_n \tag{21.12}$$

Let us assume that:

1. The heat exchange areas of the units are equal, i.e., $A_1 = A_2 = A_3 = \ldots = A_n = A$
2. The coefficients of heat transfer of the units are equal, i.e., $U_1 = U_2 = U_3 = \ldots = U_n = U$.

The total evaporation capacity is, therefore:

$$V_{Total} = \frac{UA(\sum \Delta T)}{\lambda} \tag{21.13}$$

Note that the number of effects, n, does not appear in Eq. (21.13). This means that the evaporation capacity of a multiple-effect system with n units of area A is the same as the evaporation capacity of a single-effect evaporator of area A, provided that the same total temperature drop is available in both cases. The economic consequences of this conclusion are also obvious: in a multiple-effect system with a given evaporation capacity and operating at a given total temperature drop, *the total heat exchange area is proportional to the number of effects*.

Remembering that the capital cost of evaporators increases linearly with the total heat transfer area, while steam and cooling water constitute the major portion of running costs, it may be concluded that:

1. The capital cost of a multi-effect evaporator is nearly proportional to the number of effects
2. The running cost of a multi-effect evaporator is nearly inversely proportional to the number of effects.

The total cost of evaporation must therefore have a minimum value for a given number of effects, which is the *economically optimal number* of effects (Figure 21.7). A different optimization approach, combining economics and product quality (lycopene retention in tomato paste), has been evaluated by Simpson *et al.* (2008).

Because of the simplifying assumptions made for convenience, the conclusions of the analysis above are only approximations. Thus, the assumption of equal heat transfer coefficient in all the stages may introduce a serious deviation from reality. Furthermore, multi-effect evaporators are seldom constructed with equal heat exchange area in all the stages. As the concentration increases and the liquid volume decreases accordingly, the heat transfer area must be made smaller so that the remaining liquid can still properly wet the entire surface. For accurate calculations or design purposes, more realistic assumptions have to be made (Angletti and Burton, 1983). The computation is more complicated, and often requires trial-and-error techniques (Holland, 1975).

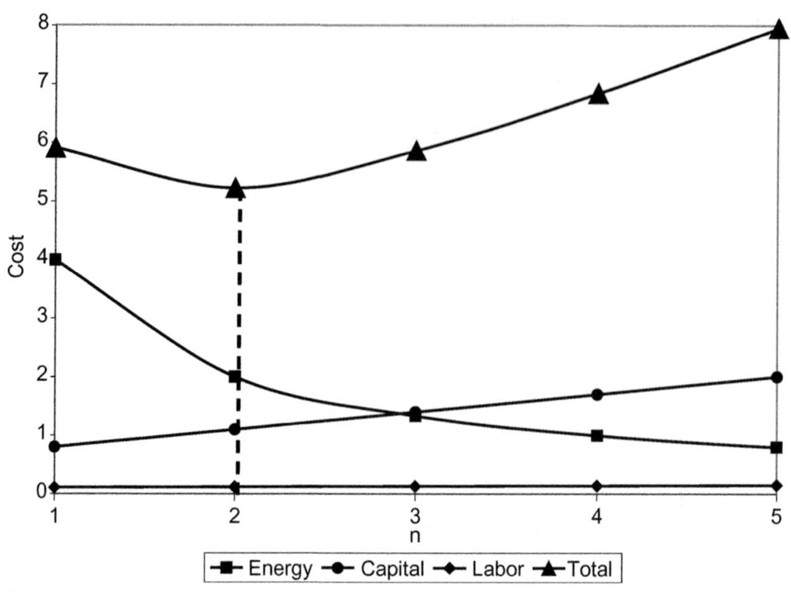

FIGURE 21.7

Economically optimal number of effects.

EXAMPLE 21.5

A backward feeding, double-effect evaporator is used for concentrating an aqueous solution from 2% to 20% soluble solids. The system, shown in Figure 21.8, consists of two identical evaporation units, each with a heat exchange surface of 100 m^2. The first effect receives saturated steam at 120°C. The second effect is connected to the condenser, where the vapor condenses at the saturation temperature of 40°C. The feed solution is introduced into the second effect at 40°C. An overall heat transfer coefficient of $2500\ W \cdot m^{-2} \cdot K^{-1}$ is assumed for both effects.

Calculate the evaporation capacity and steam consumption of the system.

Note: The solutions being relatively diluted, it will be assumed that their specific enthalpy is similar to that of water at the same temperature. BPE will be neglected. It will be also assumed that the latent heat of evaporation is the same in both effects. The error introduced by this assumption is small and can be corrected by iteration.

Solution

Let V_1 and V_2 be the rate of vapor generation in Effect 1 and Effect 2, respectively.

In Effect 2, the feed is already at the boiling temperature. Therefore heat is not needed to bring the feed to boiling and all the heat is utilized for vapor generation. Since the latent heat of evaporation is assumed to be the same in both effects, 1 kg of vapor from Effect 1 generates 1 kg of vapor in Effect 2. Therefore:

$$V_1 = V_2$$

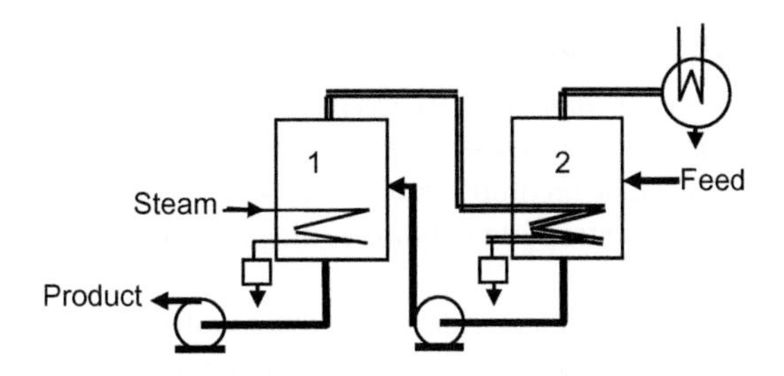

FIGURE 21.8

Double effect evaporator example.

In the first effect, heat is needed for vapor generation *and* for bringing the liquid from Effect 2 to the boiling temperature in Effect 1:

$$q_1 = UA(120 - T_1) = V_1\lambda + (F - V_2) \times (T_1 - 40) \times C_P$$
$$q_2 = UA(T_1 - 40) = V_2\lambda$$

Material balance on solute gives:

$$0.02 \times F = 0.2 \times (F - V_1 - V_2)$$

Setting $V_1 = V_2$, taking $\lambda = 2407\ kJ \cdot kg^{-1}$ and substituting the data, we obtain two equations with two unknowns, V_1 and T_1, namely:

(1) $2500 \times 100 \times (120 - T_1) = 2407\,000 \times V_1 + 1.22 \times V_1 \times (T_1 - 40) \times 4180$
(2) $2500 \times 100 \times (T_1 - 40) = 2407\,000 \times V_1$

We solve and obtain:

$$V_1 = V_2 = 3.99\ kg \cdot s^{-1}.$$

Hence, the total evaporation capacity of the system is $V = 2 \times 3.99 = 8\ kg \cdot s^{-1}$.

The temperature in the first effect is found to be 78.4°C. At this temperature, $\lambda = 2310\ kJ \cdot kg^{-1}$, about 5% less than the assumed value. A more accurate result can be obtained by correcting the calculations accordingly.

The steam consumption will be calculated from the heat transferred at Effect 1:

$$q_1 = UA(120 - T_1) = 2500 \times 100 \times (120 - 78.4) = 10400000\ W$$

The latent heat of evaporation of steam at 120°C is $2192\ kJ \cdot kg^{-1}$. Hence the steam consumption is:

$$s = \frac{10400\,000}{2192000} = 4.74\ kg \cdot s^{-1}$$

The ratio of evaporation capacity to steam consumption is 8/4.74 = 1.69, i.e., much better than that of a single-effect evaporator.

21.4.2 Vapor recompression

Compression of the vapor generated by evaporation increases its saturation temperature and therefore its usefulness as a source of heat. A hypothetical single-stage evaporator with vapor recompression is represented in Figure 21.9, together with the corresponding entropy-temperature diagram describing the thermodynamics of the process.

The vapor leaving the product side of the evaporator is compressed from P_1 (low pressure) to P_2 (high pressure). Reversible adiabatic (isentropic) compression is assumed. The compressed vapor is introduced (preferably after de-superheating) to the steam side of the same evaporator, where it condenses at the saturation temperature corresponding to P_2. Since $P_2 > P_1$ we know that $T_2 > T_1$. The temperature drop necessary for heat transfer has therefore been assured. For the system to be self-sufficient in steam, the heat absorbed for the evaporation of a given mass of water at P_1 must be exactly equal to the heat given off by the condensation of the same mass of vapor at P_2.

Vapor recompression can work in single-stage evaporation as shown, but is usually applied in combination with multi-effect evaporation. Two methods of vapor recompression are used:

1. *Mechanical recompression*. Because of the large volumes of vapor to be handled, rotating compressors (turbo-compressors, centrifugal compressors and blowers) are used. The main shortcoming of mechanical recompression is the high rate of wear, accelerated greatly by the presence of water droplets in the vapors.
2. *Thermo-recompression*. The low-pressure vapors from the evaporator can be compressed with the help of a small amount of high-pressure live steam, using a steam ejector (see Chapter 2, Section 2.4.4). The use of thermo-recompression in a multi-effect evaporator is illustrated in Figure 21.10.

Ejectors are relatively inexpensive and their cost of maintenance is low. Thermo-recompression is usually more economical than mechanical recompression. Because of the high velocities of gas flow attained at the nozzle, ejectors may emit a disturbing high-pitched noise, requiring appropriate acoustic insulation for the safety and comfort of personnel.

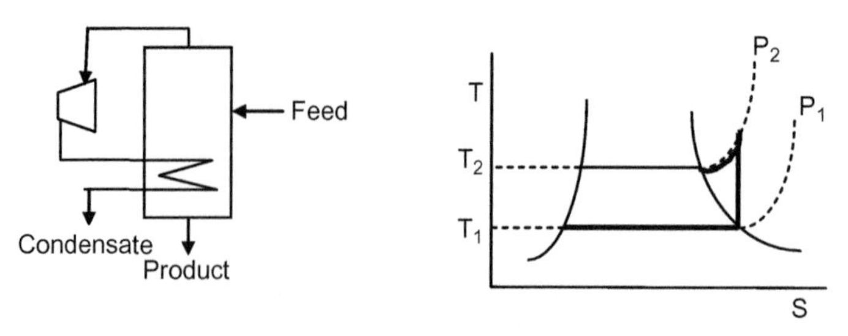

FIGURE 21.9

Single-stage evaporator with mechanical vapor recompression.

21.5 Condensers

Most evaporators in the food industry work under reduced pressure. Because of the extensive volume of vapors to be moved, maintaining reduced pressure with the help of vacuum pumps alone would require very large pumps. For this reason, the vapors are first condensed by cooling in a condenser installed between the vapor separator and the vacuum pump. The condensers are normally water-cooled. There are two types of condensers: direct (jet) condensers and indirect (surface) condensers (Figure 21.11).

In jet condensers, cooling water is directly injected into the vapors. In surface condensers, the vapors condense on the surface of coils or tubes inside which cooling water is circulated.

Because vapor/liquid separation is not perfect, the condensates of evaporator vapors are not pure distilled water but may contain a certain quantity of solutes carried over by the liquid droplets escaping the separator. The condensates are therefore unfit for reuse (say, as feed water to the steam boiler or for recycling through a cooling tower) or for discharge (because of their high BOD) without appropriate treatment. The use of membrane processes for the treatment of evaporation condensates has been discussed in Chapter 10. Direct condensers discharge large quantities of effluent (condensed vapor plus cooling water) requiring treatment. The quantity of contaminated effluent from indirect condensers is much

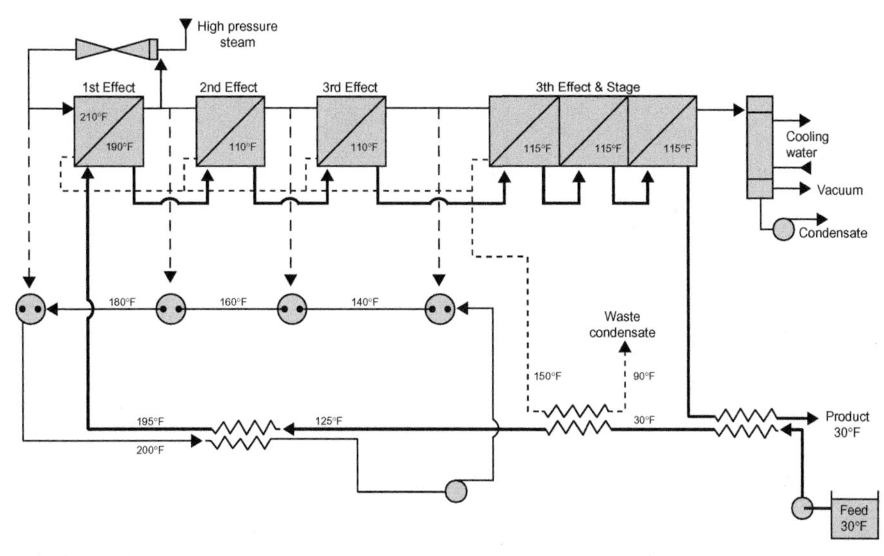

FIGURE 21.10

Quadruple-stage evaporator with thermal recompression.

Figure from Evaporator Handbook, *courtesy of APV.*

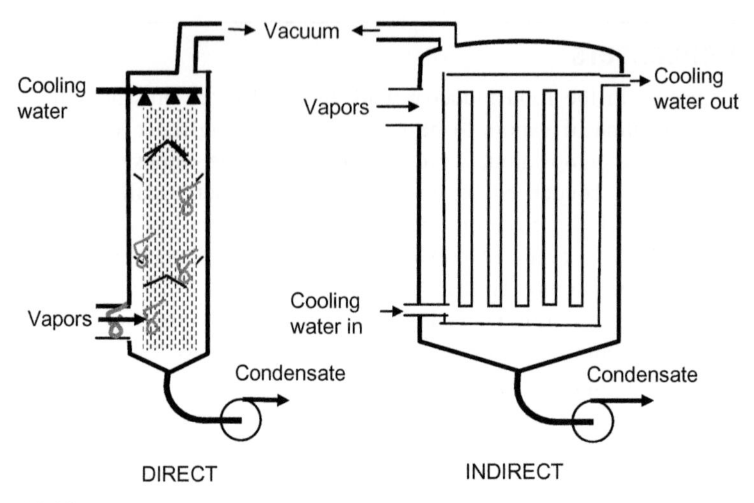

FIGURE 21.11

Direct and indirect condensers.

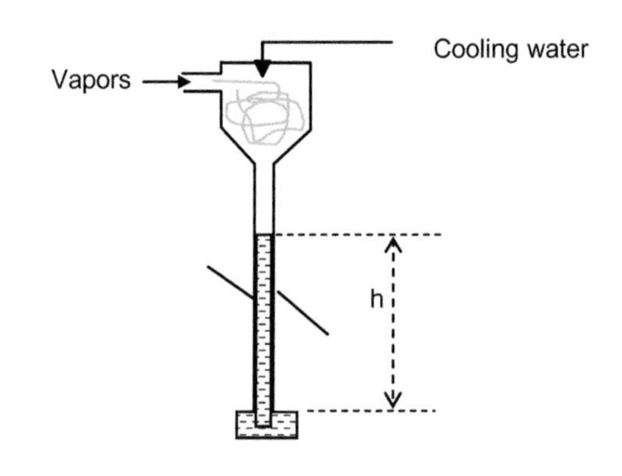

FIGURE 21.12

Barometric leg.

smaller. On the other hand, indirect condensers are more expensive, and present problems of scale deposition and corrosion.

The condenser operates under reduced pressure. Either of the two following methods can be used for the continuous discharge of the condensates to ambient pressure: mechanical pumping or a "barometric leg". The barometric leg is simply a sufficiently high vertical pipe. The hydrostatic pressure of the column of water (z) in the pipe compensates for the difference between the reduced pressure in the condenser and atmospheric pressure (Figure 21.12).

21.6 Evaporators in the food industry

The many different types of evaporators used in the food industry, have been developed or adapted in response to the peculiar needs of their particular application (Armerding, 1966; Gull, 1965; Jorge *et al.*, 2010). In this paragraph, the main types will be analyzed, following their historical evolution.

21.6.1 Open pan batch evaporator

The open pan batch evaporator is the simplest and surely the oldest type of evaporator. A batch of liquid is boiled in an open heated pan (Figure 21.13). Hemispherical, steam-jacketed pans are still widely used for small-scale batch evaporation.

21.6.2 Vacuum pan evaporator

In 1813, E.C. Howard invented the "vacuum pan", a closed, jacket-heated vessel connected to a condenser and a vacuum pump (Billet, 1989). This type of evaporator (Figure 21.14) is still used for small-scale vacuum evaporation ("finishing" tomato concentrate, production of jam, ketchup, sauces, etc.). It can be equipped with an anchor agitator and operated either in batch or in continuous mode.

21.6.3 Evaporators with internal tubular heat exchangers

The principal shortcoming of pan evaporators is their low surface/volume ratio, resulting in poor evaporation capacity and extremely long retention time. Tubular heat exchangers (Billet, 1989; Minton, 1986) represent an attempt to fit more heat-exchange area into an evaporator of a given volume. The tubular (shell-and-tube) heat exchanger can be installed either inside or outside the evaporator vessel. An early type of this group, built by N. Rillieux in 1844, had horizontal tubes inside the vessel (Figure 21.15). Having many of the disadvantages of the coil-heated

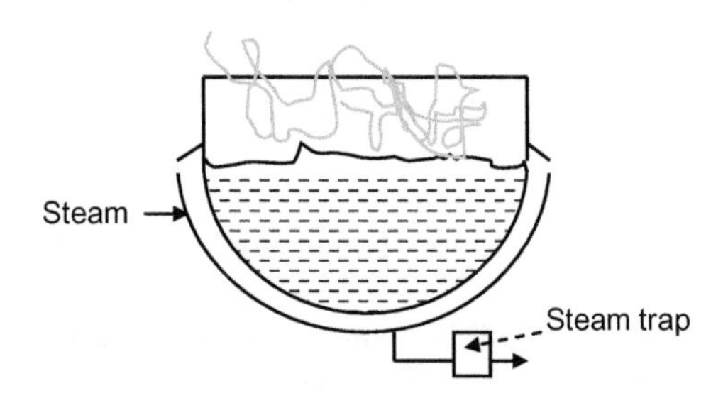

FIGURE 21.13

Open pan evaporator.

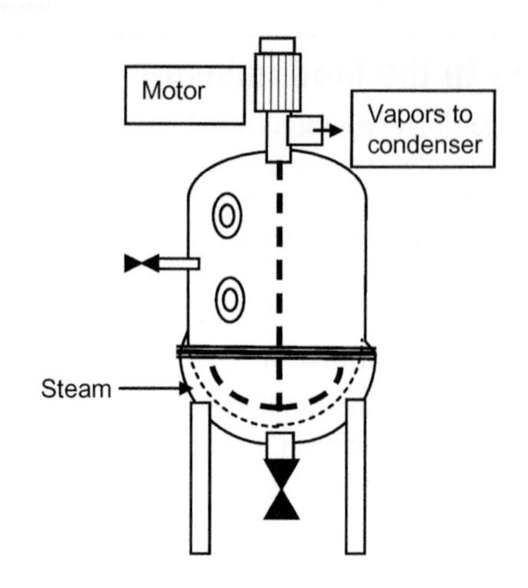

FIGURE 21.14

Batch vacuum pan with agitation.

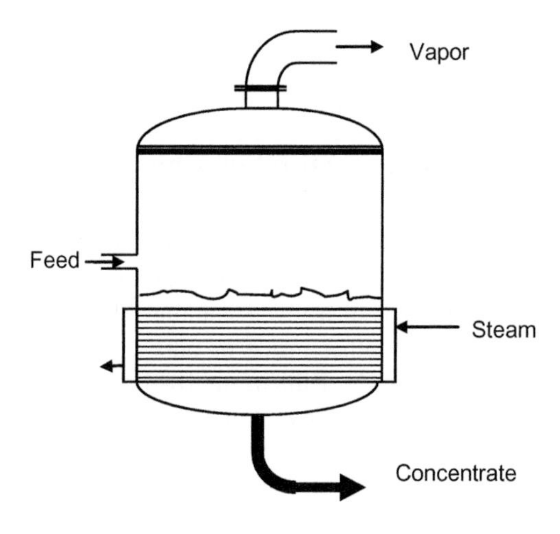

FIGURE 21.15

Evaporator body with internal horizontal heating tubes.

vessel, it was replaced by the "Robert Evaporator", equipped with vertical, short tubes (Figure 21.16).

Variations of this type can still be found, particularly in the sugar industry. In these evaporators with internal heat-exchangers, liquid movement over the exchange areas depends on natural convection. The rate of heat transfer is low, and is reduced even more as the viscosity increases. The bulk of the liquid is in

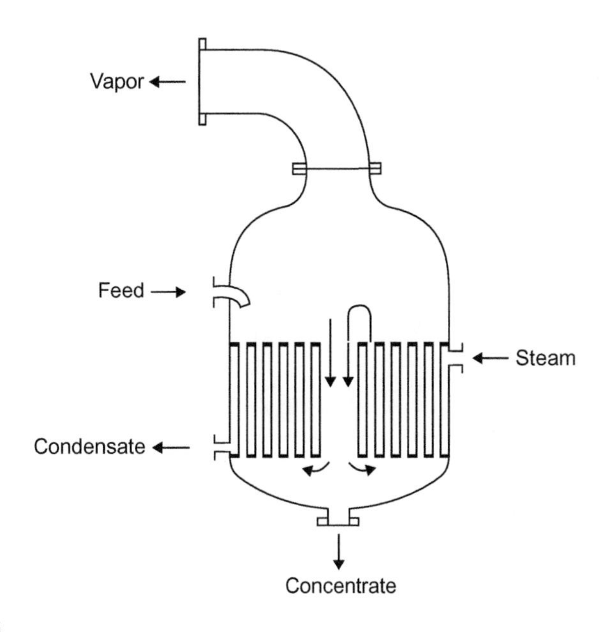

FIGURE 21.16

"Robert evaporator".

Figure courtesy of Alfa-Laval.

constant contact with the hot surface of the tubes. Evaporators of this type are therefore not suitable for concentrating heat-sensitive materials.

21.6.4 Evaporators with external tubular heat exchangers

The principle of operation of this type of evaporators was explained in Example 21.4. The heat transfer element is separated from the evaporator vessel. The liquid is circulated through the tubes of the heat exchanger (calandria) either by thermal convection (thermo-siphon) or by mechanical pumping. Sufficient pressure is maintained in the heat exchanger to prevent excessive boiling in the tubes. At a given time, the bulk of the liquid is in the vessel and only a portion of it is in contact with hot surfaces (Figure 21.17). The external heat exchanger is often positioned at an angle so as to accommodate longer tubes within limited height. Some of the high-capacity forced recirculation evaporators used for the concentration of tomato juice (Figure 21.18) belong to this type.

21.6.5 Boiling film evaporators

The shortcomings of "flooded tube" evaporators have been analyzed in Example 21.4. To overcome these disadvantages, the film evaporator concept was developed towards the end of the 19th century. In a film evaporator, the liquid flows

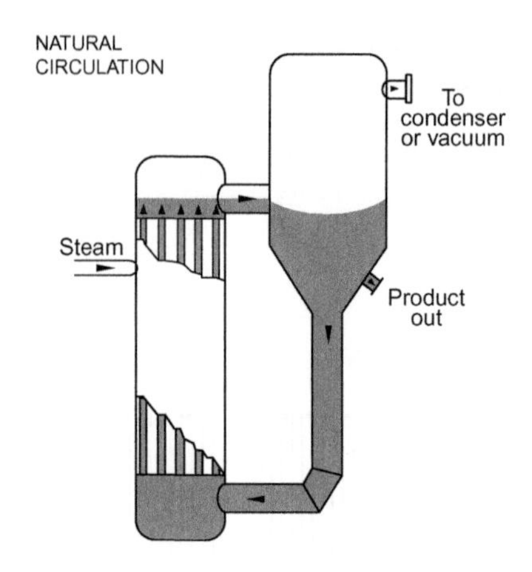

FIGURE 21.17

Evaporator with external heat exchanger, natural circulation.

Figure from Evaporator Handbook, *courtesy of APV.*

rapidly over the heating surfaces as a thin, boiling film. The heating areas may be the internal walls of tubes, the plates of a plate heat exchanger or a rotating conical surface. The liquid film is moved by gravity, by centrifugal force or by mechanical agitation, all these usually assisted by the dragging force of the vapor. The objectives are to achieve rapid evaporation, to avoid the need for recirculation and, thus, to reduce the residence time of the product in the evaporator.

1. *Climbing film evaporators*. Historically, these were the first industrial film evaporators. The liquid is introduced at the lower end of vertical tubes (Figure 21.19) and quickly reaches its boiling point. The vapor formed moves upwards and drags with it the liquid film which continues to boil and generate more vapor. As the mixture travels upwards, the volume and therefore the velocity of the vapor phase increases. The resulting increase in drag compensates for the increasing resistance of the more concentrated liquid to flowing against gravity. At the upper end of the tube, the mixture of vapor and concentrated liquid is ejected into the separator.

   Drag by the generated vapors being the only driving force, the following consequences must be kept in mind:

   - It is important to produce vapors in sufficient quantity even right at the entrance to the tubes. For this reason, a fairly high temperature difference (at least 15°C) must be applied and preheating of the feed is recommended.
   - The climbing film mechanism does not work well at very low pressures (low vapor density, less drag).
   - This type of evaporator does not perform well with highly viscous liquids.
   - The tubes must be short.

(a)

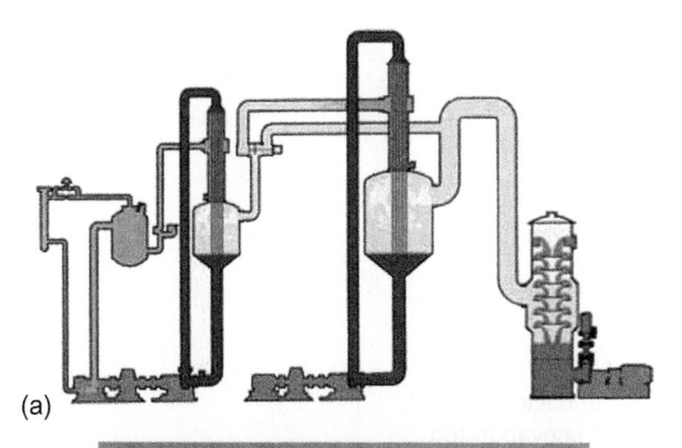

(b)

FIGURE 21.18

(a) Two-stage evaporator with forced recirculation; (b) High-capacity forced circulation evaporator in plant.

Figure and photograph courtesy of Rossi&Catelli.

2. *Falling film evaporators* (Wiegand, 1971). In this type of evaporator, the feed is introduced at the upper end of the vertical tubes, with due care taken to distribute the liquid as a regular film over the entire periphery (Figure 21.20). The film travels downwards by gravity, assisted by vapor drag. In comparison with the climbing film evaporator, thinner, faster-moving films are obtained. The tubes can be long; lengths of 20 m or more are not uncommon. High-capacity falling film evaporators are widely used for the concentration of milk and fruit juices.

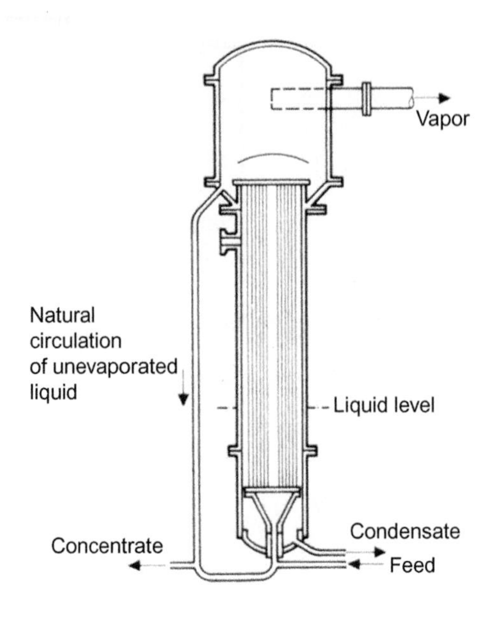

FIGURE 21.19

Climbing film evaporator.

Figure from Evaporation Handbook, *courtesy of Alfa-Laval.*

3. *Plate evaporators*. In this type of evaporator, a plate heat exchanger is used (Figure 21.21). The liquid flows over the surface of the plates. A relatively large gap between the plates is provided for vapor flow. Climbing- or falling-film set-ups are possible, and sometimes both are used in different sections of the same unit.
4. *Agitated or swept-surface evaporators*. These film evaporators (Sangrame *et al*., 2000; Stankiewicz and Rao, 1988) consist of a vertical, jacketed cylinder equipped with a central agitator (Figure 21.22). The rapidly rotating impeller projects the liquid to the heated wall as a thin film. Very rapid heat exchange and short residence time are among the advantages of this type. Its main disadvantages are high capital and maintenance costs.
5. *Centrifugal evaporators*. In centrifugal evaporators (Jebson *et al*., 2003; Malkki and Velstra, 1967) the heat surface area is a rapidly spinning, steam-heated hollow cone (Figure 21.23). The liquid film flows over this surface at a very high velocity while heating steam condensates on the other side of the surface. Contact time is very short. Centrifugal evaporators are particularly suitable for concentrating heat-sensitive, highly viscous products. They are relatively expensive for their capacity. As they are normally operated in single-effect fashion, their steam economy is also low.

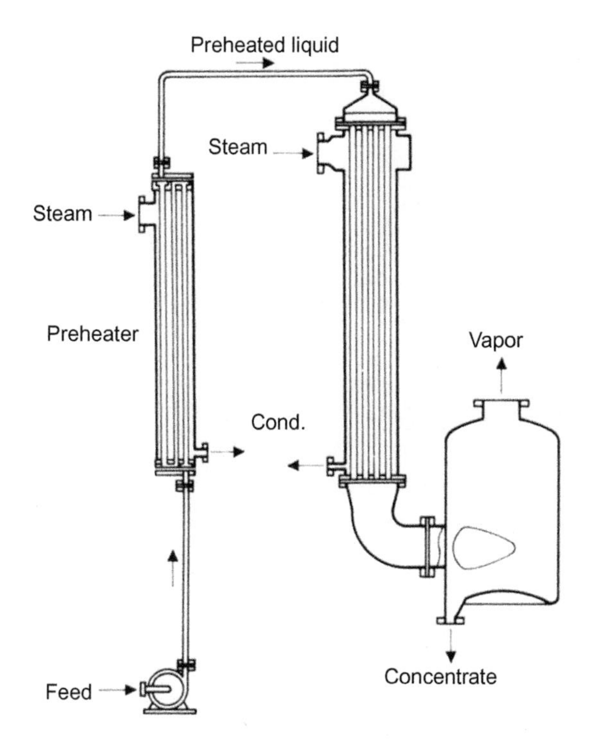

FIGURE 21.20

Falling film evaporator.

Figure from Evaporation Handbook, *courtesy of Alfa-Laval.*

21.7 Effect of evaporation on food quality

21.7.1 Thermal effects

During evaporation, foods are susceptible to thermal damage, depending on the time–temperature profile of the process. The following are some examples of types of thermal damage associated with evaporation:

- Non-enzymatic (Maillard) browning
- Induction of "cooked taste" in fruit juices and milk
- Loss of carotenoid pigments (e.g., lycopene in tomato juice)
- Protein denaturation (milk).

The rate and extent of browning discoloration is concentration-dependent. As the concentration of the food increases during evaporation, so does its sensitivity to high temperature. This type of quality loss – loss of "fresh" taste and induction of "cooked taste" – is common to most fruit juices, and particularly to tomato, citrus, apple and grape juices. Some cooked taste in tomato concentrates to be

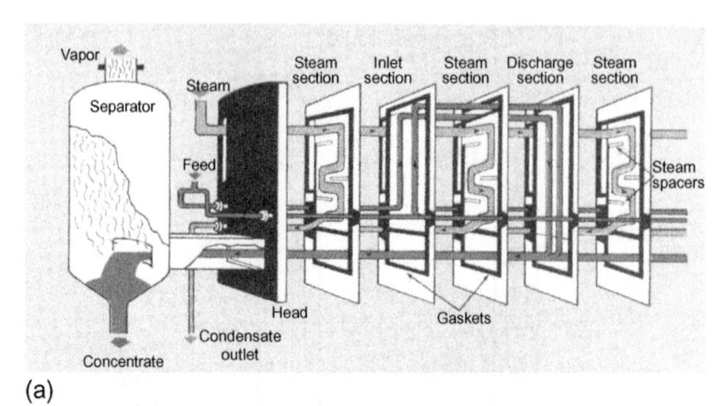

(a)

(b)

FIGURE 21.21

(a) Structure of a plate evaporator; (b) Plate evaporators in plant.

Figure and photograph from Evaporator Handbook*, courtesy of APV.*

used for cooking may not be objectionable; however, if the concentrate is to be reconstituted as tomato juice by dilution with water, cooked taste is considered a defect. Cooked taste is due to the formation of precursors of dark pigments in non-enzymatic browning, such as hydroxymethyl-furfural.

As explained in Section 21.3.2, thermal damage is drastically reduced by lowering the evaporation temperature – i.e., by operating the evaporator under vacuum. On the other hand, too low an evaporation temperature may result in longer residence time. In the case of lycopene loss in tomato juice, it has been found that high temperature–short time evaporation results in better retention of the lycopene pigment (Monselise and Berk, 1954).

21.7.2 Loss of volatile flavor components

A certain proportion of the desirable volatile components, known as the "aroma", "fragrance" or "essence", is lost when fruit juices or coffee extract are concentrated by evaporation (Johnson *et al.*, 1996; Mannheim and Passy, 1974). The extent of

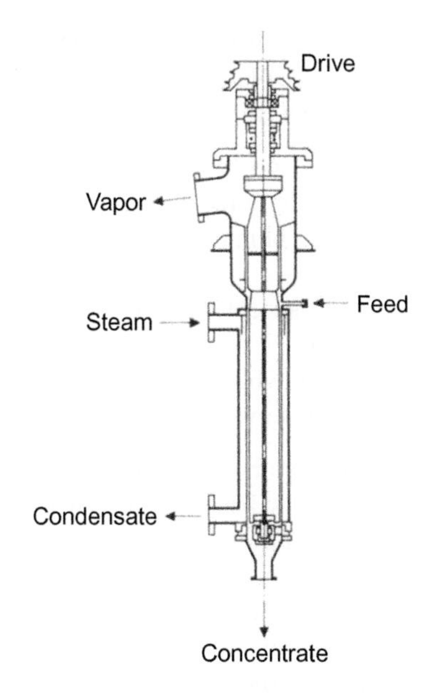

FIGURE 21.22

Swept-surface evaporator.

Figure from Evaporation Handbook, *courtesy of Alfa-Laval.*

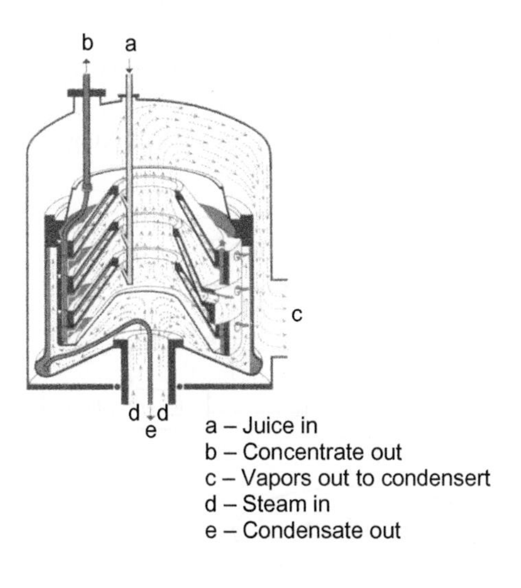

FIGURE 21.23

Centrifugal "Centritherm" evaporator.

Figure from Evaporation Handbook, *courtesy of Alfa-Laval.*

aroma loss depends on the volatility of the flavor substances in relation to that of water (see Section 13.2, on relative volatility). Karlsson and Trägårdh (1997) classify the fruit aromas into four groups, with respect to their relative volatility:

1. High-volatility aromas, such as apple aroma, which are practically lost completely when only 15% of the juice has been evaporated.
2. Medium-volatility aromas (e.g., plum, grape), lost almost completely when 50% of the juice is evaporated.
3. Low-volatility aromas (peach, apricot), lost to the extent of about 80% when 50% of the juice is evaporated.
4. Very low-volatility aromas, (strawberry, raspberry), of which 60–70% or less are lost when 50% of the juice is evaporated.

Natural aromas are complex mixtures of volatile organic compounds (alcohols, aldehydes, ketones, esters, phenolic substances, terpenes, etc.). The different components of a given aroma do not have the same volatility. During evaporation, some components are lost to a greater extent than others; consequently, the characteristic "note" of a fruit juice may be altered when concentrated by evaporation. In the case of pineapple juice, for example, 90% removal of the esters requires 80% evaporation while 90% loss of the carbonyls occurs with only 47% evaporation (Ramteke *et al.*, 1990).

Loss of volatiles is not always undesirable. The process of "deodorization", sometimes applied to milk and cream, consists of partial removal of objectionable odorous volatiles by vacuum evaporation.

The two major methods for the partial compensation of aroma loss by evaporation are *aroma recovery* and "*cut-back*" (Figure 21.24). Aroma recovery consists of the separation of the aroma from the starting material by different methods, production of the "essence" by concentration of the aroma, and admixture of the essence to the final product. In the cut-back process, applied mainly to fruit juices, part of the lost flavor is restored by mixing the concentrate with a

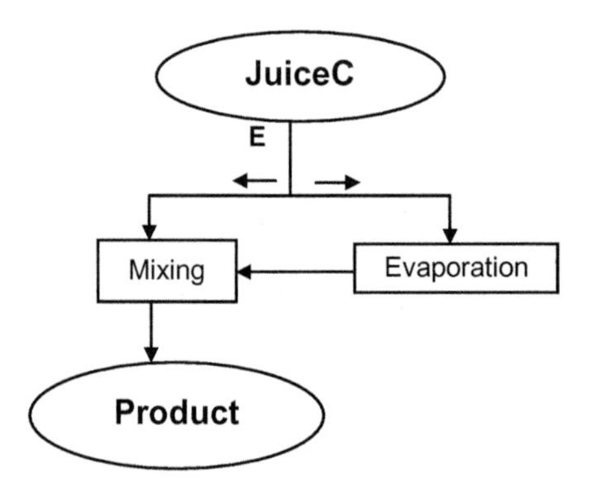

FIGURE 21.24

Schema of the "cut-back" process.

certain quantity of untreated juice. As an example, orange juice is concentrated by vacuum evaporation from 12 Bx to 65 Bx. Two parts of the concentrate (practically devoid of volatile aroma) are mixed with one part of the original juice, to obtain a 47-Bx concentrate, with some aroma.

References

Angletti, S.M., Burton, H., 1983. Modelling of multiple effect falling film evaporators. J. Food Technol. 18, 539–563.

Armerding, G.D., 1966. Evaporation methods as applied to the food industry. Adv. Food Res. 15, 303–358.

Assawarachan, R., Noomhorm, A., 2011. Mathematical models for vacuum-microwave concentration behavior of pineapple juice. J. Food Proc. Eng. 34 (5), 1485–1505.

Billet, R., 1989. Evaporation Technology. Wiley–VCH, Cambridge, UK.

Coulson, M.A., McNelly, M.J., 1956. Heat transfer in a climbing film evaporator. Trans. ChemE. 34, 247–257.

Dale, M.C., Okos, M.R., Nelson, P., 1982. Concentration of tomato products: analysis of energy saving process alternatives. J. Food Sci. 47 (6), 1853–1858.

Gull, H.C., 1965. Modern developments in the evaporator field. J. Soc. Dairy Technol. 18 (2), 98–108.

Holland, C.D., 1975. Fundamentals and Modeling of Separation Processes – Adsorption, Distillation, Evaporation and Extraction. Prentice-Hall, Englewood Cliffs, NJ.

Jebson, R.S., Chen, H., Campanella, O.H., 2003. Heat transfer coefficient for evaporation from the inner surface of a rotating cone – II. Trans. ChemE. 81 (Part C), 293–302.

Johnson, J.R., Bradock, R.J., Chen, C.S., 1996. Flavor losses in orange juice during ultrafiltration and subsequent evaporation. J. Food Sci. 61 (3), 540–543.

Jorge, L.M.M., Righetto, A.R., Polli, P.A., Santos, O.A.A., Maciel Filho, R., 2010. Simulation and analysis of a sugarcane juice evaporation system. J. Food Eng. 99 (3), 351–359.

Karlsson, H.O.E., Trägårdh, G., 1997. Aroma recovery during beverage processing. J. Food Eng. 34 (2), 159–178.

Kavak Akpınar, E., 2009. Drying of parsley leaves in a solar dryer and under open sun: modeling, energy and exergy aspects. J. Food Proc. Eng. 34 (1), 27–48.

Malkki, Y., Velstra, J., 1967. Flavor retention and heat transfer during concentration of liquids in a centrifugal film evaporator. Food Technol. 21 (9), 1179–1182.

Mannheim, C.H., Passy, N., 1974. In: Spicer, A. (Ed.), Advances in Preconcentration and Dehydration. Applied Science, London, UK, pp. 151–194.

McAdams, W.H., 1954. Heat Transmission, third ed. McGraw-Hill, New York, NY.

McNelly, M.J., 1953. A correlation of rates of heat transfer to nucleate boiling liquids. J. Imperial College Eng. Soc. 7, 18–34.

Minton, P.E., 1986. Handbook of Evaporation Technology. Noyes Publications, Saddle River, NJ.

Monselise, J.J., Berk, Z., 1954. Oxidative destruction of lycopene during the manufacture of tomato puree. Bull. Research Council of Israel 4, 188–190.

Peleg, M., Mannheim, C.H., 1970. Production of frozen orange juice concentrate from centrifugally separated serum and pulp. J. Food Sci. 35 (5), 649–651.

Perry, R.H., Green, D.W., 1997. Perry's Chemical Engineer's Handbook, seventh ed. McGraw-Hill, New York, NY.

Ramteke, R.S., Eipeson, W.E., Patwardhan, M.V., 1990. Behaviour of aroma volatiles during the evaporative concentration of some tropical juices and pulps. J. Sci. Food Agr. 50 (3), 399–405.

Rumsey, T.R., Conany, T.T., Fortis, T., Scott, E.P., Pedersen, L.D., Rose, W.W., 1984. Energy use in tomato paste evaporation. J. Food Poc. Eng. 7 (2), 111–121.

Sangrame, G., Bhagavati, D., Thakare, H., Ali, S., Das, H., 2000. Performance evaluation of a thin film scraped surface evaporator for concentration of tomato pulp. J. Food Eng. 43 (4), 205–211.

Simpson, R., Almonacid, S., López, D., Abakarov, A., 2008. Optimum design and operating conditions of multiple effect evaporators: tomato paste. J. Food Eng. 89 (4), 488–497.

Stankiewicz, K., Rao, M.A., 1988. Heat transfer in thin-film wiped-surface evaporation of model liquid foods. J. Food Proc. Eng. 10 (2), 113–131.

Stephan, K., Abdelsalam, M., 1980. Heat transfer correlation for natural convection boiling. Int. J. Heat Mass Transfer 23, 73–87.

Thijssen, H.A.C., Van Oyen, N.S.M., 1977. Analysis and economic evaluation of concentration alternatives for liquid foods – quality aspects and costs of concentration. J. Food Proc. Eng. 1 (3), 215–240.

Wiegand, J., 1971. Falling film evaporators and their application in the food industry. J. Appl. Chem. & Biotech. 21 (12), 351–358.

CHAPTER

Dehydration

22

22.1 Introduction

Drying is one of the most ancient methods of food preservation known to mankind. Preservation of meat, fish and food plants by drying in the sun or in the naturally dry air of the deserts and mountains has been practiced since prehistoric times, and is still a vital operation in the life of many rural communities.

Drying or dehydration is, by definition, the removal of water by evaporation, from a solid or liquid food, with the purpose of obtaining a solid product sufficiently low in water content. (Note: *osmotic dehydration*, where removal of water takes place by virtue of a difference in osmotic pressure and not by evaporation, is discussed at the end of this chapter.)

The main technological objectives of food dehydration are:

- Preservation as a result of depression of water activity.
- Reduction in weight and volume.
- Transformation of a food to a form more convenient to store, package, transport and use – for example, transformation of liquids such as milk, eggs, fruit and vegetable juices or coffee extract to a dry powder that can be reconstituted to the original form by addition of water (instant products).
- Imparting to a food product, a particular desirable feature, such as a different flavor, crispiness, chewiness, etc. – i.e., creating a new food (e.g., transformation of grapes to raisins).

Despite the importance of drying as an industrial operation and recent progress in drying research, the physical principles of the complex phenomena that occur in the course of dehydration and rehydration are not entirely understood. Modeling of drying is discouragingly difficult in the case of food materials. A completely satisfactory model of drying kinetics, applicable to foods, is not available at the moment. Yet, in the following discussion of drying, extensive use will be made of models. It should be remembered that these theoretical models are only approximations. Their use as an exclusive tool for process development or design of equipment is not recommended. "Food drying engineering" still largely relies on experience and experimentation.

Food Process Engineering and Technology. DOI: http://dx.doi.org/10.1016/B978-0-12-415923-5.00022-8
© 2013 Elsevier Inc. All rights reserved.

The most important engineering and technological issues in food dehydration are the following:

- *The kinetics of drying.* With some notable exceptions such as spray drying, drying is a relatively slow process. Knowledge of the factors that affect the rate of drying is essential for the optimal design and operation of drying systems.
- *Product quality.* Removal of water is not the only consequence of most drying operations. Other important quality-related changes in taste, flavor, appearance, texture, structure and nutritive value may occur in the course of drying. The extent of such changes depends on the process conditions.
- *Energy consumption.* Most common drying processes use extensive quantities of energy at relatively low efficiency. Energy-wise, drying is a wasteful water removal process, compared to other water removal operations such as evaporation or membrane separation.

The mechanism of water removal by drying involves two simultaneous processes, namely transfer of heat for the evaporation of water to the food and transport of the water vapors formed away from the food. Drying is, therefore, an operation based on *simultaneous heat and mass transfer.* As explained in subsequent sections, the rate-limiting mechanism may be superficial evaporation or internal transport of water, depending on the conditions.

Depending on the mode of heat and mass transfer, industrial drying processes can be grouped in two categories: convective drying and conductive (boiling) drying.

1. *Convective drying.* Hot and dry gas (usually air) is used both to supply the heat necessary for evaporation and to remove the water vapor from the surface of the food. Both heat and mass exchanges between the gas and the particle are essentially convective transfers, although conduction and radiation may also be involved to some extent. This widespread mode of drying is also known as *air drying.*
2. *Conductive (boiling) drying.* The moist food is brought into contact with a hot surface (or, in a particular application, with superheated steam). The water in the food is "boiled off". In essence, boiling drying is tantamount to evaporation to dryness. Vacuum drying, drum drying and drying in superheated steam are cases of this mode of drying.

Freeze drying (lyophilization) is another method of water removal based on the sublimation of water from a frozen material under high vacuum. In view of the peculiar mechanisms involved, freeze drying will be discussed in a separate chapter.

22.2 Thermodynamics of moist air (psychrometry)

22.2.1 Basic principles

In most food dehydration processes, air is the drying medium. It is therefore appropriate to review basic concepts pertaining to the thermodynamic behavior of air–water vapor mixtures (moist air) before discussing drying kinetics and drying processes.

Although dry air is in itself a mixture of gases (nitrogen, oxygen, carbon dioxide, etc.), we shall consider moist air as consisting of two components only: dry air and water vapor. Gibbs' Phase Rule (Josiah Willard Gibbs, American mathematician–physicist, 1839–1903) establishes the number of possible "degrees of freedom" in a system in equilibrium as follows:

$$F = C - P + 2 \tag{22.1}$$

where:

F = number of degrees of freedom (number of possible variables)
C = number of components
P = number of phases.

For mono-phase (homogeneous) moist air, C = 2 and P = 1. It follows that the number of independent variables is three – e.g., temperature, pressure and moisture content. Psychrometry is most commonly applied to air at atmospheric pressure. In this case, the pressure is not a variable. It follows that the state of homogeneous moist air at atmospheric pressure can be unequivocally defined by two variables: temperature and moisture content (humidity). Consequently, it is customary to represent the state of moist air graphically on a system of two axes: temperature as the abscissa and humidity as the ordinate. The resulting graph is known as the "Psychrometric Chart" (see Appendix, Figure A.2).

The state of two-phase moist air (e.g., moist air containing mist) at atmospheric pressure can be unequivocally defined by one variable only. For processes at variable pressure, one more degree of freedom has to be added.

22.2.2 Humidity

The moisture content of air (*humidity, absolute humidity*) is expressed as the mass ratio of water vapor to dry air (kg water vapor per kg dry air) and is dimensionless. At atmospheric pressure, the air–moisture mixture can be treated as an ideal mixture of gases and Dalton's Law (Chapter 13, Section 13.2) can be applied:

$$P = p_w + p_a \tag{22.2}$$

where P = total pressure (usually atmospheric) and p_w and p_a = partial pressures of water vapor and dry air, respectively.

Assuming perfect gas behavior the mass of water vapor per unit volume of mixture is:

$$m_w = \frac{18 p_w}{RT} \tag{22.3}$$

Similarly, the mass of dry air (average molecular weight = 29) per unit volume of mixture is:

$$m_a = \frac{29p_a}{RT} \tag{22.4}$$

It follows that the absolute humidity of the air, H, is:

$$H = \frac{m_w}{m_a} = \frac{18.0153p_w}{28.966p_a} = \frac{0.6219p_w}{P - p_w} \cong 0.62\frac{p_w}{P} \tag{22.5}$$

Note: The approximation $P - p_w \approx P$ is permissible only at ambient temperatures and in relatively dry air where p_w is much smaller than the total pressure P.

22.2.3 Saturation, relative humidity (RH)

Moisture-saturated air is air in equilibrium with pure liquid water at a given temperature. Consequently, the partial vapor pressure of water vapor in moisture-saturated air is equal to the vapor pressure of liquid water at the same temperature:

$$(p_w)_{sat.} = p_0 \tag{22.6}$$

The saturation humidity H_s is the maximum quantity of water vapor that air can contain at a given temperature, without phase separation.

The relative humidity (ϕ or RH) is the ratio (as percentage) of the partial pressure of water vapor in air to the vapor pressure of liquid water at the same temperature.

$$RH = \frac{p_w}{p_0} \times 100 \tag{22.7}$$

"Percent saturation" (S), often confounded with relative humidity, is:

$$S = \frac{H}{H_s} \times 100 \tag{22.8}$$

The saturation and constant percent saturation lines are indicated on the psychrometric chart.

22.2.4 Adiabatic saturation, wet-bulb temperature

If a mass of air is brought to contact with water under adiabatic conditions (no heat transfer with the exterior), the humidity of the air increases until saturation is reached. Since there is no external source of heat, water is evaporated using heat from the air itself. Consequently, the air is cooled at the same time that it is humidified. The process described is called "adiabatic saturation", and the temperature reached at saturation is called the "temperature of adiabatic saturation". It can be demonstrated that:

1. The temperature of adiabatic saturation is a sole function of the initial conditions (temperature and humidity) of the air. It follows that the temperature of adiabatic saturation is a thermodynamic property of moist air.
2. The process of adiabatic saturation follows a straight line on the psychrometric chart.

If the bulb of a mercury thermometer is wrapped in wet tissue and air is passed over the bulb, the air (and the thermometer) will be eventually cooled to the adiabatic saturation temperature. For this reason, the *wet bulb temperature* is often confounded with the adiabatic saturation temperature of the air. One of the ways to describe unequivocally the state of moist air is to specify its dry bulb (normal) and wet bulb temperatures. It should be remembered, however, that the wet bulb temperature is an empirical value (not a property) depending on the kinetics of the measurement, while the adiabatic saturation temperature is a thermodynamic property.

22.2.5 Dew point

The *dew point* is yet another property of moist air. If a mass of moist air is cooled at constant humidity, saturation is reached and liquid water is formed in the form of dew or mist. The temperature at which this occurs is called the "dew point" of the air. The dew point is one of the phenomena on which certain hygrometers (instruments for measuring humidity in air) are based. The air makes contact with a mirror that is gradually cooled, and the temperature at which the mirror becomes foggy is the dew point of the air. Based on the relationship between the water activity of a food sample and the relative humidity of the air in equilibrium contact with the food, the foggy mirror phenomenon is one of the methods for measuring water activity in the laboratory.

EXAMPLE 22.1

A mass of air at 30°C has a saturation of 60%. Find the absolute humidity, wet-bulb temperature and dew point of that air.

Solution

All the properties are obtainable from the psychrometric chart in the Appendix (Figure A.2).

The absolute humidity is $H = 0.016\ kg \cdot kg^{-1}$

The adiabatic saturation (wet bulb) temperature is 23.6°C.

The dew point is 21.2°C.

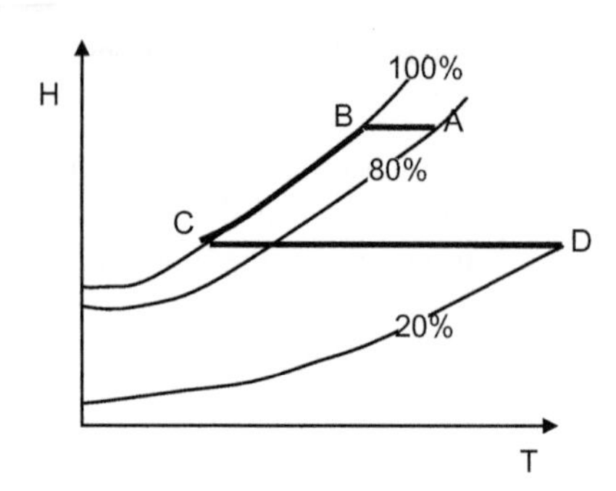

FIGURE 22.1

Dehumidification of air by cooling.

EXAMPLE 22.2

The air in the room used for packaging instant coffee must have a percent saturation of 20% or less, in order to prevent moisture sorption and caking of the product. The dehumidified air is "produced" by cooling air at 30°C and 80% saturation beyond saturation, separating the condensed water and heating the air back to 30°C.

Show the process on the psychrometric chart and find the temperature to which the air has to be cooled.

Solution

The process is shown on the general form of the psychrometric chart (Figure 22.1). Line AB represents cooling at constant humidity until saturation is reached. The section BC represents condensation at 100% saturation. The line CD represents heating of the dehumidified air. Point C shows the temperature to which air has to be cooled to perform the desired dehumidification.

The result is 5°C.

22.3 Convective drying (air drying)

A typical example of this category is tray drying. The moist food is placed on a tray. Hot and dry air is passed over the food (cross-flow drying) or through it (through-flow drying). Heat flows from the hot air to the cooler food by virtue of the temperature gradient and induces evaporation. Water vapor passes from the moist food to the dry air by virtue of the vapor pressure gradient.

22.3.1 The drying curve

As explained above, the rate of drying is of particular engineering and economic importance, mainly because it determines the production capacity of the dryer.

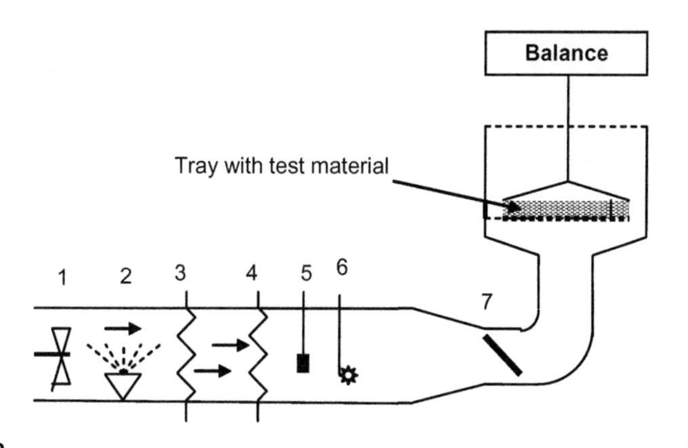

FIGURE 22.2

Schema of the set-up for the determination of drying curves. 1, Fan; 2, humidification; 3, cooling (drying); 4, heating; 5, thermometer; 6, hygrometer; 7, damper.

The drying rate is defined either as *the mass of water removed per unit time per unit mass of dry matter* (denoted as Φ) or as *the mass of water removed per unit time per unit area* (water flux, denoted by N). The factors that affect drying rate may be divided into the following two groups:

1. *Internal conditions:* variables of the material subjected to drying – shape and size, structure (e.g., porosity), moisture content, water vapor pressure as a function of composition and temperature (i.e., water vapor sorption isotherms at different temperatures).
2. *External conditions:* temperature, humidity and velocity of the air.

Drying rate data are usually represented in the form of *drying curves.* A drying curve is the plot of the drying rate Φ or N versus the remaining water content X. Water content X is expressed as kg of water per kg of dry matter.

$$\Phi = -\frac{dW}{M\,dt} = -\frac{dX}{dt} \quad \text{with} \quad X = \frac{W}{M} \tag{22.9}$$

where W = mass of water in the food and M = mass of dry matter in the food.

To build a drying curve experimentally, a thin layer of the food is placed in a drying chamber, on a tray attached to a balance (Figure 22.2). Air at known and constant temperature, humidity and velocity is passed in either a cross-flow or a through-flow direction and the weight of the tray is determined periodically and recorded (usually, air flow has to be stopped for a short time during weighing). If water is the only volatile substance in the sample, then the weight loss is taken to be equal to the loss of water mass $-\Delta W$. The mass of dry matter in the sample, M, is assumed to be constant.

A hypothetical W versus time plot is shown in Figure 22.3. From the data, the drying rate is calculated and plotted against X, to obtain the drying curve.

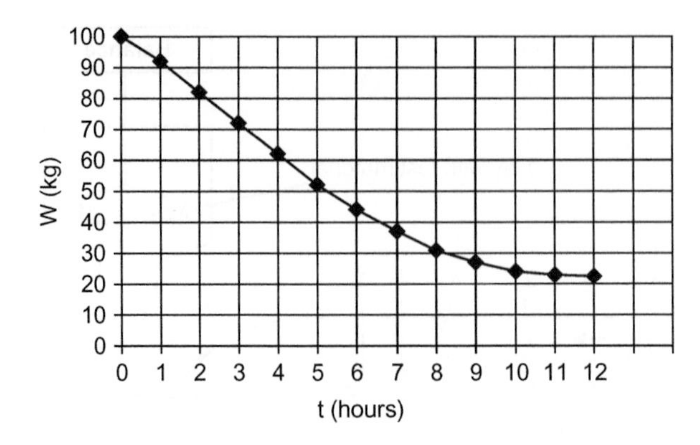

FIGURE 22.3

Theoretical W versus t curve.

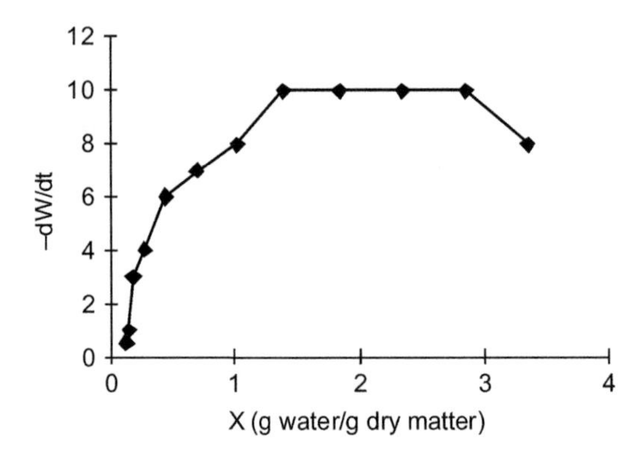

FIGURE 22.4

Drying curve corresponding to data in Figure 22.3.

A drying curve corresponding to the data of Figure 22.3 is shown in Figure 22.4. (Note: Since the moisture content X decreases during the experiment, the course of the process on the graph is from right to left.) Practically, the most sophisticated part of the experimental determination of a drying curve is the production of an air stream having precisely constant temperature, humidity and velocity.

The drying curve of Figure 22.4 is usually modeled so as to show three regions or phases:

1. Region I: phase of *rising rate*. The rate of drying increases as water is removed. Physically, this behavior is attributed to the "conditioning" of the

sample (e.g., warming up, opening the pores, etc.). This phase is usually short and is not always observed in drying experiments. It is often omitted in the calculation of drying time.

2. Region II: phase of *constant rate.* The drying rate remains nearly constant as water is removed. Truly constant rate drying may be observed when slowly drying wet sand or paper (Krischer and Kast, 1992) but seldom when drying real foods (Bimbenet *et al.*, 2002). The physical background of the constant drying rate regime will be explained later.
3. Region III: phase of *falling rate.* Below a certain moisture content, called the "critical moisture content", X_C, the drying rate drops sharply as water is removed. The probable mechanisms responsible for the falling rate will be discussed later.

22.3.2 The constant rate phase

Theory predicts that the rate of convective evaporation from a water-saturated surface is constant if the external conditions (temperature, humidity and velocity of the air) are constant.

Consider a mass of wet material subjected to convective drying by air (Figure 22.5). Heat is transferred convectively from the warm air to the cooler wet surface through the boundary layer at the interface and induces some evaporation. The water vapor formed is transported by convection from the moisture-saturated surface to the drier air.

The following transport equations can be written:

For heat transfer:

$$q = hA(T_a - T_s) \tag{22.10}$$

For mass transfer:

$$-\frac{dW}{dt} = k_g A(p_s - p_a) = k'_g A(H_s - H_a) \tag{22.11}$$

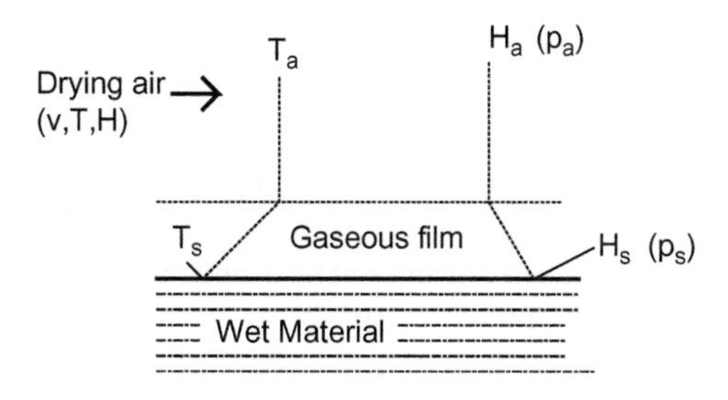

FIGURE 22.5

Heat and mass transfer in convective drying.

where:

$-dW/dt$ = rate of water transfer, $kg \cdot s^{-1}$
h = coefficient of convective heat transfer, $W \cdot m^{-2} K^{-1}$
k_g and k'_g = coefficients of convective mass transfer, $kg \cdot m^{-2} \cdot s^{-1} \cdot Pa^{-1}$ and $kg \cdot m^{-2} \cdot s^{-1}$, respectively
A = area of active transfer, m^2
p_a and p_s = partial pressure of water vapor in the air and on the wet surface, respectively, Pa
H_a and H_s = humidity in the air and on the wet surface, respectively, (dimensionless).

Assume that, at steady state, the only effect of heat transfer is evaporation of water (no sensible thermal effects). Then:

$$N = -\frac{dW}{Adt} = k'_g(H_s - H_a) = \frac{h(T_a - T_s)}{\lambda} \quad (22.12)$$

As long as the wet surface is saturated with water, T_s is the wet bulb temperature of the drying air and is constant (see Section 22.2). The humidity H_s is the adiabatic saturation humidity of the air and is also constant. It follows that the drying rate *per unit area* $N = [-dW/Adt]$ should also be constant as long as the following three conditions are satisfied:

1. The wet surface is water-saturated (i.e., behaves like the surface of pure water)
2. The air temperature, humidity and velocity are kept constant
3. Heat is transferred to the wet surface only by convection from the drying air.

To the extent that these conditions are fully satisfied, the temperature of the food in the constant rate period does not exceed the adiabatic saturation temperature of the air, even if the actual temperature (dry bulb temperature T_a) of the air is quite high.

As mentioned, the theoretical analysis above predicts a period of constant drying rate *per unit area.* In the absence of shrinkage, the area is constant and so is the area per unit mass of dry matter. In this case, the drying rate *per unit mass of dry matter* Φ, as defined in Section 22.3.1, would also be constant.

Constant rate drying prevails as long as any amount of water lost by evaporation is replaced by water transported from the interior to the surface by diffusion. In other words, the constant rate drying period can be defined as the phase of drying during which evaporation from the surface and not internal diffusion is the rate-controlling mechanism (Chu *et al.*, 1956).

In practical food dehydration these conditions are seldom met, and therefore truly constant rate drying is, at most, an approximation that can be applied only to foods with a high initial moisture content and an inert solid matrix. Real situations deviate from the ideal conditions required for constant rate drying in the following aspects:

- In real drying, heat is not transferred to the drying surface solely by convection. Conduction and radiation transfer also take place, to some extent. As a result, the temperature at the drying surface is higher than the wet bulb temperature of the air (see Section 22.3.7).
- Even in the case of foods with fairly high water content, the surface of the food does not behave like pure water because of the water vapor pressure depression due to the soluble components, among other reasons.
- The heat exchanged is not used only for evaporation. Sensible heat effects also occur.
- Usually, the "no-shrinkage" condition is not met.

As a result of these deviations from the ideal model, the drying rate declines from the very beginning of the process. Notwithstanding these deviations, the existence of a constant rate period is often assumed in the modeling of convective drying.

EXAMPLE 22.3

"Filo" dough sheets, 2 mm thick, are air-dried from both sides, from an initial moisture content of $X_0 = 1$ kg water/kg dry matter to a final moisture content of $X = 0.25$ kg water/kg dry matter. The temperature of drying air is 50°C and its percent saturation is 20%. In order to avoid mechanical stress, deformation and cracks, the moisture gradient in the dough must be maintained below 200 m^{-1}. It is known that, under the conditions of the process, the convective heat transfer coefficient is given by the correlation:

$$h = 10(v)^{0.8}$$

where h = convective heat transfer coefficient, $W \cdot m^{-2} \cdot K^{-1}$, and v = air velocity, $m \cdot s^{-1}$.

It is also known that, under the conditions of the process, the drying rate is constant.

The density of the dough is $\rho = 1030\ kg \cdot m^{-3}$, the diffusivity of water in the dough is $D = 3 \times 10^{-9}\ m^2 \cdot s^{-1}$ and the latent heat of evaporation of water is $\lambda = 2300\ kJ \cdot kg^{-1}$.

What is the maximum permissible air velocity?

Solution

During drying, moisture gradient is at its maximum at the surface. Hence the drying conditions must comply with the following condition:

$$-\left(\frac{\partial X}{\partial z}\right)_{z=0} \leq 200\ m^{-1}$$

The rate of moisture transport to the surface is, by Fick's Law:

$$\frac{dW}{AdT} = -D\rho\left(\frac{\partial X}{\partial z}\right)_{z=0}$$

Since there can be no accumulation on the interface, the rate of moisture transport to the surface must be equal to the rate of evaporation from the surface.

It is known that drying takes place at a constant rate. The rate of evaporation from the surface is then given by Eq. (22.12):

$$N = -\frac{dW}{Adt} = \frac{h(T_a - T_s)}{\lambda}$$

It follows that:

$$\frac{h(T_a - T_s)}{\lambda} = D\rho\left(\frac{\partial X}{\partial z}\right)_{z=0}$$

From the psychrometric chart, we find the adiabatic saturation temperature of the air $T_s = 28.8°C$. Substitution in the equation leads to:

$$\frac{h_{max}(50 - 28.8)}{2300000} = 3 \times 10^{-9} \times 1030 \times 200 \quad \Rightarrow \quad h_{max} = 67\ \mathrm{W \cdot m^{-2} \cdot K^{-1}}$$

But $h = 10\,(v)^{0.8}$, hence $67 = 10\,(v_{max})^{0.8}$.

$$v_{max} = 10.78\ \mathrm{m \cdot s^{-1}}.$$

22.3.3 The falling rate phase

The rate of water transfer from the interior of the particle to its surface decreases continuously as the product becomes drier. When the supply of water to the surface drops below the rate of evaporation, the moisture content of the surface begins to decrease rapidly and quickly approaches the equilibrium moisture content corresponding to the relative humidity of the air on the sorption isotherm of the material. From that moment, internal transport and not evaporation becomes the rate-limiting factor, and the falling rate period begins. The average moisture content of the food when this occurs is called the "critical moisture content", X_c. Obviously, a sharp transition from constant to falling rate can only be observed if a real constant rate period exists – i.e., when drying inert, non-shrinking materials, but not in foods. In the case of foods, the negative slope of the drying curve becomes gradually steeper but there is no well-defined point of transition. Nevertheless, the concept of critical moisture is retained for the purpose of modeling.

Since the surface of the food is no longer water saturated, the temperature of the food rises during the falling rate period and approaches asymptotically the dry bulb temperature of the air. Theoretically, drying stops (i.e., the drying rate becomes 0) when the moisture content everywhere in the food has been reduced to the equilibrium moisture content X_e.

Many different theories concerning the mechanism of internal water transfer in drying have been proposed (Bruin and Luyben, 1980; Barbosa-Canovas and Vega-Mercado, 1996). These include liquid water diffusion by virtue of concentration gradients, water vapor diffusion under the effect of vapor pressure differences, capillary transport, evaporation–condensation, etc. Probably, a number of different mechanisms are simultaneously responsible for the internal movement of water molecules during drying. Based on simultaneous heat and mass transfer functions, a very large number of mathematical, empirical and semi-empirical models for the simulation and prediction of drying curves have been proposed for specific products or for foods in general (Barati and Esfahani, 2011; Chen and Pei, 1989; Di Bonis and Ruocco, 2008; Sander and Kardum, 2009; Sereno and Madeiros, 1990).

The shape of the drying curve in the falling rate stage depends on the mechanism of internal water transfer. If molecular (Fickean) diffusion is assumed to be the principal mechanism, a linear relationship should exist between the logarithm of the residual removable water content and drying time (see Chapter 3, Section 3.5.2):

$$\frac{d[\ln(X - X_e)]}{dt} = Cons\tan t \tag{22.13}$$

Taking derivatives with respect to X on both sides and rearranging we obtain:

$$\frac{dX}{dt} = K(X - X_e) \tag{22.14}$$

Equation (22.14) indicates that the falling rate portion of the idealized drying curve should be a straight line between X_c and X_e (Figure 22.6). The theoretical curve describing the falling rate regime with the assumption of capillary transport would also be a straight line. In reality, even if the mechanism of molecular diffusion is taken for granted, the application of Eq. (22.13) to food drying is problematic. The diffusivity of water in the food is strongly temperature- and composition-dependent (Table 22.1). The concentration dependence of diffusivity is a logical consequence of the sorption isotherm (dependence of water activity on concentration). Other phenomena, such as migration of solutes, shrinkage, and thermal effects, also contribute to the deviation from the simplified Fickean model.

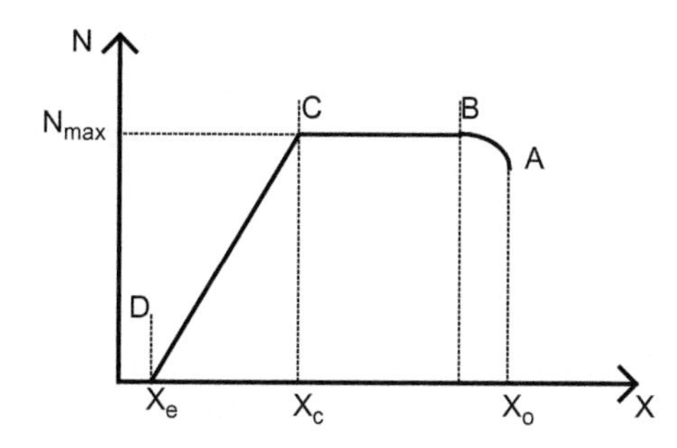

FIGURE 22.6

Theoretical drying curve model.

Table 22.1 Water Diffusion Coefficient of Skim Milk.

	Diffusion Coefficient D ($m^2 \cdot s^{-1}$)		
	Moisture Content (kg water/kg dry matter)		
Temperature °C	**0.1**	**0.3**	**0.7**
10	3×10^{-13}	1.6×10^{-11}	1.2×10^{-10}
30	1.2×10^{-12}	5×10^{-11}	2.2×10^{-10}
50	3.3×10^{-12}	1×10^{-10}	3.3×10^{-10}
70	1.3×10^{-11}	2.2×10^{-10}	7.4×10^{-10}

Calculated from a figure in Bruin and Luyben (1980).

Table 22.2 Effective Diffusivity of Water in Some Foods, Calculated from Drying Rate.

Material	*Temp. °C*	*D_{eff}($m^2 \cdot s^{-1}$)*	*Reference*
Avocado	50	2.3×10^{9}	Daudin, 1983
Sugar beet	50	5×10^{-10}	Daudin, 1983
Manioc	50	3×10^{-10}	Daudin, 1983
Apple	50	1.1×10^{-9}	Daudin, 1983
Potato	50	2×10^{-10}	Daudin, 1983
Apple, Granny Smith	76	1.15×10^{-10}	Bruin and Luyben, 1980
Pepperoni sausage, 13.3% fat	12	5.7×10^{-11}	Bruin and Luyben, 1980

An average diffusivity value can be calculated from experimental drying curves for the entire curve or parts of it. The values obtained are reported as "*effective diffusivity*". Some effective diffusivity values, calculated by different authors from experimental drying curves, are given in Table 22.2.

In some models the concave falling rate portion of the drying curve is approximated by two or more linear segments, representing so-called second, third, etc., falling rate periods (Figure 22.7).

22.3.4 Calculation of drying time

What is the time required for drying a food from an initial moisture content X_1 to a final moisture content X_2, under constant external drying conditions? Equation (22.9) can be written as follows:

$$-dw = -MdX = \Phi Mdt \tag{22.15}$$

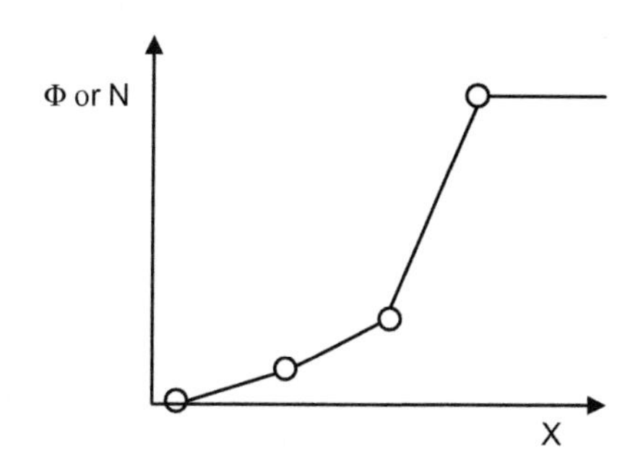

FIGURE 22.7

Drying curve approximated as a series of linear falling rate periods.

Integration gives the drying time:

$$t = -\int_{X_1}^{X_2} \frac{dX}{\Phi} \quad (22.16)$$

The integral can be calculated if $\Phi = f(X)$ is known as a drying curve or as an algebraic expression. In the absence of such information, a model of drying kinetics must be assumed. As an example, it will be assumed that the drying curve consists of a constant rate portion from X_1 to X_c, at a known constant rate Φ_0, followed by a linear falling rate portion from a known (or assumed) X_c to X_e.

If both X_1 and X_2 are larger than X_c, the entire drying period takes place at constant rate. Then:

$$t = \frac{X_1 - X_2}{\Phi_0} \quad (22.17)$$

If both X_1 and X_2 are smaller than X_c, the entire drying period takes place during the falling rate phase. Then:

$$t = \frac{X_c - X_e}{\Phi_0}\int_{X_2}^{X_1} \frac{dX}{X - X_e} = \frac{X_c - X_e}{\Phi_0} \cdot \ln\left(\frac{X_1 - X_e}{X_2 - X_e}\right) \quad (22.18)$$

If X_1 is larger than X_c and X_2 is smaller than X_c, the drying period consists of a constant rate phase and a falling rate phase. Then:

$$t = \frac{X_1 - X_c}{\Phi_0} + \frac{X_c - X_e}{\Phi_0} \cdot \ln\left(\frac{X_c - X_e}{X_2 - X_e}\right) \quad (22.19)$$

EXAMPLE 22.4

Estimate the time necessary to dry dates from 75% to 20% moisture content (on wet basis) under constant external conditions. Under the conditions of the process, a falling rate regime is known to prevail during drying. A linear relationship between drying rate and residual moisture is assumed. The initial drying rate (when the moisture content of the dates is 75%) is 0.5 kg water removed per kg of dry matter per hour. The moisture content of the dates at equilibrium with the drying air is 8% (wet basis).

Solution

We convert the "wet basis" moisture content data to "dry basis" values:

$$X_1 = \frac{75}{100-75} = 3 \quad X_2 = \frac{20}{100-20} = 0.25 \quad X_E = \frac{8}{100-8} = 0.087$$

$$t = \frac{X_c - X_e}{\Phi_0} \cdot \ln\left(\frac{X_1 - X_e}{X_2 - X_e}\right) = \frac{3-0.087}{0.5} \ln\left(\frac{3-0.087}{0.25-0.087}\right) = 5.89 \text{ h}$$

The drying time is 5.89 hours.

EXAMPLE 22.5

Calculate the time necessary to dry a food material from 80% to 20% moisture content. The food is dried on trays, from one side. Loading rate is 10 kg per m^2.

Drying air data:

Temperature DB = 70°C; WB = 30°C
Velocity, $v = 10\ m \cdot s^{-1}$
Density $\rho = 1\ kg \cdot m^{-3}$
Critical moisture of the food = 45%.
Equilibrium moisture of the food = 0.
All moisture data are w/w, wet basis. Assume that there is no shrinkage.
Assume $h = 20 \cdot G^{0.8}$ where $G = v\rho$.

Solution

Drying is at constant rate from 80% to 45% moisture, and at a falling rate from 45% to 20% moisture. A standard drying curve with a sharp change at critical moisture and linear falling rate will be assumed.

For the constant rate period, drying rate N is:

$$N = -\frac{dW}{Adt} = \frac{h(T_a - T_s)}{\lambda}$$

$$h = (v\rho)^{0.8} = (10 \times 1)^{0.8} = 6.31\ \text{W}\cdot\text{m}^{-2}\cdot\text{K}^{-1}$$

From water vapor tables, at 30°C: $\lambda = 2430\ kJ \cdot kg^{-1}$.

$$N = \frac{6.31(70-30)}{2430 \times 10^3} = 0.103 \times 10^{-3}\ \text{kg}\cdot\text{m}^{-2}\cdot\text{s}^{-1}$$

The loading rate is $10\ kg \cdot m^{-2}$ wet material at 80% moisture – i.e., $2\ kg \cdot m^{-2}$ dry material. Converting N to Φ:

$$\Phi = N \times 2 = 0.206 \times 10^{-3}\ \text{kg}\cdot\text{kg}^{-1}\cdot\text{s}^{-1}$$

The moisture content data are converted to dry basis:

$$X_1 = \frac{80}{100-80} = 4 \quad X_2 = \frac{20}{100-20} = 0.25 \quad X_C = \frac{45}{100-45} = 0.82 \quad X_e = 0$$

For drying time, Eq. (22.19) is applied:

$$t = \frac{X_1 - X_c}{\Phi_0} + \frac{X_c - X_e}{\Phi_0} \cdot \ln\left(\frac{X_c - X_e}{X_2 - X_e}\right)$$

$$t = \frac{4 - 0.82}{0.206 \times 10^{-3}} + \frac{0.82}{0.206 \times 10^{-3}} \ln\left(\frac{0.82}{0.25}\right) = 20.17 \times 10^3 \text{ s} = 5.6 \text{ h}$$

The total drying time is 5.6 hours.

22.3.5 Effect of external conditions on the drying rate

1. *Air velocity*. Increasing the air velocity accelerates both heat and mass transfer at the interface and therefore increases the drying rate *as long as the rate-controlling mechanism is evaporation at the surface*. The following empirical correlation gives approximate values of the convective heat transfer coefficient as a function of air flow:

$$h = 20.G^{0.8} \tag{22.20}$$

where:
h = coefficient of convective heat transfer, $W \cdot m^{-2} K^{-1}$
G = vρ = superficial mass flow-rate of the air, $kg \cdot m^{-2} \cdot s^{-1}$ (to obtain G, multiply air velocity by the density of the air at the prevailing temperature and pressure).

Air velocity has no direct effect on the internal water transport, and therefore should not affect drying rate at the falling rate period. For the entire drying period, the following approximate correlation may be used:

$$h = 20 \cdot G^{0.5} \tag{22.21}$$

2. *Air temperature*. According to Eq. (22.12), drying rate in the constant rate phase is proportional to the wet bulb temperature depression $T - T_s$. In the falling rate period the effect of air temperature is indirect, mainly through its influence on the diffusivity of water.
3. *Air humidity*. In the constant rate period, drying rate is proportional to the difference H_S-H between the adiabatic saturation humidity of the air and its actual humidity. Theoretically the humidity of the air has no effect on the falling rate phase, except for its obvious relation with X_e through the sorption isotherm of the food.

In summary, all the external conditions affect the shape of the drying curve. Theoretically, every external factor that increases the initial drying rate (higher air temperature, lower air humidity, higher air velocity) is liable to increase the critical moisture content and thus anticipate the onset of the falling rate, as shown schematically in Figure 22.8. Too rapid initial drying is one of the reasons

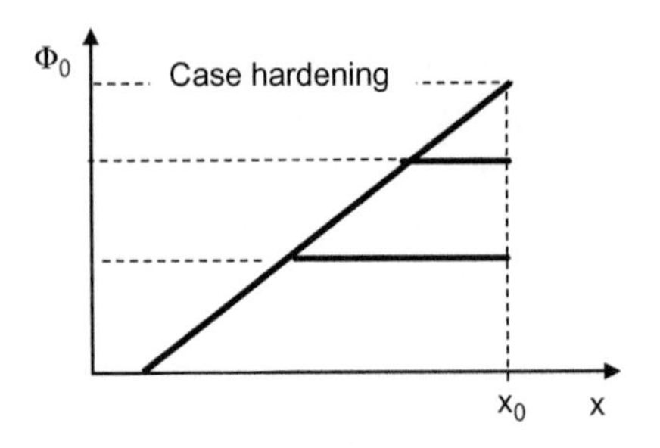

FIGURE 22.8

Conditions for case hardening.

responsible for the phenomenon known as "*case hardening*", whereby a dry, hard crust impermeable to moisture forms on the surface while the water content at the center is still quite high. Case-hardening is generally undesirable in most food dehydration processes but may be desirable in other cases, such as bread baking and frying.

22.3.6 Relationship between film coefficients in convective drying

As stated before, the convective heat and mass transfer coefficients in air drying are interrelated.

The relationship between the mass transfer coefficients k_g and k'_g is given by the following approximate formula:

$$k'_g/kg = 1600 \text{ kPa} \tag{22.22}$$

The heat and mass transfer coefficients are interrelated as follows:

$$h/k'_g = C_H \text{ J}\cdot\text{kg}^{-1} \tag{22.23}$$

where C_H is the specific heat of the humid air.

The following is another practical correlation between coefficients (Bimbenet *et al.*, 2002):

$$h/k_g \approx 65\lambda \text{J}\cdot\text{Pa}\cdot\text{K}^{-1}\cdot\text{kg}^{-1} \tag{22.24}$$

where λ represents the latent heat of evaporation of water at the wet bulb temperature of the air.

22.3.7 Effect of radiation heating

In practice, a certain proportion of the heat is transferred to or from the drying surface by radiation, in addition to convection from the air. In some dryers, radiating bodies are installed to accelerate drying; in this case, the proportion of heat transferred by radiation may be considerable.

Combined convection–radiation heat transfer was discussed in Chapter 3. A "pseudo" combined heat transfer coefficient, h_r, was defined. Thus, for calculation of the constant drying rate Eq. (22.12) is rewritten, adding the portion of the heat transferred by radiation:

$$-\frac{dW}{Adt} = K'_g(H_s - H_a) = \frac{h(T_a - T_s)}{\lambda} + \frac{h_r(T_r - T_s)}{\lambda} \tag{22.25}$$

where T_r is the temperature (in °C) of the radiating hot surface.

However, h_r depends on the temperatures of the surfaces emitting and receiving radiation. Note that the surface temperature T_s is no longer the wet bulb temperature, but higher. Thus, the solution of Eq. (22.25) requires trial and error. A surface temperature is assumed, the corresponding H_s is found on the psychrometric chart, h_r is calculated and the process is repeated until the values of H_s, T_s and h_r satisfy Eq. (22.25). For exchange between two parallel plates behaving like black bodies, h_r is given by Eq. (3.61), which becomes in this case:

$$h_r = \sigma \frac{T_r^4 - T_s^4}{T_r - T_s} \tag{22.26}$$

The application of microwave energy in drying has been investigated (Khraisheh *et al.*, 1995; Li *et al.*, 2010; McMinn *et al.*, 2003). The combination of microwave and convective drying has been reported to give better results than microwaves alone. Microwave drying is used industrially in the production of hydrocolloid gums.

22.3.8 Characteristic drying curves

As explained in Section 22.3.5, the drying curve is affected by the external drying conditions (air temperature, moisture and velocity). If a given material is dried under different external conditions, the result is a family of drying curves. In practice, it is impossible and often undesirable to maintain constant external conditions. The conditions change by virtue of the process itself. As the food is dried, the air is humidified and cooled. The nearly constant air conditions required for the experimental determination of drying curves are obtained by maintaining a very high air/food mass ratio (thin-layer drying), so that the changes in air properties as a result of drying itself are negligible. How, then, can drying curves be used for the analysis of drying processes (calculation of drying time, estimation of the temperature and moisture distribution within the mass, etc.) where the air/food mass ratio is more realistic, such as drying a thick bed of food by air flowing through it or drying food on a long tray by air flowing longitudinally over its surface? The accepted method of calculation is to consider the bed (or the tray) as consisting of a number of finite elements (thin layers or short lengths), calculate the extent of drying in each element as drying under constant conditions, determine the changed conditions of the air emerging from each element, and calculate drying in the next element using the changed conditions of the air. This procedure requires access to a large number of experimental drying curves.

This problem led a number of researchers to define a single characteristic curve, capable of describing the drying profile of a given product under different external conditions (Daudin, 1983). The characteristic curve is constructed by applying certain transformations on the variables of the drying curve, X and Φ. In one of those empirical models (Fornell, 1979), the following transformations are made:

$$\begin{aligned} \Phi &\Rightarrow [\Phi] = \frac{\Phi}{(T_a - T_s)\cdot v^{0.5}} \\ X &\Rightarrow [X] = \frac{X}{X_0} \end{aligned} \tag{22.27}$$

If the results of drying experiments, obtained under different external conditions are plotted as [Φ] vs [X] instead of Φ vs X, all the points are supposed to fall on the same single curve. Successful grouping of the drying curves to a single characteristic curve has been reported for pieces of carrot, apple, potato and sugar beet, but the method does not seem to be universally applicable (Daudin, 1983).

22.4 Drying under varying external conditions

The process conditions in a dryer vary with location and time. Frequently, the conditions are changed intentionally for process optimization. For example, vegetables with high initial moisture content (e.g., onions) are usually dried by a multi-stage process. In the first stage, the temperature and velocity of the air are relatively high. At the final stage, during which the falling rate regime prevails, air at moderate temperature and low velocity is used. Here, however, only the changes resulting from drying itself, without external intervention, will be discussed, with the help of some examples. In all the examples, only drying under the constant rate regime (water-saturated surfaces) will be considered. This is because under the falling rate regime drying follows the course dictated by internal transport, with almost no effect of external conditions, other than temperature, on drying kinetics.

22.4.1 Batch drying on trays

Consider a tray (Figure 22.9) on which a layer of wet food is being dried with a stream of air flowing over the surface of the food (cross-flow drying). In passing over the food, the air is humidified and cooled. It follows that portions of the food are exposed to different air conditions, depending on their position on the tray. Assume that the rate-limiting mechanism, everywhere on the tray, is

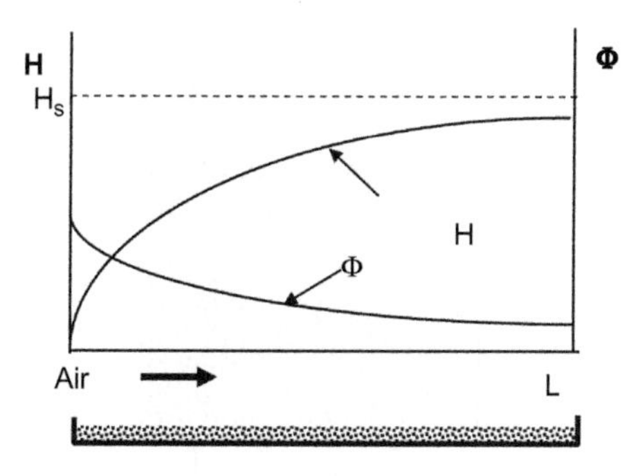

FIGURE 22.9

Drying rate and air humidity distribution in tray drying.

evaporation – i.e., nowhere on the tray is the moisture content of the food below the critical moisture level. It will be also assumed that the system is adiabatic.

Let dH be the increase in the humidity of the air as it passes over a segment of the tray of length dL. A material balance can be written, equating the moisture gained by the air with the water lost by the food on that tray segment:

$$G.dH = k_g'(a \cdot dL)(H_s - H) = \left(-\frac{dW}{Adt}\right) a \cdot dL \tag{22.28}$$

where:

G = mass flow-rate of the air (dry basis), kg · $^{-1}$
H = local humidity of the air (dimensionless)
H_s = adiabatic saturation humidity of the air (dimensionless)
a = width of the tray, m.

Separation of the variables and integration gives:

$$\int_{H_0}^{H} \frac{dH}{H_s - H} = \int \frac{k_g' a \cdot dL}{G} \quad \Rightarrow \quad \ln \frac{H_s - H_0}{H_s - H} = \frac{k_g' a \cdot L}{G} \tag{22.29}$$

Equation (22.29) indicates that the dependence of (H_s − H) of L is logarithmic (Figure 22.10). However, since drying is assumed to take place from water-saturated surfaces everywhere on the tray, the rate of drying is proportional to (H_s − H). Hence:

$$N_L = N_0 \exp\left(-\frac{k_g' aL}{G}\right) \tag{22.30}$$

where $N_L = (-dW/Adt)_L$ = local drying rate per unit area at location L on the tray, and $N_0 = (-dW/Adt)_0$ = local drying rate per unit area at location 0 on the tray.

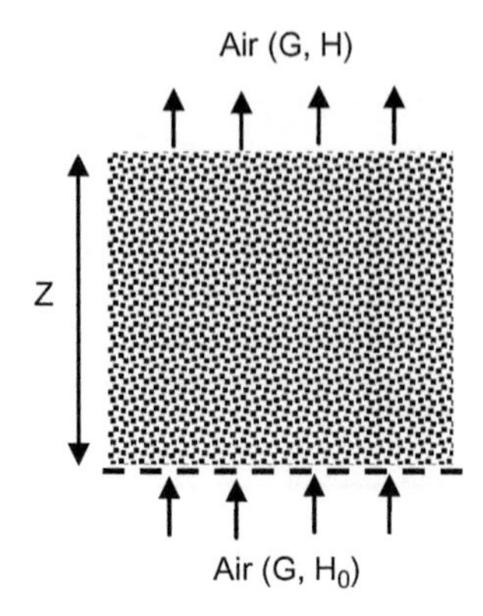

FIGURE 22.10

Through-flow drying.

It follows that the drying rate, although constant with time, varies with location. At any time, the food closest to the entrance of the air will also be the driest, and *vice versa.* The residual moisture content of the food will be a function of location and time.

22.4.2 Through-flow batch drying in a fixed bed

Consider a bed of humid food particles, dried with air flowing through the bed (Figure 22.10).

Assume, as in the previous example, that the moisture content of every particle in the bed is still above the critical moisture and that the bed is adiabatic. Let dH be the increase in the humidity of the air as it passes through a bed thickness of dz, and let dA be the total area of the particles confined in the bed thickness dz. If the particles are assumed to be non-shrinking spheres of equal diameter d_p, one can write:

$$dA = \frac{S(1-\varepsilon)dz}{\pi\, d_p^3/6} \times \pi\, d^2 = \frac{6S(1-\varepsilon)dz}{d_p} \tag{22.31}$$

where S is the cross-sectional area of the bed.

A material balance on water is written as before:

$$G\, dH = k'_g \cdot dA \cdot (H_s - H) = \left[\frac{6S(1-\varepsilon)k'_g}{d_p}\right](H_s - H)dz \tag{22.32}$$

Denoting as K the group of invariables in the bracket, integration gives:

$$\int_{H_0}^{H_z} \frac{dH}{H_s - H} = \frac{K}{G} \cdot \int_0^Z dz \;\Rightarrow\; \ln\left(\frac{H_s - H_0}{H_s - H}\right) = \frac{KZ}{G} \tag{22.33}$$

Again, a linear relationship is found between ln $(H_s - H)$ and the distance Z.

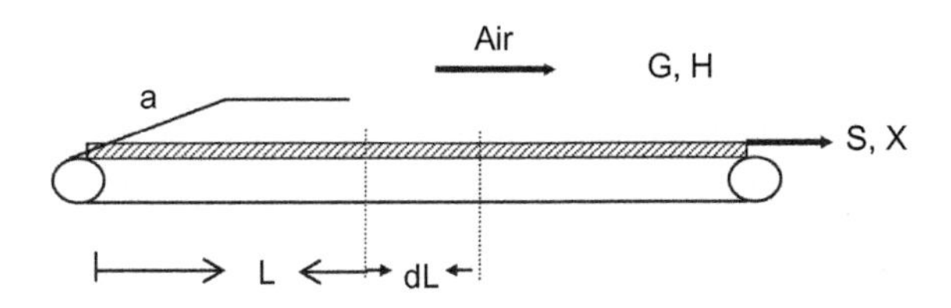

FIGURE 22.11

Continuous drying on a belt.

22.4.3 Continuous air drying on a belt or in a tunnel

Consider a belt conveyor (Figure 22.11) carrying food at the rate of F kg $\cdot^{-1}$ (dry basis). Air, at a flow-rate of G, flows over the surface of the food.

Material balance on water, for a belt segment dL gives:

$$-F\,dX = N \cdot a \cdot dL \tag{22.34}$$

N is the local drying rate. Assuming again drying from a water-saturated surface, N can be written as a function of the position L, as formulated in Eq. (22.30). Combining Eq. (22.30) and Eq. (22.34) and integrating between L = 0 and L = L, we get:

$$X_0 - X_L = \pm \frac{G}{F} \cdot \frac{N_0 - N_L}{k'_g} \tag{22.35}$$

The plus and minus signs apply to co-current and countercurrent operation, respectively.

22.5 Conductive (boiling) drying

22.5.1 Basic principles

Drying by boiling, also known as *contact drying* or *conductive drying*, is a process whereby heat is transferred to the wet material from a heated surface in contact with it. The principal mechanism of heat transfer in boiling drying is conduction, with the exception of superheated steam drying where heat is transported by convection. Boiling drying is applied most commonly (but not exclusively) to foods in liquid or slurry form. Unlike air drying, the temperature of the material during most of the drying period is at or above the boiling point of the liquid at the prevailing pressure, hence the name of "drying by boiling". As stated in the introduction to this chapter, boiling drying is similar to evaporation. The main difference between the two operations is in that boiling drying is continued to a much lower final moisture content. As in evaporation, the rate-controlling factor is heat transfer.

In order to transfer heat to the boiling liquid at a sufficiently high rate, the heated surface must be at a temperature much higher than the boiling point at the process pressure. Consequently, the temperature of the food in contact with the surface may reach very high temperatures, particularly towards the end of drying. One or both of the two following measures are taken in order to limit product overheating:

1. The material (liquid, slurry, paste) is applied on the heated surface as a very thin layer so that drying time is fairly short.
2. Drying takes place under reduced pressure (vacuum) so as to lower the boiling point of the material.

Regardless of these measures, thermal damage is generally more serious in the case of contact drying. On the positive side, contact drying is more economical in energy expenditure because heat is transferred from the primary source of energy (steam, electricity) directly to the food, without air as intermediate transfer medium.

22.5.2 Kinetics

The kinetics of contact drying, like that of air drying, shows a pattern consisting of three stages (Figure 22.12):

Stage 1: This is the stage of heating the feed up to the boiling point. Only a negligible proportion of the water is evaporated during this stage.
Stage 2: The temperature of the material is maintained at the boiling point which rises slightly as the product becomes more concentrated. The viscosity of the material increases. Most of the water is evaporated during this stage.

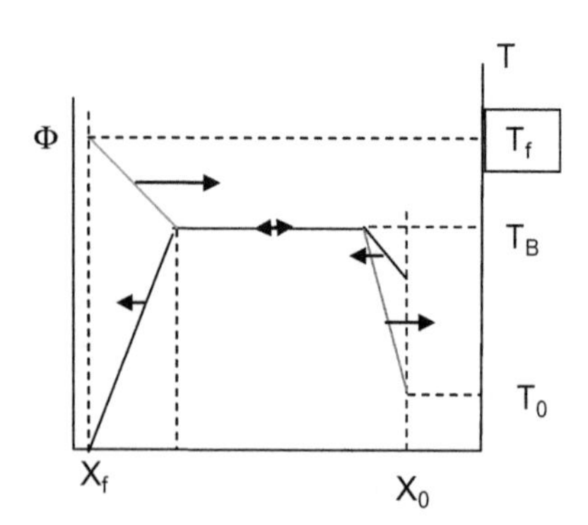

FIGURE 22.12

Stages in contact drying.

The rate of drying is governed by the rate of heat transfer, as shown in Eq. (22.36):

$$\frac{-dW}{dt} = \frac{-M.dX}{dt} = \frac{A.U.(T_h - T_B)}{\lambda} \tag{22.36}$$

where:
A = area of the contact surface, m^2
U = coefficient of contact heat exchange, $W \cdot m^{-2} K^{-1}$
T_h = temperature of the heated surface, °C
T_B = boiling point of the material, °C
λ = latent heat of evaporation of water, $J \cdot kg^{-1}$.

Assuming constant boiling temperature and constant heat transfer coefficient, integration of Eq. (22.36) yields:

$$t = \left(\frac{\lambda}{U.\Delta T}\right)\left(\frac{M}{A}\right)(X_0 - X) \tag{22.37}$$

The time t required to reduce the water content of the food from X_0 to X during Stage 2 is therefore proportional to the temperature difference ΔT between the heated surface and the boiling liquid, and to the "loading factor" M/A, i.e., the mass of the feed (dry basis) per unit area of heated surface – hence the thickness of the food layer.

Stage 3: This is the stage of falling rate of drying. Because of the high viscosity of the concentrated liquid, the overall coefficient of heat transfer U is reduced and the rate of heat dissipation through the drying layer is slowed down. The temperature of the product now rises and tends asymptotically to that of the contact surface. Some porosity may be created in the drying material, causing a further decrease in the rate of heat transfer but improving mass transfer somewhat. The simultaneous drop in ΔT and U results in rapid decay of the drying rate. At this stage, most of the residual water in the product is adsorbed onto the dry matter; therefore, removal of water now takes place by desorption until the desired final dryness is reached.

22.5.3 Systems and applications

22.5.3.1 Drum drying

In drum drying, the heated surface is the envelope of a rotating horizontal metal cylinder. The cylinder is heated by steam condensing inside, at a pressure in the range of 200–500 kPa, bringing the temperature of the cylinder wall to 120–155°C. The wet material is applied on the drum surface as a relatively thin layer by a variety of different methods to be described later. The dried product is removed from the drum with the help of a blade (Figure 22.13). In vacuum drum drying, applied to materials highly sensitive to heat, the drum and its accessories are enclosed in a vacuum chamber. Drum drying is extensively used in the

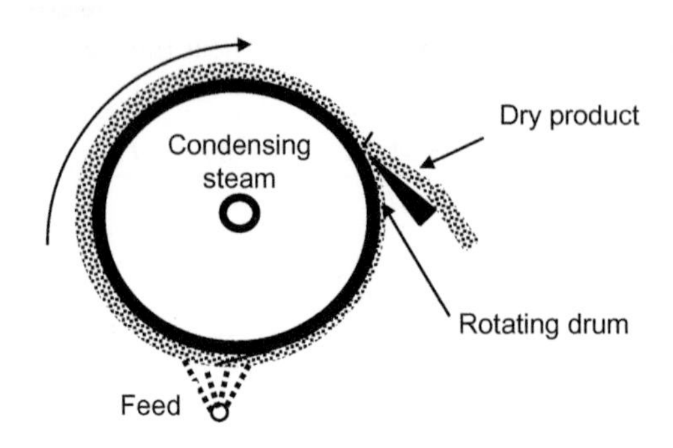

FIGURE 22.13

Principle of drum drying.

production of instant mashed potatoes, pre-cooked cereals, soup mixtures and low-grade milk powder.

22.5.3.2 Belt drying

In this application, the heated surface is a metal belt conveyor heated by contact or radiation by hot elements installed on both sides. In one case, known as "foam-mat drying", the belt is used for drying concentrated juices. Because of the high viscosity of the feed material, mass transfer is important right from the start of the process. In order to improve mass transfer, the concentrates are first foamed and placed on the belt as a porous mat. Some of the tomato powders available commercially are made by foam-mat drying of tomato paste. Belt drying can also be carried out under vacuum (Figure 22.14).

22.5.3.3 Vacuum tray drying

Drying under vacuum is particularly intended for heat-sensitive materials. Industrial vacuum tray drying is essentially a scaled up version of the laboratory vacuum dryer. The material to be dried is spread on heat-conducting trays, and the trays are placed on heated shelves enclosed in a vacuum chamber. This is a batch process, seldom used for industrial-scale dehydration of foods. A drying method, consisting of a first stage of convective drying and a second stage of vacuum drying, was applied to mushrooms and parsley and found to result in shorter total drying time and superior product quality (Zecchi *et al.*, 2011).

22.5.3.4 Drying with superheated steam

In this method of dehydration, the heating medium is superheated steam Chu *et al.* (1953). The material to be dried is brought into contact with water vapors at a temperature considerably higher than the saturation temperature at the prevailing pressure. Heat is transferred to the wet material by convection. Water evaporates, and the vapors released mix with the heating medium. Since both consist of water vapor, there is no diffusional resistance to mass transfer at the surface and the

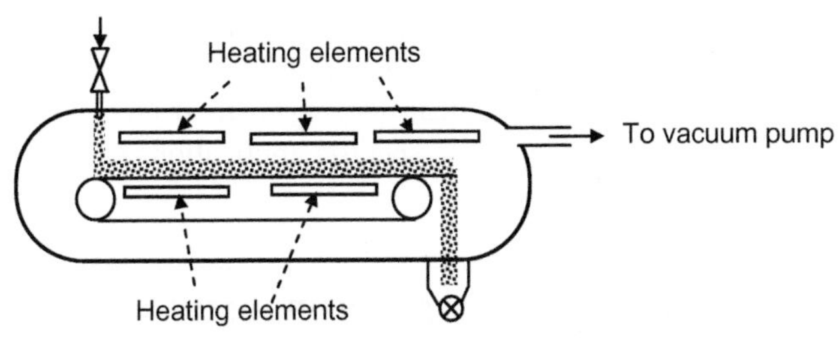

FIGURE 22.14

Vacuum belt dryer.

vapors from the product simply flow into the medium by virtue of a pressure difference. In the case of solid foods and highly viscous fluids, mass transfer considerations apply to internal moisture transport (Sa-adchom *et al.*, 2011). The temperature of the product rises to the boiling point at the prevailing pressure, and is maintained at that level as long as the water content is high. Thereafter, the temperature of the product rises above the boiling point and tends towards the temperature of the superheated steam. In a system operating at atmospheric pressure, this may result in final product temperatures in the range of 120–150°C. Although saturated steam drying can also be performed at reduced pressure (vacuum) to prevent product overheating, high-pressure operation is preferred because of the greatly increased heat transfer coefficient (Svensson, 1980). Consequently, superheated steam drying is mainly used for the dehydration of materials that are not particularly prone to thermal damage, such as cooked meat, wood, paper, and cellulose pulp (Mujumdar, 1992; Sa-adchom *et al.*, 2011; Svensonn, 1980). Alternatively, a combined process consisting of steam drying followed by convective drying may be applied to foods with high initial moisture content (Somjai *et al.*, 2009). The main industrial application in the food arena to date is in the dehydration of spent beet pulp in the production of sugar (Jensen *et al.*, 1987).

The main advantage of steam drying is its excellent energy economy. Unlike air drying, all the heat supplied to the system, which is part of the excess enthalpy of superheat, is utilized for bringing the product to the boiling temperature and for evaporation. The gas leaving the dryer is live steam that can be used for heating duties elsewhere in the plant. Thus, in the case of the sugar industry, the "spent" steam leaving the pulp dryer can be used for heating the evaporators, with or without recompression (see Chapter 19). The energy cost of steam drying has been reported to be 50% of that of air drying, provided that the spent steam is utilized (Svensson, 1980). In addition, steam drying reduces the risk of oxidation since there is no contact with air. This is probably the reason for the higher phenol content found in mate leaves dried by superheated steam in comparison to air-dried leaves (Zanoelo *et al.*, 2006). On the other hand, the capital cost of steam drying is considerably higher than that of air drying. Steam drying may have additional advantages in specific applications in the food industry – for example, where sterilization of the product is an objective. In summary, although the present utilization of superheated steam drying in food processing is limited, the potential advantages of the method in specific applications cannot be ignored.

22.6 Dryers in the food processing industry

The numerous types of dryers in use in the food industry may be classified according to different criteria:

- Method of operation – batch, continuous
- Mechanism of heat transfer – convection (air), convection (steam), conduction (contact), radiation (infra-red, microwave, sun)
- Physical state of the feed material – solid, liquid, paste
- Movement of the material during drying – static, moving, fluidized.
- Pressure of operation – atmospheric, vacuum, high pressure.

Table 22.3 lists the main types of dryers used by the food industry.

22.6.1 Cabinet dryers

Cabinet dryers are used for batch drying of solid foods at a small to moderate scale (say, 2000–20 000 kg per day). They are inexpensive and simple to construct. Cabinet dryers consist of a closed compartment in which trays containing the food to be dried are placed (Figure 22.15).

The trays rest on shelves with adequate spacing between them. Heated dry air circulates between the shelves. Very often, tray bottoms are slatted or perforated in order to provide some air flow through the trays. As illustrated in the example in Section 22.4.1, the drying rate, and hence the moisture content of the material, depends on its position on the tray. The material located closest to the entrance of dry air has the lowest moisture content. In order to secure more uniform drying, the direction of air flow may be reversed or the trays may be rotated periodically. The cabinet is usually equipped with movable baffles, adjusted so as to have uniform distribution of the drying air throughout the cabinet. Cabinet driers are frequently found in rural installations, where they are used for drying fruits (grapes, dates, apples), vegetables (onion, cabbage) and herbs (parsley, basil, mint, dill). Air inlet temperatures are usually in the range of 60–80°C. Air velocity is a few meters per second, and must be adjusted according to the size, shape and density of the food particles so as to avoid entrainment of dry particles with the wind. Depending on the product and the conditions, the duration of a batch is typically 2–10 hours. Most cabinet dryers feature means for adjustable recirculation of the air. The rate of recirculation is increased as drying progresses, when the air exiting the cabinet is warmer and less humid. Recirculation results in considerable saving in energy costs.

Table 22.3 Principal Types of Dryers in the Food Industry.

Dryer Type	*Operation*	*State of Feed*	*Movement of Bulk*	*Product Examples*
Cabinet	B	S	0	Fruit, vegetables, meat, fish
Tunnel	C	S	0	Fruit, vegetables
Belt	C	S, P	0	Fruit, vegetables, tomato
Belt-trough	C	S	M	Vegetables
Rotary	C	S	M	Animal feed, waste
Bin	B	S	0	Vegetables
Grain dryers	B, C	S	0, M	Grain
Spray	C	L, P	M	Milk, coffee, tea
Fluid bed	B,C	S	F	Vegetables, grain, yeast
Pneumatic	C	S	M	Flour
Drum	C	L, P	0	Mashed potato, soup
Screw conveyor	C	S, P	M	Grain, waste
Mixer	B	S	M	Particles, powders
Solar	B,C	All	All	All
Sun drying	B	S	0	Fruit, vegetables, fish

B, batch; C, continuous; S, solid; L, liquid; P, paste; 0, static; M, moving; F, fluidized.

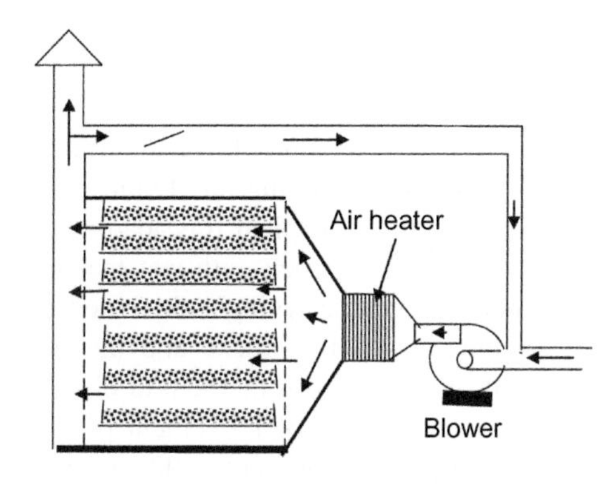

FIGURE 22.15

Cabinet dryer.

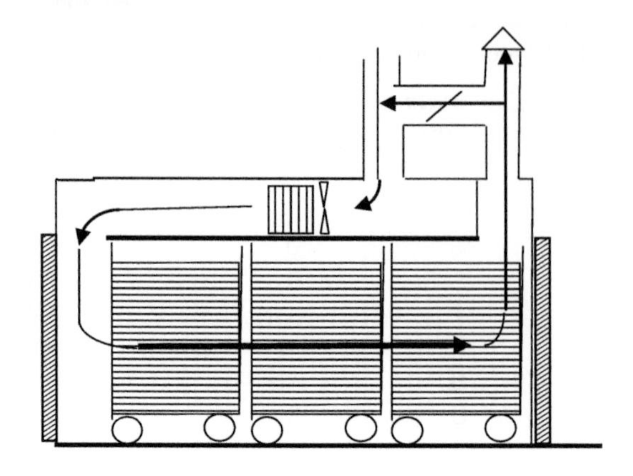

FIGURE 22.16

Short tunnel dryer.

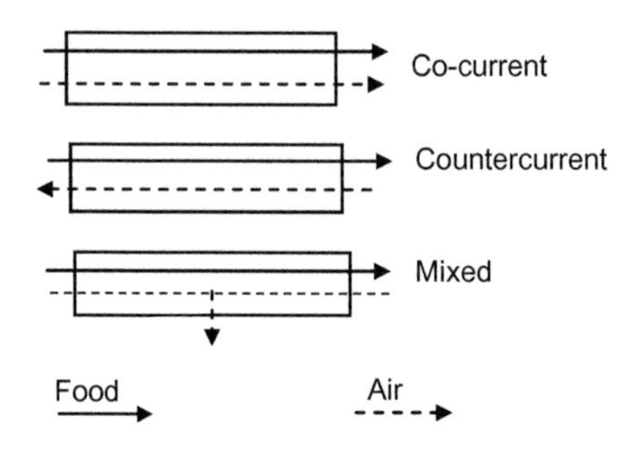

FIGURE 22.17

Flow patterns in tunnel dryers.

22.6.2 Tunnel dryers

Tunnel dryers consist of long tunnels through which trucks carrying stacks of trays travel with or against a stream of drying air (Figure 22.16). The material to be dried is evenly spread on the trays. Typical tray loading for wet vegetables is in the order of 10–30 kg per m^2. As a truck with wet material is introduced into the tunnel at one end, another truck, carrying dehydrated product, exits at the other end. Depending on the size of the trucks and the tunnel, the trucks are moved manually or mechanically (e.g., with the help of chains).

With respect to the relative direction of movement of the air and the trucks, tunnel dryers operate in co-current, countercurrent or mixed current fashion (Figure 22.17). In the case of the co-current tunnel, air with the highest temperature and lowest humidity meets the food with the highest humidity and lowest temperature. This provides the highest "driving force" for drying, and therefore the most rapid rate of water transfer at the entrance to the tunnel. If the feed material is sufficiently humid, its temperature remains low despite the contact with hot air. The "driving force", however, diminishes as the food travels towards the exit. The air at the exit end of the tunnel is the most humid and the coolest. Consequently, the final residual moisture content of the product may not be as low as desired. The converse occurs in the case of countercurrent tunnels. The starting rate of drying is lower, but it is possible to dehydrate the product to the lower final moisture content. The mixed flow, central exhaust tunnel functions as two tunnels in series. Its first portion is co-current, and provides the desired high initial drying rate; its last portion is countercurrent, and gives the desired finishing effect. Unlike cabinet drying, tunnel drying provides the possibility of exposing the product to a changing profile of external conditions. In addition to air temperature and humidity, it is possible to vary air velocity.

In one model used for drying fruit, the tunnel is designed as two units in series, with a small cross-section resulting in higher air velocity in the first unit (Figure 22.18).

22.6.3 Belt dryers

Belt dryers are among the most versatile continuous dryers for solid foods (Kiranoudis, 1998). They are used extensively, mainly for large-scale dehydration of vegetables. Essentially, a belt dryer functions like a tunnel dryer with the difference that trays and trucks have been replaced by belt conveyors. These dryers can be operated in cross-flow or through-flow mode, or a combination of both. In dryers for through-flow operation, the belt is made of metal mesh to allow air circulation through the bed. Belt dryers can be single-stage or multi-stage. Multi-stage dryers consist of a number of belts in series.

The advantages of belt dryers over tunnel dryers include the following:

- Continuous feeding and discharging of the belts is much easier and less labor-intensive than loading and unloading trays.

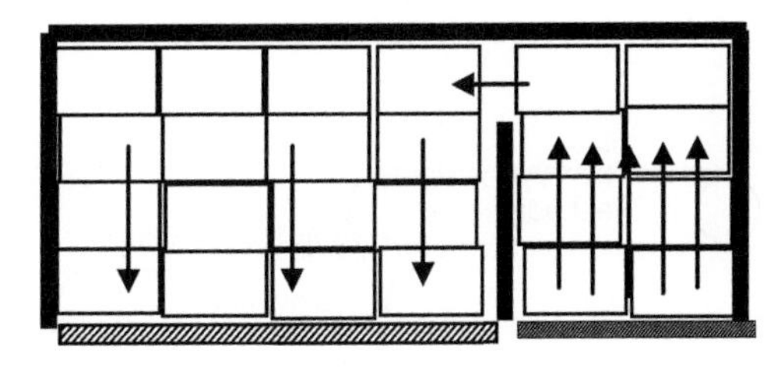

FIGURE 22.18

Two-stage tunnel dryer.

- In tray dryers (tunnel or cabinet), the bed is static during the entire process. In the absence of mixing, the moisture content of the food on a tray is uneven – the top surface of the bed may be quite dry while the material at half-depth is still wet. In multi-stage belt dryers, the bed is refilled and mixed at each transfer from one belt to another (Figure 22.19).
- Most food materials shrink upon drying. In tray dryers, shrinking results in a diminishing rate of utilization of the tray surface as drying progresses. In multi-stage belt dryers, it is possible to maintain adequate bed thickness by moving each stage at a slower speed than the previous one (Figure 22.20). Thus, higher loading per m^2 can be achieved in belt dryers.

The exact dimensions and operating conditions of conveyor-belt dryers are optimized for the most economic performance. Product quality parameters, such as color, may be included in the optimization (Kiranoudis and Markatos, 2000).

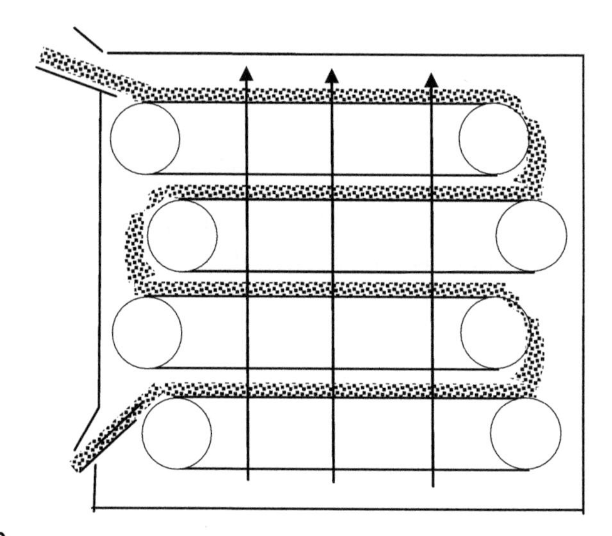

FIGURE 22.19

Multi-stage belt dryer.

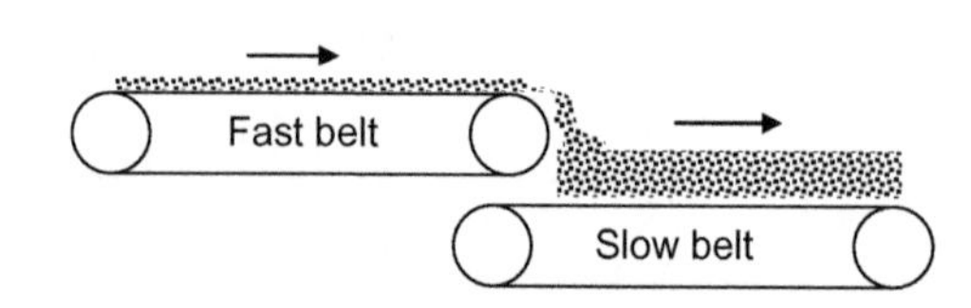

FIGURE 22.20

Effect of varying belt speed on bed thickness.

22.6.4 Belt-trough dryers

The belt-trough dryer is a special type of dryer designed for the initial drying of vegetables in small pieces. The unit consists of a wide mesh belt, freely hanging over two cylindrical rollers so as to form a trough. The wet material is fed onto the belt at one end of the trough and forms a fairly thick bed. The belt is driven slowly by rotating the supporting rollers. The movement of the bed causes gentle and continuous mixing of the bed, while a vertical blast of hot air circulates through the material. The blast is sufficiently strong to slightly expand the bed without fluidizing it. Evaporation is rapid, and most of the water in the wet material is removed – typically in less than 1 hour. The trough assembly is tilted towards one end, causing the material to move slowly down the slope towards the discharge. The rate of discharge, and hence the hold-up of the unit, is regulated by an adjustable weir placed at the discharge end.

22.6.5 Rotary dryers

Rotary dryers are mainly used in the chemical and mineral industries. In the area of food, their most common applications are for dehydrating waste materials (citrus peels, vegetable trimmings) and animal feedstuffs (alfalfa). Rotary dryers consist of a metal cylinder with internal flights or louvers (Figure 22.21). The cylinder is slightly inclined, and the material is fed in at the high end and discharged at the low end. Hot air is blown in a co-current or countercurrent direction. As the cylinder rotates, the material climbs in the direction of rotation. When it reaches a position where its angle of repose has been exceeded, the material falls back to the bottom of the cylinder (Figure 22.21). Most of the drying takes place while the material falls through the air blast. Using very hot air or combustion gases, rotary dryers can also function as roasters for nuts, sesame seeds and cocoa beans. A detailed method for the design of rotary dryers, based on a heat exchange approach, has been described by Nonhebel (1971).

22.6.6 Bin dryers

Tray, belt and belt-trough dryers are quite efficient in removing most of the water of high-moisture materials in the initial stage of drying. In the latter stages of the falling rate period, however, removal of the residual moisture takes a long time and external turbulence and mixing cannot accelerate the process. In the case of vegetables, using the more expensive types of dryers for reducing the moisture content below 15–20% would be uneconomical. Bin dryers provide the ideal solution for "finishing" the process to the desired final moisture content of 3–6%.

As their name implies, bin dryers consist of simple containers with a perforated or mesh bottom (Figure 22.22). The partially dehydrated material is placed in the bin as a deep bed, and a slow current of air at moderate temperature is

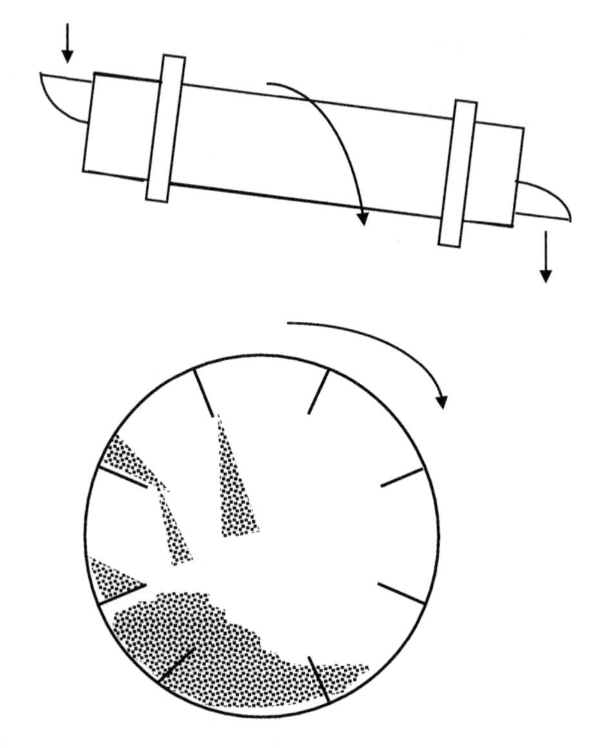

FIGURE 22.21

Generalized schema of the rotary dryer.

passed through the bed until the desired final dryness is reached. Another function of the bin dryer is moisture equilibration. As explained previously, at the exit from the more "rapid" dryers the moisture content distribution between and within particles is not uniform. The long residence in the bins allows moisture equilibration in the product as a result of internal transport.

22.6.7 Grain dryers

Cereal grains and oilseeds are often harvested with a moisture content that does not warrant safe storage for a long period. In this case, it is necessary to dry the grains before storage. In terms of tonnage, grain drying is probably the largest drying operation in the food industry and agriculture. Grain drying is also called "conditioning". The initial moisture content depends on the climatic conditions during the harvest season. Typically, grain may contain 25% moisture when harvested and must be dried to 12–15% moisture content before storage. Because of the relatively low moisture content throughout, drying takes place entirely under the falling rate regime and takes a long time (e.g., several days).

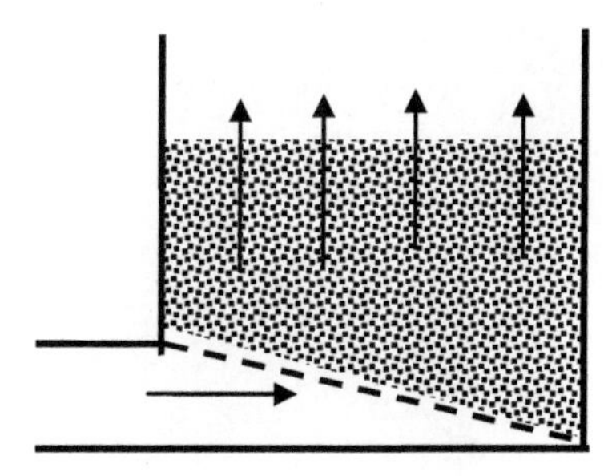

FIGURE 22.22

Bin dryer.

Air temperatures used in grain drying are, typically, 50–70°C, the lower temperatures being applied to grain for seed and the upper temperatures to grain for processing.

Grain dryers may be batch or continuous. Batch dryers function as very large bin dryers, except that special precautions must be taken for the proper distribution through the bed, which can be many meters high. The wet grain enters through the top. Often, the upper section of the bin or tower is separated from the rest. There, the wet grain is pre-dried and then dumped to the section below. In order to avoid heat damage, dried grain is usually cooled by passing unheated air through the bed. Frequently, the batch dryer also serves as a long-term storage silo. In continuous grain dryers, hot air is passed through continuously flowing or agitated grain. After drying, the grain is continuously cooled with air before being discharged. Drying rates are high. Precise control is required in order to avoid over-drying and over-heating. A period of "tempering" is provided for moisture equilibration.

22.6.8 Spray dryers

Spray dryers are used for drying liquid solutions and suspensions, with the objective of producing light, porous powders. Spray drying is the standard method for the production of milk and whey powders, coffee creamers, cheese powder, dehydrated yeast extract, instant coffee and tea, isolated soybean protein, enzymes, maltodextrin, egg powder and many other products in powder form. (Filkova and Mujumdar, 1995). Spray drying is also one of the methods used for microencapsulation.

The liquid is dispersed (atomized) as a spray of fine droplets into very hot air inside a large chamber. Because of their small size and the high temperature of the air, the droplets are dried in a matter of seconds and transformed into particles of solid powder. At the exit from the chamber, the solid particles are separated

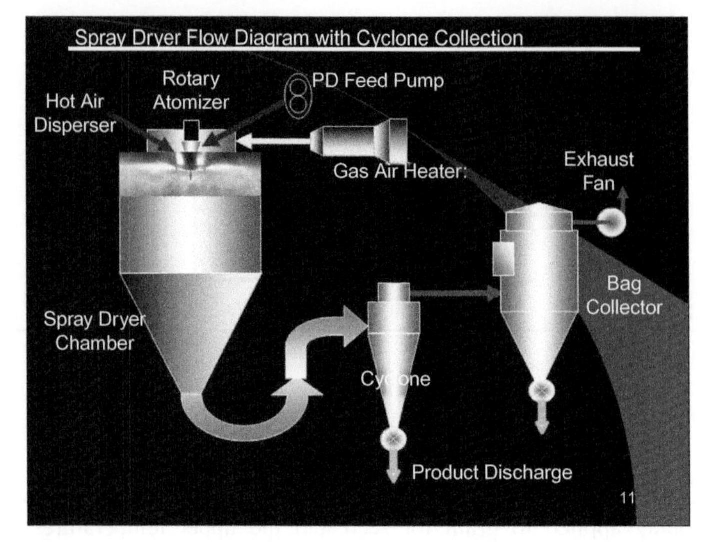

FIGURE 22.23

Spray dryer.

Figure courtesy of GEA-Niro.

from the humid air. A spray dryer system consists of the following elements (Figure 22.23):

- An air heater
- A device for forming the spray (nozzle or atomizer)
- A pump for feeding the liquid to the atomizer or nozzle
- A drying chamber
- Solid–gas separators (cyclones)
- Fans for moving the air through the system
- Control and measurement instruments.

1. *Air heater*. Inlet air temperatures in spray drying are in the order of 200–250°C and sometimes higher. Usually, steam cannot be used for heating the air because of the high pressure that would be required to deliver heat at such high temperatures. Electrical heating is practiced in small installations. Combustion gases are the preferred source of heat at industrial scale. In direct heating, the combustion gases at 400–500°C are mixed directly with fresh air to produce a gas mixture at the desired temperature. In indirect heating, air is heated by combustion gases in heat exchangers, without direct contact between the two streams. Direct heating is simple, less expensive and provides excellent energy utilization, but the direct contact between the food and combustion gases may be problematic.

FIGURE 22.24

Rotary atomizer.

Photograph courtesy of GEA-Niro.

2. *Formation of the spray*. There are five methods for dispersing the feed as a spray: centrifugal (turbine) atomizers, pressure nozzles, bi-fluid nozzles, hot air dispersers (e.g., Leaflash dryers) and ultrasonic atomizers.
   a. *Centrifugal (rotary) atomizers* (Figure 22.24) consist of a wheel, similar to a closed impeller in a centrifugal pump. The atomizer, which can be motor-driven or moved by compressed air, spins at high speed, corresponding to tip velocities in the order of $10^2\ m \cdot s^{-1}$. The liquid is fed to the center of the wheel, flows through the channels to the periphery and is ejected as a film that disintegrates to droplets. Centrifugal atomizers form sprays with fairly narrow droplet size distribution. Uniformity of drop size is important for even drying and for producing a powder with uniform particle size distribution. The mean Sauter diameter (see Section 6.2.1) of the droplets depends on the wheel diameter D, speed of rotation N, mass flow-rate G, liquid viscosity μ, density ρ and surface tension σ, according to the following approximate empirical equation (Masters, 1991):

$$\overline{dsv} = k(N)^{-0.8}(G)^{0.2}(D)^{-0.6}(\mu)^{0.2}(\rho)^{0.5}(\sigma)^{0.2} \qquad (22.38)$$

   Equation (22.38) indicates that viscosity, surface tension and feed flow-rate have little influence on the drop size. Uniformity of drop size is the main advantage of centrifugal atomizers. Their disadvantages are high capital and maintenance costs, and high energy consumption.
   b. *Pressure nozzles* (Figure 22.25). The liquid is fed to a narrow nozzle at high pressure. At the exit from the nozzle, the liquid jet disintegrates to form the spray. Nozzles that deliver a conical spray are selected. The droplets are relatively large and size distribution is wide.

   The following approximate relation expresses the dependence of mean droplet size on pressure drop ΔP and other variables (Masters, 1991). Note the strong dependence on viscosity.

$$\overline{d_{VS}} = k(G/\rho)^{0.25}\mu(\Delta P)^{-0.5} \qquad (21.39)$$

FIGURE 22.25

Pressure nozzle.

Photograph courtesy of GEA-Niro.

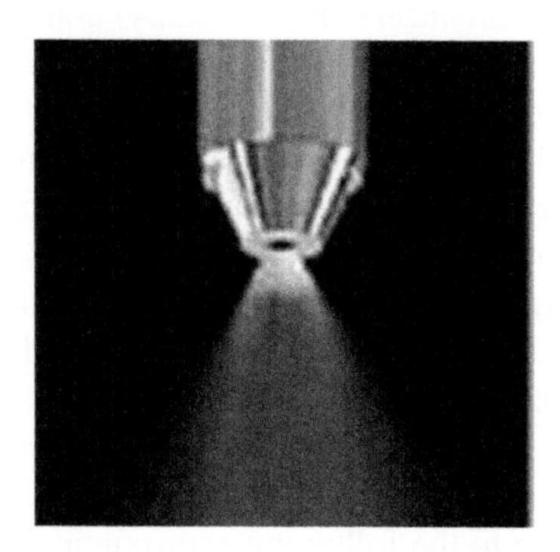

FIGURE 22.26

Bi-fluid nozzle.

Photograph courtesy of GEA-Niro.

c. *Bi-fluid (kinematic) nozzles*. The liquid and the pressurized air are supplied separately to the nozzle (Figure 22.26). The two fluids meet at the exit from the nozzle. The liquid is disintegrated by the high velocity air jet. The droplets are finer and more uniform than those in sprays delivered by pressure nozzles.

d. *"Leaflash" atomization*. "Leaflash" is a relatively novel spray-drying system with a special atomization mechanism. In the atomizer head of this system, very hot air (typically at 300–400°C; Bhandari *et al.*, 1992) at high velocity tears a film of fluid into droplets. The air that induces atomization is also the air used for drying. Because of the intimate contact with very hot air, the droplets are dried even more rapidly than in classical spray drying. Consequently, the residence time of the product in the dryer is particularly short.

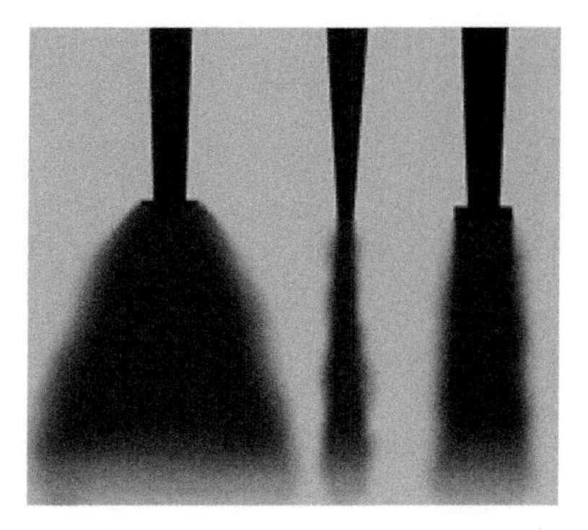

FIGURE 22.27

Different types of sprays obtained by ultrasonic atomization.

Photograph courtesy of SonoTek Corporation.

  e. *Ultrasonic atomization*. If a surface (such as a rod or a disc) wetted with liquid is set to vibrations at ultrasound frequency and appropriate amplitude, the liquid film is atomized as very fine droplets. This is the operating principle of the ultrasonic nozzle (Figure 22.27), used for spray-drying and micro-encapsulation (Bittner and Kissel, 1999; Yeo and Park, 2004).

3. *Feed pump*. The starting materials for spray drying are often highly viscous solutions or suspensions. Frequently, the feed has to be pumped to a considerable height, from ground floor to the top of the spray dryer. If the spraying devices are pressure nozzles, the liquid must be delivered to the nozzle at MPa range pressure. For all these reasons, positive displacement pumps are used as feed pumps. Furthermore, the controlled variable in spray drying is the feed flow-rate (see below). Therefore, feeding pumps must be equipped with automatically controlled variable speed drive.
4. *Drying chamber*. The drying chamber consists, most commonly, of a vertical cylindrical section with a conical bottom. Industrial spray dryers are very large in diameter and in height, and often occupy an entire building or must be installed outdoors. The large volume of the drying chamber is required in order to provide the residence time necessary for complete drying. In spray dryers with centrifugal atomizers, the droplets are ejected in radial direction. In this case, a large diameter is required in order to prevent the droplets from reaching the chamber walls while they are still wet and sticky. Dryers with pressure or bi-fluid nozzles are tall but less wide, because of the narrower spray angle. The conical bottom serves to collect the major part of the dried product but adds to the height of the dryer. A spray dryer of a different model has a nearly flat bottom and uses a rotating suction duct (air broom) for collecting the product.

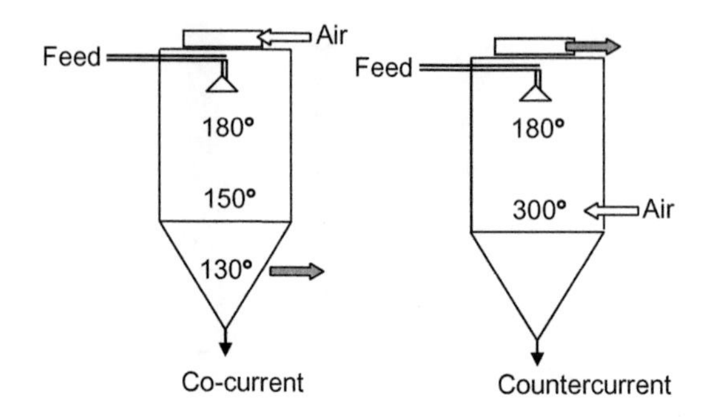

FIGURE 22.28

Typical temperature distribution in co-current and countercurrent spray dryers (inlet air temperature in both cases is 350°C).

Adapted from Masters (1991).

5. *Air flow*. Air is driven through the heater and the dryer by low-pressure, high-capacity blowers. Air movement may be co-current with or countercurrent to the direction of the product. The typical temperature profile in the dryer for the two types of flow pattern is shown in Figure 22.28. A certain portion of the dry particles and particularly the fines are carried away with the exhaust air and recovered in cyclones, but a sizable portion of powder is deposited on the walls of the drying chamber. The flow pattern of air and particles, the drying kinetics and the rate of deposition of powder on the chamber walls have been studied by Roustapour *et al.* (2009).

   For environmental reasons, it is often necessary to pass the exhaust air through a scrubber before discharge to the atmosphere.
6. *Control*. Because of the rapidity of the process, it is difficult to control exactly the residual moisture content of the product and to adjust operating conditions accordingly. The approximate method of control most commonly applied assumes that the temperature of exhaust air is related to the residual moisture content of the product. For foods, exhaust temperatures in the range of 90–110°C are specified. If the exhaust temperature is too low, it is assumed that the product is too moist. In this case, the controller reduces the feed rate by regulating the speed of the feed pump. Contrarily, high exhaust temperature indicates poor utilization of the drying power of the air, with possible thermal damage to the product. The controller increases the feed rate accordingly (Figure 22.29). An additional independent control loop regulates the inlet temperature of the air.

Due to the difficulty of achieving precise control of the final moisture content, it is often necessary to include a second stage of drying, usually in a fluidized bed dryer (Figure 22.30). This second stage may also serve for agglomeration.

The relationship between product quality and process conditions in spray drying is discussed in Section 22.7.

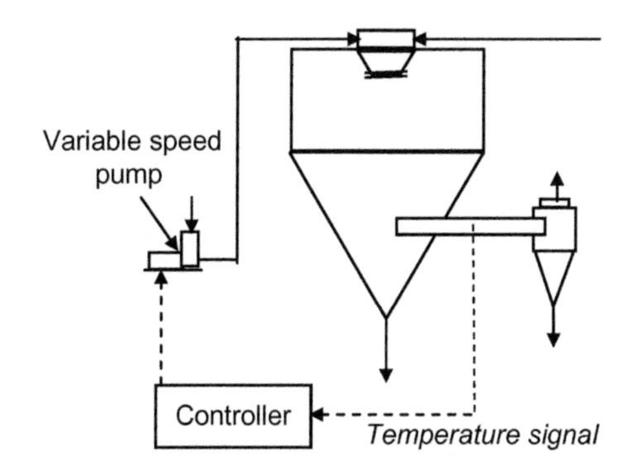

FIGURE 22.29

Control of a spray dryer.

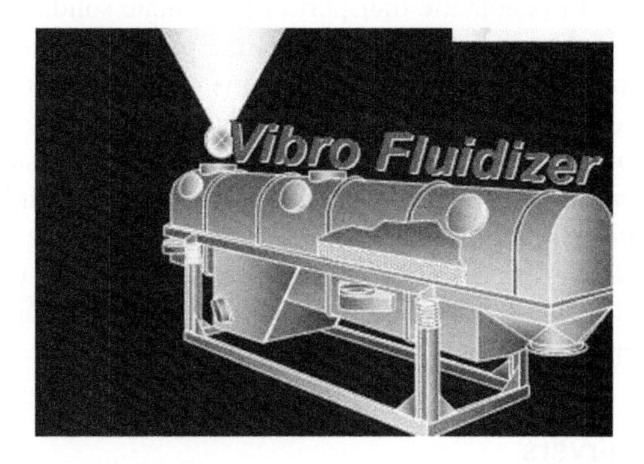

FIGURE 22.30

Vibrating belt fluidized bed after-dryer.

22.6.9 Fluidized bed dryer

The properties of fluidized beds and the conditions for fluidization have been discussed in Section 2.5.3. In fluidized bed dryers, hot and dry air is used both for fluidization and for drying. Fluidized bed drying can be applied to particulate, non-sticky foods with particle size within the range of 0.05–10 mm, depending on the density (Cil and Topuz, 2010). Fluidized bed drying can be in batches or continuous (Figure 22.31). Due to the efficient heat and mass transfer, the product is dried rapidly. Part of the hot air may be recirculated, as in a cabinet dryer. Sticking and product accumulation in continuous dryers is largely prevented by vibrating the fluid bed. Fluidized bed dryers are also used for powder agglomeration and coating of particles (Lin and Krochta, 2006).

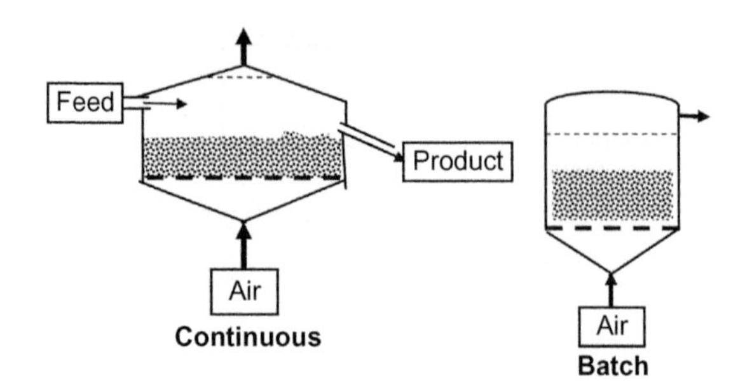

FIGURE 22.31

Schema of fluidized bed drying.

22.6.10 Pneumatic dryer

The basic principles of pneumatic transport of particulate solids were discussed in Chapter 2, Section 2.5.4. In pneumatic dryers, the particles are dried while being moved in a stream of hot, dry air. This method is often used for removing "free" moisture in the constant drying rate period. High drying rates can be achieved (hence the name of "flash dryer"), but the residence time is too short for complete drying. For this reason, pneumatic drying is often used as a method of pre-drying, followed by another type of drying. It is also possible to recycle part of the product in order to achieve the desired final moisture content. The use of pneumatic drying in the food industry is limited to the drying of flours, starch, gluten powder, casein powder, etc.

22.6.11 Drum dryers

The basic principle of operation of the drum dryer was discussed in Section 22.5.3. The different types of drum dryers vary in the method used for applying the wet material onto the surface of the drum (Figure 22.32).

Drum dryers are classified into two types, namely single-drum and double-drum dryers. Double-drum dryers consist of two drums rotating in opposite directions, with a narrow, adjustable gap between the two. The so-called "twin" drum dryer consists, in fact, of two independent co-rotating single drums sharing some of the accessory devices.

The simplest method of application is dip feeding. The drum is partially immersed in the feed fluid contained in a tray, and a film of fluid adheres to the immersed segment of the drum. Fresh material is continuously supplied to the tray. Because the mass of material in the tray is heated by the drum, this method is less suitable for heat-sensitive products. In the double-drum dryer with nip feed, the feed fluid is introduced into the pool formed between the two drums. The thickness of the adhering film is controlled by adjusting the gap. This type of feeding is used with less viscous materials, such as milk and other dairy products.

FIGURE 22.32

Drum dryer.

Photograph courtesy of the Department of Biotechnology and Food Engineering, Technion, I.I.T.

Applicator rolls are used for spreading viscous liquids, purees and pastes on the drum surface. Single-drum dryers with applicator rolls are used extensively in the production of instant mashed potato. Multiple applications resulting in thicker layers are made possible by mounting a number of applicator rolls on the drum periphery (Figure 22.33). Application rolls are also useful for pressing the film upon the drum, to restore good contact and to reduce porosity.

Removal of the water vapors from the vicinity of the drums is essential in order to prevent moisture adsorption by the dry product. To this end, drum dryer installations are usually equipped with venting hoods of adequate size.

22.6.12 Screw conveyor and mixer dryers

Screw conveyors with jacketed troughs and hollow augers are used for drying slurries and wet particulate solids (Waje *et al.*, 2006). The trough and screw auger are steam heated. Heat transfer is mainly by conduction. These dryers, also known as *hollow flight dryers*, are frequently used for drying waste materials and biomass (sludge) from wastewater treatment processes.

Dryers built on the principle of ribbon, double-cone or V-tumbler mixers (Chapter 7, Section 7.5.3) are used for small-scale batch drying of solids. The mixers are jacketed for heating with steam or hot water. Some models operate under vacuum.

22.6.13 Sun drying, solar drying

Sun drying (Bansal and Garg, 1987; Ekechukwu and Norton, 1999) refers to the dehydration of foods by direct exposure to radiation from the sun. Important quantities of fruits, vegetables, grains and fish are dried by this method. Sun-dried

SINGLE DRUM DRYER WITH APPLICATOR ROLLS
The wet product is applied to the drum by means of applicator rolls. According to the number of the applicator rolls used the layer formed on the drying drum is thicker or thinner. This arrangement is suitable for the processing of pulpy or pasty products.
Typical applications are: Cereal based breakfast foods, Babyfood products, Pre-gelatinised starches, Fruit pulp and pastes, Potato flakes.

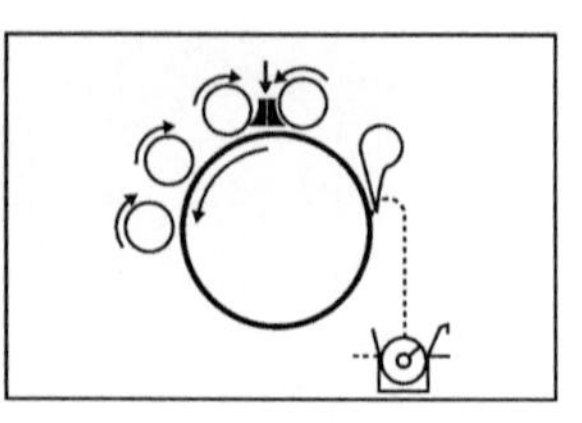

DOUBLE DRUM DRYER WITH FEED ROLLS
This is a special'Hybrid' system used to impart certain physical properties, particularly density, to specific products
Prinicple applications are: Cereal based breakfast foods, Baby food products, Fruit pulp and pastes.

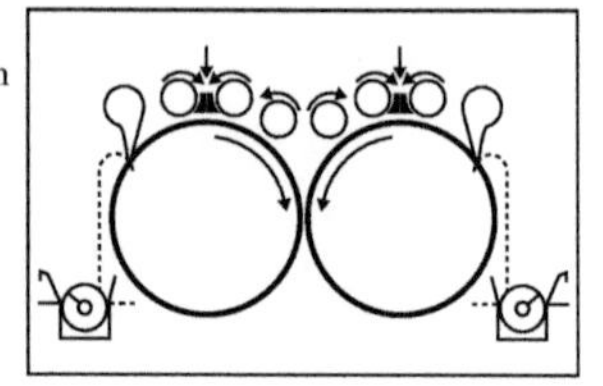

TWIN CYLINDER DRYER WITH NIP FEED
The material to be dried is pumped, either directlyor through spray nozzles, into the nip formed between two drying drums. This is the oldest and simplest form of drum dryer. The thickness of the product film may be varied by adjustment of the gap between the drying drums or cylinders.
Typical applications are: Drying of Yeast, Milk proudcts, Detergents, Dyestuff manufacture.

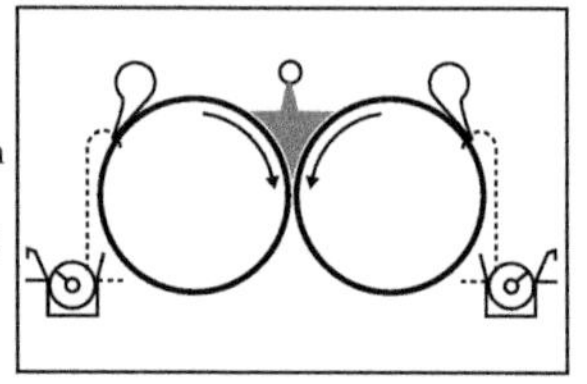

SINGLE DRUM DRYER WITH APPLICATOR ROLLS
This is a more specialised type of drying, for example in the Chemical Inudstry. The applicator roll is located underneath the drum dryer and dips into the product. A liquid film is then transferred to the drying drum.
Typical applications are: Animal based glu, Gelatin, Pesticides.

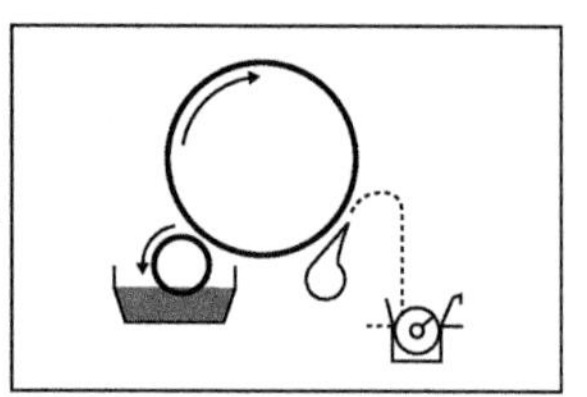

DIP FEED DRUM DRYER
One of the most basic forms of the drum dryer, where a film of the proudct to be dried is picked up on the surface of the dryer drum as it rotates through a feed tray mounted below. The feed tray may be cooled or fitted with a recirculation system to prevent overheating or settling out of product from suspension.
Typical applications are: Drying of cereals, Spent Yeast.

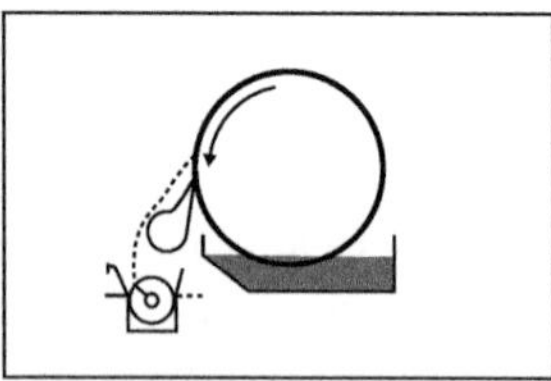

FIGURE 22.33

Drum drying feeding methods.

Figure courtesy of Simon Dryers.

tomatoes are a specialty product of increasing popularity. A large proportion of the raisins and practically all of the dried apricots and figs produced in the world are sun dried. Sun drying of fish at village level is common in tropical regions. The term "solar dryer" is reserved for a large variety of convective dryers wherein

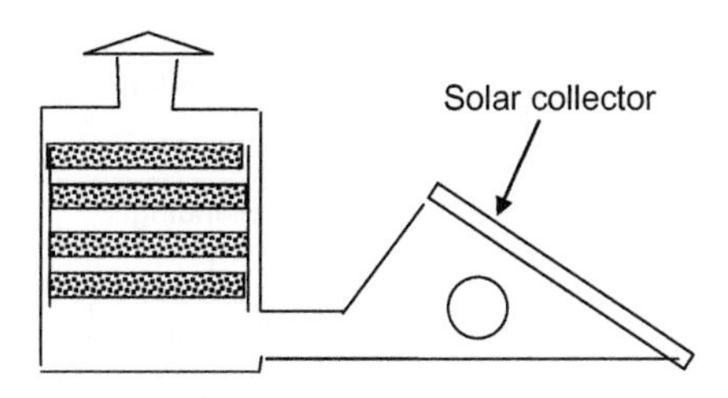

FIGURE 22.34

A simple solar dryer.

the products are not exposed to sun but dried indirectly by air heated by solar energy (Kadam *et al.*, 2011; Kavak Akpinar, 2009). An example of solar dryer is shown in Figure 22.34. Obviously, both types of drying are feasible only if sufficient and reliable insolation can be assumed.

22.7 Issues in food drying technology

22.7.1 Pre-drying treatments

Most vegetables are blanched before drying. The objectives of blanching, explained in the chapter on freezing, are also applicable to drying. In addition, due to its effect on the permeability and integrity of cell membranes, blanching often has a beneficial effect on the drying rate. Ramirez *et al.* (2012) studied the effect of four types of pre-treatment (immersion in boiling water, vacuum impregnation, freeze–thaw cycles, and compression) on the microstructure and drying rate of apple. Immersion in boiling water, freezing–thawing, and compression resulted in increased drying rate and diffusivity. In order to preserve their characteristic pungency, onions are usually not blanched.

Fruits with tough skins, such as prunes, figs and grapes, are dipped in hot water or a hot solution of potassium carbonate. This treatment (checking) induces small cracks, removes the waxy layer covering the skin and accelerates drying. Dipping in a solution of sodium bisulfite is often practiced in the case of grapes (bleached raisins), apricots and some vegetables. Some producers dip carrot pieces in a hot solution of starch. Upon drying, the starch forms a protective coating that retards oxidation of the carotenoid pigments (Zhao and Chang, 1995). Dipping in solutions of ascorbic acid, citric acid or a mixture of both is effective in preventing browning discoloration in vegetables.

The time–temperature profile of the product during drying is usually insufficient for the destruction of microorganisms. Therefore, thermal processing before drying is required in critical cases such as milk and dairy products.

The most common treatment given to liquid foods before drying is preconcentration. As explained in Section 22.8, removal of water by drying is more

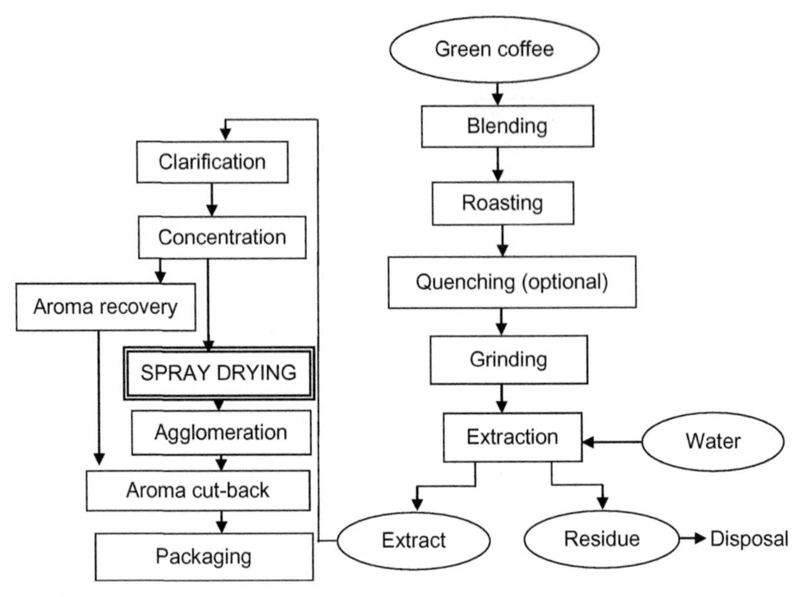

FIGURE 22.35

Process flow diagram for spray-dried instant coffee.

costly than by evaporation. Milk and coffee extract are pre-concentrated by evaporation, freeze concentration or membrane processes before spray drying (see the process flow diagram for spray-dried instant coffee, Figure 22.35). The upper limit of concentration before drying is dictated by the viscosity of the concentrate that can still be successfully spray dried.

22.7.2 Effect of drying conditions on quality

Heat and mass transport phenomena may have profound effects on the quality of dehydrated foods (Patel and Chen, 2005). Texture, structure, appearance, color, flavor, taste and nutritional value are all subject to change as a result of the drying process.

Food structure is changed by drying (Aguillera *et al.*, 2003; Sansiribhan *et al.*, 2010; Wang and Brennan, 1995). The most general and the most evident effect, as well as probably the most difficult to avoid, is shrinkage. In foods consisting of soft, pliable, hydrophilic gels not containing rigid fillers, the decrease in volume corresponds roughly to the volume of the removed water, and shrinkage is isotropic. The shape of the particle is therefore not affected. This case is, however, not very frequent in solid foods. Usually the solid matrix has some resilience, and therefore shrinkage is less than the volume of removed water. Mechanical stresses are induced and deformation occurs (Bar *et al.*, 2002; Earle and Ceaglske, 1949; Eichler *et al.*, 1998). The dried product is porous (Krokida and Maroulis, 1997).

During drying, solutes are transported along with water. As a result, the distribution of components in the dried product may be different from that of the starting material. The concentration of solutes such as sugars and proteins may be higher at the surface of the dehydrated product (Fäldt, 1995). Similarly, flow of liquid fat may result in higher fat content at the surface of fat-containing powders obtained by spray-drying (Kim *et al.*, 2008). The initial increase of concentration in the liquid may result in the crystallization of the solutes. Solute crystallization may be enhanced by increasing the drying temperature and air humidity (Islam *et al.*, 2010).

The most common thermal effects are non-enzymatic browning, denaturation of proteins, and thermal destruction of heat-sensitive vitamins and pigments. Protein denaturation is the main cause of decrease in the dispersibility/solubility of milk powder obtained by drying at high temperature.

With respect to the effect of drying on product quality, spray drying occupies a peculiar position. Because of the extremely fast drying rate, short residence time of the food in the dryer and the relatively low temperature during most of the drying period, thermal damage in spray drying is remarkably low. Even when suspensions of live cells (starter cultures, yeast) are spray dried, the survival rate of the cells in the process is remarkably high (Bonazzi *et al.*, 1996).

The retention of volatile aromas in spray drying is also much better than might be expected from the volatility of the aroma relative to water (Hecht and King, 2000; Senoussi *et al.*, 1995). The retention of organic trace components in drying in general, and in spray drying in particular, has been extensively investigated (Bangs and Reineccius, 1990; Coulter and Reineccius, 1969; Etzel and King, 1984; Rulkens, 1973; Thijssen, 1975). Thijssen proposed the "Selective Diffusion Theory" to explain the retention of volatiles in spray drying. According to this theory, as the concentration of carbohydrates increases during drying, the diffusivity of the organic trace components decreases faster than the diffusivity of water. One of the practical consequences of his theory is that the retention of volatile aroma increases as the solids content of the feed solution is increased.

22.7.3 Post-drying treatments

As a result of thermal damage during drying, the dried product may contain defective particles. These are removed by visual or automatic inspection and sorting. Automatic color sorters are used in large-scale production of dehydrated vegetables.

Some dehydrated products in powder form (e.g., milk powder, infant formulae, soluble coffee) are agglomerated. Agglomeration as a unit operation will be discussed later in this section.

Considerable quantities of dehydrated herbs and spices are irradiated with ionizing radiations in order to meet microbiological standards. Some dried products, such as soluble coffee, are highly hygroscopic, and must be protected against adsorption of moisture immediately after drying. For the intermediate storage and packaging of such products, air-conditioned rooms kept at low relative humidity must be provided. Proper packaging and adequate storage conditions are, of course, the most important factors in post-drying handling of all dehydrated products.

22.7.4 Rehydration characteristics

Dehydrated products may be classified into those that are used as such (e.g., most dried fruits) and those that must be "reconstituted" by rehydration (e.g., milk powder, mashed potato flakes, most dried vegetables). Ideally, dehydrated foods should be able to regain their original moisture content, volume, shape and quality when rehydrated. This, however, seldom occurs. A number of indices are used to define the rehydration characteristics of dehydrated foods (Lewicki, 1998). The "rehydration ratio" is the ratio of the mass of rehydrated and drained food to the mass of the original material. It is usually expressed as a percentage. The "rehydration rate" or, conversely, "rehydration time" refers to the kinetics of rehydration. The "rehydratability" of dehydrated foods depends largely on the conditions of drying (Bilbao-Sǎinz *et al.*, 2005; Saravacos, 1967). High porosity seems to be among the most important requisites for complete and fast rehydration (Marabi and Saguy, 2004). The rate-limiting factor is adsorption and internal transport of the water. Consequently, agitation has only a slight effect on rehydration kinetics, except if the rehydration medium is highly viscous (Marabi *et al.*, 2004).

Rehydration of dried foods involves a sequence of events: wetting of the surface, penetration of the water into the pores, adsorption on the surface of the matrix, diffusion into the solid matrix, and equilibration. It should be kept in mind that, simultaneously with the penetration of water, there may be leaching of soluble components out of the food particle. In the case of spray-dried "instant" powders, the rehydration characteristics of the product are evaluated in terms of the following properties (Barbosa-Cánovas and Vega-Mercado, 1996):

- *Wettability*. This is the ability of the powder to absorb water on its surface. This property can be evaluated quantitatively in terms of surface tension and wetting angle, or empirically by measuring the rate of penetration of water into a column of powder.
- *Sinkability*. For rapid reconstitution, a quantity of powder placed on the surface of water should sink down into the water as quickly as possible.
- *Dispersibility*. This property represents the ability of the powder to disperse quickly and evenly into the water, without forming lumps.
- *Solubility*. This property is a function of the chemical composition but may be affected by drying (e.g. denaturation of proteins).

22.7.5 Agglomeration

Agglomeration (Dumoulin, 2008; Schubert, 1981; Schubert, 1993; Ormos, 1994) is a "size augmentation" process applied to powders. The most commonly applied process ("rewet" agglomeration) consists of wetting the powder with water or with an aqueous solution, followed by re-drying under agitation. Wetting creates liquid bridges between powder particles. These bridges contain solutes, either dissolved from the powder or added previously to the wetting liquid. Upon re-drying the liquid bridges are converted to solid bridges, binding powder particles together. Agitation during drying is essential for controlling the ultimate size of the agglomerates.

The objectives of agglomeration are:

- To produce a rigid porosity, and thus prevent lump formation and improve the rehydration properties of instant powders (Schubert, 1983)
- To prevent segregation of components in composite powders
- To reduce the proportion of fines
- To control the bulk density of the powder
- To improve flow characteristics.

The methods of agglomeration differ from each other, mainly regarding the manner in which agitation is applied. Thus, agglomeration can be performed in a rotating pan, in a tumbler mixer or in a fluidized bed. An example of a fluidized bed method for powder agglomeration is shown in Figure 22.36.

22.8 Energy consumption in drying

Drying processes usually consume considerable quantities of energy. A number of different sources of energy are used in drying processes: fuel, steam, hot water, electric current, solar heat etc. The main parameters used to characterize the energy consumption of drying processes are:

1. *The specific energy consumption.* This is the energy consumed per unit mass of product. This parameter has economic significance.
2. *The efficiency of energy usage.* This is the proportion of the energy used for the evaporation of water only. This parameter has similarities with the steam economy of evaporators. It is related to the engineering aspects of the process.

The energy usage efficiency of conduction (boiling) drying is similar to that of single-effect evaporation without vapor recompression. Recompression of the water vapor released in drying and its recycling as heating steam is theoretically possible in some cases, but has not been applied commercially. The energy efficiency of drying with superheated steam is particularly high, but the use of this method for drying foods has limited possibilities, as explained previously.

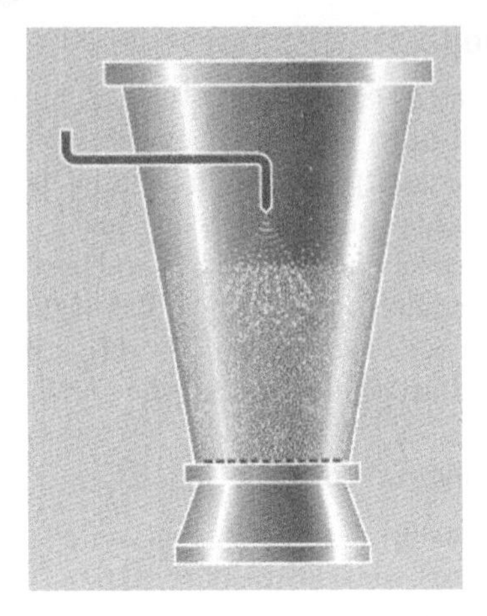

FIGURE 22.36

Top-spray fluidized bed agglomeration.

Figure courtesy of Niro.

The energy usage efficiency of convective (air) drying is significantly lower, mainly for two reasons:

1. While, in conductive drying, heat is supplied directly to the material being dried, in convective drying, heat is supplied to air. The heated air in turn transfers heat to the material being dried. Air serves here as an intermediate heat-transport agent. Any time an intermediate agent is used to transfer energy, the efficiency of the transfer is reduced.
2. In order to maintain the necessary driving force for drying, the air is not allowed to reach saturation. "Spent air" leaving the dryer is still hot and only partially humidified – i.e., it has un-utilized drying capacity. Spent air is usually released to the environment, and represents considerable loss of energy.

One of the methods used to save energy in drying liquid foods is to remove most of the water by pre-concentrating the feed through an energy-efficient method (e.g., multi-effect evaporation) before drying. Such combined processes are extensively used in drying milk and other dairy products, such as instant coffee, etc.

EXAMPLE 22.6

It is necessary to compute the specific energy consumption and the energy consumption efficiency of a spray dryer for milk, according to the following data.

Feed: pre-concentrated milk, 35% w/w dry matter, 30°C, 3600 $kg \cdot h^{-1}$
Product: milk powder, 2.5% residual moisture, 90°C
Drying air: ambient air at 22°C, 40% RH is heated indirectly by fuel burners to 180°C

Spent air: leaves the dryer at 90°C
Specific heat of the feed and the product: 3.1 and 1.2 $kJ \cdot kg^{-1} \cdot K^{-1}$. respectively

Heat losses are negligible.

Solution

In: 3600 $kg \cdot h^{-1} = 1\ kg \cdot s^{-1}$ milk at 35% solids. This yields 0.36 $kg \cdot s^{-1}$ milk powder at 2.5% moisture. $1 - 0.36 = 0.64$ kg of water evaporated.

The air is heated at constant moisture (H_1) from 22°C to 180°C, then adiabatically humidified and cooled to 95°C. At this point its moisture content is H_2.

H_1 and H_2 are read from the psychrometric chart:

$H_1 = 0.007\ kg \cdot kg^{-1}$
$H_2 = 0.046\ kg \cdot kg^{-1}$.

(Note: Because the temperature of 180°C is outside the range of the psychrometric chart (see Appendix, Figure A.2), H_2 is found by extrapolation. Alternatively, the chart at www.engineeringtoolbox.com/psycrometric-chart-d_251.html that covers high temperatures may be consulted.)

The mass flow-rate of air is then:

$$G_a = \frac{0.64}{0.046 - 0.007} = 16.4\ kg \cdot s^{-1}$$

At an average C_p of 1 $kJ \cdot kg^{-1} \cdot K^{-1}$, the heat transfer rate to the air is:

$$q_{in} = 16.4 \times 1 \times (180 - 22) = 2591\ kW$$

The energy spend for evaporation (at an average latent heat of 2270 $kJ \cdot kg^{-1}$) is:

$$q_{eva} = 0.64 \times 2270 = 1453\ kJ \cdot kg^{-1}$$

The specific energy expenditure (per kg of product is):

$$E = \frac{2591}{0.36} = 7197\ kJ \cdot kg^{-1}$$

The energy utilization efficiency is:

$$\eta_E = \frac{1453}{2591} \times 100 = 56\%$$

Energy of 7197 kJ is consumed for each kg of milk powder leaving the dryer. Even this high figure is an underestimation, because heat losses and inefficiencies in the indirect heating of the air have not been considered. Only 56% of this energy is utilized for the removal of water. Most of the rest is ejected with the spent air.

22.9 Osmotic dehydration

Osmotic dehydration is the removal of water by immersing the food in a solution of salt or sugars of high osmotic pressure. Water is transferred from the food to the solution by virtue of the difference in osmotic pressure.

Essentially, the process of partial removal of water by osmosis is an operation that has been known and practiced for centuries. Salting fish and candying fruit are examples of long established food processing techniques where removal of water takes place together with solute penetration. Nevertheless, the process, now named "osmotic dehydration" or "dewatering–impregnation soaking in concentrated solutions" (Raoult-Wack *et al.*, 1991, 1992), has recently attracted considerable research interest, and a voluminous literature on the subject has accumulated. (Seguí, 2012; Torreggiani, 1995). The commercial application of the process, outside the classical processes mentioned above, has been rather limited.

Physically, the process of osmotic dehydration is simple. The prepared (peeled, sliced or cut, etc.) material is immersed in the "osmotic solution", a relatively concentrated solution of sugars (glucose, sucrose, trehalose, etc.) or salt, or both. Water and some of the natural solutes of the food pass to the osmotic solution, while at the same time a certain amount of the "osmotic solute" penetrates the food. Selection of the osmotic solution composition and the process conditions aims at maximizing the water removal and minimizing all other transports. The osmotic solution is recycled, after concentration by evaporation. Energy consumption per kg of water removed is, of course, much lower than in ordinary drying, particularly if the osmotic solution is concentrated by multi-effect evaporation or reverse osmosis. Continuous operation has been tried at pilot-plant level (Barbosa-Cánovas and Vega-Mercado, 1996).

The rate of water removal is fast in the beginning, but slows down considerably as the osmotic pressure difference becomes gradually smaller. Consequently, water removal by osmosis cannot be carried too far. Osmotically dehydrated products still contain too much moisture to be shelf-stable. It is suggested that these materials can actually serve as intermediate products, intended for further processing such as drying, freezing or thermal processing (Raoult-Wack *et al.*, 1989).

The transport phenomena in osmotic drying of cellular tissue cannot be explained solely in terms of Fickean diffusion and effective diffusivity coefficients (Seguí *et al.*, 2012). The microstructure of cellular tissue contains elements (such as membranes) that interfere with mass transfer (Ferrando and Spiess, 2002). At the same time, the microstructure itself is affected by osmotic dehydration (Nieto *et al.*, 2004).

References

Aguillera, J.M., Chiralt, A., Fito, P., 2003. Food dehydration and product structure. Trends Food Sci. Technol. 14 (10), 432–437.

Bangs, E.W., Reineccius, G.A., 1990. Prediction of flavor retention during spray drying: an empirical approach. J. Food Sci. 55 (6), 1683–1685.

Bansal, N.K., Garg, H.P., 1987. Solar crop drying. In: Mujumdar, A.S. (Ed.), Recent Developments in Solar Drying. Hemisphere, New York, NY.

Bar, A., Ramon, O., Cohen, Y., Mizrahi, S., 2002. Shrinkage behavior of hydrophobic hydrogels during dehydration. J. Food Eng. 55 (3), 193–199.

Barati, E., Esfahani, J.A., 2011. A new solution approach for simultaneous heat and mass transfer during convective drying of mango. J. Food Eng. 102 (4), 302–309.

Barbosa-Cánovas, G.V., Vega-Mercado, H., 1996. Dehydration of Foods. Chapman & Hall, New York, NY.

Bhandari, B.R., Dumoulin, E.D., Richard, H.M.J., Noleau, I., Lebert, A.M., 1992. Flavor encapsulation by spray drying: application to citral and linalyl acetate. J. Food Sci. 57 (1), 217–221.

Bilbao-Sáinz, C., Andrés, A., Fito, P., 2005. Hydration kinetics of dried apple as affected by drying conditions. J. Food Eng. 68 (3), 369–376.

Bimbenet, J.J., Duquenoy, A., Trystram, G., 2002. Génie des Procédés Alimentaires. Dunod, Paris, France.

Bittner, B., Kissel, T., 1999. Ultrasonic atomization for spray drying: a versatile technique for the preparation of protein loaded biodegradable microspheres. J. Microencapsul. 16 (3), 325–341.

Bonazzi, C., Dumoulin, E., Raoult-Wack, A.L., Berk, Z., Bimbenet, J.J., Courtois, F., et al., 1996. Food drying and dewatering. Dry. Technol. 14 (9), 2135–2170.

Bruin, S., Luyben, K.Ch., 1980. Drying of food materials. A review of recent developments. In: Mujumdar, A.S. (Ed.), Advances in Drying. Hemisphere, New York, NY.

Chen, P., Pei, D.L.T., 1989. A mathematical model of drying processes. Intl. J. Heat Mass Trans. 32 (2), 297–310.

Chu, J., Lane, A., Conklin, D., 1953. Evaporation of liquids into their superheated vapors. Ind. Eng. Chem. 53 (3), 275–280.

Cil, B., Topuz, A., 2010. Fluidized bed drying of corn, bean and chickpea. J. Food Proc. Eng. 33 (6), 1079–1096.

Coulter, S.T., Reineccius, G.A., 1969. Flavor retention during drying. J. Dairy Sci. 52 (8), 1219–1223.

Daudin, J.D., 1983. Calcul des cinétiques de séchage par l'air chaud des produits biologiques solides. Sci. Aliments 3, 1–36.

Di Bonis, M.V., Ruocco, G., 2008. A generalized conjugate model for forced convection drying based on evaporative kinetics. J. Food Eng. 89 (2), 232–240.

Dumoulin, E., 2008. From powder end use properties to process engineering. In: Gutiérrez-López, G.F., Barbosa-Cánovas, G.V., Welti-Chanes, J., Parada-Arias, E. (Eds.), Food Engineering Integrated Approach. Springer, New York, NY.

Earle, P.E., Ceagiske, H.N., 1949. Factors causing the cracking of macaroni. Cereal Chem. 26, 267–286.

Eichler, S., Ramon, O., Ladyzhinski, I., Cohen, Y., Mizrahi, S., 1998. Collapse processes in shrinkage of hydrophilic gels during dehydration. Food Res. Int. 30 (9), 719–726.

Ekechukwu, O.V., Norton, B., 1999. Review of solar energy drying systems II: an overview of solar drying technology. Energy Conserv. Manage. 409 (6), 615–655.

Etzel, R.M., King, C.J., 1984. Loss of volatile trace organic components during spray drying. Ind. Eng. Chem. Process Res. Dev. 23, 705–710.

Fäldt, P., 1995. Surface Composition of Spray-Dried Emulsions, Doctoral Thesis, Lund University, Lund, Sweden.

Ferrando, M., Spiess, W.E.I., 2002. Transmembrane mass transfer in carrot protoplasts during osmotic treatment. J. Food Sci. 67 (7), 2673–2680.

Filkova, I., Mujumdar, A.S., 1995. Industrial spray drying systems. In: Mujumdar, A.S. (Ed.), Handbook of Industrial Drying, second ed. Marcel Decker, New York, NY.

Fornell, A., 1979. Séchage Des Produits Biologiques Par l'Air Chaud, Doctoral Thesis, ENSIA, Massy, France.

Hecht, J.P., King, C.J., 2000. Spray drying: influence of drop morphology on drying rates and retention of volatile substances I; Single drop experiments. Ind. Eng. Chem. Res. 39 (6), 1756–1765.

Islam, M.I.U., Langrish, T.A.G., Chiou, D., 2010. Particle crystallization during spray drying in humid air. J. Food Eng. 99 (1), 55–62.

Jensen, S.A., Borreskov, J., Dinesen, D.K., Madsen, R.F., 1987. Beet pulp drying in superheated steam under pressure. Zuckerind 112 (10), 886–891.

Kadam, D.M., Nangare, D.D., Singh, R., Kumar, S., 2011. Low-cost greenhouse technology for drying onion (Allium cepa L.) slices. J. Food Proc. Eng. 34 (1), 67–82.

Kavak Akpinar, E., 2009. Drying of parsley leaves in a solar dryer and under open sun: modeling, energy and exergy aspects. J. Food Proc. Eng. 34 (1), 27–48.

Khraisheh, M.A.M., Cooper, T.J.R., Magee, T.R.A., 1995. Investigation and modeling of combined microwave and air drying. Food Bioprod. Proc. 73, 121–126.

Kim, E.H.J., Chen, X.D., Pearce, D., 2008. Surface composition of spray dried milk powder 2. Effect of spray-drying conditions on the surface composition. J. Food Eng. 94 (2), 169–181.

Kiranoudis, C.T., 1998. Design and operational performance of conveyor-belt drying equipment. Chem. Eng. J. 69 (1), 27–38.

Kiranoudis, C.T., Markatos, N.C., 2000. Pareto design of conveyor-belt dryers. J. Food Eng. 46 (3), 145–155.

Krischer, O., Kast, W., 1992. Die Wissenschaftlischen Grundlagen Der Trockningstechnik, third ed. Springer Verlag, Berlin, Germany.

Krokida, M.N.K., Maroulis, Z.E., 1997. Effect of drying method on shrinkage and porosity. Dry. Technol. 10 (5), 1145–1155.

Lewicki, P.P., 1998. Some remarks on rehydration of dried foods. J. Food Eng. 36 (1), 81–87.

Li, Z., Raghavan, G.S.V., Orsat, V., 2010. Temperature and power control in microwave drying. J. Food Eng. 97 (4), 478–483.

Lin, S.Y., Krochta, J.M., 2006. Fluidized-bed system for whey protein film coating of peanuts. J. Food Proc. Eng. 29 (5), 532–546.

Marabi, A., Saguy, I.S., 2004. Effect of porosity on rehydration of dry food particulates. J. Food Sci. Agric. 84 (10), 1105–1110.

Marabi, A., Jacobson, M., Livings, S.J., Saguy, I.S., 2004. Effect of mixing and viscosity on rehydration of dry food particulates. Eur. Food Res. Technol. 218 (4), 339–344.

Masters, K., 1991. Spray Drying Handbook. Longman Scientific and Technical, Harlow, UK.

McMinn, W.A.M., Khraisheh, M.A.M., Magee, T.R.A., 2003. Modeling the mass transfer during convective, microwave and combined microwave-convective drying of solid slabs and cylinders. Food Res. Intl. 36 (9–10), 977–983.

Mujumdar, A.S., 1992. Superheated steam drying of paper: principles, status and potential. In: Mujumdar, A.S. (Ed.), Drying of Solids. International Science Publisher, Enfield, UK.

Nieto, A.B., Salvatori, D.M., Castro, M.A., Alzamora, S.M., 2004. Structural changes in apple tissue during glucose and sucrose osmotic dehydration: shrinkage, porosity, density and microscopic features. J. Food Eng. 61 (2), 269–278.

Nonhebel, G., 1971. Drying of Solids in the Chemical Industry. Butterworth, London, UK.

Ormos, Z.D., 1994. Granulation and coating. In: Chulia, D., Deleuil, M., Pourcelot, Y. (Eds.), Handbook of Powder Technology, vol. 9. Elsevier, Amsterdam, The Netherlands.

Patel, K.P., Chen, X.D., 2005. Prediction of spray-dried product quality using two simple drying kinetics models. J. Food Process Eng. 28 (6), 567–594.

Ramirez, C., Troncoso, E., Muñoz, J., Aguilera, J.M., 2012. Microstructure analysis on pre-treated apple slices and its effect on water release during air drying. J. Food Eng. 106 (3), 253–261.

Raoult-Wack, A.L., Lafont, F., Rios, G., Guilbert, S., 1989. Osmotic dehydration. Study of mass transfer in terms of engineering properties. In: Mujumdar, A.S., Roques, M. (Eds.), Drying '89. Hemisphere Publishing, New York, NY.

Raoult-Wack, A.L., Guilbert, S., Le Maguer, M., Rios, G., 1991. Simultaneous water and solute transport in shrinking media – Part 1. Application to dewatering and impregnation soaking process analysis (osmotic dehydration). Dry. Technol. 9 (3), 589–612.

Raout-Wack, A.L., Lenart, A., Guilbert, S., 1992. Recent advances in dewatering through immersion in concentrated solutions ("Osmotic dehydration"). In: Mujumdar, A.S. (Ed.), Drying of Solids. International Science Publisher, Enfield, UK.

Roustapour, O.R., Hosseinalipour, M., Ghobadian, B., Mohaghegh, F., Azad, M.V., 2009. A proposed numerical–experimental method for drying kinetics in a spray dryer. J. Food Eng. 90 (1), 20–26.

Rulkens, W.H., 1973. Retention of Volatile Trace Components in Drying Aqueous Carbohydrate Solutions, Doctoral thesis, Technische Hogeschool, Eindhoven, The Netherlands.

Sa-adchom, P., Swasdisevi, T., Nathakaranakule, A., Soponronnarit, S., 2011. Drying kinetics using superheated steam and quality attributes of dried pork slices for different thickness, seasoning and fibers distribution. J. Food Eng. 104 (1), 105–113.

Sander, A., Kardum, J.P., 2009. Experimental validation of thin-layer drying models. Chem. Eng. Technol. 32 (4), 590–599.

Sansiribhan, S., Devahastin, S., Soponronnarit, S., 2010. Quantitative evaluation of microstructural changes and their relations with some physical characteristics of food during drying. J. Food Sci. 75 (7), E453–E461.

Saravacos, G.D., 1967. Effect of drying method on the water sorption of dehydrated apple and potato. J. Food Sci. 32 (1), 81–84.

Schubert, H., 1981. Principles of agglomeration. Intl. Chem. Eng. 21 (3), 363–377.

Schubert, H., 1993. Instantization of powdered food products. Intl. Chem. Eng. 33 (1), 28–45.

Seguí, L., Fito, P.J., Fito, P., 2012. Understanding osmotic dehydration of tissue structured foods by means of a cellular approach. J. Food Eng. 110 (3), 240–247.

Senoussi, A., Dumoulin, E., Berk, Z., 1995. Retention of diacetyl in milk during spray drying and storage. J. Food Sci. 60 (5), 894–897, 905.

Sereno, A.M., Madeiros, G.L., 1990. A simplified model for the prediction of drying rates of foods. J. Food Eng. 12 (1), 1–11.

Somjai, T., Achariyaviriya, S., Achariyaviriya, A., Namsanguan, K., 2009. Strategy for longan drying in two-stage superheated steam and hot air. J. Food Eng. 95 (2), 313–321.

Svensson, C., 1980. Steam drying of pulp. In: Mujumdar, A.S. (Ed.), Drying '80. Hemisphere Publishers, New York, NY.

Thijssen, H.A.C., 1975. Process conditions and retention of volatiles. In: Goldblith, S.A., Rey, L., Rothmayr, S. (Eds.), Freeze *Drying and Advanced Food Technology.* Academic Press, London, UK.

Torreggiani, D., 1995. Technological aspects of osmotic dehydration in foods. In: Barbosa-Cánovas, G.V., Welti-Chanes, J. (Eds.), Food Preservation by Moisture Control. Technomics Publishing Co. Inc., Lancaster, PA.

Waje, S.S., Thorat, B.N., Mujumdar, A.S., 2006. An experimental study of the performance of a screw conveyor dryer. Dry. Technol. 24 (3), 293–301.

Wang, N., Brennan, J.G., 1995. Changes in the structure, density and porosity of potato during dehydration. J. Food Eng. 24 (1), 61–76.

Yeo, Y., Park, K., 2004. A new microencapsulation method using an ultrasonic atomizer based on interfacial solvent exchange. J. Controlled Release 100, 379–388.

Zanoelo, E.F., Cardozo-Filho, L., Cardozo-Juinior, E.L., 2006. Superheated steam-drying of mate leaves and effect of drying on the phenol content. J. Food Proc. Eng. 29 (3), 253–268.

Zecchi, B., Clavijo, L., Martínez Garreiro, J., Gerla, P., 2011. Modeling and minimizing process time of combined convective and vacuum drying of mushrooms and parsley. J. Food Eng. 104 (1), 49–55.

Zhao, Y.P., Chang, K.C.Y.P., 1995. Sulfite and starch affect color and carotenoids of dehydrated carrots (*Daucus carota*) during storage. J. Food Sci. 60 (2), 324–326.

CHAPTER

Freeze Drying (Lyophilization) and Freeze Concentration

23

23.1 Introduction

Freeze drying or *lyophilization* is the removal of water by sublimation from the frozen state (ice). In this process, the food is first frozen and then subjected to high vacuum, whereby the water ice sublimates (i.e., evaporates directly, without melting). The water vapor released is usually caught on the surface of a condenser at very low temperature. The heat of sublimation is supplied to the food by various methods, described below.

As a physical phenomenon and a laboratory technique, freeze drying was already known at the end of the 19th century. However, it did not develop into an industrial process until after World War II (Flosdorf, 1949; King, 1971, 1975). The first commercial applications were in the pharmaceutical industry (antibiotics, living cells, blood plasma, etc.), and this is still the greatest user of freeze drying (Oetjen and Haseley, 2004; Pikal, 2007; Santivarangkna, 2011). Industrial freeze drying of foods started in the late 1950s.

In the food industry, interest in commercial freeze drying arises from the superior quality of the freeze dried products as compared with foods dehydrated by other methods. Freeze drying is carried out at low temperatures, thus preserving flavor, color and appearance, and minimizing thermal damage to heat-sensitive nutrients. Since the entire process occurs in solid state, shrinkage and other kinds of structural changes are largely avoided.

Freeze drying is, however, an expensive method of dehydration. It is economically feasible only in the case of high added-value products and whenever the superior quality of the product justifies the higher production cost (Ratti, 2001).

23.2 Sublimation of water

Sublimation is the direct transition from solid state to gaseous state without melting. Sublimation occurs at a definite range of temperatures and pressures, depending on the substance in question. The phase diagram of pure water (Figure 23.1) indicates that sublimation of water ice can occur only if the vapor pressure and

Food Process Engineering and Technology. DOI: http://dx.doi.org/10.1016/B978-0-12-415923-5.00023-X
© 2013 Elsevier Inc. All rights reserved.

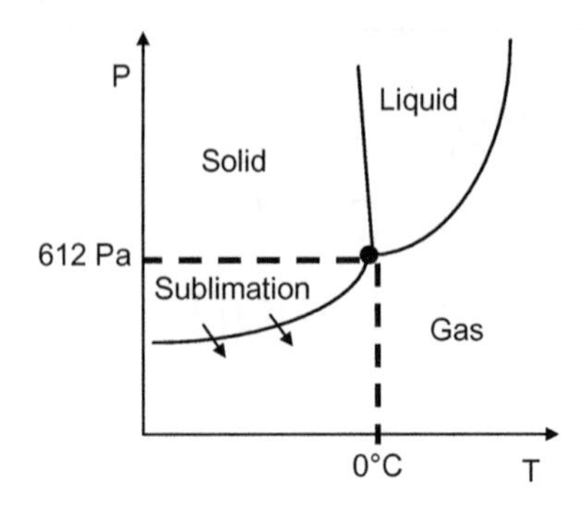

FIGURE 23.1

Phase diagram of water showing conditions for sublimation.

temperature are below those of the *triple point* of water – i.e., below 611.73 Pa and 0.01°C, respectively.

Theoretically, freeze drying should be possible at atmospheric pressure, provided that the partial pressure of water vapor is very low, i.e., the air is very dry (Boeh-Ocansey, 1984; Heldman and Hohner,1974; Karel, 1975). "Atmospheric freeze drying" occurs in nature – for example, when snow "disappears" without melting in cold, dry weather. In practice, however, freeze drying is carried out at very low *total* pressure (typically 10–50 Pa). At such low pressure, water vapor has a very large specific volume. In order to remove such large volumes of vapor in the gaseous state, the vacuum pump must have an unrealistically large displacement capacity. To overcome this problem, the vapors are condensed as ice crystals on the surface of condensers kept at an extremely low temperature (typically −40°C or less).

Freeze drying occurs in two stages (Oetjen and Haseley, 2004; Pikal *et al.*, 1990). The first stage is *sublimation drying*, in which sublimation of the frozen water (ice crystals) occurs. Normally, most of the water in the food is removed at this stage. The second stage is *desorption drying*, during which most of the unfrozen water adsorbed on the solid matrix is removed. Typically, freeze drying is carried to a final moisture content of 1–3%.

23.3 Heat and mass transfer in freeze drying

23.3.1 Heat and mass transfer mechanisms

Like any other dehydration process, freeze drying involves simultaneous heat and mass transfer (Karel, 1975). The distribution of moisture content through the material undergoing lyophilization is different from that observed in other dehydration processes (Figure 23.2). Ideally, two distinct zones, separated by a fairly

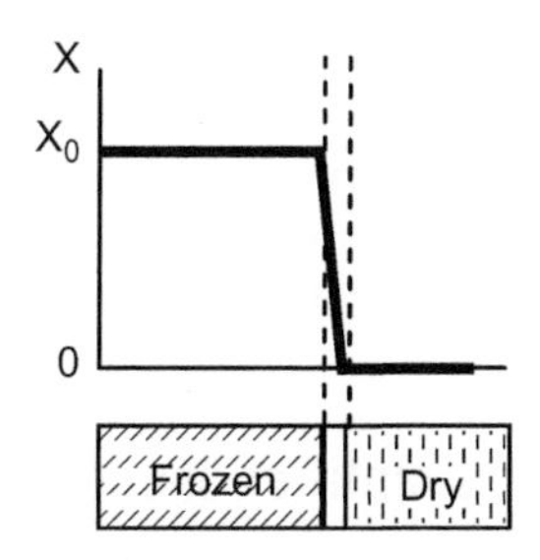

FIGURE 23.2

Moisture distribution in a material undergoing freeze drying.

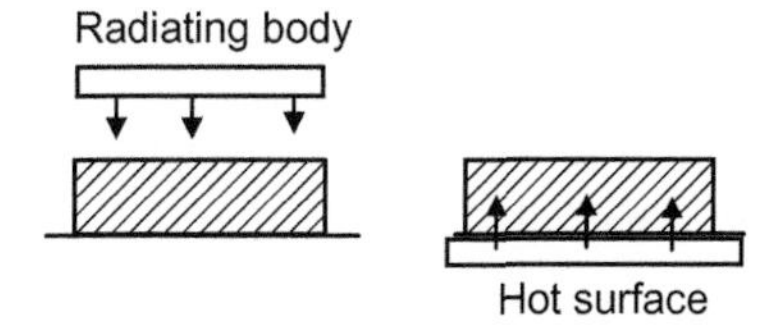

FIGURE 23.3

Modes of heat transfer in freeze drying.

narrow phase transition interface, are observed: a frozen zone and a "dry" zone. The frozen zone is at the original moisture content. The dry zone is devoid of ice crystals, and the only moisture in it is water adsorbed on the solid matrix. The proportion of adsorbed moisture is usually small, and the water content distribution across the material can be idealized as a step function. The interface between the two zones is known as the "sublimation front".

Heat can be delivered to the sublimation front by a number of mechanisms. The two most commonly applied modes of heat supply are radiation from hot surfaces, and conduction (Figure 23.3). Microwave heating is an interesting option (see below).

1. *Radiation from hot surfaces*. In this case, heat is delivered to the exposed surface of the dry zone by thermal radiation and then travels through the dry zone to the sublimation front, mainly by conduction. The water vapor released also travels through the dry zone but in the opposite direction, from the sublimation front to the surface and from there to the condenser.
2. *Contact with a hot surface*. In a common embodiment of this mode, trays containing the frozen material are placed on heated shelves. Heat is transferred to the sublimation zone by conduction through the frozen layer. Water vapor is transferred through the dry layer as before.

In practice, both modes of heat transfer take place at the same time. In a common type of freeze dryer, trays of the frozen material are placed on heated shelves. The shelves deliver heat to the tray above by conduction and to the tray below by radiation.

The quantity of heat to be transported to the sublimation front is considerable. The latent heat of sublimation is equal to the sum of the heat of evaporation and the latent heat of fusion. For water at 0°C, this amounts to approximately 3000 $kJ \cdot kg^{-1}$.

23.3.2 Drying kinetics – simplified model

Freeze dryers are expensive systems, both in capital investment and in cost of operation. Production capacity and therefore drying time are thus of great importance. The following analysis of drying time in freeze drying is based on simplifying assumptions and approximations. It does not permit prediction of the exact drying time, but it provides valuable information on the effect of processing conditions on drying rate and underlines the limitations in the selection of process variables.

Consider a homogeneous slab of frozen food, at a temperature well below melting, undergoing lyophilization (Figure 23.4). Calculation of the freeze-drying time of the slab is required. The following assumptions are made:

- The slab is dried from one side only. Both heat transfer and mass transfer are unidirectional.
- Heat is supplied to the surface of the slab by radiation from a hot surface at a distance from the slab.
- All the heat supplied is used exclusively for the sublimation of ice crystals. As long as sublimation occurs, sensible heat effects are negligible.
- There is a sharp sublimation front between the totally iceless (dry) zone and the frozen zone.
- The vapors are condensed on a cold surface (condenser) as ice. The resistance of the chamber space between the slab surface and the condenser to mass transfer is negligible because of the high vacuum. Hence, the water vapor pressure at the condenser surface is nearly equal to that measured in the chamber.

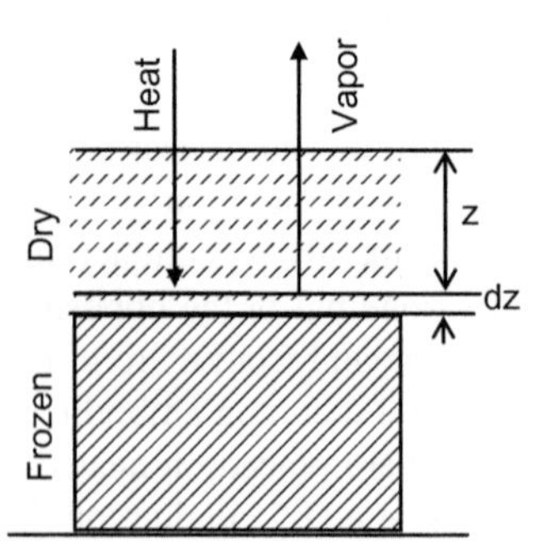

FIGURE 23.4

Kinetics of freeze drying.

- The temperature at the slab surface is kept constant, by controlling the temperature of the radiating body.
- The temperature of the frozen product and hence of the sublimation front is constant.
- The gas in the freeze-dryer chamber consists, practically, of water vapor only. The proportion of non-condensables (air) is negligible.

Assume that an incremental thickness dz of the slab is dried in a time increment dt. The rate of ice sublimation dw/dt is:

$$\frac{dw}{dt} = A\rho_i(w_i - w_f)\frac{dz}{dt} \tag{23.1}$$

where:

A = area of the slab surface, m^2
ρ_i = density of the frozen food, $kg \cdot m^{-3}$
w_i, w_f = initial and final water content, respectively, $kg \cdot kg^{-1}$.

At steady state, the rate of sublimation must be in accordance *both* with the rate of heat transfer to the sublimation front and with the rate of mass transfer from the sublimation front.

Let us consider heat transfer first. The rate of heat supply q ($J \cdot s^{-1}$) must be equal to the rate of sublimation ($\cdot s^{-1}$) multiplied by the latent heat of sublimation λ_S ($J \cdot kg^{-1}$):

$$q = A\rho_i(w_i - w_f)\lambda_s\frac{dz}{dt} \tag{23.2}$$

On the other hand, q is given by the rate of conductive transport from the slab surface to the sublimation front through the dry layer:

$$q = \frac{kA(T_0 - T_i)}{z} \tag{23.3}$$

where k = thermal conductivity of the dry layer, $W \cdot m^{-1} \cdot K^{-1}$, and T_0, T_i = temperature at the slab surface and at the sublimation front, respectively, K.

Combining Eqs (23.3) and (23.2) and integrating from z = 0 to z = Z (total thickness of the slab, m), we obtain:

$$t = \frac{Z^2(w_i - w_f)}{2(T_0 - T_i)}\left[\frac{\rho_i\lambda_s}{k}\right] \tag{23.4}$$

Let us now consider mass transfer. At steady state, the rate of sublimation must be equal to the rate of removal of the vapors by mass transfer through the dry layer:

$$A\rho_i(w_i - w_f)\frac{dz}{dt} = \Pi A\frac{p_i - p_0}{z} \tag{23.5}$$

where Π = permeability of the dry layer to water vapor, $kg \cdot s^{-1} \cdot m^{-1} \cdot Pa^{-1}$ (see Chapter 3, Section 3.3.4.4) and p_i, p_0 = water vapor pressure at the sublimation front and at the slab surface, respectively, Pa · s.

After integration we obtain:

$$t = \frac{Z^2(w_i - w_f)}{2(p_i - p_0)}\left[\frac{\rho_i}{\Pi}\right] \quad (23.6)$$

Equations (23.4) and (23.6) lead to a number of conclusions:

1. Drying time is proportional to the square of the slab thickness Z. Consequently, tray loading (mass of material per unit tray area) and the size of the particles (in the case of particulate materials) have a very strong effect on drying time. However, this conclusion is not always supported by experimentation. (Khalloufi *et al.*, 2005).
2. Drying time is proportional to the difference between initial and final water (ice) content. This is obvious.
3. Drying time strongly depends on the two independent and controllable process variables, namely the temperature at the surface T_0 and the vapor pressure at the condenser p_0 (hence condenser temperature). The potential and limitations of manipulating these parameters for the purpose of accelerating freeze drying will be discussed in point (6) below.
4. Drying time is inversely proportional to the thermal conductivity k and permeability Π of the dry layer. Both parameters depend on the porosity of the layer and on the pressure, but in opposite directions. A more porous matrix has lower thermal conductivity but higher permeability to vapor transfer. The porosity of the dry layer depends on the volumetric proportion of ice in the frozen material, hence on the initial moisture content (Simatos and Blond, 1975). Furthermore, the nature of the porosity depends on the size distribution of the ice crystals. If the food has been frozen slowly, large crystals predominate in the frozen material and the dry layer has a more open porosity, resulting in higher permeability. Pressure has been found to have a considerable effect, both on the thermal conductivity and on the permeability of the dry layer (Karel *et al.*, 1975). Thermal conductivity increases and permeability decreases with pressure, but the effect of pressure on the thermal conductivity is stronger. As a result, an increase in chamber pressure accelerates the rate of freeze drying.
5. Increasing the temperature T_i of the frozen zone would result in accelerated freeze drying, but this option is not easily practicable. As a rule, T_i is kept well below the initial freezing point of the food in order to reduce, as much as possible, the proportion of unfrozen water and to prevent melting. It remains fairly constant during the process. The water vapor pressure at the sublimation front p_i, being thermodynamically linked to T_i, also remains practically constant.

6. An expression defining the conditions for steady state is derived by equating Eqs (23.3) and (23.5):

$$\frac{p_i - p_0}{T_0 - T_i} = \frac{k}{\Pi \lambda_s} \Rightarrow T_0 = T_i + \frac{\Pi \lambda_s (p_i - p_0)}{k} \quad (23.7)$$

Stable operation of the freeze dryer is usually maintained by regulating T_0 by controlling the rate of heat supply to the surface.
Equations of the same type are obtained and the same conclusions are valid for the case of conduction heating of the slab from below. For this case, T_0 has to be replaced by T_p, the temperature of the warm surface touching the slab from below.

EXAMPLE 23.1

A slab of frozen orange juice is to be freeze dried, from a moisture content of 87% to 3% (w/w, wet basis). The slab, 1.2 cm. thick, rests on a tray and is heated by radiation from its upper surface. The source of radiation is regulated so as to maintain the surface temperature at 30°C at all times. The frozen juice is at −18°C. The latent heat of sublimation is 3000 $kJ \cdot kg^{-1}$. The thermal conductivity of the dry layer at the pressure of operation is 0.09 $W \cdot m^{-2} \cdot K^{-1}$. The density of the frozen juice is 1000 $kg \cdot m^{-3}$.

a. Estimate the drying time. Neglect sensible heat effects, and assume that heat transfer is the rate-limiting factor.
b. Using the same simplifying assumptions, estimate the condenser temperature at which the system will be at steady state. The permeability of the dry layer to water vapor at the operation conditions is 0.012×10^{-6} $kg \cdot m^{-1} \cdot Pa \cdot s$.

Solution

a. Considering that the rate-limiting factor is heat transfer, Eq. (23.4) will be applied:

$$t = \frac{Z^2(w_i - w_f)}{2(T_0 - T_i)}\left[\frac{\rho_i \lambda_s}{k}\right]$$

$$t = \frac{(0.012)^2(0.87 - 0.03)}{2(30 - (-18))}\left[\frac{1000 \times 3000000}{0.09}\right] = 0.042 \times 10^6 \text{s} = 11.67 \text{ h}$$

The freeze-drying time (not including time for desorption) is 11.67 hours.

b. To find the steady-state conditions we use Eq. (23.7):

$$\frac{p_i - p_0}{T_0 - T_i} = \frac{k}{\Pi \lambda_s} \Rightarrow p_i - p_0 = \frac{k(T_0 - T_i)}{\Pi \lambda_s} = \frac{0.09 \times 48}{0.012 \times 10^{-6} \times 3000000} = 120 \text{ Pa}$$

The vapor pressure of ice as a function of temperature is found in Table A.13 (see Appendix).
At −18°C, $p_i = 125$ Pa.
$p_0 = 125 - 120 = 5$ Pa.
The vapor pressure of ice is 5 Pa at −48°C.
The temperature of the condenser for steady-state operation is −48°C.

Note: Due to extreme simplification of the model, the results are, at most, approximations.

23.3.3 Drying kinetics – other models

More detailed models for heat and mass transfer in freeze drying have been developed and validated experimentally (Brülls and Rasmuson, 2002; George and Datta, 2002; Khalloufi *et al.*, 2005; Liapis and Bruttini, 1995; Nastaj, 1991). Some of the findings are contradictory. Thus, George and Datta (2002) determined that mass transfer is the rate-controlling factor while many authors consider heat transfer to be the controlling mechanism. Most of the simulation models consist of sets of partial differential equations that are solved by numerical methods. Usually, only the sublimation phase is included in the simulations. Inclusion of the desorption phase would require knowledge and modeling of the sorption isotherm (Khalloufi *et al.*, 2005).

23.4 Freeze drying in practice

23.4.1 Freezing

The solid slab model considered above is not common in practice. Solid materials are cut, and liquid or semi-liquid foods are frozen and then granulated before lyophilization, in order to increase the surface area. In laboratory freeze dryers, the material is frozen in the drying chamber itself by cooling the trays and/or through evaporative cooling as vacuum is applied. In industry, the material is frozen outside the freeze dryer. Liquid foods may be frozen on a refrigerated drum and scraped off as flakes. It should be remembered that all the water in the food cannot be frozen. The unfrozen, freeze-concentrated portion of the liquid may solidify as a glass. Indeed, when frozen as a slab, coffee extract and orange juice form a glassy, impermeable layer on the surface. If a canvas or metal mesh is frozen onto the surface of the slab and then pulled away before freeze drying, thus exposing a rough surface, the rate of drying is considerably improved. Another way to avoid the formation of an impermeable layer is to slush-freeze the liquid in a swept-surface freezer, then hard-freeze the slush.

The rate of freezing affects the size of ice crystals and therefore the permeability of the dry layer once the ice is removed by sublimation. Quast and Karel (1968) found that slow freezing of coffee resulted in higher permeability of the dry layer.

The effect of freezing conditions on the quality of freeze-dried coffee has been studied by Flink (1975).

23.4.2 Drying conditions

Control of the drying conditions has several objectives, including:

- Shortening the drying time
- Maximizing product quality, particularly to maximize retention of volatile aromas (Krokida and Philippopoulos, 2006)
- Preventing melting and collapse.

As explained previously, not all the water in the frozen material is in the form of ice. The non-frozen portion of the water and the water vapor adsorbed on the matrix may act as a plasticizer and soften the dry layer, if the temperature is not sufficiently low. Such softening results in collapse of the porous structure of the dry layer and in loss of all the advantages of freeze drying. The glass-transition properties of the dry layer have a profound influence on the tendency of the material to shrink and undergo collapse (Khalloufi and Ratti, 2003). Knowledge of the temperature at which collapse occurs is essential for proper control of the process.

23.4.3 Microwave freeze-drying

Considering the strong influence of heat transfer on the kinetics, the use of microwaves for heating might be expected to accelerate the rate of freeze drying considerably. It is not surprising, therefore, to find microwave freeze-drying among the first investigated food applications of microwaves (Copson, 1958; Ang *et al.*, 1977). An important advantage of microwave freeze-drying is the fact that the dry layer is practically transparent to microwaves by virtue of its relatively low loss factor (see Chapter 3, Section 3.8.1). Energy is therefore delivered directly to the frozen water where it is needed for sublimation. Excessive heating of the dry layer, and hence thermal damage to the product, is avoided (Sunderland, 1980).

Unfortunately, however, a number of factors hamper the practical application of microwave heating to freeze drying:

1. *Uneven heating*. Different components (e.g., fat) have different loss factors. Particles of different sizes absorb microwave energy at different rates. Changes in formulation (e.g., addition of salt) may affect the rate of heating (Wang *et al.*, 2010).
2. *The risk of melting*. The loss factor of liquid water is considerably higher than that of ice; thus any amount of local melting would cause rapid propagation of melting and collapse.
3. *Gas ionization, glow discharge*. Ionization of the gas in the microwave cavity and the occurrence of glow discharge are undesirable effects that may take place when operating with microwaves at high vacuum.

Avoidance of some of these problems would require special control systems, capable of matching continuously the impedance of the energy source (the microwave generator) with that of the load (the product undergoing sublimation) and monitoring mass transfer by adjusting the chamber pressure. Models describing the coupling between the electromagnetic field distribution and heat and mass transfer in sublimation have been studied (Heng *et al.*, 2007; Ma and Peltre, 1975; Tao *et al.*, 2005; Wang *et al.*, 1998). Various food materials have been subjected to microwave freeze-drying at laboratory scale (Ang *et al.*, 1977; Duan *et al.*, 2010; Wang *et al.*, 2010). However, scaling up of microwave freeze-drying to an industrial process has not, as yet, happened.

23.4.4 Freeze drying: commercial facilities

Industrial freeze drying is practiced in three types of commercial set-ups:

- Large-scale, in-house freeze drying for one product (usually coffee).
- Small to moderate in-house freeze drying for the production of freeze-dried ingredients needed in the composition of the products made by the company (e.g., meat, poultry, pasta, vegetables, herbs and condiments for dry soups).
- Custom freeze-drying facilities, dedicated to custom freeze drying for third parties. In view of the high capital cost of the equipment, the economic soundness of a custom freeze-drying facility depends strongly on the rate of utilization of the plant.

The first type of commercial set-up utilizes continuous freeze dryers. The others usually make use of batch lyophilizers.

23.4.5 Freeze dryers

A freeze dryer consists of the following basic elements (Lorentzen, 1975) (Figure 23.5):

- A drying chamber, capable of maintaining vacuum with minimal leakage
- Elements for supporting the material being lyophilized (trays, shelves, carts, etc.)

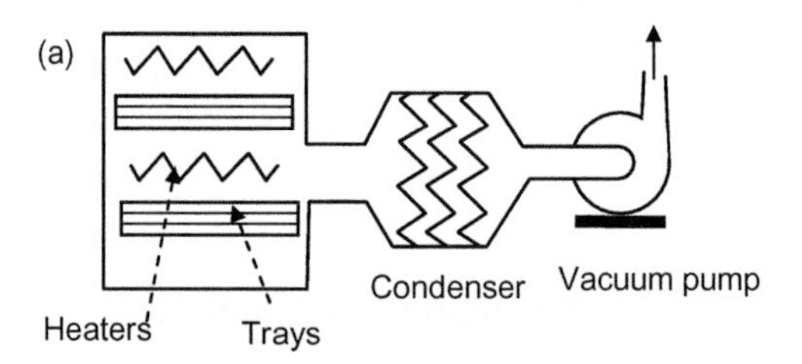

FIGURE 23.5

a. Elements of a freeze dryer. b. A batch freeze dryer.

Photograph courtesy of the Department of Biotechnology and Food Engineering, Technion.

- A source of heat (radiating hot surfaces, liquid heating medium circulating over heat-transfer surfaces, microwaves, etc.)
- A refrigerated ice condenser and its refrigeration system
- A vacuum pump capable of producing very high vacuum
- Control and measurement instruments (pressure, temperature, possibly weight).

Batch dryers with capacities ranging from 100 to 1500 kg per batch are available. The drying chamber is usually a horizontal cylinder with one door for charging and discharging. The condenser may be installed inside the chamber, or in a separate space flanged to the chamber. Bearing in mind the very large volume of vapor to be moved to the condenser, the distance between the trays and the condenser is kept as short as possible. In large units, two condensers per chamber, working alternately, are provided. While one condenser is in service, the other is isolated from the chamber and defrosted. The material to be lyophilized is spread on aluminum trays. Heat is usually supplied by circulating a heating medium through the hollow shelves supporting the trays.

Continuous dryers are horizontal cylindrical tunnels with locks for introducing and removing the material without breaking the vacuum (Figure 23.6). Several pairs of condensers, operating alternately, are installed along the tunnel. The material is placed on trays, and the stacks of trays are pushed through the tunnel. Continuous freeze dryers with tray areas of a few hundreds of square meters are available.

A method for conveying particulate materials through a continuous freeze dryer by means of vibrating supports has been proposed (Oetjen, 1999).

23.5 Freeze concentration

23.5.1 Basic principles

The basic mechanism of freeze concentration is simple. When a solution or a liquid food is frozen, water is separated from the rest of the solution as crystals of

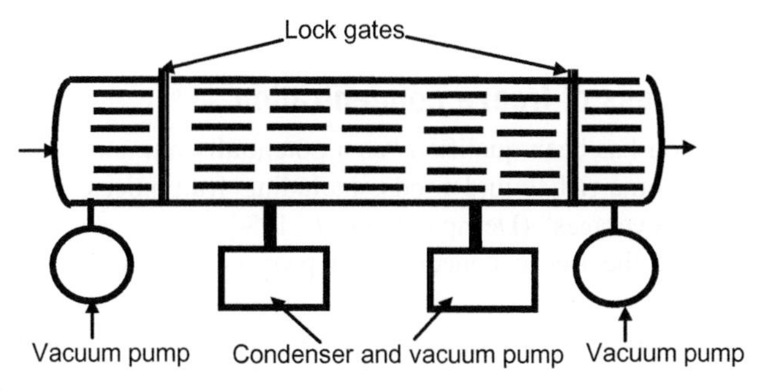

FIGURE 23.6

Schematic structure of a continuous freeze dryer.

pure water ice. Subsequently, the mixture can be separated into ice and a concentrated solution. Since freeze concentration occurs without heating and without boiling, the thermal damage and loss of volatile aroma that characterize evaporative concentration are largely avoided. Thus, freeze concentration is, in theory, particularly suitable for heat-sensitive liquid foods containing volatile aromas. Coffee extracts, fruit juices and their recovered essences would be natural candidates for freeze concentration (Aider and Halleux, 2009). The process has been also applied to skim milk (Hartel and Espinel, 1993). As we shall see, however, certain technological problems hamper the widespread application of freeze concentration in the food industry.

EXAMPLE 23.2

A "weak" wine has 10.5% (w/w) ethyl alcohol (MW = 46) and 1.5% (w/w) "dry extract" (assimilated to glucose, MW = 180). The requirement is to concentrate the wine to 13% alcohol by freeze concentration. Find the temperature to which the wine has to be cooled and the quantity of water to be removed as ice from 100 kg of wine.

Solution

Since only water is removed by freeze concentration, the "concentrated" wine will contain 13% alcohol and $1.5 \times 13/10.5 = 1.86\%$ dry extract. The molar concentration of water in the concentrated wine will be:

$$x_w = \frac{(100 - 13 - 1.86)/18}{13/46 + 1.86/180 + (100 - 13 - 1.86)/18} = 0.9417$$

The freezing point for an aqueous solution with a water molar concentration of 0.94 is found as in the Example. The result is −6.2°C.

Let W kg be the quantity of water to be removed from 100 kg of wine.

$$10.5 = 0.13 \times (100 - W) \Rightarrow W = 19.2 \text{ kg}$$

The wine has to be cooled to −6.2°C, and 19.2 kg of water must be removed as ice from each 100 kg of weak wine.

23.5.2 The process of freeze concentration

Freeze concentration has been qualified "as a potentially attractive method for the concentration of aroma-rich liquid foods, including fruit juices, coffee, tea and selected alcoholic beverages" (Deshpande *et al.*, 1984).

The principle of the freeze concentration process is shown schematically in Figure 23.7.

The system consists of two stages: freezing (crystallization) and separation (Thijssen, 1975). The stage of crystallization comprises nucleation and crystal growth, as described in Chapter 14. The technological and economic success of the process depends largely on the efficiency and quality of the separation

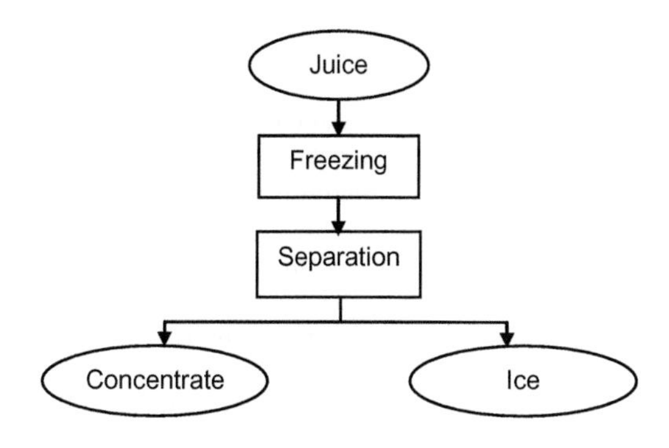

FIGURE 23.7

Flow diagram of freeze concentration.

process. Ideally, only pure water should be removed as ice, but in practice considerable quantities of solutes adhere to the surface of the crystals (Bayindirli *et al.*, 1993). The fractional loss of solutes depends on the size and shape of the crystals, their proportional volume in the mixture, and the viscosity of the concentrate. With respect to size and shape, large and near-spherical crystals are best. If the concentration ratio is substantial, the quantity of ice to be crystallized and removed and, with it, the proportional loss of solids would be very large. For example, in order to concentrate citrus fruit juices at the concentration ratios practiced by evaporative concentration (e.g., six to one), 5 kg of ice has to be formed and removed for every 6 kg of juice. The quantity of the viscous concentrate adhering to the crystals would be considerable. For this reason, freeze concentration is preferentially applied to processes requiring moderate concentration ratios (see Example 23.2). Another solution is multi-stage freeze concentration, where a small proportion of the ice is removed at each stage.

In practice, the crystallization unit is usually a swept-surface freezer and the separation unit is a centrifuge (Van Pelt, 1975). In one variation of the process, a "recrystallizer" is inserted after the freezer. This is a stirred vessel where crystal growth takes place by the Ostwald ripening mechanism discussed in Section 14.2.2. The very small crystals disappear while the larger crystals grow further.

Solid-liquid centrifuges, such as the basket centrifuges (Section 9.3.4) are the standard equipment for the separation of the ice crystals. Obviously, the crystals should be "washed" in order to recover the concentrate entrapped in the ice cake. If the centrifuge is not insulated, the water released by the melting ice participates in the washing. Another device for carrying-out the separation step after recrystallization is the "wash column" (Van Pelt, 1975), consisting of a vertical cylinder equipped with a piston at the bottom. The column is filled with "magma" from the recrystallizer. The concentrate is recovered by a combined action of pressing and washing with water obtained by melting a small proportion of the ice, at the top of the wash column.

References

Aider, M., Halleux, D., 2009. Cryoconcentration technology in the bio-food industry: Principles and applications. LWT 42 (3), 679–685.

Ang, T.K., Pei, D.C.T., Ford, J.D., 1977. Microwave freeze drying: an experimental investigation. Chem. Eng. Sci. 32 (12), 1477–1489.

Bayindirli, L., Özilgen, M., Ungan, S., 1993. Mathematical analysis of freeze concentration of apple juice. J. Food Eng. 19 (1), 95–107.

Boeh-Ocansey, O., 1984. Effects of vacuum and atmospheric freeze-drying on quality of shrimp, turkey flesh and carrot samples. J Food Sci. 49 (6), 1457–1461.

Brülls, M., Rasmuson, A., 2002. Heat transfer in vial lyophilization. Intl. J. Pharm. 46 (1), 1–6.

Copson, D.A., 1958. Microwave sublimation of foods. Food Tech. 12, 270–272.

Deshpande, S.S., Cheryan, M., Sathe, S.K., Salunkhe, D.K., 1984. Freeze concentration of fruit juices. Crit. Rev. Food Sci. Nutr. 20 (3), 173–248.

Duan, X., Zhang, M., Mujumdar, A.S., Wang, S., 2010. Microwave freeze drying of sea cucumber (*Stichopus japonicus*). J. Food Eng. 96 (4), 491–497.

Flink, J., 1975. The influence of freezing conditions on the properties of freeze dried coffee. In: Goldblith, S.A., Rey, L., Rothmayr, W.W. (Eds.), Freeze Drying and Advanced Food Technology. Academic Press, London, UK.

Flosdorf, E.W., 1949. Freeze-Drying. Reinhold, New York, NY.

George, J.P., Datta, A.K., 2002. Development and validation of heat and mass transfer models for freeze drying of vegetable slices. J. Food Eng. 52 (1), 89–93.

Hartel, R.W., Espinel, L.A., 1993. Freeze concentration of skim milk. J. Food Eng. 20 (2), 101–120.

Heldman, D.R., Hohner, G.A., 1974. An analysis of atmospheric freeze drying. J. Food Sci. 39 (1), 147–155.

Heng, S., Hongmei, Z., Haidong, F., Lie, X., 2007. Thermoelectromagnetic coupling in microwave freeze drying. J. Food Proc. Eng. 30 (2), 131–149.

Karel, M., 1975. Heat and mass transfer in freeze drying. In: Goldblith, S.A., Rey, L., Rothmayr, W.W. (Eds.), Freeze Drying and Advanced Food Technology. Academic Press, London.

Khalloufi, S., Robert, J.L., Ratti, C., 2005. Solid foods freeze-drying simulation and experimental data. J. Food Proc. Eng. 28 (2), 107–132.

King, C.J., 1971. Freeze Drying of Foods. CRC, Butterworth, London.

King, C.J., 1975. Application of freeze drying to food products. In: Goldblith, S.A., Rey, L., Rothmayr, W.W. (Eds.), Freeze Drying and Advanced Food Technology. Academic Press, London, UK.

Krokida, M.K., Philippopoulos, C., 2006. Volatility of apples during air and freeze drying. J. Food Eng. 73 (2), 135–141.

Liapis, A.I., Bruttini, R., 1995. Freeze drying. In: Mujumdar, A.S. (Ed.), Handbook of Industrial Drying, second ed. Marcel Dekker, New York, NY.

Lorentzen, J., 1975. Industrial freeze drying plants for foods. In: Goldblith, S.A., Rey, L., Rothmayr, W.W. (Eds.), Freeze Drying and Advanced Food Technology. Academic Press, London, UK.

Ma, Y.H., Peltre, P.R., 1975. Freeze dehydration by microwave energy – 1. Theoretical investigation. AIChE J. 21 (2), 335–344.

Nastaj, J., 1991. A mathematical modeling of heat transfer in freeze drying. In: Mujumdar, A.S., Filkova, I. (Eds.), Drying '91. Elsevier, London, UK.

Oetjen, G.-W., Haseley, P., 2004. Freeze Drying, second ed. Wiley-VCH, Weinheim, Germany.

Pikal, M.J., 2007. Freeze drying, third ed. Encyclopedia of Pharmaceutical Technology, vol. 3. Inform Healthcare, New York, NY.

Pikal, M.J., Shah, S., Roy, M.L., Putman, R., 1990. The secondary stage of freeze drying: drying kinetics as a function of temperature and chamber pressure. Intl J. Pharmaceutics 60, 203–217.

Quast, D.G., Karel, M., 1968. Dry layer permeability and freeze-drying rates in concentrated fluid systems. J. Food Sci. 33 (2), 170–175.

Ratti, C., 2001. Hot air and freeze-drying of high-value foods: a review. J. Food Eng. 49 (4), 311–319.

Santivarangkna, C., Aschenbrenner, M., Kulozik, U., Foerst, P., 2011. Role of glassy state on stabilities of freeze-dried probiotics. J. Food Sci. 76 (8), R152–R156.

Simatos, D., Blond, G., 1975. The porous texture of freeze dried products. In: Goldblith, S. A., Rey, L., Rothmayr, W.W. (Eds.), Freeze Drying and Advanced Food Technology. Academic Press, London, UK.

Sunderland, J.E., 1980. Microwave freeze dryng. J. Food Proc. Eng. 4 (4), 195–212.

Tao, Z., Wu, H.W., Chen, G.H., Deng, H., 2005. Numerical simulation of conjugate heat and mass transfer process within cylindrical porous media with cylindrical dielectric cores in microwave freeze-drying. Intl. J. Heat Mass. Trans. 48 (3–4), 561–572.

Thijsenn, H.A., 1975. Current developments in the freeze concentration of liquid foods. In: Goldblith, S.A., Rey, L., Rothmayr, W.W. (Eds.), Freeze Drying and Advanced Food Technology. Academic Press, London, UK.

Van Pelt, W.H.J.M., 1975. Freeze concentration of vegetable juices. In: Goldblith, S.A., Rey, L., Rothmayr, W.W. (Eds.), Freeze Drying and Advanced Food Technology. Academic Press, London, UK.

Wang, Z.H., Tao, Z., Wu, H.W., 1998. Numerical study on sublimation–condensation phenomena during microwave freeze drying. Chem. Eng. Sci. 53 (18), 3189–3197.

Wang, R., Zhang, M., Mujumdar, A.S., 2010. Effect of food ingredient on microwave freeze-drying of instant vegetable soup. LWT 43 (7), 1144–1150.

CHAPTER

Frying, Baking, Roasting 24

24.1 Introduction

Frying, baking and roasting are thermal processes of food preparation, and are three of the most basic and familiar unit operations of cookery at home and in food service. Yet, as industrial processes carried out at relatively large scale, they have also attracted the attention of researchers who have tried to elucidate the physical and chemical mechanisms involved, their effect on the sensory and nutritional quality of the product, and their food safety aspects. Engineers have developed better performing equipment systems, especially designed and optimized for large-scale industrial duty. Efforts have been made to model these processes and to optimize them (Paulus, 1984).

Roasting and baking are *cooking in hot air*. Although the same basic principles govern both operations, the term baking is usually reserved for dough products while the term roasting refers to the application of dry heat to all other types of foods, from meats to snacks. Frying, on the other hand, is *cooking in fat*. As a unit operation, it is quite distinct in the mechanisms involved, its kinetics, its nutritional and safety aspects, and the operation principles of the equipment used for its implementation.

The primary objective of baking, roasting and frying is the transformation of foods into products with improved eating quality. At the same time, these processes contribute to the short-term stability of foods by means of two mechanisms: heating and drying. Long-term preservation is not usually expected from these operations, unless they are combined with other preservation methods.

24.2 Frying

24.2.1 Types of frying

Frying is a cooking method where fat or oil is used as the heat transfer medium, in direct contact with the food (Varela *et al.*, 1988). We distinguish between different types of frying:

1. *Pan frying*. Pan frying is applied to flat, wide and relatively thin pieces of food, such as patties, fillets and omelets. The food is cooked by contact with a small amount of hot fat, usually on one side and without agitation.

Food Process Engineering and Technology. DOI: http://dx.doi.org/10.1016/B978-0-12-415923-5.00024-1
© 2013 Elsevier Inc. All rights reserved.

2. *Stir frying*. Small to medium-size food particles are rapidly cooked in a small amount of very hot fat, with constant agitation. "Sautéing" is a variation of this kind of frying.
3. *Deep frying*. Food particles are immersed in hot fat or oil, and heat transfer occurs uniformly over the entire surface of the food.

Deep-fat frying has lately become a large-scale industrial operation. Fried potato products (french fries and potato chips or crisps) are by far the most important industrial products manufactured by immersion frying. Deep frying is also used for the production of fried snacks (e.g., tortilla chips) and center-of-the-plate components of ready meals (e.g., battered and breaded fish fillets, fish sticks, vegetarian patties, poultry or veal schnitzels, etc.). Many of these products are commercialized as frozen heat-and-serve foods.

24.2.2 Heat and mass transfer in frying

Several phenomena take place simultaneously during frying:

1. *Cooking (thermal effects)*. One of the major objectives of frying is to cook the food – i.e., to induce thermal reactions such as gelatinization of starch, denaturation of proteins, hydrolysis, development of flavors, Maillard browning and caramelization, etc., known collectively as cooking.
2. *Dehydration*. Loss of moisture occurs, intentionally or unintentionally, during frying.
3. *Changes in fat content*. Oil uptake by the food occurs during deep-frying, while loss of fat may take place in pan frying (Haak *et al*., 2007; Sioen *et al*., 2006; Ufheil and Escher, 1996).
4. *Changes in texture and structure*. The most common effect is the formation of a crust.

In frying, the food is in direct contact with very hot oil, at 160–180°C. As a result, heat transfer is fast and a large temperature gradient is created at the food/oil interface. The steep temperature gradient explains some of the peculiarities of frying, such as the creation of the characteristic dual structure, with a dry, crisp, golden crust and a moist center. Coating with batter and bread crumbs (breading) has the dual purpose of creating a golden, crisp and tasty crust and protecting the moist and delicate interior (e.g., fish or seafood) from overheating and drying.

Heat and mass transfer in deep frying is complex (Baumann and Escher, 1995). Heat transfer from the oil to the product is retarded by the rapid evaporation of water, causing an insulating layer of vapor to form on the surface. The mechanism and kinetics of water loss depend on the presence or absence of a crust (Ashkenazi *et al*., 1984). Oil uptake by the product is an important issue, affecting the nutritional and organoleptic characteristics of the product as well as the process conditions and economics (Pedreschi and Moyano, 2005). Fried products may contain up to 40% oil. In the light of present knowledge regarding the health aspects of lipid intake, the fat content of fried products is a source of concern to consumers. For many years, it has been thought that the oil penetrates the fried product to fill pores left by the evaporating water. It should be considered, however, that during frying, and particularly in the initial phase of drying, there is a rapid flow of water vapor from the food to its surface, opposing the penetration of oil. It has been shown that most of the oil intake takes place during the cooling phase, after the product has been removed from the hot oil (Bouchon and Pyle, 2005; Ufheil and Escher, 1996). Oil adhering to the surface is "sucked-in" as the product cools. Thus, the nature of the surface of fried food and the viscosity of the oil have a decisive effect on oil intake. It should be remembered that frying causes dehydration of the molecules at the surface (e.g., caramelization of sugars, dextrinization of starch) and thus renders the surface more hydrophobic – i.e., more suitable for the absorption of lipids.

24.2.3 Systems and operation

Fryers can be batch or continuous (Morton and Chidley, 1988). Industrial fryers are mostly continuous (Figure 24.1). Continuous fryers for deep frying contain an oil bath through which the product is conveyed on a mesh belt. The oil is heated by combustion gases or by electric resistances. It is important to provide a large heat transfer area (less watts per unit area) in order to avoid local overheating of the oil. The oil is continuously filtered to remove particles that catalyze oxidation. At the same time, fresh oil is added to make up for the oil intake by the product.

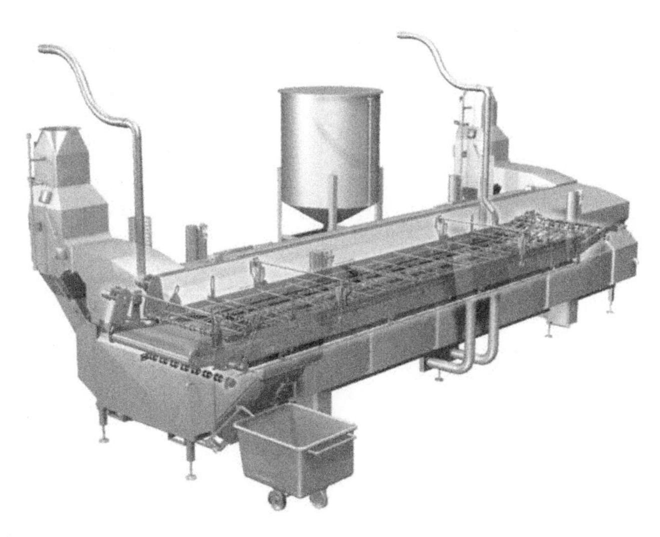

FIGURE 24.1

Continuous fryer.

Figure courtesy of FMC FoodTech.

Usually, the industrial continuous fryer is part of a production chain comprising pre-dusting, battering and breading equipment before it, and a baking oven, cooler and freezer after it (Figure 24.2).

24.2.4 Health aspects of fried foods

Various fats, both of vegetal and of animal origin, are used for frying. Prolonged heating at high temperature in the presence of moisture causes hydrolysis (increase in free fatty acid content; Kalogeropoulos, 2007), lipid oxidation, formation of cyclic fatty acids, polymerization, etc. As some of these changes have an adverse effect on food safety, it is important to reduce their occurrence by proper choice of the type of fat, frequent change of the oil in the fryer, avoiding unnecessarily high frying temperatures, and minimizing fat uptake by the product.

Recently, the health implications of frying became an actively researched subject in connection with the acrylamide issue. Acrylamide (C_3H_5NO) is a substance formed when asparagines react with reducing sugars at high temperatures. It is suspected to cause cancer. Starch-containing foods that have been subjected to high temperatures (such as fried potato products) seem to be the main source of concern (Becalski *et al.*, 2003).

24.3 Baking and roasting

Baking is defined as "cooking of food in an oven in air to which water vapor may or may not be added" (Holtz *et al.*, 1984). Roasting fits the same definition. In an oven, the product is heated simultaneously by convection, conduction and radiation, although the relative importance of each one of the heat transfer mechanisms

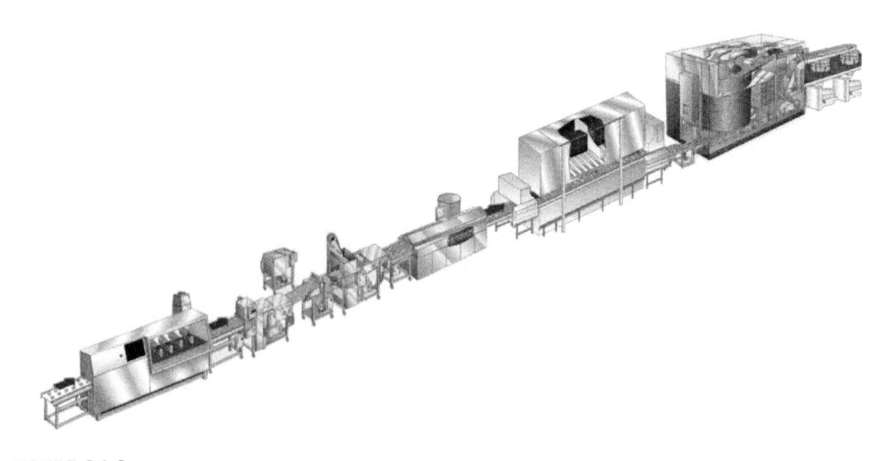

FIGURE 24.2

Continuous frying line for fried–baked–frozen products.

Figure courtesy of FMC FoodTech.

may differ at various stages of the process. Convection heating occurs by the action of hot gases. Its magnitude depends on the rate of circulation of the hot gases. Conduction takes place between the product and the hot surfaces on which it rests (trays, bricks, etc.). Radiation heating is due to radiant heat from the walls or hot bodies installed in the oven.

Depending on the method used for producing the hot gas, we distinguish between two types of oven. In direct heating, the product comes to contact with combustion gases. In order to avoid contamination of the product with residues of combustion, the "purity" of the fuel used and the efficiency of the burners must be considered. Natural gas and propane gas are among the preferred fuels. In indirect heating, a conductive separation is provided between the heating medium (steam or combustion gases) and the air.

Industrial ovens for baking bread, crackers, biscuits or pizza are continuous. The most common types of industrial ovens for large-scale baking are tunnel ovens (Figure 24.3).

In multi-section ovens, the temperature, humidity and velocity of the air can be adjusted individually at each section. Thus, in the case of bread baking, steam is injected at the earlier sections, where the objective is to "cook" the dough (to cause setting of the open structure created by the gluten, gelatinize some of the starch and produce the characteristic shiny appearance of the surface) without excessive drying. At this stage, condensation of the steam on the surface contributes to heat transfer. In subsequent sections, dry air is used to complete the cooking action and to remove some water, depending on the product. Here, the convective heat transfer coefficient from the hot gases to the product surface is in the order of $10\text{–}50\ \mathrm{W \cdot m^{-2} \cdot K^{-1}}$ (Bimbenet *et al.*, 2002). Humidity in these sections is kept low by extracting some of the air through chimneys. In the later sections, the proportion of heat supplied by radiation is increased in order to enhance heating, browning and drying of the surface, for the formation of the crust (Mälki *et al.*, 1984).

Conventionally, air in tunnel ovens flows tangentially to the product at moderate velocity. Recently, however, ovens have been developed where the hot gas flow, at

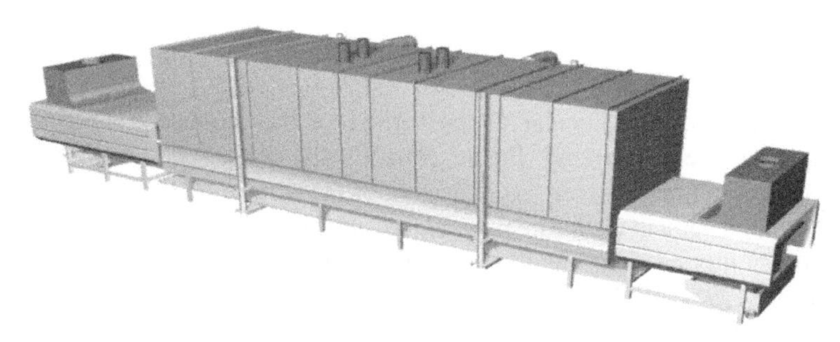

FIGURE 24.3

Tunnel oven.

Figure courtesy of FMC FoodTech.

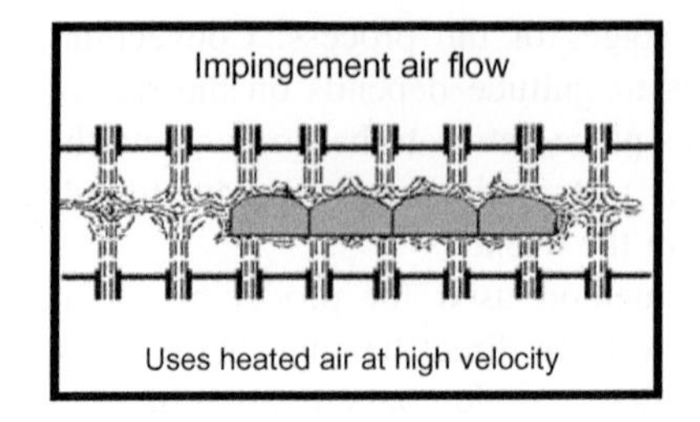

FIGURE 24.4

Principles of impingement oven.

Figure courtesy of FMC FoodTech.

FIGURE 24.5

Continuous roasting of chicken.

Photograph courtesy of FMC FoodTech.

higher velocity, is perpendicular to the surface of the product. These ovens are known as "impingement ovens" (Figure 24.4; Wählby *et al.*, 2000; Xue and Walker, 2003). Because of the relatively rapid heating of the surface by the high-velocity jet of hot gas, impingement ovens are particularly suitable for roasting meat.

The possibility of using microwaves for baking has been investigated (Ovadia and Walker, 1995). While microwaves are efficient in providing the energy for cooking the dough, they cannot cause crust formation and surface browning. For this purpose, microwave heating must be supplemented with radiation – for example, from a halogen lamp (Keskin *et al.*, 2004).

Continuous ovens are used also for large-scale roasting (Figure 24.5).

References

Ashkenazi, N., Mizrahi, S., Berk, Z., 1984. Heat and mass transfer in frying. In: McKenna, B.M. (Ed.), Engineering and Food. Elsevier Applied Science Publishers, London, UK.

Baumann, B., Esher, F., 1995. Mass and heat transfer during deep-fat frying of potato slices. I. Rate of drying and oil uptake. LWT 28 (4), 395–403.

Becalski, A., Lau, B.P.-Y., Lewis, D., Seaman, S.W., 2003. Acrylamide in foods: occurrence, sources and modeling. J. Agric. Food Chem. 51 (3), 802–808.

Bimbenet, J.J., Duquenoy, A., Trystram, G., 2002. Génie des Procédés Alimentaires. Dunod, Paris, France.

Bouchon, P., Pyle, D.L., 2005. Modelling oil absorption during post-frying cooling. I: Model development. Food Bioproduct Proc. 83 (C4), 253–260.

Haak, L., Sioen, I., Raes, K., Van Camp, J., De Smet, S., 2007. Effect of pan frying in different culinary fats on the fatty acid profile of pork. Food Chem. 102 (3), 857–864.

Holtz, E., Skjöldenbrand, C., Bognar, A., Piekarski, J., 1984. Modeling the baking process of meat products using convention ovens. In: Zeuthen, P., Cheftel, J.C., Erickson, C., Jul, M., Leniger, H., Linko, P., Varela, G., Vos, G. (Eds.), Thermal Processing and Quality of Foods. Elsevier Applied Science Publishers, London, UK.

Kalogeropoulos, N., Salta, F.N., Chiou, H., Andrikopoulos, N.K., 2007. Formation and distribution of fatty acids during deep and pan frying of potatoes. Eur. J. Lipid Sci. Technol. 109 (11), 1111–1123.

Keskin, S.Ő., Sumnu, G., Sahin, S., 2004. Bread baking in halogen lamp–microwave combination oven. Food Res. Int. 37 (5), 489–495.

Mälkki, Y., Seibel, W., Skjöldenbrand, C., Rask., Ö., 1984. Optimization of the baking process and its influence on bread quality. In: Zeuthen, P., Cheftel, J.C., Erickson, C., Jul, M., Leniger, H., Linko, P., Varela, G., Vos, G. (Eds.), Thermal Processing and Quality of Foods. Elsevier Applied Science Publishers, London, UK.

Morton, I.D., Chidley, J.E., 1988. Methods and equipment in frying. In: Varela, G., Bender, A.E., Morton, I.D (Eds.), Frying of Foods. Ellis Horwood Ltd, Chichester, UK.

Ovadia, D.Z., Walker, C.E., 1995. Microwave baking of bread. J. Microw. Power Electromagn. Energy 30 (2), 81–89.

Paulus, K.O., 1984. Modelling in industrial cooking. In: Zeuthen, P., Cheftel, J.C., Erickson, C., Jul, M., Leniger, H., Linko, P., Varela, G., Vos, G. (Eds.), Thermal Processing and Quality of Foods. Elsevier Applied Science Publishers, London, UK.

Pedreschi, F., Moyano, P., 2005. Oil uptake and texture development in fried potato slices. J. Food Eng. 70 (4), 557–563.

Sioen, I., Haak, L., Raes, K., Hermans, C., De Henauw, S., Van Camp, J., 2006. Effects of pan frying in margarine and olive oil on the fatty acid composition of cod and salmon. Food Chem. 98 (4), 609–617.

Ufheil, G., Escher, F., 1996. Dynamics of oil uptake during deep fat frying of potato slices. LWT 52 (3), 640–644.

Varela, G., Bender, A.E., Morton, I.D. (Eds.), 1988. Frying of Foods. Ellis Horwood Ltd, Chichester, UK.

Wählby, U., Skjöldebrand, C., Junker, E., 2000. Impact of impingement on cooking time and food quality. J. Food Eng. 43 (3), 179–187.

Xue, J., Walker, C.E., 2003. Humidity change and its effect on baking in an electrically heated air jet impingement oven. Food Res. Intl. 36 (6), 561–569.

CHAPTER

Chemical Preservation

25

25.1 Introduction

In its broadest sense, chemical preservation can be defined as the action of preventing or retarding food spoilage of any kind by changing the chemical composition of the food, either by the addition of substances known as preservatives and antioxidants or by promoting certain biochemical processes (e.g., fermentation) that result in the *in situ* production of preserving substances. Both kinds of chemical preservation have been practiced by humans since ancient times. Salting, smoking, the use of spices and vinegar, and curing meat with saltpeter (nitrate) are examples of ancient preservation techniques based on added preservatives. Preservation of beverages by alcoholic fermentation and of milk by lactic fermentation are examples of ancient food preservation practices based on the *in situ* creation of preserving conditions through biochemical processes.

Modern science has brought profound changes in the area of chemical preservation of foods. On one hand, advances in synthetic chemistry, while aiming primarily to serve medicine, have produced chemicals with antimicrobial or antioxidant properties, capable of being used as food preservatives. Advances in molecular biology have led, on one hand, to the discovery of antimicrobial molecules such as antibiotics and bacteriocins, and on the other hand, to better understanding of the mechanisms describing the response and resistance of microorganisms to chemical preservation. Consumer interest in natural foods has led to intensive research on the possible preserving action of spices, alone or as an additional hurdle. Detailed toxicological studies have provided, and continue to generate, valuable knowledge with respect to the safety and health aspects of chemical preservation, on which a vast volume of national and international regulation and legislation are based.

25.2 Chemical control of microbial spoilage

25.2.1 Kinetics, dose–response function

A distinction is usually made between microstatic and microcidal agents – i.e., between those that prevent growth of microorganisms and those that kill them. Depending on the class of the target microorganisms, the antimicrobials are

Food Process Engineering and Technology. DOI: http://dx.doi.org/10.1016/B978-0-12-415923-5.00025-3
© 2013 Elsevier Inc. All rights reserved.

qualified as *bacteriostatic/bactericidal* or *fungistatic/fungicidal.* According to Lück and Jager (1997), such differentiation is not justified because all antimicrobial agents actually destroy their target organism. The distinction stems from the difference in the death rate, which depends on the concentration of the preservative. If the concentration is lower than a certain value, microorganisms continue to grow; if the concentration is sufficiently high, the number of living cells declines. According to this analysis, static behavior would be observed only at the preservative concentration range at which the rate of growth is roughly equal to the rate of destruction (Figure 25.1). This view, however, is not widely accepted.

The death rate of a given microorganism in the presence of a given preservative depends on a number of factors, the most important being:

- Duration of the exposure
- Preservative concentration
- Temperature
- Environmental conditions (properties of the medium), particularly water activity and pH.

As to the effect of time, first-order kinetics is usually assumed (Lück and Jager, 1997) and occasionally confirmed experimentally (see, for example, Huang *et al.*, 2009). At constant concentration of the antimicrobial agent, first-order kinetics corresponds to the log-linear destruction model, already discussed in relation to thermal inactivation (Chapter 17, Section 17.2), and formulated in Eq. (25.1):

$$-\log\frac{N}{N_0} = kt \tag{25.1}$$

where:

N, N_0 = number of living cells at time t and time 0, respectively
t = time (duration) of contact
k = rate constant.

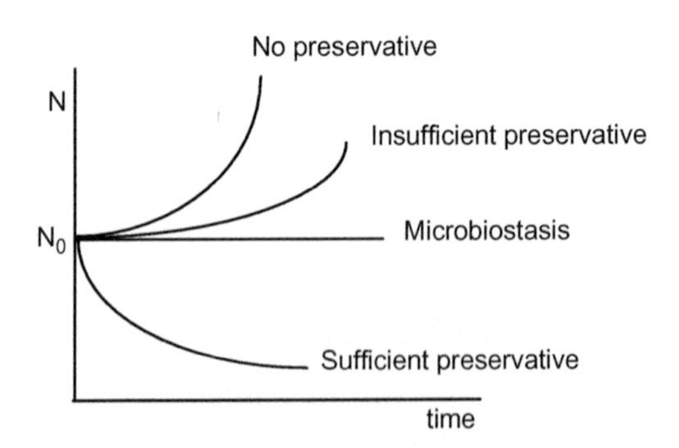

FIGURE 25.1

Effect of antimicrobial concentration on microbial population over time.

A model that apparently fits better a considerable proportion of available experimental data is the Weibullian Power Law model (Buzrul, 2009; Peleg, 2006), also discussed in Section 17.2, and formulated in Equation 25.2:

$$\log\frac{N}{N_0} = -bt^n \tag{25.2}$$

where b and n are the parameters of the model. The discrepancy between the two models depends on the magnitude of the exponent n. At $n = 1$, the Weibullian model and first-order kinetics are identical.

In practice, however, the issue of time of exposure is often disregarded because either the effect of the antimicrobial agent is immediate (as in the case of disinfectants) or the time of contact is relatively long (such as long-term stabilization with the help of stable chemicals). On the other hand, if the chemical is applied onto the surface of the food and propagates into the food by diffusion, time is of the essence (Bae *et al.*, 2011a).

The relationship between the rate constant k or the parameters b, n and the concentration of the preservative is the *dose–response function*, usually described as a *dose–response curve*. The dose–response concept, mainly applied in toxicology and pharmacology, is also used in relation to chemical preservation of foods against microbial spoilage.

Consider a single cell in a population of microorganisms exposed to a stable preservative. Only two possibilities exist: either the cell is inactivated or it is not. The microorganism will be killed if the concentration X of the preservative is greater than a limit value X_C. This critical concentration, X_C, represents the resistance of the cell to the preservative. If all the cells in the population had the same resistance, the dose–response curve would be a step function (Figure 25.2) and X_C could be taken as a measure of the resistance of the population to the preservative. Dose–response curves showing an almost step-function relationship (Figure 25.3) have often been reported.

If the population is not of uniform resistance, then a statistical model describing the distribution of X_c should be brought into the picture and a mean resistance should be defined according to the distribution model adopted. Peleg (2006) has

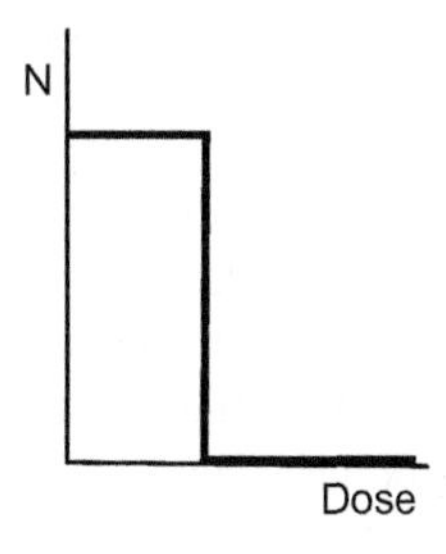

FIGURE 25.2

Dose–response curve as a step function.

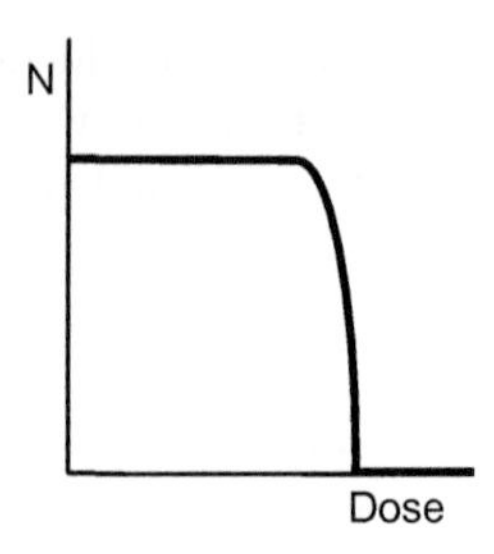

FIGURE 25.3

Typical shape of frequently observed dose–response curves.

studied two distribution models, namely a *Fermi distribution* and a *Weibull distribution*, and investigated the fit of published experimental data to dose–response curves generated using those two models (see also Peleg *et al.*, 2011).

It should be noted that microorganisms previously sensitive to the action of a preservative may acquire resistance through membrane modification or genetic transformation (Bower and Daeschel, 1999).

25.2.2 Individual antimicrobial agents

Over the years, a considerable number of chemicals have been found to possess antimicrobial properties (Jacobs, 1951). For certain periods many have been actually used as food preservatives, but national and international food laws and regulations have reduced the number of permitted chemical preservatives considerably. Some of these preservatives are discussed below. It is important to emphasize that chemical preservatives should by no means be considered a substitute for good manufacturing practice. In fact, because of the growing consumer resistance to chemicals in food, it is advisable to reduce the use of chemical preservatives to the strictly necessary minimum.

25.2.2.1 Salt

Salt is added to foods for taste or for preservation. For many centuries, common salt has been used as the most important food preservative. In recent times its importance as a preservative has somewhat declined, mainly as a result of the expansion of refrigeration and the increasing consumer concern about the effect of sodium intake on blood pressure (Doyle and Glass, 2010). Although salt is not considered a food additive, the sodium content of foods is usually declared on labels. Salt affects microbial growth, primarily, by depressing the water activity. At the customary level of use for taste alone (say, 0.5–2%), the water activity would not be sufficiently reduced for preservation. However, if the concentration of salt in the *water phase* of the food is considered, the picture could be different. For example, salted butter contains about 15% water and 1.5% salt. This means that the concentration of salt *in the aqueous phase* of salted butter is about 10%. This corresponds to a water activity level of about 0.94, at which many species of bacteria (but not most yeasts and molds) are inhibited.

The lowering of water activity alone does not explain the antimicrobial action of salt. Salt inhibits certain bacteria better than other a_w-depressing agents, such as glycerol, at equal or lower water activity levels (Lück and Jager, 1997). It is usually assumed that salt exerts specific toxicity towards microorganisms. The exact mechanism of such toxicity is not known.

According to their sensitivity to salt, microorganisms are classified as *slightly halophilic*, *moderately halophilic* or *extremely halophilic* ((Lück and Jager, 1997). A saturated solution of salt contains 34 grams of sodium chloride per 100 grams of water and has a water activity of 0.75. Highly halophilic microorganisms (halobacteria) have been detected in the water of salty lakes at near-saturation concentration.

At equal molar concentration, potassium chloride is as effective as sodium chloride as a preservative (Bidias and Lambert, 2008). However, the bitter taste of potassium chloride hampers the replacement of common salt by potassium chloride as a preservative, despite the inclination to reduce sodium intake for health reasons.

Salt, even at low concentrations, constitutes a significant hurdle in combination with other preserving factors. It intensifies the effectiveness of other antimicrobials and may increase the heat resistance of some microorganisms (Lück and Jager, 1997). Salt is used in pickling of vegetables and olives through lactic fermentation. The function of salt in this particular application is to extract from the vegetables the water and sugars necessary for fermentation, and to inhibit growth of salt-sensitive microorganisms until the more salt-tolerant lactic acid bacteria grow and produce the acidity needed for preservation.

Salting of foods is done by spreading dry salt on the surface of the food (pile salting; Corzo *et al.*, 2012), by immersion in brine (Berhimpon *et al.*, 1991), by vacuum impregnation (Larrazábal-Fuentes *et al.*, 2009) or by injection (Porsby *et al.*, 2008).

25.2.2.2 Smoke

Widely practiced since prehistoric times, smoking is among the oldest methods of preservation. It is mainly applied to meat, sausages, fish, poultry and some varieties of cheese. Basically, there are two types of smoking: cold smoking at ambient temperature, and hot smoking at 60–100°C. Artificial smoking or the application of "liquid smoke" (a purified liquid condensate of wood smoke) to the surface of foods imparts a smoky flavor and has been shown to have some (Sofos *et al.*, 1988) or no (Catte *et al.*, 1999) antimicrobial effect.

The preserving action of smoking is due mainly to three factors: the antimicrobial properties of phenolic substances in wood smoke; drying; and thermal inactivation (in the case of hot smoking). Since salt, in one form or another, is usually added in the course of smoking, the observed antimicrobial effects should be attributed to the combined action of smoke and salt (Cornu *et al.*, 2006; Porsby *et al.*, 2008). Both cold and hot smoking have been found to be effective against *Listeria monocytogenes* in the production of smoked salmon (Rørvik, 2000). Smoking is believed to inhibit aerobic bacteria due to the reduced oxygen content of the atmosphere in the smoke room. In addition, the presence of carbon monoxide and carbon dioxide in smoke may contribute to the inhibitory action against aerobic microorganisms (Kristinsson *et al.*, 2008).

25.2.2.3 Carbon dioxide

The importance of CO_2 gas as a respiration depressor in storage (controlled atmosphere, CA) or in the package (modified atmosphere packaging, MAP) is mentioned in Sections 19.2.4 and 27.3.3. The antimicrobial activity of CO_2 will be discussed here. Carbon dioxide has been found to inactivate a vast array of bacteria, yeast and molds (Daniels *et al.*, 1985; Haas *et al.*, 1989). Being a gas, CO_2 easily penetrates porous and multiphase foods such as ground meat (Bae *et al.*, 2011b). The antimicrobial effectiveness of carbon dioxide depends on the duration of contact, pressure, temperature, pH and water activity. Pressure is particularly important; hence the preponderance of *high-pressure* or *dense-phase* CO_2 treatment processes (Buzrul, 2009; Erkmen, 2000; Haas *et al.*, 1989; Huang *et al.*, 2009). In combination with low temperature, carbon dioxide has been found to preserve the freshness of shell eggs during long-term storage (Yanagisawa *et al.*, 2009). CO_2 gas seems to inactivate microorganisms directly and not indirectly by depriving cells of adequate supply of oxygen, since anaerobic microorganisms are also destroyed by carbon dioxide. According to Haas *et al.*, (1989), the effectiveness of carbon dioxide is antagonized by lowering water activity, hence the absence of antimicrobial effect when applied to dry spices. Carbon dioxide does not prevent production of botulinal toxin in meat (Moorhead and Bell, 2000).

25.2.2.4 Sulfur dioxide and sulfites

History shows that sulfur dioxide, produced by burning sulfur, was already being used as an antiseptic fumigant in ancient Egypt, Greece and China. Its use in wine making in the Middle Ages has been extensively documented.

Sulfur dioxide is a highly reactive substance. Of particular significance in food technology is its ability to link with aldehydes and reducing sugars, to produce addition compounds. Through this type of reaction, sulfur dioxide antagonizes Maillard and enzymatic browning – hence its use in the production of dried fruits and frozen potato products, and in sugar refining. Sulfur dioxide is soluble in water. In aqueous solution it can be present as free SO_2 or as the ions HSO_3^- or SO_3^-, depending on the pH. Free SO_2 is the predominant form at low pH, while the sulfite ion SO_3^{--} is most stable at high pH. Only free sulfur dioxide, not bound to aldehydes, is active. Most of the dissolved sulfur dioxide can be removed by boiling. In practice, this is an important advantage of SO_2 as a temporary preservative.

Sulfur dioxide is active against bacteria, yeasts and molds. The most effective form in solution is the undissociated molecule, hence the loss of antimicrobial activity at high pH. Sulfur dioxide and sulfites are used as antimicrobials, mainly in wine, fruit juices, concentrates and fruit pulps, but they are also used for their antioxidant and anti-browning effects in a multitude of other foods. In wine making, sulfites (including bisulfite and meta-bisulfite) are added to grape must to prevent growth of wild yeasts, bacteria and mold during alcoholic fermentation. To this end, cultured sulfite-resistant yeast species are used for fermentation. Typical levels of concentration, as free SO_2, are 30–40 ppm in moderate climates and up to 200 ppm in warm climates.

In the USA, sulfur dioxide is generally recognized as safe (GRAS) "when used in accordance with good manufacturing practice", but its use in meats and on fresh fruits and vegetables is prohibited (FDA, 2011a). The presence of the preservative must be declared on the label. Some people, and especially infants and persons with respiratory problems, are particularly sensitive to sulfur dioxide. Cases of death as a result of exposure to SO_2 in food are known. The practice of spraying salad bars in restaurants with sulfite-containing solutions in order to preserve freshness and color has been banned.

25.2.2.5 Low pH

The acidity of foods is among the principal parameters governing the growth of microorganisms. Bacteria prefer a pH of 6–7.5 but they tolerate a wider range, of 4–9 (Doores, 1983). Yeasts and molds are more acid-resistant, and some can grow at pH as low as 3.5. Traditionally, low pH has been considered a mark of safety against most food pathogens. The recent detection of the pathogen *Escherichia coli* 0157.H7 in low-pH foods, such as apple juice and mayonnaise (Zhao and Doyle, 1994), has put into serious doubt the validity of this belief.

The pH of food can be lowered by addition of acids or by *in situ* generation of lactic acid by fermentation. For artificial acidification, food acids (acetic, citric, malic, phosphoric, gluconic, etc.), some of their salts, or certain derivatives such as glucono-delta-lactone are added. Artificial acidification is widely applied in order to permit less drastic retorting conditions in heat-sensitive products such as canned mushrooms.

A very large group of bacteria, known as "lactic acid bacteria", ferment sugars and produce lactic acid, thus lowering the pH of the food (Wood, 1998). Dairy products such as yogurt, sour cream, buttermilk and some soft cheeses are stabilized by lactic fermentation (Tamime, 2006). Pickled vegetables, sauerkraut and olives are preserved by lactic fermentation (Norris and Watt, 2003). Lactic acid fermentation is applied in the production of salami-type sausages and cured ham, together with other preservatives.

25.2.2.6 Ethanol

Although no one today would view wine as "preserved grape juice" or beer as "preserved drinking water", it is possible that alcoholic fermentation was originally used by our ancestors for preservation. At sufficiently high concentration (60–70%), ethanol is used as a disinfectant and kills microorganisms of all kinds by denaturing their vital proteins. However, it has also been found to have antimicrobial effects at lower concentration of 5–20% (Davidson *et al.*, 1983). It has no effect on bacterial spores.

At present, the use of ethanol in food processing as an added preservative is extremely limited. Grape juice, preserved by addition of about 20% ethanol, is used for sweetening fortified wine (Lück and Jager, 1997). The preservation of entire berries and plums in distilled brandy, at ethanol concentration of about 20%, is practiced in small industries and homes in Europe. Tofu preserved in rice alcohol (sake) is a greatly appreciated delicacy in the Far East.

25.2.2.7 Nitrates and nitrites

Nitrates, under their popular name of saltpeter, have been used in meat products, fish and cheese for many centuries. They have been used to prevent blowing of hard cheeses by the action of gas-forming bacteria. The main reason for their inclusion in hams and sausages was their ability to fix the red color of meat, although at the same time they exerted valuable antimicrobial activity against *Clostridium botulinum*. Today, it is known that both properties are due to the nitrites formed by the reduction of nitrates, and not to the nitrates themselves (Tompkin, 1983). Nitrites conserve the red color of meat by reacting with the myoglobin in muscle tissue. The antimicrobial effectiveness of nitrites is stronger at low pH, and is apparently due to the formation of nitrous acid. The use of nitrates as a reserve for the generation of nitrites is now prohibited in some countries. In the USA (FDA, 2011b), the maximum permitted level of usage of sodium nitrite is 10 ppm in tuna (as a color fixative) and 200 ppm together with up to 500 ppm of sodium nitrate in cured meat products (as a color fixative and a preservative).

Nitrites are suspected to react with secondary and tertiary amines in the food or in the gastro-intestinal tract, to form carcinogenic nitrosamines. In response to increasing consumer opposition to nitrites in food, the industry is actively looking for alternative ways to preserve meat color (e.g., ascorbic acid, CO_2), while addressing the problem of microbial contamination by increasing the level of sanitation and the use of other preservatives such as sorbates (Sofos *et al.*, 1979).

25.2.2.8 Benzoic acid, benzoates and parabens

Benzoic acid, C_6H_5COOH, and its derivatives are widely used as food preservatives. As benzoic acid has very low solubility in water, sodium benzoate is the form commonly applied. However, only the undissociated molecule has antimicrobial properties. Therefore benzoates are active only at low pH, below pH 4.5 (Chipley, 1983). They are used, at concentration levels in the range of 500–2000 ppm, to preserve fruit juices, fruit beverages, pickled vegetables, olives and similar low pH foods against spoilage by yeasts and molds, but not against bacteria. In fact, some spoilage bacteria are fairly resistant to benzoates. The use of sodium benzoate in foods, at the concentration range mentioned, is permitted in most countries.

The disadvantage of benzoates of being effective only at low pH led to the development of the *parabens*, esters of parahydroxy-benzoic acid (Davidson, 1983). These compounds are quite active against bacteria and much more effective against molds and yeasts in a wide pH range (Warth, 1989). Actually, as parabens are not dissociable, their activity is practically independent of pH. Their effectiveness against bacteria increases with the length of the alkyl radical attached to p-hydroxybenzoic acid, probably because of the greater ability of the less polar molecule to interact with the lipids of the bacterial cell membrane (Bargiota *et al.*, 1987).

25.2.2.9 Sorbic acid and sorbates

Sorbic acid, $CH_3{-}CH = CH{-}CH = CH{-}COOH$, is a di-unsaturated, six carbon-atom fatty acid. Like benzoic acid, sorbic acid is only weakly water-soluble; therefore, the form most commonly utilized is the water-soluble potassium sorbate. Like benzoic acid, the active principle is the undissociated molecule, and consequently the activity is pH dependent. Sorbic acid and sorbates are most effective against yeasts and molds. Their action against bacteria is highly selective (Sofos and Busta, 1983). Potassium sorbate is often used together with sodium benzoate.

25.2.2.10 Propionic acid and propionates

Propionic acid, $CH_3{-}CH_2{-}COOH$ and its salts are used as preservatives in baked goods and cheese. The most common forms in usage are sodium and calcium propionates. Propionates are mainly used for their mycostatic activity. In combination with control of water activity, calcium propionate was found to be effective in suppressing aflatoxin production by *Aspergillus flavus* (Alam *et al.*, 2010). Propionates are also added to high-pH white bread in order to prevent "rope disease" (development of a sticky mass and nauseating odor a short time after baking) by *Bacillus subtilis* and related bacteria. In baked goods, propionates are added to the dough in amounts of 0.1– 0.3% of the weight of flour (Lück and Jager, 1997).

25.2.2.11 Bacteriocins

Bacteriocins are polypeptides produced by certain microorganisms. They possess antimicrobial properties against microorganisms of the same or closely related species. Although such properties characterize antibiotics, bacteriocins are not called antibiotics because of their narrow spectrum of action and their digestible proteinaceous composition excluding toxicity to humans. Bacteriocins are naturally present in food and are sometime termed biopreservatives. The list of known bacteriocins is extensive (Chen and Hoover, 2003) and continuously expanding, but *nisin* is the only purified bacteriocin approved for use in foods with certain limitations. Nisin is produced by lactic acid bacteria (Delves-Broughton, 2005), and is a short polypeptide, containing 34 amino acid residues. It belongs to the group of bacteriocins called *lantibiotics*, because it contains the unusual sulfur-containing amino acid *lanthionine*. It is heat resistant, and is known to increase the heat-sensitivity of microorganisms when present in autoclaved products. It is particularly active against gram-positive bacteria, and has no effect against yeasts and molds. Of particular interest is its effectiveness against the pathogen *Listeria monocytogenes* (Chen and Hoover, 2003; Samelis *et al.*, 2005; Schillinger *et al.*, 2001) and spore-forming pathogens such as streptococci and clostridia (Scott and Taylor, 1981a,b). It is interesting to note that nisin is more effective in preventing outgrowth of the spores than in inactivating the vegetative cell (Lück and Jager, 1997).

25.2.2.12 Spices and essential oils

Many foods naturally contain substances with some antimicrobial properties (Beuchat, 2001; Branen and Davidson, 1983). The origin of these natural preservatives can be animal (e.g., lactoperoxidase and lysozyme), microbial (e.g., alcohol, lactic acid, nisin) or plant materials (e.g., spices and essential oils). Consumer demand for all-natural foods has prompted interest in these natural antimicrobials.

A theory has been advanced claiming that, historically, spices found their way into human food not for their contribution to flavor but for their preserving action (Sherman and Billing, 1998). Regardless of the validity of this claim in explaining the evolution of human eating habits, it is a known fact that some spices and essential oils do possess some degree of inhibitory effect on some microorganisms. The number of materials tested by researchers is extensive, including cinnamon (Ceylan *et al.*, 2004), vanillin (Fitzgerald *et al.*, 2004), essential oils (Friedman *et al.*, 2002; Mosqueda-Melgar *et al.*, 2008), garlic (Rohani *et al.*, 2011), basil, clove, horseradish, marjoram, oregano, rosemary, thyme, etc. (Davidson *et al.*, 1983; Yano *et al.*, 2006). In most cases the effectiveness of the natural preservative was found to depend on the presence of an additional preservative, low temperature, low water activity or low pH. Thus, spices and essential oils can be considered as providing an additional hurdle but usually not as a single, reliable method for long-term preservation. The quantity of the spice or essential oil in food is limited by considerations of sensory effects. Furthermore, it should be kept in mind that spices may be themselves be sources of serious microbial contamination.

25.3 Antioxidants

25.3.1 Oxidation and antioxidants in food

Oxidation is one of the principal types of chemical deterioration of foods. Direct or indirect oxidation of food components is often the chief reason for the degradation of flavor, odor, taste, color and nutritional quality. Some of the deleterious oxidative processes in foods include:

- Autoxidation of lipids, resulting in bitter taste and rancid odor
- Oxidation of phenolic substances, catalyzed by polyphenol oxidases, resulting in discoloration (enzymatic browning) in fruits and vegetables
- Oxidation of aroma volatiles, resulting in loss of aroma
- Oxidation of carotenoid pigments, resulting in bleaching and loss of vitamin A activity
- Oxidation of ascorbic acid, resulting in loss of vitamin C activity
- Formation of end products with toxic properties
- Generation of carbonylic compounds, leading to non-enzymatic browning.

The principal condition for oxidation to occur is exposure to oxygen. *Singlet oxygen*, which is an excited state of molecular oxygen, is considerably more active than the normal *triplet oxygen* molecule. *Autoxidation* is a chain reaction, starting with the formation of a free radical (*initiation*), continuing with the *propagation* of the reaction with the formation of more and more free radicals and unstable molecules, and ending with reactions between the free radicals to form stable molecules (*termination*). Oxidation is accelerated by heat and light and catalyzed by some metals (e.g., Cu and Fe ions), specific enzymes and other substances, collectively termed *pro-oxidants*.

Antioxidants are substances that prevent or delay oxidative spoilage. Antioxidants may be natural or artificial (synthetic additives). The mechanism of action of antioxidants in food oxidation has been reviewed by Choe and Min (2009). Functionally, food antioxidants are classified as primary, secondary or synergistic (Rajalakshmi and Narasimhan, 1996). Primary antioxidants prevent or delay oxidation by reducing the free radicals and converting them into more stable molecules, thus interrupting the reaction chain. They are also termed "chain-breaking antioxidants". Quenching (inactivating) singlet protein is also among the protective effects of primary antioxidants (Lee and Jung, 2010). Secondary antioxidants act by decomposing active intermediate products of autoxidation, such as peroxides, and converting them to more inactive molecules. Synergistic antioxidants function by chelating pro-oxidant metals or by scavenging oxygen in limited supply (closed systems).

A number of synthetic and natural antioxidants have been found to possess interesting antimicrobial properties as well (Ahmad and Branen, 1981; Ogunrinola *et al.*, 1996; Passone *et al.*, 2007; Sun and Holley, 2012; Yousef *et al.*, 1991).

The natural antioxidant capability of foods is today one of the principal factors in health-oriented human nutrition. The claimed health benefits of antioxidants in food range from prevention of disease to delay of aging and longevity. Public interest in the subject is strong, and research activity in this area is intensive. The assessment of biological (*in vivo*) antioxidant function involves cell-based bioassays (Cheli and Baldi, 2011). Evaluation of the health-promoting capability requires epidemiological surveys and clinical testing.

25.3.2 Individual food antioxidants

25.3.2.1 Synthetic antioxidants

The principal synthetic commercial antioxidants used in foods are butylated hydroxyanisole (BHA), butylated hydroxytoluene (BHT), tertiary butyl hydroquinone (TBHQ), and esters of gallic acid, mainly propyl gallate (PG). These are phenolic substances, and function as primary antioxidants (Rajalakshmi and Narasimhan, 1996; Craft *et al.*, 2012). They are mainly used to prevent or delay lipid autoxidation in fats and fat-containing foods. In addition some may have antimicrobial activity, as mentioned above, but they are not used as antimicrobial preservatives. They are added to food at levels of hundreds of ppm. Some are fairly heat resistant and retain their activity in fried and baked foods (carry-through property). BHA is particularly heat resistant.

25.3.2.2 Natural antioxidants

The number of natural substances in foods that exhibit some antioxidant property is, logically, too large to allow listing (Brewer, 2011). These include biopolymers as well as low molecular weight compounds. Many spices contain phenolic substances that can act as potent antioxidants. The flavonoids and anthocyanins of plants are powerful antioxidants. However, only a few of the substances naturally occurring in foods are isolated, purified or synthesized and added to foods intentionally as antioxidants. *Tocopherols* (vitamin E) are present in plant materials and particularly in seed germs and nuts. One of the isomers, α-tocopherol, is also produced by chemical synthesis. It is used as a primary antioxidant in fats and fat-containing food. *Ascorbic acid* (vitamin C) is widely used in fruit and vegetable products, to prevent enzymatic browning. Its primary mechanism of action is as a reducing agent (oxygen scavenger). It also acts as a synergist with phenolic antioxidants. Ascorbic acid is water-soluble and as such cannot be used in fats. Its ester ascorbyl palmitate is more fat-soluble, and can be made more fat-dispersible by surface active agents such as lecithin and monoglycerides (Madhavi *et al.*, 1996a,b). *Citric acid* is not an antioxidant itself but it increases the effectiveness of phenolic primary antioxidants by chelating metal ions that catalyze lipid autoxidation. As such, it is widely used for the stabilization of oils. In combination with ascorbic acid, it is used for the prevention of enzymatic browning in fruit and vegetable products.

References

Ahmad, S., Branen, A.L., 1981. Inhibition of mold growth by butylated hydroxyanisole. J. Food Sci. 46 (4), 1059–1063.

Alam, S., Shah, H.U., Magan, N., 2010. Effect of calcium propionate and water activiy on growth and aflatoxin production by *Aspergillus flavus*. J. Food Sci. 75 (2), M61–M64.

Bae, Y.Y., Kim, N.H., Kim, K.H., Kim, B.C., Rhee, M.S., 2011a. Supercritical carbon dioxide as a potential intervention in ground pork decontamination. J. Food Saf. 31 (1), 48–53.

Bae, Y.Y., Choi, Y.M., Kim, M.J., Kim, B.C., Rhee, M.S., 2011b. Application of supercritical carbon dioxide for microorganism reduction in fresh pork. J. Food Saf. 31 (4), 511–517.

Bargiota, E., Rico-Muñoz, E., Davidson, P.M., 1987. Lethal effect of methyl and propyl parabens as related to *Staphylococcus aureus* lipid composition. Intl J. Food Microbiol. 4 (3), 257–266.

Berhimpon, S., Souness, R.A., Driscol, R.H., Buckle, K.A., Edwards, R.A., 1991. Salting behavior of yellowtail (*Trachurus mccullochi nichols*). J. Food Proc. Preserv. 15 (2), 101–114.

Beuchat, L.R., 2001. Control of foodborne pathogens and spoilage microorganisms by naturally occurring antimicrobials. In: Wilson, C.L., Droby, S. (Eds.), Microbial Food Contamination. CRC Press, London, UK.

Bidias, E., Lambert, J.W., 2008. Comparing the antimicrobial effectiveness of NaCl and KCl with a view to salt/sodium replacement. Intl. J. Food Microbiol. 124 (1), 98–102.

Bower, C.K., Daeschel, M.A., 1999. Resistance responses of microorganisms in food environments. Intl. J. Food Microbiol. 50 (1–2), 33–44.

Branen, A.L., Davidson, P.M. (Eds.), 1983. Antimicrobials in Foods. Marcel Dekker, New York, NY.

Brewer, M.S., 2011. Natural antioxidants: sources, compounds, mechanisms of action, and potential applications. Comprehen. Rev. Food Sci. Food Saf. 10 (4), 221–247.

Buzrul, S., 2009. A predictive model for high-pressure carbon dioxide inactivation of microorganisms. J. Food Saf. 29 (2), 208–223.

Catte, M., Gancel, F., Dzierszinski, F., Tailliez, R., 1999. Effects of water activity, NaCl and smoke concentrations on the growth of *Lactobacillus plantarum* ATCC 12315. Intl. J. Food Microbiol. 62 (1–2), 105–108.

Ceylan, E., Fung, D.Y.C., Sabah, J.R., 2004. Antimicrobial activity and synergistic effect of cinnamon with sodium benzoate or potassium sorbate in controlling *Escherichia coli* 0157:H7 in apple juice. J. Food Sci. 69 (4), 102–106 (FMS).

Cheli, F., Baldi, A., 2011. Nutrition-based health: cell-based bioassays for food antioxidant activity evaluation. J. Food Sci. 76 (9), R197–R205.

Chen, H., Hoover, D.G., 2003. Bacteriocins and their food applications. Comp. Rev. Food Sci. Food Saf. 2 (3), 82–100.

Chipley, J.R., 1983. Sodium benzoate and benzoic acid. In: Branen, A.L., Davidson, P.M. (Eds.), Antimicrobials in Foods. Marcel Dekker, New York, NY.

Choe, E., Min, D.B., 2009. Mechanisms of antioxidants in the oxidation of foods. Comprehen. Rev. Food Sci. Food Saf. 8 (4), 345–358.

Cornu, M., Beaufort, A., Rudelle, S., Laloux, L., Bergis, H., Miconnet, N., et al., 2006. Effect of temperature, water-phase salt and phenolic contents on *Listeria monocytogenes* growth rates on cold-smoked salmon and evaluation of secondary models. Intl. J. Food Microbiol. 106 (2), 159–168.

Corzo, O., Bracho, N., Rodriguez, J., 2012. Pile salting kinetics of goat sheets using Zugarramurdi and Lupin's model. J. Food Proc. Preserv. (online).

Craft, B.D., Kerrihard, A.L., Amarowicz, R., Pegg, R.B., 2012. Phenol-based antioxidants and the *in vitro* methods used for their assessment. Comprehen. Rev. Food Sci. Food Saf. 11 (2), 148–173.

Daniels, J.A., Krishnamurthi, R., Rizvi, S.S.H., 1985. A review of effects of carbon dioxide on microbial growth and food quality. J. Food Protec. 48 (6), 532–537.

Davidson, P.M., 1983. Phenolic compounds. In: Branen, A.L., Davidson, P.M. (Eds.), Antimicrobials in Foods. Marcel Dekker, New York, NY.

Davidson, P.M., Post, L.S., Branen, A.L., McCurdy, A.R., 1983. Naturally occurring and miscellaneous food antimicrobials. In: Branen, A.L., Davidson, P.M. (Eds.), Antimicrobials in Foods. Marcel Dekker, New York, NY.

Delves-Broughton, J., 2005. Nisin as a food preservative. Food Aust. 57 (12), 525–527.

Doores, S., 1983. Organic acids. In: Branen, A.L., Davidson, P.M. (Eds.), Antimicrobials in Foods. Marcel Dekker, New York, NY.

Doyle, M.E., Glass, K.A., 2010. Sodium reduction and its effect on food safety, food quality, and human health. Comprehen. Rev. Food Sci. Food Saf. 9 (1), 44–56.

Erkmen, O., 2000. Predictive modeling of *Listeria monocytogenes* inactivation under high pressure carbon dioxide. LWT 33 (7), 514–519.

FDA.,2011a. Code of Federal Regulations, 21CFR182.3862.

FDA.,2011b. Code of Federal Regulations, 21CFR172.175.

Fitzgerald, D.J., Stratford, M., Gasson, M.J., Ueckert, J., Bos, A., Narbad, A., 2004. Mode of antimicrobial action of vanillin against *Escherichia coli, Lactobacillus plantarum* and *Listeria innocuass*. J. Appl Microbiol. 97 (1), 104–113.

Friedman, M., Henika, P.R., Mandrell, R.E., 2002. Bactericidal activities of plant essential oils and some of their isolated constituents against *Campylobacter jejuni*, *Escherichia coli*, *Listeria monocytogenes*, and *Salmonella enterica*. *J Food* Prot. 65, 1545–1560.

Haas, G.J., Prescott, H.E., Dudley, E., Dik, R., Hintlian, C., Keane, L., 1989. Inactivation of microorganisms by carbon dioxide under pressure. J. Food Saf. 9 (4), 253–265.

Huang, H., Zhang, Y., Liao, H., Hu, X., Wu, J., Liao, X., et al., 2009. Inactivation of *Staphylococcus aureus* exposed to dense-phase carbon dioxide in batch systems. J. Food Proc. Eng. 32 (1), 17–34.

Jacobs, M.B., 1951. Chemical preservatives. In: Jacobs, M.B. (Ed.), The Chemistry and Technology of Food and Food Products, vol. 3. Interscience Publishers, New York, NY.

Kristinsson, H.G., Crynen, S., Yagiz, Y., 2008. Effect of a filtered wood smoke treatment compared to various gas treatments on aerobic bacteria in yellowfin tuna steaks. LWT 41 (4), 746–750.

Larrazábal-Fuentes, M.J., Escriche-Roberto, I., Camacho-Vidal, M.D.M., 2009. Use of immersion and vacuum impregnation in marinated salmon (Salmo salar) production. J. Food Proc. Preserv. 33 (5), 635–650.

Lee, J.H., Jung, M.Y., 2010. Direct spectroscopic observation of singlet oxygen quenching and kinetic studies of physical and chemical singlet oxygen quenching rate constants of synthetic antioxidants (BHA, BHT, and TBHQ) in methanol. J. Food Sci. 75 (6), C506–C513 (75).

Lück, E., Jager, M., 1997. Antimicrobial Food Additives – Characteristics, Uses, Effects, second edn. Springer-Verlag, Berlin, Germany.

Madhavi, D.L., Deshpande, S.S., Salunkhe, D.K., 1996a. Food Antioxidants. Marcel Dekker, New York, NY.

Madhavi, D.L., Singhai, R.S., Kulkarni, P.R., 1996b. Technological aspects of food antioxidants. In: Madhavi, D.L., Deshpande, S.S., Salunkhe, D.K. (Eds.), Food Antioxidants. Marcel Dekker, New York, NY.

Moorhead, S.M., Bell, R.G., 2000. Botulinal toxin production in vacuum and carbon dioxide packaged meat during chilled storage at 2 and 4°C. J. Food Saf. 20 (2), 101–110.

Mosqueda-Melgar, J., Raybaudi-Massilia, R.M., Martín-Belloso, O., 2008. Inactivation of *Salmonella enterica* ser. *Enteritidis* in tomato juice by combining of high-intensity pulsed electric fields with natural antimicrobials. J Food Sci. 73 (2), M47–M53.

Norris, L., Watt, E., 2003. Pickled. Stewart, Tabori & Chang, New York, NY.

Ogunrinola, O.A., Fung, D.Y.C., Jeon, I.J., 1996. *Escherichia coli* 0157: H7 growth in laboratory media as affected by phenolic antioxidants. J. Food Sci. 61 (5), 1017–1021.

Passone, M.A., Resnik, S., Etcheverry, M.G., 2007. Potential use of phenolic antioxidants on peanut to control growth and aflatoxin B1 accumulation by *Aspergillus flavus* and *Aspergillus parasiticus*. J. Sci. Food Agric. 87 (11), 2121–2130.

Peleg, M., 2006. Advanced Quantitative Microbiology for Foods and Biosystems. CRC Press, Boca Raton, FL.

Peleg, M., Normand, M.D., Corradini, M.G., 2011. Construction of food and water borne pathogens' dose–response curves using the expanded Fermi solution. J. Food Sci. 76 (3), R82–R89.

Porsby, C.H., Vogel, B.F., Mohr, M., Gram, L., 2008. Influence of processing steps in cold-smoked salmon production on survival and growth of persistent and presumed non-persistent *Listeria monocytogenes*. Intl. J. Food Microbiol. 122 (3), 287–296.

Rajalakshmi, D., Narasimhan, S., 1996. Food antioxidants: sources and methods of evaluation. In: Madhavi, D.L., Deshpande, S.S., Salunkhe, D.K. (Eds.), Food Antioxidants. Marcel Dekker, New York, NY.

Rohani, S.M.R., Moradi, M., Mehdizadeh, T., Saei-Dehkordi., S.S., Griffiths, M.W., 2011. The effect of nisin and garlic (*Allium sativum* L.) essential oil separately and in combination on the growth of *Listeria monocytogenes*. LWT 44 (10), 2260–2265.

Rørvik, L.M., 2000. *Listeria monocytogenes* in the smoked salmon industry. Intl. J. Food Microbiol. 62 (3), 183–190.

Samelis, J., Bedie, G.K., Sofos, J.N., Belk, K.E., Scanga, J.A., Smith, G.C., 2005. Combinations of nisin with organic acids or salts to control *Listeria monocytogenes* on sliced pork bologna stored at 4°C in vacuum packages. LWT 38 (1), 21–28.

Schillinger, U., Becker, B., Vignolo, G., Holzapfel, W.H., 2001. Efficacy of nisin in combination with protective cultures against *Listeria monocytogenes* Scott A in tofu. Intl. J. Food Microbiol. 71 (2–3), 158–168.

Scott, V.N., Taylor, S.L., 1981a. Effect of nisin on the outgrowth of *Clostridium botulinum* spores. J. Food Sci. 46 (1), 117–126.

Scott, V.N., Taylor, S.L., 1981b. Temperature, pH, and spore load effects on the ability of nisin to prevent the outgrowth of *Clostridium botulinum* spores. J. Food Sci. 46 (1), 121–126.

Sherman, W.P., Billing, J., 1998. Antimicrobial functions of spices: why some like it hot. Q. Rev. Biol. 73 (*3*), 1–47.

Sofos, J.N., Busta, F.F., 1983. Sorbates. In: Branen, A.L., Davidson, P.M. (Eds.), Antimicrobials in Foods. Marcel Dekker, New York, NY.

Sofos, J.N., Busta, F.F., Allen, C.E., 1979. *Clostridium botulinum* control by sodium nitrite and sorbic acid in various meat and soy protein formulations. J. Food Sci. 44 (6), 1662–1667.

Sofos, J.N., Maga, J.A., Boyle, D.I., 1988. Effect of ether extracts from condensed wood smokes on the growth of *Aeromonas hydrophila* and *Staphylococcus aureus*. J. Food Sci 53 (6), 1840–1843.

Sun, X.D., Holley, R.A., 2012. Antimicrobial and antioxidative strategies to reduce pathogens and extend the shelf life of fresh red meats. Comp. Rev. Food Sci. Food Saf. 11 (4), 340–354.

Tamime, A.Y. (Ed.), 2006. Fermented Milks. Blackwell Publishing, Oxford, UK.

Tompkin, R.B., 1983. Nitrite. In: Branen, A.L., Davidson, P.M. (Eds.), Antimicrobials in Foods. Marcel Dekker, New York, NY.

Warth, A.D., 1989. Relationships between the resistance of yeasts to acetic, propanoic and benzoic acids and to methyl paraben and pH. Intl. J. Food Microbiol. 8 (4), 343–349.

second ed. Wood, B.J.B. (Ed.), 1998. Microbiology of Fermented Foods, vol. 1. Blackie Academic and Professional, London, UK.

Yano, Y., Satomi, M., Oikawa, H., 2006. Antimicrobial effect of spices and herbs on *Vibrio parahaemoliticus*. Intl. J. Food Microbiol. 111 (1), 6–11.

Yanagisawa, T., Ariizumi, M., Shigematsu, Y., Koyabasi, H., Hasegawa, M., Watanabe, K., 2009. Combination of super chilling and carbon dioxide concentration techniques most effectively to preserve freshness of shell eggs during long-term storage. J. Food Sci. 75 (1), E78–E82.

Yousef, A.E., Gajewski, R.J., Marth, E.H., 1991. Kinetics of growth and inhibition of *Listeria monocytogenes* in the presence of antioxidant food additives. J. Food Sci. 56 (1), 10–13.

Zhao, T., Doyle, M.P., 1994. Fate of enterohemorrhagic *Escherichia coli* 0157:H7 in commercial mayonnaise. J. Food Protect. 57, 780–783.

CHAPTER

Ionizing Irradiation and Other Non-Thermal Preservation Processes

26

"Non-thermal processes" is a name given to a group of techniques that try to preserve foods without heating, freezing or drying, and thereby to reduce loss of quality. Most of these techniques have been developed quite recently. Some have led to more or less limited industrial application; others are still in the experimental stage or facing engineering problems that hamper their practical implementation. The first process to be described in this chapter – preservation by ionizing radiations – is, as we shall see, an exception.

26.1 Preservation by ionizing radiations

26.1.1 Introduction

Research into the potential utilization of ionizing irradiation for food preservation started in the 1940s, as an extension of studies on the effect of radiation on living cells (radiobiology) (Karel, 1971). Extensive research on irradiation preservation of food was carried out in the 1950s and 1960s, mainly under the sponsorship of the US Army. Early attempts of commercial application date from the 1960s. While considerable progress has been made in solving the technological and regulatory problems associated with the process, food irradiation does not today occupy the place it deserves in the area of commercial food preservation, for reasons to be explained later.

26.1.2 Ionizing radiations

Ionizing radiations are radiations with energy levels sufficient to cause ionization in atoms or molecules, as follows:

$$A \xrightarrow[h\nu]{} A^{+} + e^{-} \tag{26.1}$$

The ions formed are not stable and rapidly undergo changes that generate other chemical species, such as free radicals, other ions or new stable molecules. These new chemical species, and particularly the free radicals, are responsible for the biological processes of technological significance, such as the destruction of

Food Process Engineering and Technology. DOI: http://dx.doi.org/10.1016/B978-0-12-415923-5.00026-5
© 2013 Elsevier Inc. All rights reserved.

microorganisms, parasites and insects, or the inhibition of sprouting in potatoes and onions.

In practice, only two kinds of ionizing radiations are used in food processing:

- Electron beams (β rays, cathode rays)
- Electromagnetic waves (γ rays, X-rays).

Radiation is characterized by the energy of its particles or photons, usually expressed in units of MeV (million electron-volts). One MeV equals approximately 1.6×10^{-13} J.

26.1.3 Radiation sources

Sources of ionizing radiations are of two kinds: radioactive isotopes, and accelerators.

26.1.3.1 Isotope sources

The main isotope source used for food irradiation is Co^{60}. This is an artificial radioactive isotope made by bombarding stable natural Co^{59} with neutrons. Co^{60} undergoes beta decay to the stable isotope Ni^{60}, emitting one electron (beta ray) and two gamma rays. The beta radiation from Co^{60} is weak (about 0.3 MeV), while the gamma rays, with energies of 1.17 and 1.33 MeV, constitute the useful radiation from the isotope. Co^{60} has a half-life of 5.27 years.

Before its conversion to the radioactive isotope, stable cobalt metal is shaped to the required form, such as strips, rods, tubes, plates, etc. In the irradiation facility, the radioactive source is usually stored in a deep pool of water and moved by remote control to the irradiation area when needed.

The radioactivity of an isotope source is expressed in terms of the number of disintegrations per second. Its SI unit, meaning one disintegration per second, is the *becquerel*, named after Antoine Henri Becquerel (French physicist, 1852–1908, Nobel Prize in Physics, 1903), discoverer of radioactivity. The practical unit is the *curie* (Ci), named after Marie Curie (French chemist, 1867–1934, Nobel Prize in Physics, 1903, and in Chemistry, 1911), which is equal to 3.7×10^{10} becquerels.

26.1.3.2 Machine sources (electron accelerators)

Essentially, electron accelerators are devices for the production of high-energy electron beams. In these machines, electrons are accelerated to the desired energy level by various methods. The resulting high-energy electron beams can either be used as such for beta irradiation, or be directed to a target surface for the production of electromagnetic radiation (X-rays) through the Roentgen effect.

The main type of machine source used for industrial ionizing irradiation is the linear accelerator (Linac), wherein "packs" of electrons are accelerated through a linear array of plates to which high voltage, alternating in resonance with the emission of the electron bundles, is applied. The output capacity of an accelerator is expressed in terms of power.

Isotope sources are less expensive in capital cost, maintenance and energy expenditure. On the other hand, the isotope cannot be turned off and is therefore depleted even when it is not in use; it may also be seen as less safe for the environment. Machine sources can be deactivated when not in use and when access is required for maintenance. Because of the low efficiency of the conversion to X-rays, they are less economical as sources of electromagnetic radiation.

26.1.4 Interaction with matter

The principal effect of ionizing radiation on matter is, obviously, ionization. The average radiation energy needed for the ionization of an ordinary molecule is about 32 eV. As radiation travels through matter, its energy is attenuated as a result of ionization. The attenuation profile is different for different kinds of radiation.

Attenuation of an electron beam as it penetrates through matter is depicted in Figure 26.1. Electrons have a finite penetration depth, x_{max}, which depends on the energy of the electron, E, as it hits the surface, according to Eq. (26.2) ("Feather's Rule"):

$$x_{\max} = \frac{0.54E - 0.13}{\rho} \tag{26.2}$$

where:

x_{max} = maximal penetration depth, cm
E = energy of the beam at entrance, MeV
ρ = density of the matter, g^{-1}.

Attenuation of electromagnetic radiation, on the other hand, shows a logarithmic decay (Figure 26.2). There is no finite maximum penetration depth. As the radiation passes through matter, its intensity decreases asymptotically, following the well-known Beer–Lambert Law, as formulated in Eq. (26.3):

$$I = I_0 e^{-\mu x} \tag{26.3}$$

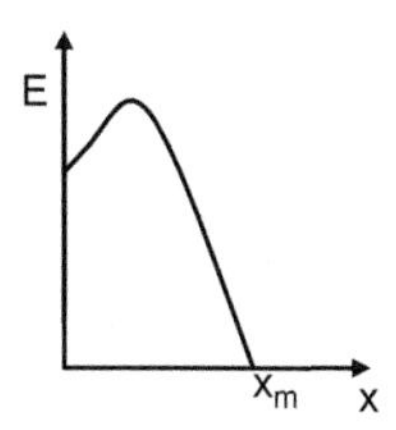

FIGURE 26.1

Attenuation of the electron beam energy through matter.

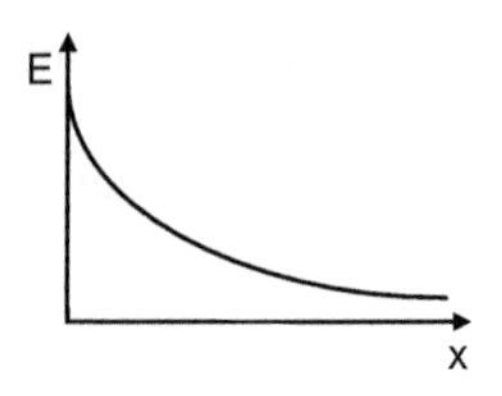

FIGURE 26.2

Attenuation of electromagnetic energy through matter.

Table 26.1 Extinction Coefficients of Different Materials to Electromagnetic Radiation

Material	Coefficient of Extinction (cm^{-1})	
	1 MeV	4 MeV
Air	0.00005	0.00004
Water	0.067	0.033
Aluminum	0.16	0.082
Steel	0.44	0.27
Lead	0.77	0.48

where I_0 and I = intensity of the radiation at depth x and depth 0, respectively, and μ = extinction coefficient.

The extinction coefficient depends on the material and the photon energy (Table 26.1). The depth of material necessary to absorb half of the incident energy is $0.693/\mu$. Extinction coefficients for different materials, at two levels of photon energy, are given in Table 26.1.

EXAMPLE 26.1

Calculate the thickness of lead needed to absorb 99.99% of the energy of gamma radiation having an incident photon energy of 4 MeV.

Solution

We apply Eq. (25.3):

$$I = I_0 e^{-\mu x}$$

For electromagnetic radiation of 4 MeV energy, μ for lead is 0.48 cm^{-1}

$$\frac{I}{I_0} = 0.0001 = e^{-0.48x}$$

x = 19.2 cm.

26.1.5 Radiation dose

The *radiation dose* is the quantity of radiation energy absorbed per unit mass of matter. The SI unit of radiation dose is the *Gray* (Gy). One Gray is equivalent to 1 joule of radiation energy absorbed per kg of matter. An older unit, the *rad*, equivalent to 100 ergs per gram, is still in use. The conversion is: 1 Gy = 100 rad.

Devices for measuring radiation doses are called "dosimeters" (Mehta and O'Hara, 2006). They fall into two categories:

1. *Primary dosimeters*. These measure dose directly, in actual terms of absorbed energy (calorimetry where the energy is measured as heat, or ionization cells in which the energy is measured in terms of electrical charges induced by ionization).
2. *Secondary dosimeters*. These measure dose indirectly, by assessing the extent of a chemical reaction induced by irradiation (e.g., oxidation of Fe^{++}, loss of methylene blue color, darkening of glass, polymer films or photographic film etc.).

Table 26.2 lists the doses necessary to induce certain effects of ionizing radiations.

EXAMPLE 26.2

What would be the temperature increase in an aluminum block used as a calorimeter, after absorbing a radiation dose of 5 kGy? The specific heat of aluminum is $900\ J \cdot kg^{-1} \cdot K^{-1}$.

Solution

$5\ kGy = 5000\ Gy = 5000\ J \cdot kg^{-1}$

$\Delta T = 5000/900 = 5.56°C.$

Table 26.2 Doses Necessary to Induce Certain Effects by Ionizing Irradiation

Effect	Dose Involved (Gy)
Medical radiography	Up to 0.01
Inhibition of sprouting in potatoes	20–200
Destruction of storage insects	50–1000
Inactivation of vegetative cells (pasteurization)	1000–3000
Inactivation of spores (sterilization)	10 000–30 000
Enzyme inactivation	>100 000

EXAMPLE 26.3

Gamma rays from a 4-MeV accelerator are to be used for irradiating frozen meat patties at a dose of 2.5 kGy. It is assumed that 75% of the radiation emitted will be absorbed by the product. What is the power (kW) of the accelerator needed to process 2000 kg patties per hour? What would be the "size" (in Ci) of a Cobalt 60 source necessary to deliver the same dose?

Solution

The energy absorbed by the product per second is:

$$E = \frac{2500 \times 2000}{3600} = 1389 \text{ W}$$

The energy emitted by the source is then:

$$E' = 1389/0.75 = 1852 \text{ W}$$

The power of the accelerator is 1.85 kW.

One disintegration in Cobalt 60 delivers a total energy of approximately 2.5 MeV; 1 MeV is equal to 1.6×10^{-13} J.

1 Ci is 3.7×10^{10} disintegrations per second. Thus, 1 Ci of Co 60 delivers a power of $3.7 \times 10^{10} \times 1.6 \times 10^{-13} = 5.92 \times 10^{-3}$ W.

The "size" C of the source needed is then:

$$C = \frac{1852}{5.95 \times 10^{-3}} = 311000 \text{ Ci}$$

The size of the radioactive source needed is 311 kilo-curies.

26.1.6 Chemical and biological effects of ionizing irradiation

As explained above, the primary action of ionizing radiation is the removal of an orbital electron and the transformation of the neutral molecule or atom to a positive ion. The ions thus formed are not stable, and they enter a series of chain reactions whereby free radicals, other ions and, finally, stable molecules are formed. The removed electron itself combines with other chemical species and starts another series of chain reactions.

The chemical and biological effects of irradiation may be the direct consequence of the ionization itself (direct effect) or the result of the action of the molecules and free radicals formed (indirect action). For example, irradiation may induce a variety of reactions, including oxidation, reduction, polymerization, depolymerization, cross-linking, hydrolysis, etc. The kinetics of alterations caused by the direct effect is independent of the composition, the temperature or the physical state of the medium, while all these factors affect the kinetics of changes induced by the indirect effect. As a rule, the destruction of microorganism cells is believed to be mainly caused by the direct effect mechanism, while chemical modifications, such as the destruction of vitamins, are usually attributed to the indirect action.

The inactivation of microorganism by ionizing irradiation is usually presented as a first-order reaction:

$$\log\frac{N}{N_0} = -kD \tag{26.5}$$

where D is the radiation dose. A "decimal reduction dose" D_{10}, similar to the "decimal reduction time" (Section 17.2.1), is defined:

$$\log\frac{N}{N_0} = -\frac{D}{D_{10}} \tag{26.6}$$

Table 26.3 lists typical values of decimal reduction doses of various microorganisms. It can be seen that, in certain cases, the medium has a considerable effect on the radio-resistance of the microorganisms.

EXAMPLE 26.4

Calculate the radiation dose needed to reduce the salmonella count in egg yolk by 7 log cycles.

Solution

$$-\frac{D}{D_{10}} = -7$$

For salmonella in egg yolk, $D_{10} = 0.8$ kGy, hence $D = 0.8 \times 7 = 5.6$ kGy.

The radiation dose needed is 5.6 kGy. This is a relatively high dose, but then the required extent of salmonella inactivation in this example is rather large.

Table 26.3 Decimal Reduction Doses of Different Microorganisms

Microorganism	Medium	D10 (kGy)
Clostridium botulinum spores	Buffer	3.3
Clostridium botulinum spores	Water	2.2
Salmonella typhimurium	Buffer	0.2
Salmonella typhimurium	Egg yolk	0.8
Escherichia coli	Buffer	0.1
Staphylococcus aureus	Buffer	0.2
Saccharomyces cereviseae	Saline	0.5
Aspergillus niger	Saline	0.5
Foot and mouth virus	Frozen	13

Data from Karel (1971).

26.1.7 Industrial applications

Theoretically, food preservation by ionizing radiations has important advantages:

1. It is a remarkably efficient method for the inactivation of microorganisms.
2. As a "cold" preservation process, it does not involve any thermal damage to quality.
3. It is a technologically and engineering-wise mature process, known for over half a century and supported by an impressive volume of research.
4. It is a permanent preservation process (like thermal processing, unlike refrigeration).
5. If applied according to the regulations, this is a safe method of food preservation, allowed by many national and international organizations.

Nevertheless, penetration of ionizing irradiation into the food process industry has not been extensive. Reasons for this include the following.

1. *Legal aspects*. As explained above, ionizing irradiation results in the formation of chemical species that were not in the food before. Therefore, legally, ionizing radiation is regarded as an "additive" rather than a "process". Demonstrating the safety of additives is a lengthy and costly endeavor.
2. *Image*. In the mind of many consumers, "irradiation" is linked to radioactivity and nuclear disasters (Eustice and Bruhn, 2006). It can be scientifically demonstrated that the probability of causing induced radioactivity in food irradiated according to the standards is practically zero – yet the consumer is persistently misinformed on this issue.
3. *Economics*. Irradiation equipment is fairly expensive. Obtaining the approval of authorities for food irradiation is costly. Gaining consumer confidence despite misinformation and campaigns with no basis in scientific evidence requires considerable investment. The risk of failure for reasons beyond the control of industry is real. Under these conditions, it is difficult to attract investment to food irradiation projects.

In recent years important progress has been achieved in the regulatory area, particularly in the USA. Applications up to doses of 3 kGy, using gamma irradiation up to 5 MeV or electrons up to 10 MeV, have been approved. For most foods, irradiation at these levels of energy and dose does not present safety problems or loss of organoleptic or nutritional quality any more than any other accepted method of preservation.

The status of food irradiation as an industrial process has been reviewed extensively, and is frequently updated (Molins, 2001; Sommers and Fan, 2006; Thorne, 1991; Wilkinson and Gould, 1996). The most important industrial applications of ionizing irradiation in the food industry today are:

- Disinfection of spices and dry condiments (up to 30 kGy)
- Control of *Salmonella* and *E. coli* in ground meat, poultry and seafood

- Pre-sterilization of packaging materials for aseptic processing
- Pre-treatment of raw materials of animal origin for pet-food production
- Surface disinfection of fresh fruits (Moreno et al., 2007) and vegetables.

Sterilization of foods by ionizing irradiation (rad-appertization) is no longer being considered, except in limited cases such as preparation of sterile food for astronauts, etc. Dairy products are not suitable for irradiation at any dose, because of undesirable effects on flavor.

26.2 High hydrostatic pressure (HHP) preservation

Preservation technologies based on the exposure of foods to high hydrostatic pressures in the range of 500–1000 MPa have been investigated in the past 30 years (Barbosa-Cánovas *et al.,* 1995; Cheftel, 1995; Knorr, 1993; Mertens and Deplace, 1993; Mertens and Knorr, 1992). These technologies are based on the observation that microorganisms are inactivated due to lethal structural and biochemical alterations caused to the cells as a result of residence under such high pressures.

In studying the kinetics of microorganism inactivation by high pressure, the "first-order" or "log-linear" model of thermal processing is adopted and a decimal reduction time is defined. Thus, Erkmen and Doğan (2004) found D values of 3.94 and 1.35 minutes at 400 and 600 MPa, respectively. Similarly, attempts have been made to define a "pressure z value", but the dependence of D on the pressure seems to be more complex. Furthermore, this dependence is found to be strongly dependent on the medium. One of the major difficulties in the application of high pressure to food preservation is the considerable variability of the piezo-resistance (resistance to high pressure) of microorganisms (Robey *et al.,* 2001). The particularly high resistance of spores to high-pressure treatment is a serious obstacle to the widespread application of HHP as a general food preservation method.

Different types of systems have been proposed for the industrial application of high-pressure processes (see, for example, Zimmerman and Bergman, 1993). Continuous processing of liquid foods in bulk does not seem to be technologically feasible. The so-called "high-pressure homogenization pasteurization" process (Maresca *et al.,* 2011) is not a HHP process. It utilizes much lower pressures, and the microbial inactivation mechanism is cell disruption due to shear rather than high pressure.

The system for processing packaged products consists of a pressure chamber and high-pressure pumps. An industrial-scale HHP system now available on the market is shown in Figure 26.3. The food in flexible packages is introduced into a chamber that is capable of withstanding high pressures. The pressure transmission medium, usually water, is introduced and pressurized by special pumps. The food is maintained under pressure for the required residence time. Since temperature affects the sensitivity of the microorganisms to high pressure, the temperature of

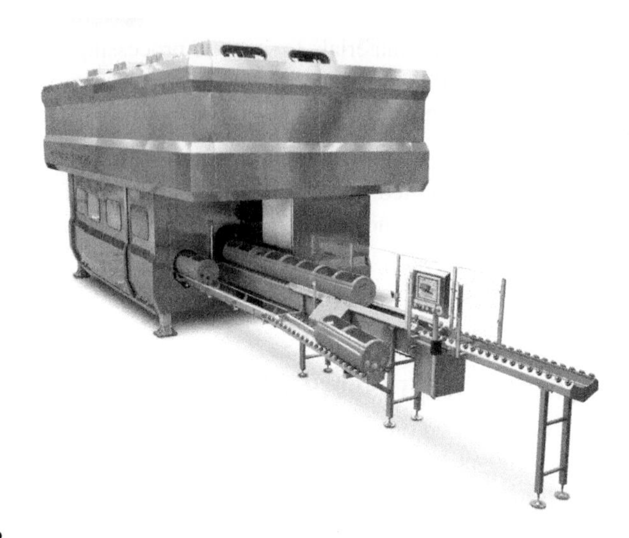

FIGURE 26.3

The Hiperbaric 420 machine for high hydrostatic pressure processing.

Figure courtesy of Hiperbaric S.A.

the process has to be controlled. Finally, the pressure is released and the packages of processed food are removed from the chamber. A color-based indicator for monitoring HHP processes has been developed (Fernández-Garcia *et al.,* 2009).

The industrial implementation of high hydrostatic pressure processing of foods faces some difficult engineering problems, such as the construction of pressure chambers capable of withstanding pressurizing–depressurizing cycles and the availability of pumps capable of delivering water at the required pressure. Advances in the design of commercial-scale HHP systems have been reviewed by Hernando-Sáiz *et al.* (2008). At any rate, the cost of the process, for the rate of production, is somewhat higher than thermal processing for equal microbial inactivation.

At present, the commercial application of high hydrostatic pressure processing is limited but expanding. It includes mainly pasteurization of some fruit juices and jams, as well as treatment of meat products (Simonin *et al.,* 2012) and cheese (Martinez-Rodriguez *et al.,* 2012) for the extension of shelf-life.

26.3 Pulsed electric fields (PEF)

Pulsed discharges of high-intensity electric fields have been used in research for the purpose of introducing genes into cells. Such discharges are used to induce reversible pore formation (electroporation) in the cell membrane. However, at higher field intensity and more prolonged exposure to the discharges, irreversible lethal effect can be achieved. The possibility of utilizing this phenomenon for the preservation of foods has been the subject of research in recent years (Barbosa-Cánovas *et al.*, 1995; Lelieveld and de Haan, 2007).

The intensity of an electric field is equal to the voltage per unit distance:

$$E = V/z \ (\mathrm{V \cdot m^{-1}}) \tag{26.7}$$

The duration of 1 pulse, τ, is given by:

$$\tau = CR \tag{26.8}$$

where C and R are the capacitance and the resistance of the circuit, respectively.

The number of pulses, n, experienced by the cell during time t is then:

$$n = t/\tau$$

In a laboratory set-up (Barbosa-Cánovas *et al.*, 1995), the food or the model material is passed through a treatment chamber. High voltage, in the order of 30–40 kV, is built up on capacitors and discharged through a gap of about 1 cm between electrodes. Pulse duration is 10–50 μs. The chamber is water-cooled.

The effect on the microorganisms depends on the field intensity, the number of pulses, the shape of the discharge pulse, the temperature, and, of course, the microorganism and the medium. The performance of pilot-scale PEF systems has been evaluated (Buckow *et al.*, 2010). The energy requirement for equal reduction of the microbial population by PEF was found to vary from one microorganism to another (Fernández-Molina *et al.*, 2006). Interestingly, a 5-minute exposure to a non-pulsed, moderate electric field (220 $\mathrm{V \cdot cm^{-1}}$) has been reported to inactivate *E. coli* (Machado *et al.*, 2010).

PEF treatment is not truly non-thermal because it involves passage of an electric current, generating heat (Sepulveda *et al.*, 2009). The term "thermization" has been applied to the combined effect of PEF and electrically induced heating (Guerrero-Beltran *et al.*, 2010). A model for the differentiation of thermal and electric field effects in PEF treatment has been proposed (Jaeger *et al.*, 2010). The effect of PEF on spores seems to be marginal. The effect of PEF on enzymes is variable (Van Loey *et al.*, 2001). PEF is mainly applied to liquid foods. Electric arc formation, resulting in intensive heating, may be a problem if the liquid contains air bubbles. After PEF treatment of the liquid in bulk, the product is sent to packaging, exposing it to recontamination unless an aseptic packaging system is used. Recently, an idea has been advanced to pre-package the product in electrically conducting plastic material prior to PEF treatment. It is claimed that a $6 \times \log 10$ inactivation can be achieved by this method (Roodenburg *et al.*, 2011).

PEF is proposed as a potential industrial process for the pasteurization of liquid foods, but this is still at the development stage (Altuntaş *et al.*, 2011). Other applications include intensification of the effect of bactericidal agents on microorganisms (Pol *et al.*, 2000).

The engineering approaches for the design of PEF treatment chambers have been reviewed by Huang and Wang (2009). The use of PEF-induced electroporation for purposes other than preservation (e.g., extraction) has been discussed elsewhere in the book.

26.4 Pulsed intense light

Exposure to short flashes of very intensive light has been found to cause lethal alterations in microorganisms. The main action seems to be DNA-related (Farkas, 1997). Usually, white broad-spectrum light (including wavelengths from UV to IR) is used. Light intensity is several $J \cdot cm^{-2}$ and the duration of the flash is measured in milliseconds. To produce the pulsed illumination, high-voltage pulses are sent to inert gas lamps.

Because of the opacity of foods, pulsed light treatment is mainly limited to the disinfection of surfaces, particularly of packaging materials and fresh fruits and vegetables (Dunn *et al.,* 1995). For in-depth effects, the optical properties of the material have to be considered. A numerical model for the prediction of microbial inactivation by intense light has been proposed (Hsu and Moraru, 2011).

References

Altuntaş, J., Akdemir-Evrendilek, G., Sangun, M., Zhang, H.Q., 2011. Processing of peach nectar by pulsed electric field with respect to physical and chemical properties and microbial inactivation. J. Food Proc. Eng. 34 (5), 1506–1522.

Barbosa-Cánovas, G.V., Pothakamury, U.R., Swanson, B.G., 1995. State of the art technologies for the stabilization of foods by non-thermal processes: physical methods. In: Barbosa-Cánovas, G.V., Welti-Chanes, J. (Eds.), Food Preservation by Moisture Control. Tecnomic Publishing Co., Lancaster, PA.

Buckow, R., Schroeder, S., Berres, P., Baumann, P., Knoerzer, K., 2010. Simulation and evaluation of pilot-scale pulsed electric field (PEF) processing. J. Food Eng. 101 (1), 67–77.

Cheftel, J.C., 1995. Review: high pressure, microbial inactivation and food preservation. Food Sci. Technol. Intl. 1, 75–90.

Dunn, J., Clark, W., Ott, T., 1995. Pulsed light treatment of food and packaging. Food Technol. 49 (9), 95–98.

Erkmen, O., Doğan, C., 2004. Kinetic analysis of *Escherichia coli* inactivation by high hydrostatic pressure in broth and foods. Food Microbiol. 21 (2), 181–185.

Eustice, R.F., Bruhn, C.M., 2006. Consumer acceptance and marketing of irradiated foods. In: Sommers, C., Fan, X. (Eds.), Food Irradiation Research and Technology. Blackwell Publishing, Oxford, UK.

Farkas, J., 1997. Physical methods of food preservation. In: Doyle, M.P., Beuchat, L.R., Montville, T.J. (Eds.), Food Microbiology: Fundamentals and Frontiers. ASM Press, Washington, DC.

Fernández García, A., Butz, P., Corrales, M., Lindauer, R., Picouet, P., Rodrigo, G., et al., 2009. A simple coloured indicator for monitoring ultra high pressure processing conditions. J. Food Eng. 92 (4), 410–415.

Fernández-Molina, J.J., Bermúdez-Aguirre, D., Altunakar, B., Swanson, B.G., Barbosa-Canovás, G.V., 2006. Inactivation of *Listeria innocua* and *Pseudomonas fluorescens* by pulsed electric fields in skim milk: energy requirements. J. Food Proc. Eng. 29 (6), 561–573.

Guerrero-Beltrán, J.A., Sepulveda, D.R., Góngora-Nieto, M.M., Swanson, B., Barbosa-Cánovas, G.V., 2010. Milk thermization by pulsed electric fields (PEF) and electrically induced heat. J. Food Eng. 100 (1), 56–60.

Hernando-Sáiz, A., Tárrago-Mingo, S., Purroy-Balda, F., Samson-Tonello, C., 2008. Advances in design for successful commercial high pressure food processing. Food Australia 60 (4), 154–156.

Hsu, L., Moraru, C.I., 2011. A numerical approach for predicting volumetric inactivation of food borne microorganisms in liquid substrates by pulsed light treatment. J. Food Eng. 105 (3), 569–576.

Huang, K., Wang, J., 2009. Designs of pulsed electric fields treatment chambers for liquid foods pasteurization process: a review. J. Food Eng. 96 (2), 227–239.

Jaeger, H., Meneses, N., Moritz, J., Knorr, D., 2010. Model for the differentiation of temperature and electric field effects during thermal assisted PEF processing. J. Food Eng. 100 (1), 109–118.

Karel, M., 1971. Physical Principles of Food Preservation. Marcel Dekker, New York, NY.

Knorr, D., 1993. Effects of high hydrostatic pressure on food safety and quality. Food Technol. 47 (6), 156–161.

Lelieveld, H.L.M., de Haas, W.H. (Eds.), 2007. Food Preservation by Pulsed Electric Fields: From Research to Application. Woodhead Publishing Co., Cambridge, UK.

Machado, L.F., Pereira, R.N., Martins, R.C., Teixeira, J.A., Vicente, A.A., 2010. Moderate electric fields can inactivate Escherichia coli at room temperature. J. Food Eng. 96 (4), 520–527.

Maresca, P., Dons, F., Ferrari, G., 2011. Application of a multi-pass high-pressure homogenization treatment for the pasteurization of fruit juices. J. Food Eng. 104 (3), 364–372.

Martínez-Rodríguez, Y., Acosta-Muñiz, C., Olivas, G.I., Guerrero-Beltrán, J., Rodrigo-Aliaga, D., Sepúlveda, D.R., 2012. High hydrostatic pressure processing of cheese. Comprehen. Rev. Food Sci. Food Safety 11 (4), 399–416.

Mehta, K., O'Hara, K., 2006. Dosimetry for food processing and research applications. In: Sommers, C., Fan, X. (Eds.), Food Irradiation Research and Technology. Blackwell Publishing, Oxford, UK.

Mertens, B., Deplace, G., 1993. Engineering aspects of high pressure technology in the food industry. Food Technol. 47 (6), 164–169.

Mertens, B., Knorr, D., 1992. Development of non thermal processes for food preservation. Food Technol. 46 (5), 124–133.

Molins, R. (Ed.), 2001. Food Irradiation – Principles and Application. Wiley-IEEE, New York, NY.

Moreno, M.A., Castel-Perez, M.E., Gomes, C., Da Silva, P.F., Kim, J., Moreira, R.G., 2007. Optimizing electron beam irradiation of "Tommy Atkins" mangoes (*Magnifera Indica* L). J. Food Proc. Eng. 30, 436–457.

Pol, I.E., Mastwijk, H.C., Bartels, P.V., Smid, E.J., 2000. Pulsed electric field treatment enhances the bactericidal action of nisin against *Bacillus cereus*. Appl. Environ. Microbiol. 66 (1), 428–430.

Robey, M., Benito, A., Hutson, R.A., Pascual, C., Park, S.F., Mackey, B.M., 2001. Variations in resistance to high hydrostatic pressure and rpoS heterogeneity in natural isolates of *Escherichia coli* 0157:H7. Appl. Environ. Microbiol. 67 (10), 4901–4907.

Roodenburg, B., De Haan, S.W.H., Ferreira, J.A., Coronel, P., Wouters, P.C., Hatt, V., 2011. Towards 6 $\log_{10}$ pulsed electric field inactivation with conductive plastic packaging materiel. J. Food Proc. Eng. (online 5 September 2011).

Sepulveda, D.R., Góngora-Nieto, M.M., Guerrero, J.A., Barbosa-Cánovas, G.V., 2009. Shelf life of whole milk processed by pulsed electric fields in combination with PEF-generated heat. LWT 42 (3), 736–739.

Simonin, H., Duranton, F., de Lamballerie, M., 2012. New insights into the high-pressure processing of meat and meat products. Comprehen. Rev. Food Sci. Food Safety 11 (3), 285–306.

Sommers, C., Fan, X. (Eds.), 2006. Food Irradiation Research and Technology. Blackwell Publishing, Oxford, UK.

Thorne, S., 1991. Food Irradiation. Elsevier Applied Science, London, UK.

Van Loey, A., Verachtert, B., Hendrickx, M., 2001. Effect of high electric field pulses on enzymes. Trends Food Sci. Technol. 12 (3–4), 94–102.

Wilkinson, V.M., Gould, G., 1996. Food Irradiation – A Reference Guide. Woodhead Publishing Co., Cambridge, UK.

Zimmerman, F., Bergman, C., 1993. Isostatic pressure equipment for food preservation. Food Technol. 47 (6), 162–163.

CHAPTER

Food Packaging 27

27.1 Introduction

Packaging technology occupies a central position in food processing (Coles *et al.*, 2003). The selection of proper packaging materials and systems is an integral part of food process and product design.

Robertson (2005) identifies four attributes of packaging: containment, protection, convenience and communication. In fact, each one of these functions comprises a number of different technological, engineering and commercial objectives.

1. *Containment*. One of the primary and obvious objectives of the package is to contain the product. This is essential for the efficient transportation, storage and distribution of the product. In addition, containment allows repartition of the product into portions of known weight or volume, and facilitates stock-keeping and merchandizing. The shape and dimensions of the package determine to a large extent the space requirement for storage, transportation and display. Shrinkable films provide a valuable medium for the compaction of product lots on pallets
2. *Protection and preservation*. In the case of foods, this is, without any doubt, the most important among the functions of packaging. By placing a more or less effective barrier between the food and the environment, the package protects the food from physical, chemical, microbial and macrobial attack from the exterior and thus has a decisive effect on the shelf-life of the product. At the same time, the package protects the environment from the food by preventing spillage, odor release, dust, etc (Johansson and Leufven, 1994). In thermal processing, the package dictates the type of processing, and *vice versa*. The package, be it a metal can, a glass jar or a plastic pouch, is expected to prevent recontamination of the thermally stabilized food inside. In recent years, packaging materials containing preserving substances have been developed, giving rise to a promising new area known as "active packaging". Finally, packaging materials with specific transport properties are the key factor in the preservation method known as the "modified atmosphere" technique.
3. *Convenience*. Convenience has long been, and continues to be, among the chief "selling" attributes of foods, and packaging contributes considerably to the convenience factor. Adapting the size of the package to the needs of particular consumer groups (family size, individual, special sizes for food

Food Process Engineering and Technology. DOI: http://dx.doi.org/10.1016/B978-0-12-415923-5.00027-7
© 2013 Elsevier Inc. All rights reserved.

service delivery, etc.) is one of the steps taken by industry to enhance product convenience through the package. Pressurized packages (for whipped cream), aerosols (for coating, flavoring, oiling, etc.), easy-open and/or resealable packages, and packages that can serve as heating utensils and as plates, cups, bowls, etc., from which the food can be eaten or drunk directly are among the convenience-driven developments of packaging technology.

4. *Communication*. The quantity of information printed on a food package has been increased constantly. In addition to text and graphics serving the purpose of product and brand identification and product promotion, the printing usually includes essential data such as a list of ingredients, net weight, nutritional data, a production date and/or a limit date for selling, the price, a barcode, and information needed for product traceability. In the future, the package of chilled foods may carry a temperature–time indicator/integrator that, by its color, will provide information on probable mishandling of the product during storage and transportation. Incidentally, packages that carry these types of monitors have been called, somewhat pretentiously, "intelligent packages".

Packaging usually consists of a number of "levels". The first level, known as the *primary package*, is the package in direct contact with the food. The primary package is the package in which a unit of the product is presented to the retail market. A can of tuna, a bag of peanuts, a jar of jam or the wrap around a chocolate candy are examples of primary packages. A number of primary packages are usually contained in an outer or *secondary package* for transportation, storage and delivery. A "case" of tuna is a carton box containing, say, 24 or 48 individual cans of tuna. A "six-pack" of beer is a package containing six bottles or cans of beer. A number of secondary packages may be collated into a "lot" contained in a *tertiary package*, and so on.

Food packaging is, by itself, a vast multidisciplinary area of studies, research and development. Entire academic departments and laboratories devoted exclusively to food packaging are in operation (Figures 27.1, 27.2). In this chapter, only some of the aspects of food packaging, related to food process engineering and technology, will be discussed. These aspects refer mainly to the packaging materials and packaging systems, to the protective function of packaging, and to some environmental issues regarding food packaging.

27.2 Packaging materials

27.2.1 Introduction

Most materials used for packaging foods belong to the following classes: metals, glass, paper and polymers. Some packaging media consist of a combination of two or more materials of the classes listed above. Enameled (lacquered) metal

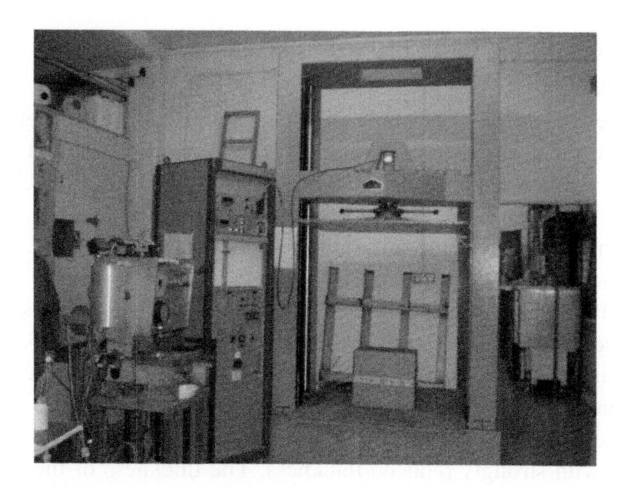

FIGURE 27.1

Press in a packaging research laboratory.

Photograph courtesy of the Department of Biotechnology and Food Engineering, Technion.

FIGURE 27.2

Vibration tables in a packaging research laboratory.

Photograph courtesy of the Department of Biotechnology and Food Engineering, Technion.

and laminates formed by binding together layers of polymer, paper and aluminum foil are common examples of such composite materials.

The chemical composition and physical properties of packaging materials determine their ability to fulfill the various functions expected from the package. The most important properties to be considered in this context are transport properties, optical properties, mechanical properties and chemical reactivity.

27.2.2 Materials for packaging foods

27.2.2.1 Metals

Metal containers offer the advantage of superior mechanical strength, impermeability to mass transfer and to light, good thermal conductivity, and resistance to relatively high temperature. The latter two properties make metal packages particularly suitable for in-package thermal processing (see Section 18.2).

Tinplate, the first material used to make metal cans and canisters, consists of a thin sheet of steel, coated with tin. The purpose of the tin coat is to reduce the risk of corrosion. The quantity of steel plate is traditionally expressed in "base box" (bb). One base box is equivalent to 112 sheets, measuring 0.356 × 0.508 m each, and weighs approximately 20–60 kg, depending on the thickness of the sheets (Hanlon *et al.*, 1998). In the past 50 years or so, advanced metallurgical processes have led to the production of steel plate with improved mechanical properties but with strongly reduced thickness. The thickness of the tin coating is quoted in nominal units of pounds per base box (lb/bb). The traditional method for coating the steel plates with tin, the "hot dip" method, has now been replaced by a process of electrolytic deposition. The electrolytic process of tinplating forms a more uniform tin coat with much less tin per unit area. Thus, both the thickness of the base plate and the weight of the tin coating per unit area of tinplate for cans have been reduced considerably, resulting in the production of lighter and less expensive cans with improved performance. For a review of processes for the production and improvement of tinplate, see Robertson (2005).

In some cases, the protection provided by tin is not sufficient for the prevention of internal or external corrosion of the can. Where the can is to face particularly severe corrosive conditions, a protective layer of polymeric lacquer or enamel is applied to the tin.

Can sizes are standardized and specified using standard denominations. In the USA, for example, cylindrical cans are specified by their diameter and their height, with both dimensions given by a three-digit code (Table 27.1).

The selection of the most suitable can for a given application involves specification of the steel base, the thickness of the tin layer, the type of enamel (where

Table 27.1 Standard Can Sizes*

Name	USA Dimensions	Capacity (l)
No. 1	211 × 400	0.30
No. 2	307 × 409	0.58
No. 2 ½	401 × 411	0.84
No. 10	603 × 700	3.07

Adapted from Lopez (1981).

applicable), and special features of can geometry. Based on their experience, can manufacturers usually supply the information for making the proper selection.

Second in importance among metal packaging materials is *aluminum*. Unlike steel, aluminum does not require the application of a protective coat because the thin film of aluminum oxide formed on the surface protects the metal against further corrosion by oxygen and mild acids, although it is attacked by alkali. Aluminum is much lighter and more ductile that tinplate, but it is more expensive. As a packaging material, aluminum is found in two forms: aluminum cans (used mainly for beer and soft beverages) and aluminum foil (as such, or in laminates). The purest form of aluminum, being the most ductile, is used for the manufacture of foil and containers.

27.2.2.2 Glass

The glass used for making containers (bottles, jars) for food packaging is soda-lime glass, containing typically 68–73% SiO_2, 12–15% Na_2O, 10–13% CaO and other oxides in lesser proportions (Robertson, 1993). The advantages of glass as a packaging material are transparence, inertness, impermeability, rigidity, thermal resistance (when properly heated), and general consumer appeal. Its disadvantages are fragility and weight. Glass containers are standardized to a much lesser degree than metal cans. In fact, most bottles and jars are tailor-made specifically for one product or one manufacturer. On the other hand, closures for glass containers are somewhat more standardized. Glass containers can be reused or recycled. Re-use is problematic, as explained in Section 28.6, but recycling (re-melting) is technically and economically feasible.

27.2.2.3 Paper

Paper products are widely used as food packages. In fact, paper, in one form or another, must have been one of the earliest food packaging materials. The main advantages of paper as a packaging material are its low cost, wide availability, low weight, printability and mechanical strength. Its most serious shortcoming is its sensitivity to moisture (Miltz, 1992). The properties of paper can be modified through the composition of the pulp, the manufacturing process, and various surface treatments. The permeability to moisture and fat can be reduced by coating with wax (waxed paper). Paper is an important component of laminated packaging materials. It is used as a primary package (boxes, wraps, pouches) and it is the principal material used for secondary packaging (corrugated cardboard boxes or cartons).

27.2.2.4 Polymers

This is, quantitatively as well as qualitatively, the most important class of packaging materials, both for food and for non-food applications (Jenkins and Harrington, 1991; Miltz, 1992). The reasons for their success and rapidly increasing share in packaging technology are numerous. Polymeric materials are fairly varied and versatile. They can be flexible or rigid, transparent or opaque, thermosetting or thermoplastic (heat-sealable), fairly crystalline or practically amorphous. They can be produced as films or as containers of many shapes and sizes. As a rule, they are much less expensive than metal or glass, and certainly much lighter. They are remarkably suitable for the application of advanced packaging technologies such as modified atmosphere (MAP), active and "intelligent" packaging (see Section 27.3).

Transport properties are the most extensively studied aspect of polymeric packaging materials for food. Unlike metal or glass, polymers are permeable to small molecules to a greater or lesser extent. Two consequences of this property are of particular interest, namely the permeability of the package to gases and vapors (particularly to oxygen and water vapor), and the migration of low molecular weight substances from the package to the food (monomers, stabilizers, plasticizers) or from the food to the packaging material and out (aroma components). Both phenomena are discussed in the next section.

With the notable exception of materials of cellulosic origin (e.g., cellophane), packaging plastics are made of synthetic polymers. Chemically, they vary in the monomers forming the polymer chain, in their molecular weight, and in the structure of the chain (linear vs branched, cross-linked, etc.). Some of the most important polymers are described below.

Polyethylene (PE) is a polymer of the olefin ethylene, $CH_2 = CH_2$. There are four kinds of polyethylene (Miltz, 1992):

1. Low density polyethylene (LDPE) is a highly branched polymer with branches consisting of short or long side chains. Short chains impart some crystallinity to the material, while long chains are responsible for the viscoelastic properties of the molten polymer. The relatively low melting range (105–115°C) allows its use as the heat-sealable layer in laminates.
2. High density polyethylene (HDPE) is a linear polymer with little branching. It is considerably more crystalline than LDPE, hence more rigid and less transparent. Its melting range is higher (128–138°C).
3. Medium density polyethylene (MDPE) has properties intermediate between those of LDPE and HDPE.
4. Linear low density polyethylene (LLDPE) is a copolymer of ethylene with small quantities of higher olefins, and has branches at regular intervals on the main chain. It is stronger than LDPE and a better heat-sealable component.

27.2.3 Transport properties of packaging materials

Glass and metals are practically impermeable to gases and vapors, so they provide an efficient barrier against material exchange between the atmosphere inside the package and the environment outside. On the other hand, polymers and paper are permeable to gases and vapors to various degrees, and their barrier properties certainly constitute the chief criterion in estimation of their suitability to serve as packaging materials in a given application. Gases and vapors may pass through packaging barriers either by molecular diffusion or by flow through holes and pores. Only the first type of transport will be discussed here.

The classical explanation of penetration assumes dissolution or adsorption of the permeant on one face of the film, molecular diffusion through the film, and desorption on the opposite face. This process of adsorption–diffusion–desorption is named "permeation", and the behavior of the penetrant–barrier couple is characterized by means of a parameter known as "permeability" or the "permeability coefficient". The concept of "permeability", Π, was developed in Section 3.3.4.4 and included in Eq. (3.16), which is reproduced here as Eq. (27.1):

$$J_G = D_G s_G \frac{p_1 - p_2}{z} = \Pi \frac{p_1 - p_2}{z} \quad (27.1)$$

where:

J_G = flux of the gas through the film of packaging material
D_G and s_G = diffusion coefficient and solubility of the gas in the film material, respectively
p_1 and p_2 = partial pressure of the gas, upstream and downstream of the transfer, respectively
z = thickness of the film, in the direction of the transfer.

Note that Eq. (27.1) applies to steady-state permeation only. In practical situations, true steady state is seldom attained. Consider, for example, the transfer of water vapor into a package containing biscuits. Even if the relative humidity of the air outside the package is maintained at a constant level (i.e., p_1 = constant), the incoming water vapor will alter p_2 at a rate that will be determined by the volume of air and the mass of biscuits in the package, as well as the sorption isotherm of the biscuits.

Note also that the permeability is the product of two fundamental properties of the permeant–barrier couple, namely, diffusivity and solubility. One notorious consequence of this fact is the high permeability of hydrophilic films (e.g., cellophane) and the low permeability of hydrophobic films (e.g., polyethylene) to water vapor.

The issue of standard units by which the permeability coefficients are expressed is problematic. According to Robertson (2005), the number of different units for permeability that appear in the literature exceeds 30. SI units of permeability are $kg \cdot m^{-2} \cdot s^{-1} \cdot Pa^{-1} \cdot m$ or $kmol \cdot m^{-2} \cdot s^{-1} \cdot Pa^{-1} \cdot m$ if the quantity of the material transferred is expressed in mass; and $m^3 \cdot m^{-2} \cdot s^{-1} \cdot Pa^{-1} \cdot m$ (equivalent to $m^2 \cdot s^{-1} \cdot Pa^{-1}$) if the quantity is given in volume. However, SI units are seldom used in the barrier permeation literature. Instead, permeability coefficients are expressed in a number of different "practical" units. The quantity transferred is usually expressed as a volume, in cm^3 (STP), for oxygen, nitrogen and carbon dioxide, and as a mass, in grams, for water vapor. The area may be given in m^2 or in cm^2. The time may be expressed in second, hours or days (24 hours). The pressure is often given in cm Hg, mm Hg or bars. The thickness of the film is, of course, never expressed in meters but in millimeters or micrometers. In addition, many practical units using grains (for water vapor), inches, square inches and mils (one-thousandth of an inch, for film thickness) are also in extensive use. The American Society for Testing and Materials (ASTM) has adopted a unit named the *barrer* (after Richard Barrer, 1910–1996). One barrer is equal to 10^{-11} cm^3 (STP) $\cdot$ cm^{-2} $\cdot$ s^{-1} $\cdot$ $(mm\ Hg)^{-1}$ $\cdot$ cm. The barrer is mainly used in connection with the permeability of contact lenses to oxygen, but much less so in the area of food packaging films. Some of the most frequently used gas permeability units and the corresponding conversion factors are given in Table 27.2.

The permeability of films to water vapor is usually expressed as the *water vapor transmission rate*, WVTR, which is the quantity of water vapor transmitted per unit area and unit time, by a film of unit thickness, under specified conditions of vapor pressure difference and temperature. Traditionally, the standard specified conditions are 90% relative humidity at 37.8°C (100°F). The units for quantity, area, time and film thickness may vary.

Typical values of permeability coefficients of a number of films to various gases and water vapor are shown in Table 27.3.

The data clearly show that some polymers (e.g., polyethylene), are excellent barriers to water vapor but quite permeable to oxygen, while the opposite is true for others (e.g., PVOH). Improved barrier properties can be achieved by binding together (laminating) films of various materials, each with a different permeability profile (Mastromatteo and Del Nobile, 2011). The permeability of the

Table 27.2 Conversion Factors for Permeability to Gases (Volumetric Flux)

Unit	Conversion Factor
1 barrer = 10^{-11} cm^2 (STP) $\cdot$ s^{-1} $\cdot$ $(mm\ Hg)^{-1}$	1
cm^3 (STP) $\cdot$ cm^{-2} $\cdot$ s^{-1} $\cdot$ $(mm\ Hg)^{-1}$ $\cdot$ cm^{-1}	10^{-11}
cm^3 (STP) $\cdot$ m^{-2} $\cdot$ d^{-1} $\cdot$ $(mm\ Hg)^{-1}$ $\cdot$ mil^{-1}	3.6
cm^3 (STP) $\cdot$ cm^{-2} $\cdot$ d^{-1} $\cdot$ bar^{-1} $\cdot$ mm	846
m^3 (STP) $\cdot$ m^{-2} $\cdot$ s^{-1} $\cdot$ Pa^{-1} $\cdot$ m [SI unit]	7.5×10^{-18}

Table 27.3 Barrier Properties of Two Polymers to Gases

Polymer	Permeability to Oxygen ($cm^3 \cdot mil^{-1} \cdot 100\ in^{-2} \cdot day^{-1}$)	Permeability to Water Vapor ($g \cdot m^{-2} \cdot day^{-1}$) at 40°C, 90% RH
Low density polyethylene	2400–3000	10–18
Polyvinyl alcohol	<0.01	200

Data from Miltz (1992).

composite laminate to a given permeant can be calculated, using the concept of "resistances in series" (see Section 3.3.4.4 and Eq. (3.13).

$$\frac{z_{total}}{\Pi_{laninate}} = \frac{z_1}{\Pi_1} + \frac{z_2}{\Pi_2} + \ldots + \frac{z_n}{\Pi_n} = \sum_1^n \frac{z_i}{\Pi_i} \tag{27.2}$$

where z_1, z_2, ..., z_n = thicknesses of individual layers and Π_1, Π_2, ..., Π_n = permeabilities of individual layers.

The integration leading to Eq. (27.1) assumes that both the solubility and the diffusivity are independent of the concentration of the permeant. The permeability of polymer films to low molecular weight gases is indeed practically independent of the concentration (partial pressure) of the permeant, but this may not be the case for condensable vapors and liquids that can alter the structure of the polymer – for example, by swelling and plasticizing. Furthermore, an "interacting" permeant may affect the permeability of the film to other permeants. Thus, the permeability of Nylon 6 to oxygen is 50 times higher at 100% relative humidity than in bone-dry gas (Ashley, 1985). On the other hand, the permeability of a hydrophobic film such as polyethylene is not affected by humidity. Similar "co-permeant" effects are exhibited also by organic vapors capable of interacting with the barrier polymer (Giacin, 1995; Johansson and Leufvén, 1994).

Obviously, the barrier properties of polymer films are dictated by their molecular structure (Giacin, 1995; Hanlon *et al.*, 1998). Complete models that can predict exactly the barrier behavior of a polymer in the light of its molecular structure are not available, but certain relationships between structural features and permeability may be established. It is known, for example, that cross-linking, higher crystallinity, high glass transition temperature and inertness to the permeant result in lower permeability (Robertson, 2005).

27.2.4 Optical properties

Some optical properties of packaging materials are of practical importance, partly because they affect the protective function of the package and partly because of their influence on its appearance and attractiveness. Transparency to light is particularly important in the case of glass and polymer films. Many deteriorative reactions are catalyzed by light in general, and ultraviolet light in particular. These include lipid oxidation, off-flavor generation, discoloration, and destruction of nutritionally important components such as riboflavin, beta-carotene, ascorbic acid and certain amino acids (Bosset *et al.*, 1994). On the other hand, transparent packages allow consumers to see the product through the package and judge on its quality by its appearance. This is the case in packaged fresh meat, poultry, fruits and vegetables, confectionery, confitures, baked goods and thermally preserved foods in glass jars (e.g., fruits in syrup, strained infant foods, etc.). A certain compromise between protection from light and transparency may be achieved by using colored plastic or glass.

The intensity of light transmitted through a thickness z of material is given by the Beer–Lambert Law, which can be written as follows:

$$T = \frac{I}{I_0} = e^{-k\,z} \tag{27.3}$$

where:

T = transmittance (fractional, may be given as a percentage)
I and I_0 = intensity of the light transmitted and incident, respectively
k = a characteristic of the material (absorbance)
z = thickness.

The characteristic absorbance parameter k depends on the wavelength of the light, and is therefore an indication of the transmitted color of the material. Considerable protection to the product can be provided by coloring the transparent packaging material (glass or plastic) with a pigment or coating it with a film of material that has a high absorbance for UV light.

Plastic packaging materials may be opaque, hazy (translucid) or transparent. Plastic materials are rendered opaque by the incorporation of very fine solid particles of white or colored pigments into the melt. Haze or cloudiness is the result of light scattering (diffraction) by the crystalline micro-regions of the polymer. Amorphous plastics such as polycarbonate are clear (transparent).

27.2.5 Mechanical properties

The ability of a package to protect its contents against external forces depends on its mechanical properties. In packaging technology, mechanical properties should be considered and evaluated at the level of the packaging material, the formed empty package, the product–package assembly and the outer packages.

The mechanical strength of cans depends on the size and structure of the can and the thickness of the tinplate. At equal tinplate thickness, cans with smaller diameters are mechanically stronger. Frequently, the side walls of the can are beaded to increase mechanical strength.

Except for the integrity and stability of the closure, mechanical strength is not an issue with glass. Relatively high output rates with minimal breakage can be achieved with adequately designed handling and conveying equipment, and with proper surface treatment to provide lubricity and prevent scratches.

Paper and particularly corrugated cardboard used for outer packaging must be tested for mechanical strength. Because the strength of paper is strongly influenced by its moisture content, paper packages must be conditioned at known humidity before testing.

27.2.6 Chemical reactivity

Of all the packaging materials, only glass can be considered to be practically inert. Almost all other packaging materials may react, to a certain extent, with the food within and the environment without. Only two of the possible interactions will be discussed here: the corrosion of tinplate, and the migration of chemical substances from the package to the food.

27.2.6.1 Corrosion of tinplate

Schematically, tinplate can be seen as consisting of three layers: the steel base, the tin coating and, between the two, a layer of Fe–Sn alloy. The tin coating is not perfect. Scratches and pores expose small areas of alloy or steel. Since the contents of the can almost always have some electrical conductivity, the system constitutes a voltaic cell. In the presence of a de-aerated acidic liquid, the iron initially acts as the anode (less "noble" than tin) and dissolves. However, the polarity is soon reversed and the tin becomes the anode with respect to the steel. Now the tin dissolves, protecting the iron. The tinplate is gradually "detinned". In both stages, dissolution of the metal generates hydrogen, which tends to polarize the cell and prevent further dissolution. If oxygen or other depolarizing agents are present, the cell is soon depolarized and dissolution continues. More tin is dissolved, more steel is exposed and more hydrogen evolves. Excessive evolution of hydrogen gas may cause swelling of the can (hydrogen swell). Detinning itself is objectionable because it imparts a metallic taste to the food and renders the internal surface of the can gray and unattractive. Tin in food is considered to be a contaminant. The regulatory tolerance in some countries is 200 mg per kg. Higher tin content is known to cause gastrointestinal disturbances, but no chronic toxicity or carcinogenicity is attributed to inorganic tin.

The type of corrosion described above is particularly severe in the cases of:

- Foods with high acidity (e.g., grapefruit, lemon, pineapple, tomato juices and concentrates)
- Tinplate with insufficient and/or porous tin coating
- Foods containing depolarizing agents such as anthocyanin pigments (red fruits)
- Cans that have not been sufficiently de-aerated or have an excessive headspace.

It should be noted, however, that mild detinning is often desirable because of the reducing effect of the hydrogen on the food. Browning reactions involving initial oxidation of ascorbic acid (e.g., in citrus products) is less severe in products packed in plain tinplate cans rather than glass or enameled tinplate.

Another type of "accidental" but severe corrosion occurs as a result of attack on the exposed iron by sulfur dioxide. Sulfur dioxide may be found as a residue in sugar. In the can it is reduced to hydrogen sulfide, which reacts with the iron to produce black iron sulfide.

Internal corrosion of tinplate may be effectively minimized by coating the tinplate with enamel, as discussed previously.

27.2.6.2 Migration of chemicals

In the past 40 years or so, the migration of low molecular weight substances from plastic packaging materials to food has been investigated intensively. The substances of interest are monomers and processing additives used in production of the plastic material.

The migration of vinyl chloride monomer from packages made of PVC (polyvinyl chloride) has attracted attention because this monomer (VCM) is a potent carcinogen. Another monomer, the presence of which in food is objectionable, is acrylonitrile monomer. Processing additives that may migrate to the food are mainly plasticizers, antioxidants and solvent residues.

While the polymer industry has invested considerable efforts to overcome the problem by technological means, research has developed increasingly sensitive methods for the detection of the contaminants. The toxicology of the substances in question is known, and regulations covering the issue are available in most countries.

27.3 The atmosphere in the package

The atmosphere surrounding the food in the package has a profound effect of the shelf life of the product. The principal techniques that make use of the in-package atmosphere for improving the preserving action of the package are vacuum packaging, controlled atmosphere packaging (CAP), modified atmosphere packaging (MAP) and the so-called active packaging. In the majority of cases, these techniques are additional "hurdles" in combined preservation processes including, almost always, refrigeration.

27.3.1 Vacuum packaging

Vacuum packaging is an old and widespread technique, applied to a variety of foods. Its main objective is to prevent oxidation reactions such as lipid oxidation, loss of certain vitamins, oxidative browning, loss of pigments, etc. Vacuum also prevents deterioration by aerobic microorganisms and, particularly, mold. Fresh meat packaged under vacuum may have a shelf life of a few weeks under refrigeration. Vacuum packaging offers additional advantages, such as reducing the volume and improving the rigidity of flexible packages. In retortable pouches, the vacuum helps press the package against the food and thus improve heat transfer.

Devices for pulling a vacuum in the package before sealing are available for cans, jars, trays and pouches.

27.3.2 Controlled atmosphere packaging (CAP)

CAP is actually controlled atmosphere storage in the package. The product is packaged in a mixture of gases, the composition of which has been precisely fixed. The packaging material is essentially impermeable to gases. True CAP is only possible if the product has no respiratory or microbial activity and is inert to the gases in the atmosphere inside the package, and there is no gas exchange across the packaging material. As these conditions are difficult to meet, in most cases CAP becomes identical to MAP (see below).

27.3.3 Modified atmosphere packaging (MAP)

Modified atmosphere packaging (MAP) has been defined as packaging of food "in an atmosphere which has been modified so that its composition is other than that of air" (Hintlian and Hotchkiss, 1986). This definition is too general, as it comprises vacuum packaging and CAP as well. A more appropriate definition would be "packaging of food in an atmosphere the composition of which is continuously modified, according to a desired profile". The product is initially packaged in a mixture of gases, the composition of which depends on the product, packaging material, expected shelf life, and storage conditions. Thereafter, modification of the atmosphere in the package is usually the result of respiration of the packaged food, selective permeability of the packaging material, and the presence of "atmosphere modifiers". Modified atmosphere packaging is applied to perishable foods such as meat (Sun and Holley, 2012) and seafood (Speranza *et al.*, 2009), and to products prone to chemical alteration, such as coffee. In the case of perishable items such as meat, fish and fresh fruits and vegetables, the product is kept under refrigeration. The packaging materials used are flexible films. The gases used for composing the initial atmosphere are carbon dioxide, nitrogen, oxygen and, occasionally, carbon monoxide. The products marketed under MAP include dairy products, bakery items, meat and poultry (Stiles, 1991), fish, and fresh fruits and vegetables.

The advantages of MAP in most cases can be attributed to the creation and maintenance of an atmosphere poor in oxygen. This, however, poses a potential danger where the development of anaerobic pathogens is possible. *Clostridium botulinum* is a strict anaerobe and *C. botulinum* type E is, in addition, psychrotropic (Skura, 1991). In such cases, the concentration of oxygen to be left intentionally in the package or the permeability of the film to oxygen is the result of a compromise between the depression of oxidative reactions and avoiding extreme anaerobiosis.

27.3.4 Active packaging

Active packaging is defined as packaging that modifies the condition of the packed food so as to extend its shelf life and improve its safety (Ahvenainen,

2003). In active packaging, active principles are included in the package or incorporated into the packaging material. Some of these principles are "atmosphere modifiers", such as oxygen absorbers, absorbers or generators of carbon dioxide, ethylene absorbers, and moisture regulators (Ooraikul and Stiles, 1991; Rooney, 1995). These are usually included in the package as a separate phase (e.g., the familiar sachets of silica used as moisture absorbers). Iron oxidation is frequently used for the removal of oxygen. Removal of ethylene in packages containing fresh fruits and vegetables is important for preventing accelerated ripening. Ethylene is adsorbed by active carbon or oxidized by potassium permanganate. Cyclodextrin has been investigated as a scavenger of undesirable compounds generated during storage (López-de-Dicastillo *et al.*, 2011). Alternatively, the active principle, which may be a preserving agent (Cooksey, 2005; Chung *et al.*, 2001; Suppakul *et al.*, 2003) or an antioxidant (Gemili *et al.*, 2010; Nerín *et al.*, 2006) is included in the packaging film and slowly released into the atmosphere of the package during storage.

A different type of active package is one that contains a microwave susceptor, for intensification of heating by microwaves. This type is widely used for packaging microwaveable popcorn.

27.3.5 Intelligent packaging

Intelligent packaging (Yam *et al.*, 2005) is a packaging system that monitors the condition of the packaged food and provides information about changes in these conditions during storage. Ideally, such systems should include built-in color-based indicators easily identified by the consumer. The monitored variables may be time–temperature, composition of the gases in the package as an indication of microbial spoilage, pH, leakage, etc.

27.4 Environmental issues

Used food packages constitute a considerable and steadily increasing proportion of solid urban waste. With the ever-rising cost of solid-waste disposal and increasing public consciousness regarding litter and environmental quality, this trend is a source of serious concern both to industry and to governing bodies. In many countries, the industry is held responsible for the problem and compelled to participate in the task of disposal.

Recycling is one of the preferred approaches in disposal. Glass, metal (particularly aluminum) and certain kinds of paper packages (e.g., corrugated carton cases) are being successfully recycled. In the case of flexible polymer films, attempts are being made to increase their degradability in nature. One of the promising directions of active research is the development of polymers degradable by the action of microorganisms (biodegradable), such as starch-based polymers (Ching *et al.*, 1993). The possibility of producing *edible polymer packaging*

(Gonzáles *et al.*, 2011; Han and Gennadios, 2005; Krochta and De Mulder-Johnston, 1997; Wang *et al.*, 2010;) would, of course, obviate the problem of disposal.

References

Ahvenainen, R. (Ed.), 2003. Novel Food Packaging Techniques. Woodhead Publishing, Cambridge, UK.

Ashley, R.J., 1985. Permeability and plastic packaging. In: Comyn, J. (Ed.), Polymer Permeability. Elsevier Applied Science, London, UK.

Bosset, J.Q., Sieber, R., Gallman, P.U., 1994. Influence of light transmittance of packaging materials on the shelf life of milk and milk products – a review. In: Mathlouthi, M. (Ed.), Food Packaging and Preservation. Blackie Academic and Professional, Glasgow, UK.

Ching, C., Kaplan, D., Thomas, E., 1993. Biodegradable Polymers and Packaging. Technomic Publishing Company Inc., Lancaster, PA.

Chung, D., Papadakis, S.E., Yam, K.L., 2001. Release of propyl paraben from polymer coting into water and food simulating solvents for antimicrobial packaging applications. J. Food Proc. Preserv. 25 (1), 71–87.

Coles, R., McDowell, D., Kirwan, M.J., 2003. Food Packaging Technology. Blackwell Publishing, Oxford, UK.

Cooksey, K., 2005. Effectiveness of antimicrobial food packaging materials. Food Addit. Contam. 22 (10), 980–987.

Gemili, S., Yemenicioğlu, A., Alsoy Altnkaya, S., 2010. Development of antioxidant food packaging materials with controlled release properties. J. Food Eng. 96 (3), 326–332.

Giacin, J.R., 1995. Factors affecting permeation, sorption and migration processes in package-product systems. In: Ackermann, P., Jägerstad, M., Ohlsson, T. (Eds.), Foods and Packaging Materials – Chemical Interactions. The Royal Society of Chemistry, Cambridge, UK.

Gonzáles, A., Strumia, M.C., Alvarez Igarzabal, C.I., 2011. Cross-linked soy protein as material for biodegradable films: synthesis, characterization and biodegradation. J. Food Eng. 106 (4), 331–338.

Han, J.H., Gennadios, A., 2005. Edible films and coatings: a review. In: Han, J.H. (Ed.), Innovations in Food Packaging. Elsevier Academic Press, New York, NY.

Hanlon, J.F., Kelsey, R.K., Forcinio, H.E., 1998. Handbook of Package Engineering, third ed. Technomic Publishing Co., Lancaster, PA.

Hintlian, C.B., Hotchkiss, J.H., 1986. The safety of modified atmosphere packaging: a review. Food Technol. 40 (12), 70–76.

Jenkins, W.A., Harrington, J.P., 1991. Packaging Foods with Plastics. Technomic Publishing Co., Lancaster, PA.

Johansson, F., Leufven, A., 1994. Food packaging polymer films as aroma vapor barriers at different relative humidities. J. Food Sci. 59 (6), 1328–1331.

Krochta, J.M., De Mulder-Johnston, C., 1997. Edible and biodegradable polymer films: challenges and opportunities. Food Technol. 51 (2), 61–74.

Lopez, A., 1981. eleventh ed. A Complete Course in Canning, vol. 1. *The Canning Trade*, Baltimore, MD.

López-de-Dicastillo, C., Catalá, R., Gavara, R., Hernández-Muñoz, P., 2011. Food applications of active packaging EVOH films containing cyclodextrins for the preferential scavenging of undesirable compounds. J. Food Eng. 104 (3), 380–386.

Mastromatteo, M., Del Nobile, M.A., 2011. A simple model to predict the oxygen transport properties of multilayer films. J. Food Eng. 102 (2), 170–176.

Miltz, J., 1992. Food packaging. In: Heldman, D.R., Lund, D.B. (Eds.), Handbook of Food Engineering. Marcel Dekker, New York, NY.

Nerín, C., Tovar, L., Djenane, D., Camo, J., Salafranca, J., Beltrán, J.A., et al., 2006. Stabilization of beef meat by a new active packaging containing natural antioxidants. J. Agric. Food Chem. 54 (20), 7840–7846.

Ooraikul, B., Stiles, M.E. (Eds.), 1991. Modified Atmosphere Packaging of Foods. Ellis Horwood, Chichester, UK.

Robertson, G.L., 1993. Food Packaging, Principles and Practice, first ed. Marcel Dekker, New York, NY.

Robertson, G.L., 2005. Food Packaging, Principles and Practice, second ed. CRC Press, New York, NY.

Rooney, M.L., 1995. Active Food Packaging. Blackie Academic & Professional, London, UK.

Skura, B.J., 1991. Modified atmosphere packaging of fish and fish products. In: Ooraikul, B., Stiles, M.E. (Eds.), Modified Atmosphere Packaging of Foods. Ellis Horwood, Chichester, UK.

Speranza, B., Corbo, M.R., Conte, A., Sinigaglia, M., Del Nobile, M.A., 2009. Microbiological and sensorial quality assessment of ready-to-cook seafood products packaged under modified atmosphere. J. Food Sci. 74 (9), M473–M478.

Stiles, M.E., 1991. Modified atmosphere packaging of meat, poultry and their products. In: Ooraikul, B., Stiles, M.E. (Eds.), Modified Atmosphere Packaging of Food. Ellis Horwood, Chichester, UK.

Sun, X.D., Holley, R.A., 2012. Antimicrobial and antioxidative strategies to reduce pathogens and extend the shelf life of fresh red eats. Comprehen. Rev. Food Sci. Food Saf. 11 (4), 340–354.

Suppakul, P., Miltz, J., Sonneveld, K., Bigger, S.W., 2003. Active packaging technologies with an emphasis on antimicrobial packaging and its applications. J. Food Sci. 68 (2), 408–420.

Wang, L., Auty, M.A.E., Kerry, J.P., 2010. Physical assessment of composite biodegradable films manufactured using whey protein isolate, gelatin and sodium alginate. J. Food Eng. 96 (2), 199–207.

Yam, K.L., Takhistov, P.T., Miltz, J., 2005. Intelligent packaging: concepts and applications. J. Food Sci. 70 (1), R1–10.

CHAPTER

Cleaning, Disinfection, Sanitation

28

28.1 Introduction

The importance of hygiene in food processing cannot be over-emphasized. In fact, hygiene is, ideally, the chief concern of all personnel dealing with food and food processing. Hygiene considerations affect all the phases of food processing, including decisions in plant and process design, equipment selection, specification and handling of raw materials and packaging, maintenance of the physical plant and its environment, selection and training of the personnel at all levels, the overall conduct of activities, and the "culture" in a food processing plant. The importance attributed to hygiene in a food processing company is reflected in the quality and safety of the products. Failure to maintain strict sanitary conditions has been the main reason for serious commercial setbacks in the food industry.

One of the problems in the application of engineering methodology to hygiene is the difficulty in expressing "dirt" and "cleanliness" in objective, exact and quantitative terms. Despite considerable progress in establishing quantitative standards and a rational approach to hygiene, subjective attitudes based on personal and collective customs, culture and esthetics still have an important part in the daily practice.

The objective of "cleaning" is the removal of "dirt". Dirt or soil has been defined as "substance in the wrong place" (Plett, 1992). Residues of a perfectly edible product on a conveyor belt, a film of milk on the walls of an empty tank, or scale on the surface of a heat exchanger are "dirt" and must be removed, just as mud on a tomato or sand on a strawberry. "Disinfection" refers to the destruction of microorganisms. "Sanitation" is the maintenance of hygienic conditions through cleaning, disinfection and preventive management (Marriott, 1985).

Cleaning operations are an integral part of the production process. Consequently, the cost of cleaning is an integral component of the cost of production. One of the purposes of rationalization of sanitation procedures is to develop *optimization* methods. Optimization of cleaning operations refers not only to their economic cost but also to their environmental impact. Thus, topics such as water- and energy-saving and recycling of solid waste are included in the objectives of optimization.

Food Process Engineering and Technology. DOI: http://dx.doi.org/10.1016/B978-0-12-415923-5.00028-9
© 2013 Elsevier Inc. All rights reserved.

In a food processing facility, the "objects" subjected to cleaning or disinfection include:

1. Raw materials
2. Vehicles used to transport raw materials and products
3. Process equipment, tools and contact surfaces
4. Packaging materials
5. Employees (personal hygiene, garments, etc.)
6. Buildings (floors, walls, windows, piping, etc.) and their surroundings
7. Storage areas
8. Water (both incoming and outgoing, including waste water and effluents)
9. Air (both incoming and outgoing) and gaseous emanations (odor abatement).

Until recently, food plants utilized enormous quantities of water for cleaning. Most of the used water was not recycled but discharged as waste. Another resource that was also used indiscriminately and in exaggerated quantity was energy. Today, the laws and economics compel the food industry to develop more rational technologies of cleaning, less wasteful in water and energy and more environmentally friendly.

28.2 Cleaning kinetics and mechanisms

Various kinetic models have bee applied to cleaning operations. The most commonly accepted model assumes first-order reaction kinetics (Loncin and Merson, 1979), as shown in Eq. (28.1):

$$-\frac{dm}{dt} = km \Rightarrow \ln\frac{m}{m_0} = -kt \tag{28.1}$$

where m = mass of soil per unit area, and k = a constant that depends on the properties of the contaminant, the surface on which it sits, the detergent, the temperature and the intensity of shearing action.

28.2.1 Effect of the contaminant

The types of soil and fouling that are of relevance in food processing are too numerous for systematic classification. Some components in food process fouled layers have been listed by Plett (1992). They include sugars (soluble in water, easily removed), fat (insoluble in water, saponified by alkali, removable with surfactants), proteins (difficult to remove with water, easily removed by alkali) and mineral salts (easily removed by water if monovalent, removed with acid if polyvalent). Soil on food, on the other hand, includes many other types: earth and mud, dust, chemical residues and, first and foremost, microorganisms as individual cells or as biofilms.

Soil is usually not uniform. If it has been deposited in successive layers, such as in the case of fouling on a heat surface area, each layer may have a different composition, structure and, certainly, different age. Furthermore, the first layer, attached directly to the clean surface, may be bound differently in comparison to subsequent layers deposited on other layers of soil. In this case, the kinetics of cleaning can be expected to change as the contaminant is removed. Bourne and Jennings (1963) studied the kinetics of removal of a fat model (tristearine) from a stainless steel strip with 0.03-M sodium hydroxide. They found that the kinetics could be best represented by two first-order processes, one fast and the other slow. They concluded that the film of lipid behaved like two species of contaminant and that the cleaning action itself consisted of two mechanisms, namely a time-dependent flow mechanism and a time-independent surface tension controlled mechanism.

First-order kinetics has also been applied to the detachment of microorganism cells from contact surfaces. Demilly *et al.* (2006) studied the removal of yeast cells from stainless steel surfaces by rinsing with a buffered solution. The percentage of cells detached by the cleaning action was found to increase with time, to attain a plateau at about 95%. The data were fitted with first-order kinetics, assuming a "saturation level" or "detachment efficiency" which is the maximum percentage of cells detached.

The kinetics of cleaning the deposit formed on a stainless steel tube fouled with a whey protein solution, using a solution of sodium hydroxide as detergent, was studied by Gillham *et al.* (1997). The thickness of the remaining deposit, determined using SEM imaging, was plotted against the time of contact with the detergent. The plot showed considerable scatter, which was interpreted by the authors as an indication of the "random nature of cleaning".

The removal of simple soluble deposits (e.g., a crystalline scale) can be described by the diffusion controlled dissolution–mass transfer model (see Chapter 14, Section 14.4.2), as shown by Davies *et al.* (1997) and Schlüsser (1976).

The adhesion of microorganisms to surfaces is of obvious relevance to health, but the subject of the attachment of microorganisms to the surface of food and food processing equipment has been also researched extensively (Flint *et al.*, 1997; Notermans *et al.*, 1991). The initial adhesion of microorganisms to various supports may be governed by non-specific Van der Waals forces or by hydrophobic bonding. Hydrophobicity of the support and of the cell surface is therefore one of the most important criteria for the initial adhesion (Kang *et al.*, 2004). Often, the adhesion of microorganisms to inert surfaces is made possible by the presence of a film of previously adsorbed organic molecules on the surface. The strength of adhesion is also affected by the morphology of the cell, with irregular cells being more strongly attached than ovoid cells (Gallardo-Moreno *et al.*, 2004). Simultaneously with adhesion, microorganisms are removed from the surface by various factors exerting shear. Here again, the strength of adhesion is decisive in determining the rate of detachment. Given favorable conditions, surface growth and accumulation of microorganisms may occur. At this stage, many types of microorganisms tend to form *biofilms* and intensify their adhesion to the surface by secreting extracellular polymeric substances (EPS). Biofilms can be viewed as micro-ecosystems with a certain degree of internal organization and communication, just as in colonies of certain microorganisms (Wolfaardt *et al.*, 1994). Once formed, biofilms constitute a type of contaminant that is most difficult to remove (Escher and Characklis, 1990). Microorganisms in a biofilm are more resistant to stress factors and disinfectants than their free (planktonic) counterparts (Joseph *et al.*, 2001; Rode *et al.*, 2007). Furthermore, biofilms are found to survive desiccation far better than free cells (Hansen and Vogel, 2011).

28.2.2 Effect of the support

Thermodynamically, adhesion of soil to a surface is exothermic and its removal is endothermic. Surface energy is therefore a critical factor in cleaning. Cleaning kinetics depends on both the nature of the surface and the geometry of the support (Bitton and Marshall, 1980). Smooth surfaces, such as stainless steel and glass, are easier to clean than rough, porous supports such as rubber and wood. Stainless steel is the preferred construction material in the food industry, not only because of its smooth surface but also because of its resistance to corrosion and to the effect of cleaning fluids. Stainless steel is available with different types of surface finishing. Opinions differ as to the effect of surface smoothness of stainless steel on the ease of cleaning, with some reports showing no effect at all and others indicating a positive correlation between smoothness and the completeness of cleaning (Plett, 1992). Curiously, Demilly *et al.* (2006) observed faster detachment of yeast cells from etched rather than from mirror-polished stainless steel.

As explained in the previous section, hydrophobicity is one of the most decisive properties of surfaces with respect to the adhesion of microorganisms. This is particularly important with polymeric surfaces. Cleaning of polymer surfaces is particularly relevant in connection with membrane processes (Kang *et al.*, 2003). Hydrophilic membranes have been found to foul less than hydrophobic ones (Ridgway, 1988). Biofilms have been found to adhere to polymeric surfaces better than to stainless steel or concrete (Joseph *et al.*, 2001).

The geometry of contact surfaces and flow channels is of utmost importance with respect to cleaning. Hard to reach zones that are hidden from inspection and cleaning are among the main sources of contamination in food plants. With respect to flow channels, the most important consideration is to avoid "dead zones" where flow is not sufficiently rapid to prevent build-up of contamination.

28.2.3 Effect of the cleaning agent

In general terms, cleaning agents remove soil through one of the following mechanisms, or by their combined effect:

1. Mechanical action
2. Physico-chemical alteration of the contaminant
3. Reducing the surface tension of water.

Mechanical removal of soil is performed in many processes, such as screening, air filtration, cyclone separation, electrostatic removal of dust, magnetic separation, washing with high-pressure water jets, etc. The classical cleaning media for this type of cleaning are water and air. Recently, cleaning systems using high-velocity jets of solid or liquid carbon dioxide have been introduced to the food industry, as a water-saving, effluent-free, environmentally friendly process.

Physico-chemical alteration mechanisms may be classified into purely physical actions and chemical reactions. Physical interactions of the cleaning agent with soil include wetting, softening (plasticizing), swelling, heating, dissolution, colloidal dispersion (e.g., emulsifying) and foaming. Water and steam are the main cleaning agents that work by this mechanism. The use of steam guns is not recommended because superheated steam tends to "fix" the soil by drying it.

Alteration of the soil by deep chemical reaction includes saponification of lipids, hydrolysis of proteins and decomposition of salts (e.g., effect of acid on calcium carbonate scale) and, of course, destruction of microorganisms by disinfectants. The most commonly used detergents are dilute NaOH and dilute acids. Where necessary, alkaline media milder than NaOH, such as sodium carbonate or trisodium phosphate, are used. Acids used for cleaning include nitric, phosphoric and carboxylic (acetic, citric, lactic, etc.) acids. Acting alone at the customary concentration, acids are not effective in removing fat and protein deposits. For cleaning fouled surfaces of evaporators and heat exchangers in dairy plants, a two-step treatment is recommended. The first step is an acid wash, resulting in the removal of the mineral layer and causing the protein to swell and detach from the support. The second step is treatment with alkali, in order to gel, hydrolyze and finally dissolve the soil. Enzymes (proteases, lipases) are used in some specific cleaning operations (e.g., cleaning drain pipes). Disinfectant action will be discussed in a subsequent section.

As explained above, removal of soil from a surface requires energy. The energy requirement can be drastically reduced by the use of surface active agents (surfactants). Surfactants are amphiphilic compounds, having one or more hydrophilic groups attached to a hydrophobic chain. They are classified according to the nature of their hydrophilic group as follows:

1. *Anionic surfactants*. Their polar end is a dissociable anionic group such as a sulfonate or the salt of a fatty acid (plain soap). They are active in alkaline to neutral media, and their activity is depressed by salts. Due to their low cost and high effectiveness, anionic surfactants are the most extensively used group.

2. *Cationic surfactants*. These are solubilized by an amine group. Quaternary ammonium derivatives are also potent bactericides. Cationic surfactans are active at low pH.
3. *Non-ionic surfactans*. These are condensates of ethylene oxide (providing the polar end) with long-chain alcohols (providing the hydrophobic end).

The cleaning action of surfactants is based on alteration of the water–soil interface. Consider the behavior of a surfactant in water, as shown in Figure 28.1.

The surfactant molecules are represented with their hydrophilic end as a head and their hydrophobic chain as a tail. At the free surface of the water, the surfactant molecules accumulate with their hydrophilic head towards the water and their hydrophobic tail away from the water. Thus, the surface of the water is rendered less polar and can wet a non-polar soil such as fat effectively, and disintegrate a film of such soil. If the concentration of the surfactant is sufficiently high (say, above 0.1%), its molecules will self-assemble in micelles around particles of soil. The soil will thus be emulsified and rinsed away.

By increasing the wetting action of the cleaning fluid, surfactants also improve its cleaning action by facilitating its penetration power into narrow passages.

28.2.4 Effect of the temperature

We refer here to the effect of temperature on the removal of soil, and not to the thermal inactivation of microorganisms. As a rule, cleaning is more effective and rapid at high temperatures. The cleaning rate constant k in Eq. (28.1) follows Arrhenius' Law "over large intervals of temperature" (Loncin and Merson, 1979). Because of the complex nature of cleaning, the energy of activation varies depending on the rate-limiting mechanism. If the rate of cleaning is controlled by mass transfer, then the energy of activation is low, corresponding to Q_{10} values (see Chapter 4, Section 4.2.3) around 1. On the other hand, if the cleaning is mainly the result of chemical reactions, then Q_{10} values of 2 or higher can be expected (Plett, 1992).

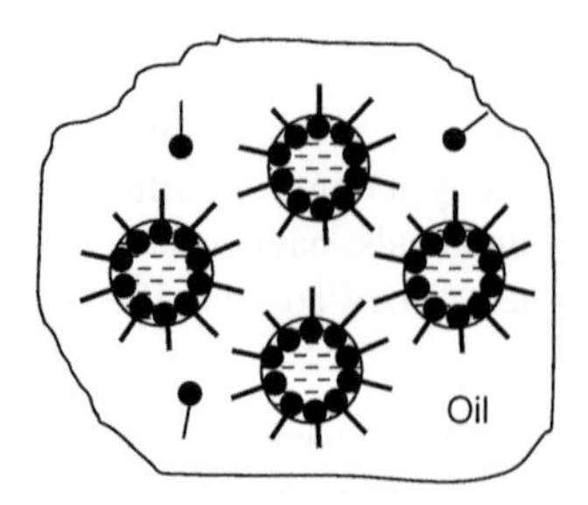

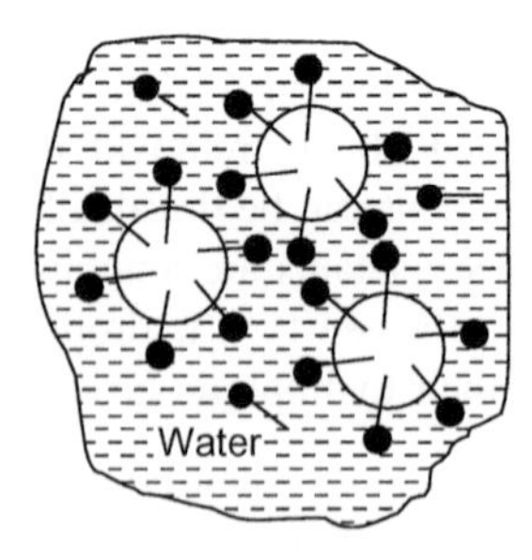

FIGURE 28.1

Mechanism of surfactant action in two media.

28.2.5 Effect of mechanical action (shear)

In cleaning, mechanical action is applies by fluid flow, high-impact fluid jets, scrapers, brushes, vibration, etc. The intensity of shear needed to remove soil is determined in a number of different experimental set-ups and expressed either as a critical shear stress at the wall or as a critical Reynolds number. Demilly *et al.* (2006), for example, found that cell detachment in their set-up did not occur at shear stress below 15 Pa. Above that critical value, the rate of cleaning increased linearly with shear stress. Loncin and Merson (1979) recommend a minimum Reynolds number of 5000–6000 while Jennings *et al.* (1957) claim that circulation cleaning is effective only above a Reynolds number of 25 000.

Tanks and other cylindrical vessels are usually cleaned by high-pressure sprays. In most installations, perforated spray balls are installed permanently inside the vessel. Removal of soil occurs as a result of perpendicular impact and tangential shear, and depends on the kinetic energy of the spray drop, hence on the mass of the drop and its velocity. Piping and tubing are cleaned by circulating the cleaning fluid through it under turbulent conditions. (circulation cleaning).

Ultrasonic vibration, an expensive technique, is used to clean small parts and tools.

28.3 Kinetics of disinfection

Microorganisms may be destroyed by heat, ionizing irradiation, ultraviolet irradiation or chemical agents. Here, we refer to disinfection as the destruction of microorganisms by chemical agents, also termed sanitizers.

The number of chemicals that could potentially serve as disinfectants is considerable. Only a few, however, are allowed as sanitizers for foods and surfaces in contact with food. Ideally, the sanitizer should be:

1. Capable of high microbial destruction against a wide range of microorganisms
2. Stable in concentrated and dilute form
3. Safe to use with food, and non-toxic
4. Non-irritant, non-corrosive
5. Able to leave residues that can be removed easily and completely by rinsing
6. Environmentally friendly, and quickly degraded in effluents
7. Easy to use, to measure and to control
8. Inexpensive and easily available.

The basic kinetics of chemical disinfection is usually assumed to obey a first-order rate equation, similar to the kinetics inactivation of microorganisms by heat or antimicrobial agents:

$$-\frac{dN}{dt} = kN \Rightarrow \ln\frac{N}{N_0} = kt \qquad (28.2)$$

The rate constant depends on the microorganism, the disinfectant, the concentration of the disinfectant, the pH, the temperature, etc. The relationship between the disinfectant concentration and the contact time necessary to achieve a determined reduction in the number of microorganisms (e.g., 95%) is given by the century-old Watson–Chick Law (Chick, 1908; Watson, 1908):

$$C^n t = K \tag{28.3}$$

where C = concentration of the disinfectant, and K and n = constants.

The power n is often near 1. With n = 1 the Chick–Watson equation is reduced to the first-order equation and the condition for reaching a given reduction ratio in the number of microorganisms is to maintain the product $C \cdot t$ constant. In other words, in order to compensate for a reduction of sanitizer concentration to half, the contact time must be doubled, and *vice versa*.

The disinfectants most commonly utilized in the food industry are chlorine and its derivatives, iodine derivatives, ozone, hydrogen peroxide, and quaternary amines.

Chlorine is among the oldest sanitizers in use. It is inexpensive and widely available. It is commercialized as a gas or a liquefied gas. When added to water it is converted to hypochlorous acid, HOCl. Hypochlorous acid and hypochlorites are the most active chlorine-based sanitizers. The active form is the undissociated acid. Chlorine is therefore more active at low pH. Chlorine is a highly reactive substance. It combines with organic substances present in the system. Therefore, it is advisable to wash away the organic impurities before sanitizing with chlorine. At any rate, the factor that must be controlled is not the total concentration of chlorine but the concentration of active chlorine. In most applications, rinsing after sanitizing with chlorine is not necessary. Automatic chlorinators that dispense the necessary quantity of chlorine to maintain the desired level of active chlorine are available. It is also possible to produce chlorine *in situ* by electrolysis of sodium chloride. Electrolytic chlorinators are used in relatively small applications. Chlorine is corrosive, and may cause pitting in stainless steel under conditions of prolonged contact. One of the classical uses of chlorine in the food industry is for sanitizing the water used for cooling cans after thermal processing.

Ozone is a very powerful sanitizer (Kim *et al.*, 1999) but is still relatively expensive. It is not available commercially as such, but must be produced *in situ* by passing a current of air through cold corona discharge, causing some of the oxygen to convert to ozone. As soon as it is formed, however, ozone starts to be converted back to oxygen. Its half-life is in, in most cases, a few minutes. Because of its overall oxidation power, ozone is also effective in destroying malodors. Ozone is applied to raw materials by contact with ozonized water (Alexandre *et al.*, 2011; Güzel-Seydim *et al.*, 2004). An interesting application is the use of ice made with ozonized water for extending the shelf life of chilled fresh fish. The factor that has limited, so far, the application of ozone disinfection in the food industry has been the unavailability of economic, large-scale ozone generators, but steady progress is being made in that area. Another potent oxidant, hydrogen peroxide, is the preferred sanitizer for the disinfection of empty packages before aseptic filling.

The use of other gaseous disinfectants (fumigants), such as methyl bromide and ethylene oxide, is now limited or prohibited by regulations.

28.4 Cleaning of raw materials

In a food plant, soil on the raw materials is among the main sources of contamination. Cleaning is usually the first operation to which the raw material is exposed. In fact, it is a good idea physically to separate the plant area where initial cleaning takes place from the rest of the plant. Some of the "dirty" areas that should be separated by physical partitions are:

- The area where fruits and vegetables are peeled, soaked, washed and inspected (usually outdoors)
- The area where eggs are broken
- The area where animals are slaughtered, skinned, de-feathered and eviscerated
- The area where fish is de-scaled, washed and gutted.

Raw material cleaning methods fall into two categories, namely wet cleaning and dry cleaning. Wet cleaning comprises soaking, washing, rinsing and sanitizing. Pre-soaking of vegetables helps loosen the soil, and serves to remove stones and some of the foreign materials that may damage the equipment. The soaker-washer (Figure 28.2) is often the first piece of equipment in a fruit or vegetable processing line. It consists of a soaking tank, followed by an elevator equipped with water sprays. It is often followed by a rotary drum washer (Figure 28.3) and an inspection table. It is important to replace a certain proportion of the dirty soak water in the tank of the soaker-washer with clean water in order to prevent recontamination. It is often observed, for example, that soaking may increase the mold count of tomatoes if an adequate rate of water change is not maintained.

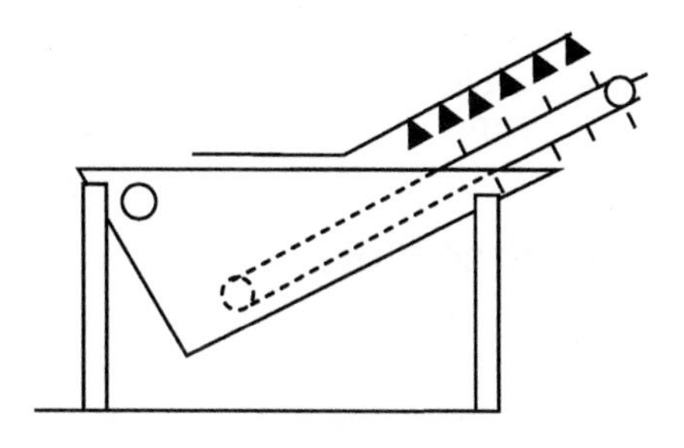

FIGURE 28.2

Schema of a soaker-washer.

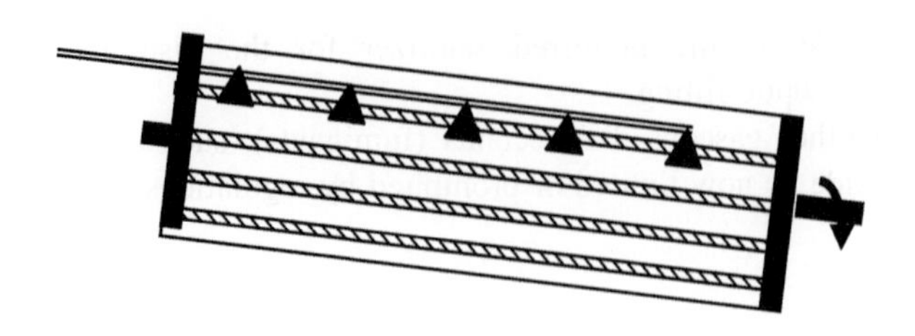

FIGURE 28.3

Schema of a rotary washer.

Washing vegetables employs very large quantities of water and produces a large volume of effluents. The obvious approach for reducing usage of clean water is reuse of the effluents after sedimentation, filtration and sanitation. Chlorine is the most common sanitizer used. The countercurrent principle is applied – i.e., the cleanest water is used to wash the cleanest material.

It should be realized that washing and sanitizing will be the final and only process of preservation applied to some foods, such as pre-packaged fresh produce. Insufficient cleaning and sanitizing of such product may therefore result in serious food safety deficiency. Thorough cleaning of leaf vegetables is particularly difficult because of dirt and biofilms that may be hidden in hard-to-reach places.

Dry cleaning is applied primarily to grain products, and is based mainly on differences in density and/or particle size. One application is the separation of dust and chaff from grain with a vertical current of air, or the removal of leaf fragments from olives or other fruits. Dry cleaning also includes removal of tramp metal and other foreign materials. The destruction of insects and their eggs by impact mills (Entoleters) may be considered a dry-cleaning operation.

28.5 Cleaning of plants and equipment

The establishment of a detailed, clear and precise program of plant and equipment cleaning procedures and its strict application should be the first condition for the management of hygiene in a food processing facility.

Cleaning strategies may be classified into two groups: cleaning out-of-place (COP) and cleaning in-place (CIP).

28.5.1 Cleaning out-of-place (COP)

In a COP operation, the equipment is disassembled as much as needed to expose all possibly soiled surfaces. The parts are rinsed, cleaned, sanitized and reassembled. The actual cleaning operations may be manual or machine-assisted. Tubing and equipment parts are cleaned by the action of turbulent fluid flow and brushes. Permanent equipment, such as large tanks, elevators, conveyors, etc., is

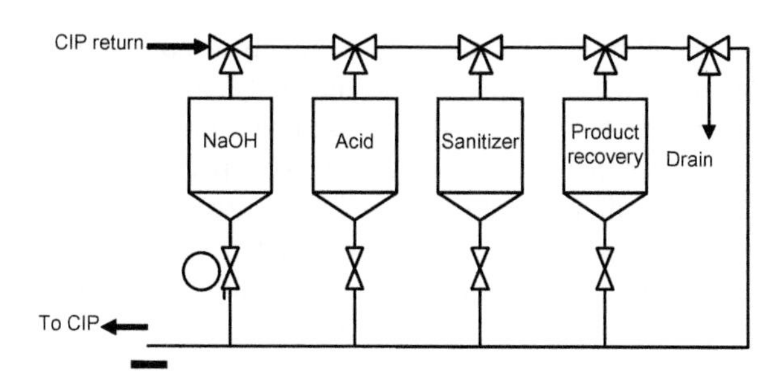

FIGURE 28.4

Schema of a reuse CIP system.

cleaned individually, using high-pressure nozzles for hot water, detergent and sanitizer. The COP units contain reservoirs for cleaning solutions, recirculation pumps, nozzles (hot and cold liquid guns), and devices for heating the cleaning solution. To be applicable, the COP strategy requires, in the first place, that the production line can be disassembled into relatively small and numerous parts. This condition has obvious negative economic consequences both in capital investment and in labor cost.

28.5.2 Cleaning in-place (CIP)

Today, in many food industries, CIP is not only a cleaning method but also a strategic decision built into the design of the plant and its individual parts. Equipment manufacturers today offer units that are specifically designed and built for CIP operation. Basically, CIP is a hydrodynamic cleaning process where rinsing, cleaning and sanitizing fluids circulate along the path of the product and provide the detergency as well as the mechanical action needed for the removal of soil, without dismantling the line. We distinguish between two main types of CIP systems: single-use and reuse systems.

In single-use CIP, the water, detergent solutions and sanitizers are recycled through the system in a pre-determined sequence for only one cleaning cycle and then discharged. In order to attain the degree of turbulence required for efficient cleaning, CIP systems operate at a very high flow-rate, resulting in high water usage and a large effluent volume. Some degree of water saving may be achieved by collecting the water from the previous rinsing and using it for the pre-rinse of the next cycle. Single-use systems, however, are less expensive and require little space.

In reuse CIP, the cleaning fluids from one cycle are stored and reused in subsequent cycles, applying, as much as possible, the countercurrent principle by judicious programming of the sequences. Reuse CIP systems are more expensive and require more floor space (Figure 28.4).

A basic CIP operation can be run by a relatively simple system consisting of storage and recirculation tanks, pumps and manually operated valves. Some systems are indeed simple. They are sold pre-assembled on a skid, ready to be connected. In larger, more sophisticated plants, CIP systems are much more complex. They usually feature a digitally programmable control system that governs and records the entire operation, air- or power-actuated control valves, measuring instruments for evaluating the quality of the cleaning fluids and the effluents, locks and alarms to prevent wrong types of flow (e.g., flow of detergent into the product), etc.

The cleaning efficiency of CIP is considerably enhanced by applying a pulsed rather than a steady flow (Gillham *et al.*, 2000).

28.6 Cleaning of packages

Washing used and returned bottles was once a major cleaning problem in food plants. The bottles had to be washed in hot alkali, then rinsed and inspected. The cost of collection, classification, transport and cleaning, added to breakage and zero-tolerance inspection in order to prevent faulty cleaning, made the reuse of glass containers an almost unacceptable option. New glass is usually cleaned with an air blast, often followed with a spray of clean water in the inverted position, draining, and inspection. Empty metal cans are cleaned by an essentially similar procedure.

Cleaning of packages intended for aseptic filling has been discussed in Chapter 12, Section 12.4.

28.7 Odor abatement

The laws compel the industry to clean the air and gaseous emanations released to the atmosphere from dust, dirt, toxic components and odors. Note that the term "odor" has been used and not "malodor", because an odor that is usually considered to be "pleasant" (e.g., coffee) will still be treated as a "nuisance" by a court of law.

The sources of odor from the food industry are of three types:

1. Odors generated in open areas, such as the court, floors, waste-water treatment basins, and released to the open air. The obvious way to prevent this kind of odor is better hygiene.
2. Odors emanating from materials, products and by-products. Examples include untreated poultry offal and spent coffee pulp from extractors. This problem is alleviated by containment and rapid removal of the objectionable materials from the plant.

3. Gaseous emanations from localized sources such as exhaust pipes, chimneys, etc. As an example, consider the emanations from the autoclave used to cook offal in a rendering plant.

There are several methods for eliminating odors of the latter type:

- *Scrubbing*. The gaseous emanations are continuously scrubbed with water or another liquid. Typically, a scrubber is a vertical adsoption column in which the odorous gas is washed in countercurrent fashion. The scrubbing liquid may be water (for water-soluble odors), an alkali or an acid (for odors of acid or alkaline nature, respectively), or an oxidizing agent.
- *Catalytic combustion*. The gas is heated ad passed over a catalytic converter, where the odorous organic substances are burned.
- *Biofilters*. Biofilters are porous supports on which a flora of microorganisms has been cultivated. The microorganisms use the odorous organic matter as a substrate, and decompose it. Biofilms are not suitable for treating large quantities of gaseous emanations heavily contaminated with odorous components.

References

Alexandre, E.M.C., Santos-Pedro, D.M., Brandão, T.R.S., Silva, C.L.M., 2011. Influence of aqueous ozone, blanching and combined treatments on microbial load of red bell peppers, strawberries and watercress. J. Food Eng. 105 (2), 277–282.

Bitton, G., Marshall, K.C. (Eds.), 1980. Adsorption of Microorganisms to Surfaces. John Wiley & Sons, Inc., New York, NY.

Bourne, M.C., Jennings, W.G., 1963. Existence of two soil species in detergengy investigations. Nature 197, 1003–1004.

Chick, H., 1908. An investigation of the laws of disinfection. J. Hyg. 8, 92–158.

Davies, M.J., Procter, N.J.L., Wilson, D.I., 1997. Cleaning of crystalline scale from stainless steel surfaces. In: Jowitt, R. (Ed.), Engineering and Food at ICEF7. Academic Press, Sheffield, UK.

Demilly, M., Bréchet, Y., Bruckert, F., Boulangé, L., 2006. Kinetics of yeast detachment from controlled stainless steel surfaces. Colloids Surf. B Biointerfaces 51 (1), 71–79.

Escher, A., Characklis, W.G., 1990. Modeling the initial events in biofilm accumulation. In: Characklis, W.G., Marshall, K.C. (Eds.), Biofilms. Wiley, New York, NY.

Flint, S.H., Bremer, P.J., Brooks, J.D., 1997. Biofilms in dairy manufacturing plant. Biofouling 11, 81–97.

Gallardo-Moreno, A.M., González-Martín, M.L., Bruque, J.M., Pérez-Giraldo, C., 2004. The adhesion strength of *Candida parapsilosis* to glass and silicone as a function of hydrophobicity, roughness and cell morphology. Colloids Surf. A Physicochem. Eng. Asp. 249 (1–2), 99–103.

Gillham, C.R., Belmar-Beiny, M.T., Fryer, P.J., 1997. The physical and chemical factors affecting of dairy process plants. In: Jowitt, R. (Ed.), Engineering and Food at ICEF7. Academic Press, Sheffield, UK.

Gillham, C.R., Fryer, P.J., Hasting, A.P.M., Wilson, D.I., 2000. Enhanced cleaning of whey protein soils using pulsed flows. J. Food Eng. 46 (3), 199–209.
Güzel-Seydim, Z.B., Greene, A.K., Seydim, A.C., 2004. Use of ozone in the food industry. LWT 37 (4), 453–460.
Hansen, L.T., Vogel, B.F., 2011. Desiccation of adhering and biofilm *Listeria monocytogenes* on stainless steel: survival and transfer to salmon products. Int. J. Food Microbiol. 146 (1), 88–93.
Jennings, W.G., McKillop, A.A., Luick, J.R., 1957. Circulation cleaning. J. Dairy Sci. 40, 1471.
Joseph, B., Otta, S.K., Karunasagar, I., Karunasagar, I., 2001. Biofilm formation by Salmonella spp. on food contact surfaces and their sensitivity to sanitizers. Intl. J. Food Microbiol. 64 (3), 367–372.
Kang, S., Agarwal, G., Hoek, E.M.V., Deshusses, M.A., 2003. Initial adhesion of microorganisms to polymeric membranes. Proceedings of the International Conference on MEMS, NANO and Smart Systems. 2003, 262.
Kim, J.G., Yousef, A.E., Dave, S., 1999. Application of ozone for enhancing the microbiological safety and quality of foods: a review. J. Food Protec. 62 (9), 1071–1087.
Loncin, M., Merson, R.L., 1979. Food Engineering, Principles and Selected Applications. Academic Press, New York, NY.
Marriott, N.G., 1985. Principles of Food Sanitation. Avi Publishing Company, Westport, CT.
Notermans, S., Dormans, J.A., Mead, G.C., 1991. Contribution of surface attachment to the establishment of microorganisms in food processing plants: a review. Biofouling 5, 21–36.
Plett, E., 1992. Cleaning and sanitation. In: Heldman, D.R., Lund, D.B. (Eds.), Handbook of Food Engineering. Marcel Dekker, New York, NY.
Ridgway, H.F., 1988. Microbial adhesion and bio-fouling of reverse osmosis membranes. In: Pakekh, B.S. (Ed.), Reverse Osmosis Technology: Applications for High Purity Water Production. Marcel Dekker, New York, NY.
Rode, T.M., Langsrud, S., Holck, A., Møretrø, T., 2007. Different patterns of biofilm formation in Staphylococcus aureus under food-related stress conditions. Intl. J. Food Microbiol. 116 (3), 372–383.
Schlüsser, H.J., 1976. Kinetics of cleaning processes at solid surfaces. Brauwissenschaft 29 (9), 263–268.
Watson, H.E., 1908. A note on the variation of the rate of disinfection with the change of the concentration of disinfectant. J. Hyg. 8, 536–542.
Wolfaardt, G.M., Lawrence, J.R., Robarts, R.D., Caldwell, S.J., Caldwell, D.E., 1994. Multicellular organization in a degradative biofilm community. Appl. Environ. Microbiol. 60, 434–446.

Appendix

Table A.1 Common Conversion Factors

To Convert From	To	Multiply By
Length		
Foot (ft)	Meter	0.3048
Inch	Meter	0.0254
Mass		
Pound (lb)	Kilogram (kg)	0.4536
Ounce	Kg	0.0283
Area		
ft^2	m^2	0.0929
$inch^2$	m^2	0.645×10^{-3}
Volume		
ft^3	m^3	0.0283
$inch^3$	m^3	16.38×10^{-6}
Density		
$lb \cdot ft^{-3}$	$kg \cdot m^{-3}$	16.018
Pressure		
Atm	Pascal (Pa)	0.1013×10^6
$lb_f \cdot inch^{-2}$ (psia)	Pa	6.894×10^3
mm Hg (torr)	Pa	133.32
Force		
lb_f	Newton (N)	4.448
dyne	N	1.000×10^{-5}
Work		
ft lb_f	Joule (J)	1.3558
erg	J	1.000×10^{-7}
Power		
ft lb_f/s	Watt (W)	1.3558
hp	W	745.7
Energy		
kcal	Joule (J)	4186.8
BTU	J	1.055×10^3
Viscosity		
poise	$Pa \cdot s$	0.100
centipoise (cP)	$Pa \cdot s$	1.000×10^{-3}
Diffusivity		
$ft^2 \cdot h^{-1}$	$m^2 \cdot s^{-1}$	25.81×10^{-6}

(Continued)

Table A.1 *Continued*

To Convert From	To	Multiply By
Thermal conductivity		
$Btu \cdot ft^{-1} \cdot h^{-1} \cdot F^{-1}$	$W \cdot m^{-1} \cdot K^{-1}$	1.731
Heat transfer coefficient		
$Btu \cdot ft^{-2} \cdot h^{-1} \cdot F^{-1}$	$W \cdot m^{-2} \cdot K^{-1}$	5.678
Ionizing radiation dose		
rad	Gray (Gy)	0.01

Table A.2 Typical Composition of Selected Foods

	Composition per 100 g Edible Portion			
Food	**Water (g)**	**Protein (g)**	**Fat (g)**	**Carbohydrate (g)**
Fruits and nuts				
Apples	84.1	0.3	0.4	14.9
Bananas	74.8	1.2	0.2	23.0
Grapes	81.6	0.8	0.4	16.7
Oranges	87.2	0.9	0.2	11.2
Peaches	86.9	0.5	0.1	12.0
Almonds	4.7	18.6	54.1	19.6
Peanuts, roasted	2.6	26.9	44.2	23.6
Peanut butter	1.7	26.1	47.8	21.0
Walnuts	3.3	15.0	64.4	15.6
Vegetables				
Artichoke	83.7	2.9	0.4	11.9
Beans, green	88.9	2.4	0.2	7.7
Broccoli	89.9	3.3	0.2	5.5
Cauliflower	91.7	2.4	0.2	4.9
Corn, sweet	73.9	3.7	1.2	20.5
Mushroom	91.1	2.4	0.3	4.0
Peas	74.3	6.7	0.4	17.7
Potatoes	77.8	2.0	0.1	19.1
Tomato	94.1	1.0	0.3	4.0

(*Continued*)

Table A.2 *Continued*

Food	Composition per 100 g Edible Portion			
	Water (g)	Protein (g)	Fat (g)	Carbohydrate (g)
Cereal products				
Wheat flour	12.0	9.2	1.0	73.8
Rice, white	12.3	7.6	0.3	79.4
Corn meal	12.0	9.0	3.4	74.5
Corn flakes	3.6	8.1	0.4	85.0
Bread, white	35.5	8.1	2.2	51.4
Dairy products				
Milk, cow, whole	87.0	3.5	3.9	4.9
Milk, cow, non-fat	90.5	3.5	0.1	5.1
Cheese, cottage	76.5	19.5	0.5	2.0
Cheese, Cheddar	37.0	25.0	32.2	2.1
Cream, heavy	59.0	2.3	35.0	3.2
Cream, light	72.5	32.9	20.0	4.0
Butter	15.5	0.6	81.0	0.4
Ice cream	62.1	4.0	12.5	20.6
Meat, poultry, eggs				
Beef, hamburger, cooked	47.0	22.0	30.0	0
Lamb, leg	63.7	18.0	17.5	0
Chicken, roaster	66.0	20.2	12.6	0
Eggs, whole	74.0	12.8	11.5	0.7
Fish				
Cod	82.6	16.5	0.4	0
Herring	67.2	18.3	12.5	0
Salmon	63.4	17.4	16.5	0
Tuna, canned	60.0	29.0	8.2	0

From Heinz Nutritional Research Division (1958), "Nutritional Data". H.J. Heinz Company, Pittsburgh, PA

Table A.3 Viscosity and Density of Gases and Liquids

Material	T (°C)	μ (Pa · s)	ρ (kg · m^{-3})	Source
Gases at 1 atm.				
Air	0	17.2×10^{-6}	1.29	1
Air	20	18.5×10^{-6}	1.205	1
Air	60	20.0×10^{-6}	1.075	1
Air	100	21.8×10^{-6}	0.95	1
CO_2	20	14.8×10^{-6}	1.84	1
Water vapor	100	12.5×10^{-6}	0.597	1
Water vapor	200	16.3×10^{-6}	0.452	1
Liquids				
Water	0	1.79×10^{-3}	999	1
Water	20	1.00×10^{-3}	998	1
Water	40	0.664×10^{-3}	992	1
Water	60	0.466×10^{-3}	983	1
Water	80	0.355×10^{-3}	972	1
Water	100	0.281×10^{-3}	958	1
Ethanol	20	1.20×10^{-3}	790	1
Glycerol	20	1.490	1261	1
Edible oils	20	0.05–0.2	920–950	1
Edible oils	100	$5–2 \times 10^{-3}$	880–900	1
Milk	20	2×10^{-3}	1032	1
Milk	70	0.7×10^{-3}	1012	1
Beer	0	1.3×10^{-3}	1000	1
Honey	25	6	1400	1
Sucrose soln. 20% w/w	20	1.945×10^{-3}	1080	2
Sucrose soln. 20% w/w	50	0.97×10^{-3}		2
Sucrose soln. 20% w/w	80	0.59×10^{-3}		2
Sucrose soln. 40% w/w	20	6.167×10^{-3}	1180	2
Sucrose soln. 40% w/w	50	2.49×10^{-3}		2
Sucrose soln. 40% w/w	80	1.32×10^{-3}		2
Sucrose soln. 60% w/w	20	58.49×10^{-3}	1290	2
Sucrose soln. 60% w/w	50	14.0×10^{-3}		2
Sucrose soln. 60% w/w	80	5.2×10^{-3}	1410	2
Sucrose soln. 80% w/w	20			2

Sources: Bimbenet, J.J., Duquenoy, A., and Trystram, G. (2002) Genie des Procedes Alimentaires. *Dunod, Paris, France. Pancoast, H.M. and Junk, W.R. (1980)* Handbook of Sugars. *The Avi Publishing Company, Westport, CT.*

Table A.4 Thermal Properties of Materials

Material	Temperature, T (°C)	Thermal Conductivity, k ($W \cdot m^{-1} \cdot K^{-1}$)	Specific Heat, C_P ($J \cdot kg^{-1} \cdot K^{-1}$)	Thermal Diffusivity, α ($m^2 \cdot s^{-1}$)
Gases at 1 atm				
Air	0	0.0240	1005	18.6×10^6
Air	20	0.0256	1006	21.2×10^6
Air	100	0.0314	1012	33.0×10^6
Hydrogen	20	0.1850	14250	153.0×10^6
Water vapor	100	0.0242	2030	19.0×10^6
Water vapor	200	0.0300	1960	32.6×10^6
Liquids				
Water	20	0.599	4180	0.143×10^6
Water	60	0.652	4180	0.161×10^6
Water	100	0.684	4210	0.170×10^6
Ethanol	20	0.167	2440	0.087×10^6
Edible oils	20	0.17	2500	0.070×10^6
Milk	20	0.56	4000	0.135×10^6
Sucrose soln. 20% w/w	20	0.54	3800	0.131×10^6
Solids				
Copper	0	370	420	105×10^6
Aluminum		230	900	95×10^6
Steel		60	460	16×10^6
Stainless steel		15	480	4×10^6
Concrete		1.1	850	0.65×10^6
Glass		0.75	800	0.35×10^6
Water ice		2.22	2100	1.14×10^6

Extracted from Bimbenet, J.J., Duquenoy, A., and Trystram, G. (2002) Génie des Procédés Alimentaires. *Dunod, Paris, France.*

Table A.5 Emissivity of Surfaces

Material	Temperature (°C)	Emissivity, ε
Aluminum, polished	220–580	0.039–0.057
Aluminum, oxidized	200–600	0.11–0.19
Steel plate, rough	30–400	0.94–0.97
Stainless steel	200–500	0.44–0.36
Concrete tiles	1000	0.63
Glass	200–550	0.95–0.85
Water	0–100	0.95–0.963
Ice	0	0.63

Adapted from Foust, A.S., Wenzel, L.A., Clump, C.W., Maus, L., and Andersen, L.B. (1960) Principles of Unit Operations. *John Wiley & Sons, Inc., New York, NY.*

Table A.6 US Standard Sieves

US Sieve No.	Sieve Opening (μm)
4	4760
6	3360
8	2380
10	2000
20	840
30	590
40	420
50	297
60	250
80	177
100	149
120	125
140	105
170	88
200	74
270	53

Adapted from Perry, J.H. (ed.) (1950) Chemical Engineers Handbook, *3rd edn. McGraw-Hill Book Company, New York, NY.*

Table A.7 Properties of Saturated Steam – Temperature Table

		Enthalpy (kJ · kg^{-1})			Entropy (kJ · kg^{-1})	
Temperature (°C)	**Pressure (kPa)**	**Liquid**	**Gas**	**λ**	**Liquid**	**Gas**
10	1.228	42	2520	2478	0.15	8.901
20	2.339	84	2538	2454	0.30	8.667
30	4.246	126	2556	2430	0.44	8.453
40	7.384	168	2574	2407	0.57	8.257
50	12.35	209	2592	2383	0.70	8.076
60	19.94	251	2610	2358	0.83	7.910
70	31.19	293	2627	2334	0.95	7.755
80	47.39	335	2644	2309	1.08	7.612
90	70.14	377	2660	2283	1.19	7.479
100	101.3	419	2676	2257	1.31	7.355
110	143.3	461	2691	2230	1.42	7.239
120	198.5	504	2706	2203	1.53	7.130
130	207.1	546	2720	2174	1.63	7.027
140	361.3	589	2734	2145	1.74	6.930
150	475.8	632	2746	2114	1.84	6.838
160	617.8	676	2758	2083	1.94	6.750
170	791.7	719	2769	2049	2.04	6.666
180	1002	763	2778	2015	2.14	6.586
190	1254	808	2786	1979	2.24	6.508
200	1554	852	2793	1941	2.33	6.432

λ = latent heat of evaporation. This is an abridged summary table. For more detailed data, please consult Keenan, J.H., Keyes, F.G., Hill, P.G., and Moore, J.G. (1969) Steam Tables: Thermodynamic Properties of Water Including Vapor, Liquid and Solid Phases. *John Wiley & Sons, New York, NY.*

Table A.8 Properties of Saturated Steam – Pressure Table

Pressure (kPa)	Temperature (°C)	Enthalpy ($kJ \cdot kg^{-1}$)			Entropy ($kJ \cdot kg^{-1} \cdot K^{-1}$)	
		Liquid	Vapor	λ	Liquid	Vapor
10	46	192	2585	2393	0.649	8.150
20	60	251	2610	2359	0.832	7.908
40	76	318	2637	2319	1.026	7.670
75	92	384	2663	2279	1.213	7.456
100	100	417	2675	2258	1.303	7.359
150	111	467	2694	2227	1.434	7.223
200	120	505	2707	2202	1.530	7.127
250	127	535	2717	2182	1.607	7.053
300	134	561	2725	2164	1.672	6.992
350	139	584	2732	2148	1.725	6.940
400	144	605	2739	2134	1.777	6.896
450	148	623	2744	2121	1.821	6.856
500	152	640	2749	2109	1.861	6.821
550	156	656	2753	2097	1.897	6.789
600	159	671	2757	2086	1.932	6.760
650	162	684	2760	2076	1.963	6.733
700	165	697	2763	2066	1.992	6.708
750	168	710	2766	2056	2.020	6.685
800	170	721	2769	2048	2.046	6.663
850	173	732	2772	2040	2.071	6.642
900	175	743	2774	2031	2.095	6.623
950	178	753	2776	2023	2.117	6.604
1000	180	763	2778	2015	2.139	6.586

λ = latent heat of evaporation. This is an abridged summary table. For more detailed data please consult Keenan, J.H., Keyes, F.G., Hill, P.G., and Moore, J.G. (1969) Steam Tables: Thermodynamic Properties of Water Including Vapor, Liquid and Solid Phases. *John Wiley & Sons, New York, NY.*

Table A.9 Properties of Superheated Steam

	Pressure (kPa)									
	100		200		300		400		500	
Temperature (°C)	h	s	h	s	h	s	h	s	h	s
100	2676	7.361								
150	2776	7.623	2769	7.279	2761	7.078	2753	6.930	2855	7.059
200	2875	7.834	2870	7.507	2866	7.311	2860	7.171	2961	7.271
250	2974	8.033	2971	7.709	2968	7.517	2964	7.379	3064	7.460
300	3074	8.216	3072	7.893	3069	7.702	3067	7.566	3272	7.794
400	3278	8.543	3277	8.222	3275	8.033	3273	7.898	3483	8.087
500	3488	8.834	3487	8.513	3486	8.325	3485	8.191	3702	7.352
600	3705	9.098	3704	8.777	3703	8.589	3702	8.456		

h = enthalpy, $kJ \cdot kg^{-1} \cdot s^{-1}$ = entropy, $kJ \cdot kg^{-1} \cdot K^{-1}$. This is an abridged summary table. For more detailed data, please consult Keenan, J.H., Keyes, F.G., Hill, P.G. and Moore, J.G. (1969) Steam Tables: Thermodynamic Properties of Water Including Vapor, Liquid and Solid Phases. *John Wiley & Sons, New York, NY.*

Table A.10 Vapor Pressure of Liquid Water and Ice Below 0°C

Temperature (°C)	p_{water} (Pa)	P_{ice} (Pa)
0	610.5	610.5
−0.5	588.7	585.9
−1	567.7	562.2
−1.5	547.3	539.3
−2	527.4	517.3
−2.5	508.3	496.2
−3	489.7	475.7
−3.5	472.0	456.2
−4	454.6	437.3
−4.5	437.8	419.2
−5	421.7	401.7
−6	390.8	368.6
−7	362.0	338.2
−8	335.2	310.1
−9	310.1	284.1
−10	286.5	260.0
−11	264.9	238.0
−12	244.5	217.6
−13	225.4	198.6
−14	208.0	181.4
−18		125.2
−23		77.3
−28		46.8
−33		33.3
−43		9.9
−53		3.8

Table A.11 Freezing Point of Ideal Aqueous Solutions

Water Activity	Freezing Temperature (°C)
1.00	0.0
0.98	−2.1
0.96	−4.2
0.94	−6.3
0.92	−8.6
0.90	−10.8

Table A.12 Vapor–Liquid Equilibrium Data for Ethanol–Water Mixtures at 1 Atm

Ethanol in Liquid (% mol or % mass)	Ethanol in Vapor	
	(% mol)	(% mass)
0	0.0	0.0
1	6.5	10.0
3	20.5	24.8
5	32.2	38.0
10	43.7	52.0
15	50.1	59.5
20	53.2	64.8
25	55.4	68.6
30	57.5	71.4
35	59.4	73.3
40	61.4	74.7
45	63.2	75.9
50	62.2	77.1
55	67.6	78.2
60	70.3	79.4
65	72.6	80.7
70	75.4	82.2
75	78.6	83.9
80	82.1	85.9
85	85.8	88.3
90	89.8	91.3
95	94.7	95.0
97	96.8	96.9
99	98.95	98.9
100	100	100
Azeotrope mixture	89.40	95.57

Table A.13 Boiling Point of Sucrose Solutions at 1 Atm

Concentration (Degrees Brix)	Boiling Temperature (°C)
33.33	100.67
50.00	101.62
60.00	102.70
66.67	103.85
71.43	105.04
75.00	106.23
77.78	107.20
80.00	108.55

Extracted from Pancoast, H.M., and Junk, W.R. (1980) Handbook of Sugars. *The Avi Publishing Company, Westport, CT.*

Table A.14 Electrical Conductivity of Some Materials

Material	C*	A** at 0°C	A** at 100°C
Na Cl solution	1	65.8	352.5
	10	63.2	335.0
	100	57.7	295.6
	200	55.6	287.0
	500	44.0	
	1000	42.5	247.0
Glucose solution	0.4	0.13 (at 20°C)	
	$\kappa \times 10^5$ at 18°C		
Butter	646–701		
Olive oil	993		

C = concentration, millimol · liter^{-1}; A** = 10^6 κ/C*; κ = specific conductance, ohm^{-1} · cm^{-1}.*
Extracted from West, C.J. ed. (1933) International Critical Tables. *National Academy of Science, New York, NY.*

Table A.15 Thermodynamic Properties of Saturated R-134a

			h ($kJ \cdot kg^{-1}$)		s ($kJ \cdot kg^{-1} \cdot K^{-1}$)	
T (°C)	P (kPa)	v (Gas) ($m^3 \cdot kg^{-1}$)	Liquid	Gas	Liquid	Gas
−50	29.4	0.61	136	368	0.743	1.782
−45	39.1	0.47	142	371	0.770	1.773
−40	51.1	0.36	148	374	0.797	1.765
−35	66.1	0.28	155	377	0.823	1.759
−30	84.3	0.23	161	381	0.849	1.752
−25	106	0.18	167	384	0.875	1.747
−20	133	0.15	174	387	0.901	1.742
−15	164	0.12	180	390	0.926	1.738
−10	201	0.10	187	393	0.951	1.734
−5	243	0.083	193	396	0.976	1.731
0	293	0.069	200	399	1.000	1.728
5	350	0.058	207	402	1.024	1.725
10	415	0.049	214	405	1.048	1.723
15	489	0.042	220	407	1.073	1.721
20	572	0.036	228	410	1.096	1.719
25	666	0.031	235	413	1.120	1.717
30	771	0.027	242	415	1.144	1.716
35	888	0.023	249	417	1.168	1.714
40	1017	0.020	257	420	1.191	1.712
45	1160	0.017	264	422	1.215	1.710
50	1319	0.015	272	424	1.238	1.709

v = specific volume, h = enthalpy, s = entropy. Note: This is an abridged summary table. For more detailed data please consult http://refrigerants.dupont.com or http://eng.sdsu.edu/testcenter/testheme.

Table A.16 Thermodynamic Properties of Superheated R134-a

	Pressure (kPa) (Saturation Temperature, °C)											
	50 (−40)		100 (−26)		200 (−10)		400 (9)		600 (21.5)		1000 (39.3)	
Temperature (°C)	h	s	h	s	h	s	h	s	h	s	h	s
−40	374	1.767										
−35	378	1.783										
−30	382	1.799										
−25	386	1.814	384	1.753								
−20	389	1.830	388	1.768								
−15	393	1.845	392	1.784								
−10	397	1.860	396	1.799								
−5	401	1.875	400	1.815	397	1.750						
0	405	1.889	404	1.830	401	1.766						
5	409	1.904	408	1.844	406	1.781						
10	413	1.919	412	1.859	410	1.797	405	1.727				
15	417	1.933	416	1.874	414	1.812	410	1.743				
20	421	1.947	421	1.888	418	1.827	414	1.759				
25	426	1.961	425	1.903	423	1.841	419	1.775	414	1.730		

(Continued)

Table A.16 *Continued*

	Pressure (kPa) (Saturation Temperature, °C)											
	50 (−40)		100 (−26)		200 (−10)		400 (9)		600 (21.5)		1000 (39.3)	
Temperature (°C)	**h**	**s**	**h**	**s**	**h**	**s**	**h**	**s**	**h**	**s**	**h**	**s**
30	430	1.976	429	1.917	427	1.856	423	1.790	419	1.746		
35	434	1.989	433	1.931	432	1.870	428	1.805	424	1.762		
40	438	2.003	438	1.945	436	1.885	433	1.820	429	1.778	420	1.715
45	443	2.017	442	1.959	441	1.890	437	1.835	434	1.793	426	1.732
50	447	2.031	447	1.973	445	1.913	442	1.849	439	1.809	431	1.749
55	452	2.044	451	1.986	450	1.927	447	1.864	443	1.823	437	1.766
60	456	2.058	455	2.000	454	1.941	451	1.878	448	1.838	442	1.782
65	461	2.072	460	2.014	459	1.954	456	1.892	453	1.853	447	1.972
70	465	2.085	465	2.027	463	1.968	461	1.906	458	1.867	452	1.813
75	470	2.098	493	2.040	468	1.981	466	1.920	663	1.881	457	1.828
80	475	2.111	474	2.054	473	19995	470	1.934	468	1.895	463	1.842

h = enthalpy, *s* = entropy. Note: This is an abridged summary table. For more detailed data please consult: http://refrigerants.dupont.com or http://eng.sdsu.edu/testcenter/testheme.

Table A.17 Properties of Air at Atmospheric Pressure

Temperature (°C)	Density ($kg \cdot m^{-3}$)	Viscosity ($Pa \cdot s$)	Thermal Conductivity ($W \cdot m^{-1} \cdot K^{-1}$)	Specific Heat ($J \cdot kg^{-1} \cdot K^{-1}$)	Prandl Number (Dimensionless)
0	1.25	17.5×10^{-6}	0.0238	1010	0.74
20	1.16	18.2×10^{-6}	0.0252	1012	0.73
40	1.09	19.1×10^{-6}	0.0265	1014	0.73
60	1.03	20.0×10^{-6}	0.0280	1017	0.72
80	0/97	20.8×10^{-6}	0.0293	1019	0.72
100	0.92	21.7×10^{-6}	0.0308	1022	0.72
120	0.87	22.6×10^{-6}	0.0320	1025	0.72
140	0.83	23.3×10^{-6}	0.0334	1027	0.72
160	0.79	24.1×10^{-6}	0.0345	1030	0.72
180	0.75	24.9×10^{-6}	0.0357	1032	0.72
200	0.72	25.7×10^{-6}	0.0370	1035	0.72

Table A.18 Surface Tension of Selected Liquids Against Air at 20°C

Liquid	Surface Tension, γ (dyn · cm^{-1})	Surface Tension, γ (N/m)
Water	73	0.073
Ethanol	22	0.022
Methanol	23	0.023
Glycerol	63	0.063
n-Hexane	18	0.018
Edible oil (typical)	30	0.030
Honey (typical)	110	0.11

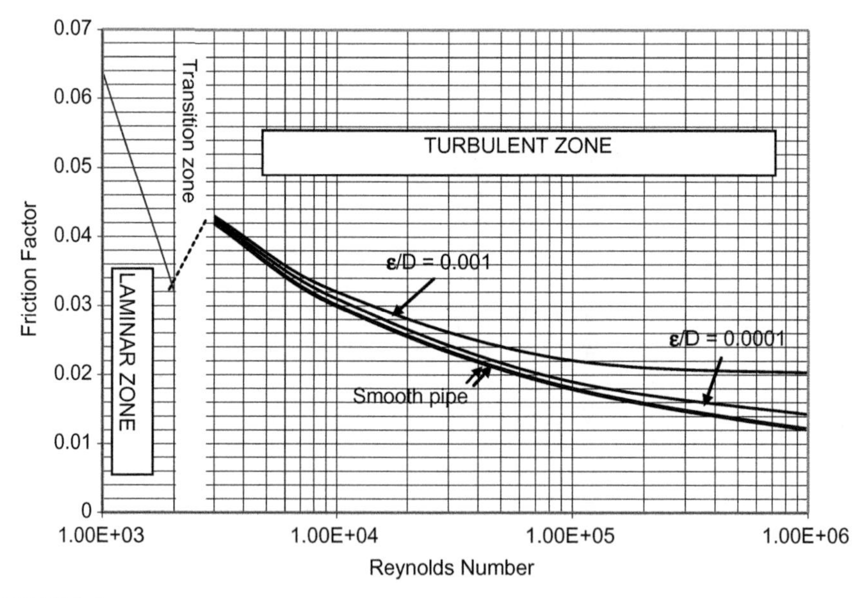

FIGURE A.1

Friction factors for flow in pipes.

Adapted from Foust, A.S., Wenzel, L.A., Clump, C.W., Maus, L., and Andersen, L.B. (1960) Principles of Unit Operations. *John Wiley & Sons, Inc., New York, NY.*

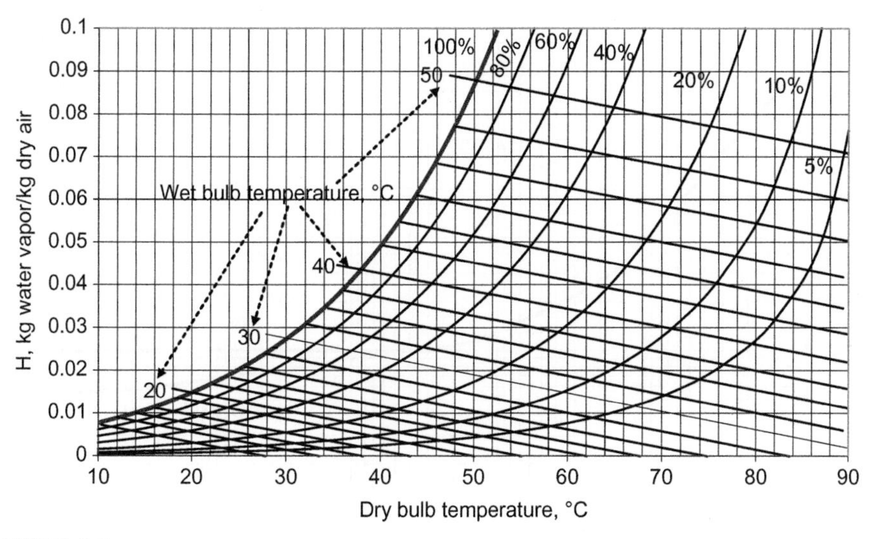

FIGURE A.2

Psychrometric chart.

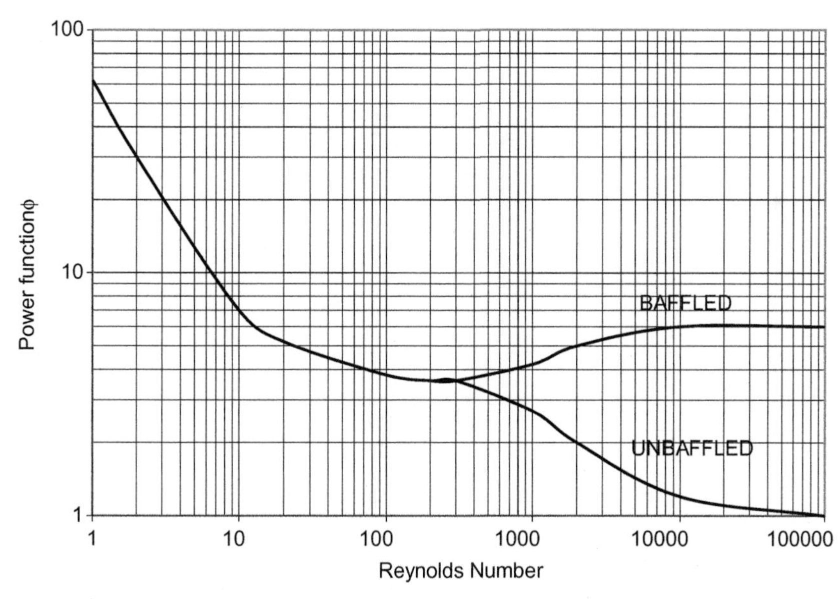

FIGURE A.3

Mixing power function, turbine impellers.

Adapted from. McCabe, W.L. and Smith J.C. (1956) Unit Operations of Chemical Engineering. *McGraw-Hill Book Company, Inc., New York, NY*

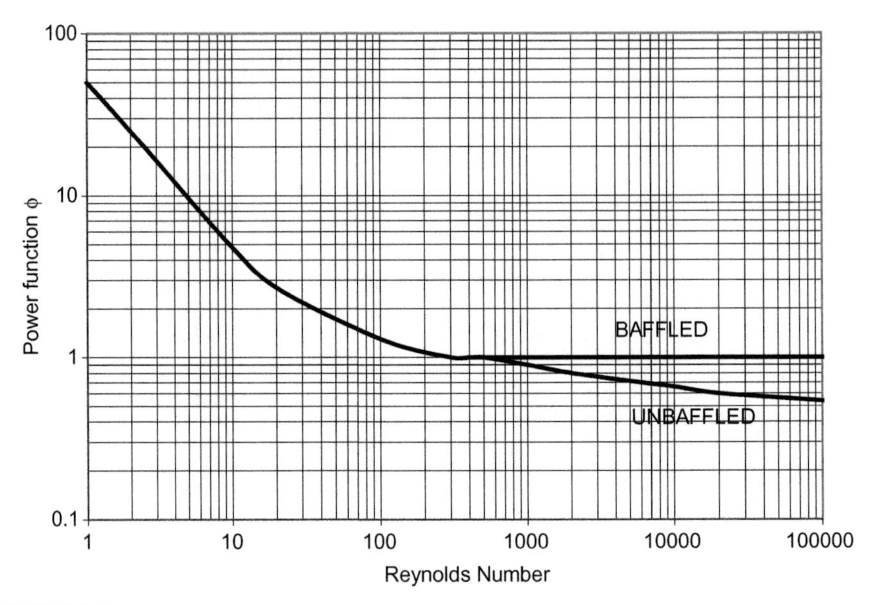

FIGURE A.4

Mixing power function, propeller impellers.

Adapted from. McCabe, W.L. and Smith J.C. (1956) Unit Operations of Chemical Engineering. *McGraw-Hill Book Company, Inc., New York, NY*

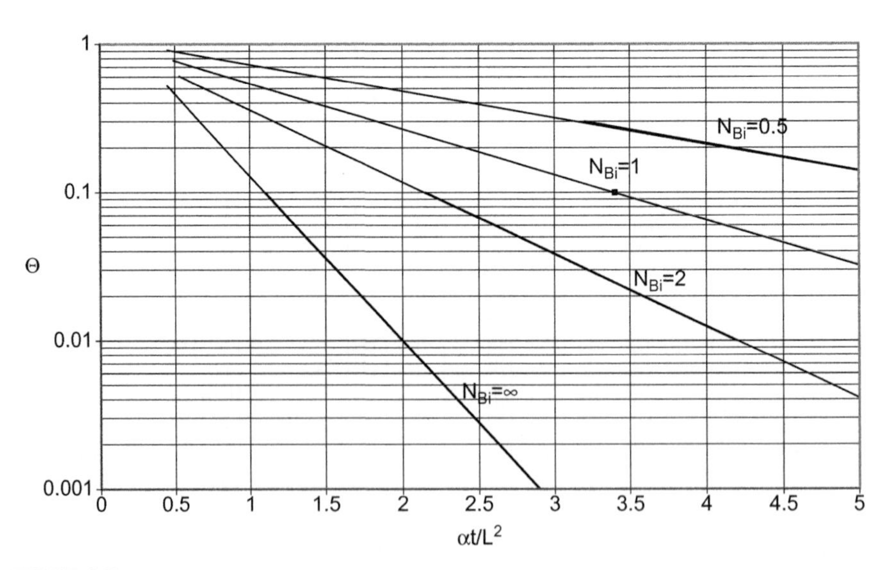

FIGURE A.5

Unsteady heat transfer in a slab: temperature at the central plane. L, half-thickness of the slab.

Adapted from Foust, A.S., Wenzel, L.A., Clump, C.W., Maus, L., and Andersen, L.B. (1960) Principles of Unit Operations. *John Wiley & Sons, Inc., New York, NY.*

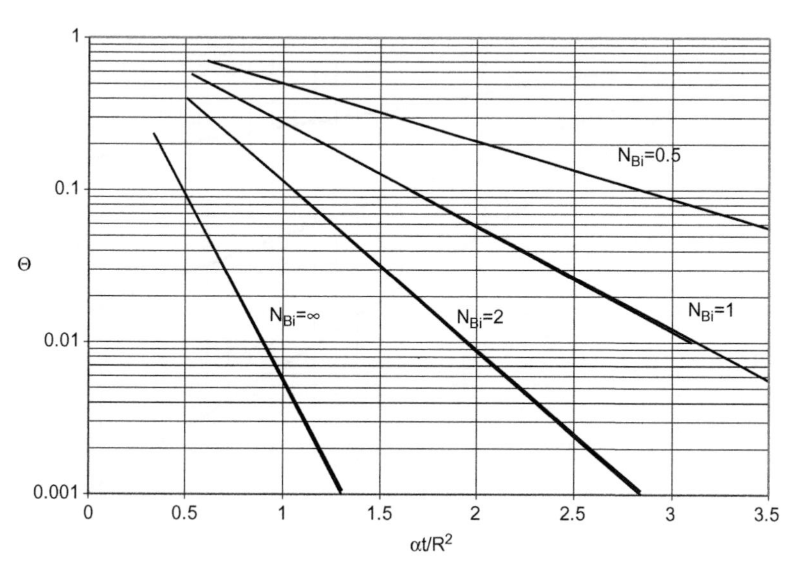

FIGURE A.6

Unsteady heat transfer in an infinite cylinder: temperature at the central axis. R, radius of the cylinder.

Adapted from Foust, A.S., Wenzel, L.A., Clump, C.W., Maus, L., and Andersen, L.B. (1960) Principles of Unit Operations. *John Wiley & Sons, Inc., New York, NY.*

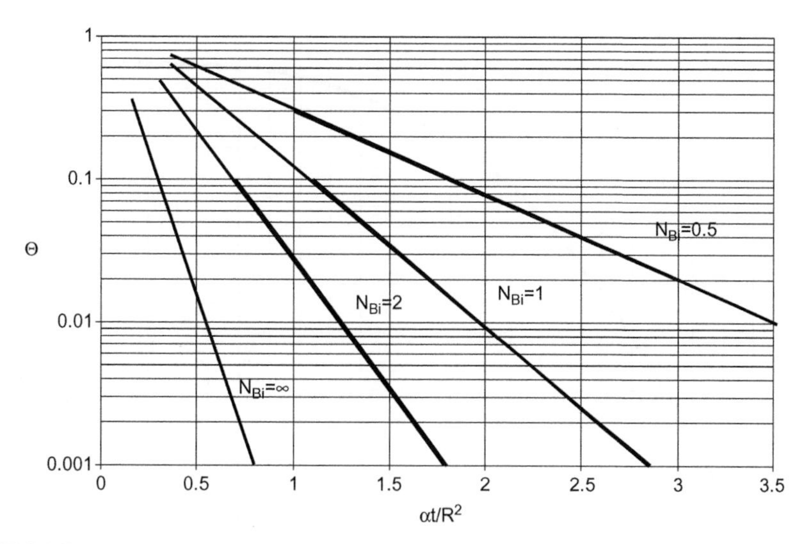

FIGURE A.7

Unsteady heat transfer in a sphere: temperature at the geometric center. R, radius of the sphere.

Adapted from Foust, A.S., Wenzel, L.A., Clump, C.W., Maus, L., and Andersen, L.B. (1960) Principles of Unit Operations. *John Wiley & Sons, Inc., New York, NY.*

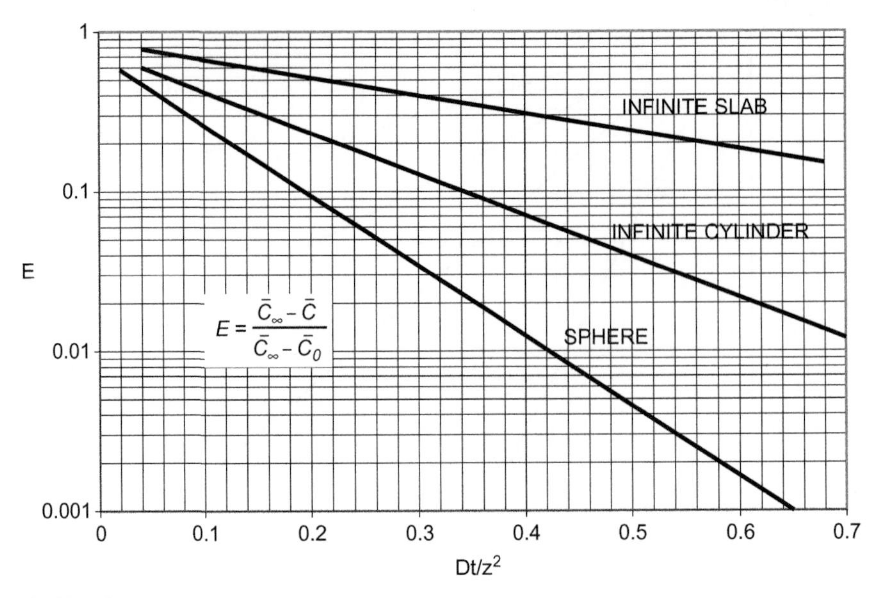

FIGURE A.8

Unsteady state mass transfer: mass-average concentration. Z, half-thickness of slab, radius of cylinder or sphere.

Adapted from Treybal, R.E. (1968) Mass Transfer Operations, *2nd edn. McGraw-Hill Book Company Inc., New York, NY.*

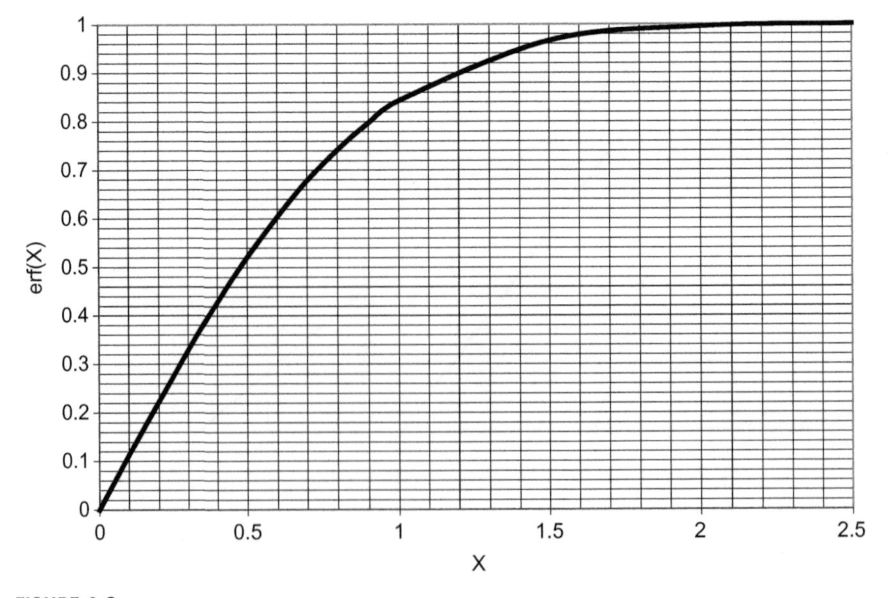

FIGURE A.9

Error function.

Food Science and Technology: International Series List

Amerine, M.A., Pangborn, R.M., and Roessler, E.B., 1965. Principles of Sensory Evaluation of Food.
Glicksman, M., 1970. Gum Technology in the Food Industry.
Joslyn, M.A., 1970. Methods in Food Analysis, Second Ed.
Stumbo, C.R., 1973. Thermobacteriology in Food Processing, Second Ed.
Altschul, A.M. (Ed.), New Protein Foods: Volume 1, Technology, Part A – 1974. Volume 2, Technology, Part B – 1976. Volume 3, Animal Protein Supplies, Part A – 1978.
Volume 4, Animal Protein Supplies, Part B – 1981. Volume 5, Seed Storage Proteins – 1985.
Goldblith, S.A., Rey, L., and Rothmayr, W.W., 1975. Freeze Drying and Advanced Food Technology.
Bender, A.E., 1975. Food Processing and Nutrition.
Troller, J.A., and Christian, J.H.B., 1978. Water Activity and Food.
Osborne, D.R., and Voogt, P., 1978. The Analysis of Nutrients in Foods.
Loncin, M., and Merson, R.L., 1979. Food Engineering: Principles and Selected Applications.
Vaughan, J.G. (Ed.), 1979. Food Microscopy.
Pollock, J.R.A. (Ed.), Brewing Science, Volume 1 –1979. Volume 2–1980. Volume 3–1987.
Christopher Bauernfeind, J. (Ed.), 1981.Carotenoids as Colorants and Vitamin A Precursors: Technological and Nutritional Applications.
Markakis, P. (Ed.), 1982. Anthocyanins as Food Colors.
Stewart, G.G., and Amerine, M.A. (Eds.), 1982. Introduction to Food Science and Technology, Second Ed.
Iglesias, H.A., and Chirife, J., 1982. Handbook of Food Isotherms: Water Sorption Parameters for Food and Food Components.
Dennis, C. (Ed.), 1983. Post-Harvest Pathology of Fruits and Vegetables.
Barnes, P.J. (Ed.), 1983. Lipids in Cereal Technology.
Pimentel, D., and Hall, C.W. (Eds.), 1984. Food and Energy Resources.
Regenstein, J.M., and Regenstein, C.E., 1984. Food Protein Chemistry: An Introduction for Food Scientists.
Gacula Jr. M.C., and Singh, J., 1984. Statistical Methods in Food and Consumer Research.
Clydesdale, F.M., and Wiemer, K.L. (Eds.), 1985. Iron Fortification of Foods.
Decareau, R.V., 1985. Microwaves in the Food Processing Industry.
Herschdoerfer, S.M. (Ed.), Quality Control in the Food Industry, Second Ed. Volume 1 –1985. Volume 2 –1985. Volume 3–1986. Volume 4–1987.
Urbain, W.M., 1986. Food Irradiation.
Bechtel, P.J., 1986. Muscle as Food.
Chan, H.W.-S., 1986. Autoxidation of Unsaturated Lipids.
Cunningham, F.E., and Cox, N.A. (Eds.), 1987. Microbiology of Poultry Meat Products.
McCorkle Jr. C.O., 1987. Economics of Food Processing in the United States.

Japtiani, J., Chan Jr., H.T., and Sakai, W.S., 1987. Tropical Fruit Processing.
Solms, J., Booth, D.A., Dangborn, R.M., and Raunhardt, O., 1987. Food Acceptance and Nutrition.
Macrae, R., 1988. HPLC in Food Analysis, Second Ed.
Pearson, A.M., and Young, R.B., 1989. Muscle and Meat Biochemistry.
Penfield, M.P., and Campbell, A.M., 1990. Experimental Food Science, Third Ed.
Blankenship, L.C., 1991. Colonization Control of Human Bacterial Enteropathogens in Poultry.
Pomeranz, Y., 1991. Functional Properties of Food Components, Second Ed.
Walter, R.H., 1991. The Chemistry and Technology of Pectin.
Stone, H., and Sidel, J.L., 1993. Sensory Evaluation Practices, Second Ed.
Shewfelt, R.L., and Prussia, S.E., 1993. Postharvest Handling: A Systems Approach.
Nagodawithana, T., and Reed, G., 1993. Enzymes in Food Processing, Third Ed.
Hoover, D.G., and Steenson, L.R., 1993. Bacteriocins.
Shibamoto, T., and Bjeldanes, L., 1993. Introduction to Food Toxicology.
Troller, J.A., 1993. Sanitation in Food Processing, Second Ed.
Hafs, D., and Zimbelman, R.G., 1994. Low-Fat Meats.
Phillips, L.G., Whitehead, D.M., and Kinsella, J., 1994. Structure–Function Properties of Food Proteins.
Jensen, R.G., 1995. Handbook of Milk Composition.
Roos, Y.H., 1995. Phase Transitions in Foods.
Walter, R.H., 1997. Polysaccharide Dispersions.
Barbosa-Canovas, G.V., Marcela Góngora-Nieto, M., Pothakamury, U.R., and Swanson, B.G., 1999. Preservation of Foods with Pulsed Electric Fields.
Jackson, R.S., 2002. Wine Tasting: A Professional Handbook.
Bourne, M.C., 2002. Food Texture and Viscosity: Concept and Measurement, Second Ed.
Caballero, B., and Popkin, B.M. (Eds.), 2002. The Nutrition Transition: Diet and Disease in the Developing World.
Cliver, D.O., and Riemann, H.P. (Eds.), 2002. Foodborne Diseases, Second Ed.
Kohlmeier, M., 2003. Nutrient Metabolism.
Stone, H., and Sidel, J.L., 2004. Sensory Evaluation Practices, Third Ed.
Han, J.H., 2005. Innovations in Food Packaging.
Sun, D.-W. (Ed.), 2005. Emerging Technologies for Food Processing.
Riemann, H.P., and Cliver, D.O. (Eds.), 2006. Foodborne Infections and Intoxications, Third Ed.
Arvanitoyannis, I.S., 2008. Waste Management for the Food Industries.
Jackson, R.S., 2008. Wine Science: Principles and Applications, Third Ed.
Sun, D.-W. (Ed.), 2008. Computer Vision Technology for Food Quality Evaluation.
David, K., and Thompson, P., (Eds.), 2008. What Can Nanotechnology Learn From Biotechnology?
Arendt, E.K., and Bello, F.D. (Eds.), 2008. Gluten-Free Cereal Products and Beverages.
Bagchi, D. (Ed.), 2008. Nutraceutical and Functional Food Regulations in the United States and Around the World.
Singh, R.P., and Heldman, D.R., 2008. Introduction to Food Engineering, Fourth Ed.
Berk, Z., 2009. Food Process Engineering and Technology.
Thompson, A., Boland, M., and Singh, H. (Eds.), 2009. Milk Proteins: From Expression to Food.

Florkowski, W.J., Prussia, S.E., Shewfelt, R.L. and Brueckner, B. (Eds.), 2009. Postharvest Handling, Second Ed.

Gacula Jr., M., Singh, J., Bi, J., and Altan, S., 2009. Statistical Methods in Food and Consumer Research, Second Ed.

Shibamoto, T., and Bjeldanes, L., 2009. Introduction to Food Toxicology, Second Ed.

BeMiller, J. and Whistler, R. (Eds.), 2009. Starch: Chemistry and Technology, Third Ed.

Jackson, R.S., 2009. Wine Tasting: A Professional Handbook, Second Ed.

Sapers, G.M., Solomon, E.B., and Matthews, K.R. (Eds.), 2009. The Produce Contamination Problem: Causes and Solutions.

Heldman, D.R., 2011. Food Preservation Process Design.

Tiwari, B.K., Gowen, A. and McKenna, B. (Eds.), 2011. Pulse Foods: Processing, Quality and Nutraceutical Applications.

Cullen, PJ., Tiwari, B.K., and Valdramidis, V.P. (Eds.), 2012. Novel Thermal and Non-Thermal Technologies for Fluid Foods.

Stone, H., Bleibaum, R., and Thomas, H., 2012. Sensory Evaluation Practices, Fourth Ed.

Kosseva, M.R. and Webb, C. (Eds.), 2013. Food Industry Wastes: Assessment and Recuperation of Commodities.

Morris, J.G. and Potter, M.E. (Eds.), 2013. Foodborne Infections and Intoxications, Fourth Ed.

Index

Note: Page numbers followed by "*f*", "*t*" and "*b*" denote figures, tables and boxes, respectively.

A

B

C

F

G

H

I

J

K

L

M

N

O

P

R

S

T

U

V

W

Y

Printed and bound by CPI Group (UK) Ltd, Croydon, CR0 4YY
08/05/2025
01864880-0002